Finite Mathematics

Finite Mathematics

Frank C. Wilson
Green River Community College

HOUGHTON MIFFLIN COMPANY
Boston New York

Vice President and Publisher: Jack Shira
Sponsoring Editor: Leonid Tunik
Associate Editor: Jennifer King
Senior Project Editor: Tamela Ambush
Editorial Assistant: Sage Anderson
Senior Manufacturing Buyer: Renee Ostrowski
Executive Marketing Manager: Brenda Bravener-Greville
Senior Marketing Manager: Jennifer Jones

Cover photograph: © George Simhoni/Masterfile, © Bilderberg/Photonica, © Steve Edson/Photonica, © Photodisc Photography

Photo Credits: Chapter 1: p. 1, © Lou Dematteis/Reuther/CORBIS; p. 43, © Richard Cummins/CORBIS; **Chapter 2:** p. 53, © Wolfgang Kaehler/CORBIS; p.57, Author; **Chapter 3:** p. 111, © Larry Williams and Associates/CORBIS; p. 132, © MacDuff/Everton/CORBIS; **Chapter 4:** p.171, © Brownie Harris Productions/CORBIS; p. 215, © Skjold Productions/PhotoEdit, Inc.; p. 265, © Connie Gardner AFP Getty Images; **Chapter 5:** p. 281, © Frederic J. Brown/Getty Images; p.330, © Reuters/CORBIS; p. 382, Bureau of Labor Statistics; **Chapter 6:** p. 383, © Tom and Dee Ann McCarthy/CORBIS; p. 418, © Getty Images News; **Chapter 7:** p. 440, © Strauss/Curtis/CORBIS; p. 458, © Chris Trotman/Duomo/CORBIS; **Chapter 8:** p. 494, © Darren Modricker/CORBIS; p. 518, © Digital Images/Getty Images; p. 536, © Animals/Animals.

Printed in the U.S.A.

Library of Congress Control Number: 2005934026

Instructor's exam copy:
ISBN 13: 978-0-618-73266-1
ISBN 10: 0-618-73266-7

For orders, use student text ISBNs:
ISBN 13: 978-0-618-33293-9
ISBN 10: 0-618-33293-6

1 2 3 4 5 6 7 8 9 –CRK– 10 09 08 07 06

Contents

Preface

To the Student

Have you ever asked, "When am I ever going to use this?" or "Why do I care?" after learning a new mathematical procedure? Many students have. This book seeks to answer those questions by teaching mathematical skills which, when applied, can improve your quality of life. Whether calculating how long it will take to pay off a car loan or predicting how much tuition will be next year, examples and exercises in this book are based on interesting and engaging real-life data. Consequently, when you find a mathematical solution, you will learn something new about the world in which we live. To make real-life data analysis even more meaningful, you will become skilled in collecting and analyzing data from your own life through the *Make It Real* projects. The analytical skills learned through these projects will remain with you long after you have left this course.

This book is written in a casual, reader-friendly style. Although key mathematical terms and concepts are appropriately addressed, the focus is conceptual understanding not mathematical jargon. If you are pursuing a degree in business, social science, or a similar field, this book is written specifically for you.

You will find that your understanding of key concepts will be enhanced through the use of a graphing calculator. The TI-83 Plus (or the TI-83 or TI-84) is the ideal calculator for this course. Because learning how to use the calculator is a challenge for many students, Technology Tips are scattered throughout the text allowing you to learn a new calculator technique when it is needed. These tips detail how to graph a function, solve an equation, find the maximum value of a function, and so on. Rather than giving a broad overview of a procedure, the tips take you through the actual keystrokes on the TI-83 Plus and show you calculator screenshots so that you can verify that you're doing each step correctly.

I have learned much about teaching from my students and deeply value their input. Likewise, I'm interested in hearing from you. Let me know how this book works for you and feel free to share any feedback you may have on how to improve this text. I may be reached by e-mail by clicking the Contact the Author icon on the text's website: **college.hmco.com/pic/wilsonFM1e.**

To the Instructor

Thank you for selecting this book. I believe you will find its approach refreshing and its content interesting to you and your students. It is written specifically for students pursuing business, social science, or a related field. Many of these students do not enjoy mathematics and are taking this course only because it is required for their major. (Many business schools require a finite math—applied calculus—statistics sequence.) Several features are included in this text to make the course content more accessible to these students including:

- an informal writing style that addresses key concepts appropriately without becoming bogged down in mathematical jargon

- examples and exercises throughout the text based on interesting and engaging real-life data
- *Make It Real* projects which allow students to collect and analyze data relevant to their personal lives
- detailed *Technology Tips* (including screenshots) which teach students how to use the TI-83 Plus calculator as a tool to analyze real-life data
- a bare-bones approach to course content. Topics covered in detail include those that are most relevant to everyday people

I believe that candid feedback from colleagues is helpful in enhancing teaching effectiveness. Please feel free to contact me with any recommendations, comments, or other feedback that you feel will enhance the effectiveness of future editions of this text. I may be reached by e-mail by clicking the Contact the Author icon on the text's website: **college.hmco.com/pic/wilsonFM1e.**

Disclaimer

In this book, I have attempted to incorporate real-world data from the financial markets to the medical field. In each case, I have done my best to present the data accurately and interpret the data realistically. However, I do not claim to be an expert in financial, medical, and other similar fields. My interpretations of real-world data and my associated conclusions may not adequately consider all relevant factors. Therefore, readers are encouraged to seek professional advice from experts in the appropriate fields before making decisions related to the topics addressed herein.

Despite the usefulness of mathematical models as representations of real-world data sets, most mathematical models have a certain level of error. It is common for model results to differ from raw data set values. Consequently, conclusions drawn from a mathematical model may differ (sometimes dramatically) from conclusions drawn by looking at raw data sets. Readers are encouraged to interpret model results with this understanding.

Acknowledgments

This textbook would not have been possible without the contributions of many colleagues. I greatly appreciate all of the people who contributed time and talent to bring this book to fruition.

The feedback from the following reviewers was invaluable and helped to shape the final form of the text: Bill Ardis, *Collin County Community College, TX;* James J. Ball, *Indiana State University, IN;* Michael L. Berry, *West Virginia Wesleyan College, WV;* Marcelle Bessman, *Jacksonville University, FL;* Mike Bosch, *Iowa Lakes Community College, IA;* Emily Bronstein, *Prince George's Community College, MD;* Dean S. Burbank, *Gulf Coast Community College, FL;* Andra Buxkemper, *Bunn College, TX;* Roxanne Byrne, *University of Colorado—Denver, CO;* Scott A. Clary, *Florida Institute of Technology, FL;* David Collingwood, *University of Washington, WA;* Mark A. Crawford Jr., *Western Michigan University, MI;* Khaled Dib, *University of Minnesota Duluth, MN;* Lance D. Drager, *Texas Tech University, TX;* Klara Grodzinsky, *Georgia Institute of Technology, GA;* Lucy L. Hanks, *Virginia Polytechnic Institute and State University, VA;* Jean B. Harper, *State Univiversity of N.Y. – College at Fredonia, NY;* Kevin M. Jenerette, *Coastal Carolina University, SC;* Cynthia Kaus, *Metropolitan State University, MN;* Michael LaValle, *Rochester Community and Technical College, MN;* Roger D. Lee, *Salt Lake Community College, UT;* Lia Liu, *Univer-*

sity of Illinois Chicago, IL; Alan Mabry, *Unitersity of Texas at El Paso;* Quincy Magby, *Arizona Western College, AZ;* Mary M. Marco, *Bucks County Community College, PA;* Nicholas Martin, *Shepherd College, WV;* William C. McClure, *Orange Coast College, CA;* James McGlothin, *Lower Columbia College, WA;* Victoria Neagoe, *Goldey Beacom College;* David W. Nelson, *Green River Community College;* Ralph W. Oberste-Vorth, *Marshall University, WV;* Armando I. Perez, *Laredo Community College;* Cyril Petras, *Lord Fairfax Community College, VA;* Mihaela Poplicher, *University of Cincinnati;* John E. Porter, *Murray State University, KY;* David W. Roach, *Murray State University, KY;* R. A. Rock, *Daniel Webster College, NH;* Arthur Rosenthal, *Salem State College, MA;* Kimmo I. Rosenthal, *Union College, NY;* Sharon Mayhew Saxton, *Cascadia Community College, WA;* Edwin Shapiro, *University of San Francisco, CA;* Denise Szecsei, *Stetson University;* Abolhassan S. Taghavy, *Richard J. Daley College, IL;* Muhammad Usman, *University of Cincinnati, OH;* Jorge R. Viramontes Olivas, *University of Texas at El Paso, TX;* Beverly Vredevelt, *Spokane Falls Community College, WA;* Michael L. Wright, *Cossatot Community College, AR.*

I extend special thanks to Cindy Harvey, who provided detailed recommendations on how to improve an early draft of this text, to Helen Medley, who reviewed the final draft text for accuracy, and to Paul Lorczak, who conducted the accuracy review for the typeset manuscript. Their suggestions greatly enhanced the quality of the text.

My appreciation goes to Donnie Hallstone, who worked many of the solutions to the probability and statistics portion of the text.

I appreciate the work of my exceptional student, Jon Austin, who developed some of the mathematical models used in the text. I value the recommendations from the math instructors who class-tested this book, including:

Beverly A. Vredevelt, *Spokane Falls Community College*
Frank "Bud" Wright, *Green River Community College*
David Nelson, *Green River Community College*
Keith Alford, *Green River Community College*

Additional comments from students using a draft copy of the text at Green River Community College and Spokane Falls Community College were also helpful.

I'm thankful for Erica Carlson, who assisted with the art manuscript and for Uli Gersiek, who created most of the artwork in the text.

The Houghton Mifflin Company team was phenomenal to work with. I extend gratitude to Lauren Schultz, who first believed in the project; Jennifer King and Marika Hoe, who kept things moving; Kasey McGarrigle, who provided behind the scenes support; and Tamela Ambush and Sage Anderson, who ensured the quality of the published text. Additionally, I'm grateful for the management and support staff at Houghton Mifflin who made the production of this text a reality.

I also thank Texas Instruments for providing a way to enhance student learning by creating the TI-83 Plus calculator. Without it, the Technology Tips would not have been possible.

On a personal note, I could not have written this text without the tireless support of my wife, Shelley Wilson. She worked overtime caring for our home and children for the three years I worked on the text. Despite this challenge, she never faltered in offering me encouragement, support, and love. For this, I am eternally grateful and indebted to her.

Frank C. Wilson

Supplements for the Instructor

Eduspace®

Powerful, customizable, and interactive, Eduspace, powered by Blackboard®, is Houghton Mifflin's online learning tool. By pairing the widely recognized tools of Blackboard with quality, text-specific content from Houghton Mifflin, Eduspace makes it easy for instructors to create all or part of a course online. Homework exercises, quizzes, tests, tutorials, and supplemental study materials all come ready-to-use or can be modified by the instructor. Visit **www.eduspace.com** for more information.

HM ClassPrep™ with HM Testing (powered by Diploma™)

HM ClassPrep offers text-specific resources for the instructor. *HM Testing* offers instructors a flexible and powerful tool for test generation and test management. Now supported by the Brownstone Research Group's market-leading *Diploma*™ software, this new version of *HM Testing* significantly improves ease of use.

Blackboard®, WebCT®, and eCollege®

Houghton Mifflin can provide you with valuable content to include in your existing Blackboard, WebCT, and eCollege systems. This text-specific content enables instructors to teach all or part of their course online.

Supplements for the Student

Eduspace®

Powerful, customizable, and interactive, Eduspace®, powered by Blackboard®, is Houghton Mifflin's online learning tool for instructors and students. Eduspace is a text-specific, web-based learning environment that your instructor can use to offer students a combination of practice exercises, multimedia tutorials, video explanations, online algorithmic homework, and more. Specific content is available 24 hours a day to help you succeed in your course.

SMARTHINKING®

Houghton Mifflin's unique partnership with SMARTHINKING brings students real-time, online tutorial support when they need it most. Using state-of-the-art whiteboard technology and feedback tools, students interact and communicate with "e-structors." Tutors guide students through the learning and problem solving process without providing answers or rewriting a student's work. Visit **www.smarthinking.com** for more information.

HM mathSpace® Student Tutorial CD-ROM

For students who prefer the portability of a CD-ROM, this tutorial provides opportunities for self-paced review and practice with algorithmically generated exercises and step-by-step solutions.

Houghton Mifflin Instructional DVDs

Hosted by Dana Mosely and professionally produced, these DVDs cover topics in the text. Ideal for promoting individual study and review, these comprehensive DVDs also support students in online courses or those who have missed a lecture.

Online Learning Center and Online Teaching Center

The free Houghton Mifflin websites at **college.hmco.com/pic/wilsonFM1e** contain an abundance of resources.

Student Solutions Guide (ISBN 0-618-33300-2) contains solutions to the odd-numbered exercises from all Exercise Sets in the book.

Excel Guide for Finite Mathematics and Applied Calculus by Revathi Narasimhan (ISBN 0-618-67691-0) provides lists of exercises from the text that can be completed after each stepped out Excel example. No prior knowledge of Excel is necessary.

Chapter 1

Functions and Linear Models

Mathematical functions are a powerful tool used to model real-world phenomena. Whether simple or complex, functions give us a way to forecast expected results. Remarkably, anything that has a constant rate of change may be accurately modeled with a linear function. For example, the cost of filling your car's gas tank is a linear function of the number of gallons purchased.

1.1 **Functions**
- Distinguish between functions and nonfunctions in tables, graphs, and words
- Use function notation
- Graph functions using technology
- Determine the domain of a function

1.2 **Linear Functions**
- Calculate and interpret the meaning of the slope of a linear function
- Interpret the physical and graphical meaning of *x*- and *y*-intercepts
- Formulate the equation of a line given two points
- Recognize the slope-intercept, point-slope, and standard forms of a line

1.3 **Linear Models**
- Use technology to model linear and near-linear data
- Use a linear equation to describe the relationship between directly proportional quantities
- Recognize and model naturally occurring linear, near-linear, and piecewise linear relationships

1.1 Functions

- Distinguish between functions and nonfunctions in tables, graphs, and words
- Use function notation
- Graph functions using technology
- Determine the domain of a function

GETTING STARTED Our society is a complex system of relationships between people, places, and things. Many of these relationships are interconnected. In mathematics, we often model the relationship between two or more interdependent quantities by using a **function.** In this section, we will show how to distinguish between functions and nonfunctions and will practice using function notation. We will also demonstrate how to use technology to draw a function graph, and discuss how to find the domain of a function.

DEFINITION: FUNCTION

A function is a rule that associates each input with *exactly one* output.

Often the rule is represented in a table of data with the inputs on the left-hand side and the outputs on the right-hand side. For example, the amount of money we pay to fill up our gas tank is a function of the number of gallons pumped (see Table 1.1).

TABLE 1.1

Gallons Pumped	Total Fuel Cost
10	$15.99
15	$23.99
20	$31.98

In this case, the input to the function is *gallons pumped* and the output of the function is *total fuel cost.* Fuel cost is a function of the number of gallons pumped because each input has exactly one output.

Similarly, the weekly wage of a service station employee is a function of the number of hours worked. The *number of hours worked* is the input to the function, and the *weekly wage* is the output of the function. Since weekly wage is a function of the number of hours worked, an employee who works 40 hours expects to be paid the same wage each time she works that amount of time.

EXAMPLE 1 **Determining If a Table of Data Represents a Function**

A car dealer tracks the number of blue cars in each of three shipments and records the data in a table (see Table 1.2). Is the number of blue cars a function of the number of cars in the shipment?

TABLE 1.2

Number of Cars in the Shipment	Number of Blue Cars
22	6
24	7
24	5

same input (24, 24) — different outputs (7, 5)

SOLUTION According to the definition of a function, each input must have exactly one output. The input value 24 has two different outputs: 5 and 7. Since the input 24 has more than one output, the number of blue cars is *not* a function of the number of cars in the shipment.

Function Notation

When we encounter functions in real life, they are often expressed in words. To make functions easier to work with, we typically use symbolic notation to represent the relationship between the input and the output. Let's return to the fuel cost table introduced previously (Table 1.3).

TABLE 1.3

Gallons Pumped	Total Fuel Cost
10	$15.99
15	$23.99
20	$31.98

Observe that the fuel cost is equal to $1.599 times the number of gallons pumped. We represent this symbolically as $C(g) = 1.599g$, where $C(g)$ represents the total fuel cost when g gallons are pumped. [$C(g)$ is read "C of g"]. The letter C is used to represent the name of the rule, and the letter g in the parentheses indicates that the rule works with different values of g (see Figure 1.1).

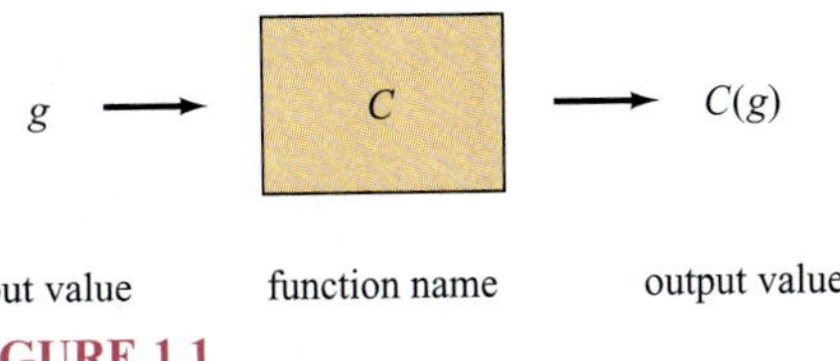

FIGURE 1.1

We call the output variable of a function the **dependent variable** because the value of the output variable *depends* upon the value of the input variable. The input variable is called the **independent variable.** (One way to remember the meaning of the terms is to observe that both ***in**put* and ***in**dependent* begin with ***in**.*) From the table, we see that

$$C(10) = 15.99$$
$$C(15) = 23.99$$
$$C(20) = 31.98$$

For this function, the independent variable took on the values 10, 15, and 20, and the dependent variable assumed the values 15.99, 23.99, and 31.98.

EXAMPLE 2 Determining a Linear Model from a Verbal Description

An electronics store employee earns \$8.50 per hour. Write an equation for the employee's earnings as a function of the hours worked. Then calculate the amount of money the employee earns (in dollars) by working 30 hours.

SOLUTION Since the employee earns \$8.50 for each hour worked, the employee's total earnings are equal to \$8.50 times the total number of hours worked. That is,

$$E(h) = 8.50h$$

where E is the employee's earnings (in dollars) and h is the number of hours worked. To calculate the amount of money earned by working 30 hours, we evaluate this function at $h = 30$.

$$\begin{aligned} E(30) &= 8.5(30) \\ &= 255 \end{aligned}$$

The employee earns \$255 for 30 hours of work.

Function notation is extremely versatile. Suppose we are given the function $f(x) = x^2 - 2x + 1$. We may evaluate the function using either numerical values or nonnumerical values. For example,

$$\begin{aligned} f(2) &= (2)^2 - 2(2) + 1 \\ &= 4 - 4 + 1 \\ &= 1 \end{aligned} \qquad f(\triangle) = (\triangle)^2 - 2(\triangle) + 1$$

$$\begin{aligned} f(a + 2) &= (a + 2)^2 - 2(a + 2) + 1 \\ &= (a^2 + 4a + 4) - 2a - 4 + 1 \\ &= a^2 + 2a + 1 \end{aligned}$$

In each case, we replaced the value of x in the function $f(x) = x^2 - 2x + 1$ with the quantity in the parentheses. Whether the independent variable value was 2, $\triangle$, or $a + 2$, the process was the same.

EXAMPLE 3 Evaluating a Function Using Function Notation

Evaluate the function $s(t) = t^3 + 4t$ at $t = 3$, $t = \triangle$, and $t = a^2$.

SOLUTION

$$\begin{aligned} s(3) &= (3)^3 + 4(3) \\ &= 27 + 12 \\ &= 39 \end{aligned} \qquad s(\triangle) = (\triangle)^3 + 4(\triangle)$$

$$\begin{aligned} s(a^2) &= (a^2)^3 + 4(a^2) \\ &= a^6 + 4a^2 \end{aligned}$$

Graphs of Functions

Functions are represented visually by plotting points on a Cartesian coordinate system (see Figure 1.2). The horizontal axis shows the value of the independent variable (in this case, x), and the vertical axis shows the value of the dependent variable (in this case, y).

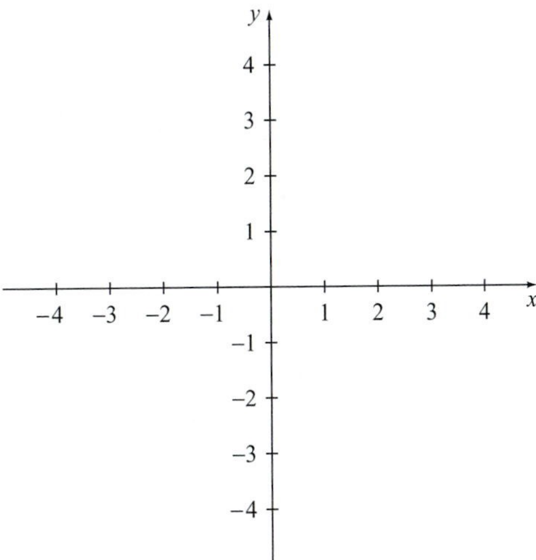

FIGURE 1.2

When using the coordinate system, y is typically used in place of the function notation $f(x)$. That is, $y = f(x)$. This is true even if the function has a name other than f.

The point of intersection of the horizontal and vertical axes is referred to as the **origin** and is represented by the ordered pair $(0, 0)$. To graph an ordered pair (a, b), we move from the origin $|a|$ units horizontally and $|b|$ units vertically and draw a point. If $a > 0$, we move to the right. If $a < 0$, we move to the left. Similarly, if $b > 0$, we move up, and if $b < 0$, we move down. For example, consider the table of values with its associated interpretation in Table 1.4 and the graph in Figure 1.3.

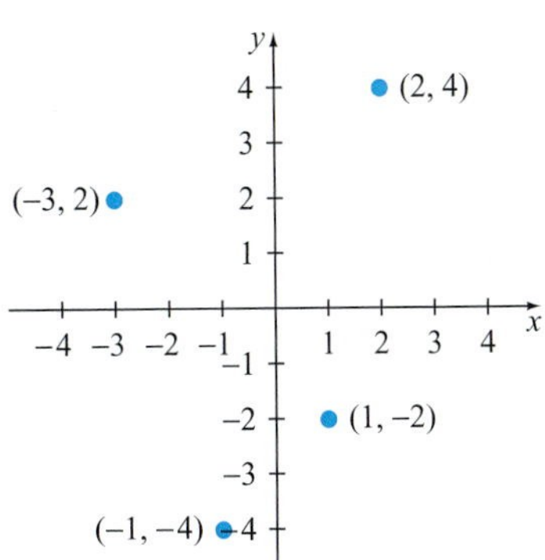

FIGURE 1.3

TABLE 1.4

x	y
-3	2
-1	-4
1	-2
2	4

Horizontal	Vertical
left 3	up 2
left 1	down 4
right 1	down 2
right 2	up 4

When we are given the equation of a function, we can generate a table of values and then plot the corresponding points. Once we have drawn a sufficient number of points to be able to determine the basic shape of the graph, we typically connect the points with a smooth curve. For example, the function $y = x^3 - 9x$ has the table of values and graph shown in Figure 1.4.

x	y
-4	-28
-3	0
-2	10
-1	8
0	0
1	-8
2	-10
3	0
4	28

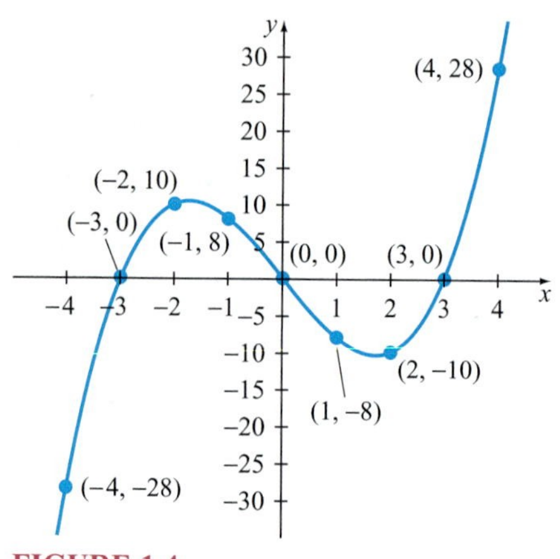

FIGURE 1.4

EXAMPLE 4

Estimating Function Values from a Graph

Estimate $f(-3)$ and $f(2)$ using the graph of $f(x) = x^3 - 16x$ shown in Figure 1.5.

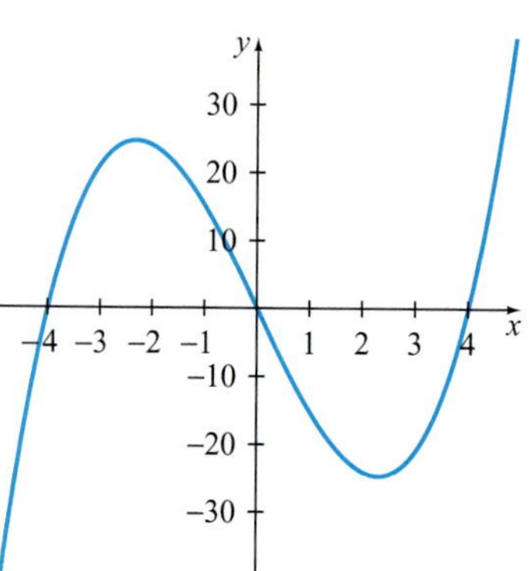

FIGURE 1.5

FIGURE 1.6

SOLUTION It appears from Figure 1.6 that $f(-3) \approx 20$ and $f(2) \approx -25$. Calculating these values with the algebraic equation, we see that

$$\begin{aligned} f(-3) &= (-3)^3 - 16(-3) \\ &= -27 + 48 \\ &= 21 \end{aligned} \quad \text{and} \quad \begin{aligned} f(2) &= (2)^3 - 16(2) \\ &= 8 - 32 \\ &= -24 \end{aligned}$$

One drawback of using a graph to determine the values of a function is that it is difficult to be precise. For this reason, algebraic methods are typically preferred when precision is important.

Not all data sets represent functions. If any value of the independent variable is associated with more than one value of the dependent variable, the table of data and its associated graph will not represent a function. For example, consider $y = \pm\sqrt{2x}$ (see Figure 1.7).

x	y
0	0
0.5	−1
0.5	1
2	−2
2	2
4.5	−3
4.5	3
8	−4
8	4

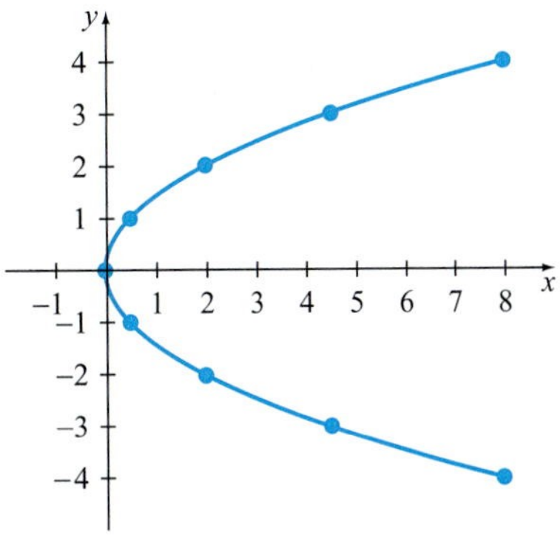

FIGURE 1.7

Each positive value of x is associated with two different values of y. We can easily determine this from the graph by observing that a vertical line drawn through any positive value of x will cross the graph twice. This observation leads us to the Vertical Line Test.

VERTICAL LINE TEST

If every vertical line drawn on a graph intersects the graph in at most one place, then the graph is the graph of a function. Otherwise, the graph is not the graph of a function.

EXAMPLE 5

Determining If a Graph Represents a Function

The graph of $y = x^2$ is shown in Figure 1.8. Does the graph represent a function?

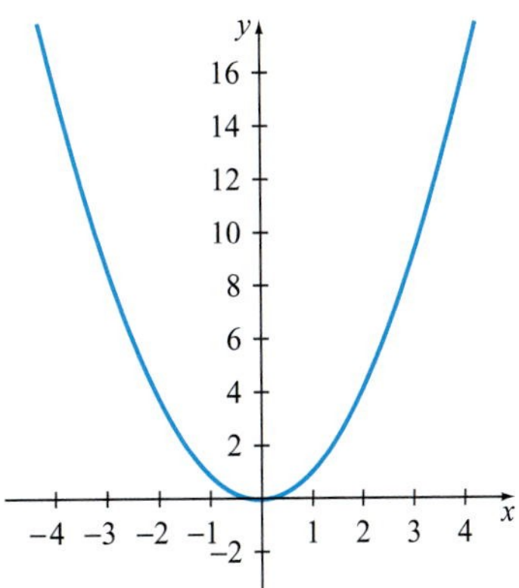

FIGURE 1.8

SOLUTION

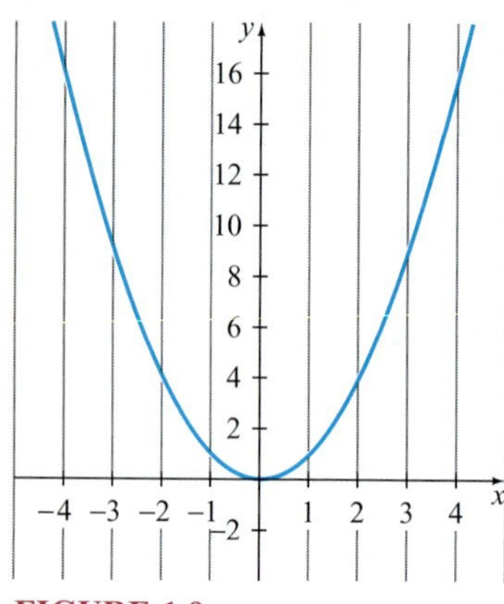

FIGURE 1.9

Each of the vertical lines drawn crosses the graph in exactly one place, as shown in Figure 1.9. Therefore, the graph passes the Vertical Line Test and y is a function of x.

EXAMPLE 6

Determining If a Graph Represents a Function

The graph of $y = \pm\sqrt{x}$ is shown in Figure 1.10. Does the graph represent a function?

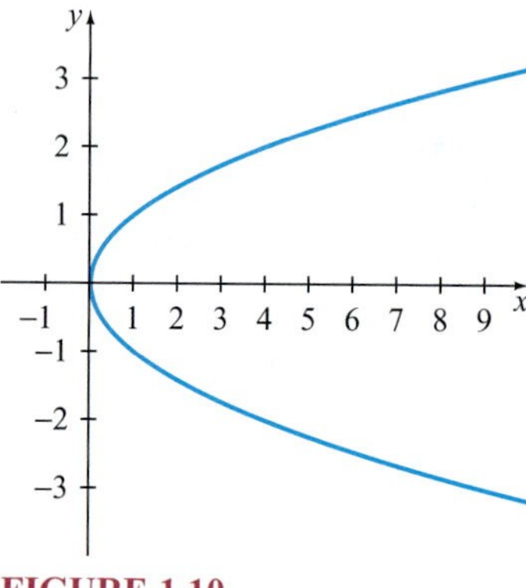

FIGURE 1.10

SOLUTION Notice that many of the vertical lines in Figure 1.11 cross the graph in two places. The graph does not pass the Vertical Line Test, so y is not a function of x. Each positive value of x corresponds with two values of y.

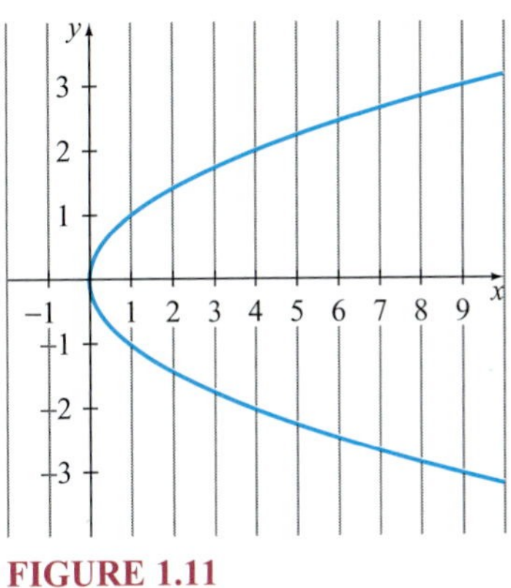

FIGURE 1.11

The graphs shown thus far have been continuous. That is, each graph could be drawn without lifting the pencil. However, some function graphs are *discontinuous*. A discontinuous graph has a break in the graph. The break in the graph is referred to as a **discontinuity.**

EXAMPLE 7

Determining If a Discontinuous Graph Represents a Function

The graph of $h(x) = \begin{cases} x \text{ if } x \le 1 \\ 2 \text{ if } x > 1 \end{cases}$ is shown in Figure 1.12. Determine if the graph represents a function.

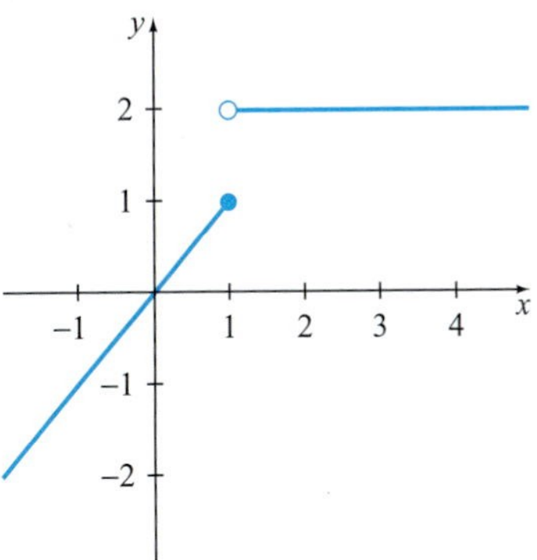

FIGURE 1.12

SOLUTION The graph is discontinuous at $x = 1$. The value of the function at $x = 1$ is 1. [The open dot at $(1, 2)$ indicates that the point $(1, 2)$ is not part of the graph.] Although the graph is discontinuous, each value of the independent variable is associated with exactly one value of the dependent variable. Therefore, the graph represents a function.

Using Technology to Graph a Function

Graphing calculators, such as the TI-83 Plus, can quickly draw the graph of a function as shown in the following Technology Tip.

TECHNOLOGY TIP

Graphing a Function

1. Bring up the graphing list by pressing the Y= button.

```
Plot1 Plot2 Plot3
\Y1=
\Y2=
\Y3=
\Y4=
\Y5=
\Y6=
\Y7=
```

2. Type in the function using the X,T,θ,n button for the variable and the ^ button to place an expression in an exponent. Make sure you use parentheses as needed.

```
Plot1 Plot2 Plot3
\Y1=X^3-16X
\Y2=
\Y3=
\Y4=
\Y5=
\Y6=
\Y7=
```

3. Specify the size of the viewing window by pressing the WINDOW button and editing the parameters. The `Xmin` is the minimum x value, `Xmax` is the maximum x value, `Ymin` is the minimum y value, and `Ymax` is the maximum y value. The `Xscl` and `Yscl` are used to specify the spacing of the tick marks on the graph.

```
WINDOW
 Xmin=-5
 Xmax=5
 Xscl=1
 Ymin=-45
 Ymax=45
 Yscl=5
 Xres=1
```

4. Draw the graph by pressing the GRAPH button.

Domain and Range

We are often interested in the set of all possible values of the independent variable and the set of all possible values of the dependent variable.

DOMAIN AND RANGE

The set of all possible values of the independent variable of a function is called the **domain.** The set of all possible values of the dependent variable of a function is called the **range.**

It is easy to remember the meaning of the terms if we observe that ***input, independent variable,*** and ***domain*** all contain the word *in.*

Consider a common kitchen blender. If we put an orange into a blender and turn the blender on, the orange is transformed into orange juice. We say that *orange* is in the domain of the blender function and *orange juice* is in the range of the blender function. On the other hand, if we put a rock into the blender and turn it on, the blender will self-destruct. We say that *rock* is not in the domain of the blender function, since blenders are unable to process rocks.

Finding the domain of a function frequently involves solving an equation or an inequality. Recall that *solving an equation* means finding the value of the variable that makes the equation a true statement.

The domain of most frequently used mathematical functions is the set of all real numbers. However, there are three common situations in which the domain of a function is restricted to a subset of the real numbers. They are:

1. A zero in the denominator
2. A negative value under a square root symbol (radical)
3. The context of a word problem

Let's look at an example for each of the situations.

EXAMPLE 8

Determining the Domain of a Function

What is the domain of $g(x) = \frac{3x - 1}{2x + 6}$?

SOLUTION We know that the value in the denominator must be nonzero, since division by zero is undefined. That is, $2x + 6 \neq 0$. To find the value of x that must be excluded from the domain, we must solve the following equation:

$$2x + 6 = 0$$

$$2x + 6 - 6 = 0 - 6 \qquad \text{Subtract 6 from both sides}$$

$$2x = -6$$

$$\frac{2x}{2} = \frac{-6}{2} \qquad \text{Divide both sides by 2}$$

$$x = -3$$

The domain of the function is all real numbers except -3. We may rewrite the function equation with the domain restriction as follows:

$$g(x) = \frac{3x - 1}{2x + 6}, \qquad x \neq -3$$

EXAMPLE 9

Determining the Domain of a Function

What is the domain of $y = \sqrt{x - 3}$?

SOLUTION We know that the value underneath the radical must be nonnegative, since the square root of a negative number is undefined in the real number system. Therefore, $x - 3 \geq 0$. Solving for x, we get

$$x - 3 \geq 0$$

$$x - 3 + 3 \geq 0 + 3 \quad \text{Add 3 to both sides}$$

$$x \geq 3$$

The domain of the function is all real numbers greater than or equal to 3.

EXAMPLE 10

Determining the Domain of a Function

The revenue R from the sale of x gallons of gasoline is given by the equation

$$R(x) = 1.379x$$

What is the domain of the function?

SOLUTION In the context of this problem, it doesn't make sense to talk about selling a negative number of gallons of gas. So $x \geq 0$. The domain of the function is all nonnegative real numbers.

1.1 Summary

In this section, you learned to distinguish between functions and nonfunctions, and you practiced using function notation. You learned how to use technology to graph a function, and you discovered how to find the domain of a function. Mastering each of these techniques will help you understand the subsequent concepts covered in this chapter.

1.1 Exercises

In Exercises 1–6, determine whether the output is a function of the input.

1. $W(a)$ = your weight in pounds when you were a years old.

2. $S(n)$ = your score on test number n in a finite math course. (Assume you could only take the test once.)

3.

Time of Day	Temperature (°F)
11:00 a.m.	68
1:00 p.m.	73
3:00 p.m.	75

4.

Time of Day	Temperature (°F)
10:00 a.m.	66
12:00 p.m.	74
2:00 p.m.	74

5.

Fish Caught	Salmon in Catch
169	24
182	32
182	47

6.

Fish Caught	Salmon in Catch
252	74
276	74
301	92

In Exercises 7–13, calculate the value of the function at the designated input and interpret the result.

7. $C(x) = 39.95x$ at $x = 4$, where $C(x)$ is the cost of buying x pairs of shoes.

8. $C(x) = 29.95x + 200$ at $x = 300$, where $C(x)$ is the cost of making x pairs of shoes.

9. $H(t) = -16t^2 + 120$ at $t = 2$, where $H(t)$ is the height of a cliff diver above the water t seconds after he jumped from a 120-foot cliff.

10. **Revenue** Find $R(t)$ at $t = 4$, where $R(t)$ is the quarterly revenue for an international tortilla producer t quarters after December 1999 as shown below.

Quarter	Revenue (in millions of dollars)
1	460.0
2	452.0
3	474.0
4	483.8
5	466.0

Source: Gruma, S.A. de C.V.

11. **Earnings per Share** Find $E(t)$ at $t = 4$, where $E(t)$ is the quarterly earnings per share for an international tortilla producer t quarters after December 1999.

Quarter	Earnings per Share (in dollars)
1	0.13
2	−0.11
3	0.15
4	0.06
5	0.03
6	0.06

Source: Gruma, S.A. de C.V.

12. **Stock Price** Find $P(t)$ at $t = 3$, where $P(t)$ is the stock price of an e-learning company at the end of the day and t is the number of days after November 4, 2001.

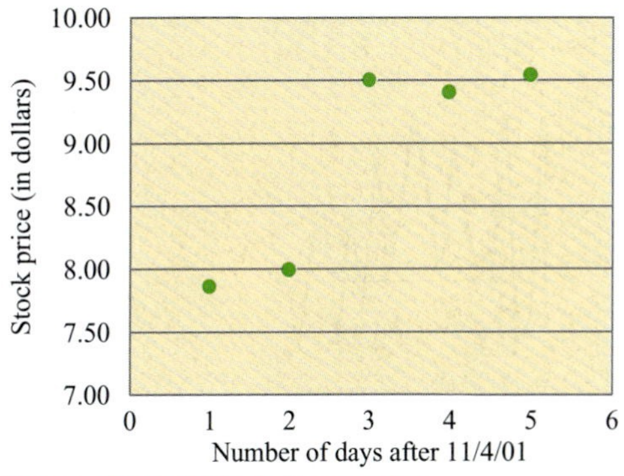

Source: Digital Think Corporation.

13. **Stock Price** Find $P(t)$ at $t = 4$, where $P(t)$ is the stock price of a computer company at the end of the day and t is the number of days after November 4, 2001.

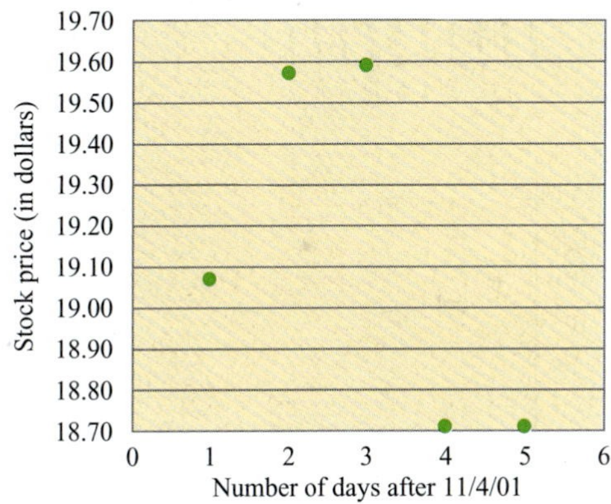

Source: Apple Computer Corporation.

In Exercises 14–19, determine whether the graphs represent functions by applying the Vertical Line Test.

14. 15.

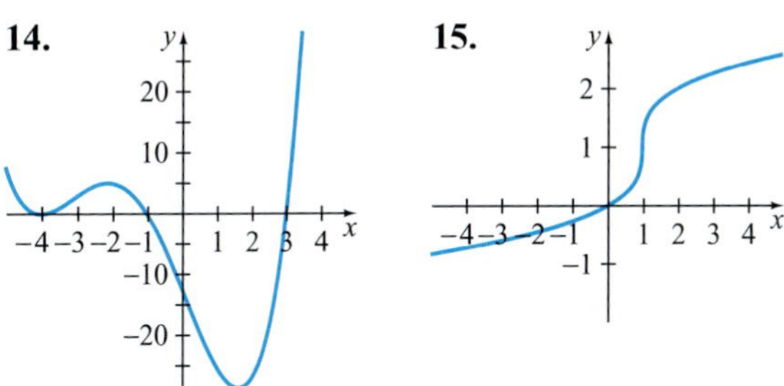

16.

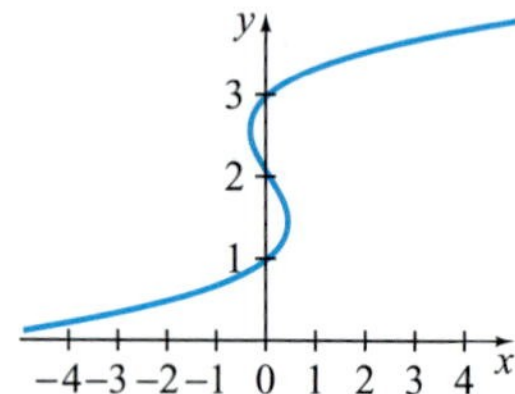

17.

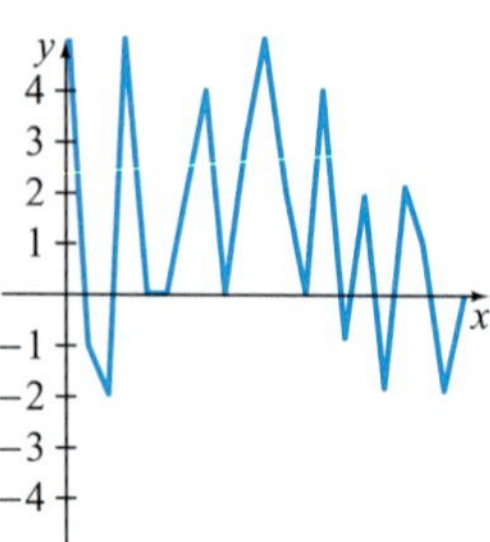

18.

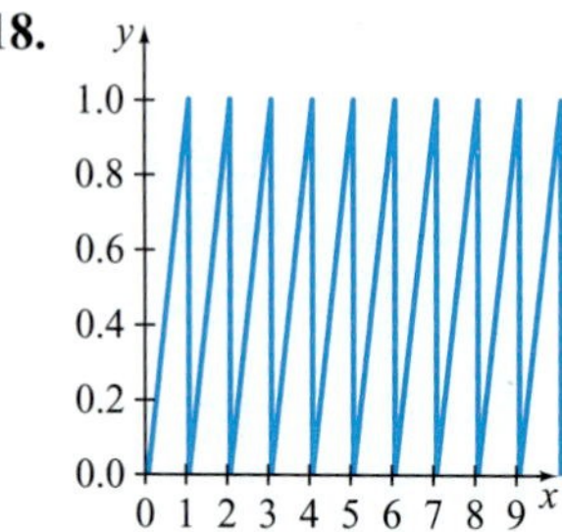

19.

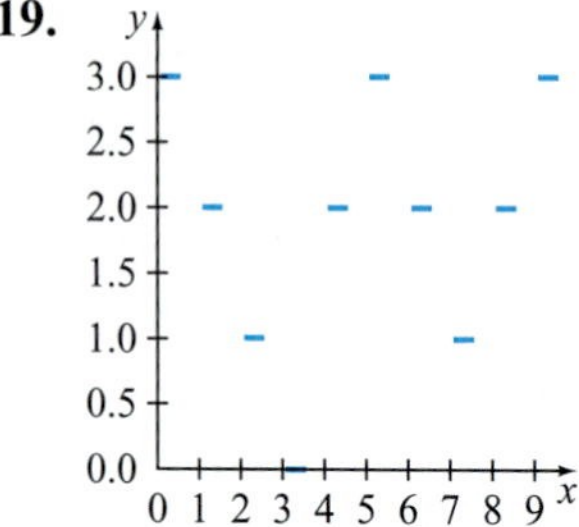

In Exercises 20–23, graph the function on your graphing calculator, using the specified viewing window. Note that $a \le x \le b$ means x min $= a$ and x max $= b$. Similarly, $c \le y \le d$ means y min $= c$ and y max $= d$.

 **20.** $y = x^2 - 5x$; $-3 \le x \le 7$, $-8 \le y \le 8$

 **21.** $y = -x^2 - 4$; $-3 \le x \le 3$, $-10 \le y \le 1$

 **22.** $y = -x + 2$; $-3 \le x \le 3$, $-2 \le y \le 6$

23. $y = -2x^2 + 1$; $-2 \le x \le 2$, $-3 \le y \le 2$

In Exercises 24–27, use the graph to estimate the value of the function at the indicated x value. Then calculate the exact value algebraically.

24. $y = 0.25(x + 1)^2(x - 3)$; $x = 4$

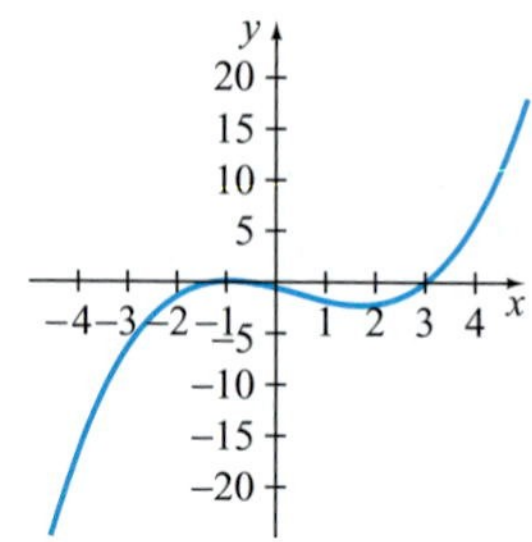

25. $y = 5|x| - x^2$; $x = 1$

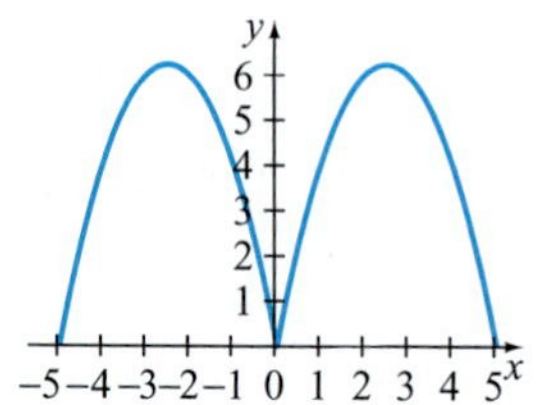

26. $y = \dfrac{|x + 2|}{x^2 + 1}$; $x = 3$

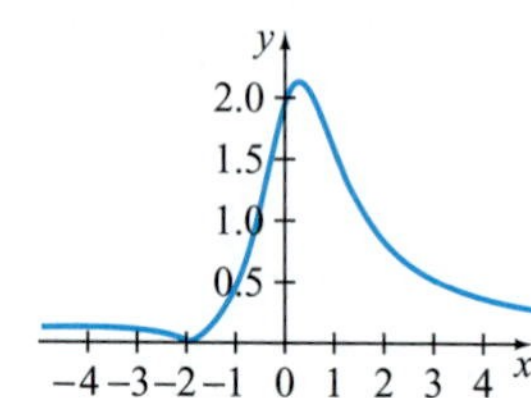

27. $y = -\dfrac{4(x^2 - 4)}{x^3 + 2x^2 + 3x + 6}$; $x = -2$

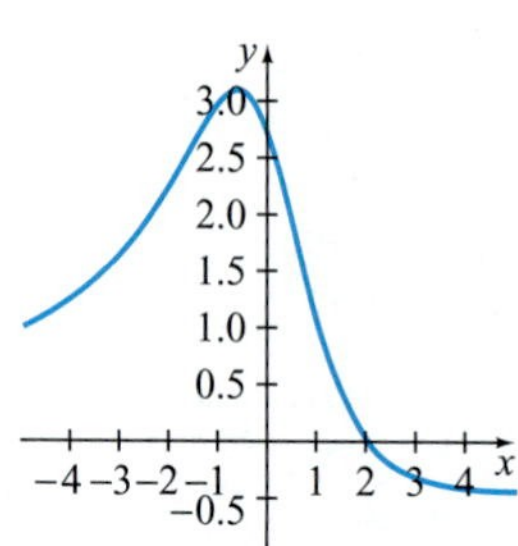

In Exercises 28–41, determine the domain of the function.

28. $f(x) = 5x^3 - 5x$

29. $f(p) = p^2 - 2p + 1$

30. $C(r) = \dfrac{r^2 - 1}{r + 1}$

31. $h(r) = \dfrac{r^2 - 1}{r^2 + 1}$

32. $S(t) = \dfrac{t - 3}{4t + 1}$

33. $g(t) = \dfrac{3t - 3}{4t - 4}$

34. $H(x) = \sqrt{x - 5}$

35. $f(a) = a^3 - \sqrt{a + 1}$

36. $f(x) = \dfrac{\sqrt{x + 5}}{x + 2}$

37. $f(x) = \dfrac{\sqrt{2x + 6}}{x^2 + 3}$

38. $C(n)$ = the cost of buying n apples

39. $P(n)$ = the profit earned from the sale of n bags of candy

40. $W(n)$ = the average birth weight of someone born in year n

41. $H(n)$ = the average height of someone born in year n

Exercises 42–46 are intended to challenge your understanding of functions and graphs.

42. Graph the function $f(x) = \frac{x - 1}{x^2 - 1}$ on your calculator. What do you think is the domain of f?

43. Determine the domain of $f(x) = \frac{x - 1}{x^2 - 1}$ algebraically and compare your answer to the solution of Exercise 42.

44. Are all lines functions? Explain.

45. Does the table of data represent a function? Explain.

x	y
0	5
1	6
1	6
3	0
4	2

46. For a given date and location, is air temperature a function of the time of day? Explain.

1.2 Linear Functions

- Calculate and interpret the meaning of the slope of a linear function
- Interpret the physical and graphical meaning of x- and y-intercepts
- Formulate the equation of a line given two points
- Recognize the slope-intercept, point-slope, and standard forms of a line

GETTING STARTED The cost of filling our car's gas tank, the number of calories we consume by eating a few bags of fruit snacks, and the amount of sales tax we pay when we buy new clothes are examples of linear functions. In this section, we will show how to calculate the slope of a linear function. We will discuss how to interpret the physical and graphical meaning of slope, x-intercept, and y-intercept. We will also show three different ways to write a linear equation and demonstrate how to find the equation of a linear function from a data set.

DEFINITION: GRAPH OF A LINEAR FUNCTION

The line passing through any two points (x_1, y_1) and (x_2, y_2) with $x_1 \neq x_2$ is referred to as the **graph of a linear function.**

Linear functions are characterized by a constant rate of change. That is, increasing a domain value by one unit will always change the corresponding range value by a constant amount. The converse is also true. Any table of data with a constant rate of change represents a linear function. The constant rate of change is referred to as the **slope** of the linear function.

DEFINITION: SLOPE

The slope of a linear function is the change in the output that occurs when the input is increased by one unit. The slope m may be calculated by dividing the difference of two outputs by the difference in the corresponding inputs. That is,

$$m = \frac{y_2 - y_1}{x_2 - x_1}$$

where (x_1, y_1) and (x_2, y_2) are data points of the linear function.

The number of calories C in n bags of fruit snacks is shown in Table 1.5.

TABLE 1.5

Bags (n)	Calories (C)
1	100
2	200
3	300

Source: Fruit Smiles 9-oz. box label

Notice that the calorie count increases by 100 calories for each additional bag. Since the dependent variable C is changing at a constant rate (100 calories per bag), the data may be modeled by a linear function. In this case, the linear function is $C = 100n$.

To calculate the slope of the function, we may use any two data points. Using the data points $(x_1, y_1) = (1, 100)$ and $(x_2, y_2) = (3, 300)$, the slope is

$$\begin{aligned} m &= \frac{300 - 100 \text{ calories}}{3 - 1 \text{ bags}} \\ &= \frac{200}{2} \frac{\text{calories}}{\text{bags}} \\ &= 100 \text{ calories per bag} \end{aligned}$$

A one-bag increase in the input results in a 100-calorie increase in the output.

Will we get the same result if we use different points? Let's check using (1, 100) and (2, 200) (see Figure 1.13).

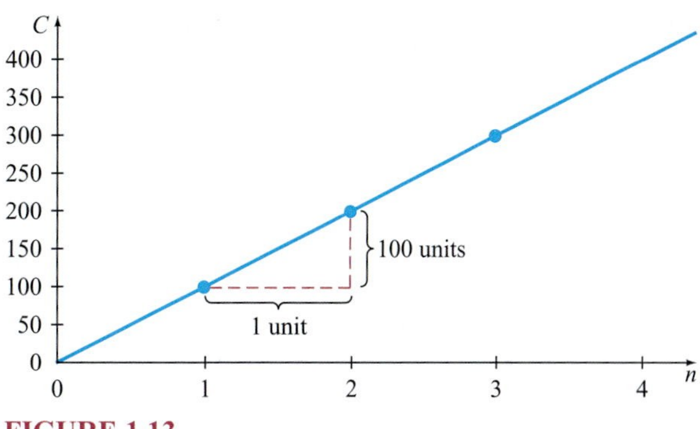

FIGURE 1.13

UNITS

$$m = \frac{100 - 200 \text{ calories}}{1 - 2 \text{ bags}}$$
$$= \frac{-100}{-1} \frac{\text{calories}}{\text{bag}}$$
$$= 100 \text{ calories per bag}$$

The result is the same! With linear functions, we may use any two points to calculate the slope.

Not all lines are functions. Vertical lines fail to pass the Vertical Line Test, so they are not functions. As shown in Example 1, vertical lines have an undefined slope.

EXAMPLE 1

Determining the Slope of a Line from Two Points on the Line

What is the slope of the line going through (2, 4) and (2, 8) (Figure 1.14)?

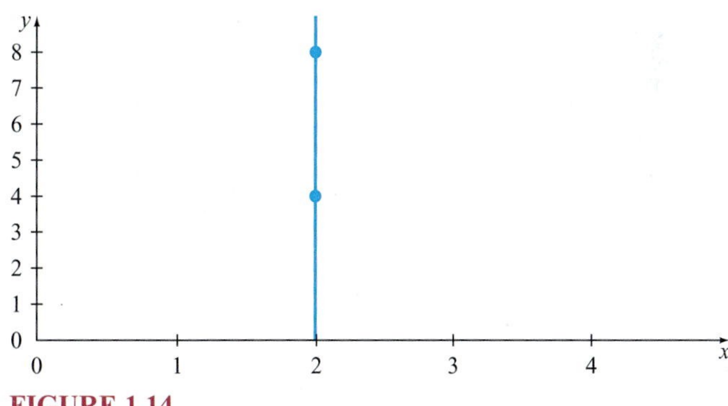

FIGURE 1.14

SOLUTION

$$m = \frac{8 - 4}{2 - 2}$$
$$= \frac{4}{0}$$
$$= \text{undefined}$$

Since division by zero is not defined, the line has an undefined slope. Vertical lines are the only lines that have an undefined slope.

EXAMPLE 2

Determining the Slope of a Line from Two Points on the Line

What is the slope of the line going through (2, 4) and (5, 4) in Figure 1.15?

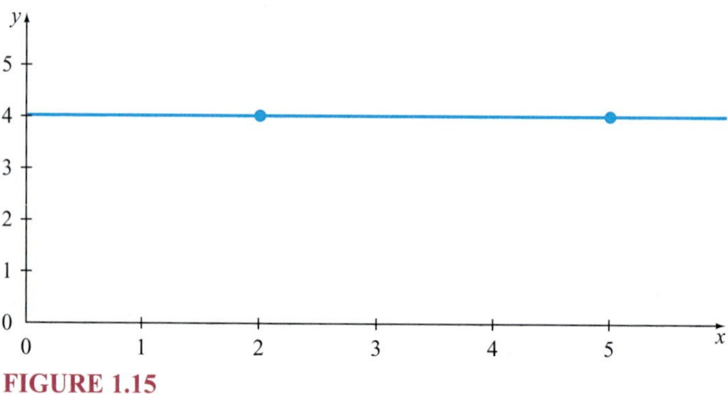

FIGURE 1.15

SOLUTION

$$m = \frac{4 - 4}{5 - 2}$$
$$= \frac{0}{3}$$
$$= 0$$

Any line with a zero slope is a horizontal line.

The absolute value of the slope is referred to as the **magnitude** of the slope. In a general sense, slope is a measure of steepness: The greater the magnitude of the slope, the greater the steepness of the line. The graph of a line with a negative slope falls as the independent variable increases. The graph of a line with a positive slope rises as the independent variable increases.

EXAMPLE 3

Determining the Sign and Magnitude of a Line's Slope from a Graph

Determine from the graph in Figure 1.16 which lines have a negative slope and which lines have a positive slope. Then identify the line whose slope has the greatest magnitude.

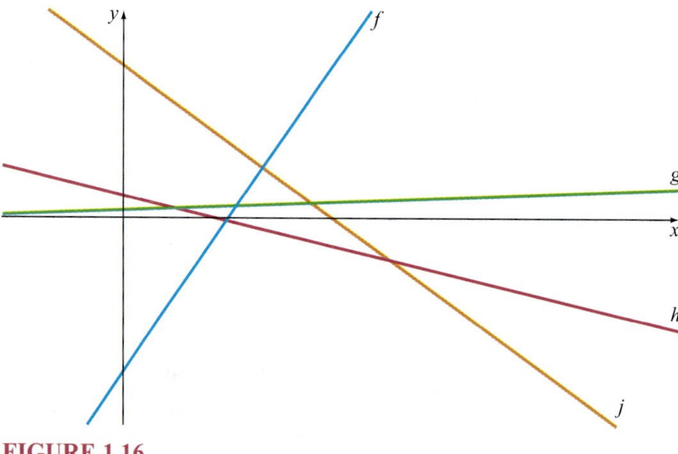

FIGURE 1.16

SOLUTION Since lines h and j fall as the value of x increases, they have negative slopes. Since lines f and g rise as the value of x increases, they have positive slopes.

The steepest line is the line whose slope has the greatest magnitude. Although both f and j are steep, line f is steeper. Therefore, f has the slope with the greatest magnitude.

Intercepts

In discussing linear functions, it is often useful to talk about where the line crosses the x and y axes. Knowing these **intercepts** helps us to determine the equation of the line.

DEFINITION: y-INTERCEPT

The y-intercept is the point on the graph where the function intersects the y axis. It occurs when the value of the independent variable is 0. It is formally written as an ordered pair $(0, b)$, but b itself is often called the y-intercept.

EXAMPLE 4

Finding the y-intercept of the Graph of a Linear Function

What is the y-intercept of the linear function $y = 3x + 5$?

SOLUTION At the y-intercept, the x coordinate is 0.

$$\begin{aligned} y &= 3(0) + 5 && \text{Substitute 0 for } x \\ &= 5 \end{aligned}$$

So the y-intercept of the function is $(0, 5)$.

DEFINITION: x-INTERCEPT

The x-intercept is the point on the graph where the function intersects the x axis. It occurs when the value of the dependent variable is 0. It is formally written as an ordered pair $(a, 0)$, but a itself is often called the x-intercept.

EXAMPLE 5

Finding the x-intercept of the Graph of a Linear Function

What is the x-intercept of the linear function $y = 3x + 5$?

SOLUTION At the x-intercept, the y coordinate is 0.

$$\begin{aligned} y &= 3x + 5 \\ 0 &= 3x + 5 && \text{Substitute 0 for } y \\ -3x &= 5 \\ x &= -\frac{5}{3} \end{aligned}$$

So the x-intercept of the function is $\left(-\frac{5}{3}, 0\right)$.

EXAMPLE 6

Determining the Slope and Intercepts of the Graph of a Linear Function

Determine the slope, the y-intercept, and the x-intercept of the linear function from its graph (Figure 1.17).

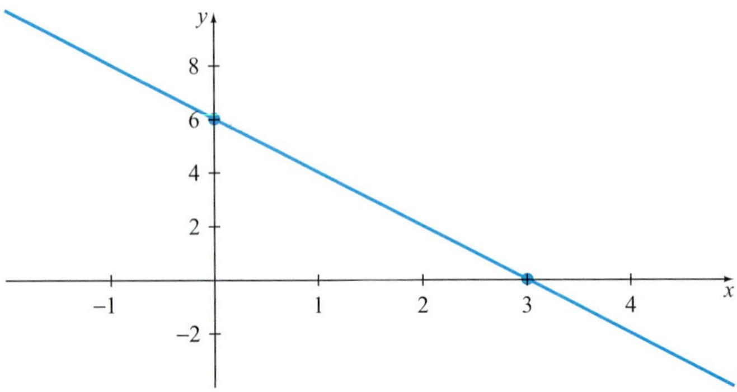

FIGURE 1.17

SOLUTION Since the graph of the function crosses the y axis at $(0, 6)$, the y-intercept is $(0, 6)$. Since the graph crosses the x axis at $(3, 0)$, the x-intercept is $(3, 0)$.

Since the line is falling as x increases, the slope will be negative. The slope of the line is

$$\begin{aligned} m &= \frac{y_2 - y_1}{x_2 - x_1} \\ &= \frac{6 - 0}{0 - 3} \\ &= \frac{6}{-3} \\ &= -2 \end{aligned}$$

Linear Equations

The graph of any line may be represented by a linear equation. Vertical and horizontal lines have the simplest linear equations.

EQUATIONS OF VERTICAL AND HORIZONTAL LINES

- The equation of a vertical line passing through a point (a, b) is $x = a$.
- The equation of a horizontal line passing through a point (a, b) is $y = b$.

If we know the slope and y-intercept of a linear function, we are able to easily determine the slope-intercept form of the linear equation.

SLOPE-INTERCEPT FORM OF A LINE

A linear function with slope m and y-intercept $(0, b)$ has the equation

$$y = mx + b$$

EXAMPLE 7

Determining the Slope-Intercept Form of a Line

What is the slope-intercept form of the linear function with slope 5 and y-intercept (0, 4)?

SOLUTION Since $m = 5$ and $b = 4$,

$$y = 5x + 4$$

is the slope-intercept form of the line.

EXAMPLE 8

Finding a Linear Model from a Verbal Description

For breakfast, we decide to eat an apple containing 5.7 grams of dietary fiber and a number of servings of Cheerios™, each containing 3 grams of dietary fiber. (**Source:** Cheerios box label.) Write the equation for our dietary fiber intake as a function of number of servings of cereal.

SOLUTION We know that if we don't eat any cereal, we will consume 5.7 grams of dietary fiber (from the apple). So the y-intercept of the fiber function is (0, 5.7). Since each serving of Cheerios contains the same amount of fiber, we know that the function is linear. The slope of the function is 3 grams per serving. So the equation of the fiber function is

$$F(n) = 3n + 5.7$$

where $F(n)$ is the amount of dietary fiber (in grams) and n is the number of servings of cereal.

To check our work, we directly calculate the number of grams of dietary fiber in a breakfast containing two servings of cereal and one apple.

$$\text{Fiber} = 3 + 3 + 5.7 = 11.7 \text{ grams}$$

Using the fiber function formula, we get

$$\begin{aligned} F(2) &= 3(2) + 5.7 \\ &= 11.7 \text{ grams} \end{aligned}$$

The results are the same, so we are confident that our formula is correct.

EXAMPLE 9

Determining a Linear Model from a Table of Data

The amount of sales tax paid on a clothing purchase in Seattle is a function of the sales price of the clothes, as shown in Table 1.6.

TABLE 1.6

Sales Price (p)	Tax (T)
\$20.00	\$1.72
\$30.00	\$2.58
\$40.00	\$3.44

Source: www.cityofseattle.net.

If the function is linear, write the equation for the sales tax as a function of the sales price.

SOLUTION We must first determine if the function is linear. Since each \$10 increase in sales price increases the sales tax by a constant \$0.86, we conclude that the function is linear. The slope of the function is

$$m = \frac{2.58 - 1.72 \text{ dollars}}{30 - 20 \text{ dollars}}$$
$$= \frac{0.86}{10} \frac{\text{dollar}}{\text{dollars}}$$
$$= 0.086 \text{ tax dollar per sales price dollar}$$

(In other words, for each dollar increase in the sales price, the sales tax increases by 8.6¢.) Since a sale of \$0 results in \$0 sales tax, we know that the y-intercept is (0, 0). The equation of the function is

$$T = 0.086p$$

To check our work, we substitute the point (40, 3.44) into the equation.

$$3.44 = 0.086(40) \qquad p = 40 \text{ and } T = 3.44$$
$$3.44 = 3.44$$

The statement is true, so we are confident that our equation is correct.

EXAMPLE 10 Finding a Linear Model from the Graph of a Linear Function

The graph in Figure 1.18 shows the balance of a checking account as a function of the number of ATM withdrawals from the account.

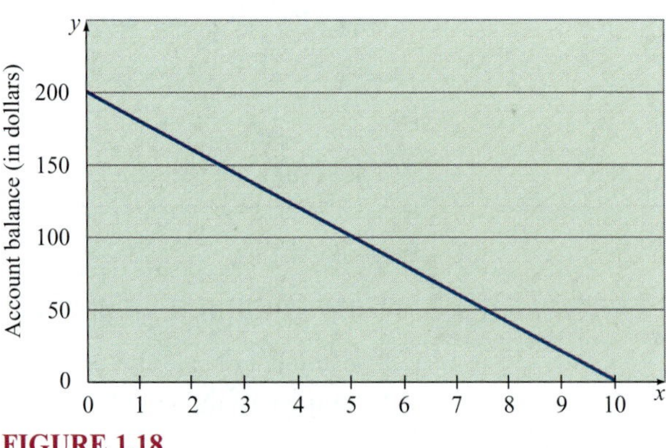

FIGURE 1.18

(a) Identify the x-intercept and y-intercept and interpret their physical meaning.

(b) Write the linear equation for the function.

SOLUTION

(a) The x-intercept is (10, 0). When there have been 10 withdrawals, the account balance is \$0. The y-intercept is (0, 200). When there have been no withdrawals, the account balance is \$200.

(b) The slope of the function is

$$m = \frac{200 - 0 \text{ dollars}}{0 - 10 \text{ withdrawals}}$$
$$= -\$20 \text{ per withdrawal}$$

So the slope-intercept form of the linear equation is

$$B = -20w + 200$$

where B is the account balance and w is the number of withdrawals.

Finding the Equation of a Line

As demonstrated in Example 10, to find the slope-intercept form of a line from two points, we need to calculate the slope and the y-intercept of the function. To do this, we proceed as follows:

1. Calculate the slope
2. Write the function in slope-intercept form, substituting the slope for m.
3. Select one of the points and substitute the output value for y and the input value for x.
4. Solve for b.
5. Write the function in slope-intercept form, substituting the y-intercept for b.

EXAMPLE 11

Finding the Slope-Intercept Form of a Line from Two Points

Find the equation of the line passing through the points (3, 5) and (7, 1).

SOLUTION

$$m = \frac{5 - 1}{3 - 7}$$
$$= \frac{4}{-4}$$
$$= -1$$

So the slope is -1. Substituting in the slope and the point (3, 5), we get

$$y = -1 \cdot x + b$$
$$5 = -1(3) + b$$
$$b = 8$$

The y-intercept is (0, 8).

The slope-intercept form of the line is $y = -1 \cdot x + 8$ or $y = -x + 8$.

Other Forms of Linear Equations

There are two additional forms of linear equations that are commonly used: standard form and point-slope form.

STANDARD FORM OF A LINE

A linear equation may be written as

$$ax + by = c$$

where a, b, and c are real numbers. If $a = 0$, the graph of the equation is a horizontal line. If $b = 0$, the graph of the equation is a vertical line. In the equation, a and b cannot both be zero.

The standard form of a linear equation is extremely useful when working with systems of equations or solving linear programming problems. We will discuss the standard form of a line further when we introduce these topics.

The point-slope form of a line is especially useful when we know a line's slope and the coordinates of a point on the line.

POINT-SLOPE FORM OF A LINE

A linear function written as

$$y - y_1 = m(x - x_1)$$

has slope m and passes through the point (x_1, y_1).

EXAMPLE 12

Finding a Linear Model from a Verbal Description

Based on data from 1980 to 1999, per capita consumption of milk as a beverage has been decreasing by approximately 0.219 gallon per year. In 1997, the per capita consumption of milk as a beverage was 24.0 gallons. (**Source:** *Statistical Abstract of the United States, 2001*, Table 202, p 129.) Find an equation for the per capita milk consumption as a function of years since 1980.

SOLUTION The slope of the line is $m = -0.219$. The point $(17, 24.0)$ lies on the line, since 1997 is 17 years after 1980. The point-slope form of the line is

$$y - 24.0 = -0.219(t - 17)$$

If preferred, the equation may be rewritten in slope-intercept form,

$$y = -0.219(t - 17) + 24.0$$
$$y = -0.219t + 3.723 + 24.0$$
$$y = -0.219t + 27.723$$

or standard form,

$$0.219t + y = 27.723$$

Graphing Linear Functions

To graph a linear function, we first generate a table of values by substituting different values of x into the equation and calculating the corresponding value of y. For example, if we are given the linear equation $y = 4x - 8$, we may choose to evaluate the function at $x = -1$, $x = 0$, $x = 1$, $x = 2$, and $x = 3$, as shown in Table 1.7.

TABLE 1.7

x	y
-1	-12
0	-8
1	-4
2	0
3	4

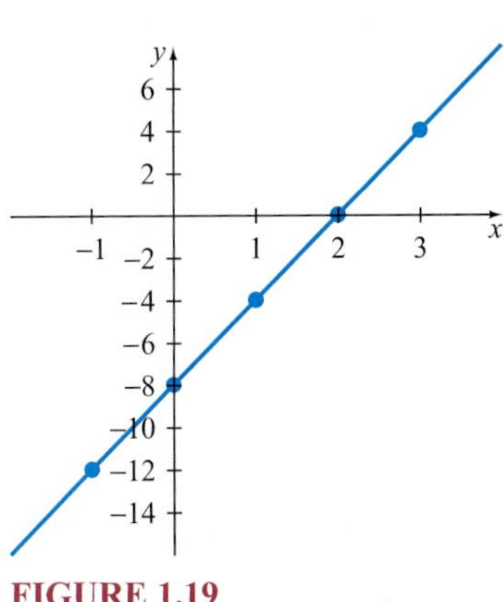

FIGURE 1.19

We then plot the points and connect them with a straight line, as shown in Figure 1.19. (Although we plotted multiple points, only two points are necessary to determine the line.)

EXAMPLE 13

Graphing a Linear Function

Graph the function $y = 2x - 3$.

SOLUTION From the equation, we see that the y-intercept is $(0, -3)$. We need to find only one more point. Evaluating the function at $x = 2$ yields $y = 1$, so $(2, 1)$ is a point on the line. We plot each point and connect the points with a straight line, as shown in Figure 1.20.

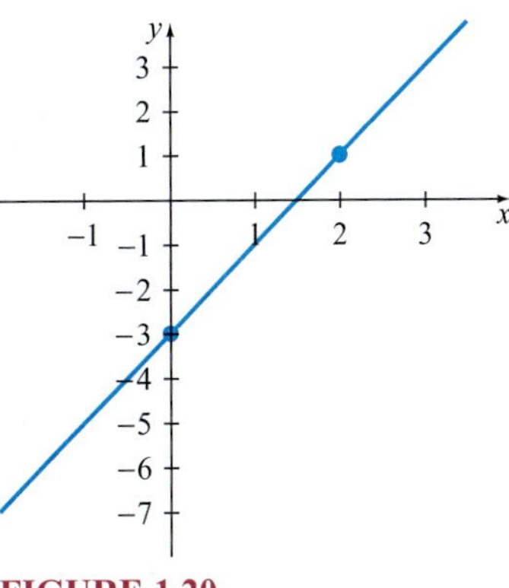

FIGURE 1.20

Recall that the standard form of a line is $ax + by = c$. Graphing a line from standard form is remarkably easy. The y-intercept of the graph occurs when the x value is equal to zero. Similarly, the x-intercept of the graph occurs when the

y value is equal to zero. Using these facts, we can quickly determine the x- and y-intercepts of the function. To find the x-intercept, we set $y = 0$.

$$\begin{aligned} ax + by &= c \\ ax + b(0) &= c \\ ax &= c \\ x &= \frac{c}{a} \end{aligned}$$

Notice that the x coordinate of the x-intercept is the constant term divided by the coefficient on the x term.

To find the y-intercept, we set $x = 0$.

$$\begin{aligned} ax + by &= c \\ a(0) + by &= c \\ by &= c \\ y &= \frac{c}{b} \end{aligned}$$

Notice that the y coordinate of the y-intercept is the constant term divided by the coefficient on the y term. Using this procedure to find intercepts will allow us to graph linear equations in standard form quickly by hand.

EXAMPLE 14

Graphing a Linear Function

Graph the linear function $2x + y = 4$.

SOLUTION The x-intercept is found by dividing the constant term by the coefficient on the x term.

$$\begin{aligned} x &= \frac{4}{2} \\ &= 2 \end{aligned}$$

The point $(2, 0)$ is the x-intercept.

The y-intercept is found by dividing the constant term by the coefficient on the y term.

$$\begin{aligned} y &= \frac{4}{1} \\ &= 4 \end{aligned}$$

The point $(0, 4)$ is the y-intercept. We graph the x- and y-intercepts and then draw the line through the intercepts, as shown in Figure 1.21.

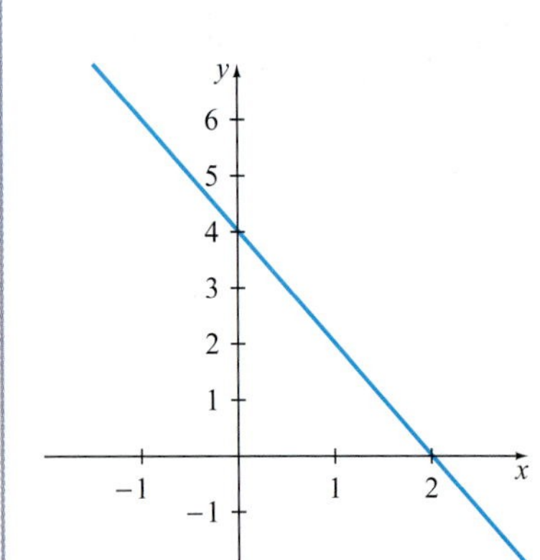

FIGURE 1.21

1.2 Summary

In this section, you learned how to calculate the slope of a linear function and how to interpret the physical and graphical meaning of slope, x-intercept, and y-intercept. You also discovered how to calculate the equation of a linear function from a set of data and learned three different ways to write a linear equation.

1.2 Exercises

In Exercises 1–6, calculate the slope of the linear function passing through the points.

1. (2, 5) and (4, 3)

2. (−3, 4) and (0, −2)

3. (1.2, 3.4) and (2.7, 3.1)

4. (7, 11) and (9, 2)

5. (2, 2) and (5, 2)

6. (4, 3) and (4, 7)

In Exercises 7–12, find the x-intercept and y-intercept of the linear function.

7. $y = 5x + 10$

8. $y = -3x + 9$

9. $y = 2x + 11$

10. $y = \frac{1}{2}x - 2$

11. $3x - y = 4$

12. $4x - 2y = 5$

In Exercises 13–18, write the equation of the linear function passing through the points in slope-intercept form, point-slope form, and standard form.

13. (2, 5) and (4, 3)

14. (−3, 4) and (0, −2)

15. (1.2, 3.4) and (2.7, 3.1)

16. (7, 11) and (9, 2)

17. (−2, 2) and (5, 2)

18. (−3.1, 4.5) and (2.1, −3.4)

In Exercises 19–25, graph the line.

19. $y = 4x - 2$

20. $x - y = 3$

21. $y - 4 = 0.5(x - 2)$

22. $y = -5x + 10$

23. $2x - 3y = 5$

24. $y - 9 = -3(x + 2)$

25. $y = -\frac{2}{3}x + \frac{4}{3}$

In Exercises 26–30, use the graph to determine the equation of the line.

26.

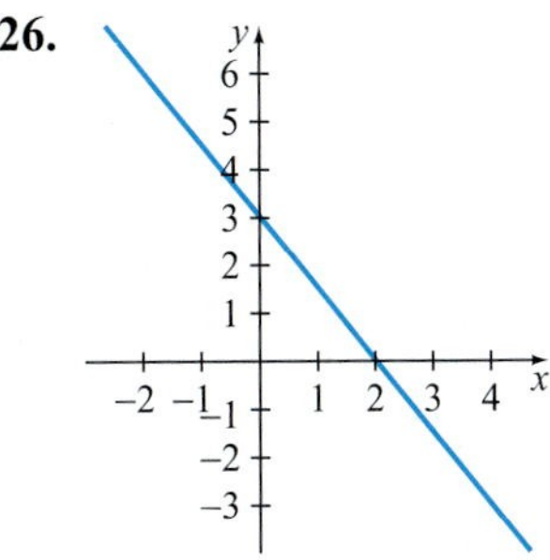

27.

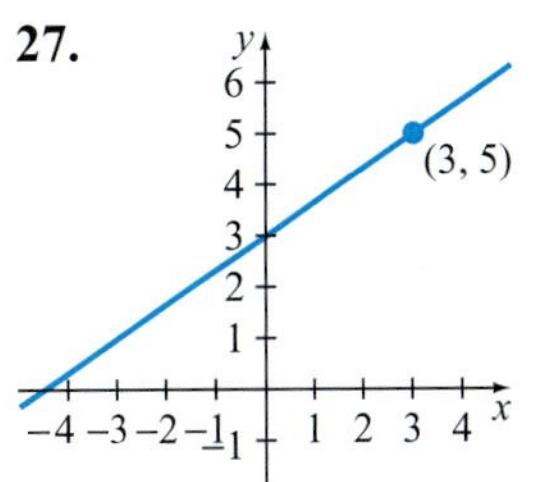

28.

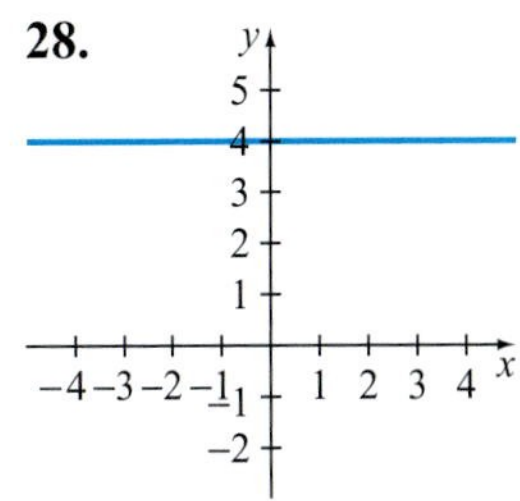

29.

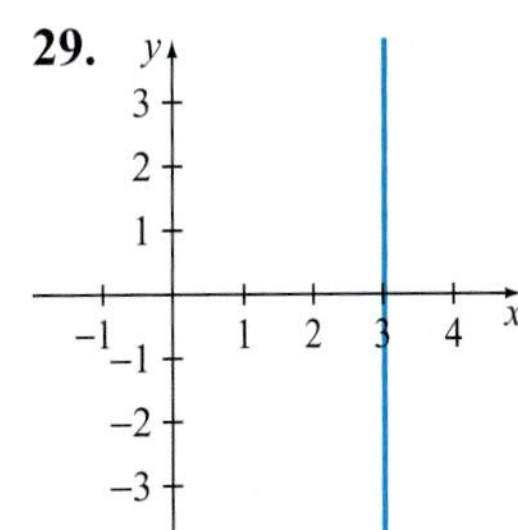

30.

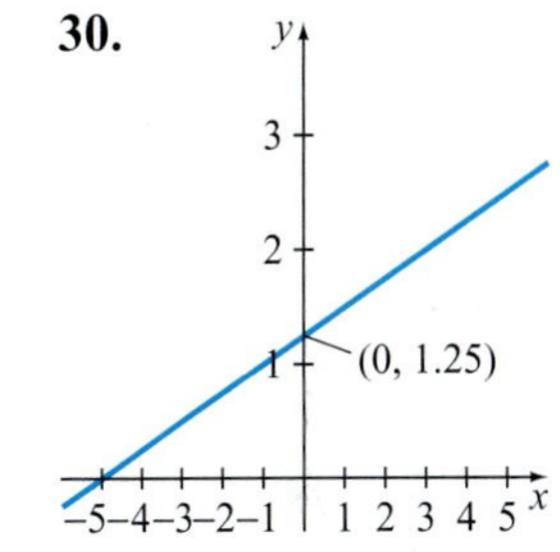

In Exercises 31–35, determine if the table of data represents a linear function. If so, calculate the slope and interpret its real-life significance.

31. **Average Personal Income**

Year	U.S. Average Personal Income (in terms of year 2000 dollars)
1989	18,593
1999	28,525

Source: www.census.gov.

32. **Internet Access**

Year	Number of People with Internet Access at Home or at Work (in thousands)
1997	46,305
1998	62,273
1999	83,677
2000	112,949

Source: www.census.gov.

33.  **Take-Home Pay**

Months (since Sept. 2001)	Take-Home Pay (dollars)
0	3,167.30
1	4,350.31

Source: Employee pay stubs.

34.  **Museum Admission Cost**

People in Group	Total Cost of Admission for a Group Visiting the Experience Music Project
3	\$59.85
4	\$79.80
5	\$99.75
6	\$119.70

Source: www.emplive.com.

35.  **Solid Waste Disposal**

Clean Wood (pounds)	Cost to Dispose of Clean Wood at Enumclaw Transfer Station
500	\$18.75
700	\$26.25
900	\$33.75
1,000	\$37.50

Source: www.dnr.metrokc.gov.

In Exercises 36–38, find the linear equation that models the data and answer the specified questions.

36. **Nutrition** The Recommended Daily Allowance for dietary fiber is 25 grams for a person on a 2,000-calorie-per-day diet. A $\frac{3}{4}$-cup serving of Post® Fruity Pebbles® contains 0.2 gram of dietary fiber. (**Source:** Package labeling.) A 1-cup serving of 2 percent milk fortified with vitamin A doesn't contain any dietary fiber. A large banana $\left(8 \text{ to } 8\frac{7}{8} \text{ inches long}\right)$ contains 3.3 grams of dietary fiber. (**Source:** www.nutri-facts.com.) How many servings of Fruity Pebbles with milk would you have to eat along with a large banana to consume 8 grams of dietary fiber? (Round up to the nearest number of servings.)

37. **Nutrition** The Recommended Daily Allowance for dietary fiber is 25 grams for a person on a 2,000-calorie-per-day diet. A $\frac{3}{4}$-cup serving of General Mills Wheaties® contains 2.1 grams of dietary fiber. (**Source:** Package labeling.) A 1-cup serving of 2 percent milk fortified with vitamin A doesn't contain any dietary fiber. A large $\left(8 \text{ to } 8\frac{7}{8} \text{ inches long}\right)$ banana contains 3.3 grams of dietary fiber. (**Source:** www.nutri-facts.com.) How many servings of Wheaties with milk would you have to eat along with a large banana to consume 8 grams of dietary fiber? (Round up to the nearest number of servings.)

38. **Nutrition** The Recommended Daily Allowance for dietary fiber is 30 grams for a person on a 2,500-calorie-per-day diet. A large apple $\left(3\frac{1}{4} \text{ inches in diameter}\right)$ contains 5.7 grams of dietary fiber. A large banana (8 to $\left(8\frac{7}{8} \text{ inches long}\right)$ contains 3.3 grams of dietary fiber. (**Source:** www.nutri-facts.com.) How many apples would you have to eat along with a banana to consume 30 grams of dietary fiber?

Exercises 39–51 are intended to challenge your understanding of linear equations and their graphs.

39. What is the equation of the line that passes through $(4, 7)$ and $(4, -1)$?

40. Do the two equations represent the same line? Explain.

$$3x + 4y = 12;\ y + 3 = -0.75(x - 8)$$

41. Explain why the equation of a vertical line cannot be written in slope-intercept or point-slope form.

42. **Alaska Population** The U.S. Census Bureau projects that the population of Alaska will grow at an approximate rate of 9,321 people per year between 1995 and 2025. It estimates that 791,000 people will be living in Alaska in 2025. (**Source:** www.census.gov.) What is the projected population of Alaska in 2035?

43. Are all lines functions? Explain.

44. Two lines are said to be parallel if they do not intersect. If $f(x) = mx + b$ and $g(x) = nx + c$ are parallel lines, what conclusions can you draw about b, c, m, and n?

45. Two lines are perpendicular if the slope of one line is the negative reciprocal of the slope of the other line. Find the equation of the line perpendicular to $y = 4$ that passes through the point $(3, -2)$.

46. **World Population** Based on data from 1980 to 2001, the population of the world may be modeled by $P = (81.3t + 4460)$ million people, where t is the number of years since 1980. (**Source:** Modeled from *Statistical Abstract of the United States, 2001,* Table 1324, p. 829.) According to the model, how quickly will the world population be increasing in 2010?

47. The equation of a nonvertical line is given by $ax + by = c$. What are the slope, x-intercept, and y-intercept of the line?

48. Is the function shown in the table linear? Explain.

x	y
2	11
5	17
8	23
10	29
14	35

49. In order for the function shown in the table to be linear, what must be the value of b?

x	y
2	7
5	13
8	19
10	b
14	31

50. In order for the function shown in the table to be linear, what must be the value of a?

x	y
2	6
5	12
8	18
a	26
14	30

51. What is the point of intersection of the lines $x = a$ and $y = b$?

1.3 Linear Models

- Use technology to model linear and near-linear data
- Use a linear equation to describe the relationship between directly proportional quantities
- Recognize and model naturally occurring linear, near-linear, and piecewise linear relationships

GETTING STARTED The amount of money deducted from a paycheck for Social Security, the annual income of a sales representative, and the profit from a club car wash may be modeled by linear functions. In this section, we will look at a wide variety of information encountered in everyday life and learn how to determine if the data are linear, near-linear, or piecewise linear. We will also demonstrate how to use technology to determine the linear equation that best fits a data set. Additionally, we will show how to find linear equations to model real-life phenomena.

Direct Proportions

Two quantities are said to be **directly proportional** if the ratio of the output to the input is a constant. Quantities that are directly proportional to each other may be modeled by a linear equation of the form

$$y = kx$$

The constant $k = \frac{y}{x}$ and is called the **constant of proportionality.**

EXAMPLE 1

Determining If a Data Set Is Directly Proportional

For the data in Table 1.8, determine whether the employee's weekly income is directly proportional to the number of hours worked. If so, find a linear function that models the data.

TABLE 1.8

Hours Worked (h)	Income (I)
35.0	\$210.00
38.5	\$231.00
42.3	\$253.80

SOLUTION Calculate the ratio $\frac{I}{h}$ for each ordered pair.

$$\frac{\$210}{35 \text{ hours}} = \$6/\text{hr}$$

$$\frac{\$231}{38.5 \text{ hours}} = \$6/\text{hr}$$

$$\frac{\$253.80}{42.3 \text{ hours}} = \$6/\text{hr}$$

Since $\frac{I}{h} = 6$ for each ordered pair, the weekly income is directly proportional to the hours worked with a constant of proportionality $k = 6$. The linear function is

$$I = 6h$$

EXAMPLE 2

Determining If a Data Set Is Directly Proportional

Determine whether the cost of filling a gas tank as given in Table 1.9 is directly proportional to the number of gallons put into the tank. If so, find a linear function that models the data.

Number of Gallons (g)	Cost (C)
3.50	\$5.25
6.28	\$9.41
17.34	\$25.99

SOLUTION Calculate the ratio $\frac{C}{g}$ for each ordered pair.

$$\frac{\$5.25}{3.50 \text{ gallons}} = 1.500 \text{ dollars per gallon}$$

$$\frac{\$9.41}{6.28 \text{ gallons}} \approx 1.498 \text{ dollars per gallon}$$

$$\frac{\$25.99}{17.34 \text{ gallons}} \approx 1.499 \text{ dollars per gallon}$$

The ratios are close to constant, but they are different. But wait! Although gas prices are given to the tenth of a cent (e.g., \$1.599), the cost of buying gasoline is rounded to the nearest cent. Could the difference in ratios be due to round-off error? Let's see. It looks as if $k \approx 1.499$. The associated linear model is

$$C = 1.499g$$

How good is the model?

TABLE 1.10

Number of Gallons	Cost	Model Estimate
3.50	\$5.25	\$5.2465
6.28	\$9.41	\$9.4137
17.34	\$25.99	\$25.993

As we see in Table 1.10, if we round to the nearest cent, the model fits the data perfectly. From our personal experience purchasing gasoline, we know that the cost of filling the gas tank is directly proportional to the number of gallons we put in. Our personal experience validates the accuracy of the model.

Determining If a Data Set Is Directly Proportional

A teacher's salary may fluctuate over the course of the year. (Teaching contracts typically cover a nine-month period.) A teacher recorded his paycheck amount and the OASI (Social Security) deduction over a three-month period in Table 1.11.

TABLE 1.11

Paycheck Amount (p)	OASI Deduction (S)
\$1,474.77	\$91.44
\$2,386.17	\$145.58
\$2,795.74	\$170.98

Is the OASI deduction directly proportional to the paycheck amount? If so, find a linear function that models the data.

SOLUTION Calculate the ratio $\frac{S}{p}$ for each ordered pair.

$$\frac{91.44}{1474.77} \approx 0.06200 \text{ OASI dollar per paycheck dollar}$$

$$\frac{145.58}{2386.17} \approx 0.06101 \text{ OASI dollar per paycheck dollar}$$

$$\frac{170.98}{2795.74} \approx 0.06116 \text{ OASI dollar per paycheck dollar}$$

The ratios are not constant, so, technically, we may conclude that the data are not directly proportional. (The tax law requires that 6.2 percent of the gross pay amount reduced by the cost of medical premiums be paid into OASI. In the first check, no amount of money was paid toward the teacher's medical plan. For the second two checks, the gross pay amount was reduced by the cost of the medical premiums before calculating the 6.2 percent OASI amount.)

Although the data in Example 3 are not linear, the fact that the ratios are nearly constant indicates that the data are near-linear. It looks as if $k \approx 0.0615$. A linear model for the data is

$$S = 0.0615p$$

How good is the model? Let's check.

$$S = 0.0615(1474.77) = 90.70$$
$$S = 0.0615(2386.17) = 146.75$$
$$S = 0.0615(2795.74) = 171.94$$

Table 1.12 summarizes the results.

TABLE 1.12

Paycheck Amount (p)	OASI Deduction (S)	Model Estimate (M)	Error ($S - M$)
$1,474.77	$91.44	$90.70	0.74
$2,386.17	$145.58	$146.75	−1.17
$2,795.74	$170.98	$171.94	−0.96

The error represents the difference between the actual data and the model estimate. The closer the magnitude of the error is to 0, the better the model fits the data. Is there a linear equation that better fits the data? Using technology, we can find the linear equation that best fits the data.

Linear Regression and the Line of Best Fit

Any number of lines may be used to model a near-linear set of data. We want to find the **line of best fit,** the line that best fits the data. To do this, we minimize the sum of the squares of the errors. Figure 1.22 shows three data points and a line used to model the data.

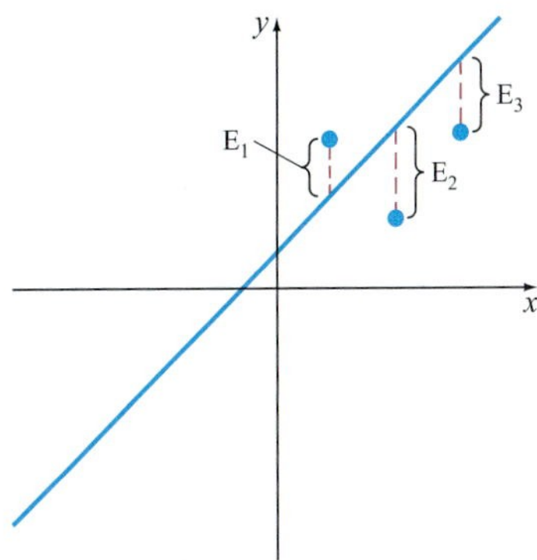

FIGURE 1.22

The vertical lines show the difference between the data output and the model estimate. The difference between the actual value and the estimated value of the individual data points (the error) is designated by $E1$, $E2$, and $E3$. To find the line of best fit, we square each of the errors, add the results together, and then minimize the sum using advanced calculus techniques. That is, we minimize $E = (E_1)^2 + (E_2)^2 + (E_3)^2$. This process is called **linear regression.** Fortunately, the TI-83 Plus calculator has a linear regression feature that does the necessary calculations automatically.

TECHNOLOGY TIP

Linear Regression

1. Press [2nd] then [0], scroll to `DiagnosticOn`, and press [ENTER] twice. This will ensure that the correlation coefficient r and the coefficient of determination r^2 will appear.

```
CATALOG
 DependAuto
 det(
 DiagnosticOff
▶DiagnosticOn
 dim(
 Disp
 DispGraph
```

2. Bring up the Statistics Menu by pressing the [STAT] button.

```
EDIT CALC TESTS
1:Edit...
2:SortA(
3:SortD(
4:ClrList
5:SetUpEditor
```

3. Bring up the List Editor by selecting `1:Edit...` and pressing [ENTER].

```
L1      L2      L3    2
1       9       ------
2       7
3       5
------
L2(4) =
```

(Continued)

4. Clear the lists. If there are data in the list, use the arrows to move the cursor to the list heading L1. Press the CLEAR button and press ENTER. This clears all of the list data. Repeat for each list with data. (*Warning:* Be sure to use CLEAR instead of DELETE. DELETE removes the entire column.)

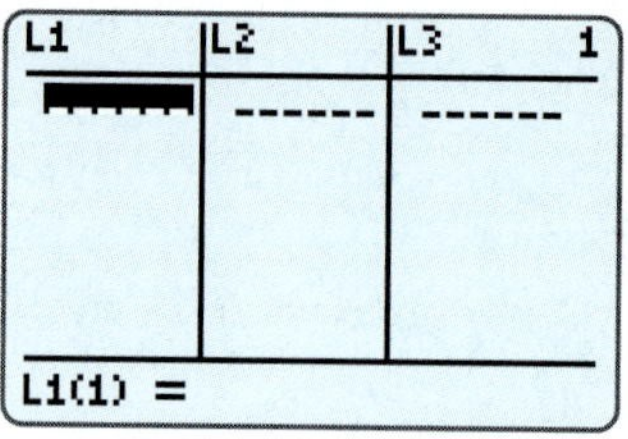

5. Enter the numeric values of the *inputs* in list L1, pressing ENTER after each entry.

6. Enter the numeric values of the *outputs* in list L2, pressing ENTER after each entry.

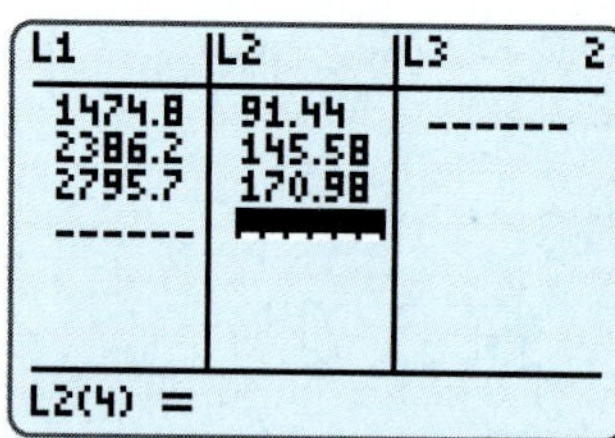

7. Return to the Statistics Menu by pressing the STAT button.

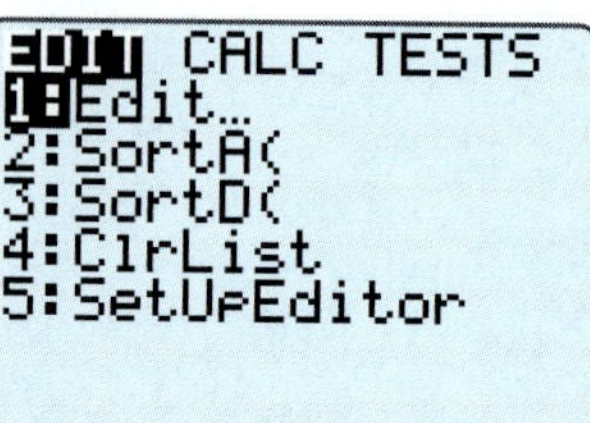

8. Bring up the Calculate Menu by using the arrows to select CALC.

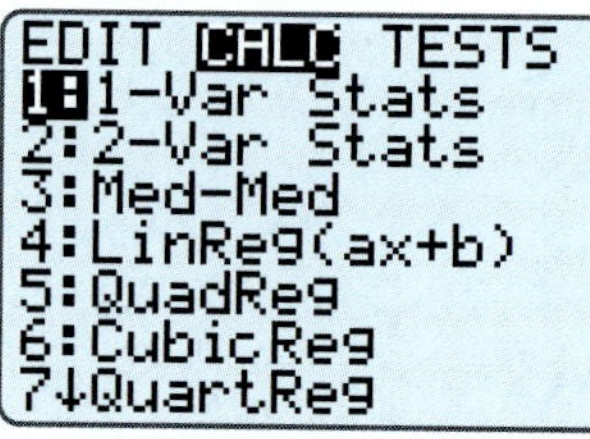

9. Calculate the linear equation of the model by selecting 4:LinReg(ax+b) and pressing ENTER twice. The line of best fit is $y = 0.06008x + 2.6928$ and has correlation coefficient $r = 0.9999$.

LinReg
y=ax+b
a=.060078251
b=2.692769342
r²=.9998949304
r=.9999474638

The correlation coefficient is a number r between -1 and 1 that measures the degree to which two variables are linearly related. If the line of best fit has a positive slope, $r > 0$. If the line of best fit has a negative slope, $r < 0$. If there is a perfect linear relationship between the two variables, $|r| = 1$. The more closely related the variables, the closer the correlation coefficient will be to ± 1. If the correlation coefficient is 0, the variables are not linearly related.

The coefficient of determination is equal to r^2. For linear functions, the coefficient of determination is the square of the correlation coefficient. (The formulas for calculating the correlation coefficient and coefficient of determination will not be shown here but may be found in many college-level statistics textbooks.)

In Example 3, the correlation coefficient was $r = 0.9999474638$. This number is extremely close to 1, so the line of best fit fits the data well. The linear function that best fits the data from Example 3 is $S = 0.06008p + 2.6928$. We can readily see from Table 1.13 that the model we found using linear regression fits the data better than the first model.

TABLE 1.13

Paycheck Amount (p)	OASI Deduction (S)	First Model $S = 0.0615p$	Linear Regression Model $S = 0.06008p + 2.6928$
\$1,474.77	\$91.44	\$90.70	\$91.30
\$2,386.17	\$145.58	\$146.75	\$146.05
\$2,795.74	\$170.98	\$171.94	\$170.66

Sometimes it is useful to graph the model. One method of doing this is detailed in the following Technology Tip.

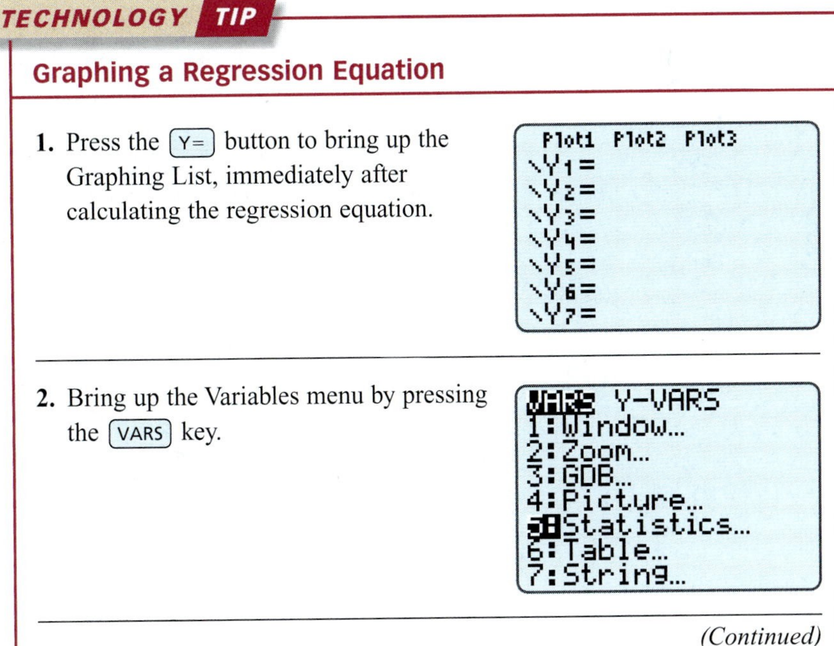

TECHNOLOGY TIP
Graphing a Regression Equation
1. Press the Y= button to bring up the Graphing List, immediately after calculating the regression equation.
Plot1 Plot2 Plot3
\Y1=
\Y2=
\Y3=
\Y4=
\Y5=
\Y6=
\Y7=
2. Bring up the Variables menu by pressing the VARS key.
VARS Y-VARS
1:Window...
2:Zoom...
3:GDB...
4:Picture...
5:Statistics...
6:Table...
7:String...
(Continued)

3. Bring up the Variables: Statistics menu by selecting `5:Statistics...` and pressing ENTER.

```
XY Σ EQ TEST PTS
1:n
2:x̄
3:Sx
4:σx
5:ȳ
6:Sy
7↓σy
```

4. Bring up the Equation menu by moving the cursor to `EQ`.

```
XY Σ EQ TEST PTS
1:RegEQ
2:a
3:b
4:c
5:d
6:e
7↓r
```

5. Select the regression equation by choosing `1:RegEQ` and pressing ENTER. This pastes the regression equation into the Graphing List.

```
Plot1 Plot2 Plot3
\Y1=.06007825101
623X+2.692769341
76
\Y2=
\Y3=
\Y4=
\Y5=
```

6. Graph the function by pressing the GRAPH button. You may need to adjust the viewing window to see the graph.

The graph of a set of discrete data is called a **scatter plot.** We can draw a scatter plot by hand, using graph paper, or we can use a calculator to draw the scatter plot.

TECHNOLOGY TIP

Drawing a Scatter Plot

1. Bring up the Statistics Plot menu by pressing the 2nd button, then the Y= button.

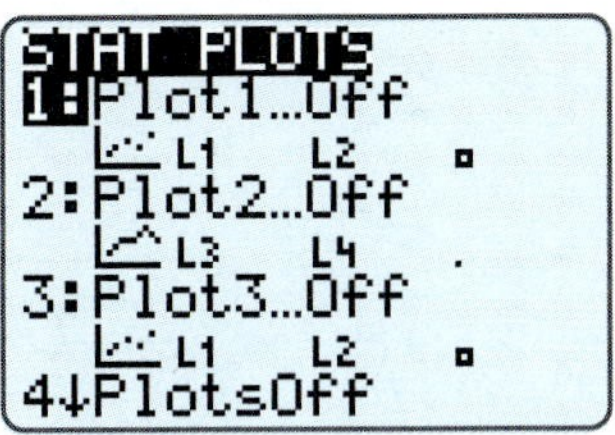

2. Open Plot1 by pressing ENTER.

3. Turn on Plot1 by moving the cursor to `On` and pressing ENTER. Confirm that the other menu entries are as shown.

4. Graph the scatter plot by pressing ZOOM and scrolling to `9:ZoomStat`. Press ENTER. This will graph the entire scatter plot along with any functions in the Graphing List. The ZoomStat feature automatically adjusts the viewing window so that all of the data points are visible.

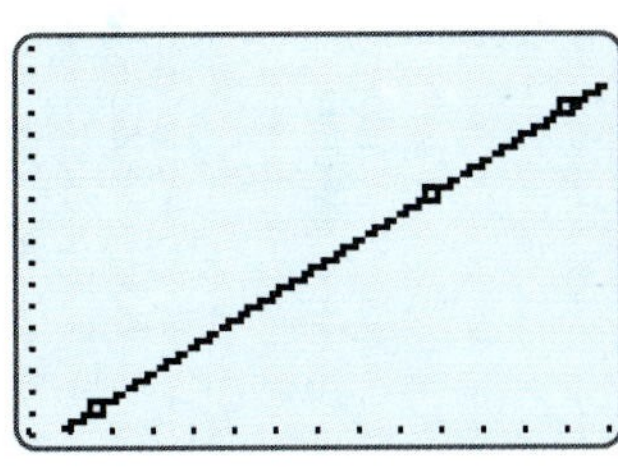

The graph shows the data points along with the linear model. We see that the linear model fits the data well.

EXAMPLE 4 Using the Line of Best Fit to Forecast Sales

The annual sales of McDonalds Corporation are shown in Table 1.14.

TABLE 1.14

Years Since 1990 (t)	Franchised Sales (S) (millions of dollars)
1	12,959
2	14,474
3	15,756
4	17,146
5	19,123
6	19,969
7	20,863
8	22,330
9	23,830
10	24,463
11	24,838

Source: www.mcdonalds.com.

(a) Draw a scatter plot of the data. Will a linear function fit the data well?
(b) Find the equation of the linear model that best fits the data.
(c) Graph the linear model simultaneously with the scatter plot. How well does the model predict the data?
(d) What does the slope of the function tell us about the company's sales?

SOLUTION

(a) Figure 1.23 shows a scatter plot of the data.

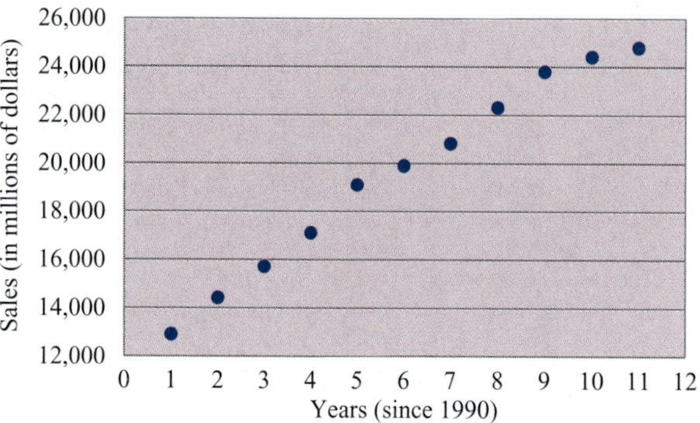

FIGURE 1.23

The sales revenue increases every year, but the amount by which it increases varies from year to year. Since the rate of increase is not constant, a linear function won't fit the data perfectly. However, the data do appear to be near-linear, so a linear model may fit the data well.

(b) Using linear regression, we determine that the linear model for sales is $S = 1233.5t + 12{,}213$ million dollars where t is the number of years since the end of 1990.

(c)

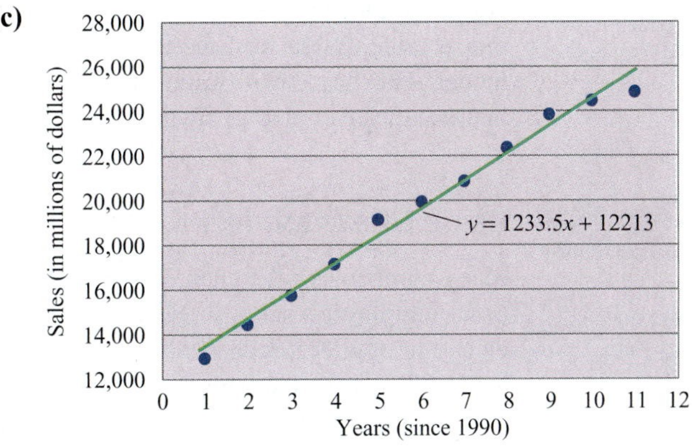

FIGURE 1.24

As shown in Figure 1.24 and Table 1.15, the linear model fits the data fairly well.

TABLE 1.15

Years Since 1990 (t)	Franchised Sales (S) (millions of dollars)	Estimated Sales (E) (millions of dollars)	Error ($S - E$) (millions of dollars)	% Error $\left(\frac{S-E}{S}\right)$
1	12,959	13,447	−488	−3.76%
2	14,474	14,680	−206	−1.42%
3	15,756	15,914	−158	−1.00%
4	17,146	17,147	−1	−0.01%
5	19,123	18,381	743	3.88%
6	19,969	19,614	355	1.78%
7	20,863	20,848	16	0.07%
8	22,330	22,081	249	1.12%
9	23,830	23,315	516	2.16%
10	24,463	24,548	−85	−0.35%
11	24,838	25,782	−944	−3.80%

Although the actual sales and the estimated sales from the model differed by as much as 944 million dollars, the model projections were within 3.88 percent of the actual values.

(d) The slope of the sales function is 1233.5 million dollars per year. According to the model, we expect that, on average, the sales will increase by about 1233.5 million dollars per year. We may use this information to help us determine whether to buy stock in the company. We should be aware that past performance is no guarantee of future results. Before making a stock purchase, we would want to look at other indicators, such as earnings per share, recent news, analyst ratings, and so on.

Finding Linear Functions to Model Real-Life Phenomena

Not all of the data we encounter in the real world are laid out nicely in an input-output table. Often we have to ferret out the information needed to formulate a model. The next two examples will illustrate real-life situations where using mathematics can give us financial leverage.

EXAMPLE 5

Using Linear Models to Make Business Decisions

Many companies offer their salespeople a base salary plus a commission based on the employee's sales. In November 2001, a Nebraska firm posted an ad on an electronic bulletin board advertising a sales representative position paying a \$36,000 base salary plus commission. (**Source:** www.dice.com.) Based on her past sales experience, a saleswoman estimates that she can generate \$600,000 in sales annually. To maintain her standard of living, she needs to earn a total of \$63,000 annually.

(a) What commission rate (as a percentage of sales) must she earn to maintain her standard of living?

(b) Write the equation for her annual income as a function of her annual sales dollar volume.

(c) If she wants to increase her annual income to \$75,000, by how much will she have to increase her annual sales?

(d) If she wanted to increase her annual income to \$75,000 without increasing her sales volume, what commission rate must she receive?

(e) How could the answers to the previous questions help her during salary negotiations?

SOLUTION

(a) We have

$$I = ms + b$$

where I is the annual income (in thousands of dollars) and s is the annual sales (in thousands of dollars). If she doesn't sell anything (assuming she doesn't get fired), she will still earn \$36,000, so (0, 36) is the y-intercept. Thus

$$I = ms + 36$$

When $s = 600$, $I = 63$, so

$$\begin{aligned} 63 &= m \cdot 600 + 36 \\ 27 &= 600m \\ m &= 0.045 \end{aligned}$$

She must earn a 4.5 percent commission to maintain her standard of living.

(b) The equation for her annual income is $I = 0.045s + 36$ thousand dollars, where s is the sales volume (in thousands of dollars).

(c) Setting $I = 75$ and solving for s, we get

$$\begin{aligned} 75 &= 0.045s + 36 \\ 39 &= 0.045s \\ s &= 866.667 \end{aligned}$$

She must increase her annual sales from \$600,000 to \$866,667 to increase her annual income to \$75,000.

(d) Letting m be the variable and setting I equal to 75, we get

$$\begin{aligned} 75 &= m \cdot 600 + 36 \\ 39 &= 600m \\ m &= 0.065 \end{aligned}$$

So to increase her income to \$75,000 without increasing sales, she must earn a 6.5 percent commission.

(e) We know that her minimum acceptable commission rate is 4.5 percent. To increase her salary without having to increase her workload, she will have to convince the employer to increase her commission rate to 6.5 percent. If the employer is unwilling to raise the commission rate, she will have to bring in an additional \$266,667 in sales to earn \$75,000. This information can give her leverage in salary negotiations.

EXAMPLE 6

Using Linear Models to Make Business Decisions

In 2001, BeautiControl Cosmetics sold a new consultant a demonstration package for \$119.82. (**Source:** BeautiControl Cosmetics consultant.) The consultant's director told her that the average revenue from each in-home party is \$250. The consultant's cost for the cosmetics is 50 percent of the sales price.

(a) Find the linear model for the consultant's cost as a function of the dollar volume of the cosmetics sold.

(b) Find the linear model for the consultant's revenue as a function of the dollar volume of cosmetics sold.

(c) Find the linear model for the consultant's profit as a function of the dollar volume of cosmetics sold.

(d) How many dollars of cosmetics must the consultant sell before she begins to make a profit?

(e) How many in-home parties must the consultant hold in order to begin making a profit?

SOLUTION

(a) Let x be the sales volume in dollars. Then the cost function C is

$$C(x) = 0.5x + 119.82$$

The cost of the demonstration package, \$119.82, is the fixed cost. The consultant's variable cost, $0.5x$ dollars, is dependent upon her sales dollar volume.

(b) The revenue function R is

$$R(x) = x$$

The consultant's revenue is equal to her sales dollar volume.

(c) The profit function is the difference between the revenue and cost functions.

$$\begin{aligned} P(x) &= R(x) - C(x) \\ &= x - (0.5x + 119.82) \\ &= 0.5x - 119.82 \end{aligned}$$

(d) We need to find the break-even point (the sales dollar volume that makes the profit equal zero). Setting the profit equal to zero, we get

$$\begin{aligned} 0 &= 0.5x - 119.82 \\ 119.82 &= 0.5x \\ x &= 239.64 \end{aligned}$$

So to break even, she must sell \$239.64 worth of cosmetics.

(e) Dividing the sales dollar volume needed to break even by the amount of revenue per party, we get

$$\frac{239.64 \text{ dollars}}{250.00 \frac{\text{dollars}}{\text{party}}} = 0.95856 \text{ parties}$$

It doesn't make sense to talk about fractions of a party, so we round up. The consultant should begin to make a profit at her first party.

Piecewise Linear Models

A function of the form

$$f(x) = \begin{cases} ax + b & x \leq c \\ dx + g & x > c \end{cases}$$

is called a **piecewise linear function.** The rule used to calculate $f(x)$ depends upon the value of x. In this case, if x is less than or equal to c, the rule is $f(x) = ax + b$. If x is greater than c, the rule is $f(x) = dx + g$.

A common piecewise linear function is the absolute value function $f(x) = |x|$. This function makes negative numbers positive and leaves positive numbers unchanged. In practical terms, it "erases" the negative sign of any negative number. The absolute value function is formally defined as a piecewise linear function.

$$\begin{aligned} f(x) &= |x| \\ &= \begin{cases} x & x \geq 0 \\ -x & x < 0 \end{cases} \end{aligned}$$

What is $|-5|$? Since $-5 < 0$, we apply the second rule of the piecewise function.

$$\begin{aligned} |-5| &= -(-5) \\ &= 5 \end{aligned}$$

Piecewise linear functions are very common in a variety of real-life situations, such as the ones given in Examples 7 and 8.

EXAMPLE 7

Using a Piecewise Linear Model to Forecast Monthly Payments

An orthodontist and the parents of a girl scheduled to receive braces agreed upon the following payment plan: a \$550 down payment when treatment begins and \$100 per month until the treatment is paid in full. The total cost for the treatment is \$1250, and the treatment is expected to take 18 months to complete. (**Source:** Author's personal payment plan.)

(a) How many months after treatment begins will the treatment be paid in full?

(b) Write the equation for the balance due as a function of the number of months since treatment began.

SOLUTION

(a) The patient's parents pay \$550 when treatment begins, so \$700 remains for the balance to be paid in full. Since \$100 is paid each month, the treatment will be paid in full 7 months later.

(b)

$$B(t) = \begin{cases} -100t + 700 & 0 \le t \le 7 \\ 0 & 7 < t \le 18 \end{cases}$$

where $B(t)$ is the balance of the account t months after treatment begins. Since payments are made monthly, the function is defined for whole number values of t.

We initially use a linear function for $B(t)$, since the account balance starts at \$700 and is reduced by a constant \$100 per month. However, after 7 months, the account balance has reached \$0. If we continued to use the linear function, the balance would end up being a negative number, which doesn't make sense in the context of the problem. Therefore, no further payments are made after the seventh month, and the account balance remains at \$0 from the end of the seventh month through the eighteenth month.

Using a Piecewise Linear Model to Make Spending Decisions

Experience Music Project, an interactive music museum in Seattle, charges adult visitors a \$19.95 admission fee. Groups of 15 or more may enter for \$14.50 per person. (**Source:** www.emplive.com.)

(a) Draw a scatter plot for group cost as a function of group size.

(b) Write the total cost of admission as a function of the number of people in a group. Confirm that the function and the scatter plot are in agreement.

(c) Calculate the cost of admitting a group of 14 people and the cost of admitting a group of 15 people.

(d) Determine the largest group we could bring in and still remain below the cost of a 14-person group.

SOLUTION

(a)

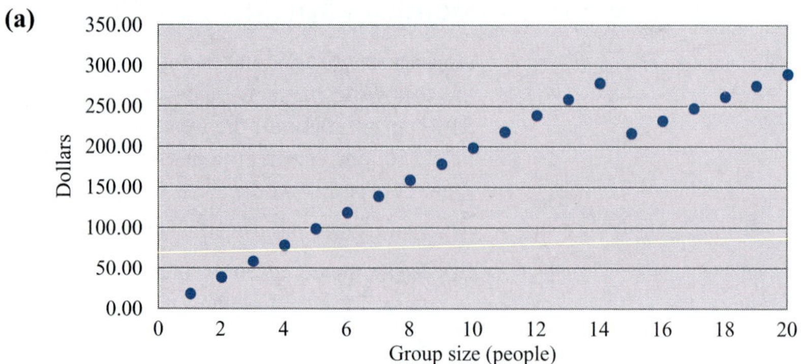

FIGURE 1.25

(b)

$$C(x) = \begin{cases} 19.95x & x < 15 \\ 14.5x & x \geq 15 \end{cases}$$

where x is the number of people in the group and C is the total cost in dollars.

(c) $C(14) = 19.95(14) = \$279.30$ is the cost for a group of 14 people.

$C(15) = 14.5(15) = \$217.50$ is the cost for a group of 15 people.

(d) From the scatter plot, it looks as if the maximum number of people we could bring in and still remain below the 14-person group price is 19 people. We can confirm the result algebraically.

$$279.30 = 14.5x$$
$$x \approx 19.26$$

Since it doesn't make sense to talk about 0.26 of a person, we round down to 19.

$$\begin{aligned} C(19) &= 14.5(19) \\ &= 275.50 \end{aligned}$$

The cost for a 19-person group is \$275.50. This is slightly less than the \$279.30 cost for a 14-person group.

Determining Which Modeling Method to Use

Although linear regression is a powerful tool for creating linear models, many linear models can be constructed algebraically, as shown in Examples 5 through 8. It is good practice to first attempt to find the linear model algebraically. A common pitfall of beginning students is to use the calculator to find the linear model when it is actually easier to find the model algebraically.

Whenever the data set contains exactly two points, we should find the linear model algebraically. In addition, there are other key phrases that alert us to the fact that a linear model may be constructed algebraically. Table 1.16 details how to interpret the mathematical meaning of some commonly occurring phrases.

TABLE 1.16

Phrase	Mathematical Meaning
constant rate of change	slope
increasing at a rate of 20 people per year	The slope of the linear function is $20\frac{\text{people}}{\text{year}}$.
tickets cost \$37 per person	The slope is $37\frac{\text{dollars}}{\text{person}}$.
sales increase by 100 units for every \$1 decrease in price	The slope is $m = \frac{100 \text{ units}}{-1 \text{ dollar}} = -100$ units per dollar.
initial price of \$1.25	The y-intercept is 1.25.
There are 350 chairs today. The number of chairs is increasing by 10 chairs per month.	The y-intercept is 350 and the slope is 10.
The price is \$2.25 and is decreasing at a constant rate of \$0.02 per day.	The y-intercept is 2.25 and the slope is -0.02.

By applying the techniques demonstrated in the table, you will increase your comfort level in working with real-world problems.

1.3 Summary

In this section, you looked at a wide variety of information encountered in everyday life, and you learned how to model linear, near linear, and piecewise linear data sets. You also learned how to use technology to determine the line of best fit to model real-life phenomena.

1.3 Exercises

In Exercises 1–5, do the following:

(a) Draw the scatter plot.
(b) Find the equation of the line of best fit.
(c) Interpret the meaning of the slope and the y-intercept of the model.
(d) Explain why you do or do not believe that the model would be a useful tool for businesses and/or consumers.

 **1.**  **Share Price**

Month in 2000	Harbor Capital Appreciation Fund Share Price on Last Day of Month
10	\$48.16
11	\$41.50
12	\$35.58

Source: www.quicken.com.

 2. Share Price

Month in 2001	CREF Stock Retirement Annuity Share Price on Last Day of Month
3	$164.5760
6	$173.1588
9	$146.4102

Source: TIAA-CREF statement.

 3.  University Enrollment

Washington State Public University Enrollment	
Years Since 1990 (t)	Students (S)
0	81,401
1	81,882
2	83,052
3	84,713
4	85,523
5	86,080
6	87,309
7	89,365
8	90,189
9	91,543
10	92,821

Source: Washington State Higher Education Coordinating Board.

 4.  Per Capita Personal Income

Per Capita Personal Income—Utah	
Years Since 1993 (t)	Personal Income (P) (dollars)
0	16,830
1	17,638
2	18,508
3	19,514
4	20,613
5	21,594
6	22,305
7	23,436

Source: Bureau of Economic Analysis (www.bea.gov).

 5.  Per Capita Income Ranking

Years since 1995	North Carolina Per Capita Income Ranking (out of 50 states)
0	42
1	34
2	32
3	28
4	28

Source: http://www.census.gov/statab.

In Exercises 6–16, formulate the linear or piecewise linear model for the real-life scenario. If a linear model represents a directly proportional relationship between the independent and dependent variables, so state.

6.  **Personal Income** In November 2001, Hall Kinion advertised an Account Executive position in Savannah, Georgia, for a salesperson with experience selling Internet services. The job advertised a $35K base salary and $24K to $48K in commissions in the first year. (**Source:** www.dice.com.)

(a) Assuming a 20 percent commission rate, how many dollars in sales would you have to generate to earn $64K annually?

(b) If you were in the sales field, why would this type of information be valuable to you?

7. **Solid Waste Disposal** King County charges \$82.50 per ton to dispose of garbage at area transfer stations. A minimum disposal fee of \$13.72 per entry is charged for vehicles disposing of trash. In addition to the disposal fee, a moderate risk waste (MRW) fee of \$2.61 per ton is assessed. A minimum MRW fee of \$1 is charged on all transactions. Tax is charged on the combined price of the disposal and moderate risk waste fees. The tax rate is 3.6 percent. The weight of the trash is rounded to the nearest 20 pounds before the fees are calculated. (**Source:** www.dnr.metrokc.gov.)

(a) Write the equation for disposal cost as a function of the number of pounds of refuse.
(b) How much will it cost to drop off 230 pounds of trash?
(c) How much will it cost to drop off 513 pounds of trash?
(d) If you ran a construction company, how could you use the results of this exercise?

8. **Cellular Phone Plan** In 2002, Qwest offered a 150-minute cellular plan including free long distance and free roaming to its Seattle-area customers for \$29.99 per month. Additional time cost \$0.35 per minute or portion of a minute. (**Source:** Qwest customer monthly statement.)

(a) Write the cellular phone cost equation as a function of the number of minutes used.
(b) How could you use the cost equation if you were enrolled in the plan?

9. **Cellular Phone Plan** Sprint PCS offered a service plan to its Seattle-area customers for \$29.99 per month that included 200 anytime minutes. The plan also included free long distance. Additional minutes or portions of a minute were \$0.40 each. (**Source:** www1.sprintpcs.com.)

(a) Determine the cost equation for the Sprint PCS plan as a function of the number of anytime minutes used.
(b) Is the Sprint plan or the Qwest plan (from Exercise 8) a better deal for a consumer who uses 300 anytime minutes monthly? Explain.

10. **Solid Waste Disposal** The Enumclaw Transfer/Recycling Station accepts clean wood (stumps, branches, etc.) in addition to household garbage. The station charges \$75 per ton with a minimum fee of \$12.75 per entry for vehicles disposing of clean wood. (**Source:** www.dnr.metrokc.gov.)

(a) How many pounds of clean wood could be dropped off at the station without exceeding the \$12.75 minimum fee?
(b) Write the equation for the disposal cost as a function of the number of pounds of clean wood.
(c) If you ran a landscaping business, how could you use the results of parts (a) and (b)?

11. **Used Car Value** In 2001, the average retail price of a 2000 Toyota Land Cruiser 4-Wheel Drive was \$43,650. The average retail price of a 1995 Toyota Land Cruiser 4-Wheel Drive was \$21,125. (**Source:** www.nadaguides.com.)

(a) Find a linear model for the value of a Toyota Land Cruiser in 2001 as a function of its production year.
(b) Use your linear model to predict the value of a 1992 Toyota Land Cruiser in 2001.
(c) A 1992 Toyota Land Cruiser had an average retail price of \$14,325 in 2001. How good was your linear model at predicting the value of the vehicle?

12. **Nutrition** The Recommended Daily Allowance (RDA) for fat for a person on a 2,000-calorie-per-day diet is less than 65 grams. A McDonalds Big Mac® sandwich contains 34 grams of fat. A large order of French fries contains 26 grams of fat. (**Source:** www.mcdonalds.com.)

(a) Write the equation for fat grams consumed as a function of large orders of French fries.
(b) Write the equation for fat grams consumed as a function of Big Macs.
(c) How many large orders of French fries can you eat without exceeding the RDA for fat?
(d) How many Big Macs can you eat without exceeding the RDA for fat?
(e) How many combination meals (Big Mac and French fries) can you eat without exceeding the RDA for fat?

13. **Nutrition** The Recommended Daily Allowance (RDA) for fat for a person on a 2,500-calorie-per-day diet is less than 80 grams. A McSalad Shaker Chef Salad® contains 8 grams of fat. A package (44.4 ml) of ranch salad dressing contains 18 grams of fat. (**Source:** www.mcdonalds.com.)

(a) Write the equation for fat grams as a function of Chef Salads.

(b) How many salads can you eat and remain below the RDA for fat?

(c) If you use one package of ranch salad dressing, how many salads can you eat and remain below the RDA for fat?

(d) The RDA for fat for a person on a 2000-calorie diet is 65 grams. If a person on a 2,000-calorie diet uses one package of ranch salad dressing, how many salads can that person eat and remain below the RDA for fat?

14. **Sales Tax** The Picture People photography studio in Tacoma, Washington, charges $40 for one of its promotional photo packages. The total purchase price for the promotional package, including tax, is $43.20. (**Source:** Picture People sales receipt.)

(a) Write the equation for Tacoma sales tax as a function of the pretax purchase price.

(b) A family has $60 cash to pay for pictures. What is the maximum pretax purchase price that the family can afford?

15. **Airline Ticket Cost** The total cost of an airline ticket includes the published fare, a federal flight segment tax, a federal security fee, and an airport Passenger Facility Charge (PFC). The flight segment tax is $3 on each segment of the itinerary. (A segment is a takeoff and landing.) The PFC varies from airport to airport; it ranges from $3 to $4.50, and it is imposed by an airport on enplaning passengers. When a passenger is departing from or connecting at any airport, that airport's PFC will apply in addition to the fare. Likewise, a $2.50 security fee is imposed on all enplaning passengers.

A Southwest Airlines refundable airline ticket from Seattle to Phoenix is advertised at $222. Write the maximum total cost of a Southwest (**Source:** www.southwest.com.) airline ticket from Seattle to Phoenix as a function of the number of segments. (Assume that a passenger travels exactly one segment per airplane. That is, if the plane stops, the passenger changes planes.)

16. **Airline Ticket Cost** The total cost of an airline ticket includes the published fare, a federal flight segment tax, a federal security fee, and an airport Passenger Facility Charge (PFC). The flight segment tax is $3 on each segment of the itinerary. (A segment is a takeoff and landing.) The PFC varies from airport to airport; it ranges from $3 to $4.50, and it is imposed by an airport on enplaning passengers. When a passenger is departing from or connecting at any airport, that airport's PFC will apply in addition to the fare. Likewise, a $2.50 security fee is imposed on all enplaning passengers.

A Southwest Airlines refundable round-trip airline ticket from Seattle to Phoenix is advertised at $444. (**Source:** www.southwest.com.) Write the maximum total cost of a round-trip airline ticket from Seattle to Phoenix as a function of the number of segments.

Exercises 17–25 are intended to challenge your understanding of linear models.

17. The standard form of a linear function in three dimensions is given by $ax + by + cz = d$, where x, y, and z are variables and a, b, c, and d are constants. A business determines that the ***revenue*** generated by selling x cups of coffee, y bagels, and z muffins is given by

$$R = ax + by + cz$$

What is the practical meaning of the constants a, b, and c?

18. A business determines that the ***cost*** of producing x cups of coffee, y bagels, and z muffins is given by

$$C = ax + by + cz + d$$

What is the practical meaning of the constants a, b, c, and d?

19. The equation of the line of best fit for the data in the table is $y = 0$.

x	y
0	-1
1	1
2	0
3	0
4	1
5	-1
6	0

This model passes through three of the seven points shown in the table. Is this model a good fit for the data? Justify your answer.

20. Are production costs directly proportional to the number of items produced for most businesses? Explain.

21. Does the fact that a linear model passes through the point $(0, 0)$ imply that the dependent variable of the original data is directly proportional to the independent variable of the original data? Explain.

22. A linear model has a correlation coefficient of $r = -1$. Does the fact that the linear model passes through the point $(0, 0)$ imply that the output of the original data is directly proportional to the input of the original data? Explain.

23. Use your calculator to find the line of best fit for the data in the table.

x	y
0	1
1	1
2	1
3	1

Why do you think the correlation coefficient is undefined? Does this mean that the model doesn't fit the data? Justify your answer.

24. Find the piecewise linear function that best fits the data in the table.

x	y
0	2
1	4
2	6
3	7
4	8
5	9
6	10

25. Find the piecewise linear function that best fits the data in the table.

x	y
-2.0	14
1.0	8
2.5	5
4.0	2
8.0	4
10.0	5

Chapter 1 Review Exercises

Section 1.1 *In Exercises 1–3, calculate the value of the function at the designated input and interpret the result.*

1. $C(x) = 49.95x$ at $x = 2$, where C is the cost (in dollars) of buying x pairs of shoes.

2. $C = 9.95x + 1200$ at $x = 400$, where C is the cost of making x pairs of shoes.

3. $H = -16t^2 + 100$ at $t = 2$, where H is the height of a cliff diver above the water t seconds after he jumped from a 100-foot cliff.

In Exercises 4–7, determine the domain of the function.

4. $f(x) = 4x^3 - 10x$

5. $f(p) = p^2 - 9p + 15$

6. $C = \dfrac{r^2 - 1}{2r + 1}$ **7.** $C = \dfrac{r^2 + 1}{r^2 - 1}$

In Exercises 8–9, interpret the meaning of the indicated point on the graph.

8. **Stock Price** Find P at $t = 4$, where P is the stock price of an e-learning company at the end of the day and t is the number of days after December 16, 2001.

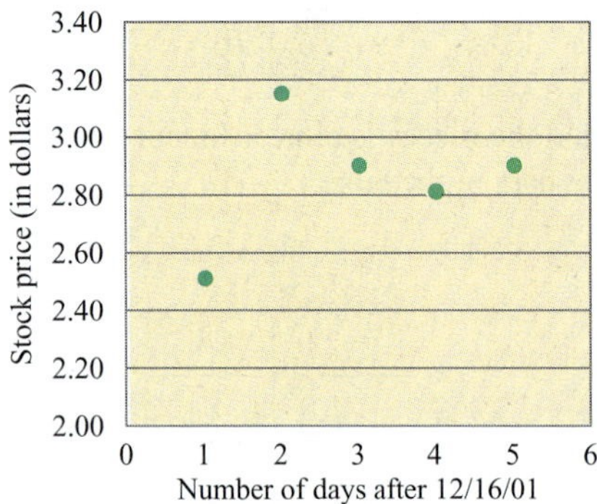

Source: Click2Learn Corporation.

9. **Stock Price** Find P at $t = 2$, where P is a computer company's stock price at the end of the day, t days after December 16, 2001.

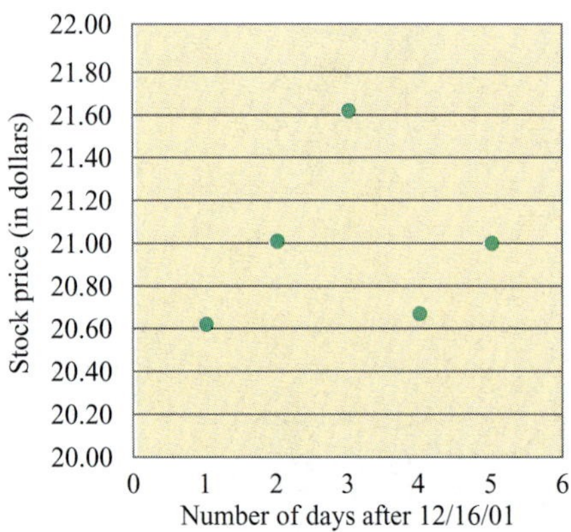

Source: Apple Computer Corporation.

In Exercises 10–11, determine whether the graphs represent functions by applying the Vertical Line Test.

10.

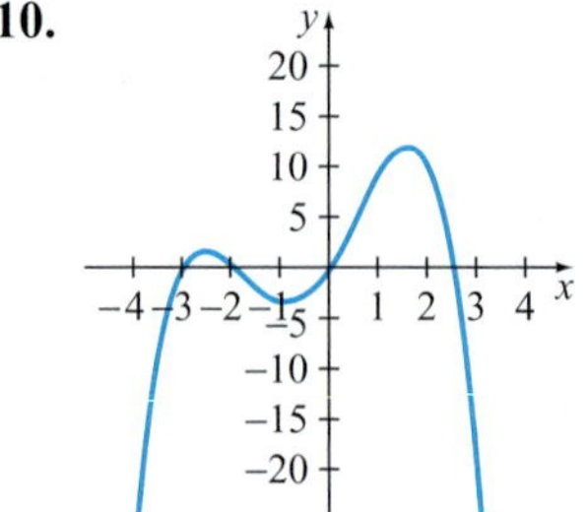

11.

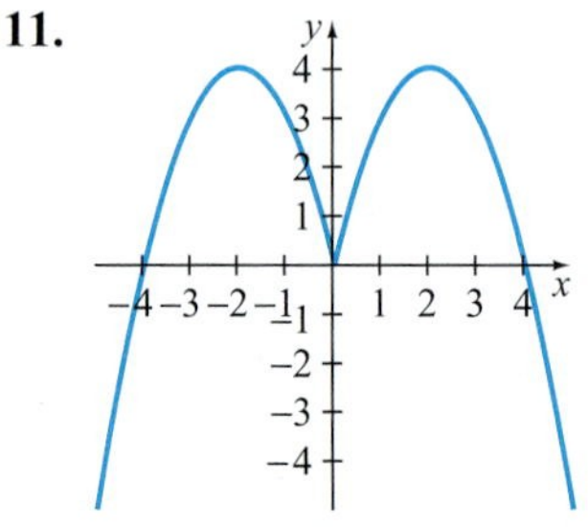

Section 1.2 *In Exercises 12–15, calculate the slope of the linear function passing through the points.*

12. (2, 9) and (4, 3)

13. (−3, −4) and (0, −2)

14. (1.3, 3.5) and (3.5, 1.3)

15. (7, 11) and (8, 0)

In Exercises 16–18, find the x-intercept and the y-intercept of the linear function.

16. $y = -5x + 10$ **17.** $y = -3x + 18$

18. $y = 2x - 12$

In Exercises 19–20, write the equation of the linear function passing through the points in slope-intercept form and in standard form.

19. (2, 5) and (4, 3) **20.** (−3, 4) and (0, −2)

Section 1.3 *In Exercises 21–23, find a model for the data and answer the given questions.*

21. **Nutrition** The Recommended Daily Allowance (RDA) for fat for a person on a 2,000-calorie-per-day diet is less than 65 grams.

A McDonalds Big N' Tasty® sandwich contains 32 grams of fat. A super size order of French fries contains 29 grams of fat. **(Source:** www.mcdonalds.com.)

(a) Write the equation for fat grams consumed as a function of large orders of French fries eaten.
(b) Write the equation for fat grams consumed as a function of Big N' Tasty sandwiches eaten.
(c) How many combination meals (Big N' Tasty sandwich and super size order of French fries) can you eat without exceeding the RDA for fat?

22. **Solid Waste Disposal** The Enumclaw Transfer/Recycling Station accepts clean wood (stumps, branches, etc.) in addition to household garbage. The station charges \$75 per ton with a minimum fee of \$12.75 per entry for vehicles disposing of clean wood. **(Source:** www.dnr.metrokc.gov.)

(a) Write a piecewise linear function to model the cost of disposing of x pounds of clean wood. (Assume that the entire amount of wood is delivered in a single load.)
(b) How much does it cost to dispose of 300 pounds of clean wood? 700 pounds?

23. **Used Car Value** In 2001, the average retail price of a 1998 Mercedes-Benz Roadster two-door SL500 was \$51,400. The average retail price of a 2000 Mercedes-Benz Roadster two-door SL500 was \$66,025. **(Source:** www.nadaguides.com.)

(a) Find a linear model for the value of a Mercedes-Benz Roadster two-door SL500 in 2001 as a function of its production year.
(b) Use your linear model to predict the 2001 value of a 1999 Mercedes-Benz Roadster two-door SL500.
(c) A 1999 Mercedes-Benz Roadster two-door SL500 had an average retail price of \$58,500 in 2001. How good was your linear model at predicting the value of the vehicle?

Make It Real

PROJECT 1

What to do

1. Find a set of at least six data points from an area of personal interest.
2. Draw a scatter plot of the data and explain why you do or do not believe that a linear model would fit the data well.
3. Find the equation of the line of best fit for the data.
4. Interpret the physical meaning of the slope and y-intercept of the model.
5. Use the model to predict the value of the function at an unknown point and explain why you do or do not think the prediction is accurate.
6. Explain how a consumer and/or a businessperson could benefit from the model.
7. Present your findings to the class and defend your conclusions.

Where to look for data

Box Office Guru
www.boxofficeguru.com
Look at historical data on movie revenues.

Nutri-Facts
www.nutri-facts.com
Compare the nutritional content of common foods based on serving size.

Quantitative Environmental Learning Project
www.seattlecentral.org/qelp
Look at environmental information in easy-to-access charts and tables.

U.S. Census Bureau
www.census.gov
Look at data on U.S. residents ranging from Internet usage to family size.

Local Gas Station or Supermarket
Track an item's price daily for a week.

School Registrar
Ask for historical tuition data.

Utility Bills
Look at electricity, water, or gas usage.

Employee Pay Statements
Look at take-home pay or taxes.

Body Weight Scale
Track your body weight for a week.

Chapter 2

Systems of Linear Equations

New companies often lose money for a few years before becoming profitable. Business owners may forecast profitability by modeling business costs and revenue with mathematical functions. Systems of linear equations are commonly used to determine when a linear revenue function is first expected to exceed a linear cost function. That is, systems of equations may be used to forecast the profitability of a company.

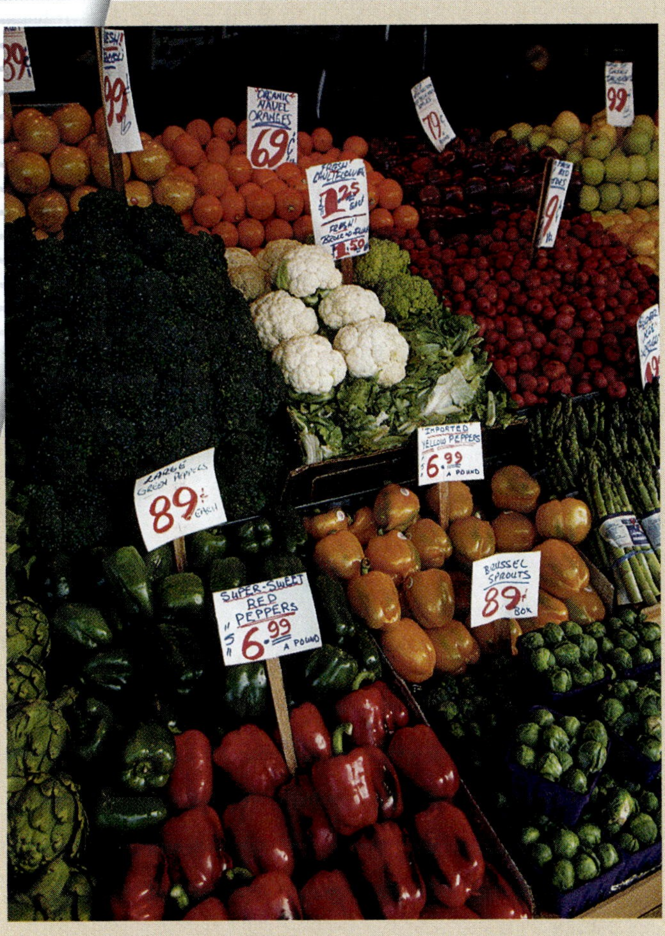

2.1 Systems of Linear Equations

- Graphically find the point of intersection of two lines
- Solve systems of linear equations using algebraic methods
- Interpret the meaning of solutions to linear systems

2.2 Using Matrices to Solve Linear Systems of Equations

- Represent systems of linear equations with augmented matrices
- Use row operations to put augmented matrices in reduced row echelon form

2.3 Linear System Applications

- Formulate linear systems from real-life data
- Interpret the meaning of solutions to linear systems

2.1 Systems of Linear Equations

- Graphically find the point of intersection of two lines
- Solve systems of linear equations using algebraic methods
- Interpret the meaning of solutions to linear systems

GETTING STARTED In order to be profitable, a company's revenue must exceed its costs. Analysts often try to predict when a new company will begin to turn a profit. When both revenue and expenditures may be modeled by linear functions, a system of linear equations may be used to determine the **break-even point,** the point at which revenue and costs are equal. In this section, we will show how to find the break-even point of a company. Additionally, we will demonstrate how to find the point of intersection of any two lines graphically and algebraically.

SYSTEM OF LINEAR EQUATIONS IN TWO VARIABLES

A **system of linear equations in two variables** is a collection of two or more linear equations. The solution to the system of equations, if it exists, is the point (or points) of intersection of the graphs of the equations. In other terms, a solution is an ordered pair that satisfies each of the equations in the system.

In reality, systems of linear equations may actually contain hundreds of variables and hundreds of equations. However, increasing the number of variables beyond three makes a graphical interpretation of the solution impossible. Even a three-variable system of equations may be difficult to represent graphically. For this reason, our graphical discussion will focus on systems of linear equations in two variables.

The Graphical Method

An estimate of the solution to a system of linear equations may easily be obtained using the graphical method.

EXAMPLE 1

Finding a Solution to a System of Linear Equations Using Technology

Find the solution to the system of linear equations. (That is, find the point of intersection of the graphs of the two lines.)

$$y = -1.5x + 8$$
$$y = x + 1$$

SOLUTION Plotting the linear equations yields the graph in Figure 2.1.

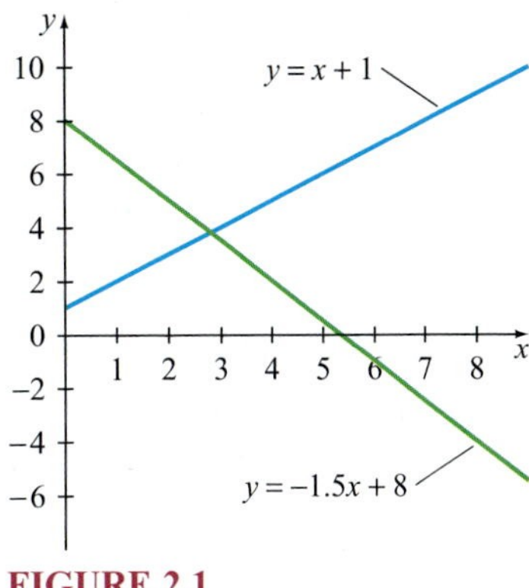

FIGURE 2.1

It appears from the graph that the lines intersect at or near (3, 4). One of the limitations of graphical analysis is that it is difficult to be precise; however, the TI-83 calculator has a built-in tool that calculates the point of intersection of two graphs. Using the technique demonstrated in the following Technology Tip, we can determine that the point of intersection of the two lines is (2.8, 3.8).

When using the calculator to determine the point of intersection, it is important to note that the calculator may not give the *exact* answer. It will, however, return the best possible decimal approximation of the exact answer. For example, if the point of intersection is $(\sqrt{3}, \sqrt{2})$, the calculator will return $(1.732050808, 1.414213562)$.

TECHNOLOGY TIP

Finding a Point of Intersection

1. Simultaneously graph both functions. Adjust the window as necessary to make the point of intersection visible. (If you've forgotten how to do this, reread the "Graphing a Function" technology tip in Section 1.1.)

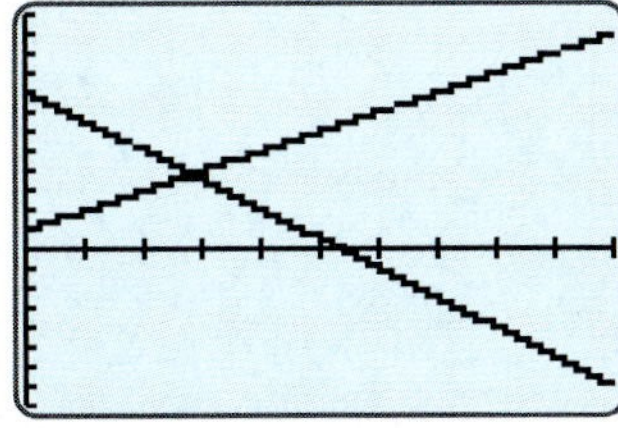

2. Press 2nd TRACE to bring up the Calculate Menu.

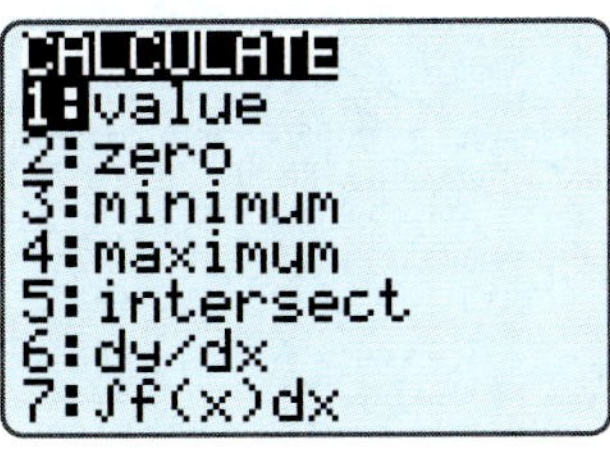

(Continued)

3. Select `5:intersect`. The calculator asks "`First curve?`" Use the blue arrow buttons to select either curve, then press ENTER. The calculator asks "`Second curve?`" Use the arrow buttons to select the second line and press ENTER.

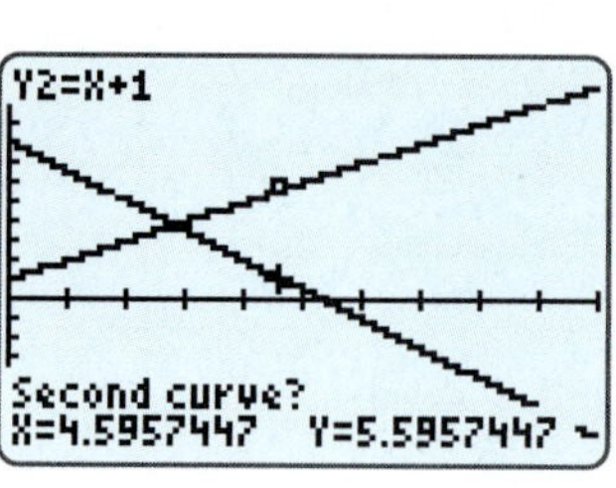

4. The calculator asks "`Guess?`" If there is more than one point of intersection, move the cursor near the desired point of intersection and press ENTER. Otherwise, just press ENTER. The point of intersection is highlighted on the graph and its coordinates are displayed.

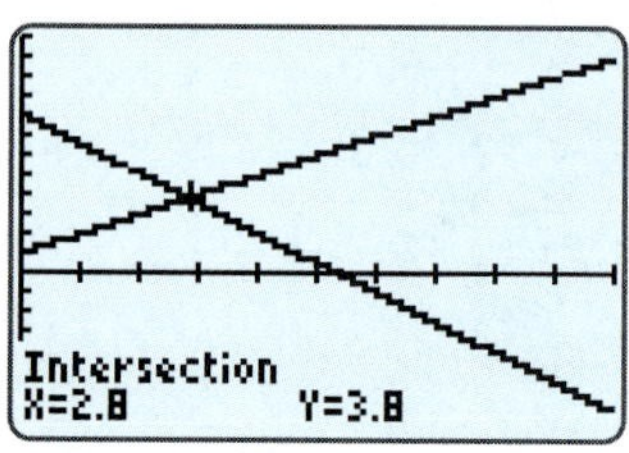

EXAMPLE 2

Finding a Solution to a System of Linear Equations Graphically

Solve the system of equations graphically.

$$2x - 3y = 6$$
$$3x + 2y = 48$$

SOLUTION We begin by writing each of the lines in slope-intercept form.

$$\begin{aligned} 2x - 3y &= 6 \\ -3y &= -2x + 6 \\ y &= \frac{2}{3}x - 2 \end{aligned} \qquad \begin{aligned} 3x + 2y &= 48 \\ 2y &= -3x + 48 \\ y &= -\frac{3}{2}x + 24 \end{aligned}$$

Then we graph the functions, as shown in Figure 2.2.

It appears that the graphs intersect at $(12, 6)$. We can verify this by using the graphing calculator or by substituting the values of x and y into each of the original equations. We will do the latter.

$$\begin{aligned} 2x - 3y &= 6 \\ 2(12) - 3(6) &= 6 \qquad \text{Substituting in } (12, 6) \\ 24 - 18 &= 6 \\ 6 &= 6 \end{aligned}$$

$$\begin{aligned} 3x + 2y &= 48 \\ 3(12) + 2(6) &= 48 \qquad \text{Substituting in } (12, 6) \\ 36 + 12 &= 48 \\ 48 &= 48 \end{aligned}$$

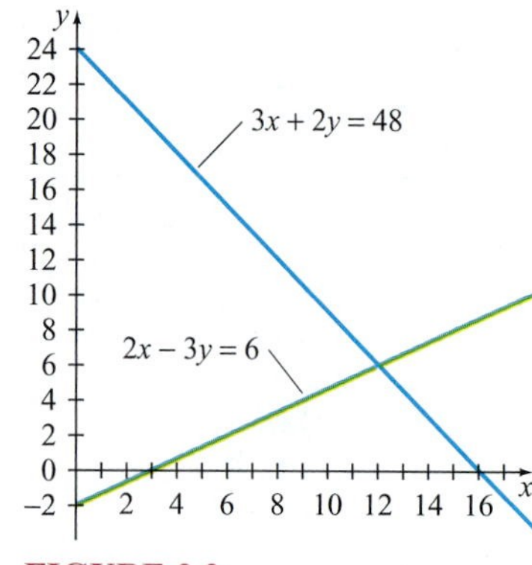

FIGURE 2.2

Since the point $(12, 6)$ satisfies both equations, $x = 12$ and $y = 6$ is the solution to the system of equations.

Business analysts are often interested in the profit, revenue, and costs of a company. In evaluating a business plan, it is important to know at what sales level revenue and costs are expected to be equal. This sales level is referred to as the **break-even point** and is the point at which the company begins to turn a profit.

Any business has two types of costs: **fixed costs** and **variable costs.** Fixed costs are those that remain constant independent of production levels. For example, building rent, product research and development, and advertising costs are often viewed as fixed costs. Variable costs are those that vary depending upon the level of production. For example, raw materials and production-line workers' wages may be viewed as variable costs. If no items are produced, no raw materials are purchased and no production-line workers are needed. As the number of items produced increases, the raw material and production-line labor costs also increase.

Break-even analysis may be applied to large companies or small home-based businesses. An example of one such business is given in Example 3.

EXAMPLE 3

Using a System of Linear Equations to Find the Break-Even Point

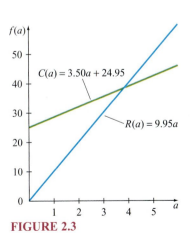

An artisan is considering selling her handmade craft angels online. She estimates her material cost for each angel to be \$3.50. An online merchant, craftmall.com, offers to advertise her product for a \$24.95 monthly fee. Comparing her craft to similar crafts on the market, she estimates that she can sell the craft angel for \$9.95. How many angels will she have to sell each month to break even? At that production level, what will be her production cost, revenue, and profit?

SOLUTION The cost equation for the craft angels is the sum of the variable cost, \$3.50 per angel, and the fixed cost, \$24.95. Let a be the number of angels sold. The cost equation is

$$y = 3.50a + 24.95 \text{ dollars}$$

The revenue equation is

$$y = 9.95a \text{ dollars}$$

She must determine when her revenue will equal her cost. In other words, she must solve the following system of equations.

$$y = 3.50a + 24.95$$
$$y = 9.95a$$

Graphing the two functions simultaneously results in the graph shown in Figure 2.3.

It appears that the graphs intersect near $(4, 38)$. At the intersection point, the revenue and cost functions are equal. We can find the exact point of intersection by setting the revenue and cost expressions equal to each other and solving algebraically.

$$9.95a = 3.50a + 24.95 \quad \text{Revenue expression = cost expression}$$
$$6.45a = 24.95$$
$$a = \frac{24.95}{6.45}$$
$$a \approx 3.868$$
$$a \approx 4 \text{ angels}$$

FIGURE 2.3

The break-even point is (3.868, 38.49); however, in the context of the problem, it doesn't make sense to talk about 3.868 angels. We conclude that she must sell four craft angels each month in order to cover her costs.

The cost to produce four angels is

$$\begin{aligned} y &= 3.50(4) + 24.95 \\ &= \$38.95 \end{aligned}$$

The revenue from the sale of four angels is

$$\begin{aligned} y &= 9.95(4) \\ &= \$39.80 \end{aligned}$$

She will have a profit of \$0.85 if she sells four angels.

When solving an application problem (such as the one shown in Example 3), it is essential to make sure that the solution makes sense in the context of the problem. A common error among students is to accept a mathematically correct answer [i.e., (3.868, 38.49)] without verifying that it makes sense in the context of the problem.

In Example 3, we found the intersection of two lines algebraically. Two common methods used to solve systems of linear equations algebraically are the substitution method and the elimination method.

HOW TO The Substitution Method

To solve a system of linear equations

$$\begin{aligned} ax + by &= c \\ dx + ey &= f \end{aligned}$$

do the following:

1. Solve the first equation for y in terms of x.
2. Replace y in the second equation with the result from Step 1.
3. Solve the second equation for x.
4. Substitute the value of x back into the first equation and solve for y.

If the system of equations models real-life data, you should also complete Steps 5 and 6.

5. Ask yourself if the mathematical solution makes sense in its real world context.
6. Reevaluate the functions, as necessary, and verbally express the real-world meaning of the result.

Although we stated that you should solve the first equation for y in terms of x, you may equivalently solve for x in terms of y and obtain the same result.

EXAMPLE 4

Solving a System of Linear Equations with the Substitution Method

Solve the system of equations.

$$5x + 2y = 12$$
$$4x - y = 7$$

SOLUTION We may solve either equation for y. Since the second equation looks easier, we will solve it for y.

$$4x - y = 7$$
$$4x = y + 7$$
$$y = 4x - 7$$

Substituting this result in for y in the equation $5x + 2y = 12$ yields

$$5x + 2(4x - 7) = 12$$
$$5x + 8x - 14 = 12$$
$$13x = 26$$
$$x = 2$$

Since $y = 4x - 7$,

$$y = 4(2) - 7 \qquad \text{Substitute } x = 2$$
$$y = 1$$

Therefore, the solution to the system of equations is $x = 2$ and $y = 1$. We can verify the accuracy of this answer by substituting these values back into each of the original equations.

$$5x + 2y = 12 \qquad 4x - y = 7$$
$$5(2) + 2(1) = 12 \qquad 4(2) - (1) = 7$$
$$12 = 12 \qquad 7 = 7$$

EXAMPLE 5

Solving a System of Linear Equations with the Substitution Method

Solve the system of equations.

$$2x - 7y = -4$$
$$x + 12y = 29$$

SOLUTION Since the coefficient on the x term in the second equation is a 1, it will be easier to solve the equation for x instead of y.

$$x + 12y = 29$$
$$x = -12y + 29$$

Substituting this value of x into the first equation yields

$$\begin{aligned} 2x - 7y &= -4 \\ 2(-12y + 29) - 7y &= -4 \\ -24y + 58 - 7y &= -4 \\ -31y &= -62 \\ y &= 2 \end{aligned}$$

To find the corresponding value of x, we back-substitute $y = 2$.

$$\begin{aligned} x &= -12y + 29 \\ x &= -12(2) + 29 \\ &= 5 \end{aligned}$$

The solution to the system of equations is $x = 5$ and $y = 2$.

Solving a system of equations by substitution is especially easy when the linear equations are written in slope-intercept form, $y = mx + b$. Since y is already written as a function of x, we can proceed directly to Step 2 of the substitution process.

EXAMPLE 6

Solving a System of Linear Equations with the Substitution Method

Based on data from 1992 to 2000, the annual value of printing products shipped in the United States may be modeled by

$$V(t) = 3203.0t + 80{,}732 \text{ million dollars}$$

and the annual value of furniture and related products shipped in the United States may be modeled by

$$V(t) = 3511.7t + 48{,}294 \text{ million dollars}$$

where t is the number of years since the end of 1992. According to the models, when will the annual value of products shipped by the printing industry and the furniture industry be the same?

SOLUTION We will replace the V in the second equation with the value of V from the first equation and solve for t.

$$\begin{aligned} 3203.0t + 80{,}732 &= 3511.7t + 48{,}294 \\ 32{,}438 &= 308.7t \\ t &\approx 105.08 \end{aligned}$$

We solve for the value of $V(t)$ by substituting $t = 105.08$ into either equation.

$$\begin{aligned} V(105.08) &= 3203.0(105.08) + 80{,}732 \\ &= 417{,}300 \text{ (accurate to five significant digits)} \end{aligned}$$

According to the model, at the end of 2097 ($t = 105$), the annual value of products shipped by the printing and furniture industries will be approximately equal ($417,300 million). Although we found a mathematical solution, we are a bit skeptical of the real-life accuracy of the result, since it is so far outside of the original data set.

EXAMPLE 7

Solving a System of Linear Equations with No Solution

Find the solution to the system of linear equations.

$$4x - 5y = 20$$
$$8x - 10y = 30$$

SOLUTION We begin by solving the first equation for y.

$$\begin{aligned} 4x - 5y &= 20 \\ 5y &= 4x - 20 \\ y &= 0.8x - 4 \end{aligned}$$

We substitute this value of y into the second equation and solve for x.

$$\begin{aligned} 8x - 10y &= 30 \\ 8x - 10(0.8x - 4) &= 30 \\ 8x - 8x + 40 &= 30 \\ 40 &= 30 \end{aligned}$$

But $40 \neq 30$. Since the system of equations led to a contradiction, the system of equations does not have a solution. This can readily be seen by writing both equations in slope-intercept form.

$$y = 0.8x - 4$$
$$y = 0.8x - 3$$

The lines are parallel, since they have the same slope but different y-intercepts (see Figure 2.4). Therefore, the two lines do not intersect and the system of equations does not have a solution.

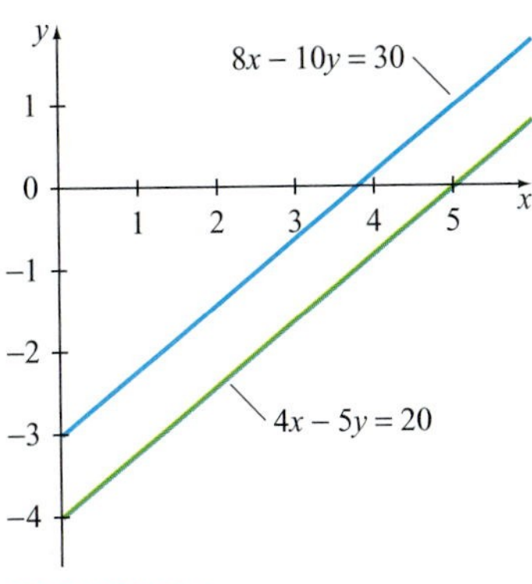

FIGURE 2.4

EXAMPLE 8

Solving a System of Linear Equations with Infinitely Many Solutions

Find the solution to the system of linear equations.

$$12x - 15y = 60$$
$$8x - 10y = 40$$

SOLUTION We begin by solving the first equation for y.

$$12x - 15y = 60$$
$$15y = 12x - 60$$
$$y = 0.8x - 4$$

We substitute this value of y into the second equation and solve for x.

$$8x - 10y = 40$$
$$8x - 10(0.8x - 4) = 40$$
$$8x - 8x + 40 = 40$$
$$40 = 40$$

As in Example 7, the x was eliminated; however, in this case, the resulting statement was true. Therefore, the system of equations has infinitely many solutions. The slope-intercept form of each equation is $y = 0.8x - 4$. Any point on this line is a solution to the system of equations.

Systems of linear equations will have 0, 1, or infinitely many solutions. A system of equations without a solution is said to be **inconsistent.** A system of equations with infinitely many solutions is said to be **dependent** (see Figure 2.5).

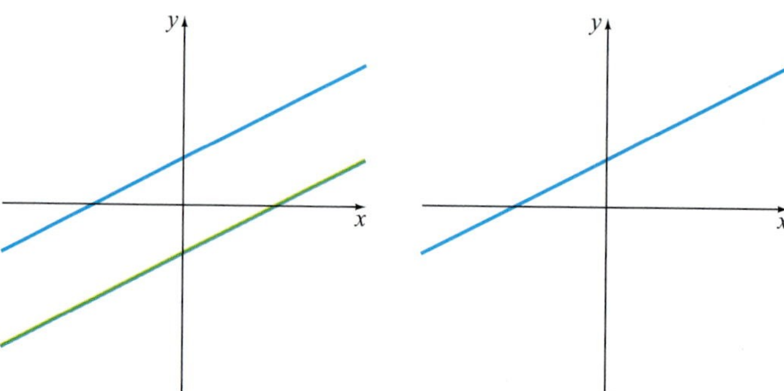

inconsistent system
two parallel lines
no solution

dependent system
both linear equations give the same line
infinitely many solutions

FIGURE 2.5

In order for a system of three or more linear equations to have a solution, all of the lines must intersect at the same point. The fact that two lines intersect at a point (a, b) does not ensure that a third line will intersect the first two lines at the same point.

EXAMPLE 9

Solving a System of Equations with More Equations than Unknowns

Solve the system of equations.

$$2x - y = 5$$
$$3x + 2y = 11$$
$$4x - 4y = 8$$

SOLUTION We will find the point of intersection of the first two lines and then check to see if the point is a solution to the third equation. The first equation may be written as $y = 2x - 5$. Substituting this value in for y in the second equation and solving for x yields

$$\begin{aligned} 3x + 2y &= 11 \\ 3x + 2(2x - 5) &= 11 \quad \text{Substitute } y = 2x - 5 \\ 3x + 4x - 10 &= 11 \\ 7x &= 21 \\ x &= 3 \end{aligned}$$

Since $y = 2x - 5$,

$$\begin{aligned} y &= 2(3) - 5 \\ &= 1 \end{aligned}$$

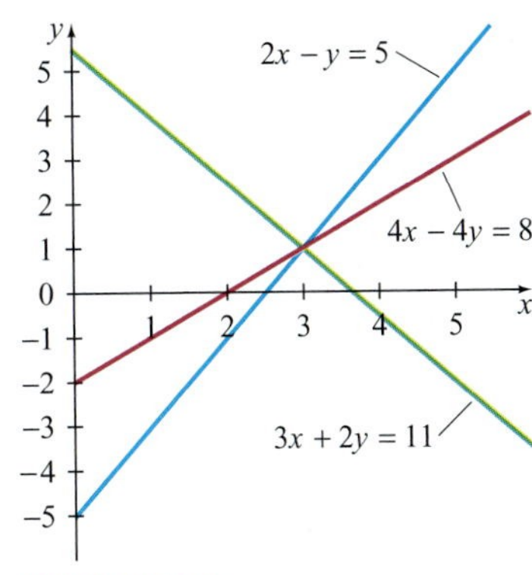

FIGURE 2.6

The point of intersection of the first two lines is (3, 1). We will check to see if this point satisfies the third equation.

$$\begin{aligned} 4x - 4y &= 8 \\ 4(3) - 4(1) &= 8 \quad \text{Substitute } x = 3 \text{ and } y = 1 \\ 12 - 4 &= 8 \\ 8 &= 8 \end{aligned}$$

Since the resulting statement was true, the solution to the system of equations is $x = 3$ and $y = 1$. (A false statement would have implied that the system was inconsistent.) We graphically confirm the result in Figure 2.6.

EXAMPLE 10

Solving an Inconsistent System of Linear Equations

Solve the system of equations.

$$\begin{aligned} 5x - 3y &= -2 \\ -x + y &= 2 \\ -3x - 2y &= -19 \end{aligned}$$

SOLUTION We will find the point of intersection of the first two lines and then check to see if the point is a solution to the third equation. The second equation may be written as $y = x + 2$. Substituting this value in for y in the first equation and solving for x yields

$$\begin{aligned} 5x - 3y &= -2 \\ 5x - 3(x + 2) &= -2 \quad \text{Substitute } y = x + 2 \\ 5x - 3x - 6 &= -2 \\ 2x &= 4 \\ x &= 2 \end{aligned}$$

Back-substituting $x = 2$ into $y = x + 2$ yields

$$\begin{aligned} y &= (2) + 2 \\ &= 4 \end{aligned}$$

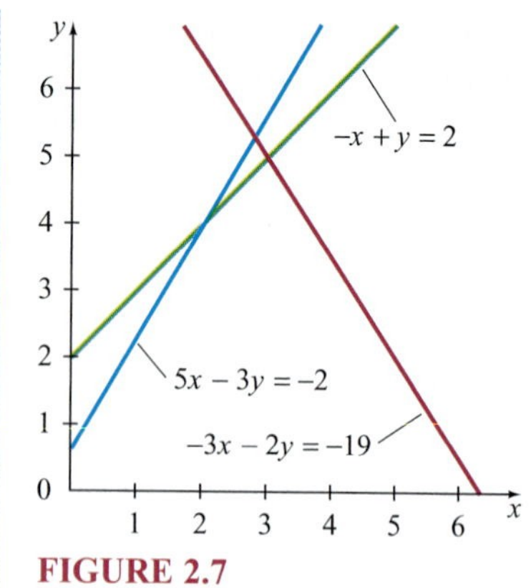

FIGURE 2.7

The point of intersection of the first two lines is $(2, 4)$. We will now determine if this point is on the third line.

$$\begin{aligned} -3x - 2y &= -19 \\ -3(2) - 2(4) &= -19 \qquad \text{Substitute } x = 2 \text{ and } y = 4 \\ -6 - 8 &= -19 \\ -14 &= -19 \end{aligned}$$

Since $-14 \neq -19$, the statement is false and $(2, 4)$ is not on the third line. Therefore the system is inconsistent. Graphing the lines confirms our result, as shown in Figure 2.7.

The Elimination Method

Although the elimination method is most commonly used when a system of linear equations contains three or more variables, it may also be used to solve two-variable systems. This method uses the properties of equivalent systems of equations to find the solution.

EQUIVALENT SYSTEMS OF EQUATIONS

The following operations on a system of equations result in a system of equations with the same solution as the original system:

1. Interchange (change the position of) two equations.
2. Multiply an equation by a nonzero number.
3. Add a nonzero multiple of one equation to a nonzero multiple of another equation and replace either equation with the result.

EXAMPLE 11

Finding an Equivalent System of Equations

Given the system of equations

$$\begin{aligned} 2x + 3y - z &= 6 \\ 3x - 6y + 2z &= 12 \end{aligned}$$

find an equivalent system of equations by doing the following:

1. Multiply the first equation by 2 and write the result as the first equation.
2. Add the first equation to the second and write the result as the second equation.

SOLUTION We first multiply the first equation $2x + 3y - z = 6$ by two.

$$\begin{aligned} 4x + 6y - 2z &= 12 \\ 3x - 6y + 2z &= 12 \end{aligned}$$

We add the two equations together and write the result as the second equation.

$$4x + 6y - 2z = 12$$
$$7x + 0y + 0z = 24$$

HOW TO The Elimination Method

To solve a system of equations using the elimination method, do the following:

1. Write each equation in standard form ($Ax + By + Cz + \cdots = K$).
2. Vertically align the variables in each equation with the corresponding variables in the other equations.
3. Use the equivalency operations to eliminate the x variable in all but the first equation.
4. Use the equivalency operations to eliminate the y variable in all but the second equation.
5. Continue to simplify the system in a like manner for any remaining variables.
6. Simplify the system to the form

$$\begin{aligned} x &= a \\ y &= b \\ z &= c \\ &\vdots \end{aligned}$$

EXAMPLE 12 Solving a System of Linear Equations with the Elimination Method

Find the solution to the system of linear equations.

$$2x - 9y = 29$$
$$3x + 8y = 22$$

SOLUTION If we multiply the first equation by 3 and the second equation by -2, we get an equivalent system.

$$6x - 27y = 87 \qquad \text{Multiply by 3}$$
$$-6x - 16y = -44 \qquad \text{Multiply by } -2$$

Now that the x terms have the opposite coefficients, we can eliminate the x terms by adding the second equation to the first equation.

$$\begin{array}{r} 6x - 27y = 87 \\ +(-6x - 16y = -44) \\ \hline 0x - 43y = 43 \end{array}$$

We place this result in the position of the second equation.

$$\begin{aligned} 6x - 27y &= 87 \\ -43y &= 43 \end{aligned}$$

Multiplying the first equation by $\frac{1}{3}$ and the second equation by $-\frac{1}{43}$ yields

$$\begin{aligned} 2x - 9y &= 29 \qquad \text{Multiply by } \tfrac{1}{3} \\ y &= -1 \qquad \text{Multiply } -\tfrac{1}{43} \end{aligned}$$

We want to eliminate the y term in the first equation. If we add nine times the second equation to the first equation, we will eliminate the y term in the first equation.

$$\begin{array}{rr} & 2x - 9y = 29 \\ + & 9y = -9 \\ \hline & 2x + 0y = 20 \end{array}$$

We place this result in the position of the first equation.

$$\begin{aligned} 2x &= 20 \\ y &= -1 \end{aligned}$$

We finish by multiplying the first equation by $\frac{1}{2}$.

$$\begin{aligned} x &= 10 \qquad \text{Multiply by } \tfrac{1}{2} \\ y &= -1 \end{aligned}$$

The solution is $x = 10$ and $y = -1$.

EXAMPLE 13

Solving a System of Linear Equations with the Elimination Method

Find the solution to the system of linear equations.

$$\begin{aligned} 3x - 2y + z &= -2 \\ x + 2y - z &= 6 \\ -x + y + z &= 0 \end{aligned}$$

SOLUTION A system of linear equations in two variables may be interpreted graphically as a group of lines drawn in two dimensions. A system of linear equations in three variables may be interpreted graphically as a group of planes drawn in three dimensions. The three-dimensional graph of the given system is shown in Figure 2.8.

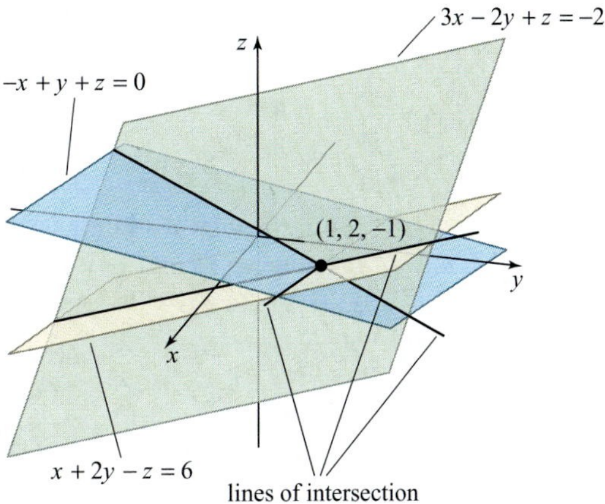

FIGURE 2.8

Notice that any two planes intersect in a line. The point where all three planes intersect will be the solution of the system of equations. In solving linear systems in three variables, we rarely draw the graphs because of the difficulty of three-dimensional graphing. However, a graphical understanding of what is going on helps to add meaning to the numerical solution.

We will proceed to solve the system algebraically.

$$\begin{aligned} &(1) \quad 3x - 2y + z = -2 \\ &(2) \quad x + 2y - z = 6 \\ &(3) \quad -x + y + z = 0 \end{aligned}$$

We have labeled the equations (1), (2), and (3) to make it easier to keep track of our work. Our strategy is to eliminate variables by adding a multiple of one equation to a multiple of another equation in such a way that one or more variables are eliminated. We will place the resultant equation in the position of either of the original equations. It is computationally convenient to have the coefficient of the x variable in the first equation equal to 1. We could divide equation (1) by 3 to achieve this result; however, that would create fractional coefficients on the y and z variables. We will instead interchange equations (1) and (2). After making the switch, we relabel the result.

$$\begin{aligned} &(1) \quad x + 2y - z = 6 \\ &(2) \quad 3x - 2y + z = -2 \\ &(3) \quad -x + y + z = 0 \end{aligned}$$

We observe that if we multiply equation (1) by -3 and add it to equation (2), the x variable in equation (2) will be eliminated. We will place the result back in equation (2). Symbolically, we can represent the operation as $-3E_1 + E_2 \rightarrow E_2$, where E_1 represents equation (1) and E_2 represents equation (2). We multiply each term in equation (1) by -3 and add it to the corresponding term in equation (2).

$$\begin{array}{r} -3x - 6y + 3z = -18 \\ + \quad (3x - 2y + z = -2) \\ \hline 0x - 8y + 4z = -20 \end{array}$$

We write the new equation back in (2).

$$\begin{array}{lrl} (1) & x + 2y - z = 6 & \\ (2) & -8y + 4z = -20 & -3E_1 + E_2 \to E_2 \\ (3) & -x + y + z = 0 & \end{array}$$

We next want to eliminate the x in the third equation. We observe that adding equations (1) and (3) will achieve the desired result. Symbolically, we write $E_1 + E_3 \to E_3$.

$$\begin{array}{r} x + 2y - z = 6 \\ +\ (-x + y + z = 0) \\ \hline 0x + 3y + 0z = 6 \end{array}$$

We write the new equation back in (3).

$$\begin{array}{lrl} (1) & x + 2y - z = 6 & \\ (2) & -8y + 4z = -20 & \\ (3) & 3y = 6 & E_1 + E_3 \to E_3 \end{array}$$

Observe that we can solve for y by multiplying (3) by $\frac{1}{3}$. Symbolically, we write $\frac{1}{3}E_3 \to E_3$.

$$\begin{array}{lrl} (1) & x + 2y - z = 6 & \\ (2) & -8y + 4z = -20 & \\ (3) & y = 2 & \frac{1}{3}E_3 \to E_3 \end{array}$$

We will interchange equations (2) and (3) so that the middle equation is of the form $y = b$. Symbolically, we write $E_2 \leftrightarrow E_3$.

$$\begin{array}{lrl} (1) & x + 2y - z = 6 & \\ (2) & y = 2 & E_2 \leftrightarrow E_3 \\ (3) & -8y + 4z = -20 & E_2 \leftrightarrow E_3 \end{array}$$

We will now eliminate the y term in equation (3). Observe that $8E_2 + E_3 \to E_3$ will do this.

$$\begin{array}{r} 0x + 8y + 0z = 16 \\ +(0x - 8y + 4z = -20) \\ \hline 0x + 0y + 4z = -4 \end{array}$$

We write the new equation back in equation (3).

$$\begin{array}{lrl} (1) & x + 2y - z = 6 & \\ (2) & y = 2 & \\ (3) & 4z = -4 & 8E_2 + E_3 \to E_3 \end{array}$$

Using $\frac{1}{4}E_3 \to E_3$, we solve for z.

$$\begin{array}{lrl} (1) & x + 2y - z = 6 & \\ (2) & y = 2 & \\ (3) & z = -1 & \frac{1}{4}E_3 \to E_3 \end{array}$$

Using $E_1 + E_3 \to E_1$, we eliminate the z term in equation (1).

$$\begin{array}{r} x + 2y - z = 6 \\ +(0x + 0y + z = -1) \\ \hline x + 2y + 0z = 5 \end{array}$$

We write the new equation back in equation (1).

$$\begin{array}{llr} (1) & x + 2y = 5 & E_1 + E_3 \to E_1 \\ (2) & y = 2 & \\ (3) & z = -1 & \end{array}$$

Using $E_1 - 2E_2 \to E_1$, we eliminate the y term in equation (1).

$$\begin{array}{r} x + 2y + 0z = 5 \\ -(0x + 2y + 0z = 4) \\ \hline x + 0y + 0z = 1 \end{array}$$

We write the new equation back in equation (1).

$$\begin{array}{llr} (1) & x = 1 & E_1 - 2E_2 \to E_1 \\ (2) & y = 2 & \\ (3) & z = -1 & \end{array}$$

The solution to the system of equations is $x = 1$, $y = 2$, and $z = -1$.

To be certain we're correct, we will check our work by plugging our solution into each of the original equations.

$$\begin{array}{lr} (1) & 3(1) - 2(2) + (-1) = -2 \\ (2) & (1) + 2(2) - (-1) = 6 \\ (3) & -1(1) + (2) + (-1) = 0 \end{array}$$

Simplifying, we get

$$\begin{array}{lr} (1) & -2 = -2 \\ (2) & 6 = 6 \\ (3) & 0 = 0 \end{array}$$

Each statement is true so we are certain that our solution is correct.

EXAMPLE 14 Solving a System of Linear Equations with the Elimination Method

Solve the system of equations.

$$\begin{aligned} 4x + 5y - 2z &= 30 \\ 3x - 2y + 7z &= -43 \\ -2x + 4y + 3z &= 1 \end{aligned}$$

SOLUTION

$$\begin{array}{lr} (1) & 4x + 5y - 2z = 30 \\ (2) & 3x - 2y + 7z = -43 \\ (3) & -2x + 4y + 3z = 1 \end{array}$$

We want to make the coefficient on the x term in equation (1) equal 1. The operation $E_1 - E_2 \to E_1$ will achieve this objective. The new system is

$$\begin{array}{lrl} (1) & x + 7y - 9z = 73 & E_1 - E_2 \to E_1 \\ (2) & 3x - 2y + 7z = -43 & \\ (3) & -2x + 4y + 3z = 1 & \end{array}$$

Next we will eliminate the x term in equation (2) with the operation $3E_1 - E_2 \to E_2$.

$$\begin{array}{lrl} (1) & x + 7y - 9z = 73 & \\ (2) & 23y - 34z = 262 & 3E_1 - E_2 \to E_2 \\ (3) & -2x + 4y + 3z = 1 & \end{array}$$

Next, we will eliminate the x term in equation (3) with the operation $2E_1 + E_3 \to E_3$.

$$\begin{array}{lrl} (1) & x + 7y - 9z = 73 & \\ (2) & 23y - 34z = 262 & \\ (3) & 18y - 15z = 147 & 2E_1 + E_3 \to E_3 \end{array}$$

We now want to eliminate the y term in equation (3). We will do this with the operation $18E_2 - 23E_3 \to E_3$. Since this computation is somewhat complex, we will show each step.

$$\begin{array}{r} 18(23y - 34z = 262) \\ \underline{-23(18y - 15z = 147)} \\ \\ 414y - 612z = 4716 \\ \underline{-(414y - 345z = 3381)} \\ 0y - 267z = 1335 \end{array}$$

We write the new equation back in equation (3).

$$\begin{array}{lrl} (1) & x + 7y - 9z = 73 & \\ (2) & 23y - 34z = 262 & \\ (3) & -267z = 1335 & 18E_2 - 23E_3 \to E_3 \end{array}$$

Solving equation (3) for z with the operation $-\frac{1}{267}E_3 \to E_3$ yields

$$\begin{array}{lrl} (1) & x + 7y - 9z = 73 & \\ (2) & 23y - 34z = 262 & \\ (3) & z = -5 & -\frac{1}{267}E_3 \to E_3 \end{array}$$

We will now eliminate the z term in equation (2) with $E_2 + 34E_3 \to E_2$.

$$\begin{array}{lrl} (1) & x + 7y - 9z = 73 & \\ (2) & 23y \quad = 92 & E_2 + 34E_3 \to E_2 \\ (3) & z = -5 & \end{array}$$

Then we eliminate the z term in equation (1) with $E_1 + 9E_3 \to E_1$.

$$\begin{array}{lrrl} (1) & x + 7y & = 28 & E_1 + 9E_3 \to E_1 \\ (2) & 23y & = 92 & \\ (3) & z & = -5 & \end{array}$$

Solving equation (2) for y with the operation $\frac{1}{23}E_2 \to E_2$ yields

$$\begin{array}{lrrl} (1) & x + 7y & = 28 & \\ (2) & y & = 4 & \frac{1}{23}E_2 \to E_2 \\ (3) & z & = -5 & \end{array}$$

We now eliminate the y term from equation (1) with $E_1 - 7E_2 \to E_1$.

$$\begin{array}{lrrl} (1) & x & = 0 & E_1 - 7E_2 \to E_1 \\ (2) & y & = 4 & \\ (3) & z & = -5 & \end{array}$$

The solution to the system of equations is $x = 0$, $y = 4$, and $z = -5$.

EXAMPLE 15

Solving a Dependent System of Equations with the Elimination Method

Solve the system of equations.

$$\begin{aligned} 2x + 6y - 4z &= 42 \\ 4x - 3y - 2z &= -42 \\ 6x + 3y - 6z &= 0 \end{aligned}$$

SOLUTION We first obtain a coefficient of 1 on the x term in equation (1) by using the operation $\frac{1}{2}E_1 \to E_1$.

$$\begin{aligned} x + 3y - 2z &= 21 \\ 4x - 3y - 2z &= -42 \\ 6x + 3y - 6z &= 0 \end{aligned}$$

We then eliminate the x terms in equations (1) and (2) with the operations $4E_1 - E_2 \to E_2$ and $6E_1 - E_3 \to E_3$.

$$\begin{aligned} x + 3y - 2z &= 21 \\ 15y - 6z &= 126 \\ 15y - 6z &= 126 \end{aligned}$$

Observe that equations (2) and (3) are identical. In fact, the operation $E_2 - E_3 \to E_3$ eliminates all variables from the third equation.

$$\begin{aligned} x + 3y - 2z &= 21 \\ 15y - 6z &= 126 \\ 0 &= 0 \end{aligned}$$

When the number of equations equals the number of variables, a unique solution may exist. However, if one of the equations "drops out" (as shown in this example), the system is either dependent or inconsistent. We will proceed to eliminate the y from equation (1) with the operation $5E_1 - E_2 \rightarrow E_1$ before discussing the final solution.

$$\begin{aligned} 5x \quad\quad - 4z &= -21 \\ 15y - 6z &= 126 \\ 0 &= 0 \end{aligned}$$

We will now solve each equation for x or y in terms of z.

$$\begin{aligned} 5x - 4z &= -21 \\ 5x &= 4z - 21 \\ x &= 0.8z - 4.2 \end{aligned} \qquad\qquad \begin{aligned} 15y - 6z &= 126 \\ 15y &= 6z + 126 \\ y &= 0.4z + 8.4 \end{aligned}$$

If we let $z = t$ (where t is any real number), then $x = 0.8t - 4.2$ and $y = 0.4t + 8.4$. The system is a dependent system of equations and thus has infinitely many solutions. Each solution is of the form $x = 0.8t - 4.2$, $y = 0.4t + 8.4$, and $z = t$. For example, if we let $t = 0$, then the solution is $(-4.2, 8.4, 0)$. If we let $t = 1$, the solution is $(-3.4, 8.8, 1)$.

In graphical terms, a dependent system of three linear equations with three variables is a set of three planes that intersect in a common line instead of a single point. The graph for the system of equations given in this example is shown in Figure 2.9.

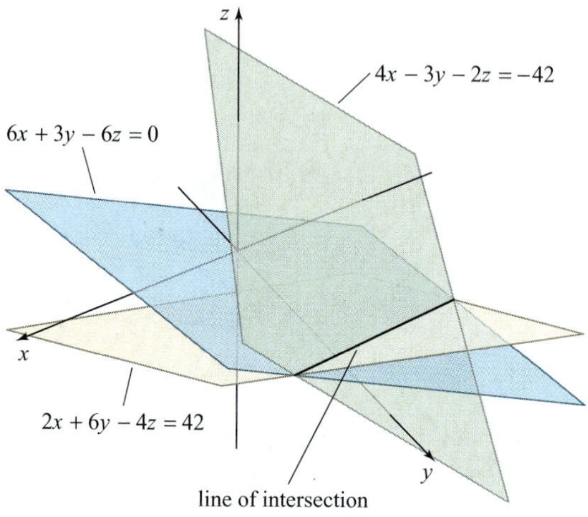

FIGURE 2.9

In the next section, we will introduce a process for solving systems of equations with three or more variables that, when mastered, is easier than the approach shown in Example 15.

2.1 Summary

In this section, you learned how to find the solution to a system of linear equations both graphically and algebraically. You learned two algebraic methods for solving linear systems of equations: the substitution method and the elimination method.

2.1 Exercises

In Exercises 1–10, use your graphing calculator to find the point(s) of intersection of the lines. If the lines do not intersect at a common point, state that a solution does not exist.

1. $y = 5x - 9$
$y = 2x - 3$

2. $-5x + 2y = 8$
$y = -3x + 26$

3. $y = 5$
$y = -3.21x + 4.32$

4. $-26x + 10y = 112$
$21x + 100y = 1000$

5. $-98x + 10y = -76$
$-98x + 10y = -28$

6. $y = 0.5x - 9$
$y = 0.5x + 10$

7. $y = 2x - 9$
$y = -3x - 9$
$y = \frac{1}{2}x - 9$

8. $y = 0.25x + 1$
$4x + y = 18$
$y = 2$

9. $y = 4.8x + 6$
$y = 1.9x - 2.1$
$y = 2x + 1.6$

10. $y = 11x - 12$
$y = 15x - 21$
$y = 15x - 12$

In Exercises 11–20, solve the system of equations algebraically. Compare your solutions to those obtained in Exercises 1–10.

11. $y = 5x - 9$
$y = 2x - 3$

12. $-5x + 2y = 8$
$y = -3x + 26$

13. $y = 5$
$y = -3.21x + 4.32$

14. $-26x + 10y = 112$
$21x + 100y = 1000$

15. $-98x + 10y = -76$
$-98x + 10y = -28$

16. $y = 0.5x - 9$
$y = 0.5x + 10$

17. $y = 2x - 9$
$y = -3x - 9$
$y = \frac{1}{2}x - 9$

18. $y = 0.25x + 1$
$4x + y = 18$
$y = 2$

19. $y = 4.8x + 6$
$y = 1.9x - 2.1$
$y = 2x + 1.6$

20. $y = 11x - 12$
$y = 15x - 21$
$y = 15x - 12$

In Exercises 21–30, determine the solution by setting up and solving a linear system of equations.

21. **Diabetes** The incidence of diabetes in men and women aged 45–64 years increased between 1994 and 1999. The incidence for women may be modeled by

$$F(t) = 3.75t + 131 \text{ people per thousand}$$

and the incidence for men may be modeled by

$$M(t) = 20t + 109 \text{ people per thousand}$$

where t is the number of years after 1994. (**Source:** Modeled from *Statistical Abstract of the United States, 2001,* Table 159, p. 109.) According to the models, when will the incidence for men and women be the same?

22. **Gasoline Prices** Based on data from 1990 to 2000, the retail cost of unleaded regular gasoline may be modeled by

$$R(t) = 0.035t + 1.16 \text{ dollars per gallon}$$

and the retail cost of unleaded premium gasoline may be modeled by

$$P(t) = 0.034t + 1.35 \text{ dollars per gallon}$$

where t is the number of years after 1990. (**Source:** Modeled from *Statistical Abstract of the United States, 2001,* Table 704, p. 467.) According to the models, when will the cost of unleaded regular and unleaded premium gasoline be the same? Based on your experience of purchasing gasoline, does your solution seem reasonable? Explain.

23. **Weekly Food Cost** Based on data from 1990 and 2000, the weekly food cost for a 15- to 19-year-old male in a four-person family t years after 1990 may be modeled by one of the following functions based on the family's food-spending behavior.

Thrifty Plan:

$$T(t) = 0.56t + 21.40 \text{ dollars}$$

Moderate–Cost Plan:

$$M(t) = 1.06t + 36.80 \text{ dollars}$$

Liberal Plan:

$$L(t) = 1.21t + 42.60 \text{ dollars}$$

(**Source:** Modeled from *Statistical Abstract of the United States, 2001,* Table 705, p. 467.)

According to the models, will the food cost of all three plans ever be the same? Explain.

24. **Weekly Food Cost** Based on data from 1990 and 2000, the weekly food cost for a 12- to 19-year-old female in a four-person family t years after 1990 may be modeled by one of the following functions based on the family's food-spending behavior. (**Source:** Modeled from *Statistical Abstract of the United States, 2001,* Table 705, p. 467.)

Thrifty Plan:

$$T(t) = 0.55t + 20.80 \text{ dollars}$$

Moderate–Cost Plan:

$$M(t) = 0.85t + 30.10 \text{ dollars}$$

Liberal Plan:

$$L(t) = 1.04t + 36.30 \text{ dollars}$$

According to the models, will the food cost of all three plans ever be the same? Explain.

25. **Bread and Gasoline Prices** Based on data from 1990 to 2000, the retail cost of unleaded regular gasoline may be modeled by

$$R(t) = 0.035t + 1.16 \text{ dollars per gallon}$$

where t is the number of years since 1990. Based on data from 1993 to 2000, the price of a loaf of whole wheat bread may be modeled by

$$C(t) = 0.0404t + 0.991$$

where t is the number of years since 1990. (**Source:** Modeled from *Statistical Abstract of the United States, 2001,* Tables 704 and 706, pp. 467–468.) Rounded up to the nearest year, when will a loaf of whole wheat bread cost more than a gallon of unleaded regular gasoline?

26. **Entertainment Revenues** Based on data from 1998 and 1999, the amount of revenue brought in by amusement and theme parks may be modeled by

$$A(t) = 177t + 7335 \text{ million dollars}$$

where t is the number of years since the end of 1998. The amount of revenue brought in by racetracks may be modeled by

$$R(t) = 507t + 4599 \text{ million dollars}$$

According to the models, in what year will racetrack revenues surpass amusement park revenues? (**Source:** Modeled from *Statistical Abstract of the United States, 2001,* Table 1231, pp. 753–754.)

27. **Artist and Athlete Earnings** In 1999 the number of paid employees in the independent artist, writer, and performer industry surpassed the number of paid employees in the sports team industry; however, the annual payroll for the artist industry still trailed that for the sports team industry by \$4.1 billion. Based on data from 1998 and 1999, the payroll for the sports team industry may be modeled by

$$S(t) = 990t + 5718 \text{ million dollars}$$

and the payroll for the independent artist industry may be modeled by

$$A(t) = -35t + 3494 \text{ million dollars}$$

where t is the number of years since 1998. (**Source:** Modeled from *Statistical Abstract of the United States, 2001,* Table 1232, p. 754.) According to the models, when were the payrolls for the two industries equal? Does this solution seem reasonable? Explain.

28. **High School Athletics** From 1971 to 1999, participation in high school athletics among women increased by 809 percent. In contrast, participation among men increased by a mere 5 percent. Based on data from 1991 to 2000, the number of women in high school sports may be modeled by

$$F(t) = 0.104t + 1.83 \text{ million}$$

and the number of men in high school sports may be modeled by

$$M(t) = 0.0606t + 3.33 \text{ million}$$

where t is the number of years since the 1990–1991 school year. (**Source:** Modeled from *Statistical Abstract of the United States, 2001,* Table 1247, p. 764.) According to the models, in what school year will the number of female athletes surpass the number of male athletes?

29. **Chicken and Fish Consumption** Using data from 1970 to 1999, the per capita consumption of chicken may be modeled by

$$C(t) = 0.982t + 23.7 \text{ pounds}$$

and the per capita consumption of fish may be modeled by

$$F(t) = 0.137t + 11.7 \text{ pounds.}$$

where t is the number of years since 1970. (**Source:** Modeled from *Statistical Abstract of the United States, 2001,* Table 202, p. 129.) According to the models, will the per capita consumption of fish ever exceed the per capita consumption of chicken? Explain.

30. **Pork and Beef Consumption** Using data from 1981 to 2001, the demand for pork may be modeled by

$$q = -21.2p + 95.2 \text{ pounds per person}$$

where p is the constant-demand price of pork in dollars per pound. The demand for beef over the same period may be modeled by

$$q = -18.1p + 128 \text{ pounds per person}$$

where p is the constant-demand price of beef in dollars per pound. (**Source:** Modeled from the Research Institute on Livestock Pricing.) Is there a price at which the demand for beef and pork will be the same?

In Exercises 31–40, solve the system of equations using the elimination method. If the system of equations is inconsistent or dependent, so state.

31. $\begin{aligned} x - y + z &= 6 \\ x + y + z &= 8 \\ x - y - z &= -4 \end{aligned}$

32. $\begin{aligned} 3x - y &= 4 \\ 2x + z &= 5 \\ x + y + z &= 5 \end{aligned}$

33. $\begin{aligned} 6x - 3y + 3z &= 6 \\ x + 5y + 10z &= 1 \\ 4x - 5y - z &= 4 \end{aligned}$

34. $\begin{aligned} -2x - y + 4z &= 8 \\ 2x + y - 4z &= -8 \\ x + y + z &= 6 \end{aligned}$

35. $\begin{aligned} 2x + 2y + 2z &= 0 \\ 3x - 5y + 4z &= -90 \\ 5x - 3y + 6z &= -90 \end{aligned}$

36. $\begin{aligned} 3x - y &= 2 \\ x + z &= 2 \\ 2x - y - z &= 2 \end{aligned}$

37. $\begin{aligned} 3x - 3y + z &= 5 \\ x + y + z &= 11 \\ 2x + 2z &= 10 \end{aligned}$

38. $\begin{aligned} -2x - 7y + 3z &= 0 \\ 5x + 4y - 2z &= 0 \\ 4x + 3y + 9z &= 0 \end{aligned}$

39. $\begin{aligned} 9x - 8y - 7z &= 16 \\ 7x - 5y + 3z &= 4 \\ 5x - 3y - z &= 6 \end{aligned}$

40. $\begin{aligned} 10x - 5y &= 0 \\ 20x + 5y + z &= 0 \\ 30x + z &= 0 \end{aligned}$

Exercises 41–45 are intended to challenge your understanding of systems of linear equations.

41. Solve the system of linear equations.

$$\begin{aligned} x + y + z + w &= 2 \\ x - y + z - w &= 2 \\ x - y - z + w &= 0 \\ x - y - z - w &= 0 \end{aligned}$$

42. Solve the system of linear equations.

$$\begin{aligned} x + y + z + w &= 1 \\ x - y + z - w &= 2 \\ x - y - z + w &= 3 \\ x - y - z - w &= 4 \end{aligned}$$

43. A failing business sells three colors of hair dye: fluorescent green, magenta, and sky blue. Its supplier charges \$3 per bottle of fluorescent green, \$2 per bottle of magenta, and \$1 per bottle of sky blue. Regardless of the color, the business sells a bottle of hair dye to consumers for \$5. The business has budgeted \$300 to spend on hair dye and hopes to generate sufficient income from hair dye sales to pay its hair dye supplier and pay the \$900 lease on the building that houses the store. It has sufficient space to store 250 bottles. Is it mathematically possible for the business to achieve its goal? Explain.

44. A system of equations has more equations than it has variables. Is it possible for such a system to have exactly one solution? Explain.

45. A system of equations has more variables than it has equations. Is it possible for such a system to have exactly one solution? Explain.

2.2 Using Matrices to Solve Linear Systems of Equations

- Represent systems of linear equations with augmented matrices
- Use row operations to put augmented matrices in reduced row echelon form

GETTING STARTED In the previous section, we solved linear systems of equations in two and three variables using the graphical, substitution, and elimination methods. For equations containing more than three variables, the elimination method is the procedure of choice. In this section, we will introduce matrix notation and demonstrate how matrices are used to represent and solve systems of linear equations. We will also show how to use technology to simplify an augmented matrix to reduced row echelon form.

Matrix Notation

AN $m \times n$ MATRIX

An $m \times n$ matrix A is an array of numbers with m rows and n columns. The plural of matrix is *matrices.*

Capital letters are typically used to represent matrices. The following examples show matrices of various dimensions.

$$A = \begin{bmatrix} 2 & 4 & 6 \\ 5 & 3 & 1 \end{bmatrix} \qquad B = \begin{bmatrix} 3 \\ -1 \\ 6 \end{bmatrix} \qquad C = \begin{bmatrix} 2.1 & 1.9 \\ 0.1 & 0.8 \end{bmatrix} \qquad D = \begin{bmatrix} 2 & 7 \end{bmatrix}$$

Matrix A is a 2×3 matrix because it has two rows and three columns. Similarly, B is a 3×1 matrix, C is a 2×2 matrix, and D is a 1×2 matrix.

A matrix that consists of a single row of numbers is called a **row matrix.** Matrix D is a row matrix. A matrix that consists of a single column of numbers is called a **column matrix.** Matrix B is a column matrix. A matrix that has the same number of rows as columns is called a **square matrix.** Matrix C is a square matrix.

The numbers (or variables) inside the matrix are called the **entries** of the matrix. The entries of an $m \times n$ matrix A may be represented as

$$A = \begin{bmatrix} a_{11} & a_{12} & \cdots & a_{1n} \\ a_{21} & a_{22} & \cdots & a_{2n} \\ \vdots & \vdots & \vdots & \vdots \\ a_{m1} & a_{m2} & \cdots & a_{mn} \end{bmatrix}$$

The subscripts of each entry indicate its row and column position. An entry a_{ij} is the term in row i and column j. For example, a_{21} refers to the entry in the second row and first column.

EXAMPLE 1

Determining the Dimensions and Entry Values of a Matrix

Determine the dimensions of the matrix A and the value of the entries a_{12}, a_{21}, and a_{24}.

$$A = \begin{bmatrix} 1 & 2 & 3 & 4 \\ 5 & 6 & 7 & 8 \\ 9 & 10 & 11 & 12 \end{bmatrix}$$

SOLUTION The matrix has three rows and four columns, so it is a 3×4 matrix. For this matrix, $a_{12} = 2$, $a_{21} = 5$, and $a_{24} = 8$.

Representing Systems of Linear Equations with Augmented Matrices

A system of linear equations may be represented with an augmented matrix. For example, the system

$$\begin{aligned} 2x + 3y &= 8 \\ 4x - y &= 2 \end{aligned} \quad \text{is written as} \quad \left[\begin{array}{cc|c} 2 & 3 & 8 \\ 4 & -1 & 2 \end{array}\right]$$

The matrix is called an **augmented matrix** because the coefficient matrix, $\begin{bmatrix} 2 & 3 \\ 4 & -1 \end{bmatrix}$, is augmented (added onto) with the column matrix, $\begin{bmatrix} 8 \\ 2 \end{bmatrix}$. (Recall that the term *coefficient* refers to the numeric value written in front of a given variable. For example, the coefficient of the term $2x$ is 2.) The vertical bar between the last two columns of numbers indicates that the matrix is an augmented matrix. The first column of the matrix contains the coefficients of the x terms. In this case, the x terms $2x$ and $4x$ have coefficients 2 and 4, respectively. The second column contains the coefficients of the y terms. In this case, the y terms $3y$ and $-y$ have coefficients 3 and -1, respectively.

EXAMPLE 2

Writing a Systems of Equations as an Augmented Matrix

Rewrite the system of equations as an augmented matrix.

$$\begin{aligned} x - 4y - z &= -5 \\ 3y + z &= 7 \\ 2x + y &= 10 \end{aligned}$$

SOLUTION Before converting a system of equations to an augmented matrix, it is useful to write the equations so that each of the variables lines up vertically. That is, we rewrite the system as

$$\begin{aligned} x - 4y - z &= -5 \\ 3y + z &= 7 \\ 2x + y \phantom{{}+z} &= 10 \end{aligned} \quad \text{The resultant augmented matrix is} \quad \left[\begin{array}{ccc|c} 1 & -4 & -1 & -5 \\ 0 & 3 & 1 & 7 \\ 2 & 1 & 0 & 10 \end{array}\right]$$

The first column represents the x variable, the second column represents the y variable, and the third column represents the z variable. Since the second equation didn't contain an x variable, we concluded that the coefficient on the second equation's x variable was zero. Similarly, the coefficient on the z variable in the third equation was zero.

Row Operations

When we introduced the elimination method, we identified three operations that yielded an equivalent system of equations. The operations were

1. Interchange (change the position of) two equations.
2. Multiply an equation by a nonzero number.
3. Add a nonzero multiple of one equation to a nonzero multiple of another equation.

Similar operations, called **row operations**, may be applied to augmented matrices.

ROW OPERATIONS

For any augmented matrix of a system of equations, the following row operations yield an augmented matrix of an equivalent system of equations:

1. Interchange (change the position of) two rows.
2. Multiply a row by a nonzero number.
3. Add a nonzero multiple of one row to a nonzero multiple of another row and replace either row with the result.

We will use the following notation to specify which operation we are applying.

Operation	Stand Alone Notation	Matrix Notation
Interchange Row 1 and Row 2	$R_1 \leftrightarrow R_2$	$\begin{bmatrix} a_{11} & a_{12} & a_{13} \\ a_{21} & a_{22} & a_{23} \\ a_{31} & a_{32} & a_{33} \end{bmatrix} \begin{matrix} R_2 \\ R_1 \\ \end{matrix}$
Multiply Row 1 by a and place back in Row 1	$aR_1 \rightarrow R_1$	$\begin{bmatrix} a_{11} & a_{12} & a_{13} \\ a_{21} & a_{22} & a_{23} \\ a_{31} & a_{32} & a_{33} \end{bmatrix} \begin{matrix} aR_1 \\ \\ \end{matrix}$
a times Row 1 is added to b times Row 2 and placed in Row 2	$aR_1 + bR_2 \rightarrow R_2$	$\begin{bmatrix} a_{11} & a_{12} & a_{13} \\ a_{21} & a_{22} & a_{23} \\ a_{31} & a_{32} & a_{33} \end{bmatrix} \begin{matrix} \\ aR_1 + bR_2 \\ \end{matrix}$

In matrix notation, the vertical position of the operation indicates how the entries of the adjacent row were calculated.

The process of simplifying the matrix is almost identical to the elimination method illustrated in the previous section. We will first seek to obtain a 1 in the a_{11} entry of the matrix. We will then seek to obtain zeros below the 1 in the first column. We will simplify the remaining columns in a similar manner until the matrix is in **reduced row echelon form.**

REDUCED ROW ECHELON FORM

An augmented matrix A is said to be in reduced row echelon form if it satisfies each of the following conditions:

1. The leading entry (first nonzero entry) in each row is a 1.
2. The leading entry in each row is the only nonzero entry in its corresponding column.
3. The leading entry in each row is to the right of the leading entry in the row above it.
4. All rows of zeros are at the bottom of the matrix.

A matrix in reduced row echelon form is also referred to as a **reduced matrix.**

EXAMPLE 3

Determining If an Augmented Matrix Is in Reduced Row Echelon Form

Which of the augmented matrices are in reduced row echelon form? For each matrix not in reduced row echelon form, specify the operations needed to produce a reduced matrix.

$$A = \left[\begin{array}{ccc|c} 1 & 0 & 1 & 5 \\ 0 & 1 & 2 & 2 \\ 0 & 0 & 0 & 0 \end{array}\right] \qquad B = \left[\begin{array}{ccc|c} 0 & 1 & 0 & 5 \\ 1 & 0 & 0 & 7 \\ 0 & 0 & 1 & 2 \end{array}\right]$$

$$C = \left[\begin{array}{ccc|c} 1 & 0 & 0 & 12 \\ 0 & 1 & 0 & -2 \\ 0 & 0 & 6 & 12 \end{array}\right] \qquad D = \left[\begin{array}{ccc|c} 1 & 2 & 0 & 4 \\ 0 & 1 & 0 & 3 \\ 0 & 0 & 1 & 2 \end{array}\right]$$

SOLUTION Matrix A is the only matrix in reduced row echelon form.

In matrix B, the leading entry in Row 2 is to the left instead of to the right of the leading entry in Row 1. Switching Row 1 with Row 2 would result in a reduced matrix.

In matrix C, the leading entry in the third row is a 6. Multiplying Row 3 by $\frac{1}{6}$ would result in a reduced matrix.

In matrix D, the leading entry in Row 2 contains a 2 above it instead of a 0. Multiplying Row 2 by -2, adding Row 2 to Row 1, and writing the final result in Row 1 would result in a reduced matrix.

EXAMPLE 4

Rewriting an Augmented Matrix in Reduced Row Echelon Form

Rewrite the system of equations as an augmented matrix, then simplify the matrix to reduced row echelon form.

$$\begin{aligned} x - 4y - z &= -5 \\ 3y + z &= 7 \\ 2x + y \quad &= 10 \end{aligned}$$

SOLUTION

$$\left[\begin{array}{ccc|c} 1 & -4 & -1 & -5 \\ 0 & 3 & 1 & 7 \\ 2 & 1 & 0 & 10 \end{array}\right]$$

We will simplify the matrix entry by entry, beginning with the first column. Our goal is to make the matrix look like

$$\left[\begin{array}{ccc|c} 1 & 0 & 0 & a \\ 0 & 1 & 0 & b \\ 0 & 0 & 1 & c \end{array}\right]$$

We have a leading 1 in Row 1. We will use that entry to eliminate the leading entry in Row 3. If we multiply Row 1 by 2 and subtract Row 3, we can eliminate the leading 2 in Row 3. We will place the result of the operation in Row 3. Symbolically, we represent this as $2R_1 - R_3 \rightarrow R_3$.

We calculate the new values for Row 3 one entry at a time as follows:

$$a_{31} = 2(1) - 2 = 0$$
$$a_{32} = 2(-4) - 1 = -9$$
$$a_{33} = 2(-1) - 0 = -2$$
$$a_{34} = 2(-5) - 10 = -20$$

Observe that the numbers in the parentheses are the entries of Row 1 and the numbers following the minus sign are the entries of Row 3. The resultant matrix is

$$\left[\begin{array}{ccc|c} 1 & -4 & -1 & -5 \\ 0 & 3 & 1 & 7 \\ 0 & -9 & -2 & -20 \end{array}\right] \quad \begin{array}{l} \\ \\ 2R_1 - R_3 \end{array}$$

Next we will eliminate the -9 in the third row using the following operation:

$$3R_2 + R_3 \rightarrow R_3$$

We calculate the new values for Row 3 one entry at a time as follows:

$$a_{31} = 3(0) + 0 = 0$$
$$a_{32} = 3(3) + (-9) = 0$$
$$a_{33} = 3(1) + (-2) = 1$$
$$a_{34} = 3(7) + (-20) = 1$$

The resultant matrix is

$$\left[\begin{array}{ccc|c} 1 & -4 & -1 & -5 \\ 0 & 3 & 1 & 7 \\ 0 & 0 & 1 & 1 \end{array}\right] \quad \begin{array}{l} \\ \\ 3R_2 + R_3 \end{array}$$

Notice that Row 3 contains only zeros in Columns 1 and 2. This means that adding a multiple of Row 3 to any of the other rows will leave the first two columns unchanged. We will use this fact to zero out the entries a_{13} and a_{23}. Using the operations

$$R_2 - R_3 \rightarrow R_2 \qquad \text{and} \qquad R_1 + R_3 \rightarrow R_1$$

we obtain the matrix

$$\left[\begin{array}{ccc|c} 1 & -4 & 0 & -4 \\ 0 & 3 & 0 & 6 \\ 0 & 0 & 1 & 1 \end{array}\right] \quad \begin{array}{l} R_1 + R_3 \\ R_2 - R_3 \\ \end{array}$$

Next we need to get a leading 1 in Row 2. Using the operation

$$\frac{1}{3}R_2 \rightarrow R_2$$

yields the matrix

$$\left[\begin{array}{ccc|c} 1 & -4 & 0 & -4 \\ 0 & 1 & 0 & 2 \\ 0 & 0 & 1 & 1 \end{array}\right] \quad \begin{array}{l} \\ \frac{1}{3}R_2 \\ \end{array}$$

To finish, we zero out the a_{12} entry with the operation

$$R_1 + 4R_2 \rightarrow R_1$$

resulting in the matrix

$$\left[\begin{array}{ccc|c} 1 & 0 & 0 & 4 \\ 0 & 1 & 0 & 2 \\ 0 & 0 & 1 & 1 \end{array}\right] \quad R_1 + 4R_2$$

The matrix is in reduced row echelon form and yields the solution to the system of equations.

$$x = 4$$
$$y = 2$$
$$z = 1$$

In Example 4, we simplified the matrix in the following order:

1. Make Column 1 look like $\begin{bmatrix} 1 \\ 0 \\ 0 \end{bmatrix}$ and the resultant matrix look like $\left[\begin{array}{ccc|c} 1 & a & d & g \\ 0 & b & e & h \\ 0 & c & f & i \end{array}\right]$.

2. Make Row 3 look like $[0 \quad 0 \quad 1]$ and the resultant matrix look like $\left[\begin{array}{ccc|c} 1 & a & c & e \\ 0 & b & d & f \\ 0 & 0 & 1 & g \end{array}\right]$.

3. Make Column 3 look like $\begin{bmatrix} 0 \\ 0 \\ 1 \end{bmatrix}$ and the resultant matrix look like $\left[\begin{array}{ccc|c} 1 & a & 0 & c \\ 0 & b & 0 & d \\ 0 & 0 & 1 & e \end{array}\right]$.

4. Make Column 2 look like $\begin{bmatrix} 0 \\ 1 \\ 0 \end{bmatrix}$ and the resultant matrix look like $\left[\begin{array}{ccc|c} 1 & 0 & 0 & a \\ 0 & 1 & 0 & b \\ 0 & 0 & 1 & c \end{array}\right]$.

Note: The variables a through i do not represent the same values in each step.

The idea is to first create a "triangle" of zeroes below the main diagonal of the matrix and then create a "triangle" of zeroes above the main diagonal of the matrix. We believe this approach is the simplest; however, regardless of the order in which you simplify the columns and rows of the matrix, the reduced row echelon form of the matrix will be the same.

When the number of columns on the coefficient side of the augmented matrix exceeds the number of rows of the matrix, the matrix corresponds to a system of equations with more variables than equations. For this type of matrix, there will be multiple solutions to the system of equations (if a solution exists).

If the number of rows exceeds the number of columns on the coefficient side of the augmented matrix, the matrix corresponds to a system of equations with more equations than variables. If a solution to the system exists, the simplified matrix will contain at least one row of zeros. Examples 5 and 6 demonstrate these concepts.

EXAMPLE 5

Rewriting an Augmented Matrix in Reduced Row Echelon Form

Rewrite the system of equations as an augmented matrix, then simplify the matrix to reduced row echelon form. Identify the solution(s) to the system of equations.

$$\begin{aligned} x + y + 6z &= 16 \\ 2x - y &= 2 \end{aligned}$$

SOLUTION

$$\left[\begin{array}{ccc|c} 1 & 1 & 6 & 16 \\ 2 & -1 & 0 & 2 \end{array}\right]$$

The operation $2R_1 - R_2 \rightarrow R_2$ will zero out the a_{21} entry.

$$\left[\begin{array}{ccc|c} 1 & 1 & 6 & 16 \\ 0 & 3 & 12 & 30 \end{array}\right] \quad 2R_1 - R_2$$

The operation $\frac{1}{3}R_2 \rightarrow R_2$ will give us a leading 1 in the a_{22} entry.

$$\left[\begin{array}{ccc|c} 1 & 1 & 6 & 16 \\ 0 & 1 & 4 & 10 \end{array}\right] \quad \tfrac{1}{3}R_2$$

The operation $R_1 - R_2 \rightarrow R_1$ will zero out the a_{12} entry.

$$\left[\begin{array}{ccc|c} 1 & 0 & 2 & 6 \\ 0 & 1 & 4 & 10 \end{array}\right] \quad R_1 - R_2$$

The matrix is in reduced row echelon form. We have the equations

$$\begin{aligned} x + 2z &= 6 \\ y + 4z &= 10 \end{aligned}$$

Solving the equations for x and y, respectively, yields

$$\begin{aligned} x &= -2z + 6 \\ y &= -4z + 10 \end{aligned}$$

Notice that we have x as a function of z and y as a function of z. We can pick z to be any number we want and then calculate the corresponding values of x and y. To represent this idea, we set z equal to an arbitrary variable, say t. The final solution is

$$\begin{aligned} x &= -2t + 6 \\ y &= -4t + 10 \\ z &= t \end{aligned}$$

There are an infinite number of points (x, y, z) that satisfy the system of equations, so the system of equations is dependent. We can calculate a particular solution by picking a value of t. For example, let $t = 0$. Then $x = 6$, $y = 10$, and $z = 0$.

EXAMPLE 6

Rewriting an Augmented Matrix in Reduced Row Echelon Form

Rewrite the system of equations as an augmented matrix, then simplify the matrix to reduced row echelon form. Identify the solution(s) to the system of equations.

$$\begin{aligned} x + 2y &= 5 \\ 3x - y &= 11.5 \\ 5x + 3y &= 21.5 \end{aligned}$$

SOLUTION

$$\left[\begin{array}{rr:r} 1 & 2 & 5 \\ 3 & -1 & 11.5 \\ 5 & 3 & 21.5 \end{array}\right]$$

The operation $3R_1 - R_2 \rightarrow R_2$ will zero out the a_{21} entry, and the operation $5R_1 - R_3 \rightarrow R_3$ will zero out the a_{31} entry. The resultant matrix is

$$\left[\begin{array}{rr:r} 1 & 2 & 5 \\ 0 & 7 & 3.5 \\ 0 & 7 & 3.5 \end{array}\right] \quad \begin{array}{l} \\ 3R_1 - R_2 \\ 5R_1 - R_3 \end{array}$$

Observe that Row 2 and Row 3 are identical. The operation $R_2 - R_3 \rightarrow R_3$ will zero out Row 3.

$$\left[\begin{array}{rr:r} 1 & 2 & 5 \\ 0 & 7 & 3.5 \\ 0 & 0 & 0 \end{array}\right] \quad \begin{array}{l} \\ \\ R_2 - R_3 \end{array}$$

The operation $\frac{1}{7}R_2 \rightarrow R_2$ will give us a leading 1 in Row 2.

$$\left[\begin{array}{rr:r} 1 & 2 & 5 \\ 0 & 1 & 0.5 \\ 0 & 0 & 0 \end{array}\right] \quad \begin{array}{l} \\ \frac{1}{7}R_2 \\ \\ \end{array}$$

The operation $R_1 - 2R_2 \rightarrow R_1$ will zero out the a_{12} entry.

$$\left[\begin{array}{rr:r} 1 & 0 & 4 \\ 0 & 1 & 0.5 \\ 0 & 0 & 0 \end{array}\right] \quad \begin{array}{l} R_1 - 2R_2 \\ \\ \\ \end{array}$$

The matrix is in reduced row echelon form. The solution to the system of equations is $x = 4$ and $y = 0.5$.

Not every system of equations has a solution. When a system of equation has no solution, a contradiction will occur in the augmented matrix. This case is demonstrated in Examples 7 and 8.

EXAMPLE 7

Using Matrices to Solve an Inconsistent System of Equations

Rewrite the system of equations as an augmented matrix, then simplify the matrix to reduced row echelon form. Identify the solution(s) to the system of equations.

$$\begin{aligned} x + 2y &= 5 \\ 3x - y &= 2 \\ 4x + y &= 8 \end{aligned}$$

SOLUTION

$$\left[\begin{array}{rr|r} 1 & 2 & 5 \\ 3 & -1 & 2 \\ 4 & 1 & 8 \end{array}\right]$$

The operation $3R_1 - R_2 \rightarrow R_2$ will zero out the a_{21} entry, and the operation $4R_1 - R_3 \rightarrow R_3$ will zero out the a_{31} entry. The resultant matrix is

$$\left[\begin{array}{rr|r} 1 & 2 & 5 \\ 0 & 7 & 13 \\ 0 & 7 & 12 \end{array}\right] \quad \begin{array}{l} \\ 3R_1 - R_2 \\ 4R_1 - R_3 \end{array}$$

The operation $R_2 - R_3 \rightarrow R_3$ will zero out the a_{32} entry.

$$\left[\begin{array}{rr|r} 1 & 2 & 5 \\ 0 & 7 & 13 \\ 0 & 0 & 1 \end{array}\right] \quad \begin{array}{l} \\ \\ R_2 - R_3 \end{array}$$

Row 3 is equivalent to

$$\begin{aligned} 0x + 0y &= 1 \\ 0 &= 1 \end{aligned}$$

But $0 \neq 1$ so the system of equations is inconsistent (has no solution).

EXAMPLE 8

Using Matrices to Solve an Inconsistent System of Equations

Rewrite the system of equations as an augmented matrix, then simplify the matrix to reduced row echelon form. Identify the solution(s) to the system of equations.

$$\begin{aligned} 3x + 3y \phantom{{}+z} &= 9 \\ 2x + 3y + z &= 6 \\ x \phantom{{}+3y} - z &= 1 \end{aligned}$$

SOLUTION

$$\left[\begin{array}{rrr|r} 3 & 3 & 0 & 9 \\ 2 & 3 & 1 & 6 \\ 1 & 0 & -1 & 1 \end{array}\right]$$

The operation $R_1 \leftrightarrow R_3$ places a 1 in the upper left-hand corner of the matrix.

$$\left[\begin{array}{ccc|c} 1 & 0 & -1 & 1 \\ 2 & 3 & 1 & 6 \\ 3 & 3 & 0 & 9 \end{array}\right] \begin{array}{l} R_3 \\ \\ R_1 \end{array}$$

The operation $2R_1 - R_2 \to R_2$ will zero out the a_{21} entry. The operation $3R_1 - R_3 \to R_3$ will zero out the a_{31} entry.

$$\left[\begin{array}{ccc|c} 1 & 0 & -1 & 1 \\ 0 & -3 & -3 & -4 \\ 0 & -3 & -3 & -6 \end{array}\right] \begin{array}{l} \\ 2R_1 - R_2 \\ 3R_1 - R_3 \end{array}$$

The operation $R_2 - R_3 \to R_3$ will zero out the a_{32} entry.

$$\left[\begin{array}{ccc|c} 1 & 0 & -1 & 1 \\ 0 & -3 & -3 & -4 \\ 0 & 0 & 0 & 2 \end{array}\right] \begin{array}{l} \\ \\ R_2 - R_3 \end{array}$$

Row 3 is equivalent to

$$0x + 0y + 0z = 2$$
$$0 = 2$$

But $0 \neq 2$. Therefore, the system of equations is inconsistent.

In each of the examples, we have looked at systems of equations with integer coefficients and rational solutions. Although the elimination process may be used on any augmented matrix, computations may become particularly messy when the system of equations has noninteger coefficients. Fortunately, the TI-83 Plus has a powerful feature that will simplify augmented matrices to reduced row echelon form.

EXAMPLE 9

Using Technology to Find the Reduced Row Echelon Form of a Matrix

Rewrite the system of equations as an augmented matrix, then simplify the matrix to reduced row echelon form. Identify the solution(s) to the system of equations.

$$0.31x + 6.23y + 2.15z = 5.61$$
$$0.07x - 7.21y + 1.15z = 2.23$$
$$4.11x - 1.23y - 3.91z = 8.99$$

SOLUTION

$$\left[\begin{array}{ccc|c} 0.31 & 6.23 & 2.15 & 5.61 \\ 0.07 & -7.21 & 1.15 & 2.23 \\ 4.11 & -1.23 & -3.91 & 8.99 \end{array}\right]$$

Using the techniques shown in the following Technology Tips, the reduced matrix is approximately equal to

$$\left[\begin{array}{ccc|c} 1 & 0 & 0 & 4.03 \\ 0 & 1 & 0 & 0.0365 \\ 0 & 0 & 1 & 1.92 \end{array}\right]$$

The solution to the system of equations is $x \approx 4.03$, $y \approx 0.0365$, and $z \approx 1.92$.

TECHNOLOGY TIP

Entering a Matrix

1. Activate the Matrix Menu by pressing [2nd] [x^{-1}]. You may or may not have some matrices displayed in the name list.

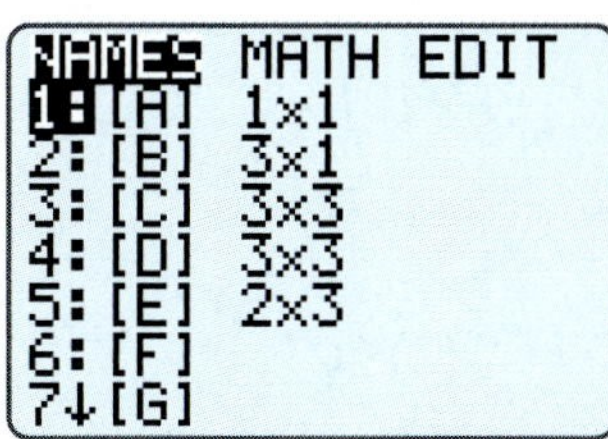

2. Create the augmented matrix A by moving the cursor to `EDIT` and pressing [ENTER].

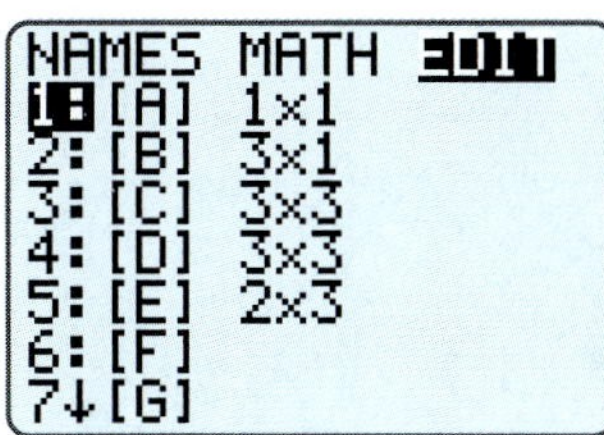

3. Enter the dimensions of the matrix. Type the number of rows and press [ENTER], then type the number of columns and press [ENTER]. The example matrix is a 3×4 augmented matrix.

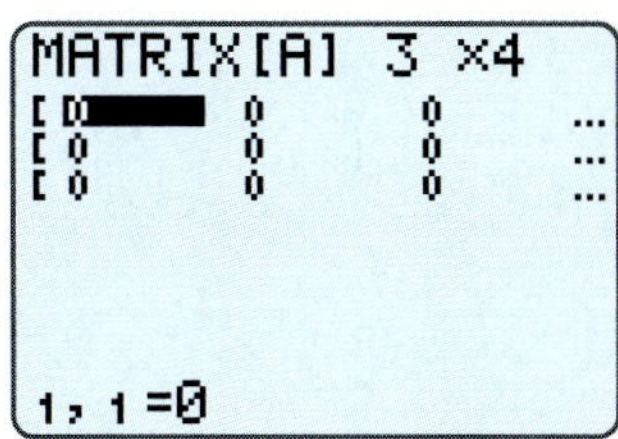

4. Enter the individual entries of the matrix. Use the blue arrow buttons to move from one entry to another. When you've entered all of the entries, press [2nd] [MODE] to exit the Matrix Editor.

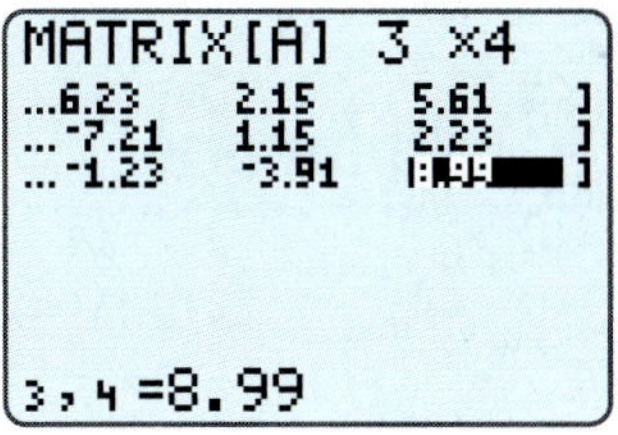

TECHNOLOGY TIP

Finding the Reduced Row Echelon Form of a Matrix *A*

1. Activate the Matrix Math Menu by pressing [2nd] [x^{-1}] and moving the cursor to the `MATH` menu item.

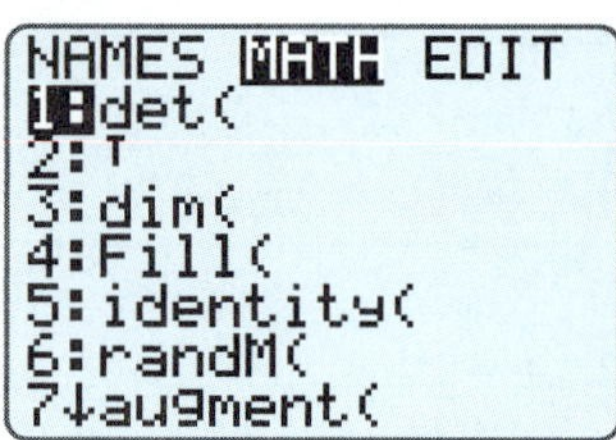

2. Select the `rref` operation by scrolling down the list to item `B:rref(`. Press [ENTER] to place the operation on the home screen. (The `rref(` function will convert a matrix to reduced row echelon form.)

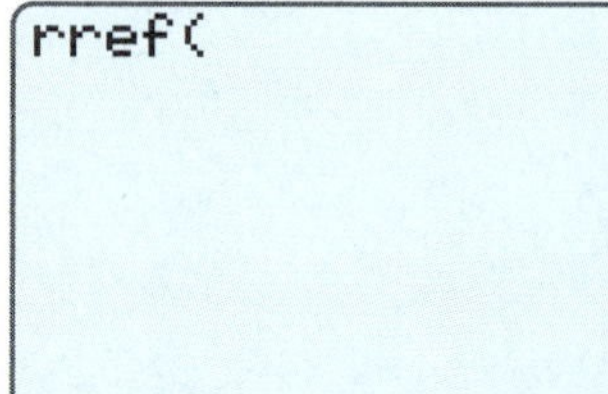

3. Select the matrix A from the Matrix Names Menu by pressing [2nd] [x^{-1}], moving the cursor to matrix A, and pressing [ENTER]. This places the matrix A on the home screen.

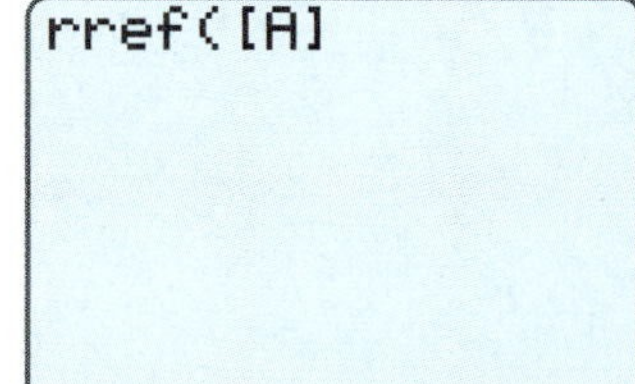

4. Calculate the reduced row echelon form of the matrix by pressing [ENTER].

```
rref([A]
[[1 0 0 4.02752...
 [0 1 0 .036500...
 [0 0 1 1.92282...
```

2.2 Summary

In this section, you learned basic matrix notation and saw how to use matrices to represent and solve systems of linear equations. You also learned how to use technology to simplify an augmented matrix to reduced row echelon form.

2.2 Exercises

In Exercises 1–5, determine the dimensions of the matrix A.

1. $A = \begin{bmatrix} 19 & 5 & 1 \\ 2 & 8 & 18 \end{bmatrix}$

2. $A = \begin{bmatrix} -10 \\ 42 \\ 17 \end{bmatrix}$

3. $A = \begin{bmatrix} 14 & 0 & -5 \end{bmatrix}$

4. $A = \begin{bmatrix} -4 & 5 & 8 & 11 \\ 11 & 5 & 4 & -6 \end{bmatrix}$

5. $A = \begin{bmatrix} 17 \end{bmatrix}$

In Exercises 6–10, write the augmented matrix as a system of linear equations, using the variables x, y, and z as appropriate.

6. $\left[\begin{array}{ccc|c} 2 & 1 & 0 & 11 \\ 1 & 0 & 4 & 0 \end{array}\right]$

7. $\left[\begin{array}{cc|c} 1 & 3 & 12 \\ 4 & 1 & 7 \\ -1 & 6 & 9 \end{array}\right]$

8. $\left[\begin{array}{ccc|c} 1 & 0 & 4 & 0 \\ 0 & 1 & 2 & 19 \\ 6 & 0 & 3 & 22 \end{array}\right]$

9. $\left[\begin{array}{ccc|c} 3 & -1 & 9 & 13 \\ -3 & 1 & 0 & -31 \end{array}\right]$

10. $\left[\begin{array}{ccc|c} 5 & 2 & 4 & 11 \\ -1 & 2 & 2 & 3 \\ 5 & 10 & 3 & 18 \end{array}\right]$

In Exercises 11–20, determine if the augmented matrix is in reduced row echelon form. If it is not reduced, specify which of the four criteria for reduced row echelon form the matrix fails to meet.

11. $\left[\begin{array}{ccc|c} 1 & 0 & 0 & 3 \\ 0 & 0 & 1 & 8 \\ 0 & 1 & 0 & -2 \end{array}\right]$

12. $\left[\begin{array}{ccc|c} 1 & 0 & 0 & 0 \\ 0 & 0 & 1 & 11 \\ 0 & 0 & 0 & 0 \end{array}\right]$

13. $\left[\begin{array}{ccc|c} 1 & 0 & 0 & 0 \\ 0 & 3 & 0 & 12 \\ 0 & 0 & 1 & 5 \end{array}\right]$

14. $\left[\begin{array}{ccc|c} 1 & 0 & 0 & 0 \\ 0 & 1 & 2 & 14 \\ 0 & 0 & 1 & 52 \end{array}\right]$

15. $\left[\begin{array}{ccc|c} 1 & 0 & 0 & 2.7 \\ 0 & 0 & 1 & 4.2 \end{array}\right]$

16. $\left[\begin{array}{ccc|c} 1 & 1 & 0 & 12.1 \\ 0 & 0 & 1 & -34.2 \end{array}\right]$

17. $\left[\begin{array}{ccc|c} 0 & 1 & 0 & 11 \\ 1 & 0 & 1 & 0 \end{array}\right]$

18. $\left[\begin{array}{ccc|c} 1 & 2 & 9 & 13 \\ 0 & 0 & 0 & 0 \end{array}\right]$

19. $\left[\begin{array}{cc|c} 1 & 0 & 12 \\ 0 & 1 & 7 \\ 0 & 1 & 9 \end{array}\right]$

20. $\left[\begin{array}{cc|c} 2 & 0 & 12 \\ 0 & 1 & 7 \\ 0 & 0 & 0 \end{array}\right]$

In Exercises 21–40, rewrite the system of equations as an augmented matrix. Then simplify the matrix to reduced row echelon form. Identify the solution(s) to the system of equations. If the system is inconsistent or dependent, so state.

21. $\begin{aligned} 2x + 5y &= 2 \\ 3x - 5y &= 3 \end{aligned}$

22. $\begin{aligned} 6x + 2y &= 10 \\ -x - 2y &= -5 \end{aligned}$

23. $\begin{aligned} -2x + 6y + 4z &= 10 \\ 4x - 12y + 2z &= -20 \\ 3x + 4y - z &= 11 \end{aligned}$

24. $\begin{aligned} 10x - y + z &= 7 \\ 9x - 2y + 4z &= 7 \\ x + 2y - 4z &= -17 \end{aligned}$

25. $\begin{aligned} 3x - 2y + z &= 6 \\ 11x - 20y - z &= 0 \\ y + z &= 3 \end{aligned}$

26. $\begin{aligned} 9x - 6y \quad &= 0 \\ 4x + 5y \quad &= 23 \\ x \quad - z &= 6 \end{aligned}$

27. $\begin{aligned} x - y &= -5 \\ 9x + y &= 25 \\ 29x + y &= 65 \end{aligned}$

28. $\begin{aligned} x - 2y &= -7 \\ 6x + 5y &= 94 \\ 10x + 3y &= 114 \end{aligned}$

29. $\begin{aligned} 2x - 4y &= 16 \\ 9x + y &= -4 \\ -3x + 6y &= -24 \end{aligned}$

30. $\begin{aligned} x - y &= 3 \\ 6x + 7y &= 44 \\ 6x - 7y &= 16 \end{aligned}$

31. $\begin{aligned} 2x - 2y &= 0 \\ 3x + y &= 4 \\ 5x - y &= 5 \end{aligned}$

32. $\begin{aligned} 8x - y &= 3 \\ x + 2y &= 11 \\ 9x + y &= 14 \end{aligned}$

33. $\begin{aligned} x + y + z &= 6 \\ 2x - y + z &= 3 \end{aligned}$

34. $\begin{aligned} 2x - 4y + 2z &= 6 \\ 5x - 5y + z &= 1 \end{aligned}$

35. $\begin{aligned} 3x + 5y + z &= 6 \\ 6x + 10y + 2z &= 10 \end{aligned}$

36. $\begin{aligned} x + y + z &= 6 \\ 2x - y + z &= 12 \end{aligned}$

37. $\begin{aligned} x + y + z &= 6 \\ 2x - y + z &= 3 \\ 4x - 2y + 3z &= 9 \end{aligned}$

38. $\begin{aligned} 5x + y + z &= 1 \\ 4x - 2y + 5z &= -2 \\ x - 7y + 6z &= -7 \end{aligned}$

39. $\begin{aligned} 3x + 2y + 3z &= 6 \\ 2x - 5y + z &= -11 \\ 4x + 2y + 3z &= 3 \end{aligned}$

40. $\begin{aligned} x + y + z &= 3 \\ 2x - y + z &= 2 \\ 4x - 2y + 3z &= 5 \end{aligned}$

In Exercises 41–50, use technology to write the matrix in reduced row echelon form. Then interpret the result.

41. $\left[\begin{array}{ccc:c} 0.1 & 0.3 & 1.2 & 2.1 \\ 0.7 & -2.8 & 3.5 & 1.2 \\ 8.2 & -7.3 & -1.1 & 3.9 \end{array}\right]$

42. $\left[\begin{array}{ccc:c} 2.6 & 0.0 & 0.2 & 9.6 \\ 3.1 & 4.0 & 7.5 & 0 \\ 9.1 & -2.1 & -0.8 & 9.9 \end{array}\right]$

43. $\left[\begin{array}{ccc:c} 0.11 & 0.99 & 1.25 & 2.61 \\ 0.97 & -2.28 & 2.15 & 11.23 \\ 5.12 & -4.43 & -1.81 & 3.29 \end{array}\right]$

44. $\left[\begin{array}{ccc:c} 1.2 & 2.3 & 3.4 & 1.0 \\ 2.3 & 3.4 & 4.5 & 2.0 \\ 3.4 & 4.5 & 5.6 & 3.0 \end{array}\right]$

45. $\left[\begin{array}{ccc:c} 0 & 2.3 & 4.4 & 8.0 \\ 2.3 & 0 & 6.5 & 6.0 \\ 4.4 & 6.5 & 0 & 4.0 \end{array}\right]$

46. $\left[\begin{array}{ccc:c} 5.7 & 7.8 & 9.4 & 8.2 \\ 1.3 & 3.0 & 0.5 & 5.9 \\ 4.7 & 0.5 & 2.0 & 3.1 \end{array}\right]$

47. $\left[\begin{array}{ccc:c} 1.2 & 3.4 & 5.6 & 7.8 \\ 0.1 & 2.3 & 4.5 & 6.7 \\ 9.8 & 7.6 & 5.4 & 3.2 \end{array}\right]$

48. $\left[\begin{array}{ccc:c} 1.7 & 2.8 & 6.4 & 8.2 \\ 1.6 & 3.6 & 2.5 & 5.9 \\ 4.9 & 0.5 & 2.0 & 3.1 \end{array}\right]$

49. $\left[\begin{array}{ccc:c} 0.50 & 0.25 & 0.25 & 1 \\ 0.25 & 0.50 & 0.25 & 1 \\ 0.25 & 0.25 & 0.50 & 1 \end{array}\right]$

50. $\left[\begin{array}{ccc:c} 1.9 & 2.8 & 3.7 & 4.6 \\ 2.5 & 3.6 & 4.7 & 5.8 \\ 3.7 & 0.3 & 2.0 & 3.3 \end{array}\right]$

2.3 Linear System Applications

- Formulate linear systems from real-life data
- Interpret the meaning of solutions to linear systems

GETTING STARTED The per capita consumption of sugar rose from 62.7 pounds in 1985 to 67.9 pounds in 1999. Over the same period, the per capita consumption of high-fructose corn syrup (used to sweeten soft drinks, juices, and other such products) rose from 45.2 pounds to 66.2 pounds. (**Source:** *Statistical Abstract of the United States, 2001,* Table 202, p. 129.) Based on the trends in sugar and corn syrup consumption, producers may want to adjust their strategic (long-range) production plan. In this section, we will show how to formulate linear systems from real-life data and use the solutions to these systems to make business decisions. We will also show some of the strengths and limitations of mathematical models.

EXAMPLE 1

Using a Linear Model to Forecast Demand for a Product

The per capita consumption of sugar and high-fructose corn syrup are shown in Table 2.1.

TABLE 2.1

Years since 1985 (t)	Per Capita Sugar Consumption (pounds) (S)	Per Capita High-Fructose Corn Syrup Consumption (pounds) (C)
0	62.7	45.2
5	64.4	49.6
10	65.5	58.4
12	66.5	62.4
13	66.9	63.8
14	67.9	66.2

(a) Use linear functions to model the consumption of each of the sweeteners.
(b) Predict the year in which the consumption of each type of sweetener will be equal.
(c) Explain how the results of your analysis could benefit a sweetener producer.

SOLUTION

(a) Using linear regression on each data set, we obtain the models

Sugar: $S = 0.339t + 62.6$ pounds
Corn syrup: $C = 1.53t + 43.8$ pounds

For both models, t is the number of years after 1985.

(b) We want to find when $S = C$. Setting $S = C$ and moving all of the variables to the left-hand side of the equal sign we obtain the system of equations

$$C - 0.339t = 62.6$$
$$C - 1.53t = 43.8$$

and the corresponding augmented matrix

$$\left[\begin{array}{cc|c} 1 & -0.339 & 62.6 \\ 1 & -1.53 & 43.8 \end{array}\right]$$

Using technology to reduce the matrix yields

$$\left[\begin{array}{cc|c} 1 & 0 & 68.0 \\ 0 & 1 & 15.8 \end{array}\right]$$

and the solution $C = 68.0$ and $t = 15.8$. That is, about 15.8 years after 1985, the per capita consumption of both sugar and high-fructose corn syrup were projected to be about 68 pounds. So in the latter part of 2001, sugar and high-fructose corn syrup consumption were projected to be equal.

(c) Based on the slopes of the linear models, we see that sugar consumption between 1985 and 1999 was increasing by 0.339 pound per year while corn syrup consumption was increasing by 1.53 pounds per year. A producer could use this information to adjust production levels of corn syrup and sugar. Recognizing that the demand for high-fructose corn syrup would surpass the demand for sugar in 2001, the producer could adjust her marketing strategy to take advantage of the increased consumer demand.

It is essential to recognize that mathematical models rarely yield exact results. Nevertheless, they are of value because they give us an idea of what may happen if current trends continue. They also offer a mechanism for making an educated guess concerning yet unknown future results.

In Example 1, we used linear regression to model the data. Linear regression is just one of the many different types of regression that may be used to find a model for a data set. In Chapter 5, we will demonstrate other types of regression, including quadratic, cubic, quartic, exponential, and logarithmic. In deciding which type of regression to use, we select the type of function that we believe will best fit the data.

EXAMPLE 2 Using a Linear Model to Forecast School Enrollment

School district facilities managers are often asked to provide information to district boards regarding projected growth in enrollment within the district. The boards use this information to determine whether the existing education infrastructure (buildings, buses, and so on) can accommodate the projected need. These projections may go out more than 25 years, but they are frequently adjusted to improve accuracy as more data become available.

In January 2002, the White River School District facilities manager published a student enrollment report including the data shown in Table 2.2.

TABLE 2.2

Years Since 1998 (t)	Projected Student Enrollment, Grade 3 (students) (T)	Projected Student Enrollment, Grade 4 (students) (F)
0	273	298
1	311	282
2	321	327
3	312	339
4	311	319
5	337	317
6	309	340
7	365	312
8	372	368
9	379	375
10	387	380

(a) Find the linear models that best fit the data for enrollment in grades 3 and 4.
(b) Use the models (not the table) to predict the years in which grade 3 enrollment is projected to exceed grade 4 enrollment.
(c) Explain why your result in part (b) is different from that shown in the table.
(d) Explain how the results of your analysis could benefit the school district.

SOLUTION

(a) Using linear regression on each data set, we obtain the models

Grade 3: $T = 9.99t + 284$ students
Grade 4: $F = 7.93t + 293$ students

For both models, t is the number of years after 1998.

We note that the correlation coefficients for the models are 0.908 and 0.830, respectively. The models don't fit the data sets as well as we would have liked. (We would prefer to see a correlation coefficient closer to 1.)

(b) We want to find when $T = F$. Setting $T = F$ and moving all of the variables to the left-hand side of the equal sign, we obtain the system of equations

$$F - 9.99t = 284$$
$$F - 7.93t = 293$$

and the corresponding augmented matrix

$$\left[\begin{array}{cc|c} 1 & -9.99 & 284 \\ 1 & -7.93 & 293 \end{array}\right]$$

Writing the matrix in reduced row echelon form and rounding to three significant digits yields

$$\left[\begin{array}{cc|c} 1 & 0 & 328 \\ 0 & 1 & 4.37 \end{array}\right]$$

and the solution $F = 328$ and $t = 4.37$. That is, about 4.37 years after 1998, enrollment in both grade 3 and grade 4 is projected to be 328 students. Looking at the slopes of our linear models, we see that grade 3 is increasing by about 10 students per year, while grade 4 is increasing by about 8 students per year. Consequently, grade 3 is expected to have a greater enrollment for the fifth and subsequent years after 1998.

(c) In the table, we see that the facilities manager projects that grade 3 will have greater enrollment 1, 5, 7, 8, 9, and 10 years after 1998. The facilities manager's projections take into consideration the fact that this year's grade 3 will be the basis for next year's grade 4, a fact not considered in our models.

(d) Our results could be used to determine if existing classrooms are sufficient to accommodate the growing third- and fourth-grade population. The results could also be used to justify hiring additional elementary school teachers.

EXAMPLE 3

Using Linear Models to Make Investment Decisions

An investor with $30,000 in an existing mutual fund IRA wants to roll over her investment into a new retirement plan with TIAA-CREF. The company offers a variety of accounts with varying returns, as shown in Table 2.3.

TABLE 2.3

CREF Variable Annuity Accounts as of 5/31/04	Unit Value	1-year Return	5-year Return	10-year Return	Return Since Inception
Bond Market	$71.01	−0.65%	6.70%	7.05%	7.63%
Equity Index	$71.94	19.16%	−0.75%	10.83%	10.89%
Global Equities	$70.45	24.50%	−0.84%	6.96%	8.52%
Growth	$55.93	16.01%	−6.72%	8.14%	8.23%
Inflation-Linked Bond	$42.14	2.11%	9.15%	—	7.53%
Money Market	$21.86	0.67%	3.16%	4.25%	4.95%
Social Choice	$101.39	11.07%	2.49%	10.01%	10.27%
Stock	$174.37	21.18%	−0.62%	9.53%	10.44%

Source: www.tiaa-cref.com.

The investor looks at the long-term performance of each account as an indicator of the likely long-term return. She would like to invest her money in the Bond Market, Equity Index, and Social Choice accounts. After analyzing her risk tolerance, she decides to invest twice as much money in the Bond Market account as in the Equity Index account. She wants to earn a 9 percent return on her investments. Assuming that she will be able to earn the 10-year average annual return, how much money should she invest in each account? [For ease of computation, round the 10-year rate to the nearest whole-number percent (e.g., 7.05 percent is rounded to 7 percent)].

SOLUTION Let x be the amount of money invested in the Bond Market account, y be the amount invested in the Equity Index account, and z be the amount invested in the Social Choice account. Since the investor has \$30,000 to invest, we know that

$$x + y + z = 30{,}000 \qquad \text{Total amount of money to be invested}$$

Since she plans to invest twice as much money in the Bond Market account as in the Equity Index account, we know that

$$x = 2y \qquad \text{The Bond Market investment is twice the Equity Index investment}$$

$$x - 2y = 0$$

Rounding to the nearest whole-number percentage, the Bond Market account has a 7 percent return rate, the Equity Index account has an 11 percent return rate, and the Social Choice Stock account has a 10 percent return rate. Since she wants to earn a 9 percent return on her total investment, we know that

$$0.7x + 0.11y + 0.10z = 0.09(30{,}000) \qquad \text{Total annual return on investment}$$

$$0.07x + 0.11y + 0.10z = 2700$$

We need to solve the following system of equations:

$$\begin{aligned} x + y + z &= 30{,}000 \\ x - 2y &= 0 \\ 0.07x + 0.11y + 0.10z &= 2700 \end{aligned}$$

The corresponding augmented matrix is

$$\left[\begin{array}{ccc|c} 1 & 1 & 1 & 30{,}000 \\ 1 & -2 & 0 & 0 \\ 0.07 & 0.11 & 0.10 & 2{,}700 \end{array}\right]$$

Entering the matrix into the TI-83 Plus and using the `rref` feature gives the results in Figure 2.10.

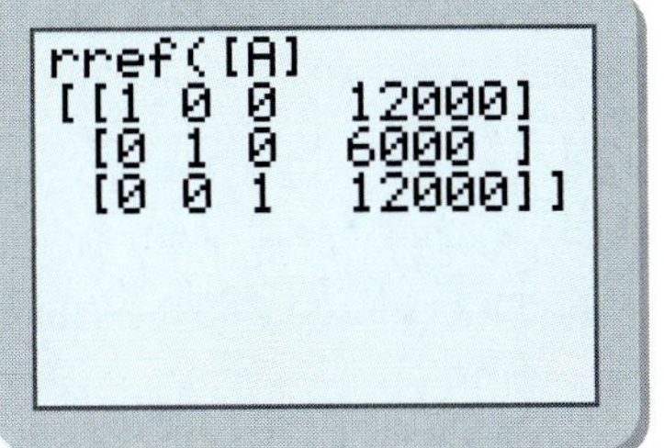

FIGURE 2.10

The reduced matrix is

$$\left[\begin{array}{ccc|c} 1 & 0 & 0 & 12{,}000 \\ 0 & 1 & 0 & 6{,}000 \\ 0 & 0 & 1 & 12{,}000 \end{array}\right]$$

and corresponds with the system of equations

$$\begin{aligned} x &= 12{,}000 \\ y &= 6000 \\ z &= 12{,}000 \end{aligned}$$

She should invest \$12,000 in the Bond Market account, \$6000 in the Equity Index account, and \$12,000 in the Social Choice account. By diversifying her investments into different types of accounts, she is able to reduce her overall risk.

EXAMPLE 4 Using a System of Linear Equations to Find a Quadratic Model

According to its 2001 annual report, the gross revenue of Johnson & Johnson and its subsidiaries increased in 1999, 2000, and 2001, as shown in Table 2.4.

TABLE 2.4 Johnson & Johnson and Subsidiaries Gross Revenue

Years Since the End of 1999 (t)	Revenue from Sales (millions of dollars) (R)
0	27,357
1	29,172
2	32,317

Source: Johnson & Johnson 2001 annual report, p. 6.

Use matrices to find a quadratic equation that models the company's gross revenue.

SOLUTION A quadratic equation will have the form $R(t) = at^2 + bt + c$, where $R(t)$ is the gross revenue (in millions of dollars) and t is the number of years since the end of 1999. Each of the data points must satisfy the quadratic equation. That is,

$$27{,}357 = a(0)^2 + b(0) + c \quad \text{Using } t = 0 \text{ and } R = 27{,}357$$
$$29{,}172 = a(1)^2 + b(1) + c \quad \text{Using } t = 1 \text{ and } R = 29{,}172$$
$$32{,}317 = a(2)^2 + b(2) + c \quad \text{Using } t = 2 \text{ and } R = 32{,}317$$

Reducing each equation yields a system of linear equations

$$\begin{aligned} c &= 27{,}357 \\ a + b + c &= 29{,}172 \\ 4a + 2b + c &= 32{,}317 \end{aligned}$$

and its corresponding augmented matrix

$$\left[\begin{array}{ccc|c} 0 & 0 & 1 & 27{,}357 \\ 1 & 1 & 1 & 29{,}172 \\ 4 & 2 & 1 & 32{,}317 \end{array}\right]$$

We can readily see that $c = 27{,}357$. We will use the TI-83 Plus to find the values of a and b. The reduced matrix

$$\left[\begin{array}{ccc|c} 1 & 0 & 0 & 665 \\ 0 & 1 & 0 & 1{,}150 \\ 0 & 0 & 1 & 27{,}357 \end{array}\right]$$

yields the corresponding solution to the system

$$a = 665$$
$$b = 1150$$
$$c = 27{,}357$$

The quadratic model for the gross revenue of Johnson & Johnson and its subsidiaries is

$$R(t) = 665t^2 + 1150t + 27{,}357 \text{ million dollars}$$

EXAMPLE 5

Using a System of Equations to Determine Business Needs

A chef has been commissioned to create a party mix containing pretzels, bagel chips, and Chex® cereal for a corporate gathering of 180 people. He estimates that on average each person will consume 1/2 cup of the party mix. The Original Chex Party Mix recipe calls for 9 cups of Chex cereal, 1 cup of pretzels, and 1 cup of bagel chips. (**Source:** www.chex.com.) However, the chef plans to modify the recipe so that there are half as many bagel chips as pretzels. How many cups of each of the ingredients will be needed to make the party mix?

SOLUTION Let x be the number of cups of pretzels, y be the number of cups of bagel chips, and z be the number of cups of Chex cereal. Since each of the 180 people is expected to consume 1/2 cup of party mix, a total of 90 cups of the party mix is needed. Therefore,

$$x + y + z = 90 \qquad \text{The quantity of the party mix is the sum of the ingredients}$$

Since there are to be half as many bagel chips as pretzels, we know that

$$\begin{aligned} y &= 0.5x \qquad \text{There are half as many bagel chips as pretzels} \\ 2y &= x \\ -x + 2y &= 0 \end{aligned}$$

No other constraints are given, so the system of equations is

$$\begin{aligned} x + y + z &= 90 \\ -x + 2y &= 0 \end{aligned}$$

with the corresponding matrix

$$\left[\begin{array}{rrr|r} 1 & 1 & 1 & 90 \\ -1 & 2 & 0 & 0 \end{array}\right]$$

Using the elimination method, we obtain

$$\left[\begin{array}{rrr|r} 1 & 1 & 1 & 90 \\ 0 & 3 & 1 & 90 \end{array}\right] \quad R_1 + R_2$$

$$\left[\begin{array}{rrr|r} 3 & 0 & 2 & 180 \\ 0 & 3 & 1 & 90 \end{array}\right] \quad 3R_1 - R_2$$

$$\left[\begin{array}{rrr|r} 1 & 0 & \frac{2}{3} & 60 \\ 0 & 1 & \frac{1}{3} & 30 \end{array}\right] \quad \begin{array}{l} \frac{1}{3}R_1 \\ \frac{1}{3}R_2 \end{array}$$

The system of linear equations is reduced to

$$x + \frac{2}{3}z = 60 \qquad \text{and} \qquad y + \frac{1}{3}z = 30$$

$$x = 60 - \frac{2}{3}z \qquad\qquad y = 30 - \frac{1}{3}z$$

Since all quantities must be nonnegative, $x \geq 0, y \geq 0$, and $z \geq 0$. Furthermore,

$$x \geq 0 \qquad\qquad y \geq 0$$

$$60 - \frac{2}{3}z \geq 0 \quad \text{and} \quad 30 - \frac{1}{3}z \geq 0$$

$$60 \geq \frac{2}{3}z \qquad\qquad 30 \geq \frac{1}{3}z$$

$$90 \geq z \qquad\qquad 90 \geq z$$

So $0 \leq z \leq 90$. The same constraints hold for each of the other ingredients as well, since the total quantity of the mix cannot exceed 90 cups.

Let $z = t$. Then the solution to the system is

$$x = 60 - \frac{2}{3}t \text{ cups of pretzels}$$

$$y = 30 - \frac{1}{3}t \text{ cups of bagel chips}$$

$$z = t \text{ cups of Chex cereal}$$

The chef has a variety of solutions. If he chooses to use 48 cups of Chex cereal, then he will need

$$x = 60 - \frac{2}{3}t \qquad\qquad y = 30 - \frac{1}{3}t$$

$$x = 60 - \frac{2}{3}(48) \qquad\qquad y = 30 - \frac{1}{3}(48)$$

$$x = 60 - 32 \qquad\qquad y = 30 - 16$$

$$x = 28 \text{ cups of pretzels} \qquad\qquad y = 14 \text{ cups of bagel chips}$$

If he chooses to use 60 cups of Chex cereal, he will need 20 cups of pretzels and 10 cups of bagel chips. In the original recipe, the quantity of cereal was nine times as much as each of the other ingredients. If the chef wants the final product to be similar to the Original Chex Party Mix, roughly 82 percent of the mix should be cereal. If 75 cups of Chex cereal are used, then 10 cups of pretzels and 5 cups of bagel chips will be required. Using this combination of ingredients, 83 percent $\left(\frac{75}{90} = 0.8333\right)$ of the mixture is cereal.

2.3 Summary

In this section, you learned how to formulate linear systems from real-life data and use the solutions to these systems to make business decisions. You also discovered some of the strengths and limitations of mathematical models.

2.3 Exercises

For Exercises 1–18, do the following:

(a) Find the linear function that best fits each data set as a function of time.

(b) Determine if and when the output of the two models will be equal.

(c) Interpret the results in part (b). Do they make sense in their real-world context?

(d) Explain how the results of your analysis could be used to improve business or society.

 **1.**  **School Enrollment**

Projected Student Enrollment

Years Since 1998 (t)	Grade 2 (students) (S)	Grade 3 (students) (T)
0	299	273
1	312	311
2	301	321
3	282	312
4	307	311
5	283	337
6	338	309
7	343	365
8	350	372
9	359	379
10	369	387

Source: White River School District.

 **2.**  **Margarine and Cheese Consumption**

Per Capita Consumption

Years Since 1980 (t)	Margarine (pounds) (M)	Mozzarella Cheese (pounds) (C)
0	11.3	3.0
5	10.8	4.6
10	10.9	6.9
15	9.2	8.1
17	8.6	8.4
18	8.3	8.8
19	8.1	9.2

Source: *Statistical Abstract of the United States, 2001,* Table 202, p. 129.

 **3.**  **Fish and Chicken Consumption**

Per Capita Consumption

Years Since 1970 (t)	Chicken (pounds) (C)	Fish and Shellfish (Boneless, Trimmed Weight) (pounds) (F)
0	27.4	11.7
5	26.4	12.1
10	32.7	12.4
15	36.4	15.0
20	42.4	15.0
25	48.8	14.9
29	54.2	15.2

Source: *Statistical Abstract of the United States, 2001,* Table 202, p. 129.

4. **Diabetes Incidence**

Incidence of Diabetes in Adults Ages 45–64

Years Since 1995 (t)	Male Diabetes Rate per 1,000 (M)	Female Diabetes Rate per 1,000 (F)
0	109	131
4	189	146

Source: *Statistical Abstract of the United States, 2001,* Table 159, p. 109.

 5.  **Baseball Game Attendance**

Major League Baseball Attendance

Years Since 1994 (t)	National League Games (millions of people) (N)	American League Games (millions of people) (A)
0	25.8	24.2
1	25.1	25.4
2	30.4	29.7
3	31.9	31.3
4	38.4	31.9
5	38.3	31.8

Source: *Statistical Abstract of the United States, 2001,* Table 1241, p. 759.

 6.  **Population of Argentina and Canada**

Population of Two Countries

Years Since 1980 (t)	Argentina (thousands) (A)	Canada (thousands) (C)
0	28,237	24,593
10	32,634	27,791
15	34,818	29,619
20	36,955	31,278
21	37,385	31,593

Source: *Statistical Abstract of the United States, 2001,* Table 1327, p. 831.

7.  **Births to Unmarried Women**

Births to Unmarried Women in the USA and UK

Years Since 1980 (t)	United States (percent of all births) (U)	United Kingdom (percent of all births) (K)
0	18	12
18	33	38

Source: *Statistical Abstract of the United States, 2001,* Table 1331, p. 836.

8. **Number of Medical Doctors**

Medical Doctors per 1,000 Population

Years Since 1990 (t)	United States (U)	United Kingdom (K)
0	2.4	1.4
7	2.7	1.7

Source: *Statistical Abstract of the United States, 2001,* Table 1332, p. 836.

 9.  **Postal Service Rates**

U.S. Postal Service Rates

Years Since 1985 (t)	1-Ounce Letter (cents) (L)	Postcard (cents) (P)
0	22	14
3	25	15
6	29	19
10	32	20
14	33	20
16	34	20

Source: *Statistical Abstract of the United States, 2001,* Table 1117, p. 697.

 10.  **Professional Athlete Salaries**

Average Major League Athlete Salaries

Years Since 1980 (t)	Baseball ($1,000s) ($B$)	Football ($1,000s) ($F$)
0	144	79
5	371	194
10	598	352
15	1,111	714
16	1,120	791
17	1,337	725
18	1,399	1,138

Source: *Statistical Abstract of the United States, 2001,* Table 1241, p. 759.

 **11.**  **Annual Earnings by Industry**

Paper Versus Publishing Industry

Years Since 1995 (t)	Paper Manufacturing Employee Earnings (dollars) (E)	Printing and Publishing Employee Earnings (dollars) (P)
0	39,458	34,539
1	40,718	35,897
2	42,129	37,427
3	43,185	39,256
4	44,900	41,083

Source: *Statistical Abstract of the United States, 2001,* Table 979, p. 622.

 **12.**  **Annual Earnings by Industry**

Plastic Versus Apparel Industry

Years Since 1995 (t)	Rubber and Plastics Manufacturing Industry Employee Earnings (dollars) (E)	Apparel and Other Textiles Manufacturing Industry Employee Earnings (dollars) (A)
0	29,867	18,800
1	30,898	19,832
2	32,237	20,838
3	33,574	22,103
4	34,508	23,255

Source: *Statistical Abstract of the United States, 2001,* Table 979, p. 622.

13.  **Clothing and Food Stores**

Retail Establishments

Year Since End of 1998 (t)	Clothing and Accessories Stores (C)	Food and Beverage Stores (F)
0	152,603	147,652
1	151,674	151,506

Source: *Statistical Abstract of the United States, 2001,* Table 1017, p. 641.

14.  **Gasoline Stations and Parts Dealers**

Retail Establishments

Year Since End of 1998 (t)	Gasoline Stations (G)	Motor Vehicle and Parts Dealers (M)
0	123,894	123,359
1	121,095	123,855

Source: *Statistical Abstract of the United States, 2001,* Table 1017, p. 641.

15.  **Specialized Clothing Stores**

Retail Establishments

Year Since End of 1998 (t)	Men's Clothing Stores (M)	Children's Clothing Stores (C)
0	11,861	5,165
1	11,445	5,333

Source: *Statistical Abstract of the United States, 2001,* Table 1017, p. 641.

16. **New and Used Vehicle Dealers**

Retail Establishments

Year Since End of 1998 (t)	New Vehicle Dealers (N)	Used Vehicle Dealers (U)
0	26,216	23,651
1	26,117	23,998

Source: *Statistical Abstract of the United States, 2001,* Table 1017, p. 641.

 **17.**  **Retail Sales**

Sales (in millions of dollars)

Year Since End of 1992 (t)	Building Materials and Supplies Stores (B)	Department Stores (D)
0	160,171	181,255
1	171,733	192,292
2	190,817	205,302
3	199,068	212,759
4	212,759	218,740
5	229,489	225,062
6	243,490	226,024
7	263,958	231,236
8	277,185	241,958

Source: *Statistical Abstract of the United States, 2001,* Table 1020, p. 644.

 **18.**   **Dining Retail Sales**

Sales (in millions of dollars)

Year Since End of 1992 (t)	Full-Service Restaurants (R)	Limited-Service Eating Places (L)
0	86,493	87,433
1	91,476	94,736
2	97,117	98,446
3	99,430	103,143
4	104,514	106,192
5	114,591	109,298
6	119,663	116,836
7	124,463	123,081
8	134,363	127,504

Source: *Statistical Abstract of the United States, 2001,* Table 1020, p. 644.

In Exercises 19–35, set up the system of linear equations and solve it using matrices.

19. **Investment Choices** The following table shows the average annual rate of return on a variety of TIAA-CREF investment accounts over a 10-year period.

CREF Variable Annuity Accounts as of 5/31/04	10-Year Average
Bond Market	7.05%
Equity Index	10.83%
Global Equities	6.96%
Growth	8.14%
Money Market	4.25%
Social Choice	10.01%
Stock	9.53%

Source: www.tiaa-cref.com.

An investor chooses to invest \$3000 in the Bond Market, Growth, and Stock accounts. He assumes that he will be able to get a return equal to the 10-year average, and he wants the total return on his investment to be 9 percent. He wants to invest three times as much money in the Bond Market account as in the Growth account. How much money should he invest in each account? [For ease of computation, round each percentage to the nearest whole-number percent (e.g., 9.53 percent = 10 percent.]

20. **Investment Choices** An investor chooses to invest \$5000 in the Global Equities, Money Market, and Social Choice accounts shown in Exercise 19. She wants to put five times as much money in the Money Market account as in the Global Equities account. She assumes that she will be able to get a return equal to the 10-year average, and she wants the total return on her investment to be 7 percent.

How much money should she invest in each account? [For ease of computation, round each percentage to the nearest whole-number percent (e.g., 6.96 percent = 7 percent).]

21. **Vehicle Sales** A company has budgeted \$970,000 to purchase new vehicles. The vehicles are offered as incentives to top salespeople and include the following models:

- 2004 Toyota Corolla 4-door sedan LE Manual, Manufacturer's Suggested Retail Price \$14,780
- 2004 Toyota Camry 4-door sedan XLE V6 Auto, Manufacturer's Suggested Retail Price \$25,405
- 2004 Prius 5-door hatchback, Manufacturer's Suggested Retail Price \$20,295

(**Source:** www.nada.com.)

The company expects to be able to negotiate lower sale prices on each of the vehicles. For budgeting purposes, the company estimates the total cost of each vehicle including tax to be as follows:

- Corolla \$14,000
- Camry \$25,000
- Prius \$20,000

Company staff anticipate that they will give away three times as many Corollas as Camrys and half as many Prius cars as Corollas. How many of each type of car should the company buy if it plans to spend the entire new vehicle budget?

22. **Vehicle Sales** Because of low sales, the company in Exercise 21 revises its ordering criteria midway through the year. The company decides to order 30 cars, including five times as many Corollas as Camrys.

Ignoring the original budget limitation, how many of each type of car should the company buy? Find three or more solutions.

23. **Johnson & Johnson Costs** According to the company's 2001 annual report, the cost of goods sold by Johnson & Johnson and subsidiaries increased in 1999, 2000, and 2001 as shown in the table.

Johnson & Johnson and Subsidiaries Cost of Goods Sold

Years Since the End of 1999 (t)	Cost of Goods Sold (millions of dollars) (C)
0	8,539
1	8,957
2	9,581

Source: Johnson & Johnson 2001 annual report, p. 6.

Use matrices to find a quadratic equation that models the company's cost of goods sold.

24. **Frito Lay Profit** According to the PepsiCo 2001 annual report, the operating profit of Frito-Lay North America increased in 1999, 2000, and 2001 as shown in the table.

Frito-Lay North America Operating Profit

Years Since the End of 1999 (t)	Operating Profit (millions of dollars) (P)
0	1,679
1	1,915
2	2,056

Source: PepsiCo 2001 annual report, pp. 23, 44.

Use matrices to find a quadratic equation that models the company's operating profit.

25. **Gatorade/Tropicana Sales** According to the PepsiCo 2001 annual report, the net sales of Gatorade/Tropicana North America increased in 1999, 2000, and 2001 as shown in the table.

Gatorade/Tropicana North America Net Sales

Years Since the End of 1999 (t)	Net Sales (millions of dollars) (R)
0	3,452
1	3,841
2	4,016

Source: PepsiCo 2001 annual report, p. 44.

Use matrices to find a quadratic equation that models the company's net sales.

26. **Gatorade/Tropicana Profit** According to the PepsiCo 2001 annual report, operating profit of Gatorade/Tropicana North America increased in 1999, 2000, and 2001 as shown in the table.

Gatorade/Tropicana North America Operating Profit

Years Since the End of 1999 (t)	Operating Profit (millions of dollars) (P)
0	433
1	500
2	530

Source: PepsiCo 2001 annual report, p. 44.

Use matrices to find a quadratic equation that models the company's operating profit.

27. **Resource Allocation: Sandwiches** A plain hamburger requires one ground beef patty and a bun. A cheeseburger requires one ground beef patty, one slice of cheese, and a bun. A double cheeseburger requires two ground beef patties, two slices of cheese, and a bun.

Frozen hamburger patties are typically sold in packs of 12; hamburger buns, in packs of 8; and cheese slices, in packs of 24.

A family is in charge of providing burgers for a neighborhood block party. It has purchased 13 packs of buns, 11 packs of hamburger patties, and 3 packs of cheese slices. How many of each type of sandwich should the family make if the goal is to use up all of the buns, patties, and cheese slices?

28. **Pet Nutrition: Food Cost** PETsMART.com sold the following varieties of dog food in June 2003. The price shown is for an eight-pound bag.

Pro Plan Adult Chicken & Rice Formula, 25 percent protein, 3 percent fiber, \$7.99
Pro Plan Adult Lamb & Rice Formula, 28 percent protein, 3 percent fiber, \$7.99
Pro Plan Adult Turkey & Barley Formula, 26 percent protein, 3 percent fiber, \$8.49
(**Source:** www.petsmart.com.)

A dog breeder wants to make 120 pounds of a mix containing 27 percent protein and 3 percent fiber. How many 8-pound bags of each dog food variety should the breeder buy? What is the breeder's food cost?

29. **Pet Nutrition: Food Cost** PETsMART.com sold the following varieties of dog food in June 2003.

Nature's Recipe Venison Meal & Rice Canine, 20 percent protein, \$21.99 per 20-pound bag
Nutro Max Natural Dog Food, 27 percent protein, \$12.99 per 17.5-pound bag
PETsMART Premier Oven Baked Lamb Recipe, 25 percent protein, \$22.99 per 30-pound bag
(**Source:** www.petsmart.com.)

A dog breeder wants to make 300 pounds of a mix containing 22 percent protein. How many bags of each dog food variety should the breeder buy? (*Hint:* Note that each bag is a different weight. Fractions of bags may not be purchased.)

30. **Utilization of Ingredients** A custard recipe calls for 3 eggs and 2.5 cups of milk. A vanilla pudding recipe calls for 2 eggs and 2 cups of milk. A bread pudding recipe calls for 2 eggs, 2 cups of milk, and 8 slices of bread. (**Source:** *Betty Crocker's Cookbook*)

A stocked kitchen contains 18 eggs, 1 gallon of milk, and 24 slices of bread. How many batches of each recipe should a chef make in order to use up all of the ingredients?

31. **First-Aid Kit Supplies** Safetymax.com sells first-aid and emergency preparedness supplies to businesses. A company that assembles first-aid kits for consumers purchases 3500 1″ × 3″ plastic adhesive bandages, 1800 alcohol wipes, and 220 tubes of antibiotic ointment from Safetymax.com.

The company assembles compact, standard, and deluxe first-aid kits for sale to consumers. A compact first-aid kit contains 20 plastic adhesive bandages, 8 alcohol wipes, and 1 tube of antibiotic ointment. A standard first-aid kit contains 40 plastic adhesive bandages, 20 alcohol wipes, and 2 tubes of antibiotic ointment. A deluxe first-aid kit contains 50 plastic adhesive bandages, 28 alcohol wipes, and 4 tubes of antibiotic ointment.

How many of each type of kit should the company assemble in order to use up all of the bandages, wipes, and antibiotics ordered?

32. **Concert Ticket Sales** On the weekend of July 23–25, 2004, the House of Blues Sunset Strip in Hollywood, California, hosted three concerts: Saves the Day, Jet, and The Bodeans. Saves the Day tickets cost \$15, Jet tickets cost \$20, and The Bodeans tickets cost \$22. (**Source:** www.ticketmaster.com.)

If concert planners expected that a total of 1200 tickets would be sold over the weekend, how many tickets for each concert needed to be sold in order to bring in $23,500? Find three different solutions.

33. **Concert Ticket Sales** On July 7, 2004, Shania Twain was scheduled to perform at the TD Waterhouse Centre in Orlando, Florida. The center offered 18,039 seats for the concert in three seating classifications: floor, lower, and upper. Based upon their location, tickets were offered at three different prices: $80, $65, and $45. (**Source:** www.ticketmaster.com.) Suppose that the average price of a floor ticket was $80, the average price of a lower ticket was $65, and the average price of an upper ticket was $45. If concert planners expected that a total of 12,000 tickets would be sold, including five times as many tickets on the lower level as on the floor, how many of each type of ticket would have to be sold in order to obtain $675,000 in ticket revenue?

34. **Concert Ticket Sales** On June 24, 2004, Madonna was scheduled to perform at Madison Square Garden in New York City. Tickets were offered at four different prices: $49.50, $94.50, $154.50, and $304.50. (**Source:** www.ticketmaster.com.) If concert planners expected that a total of 25,000 tickets would be sold, including ten times as many of the least expensive ticket as of the most expensive ticket and one-third as many $154.50 tickets as $94.50 tickets, how many of each type of ticket would need to be sold in order to earn $2,130,000 in revenue?

35. **Furniture Production** Based on data from 1991–2001, approximately 32 percent of the production value of a piece of furniture is the result of the labor cost associated with producing the piece. (**Source:** www.bls.gov.)

In June 2004, an online furniture retailer offered the following items at the indicated prices.

Teak Double Rocker: $745
Avalon Teak Armchair: $378
Teak Tennis Bench: $124
(**Source:** www.outdoordecor.com)

Suppose that the number of hours required to produce each item is as shown in the following table.

	Cut	Finish	Package
Rocker	4	7	1
Armchair	2	4	1
Bench	1	3	1

If the company has 360 labor hours available in the Cutting Department, 720 labor hours available in the Finishing Department, and 160 labor hours available in the Packaging Department, how many of each type of item should the company produce in order to use all of the available labor hours?

Exercises 36–40 are intended to challenge your understanding of linear system applications.

36. **Amusement Park Rides** Rides at an amusement park require three, four, five, or six tickets. A family purchases 75 tickets and receives two tickets from a park guest who had leftover tickets. The family wants to use all its tickets and wants the sum of five- and six-ticket rides to be twice as much as the sum of three- and four-ticket rides. What is the largest possible number of six-ticket rides that the family can go on subject to these constraints?

37. **Assortment of Coins** A coin purse contains 64 coins including pennies, nickels, dimes, and quarters. The total value of the coins is $4.94. There is the same number of nickels as dimes. How many of each type of coin is in the bag?

38. **Investments** The following table shows the average annual rate of return on a variety of TIAA-CREF investment accounts over a 10-year period.

CREF Variable Annuity Accounts (as of 5/31/04)	10 year
Bond Market	7.05%
Equity Index	10.83%
Global Equities	6.96%
Growth	8.14%
Money Market	4.25%
Social Choice	10.01%
Stock	9.53%

Source: www.tiaa-cref.com.

An investor wants to earn an 8 percent annual return on a $37,625 investment. The investor expects that the annual return on each account will be equal to the 10-year rate rounded to the nearest whole-number percentage (e.g., 10.83 percent = 11 percent).

The investor wants to invest the same amount of money in the Bond Market account as in the Global Equities account and twice as much in the Social Choice account as in the Growth account. How much money should the investor place in each account?

39. Traffic Flow The figure shows the flow of traffic at four city intersections.

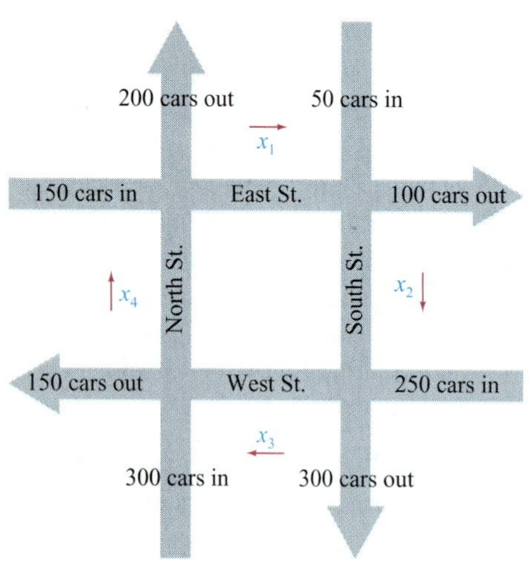

For each intersection, the number of cars entering the intersection must equal the number of cars leaving the intersection. For example, the number of cars entering the intersection of North and West Sts. is $x_3 + 300$. The number of cars leaving the intersection is $x_4 + 150$. Therefore, $x_3 + 300 = x_4 + 150$.

Find two separate set of values x_1, x_2, x_3, and x_4 that work in the traffic flow system.

40. Traffic Flow Repeat Exercise 39 for the following traffic flow diagram.

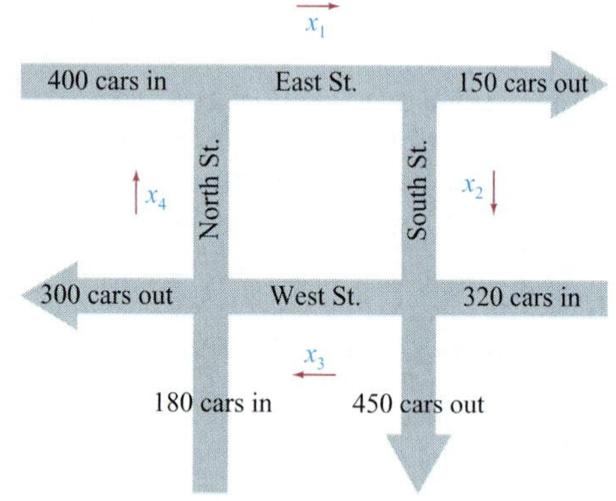

Then explain who could benefit from this type of traffic flow analysis.

Chapter 2 Review Exercises

Section 2.1 *In Exercises 1–8, use your graphing calculator to find the point(s) of intersection of the lines. If the lines do not intersect at a common point, state that a solution does not exist.*

1. $y = 3x + 4$
$y = 1.7x + 5.3$

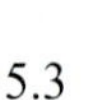

2. $y = 4.6x + 10$
$y = -10x + 44$

3. $y = 7.1x - 41.5$
$y = -1.9x + 3.5$

4. $y = 2.5x - 1$
$y = -0.25x + 10$

5. $y = 7x - 1$
$y = 7x + 1$

6. $y = 5.5x - 1.2$
$y = -2.5x + 5.5$

7. $y = 3.3x - 3.1$
$y = -3.1x + 3.3$

8. $y = -12.5x - 11.5$
$y = -6.2x + 10.4$

In Exercises 9–16, solve the system of equations algebraically. Compare your solutions to those obtained in Exercises 1–8.

9. $y = 3x + 4$
$y = 1.7x + 5.3$

10. $y = 4.6x + 10$
$y = -10x + 44$

11. $y = 7.1x - 41.5$
$y = -1.9x + 3.5$

12. $y = 2.5x - 1$
$y = -0.25x + 10$

13. $y = 7x - 1$
$y = 7x + 1$

14. $y = 5.5x - 1.2$
$y = -2.5x + 5.5$

15. $y = 3.3x - 3.1$
$y = -3.1x + 3.3$

16. $y = -12.5x - 11.5$
$y = -6.2x + 10.4$

Section 2.2

In Exercises 17–24, determine if the augmented matrix is in reduced row echelon form. If it is not reduced, specify which of the four criteria for reduced row echelon form the matrix fails to meet.

17. $\left[\begin{array}{ccc|c} 1 & 0 & 0 & 3 \\ 0 & 1 & 1 & 7 \\ 0 & 1 & 0 & -2 \end{array}\right]$

18. $\left[\begin{array}{ccc|c} 1 & 0 & 0 & 9 \\ 0 & 1 & 0 & 11 \\ 0 & 0 & 1 & 2 \end{array}\right]$

19. $\left[\begin{array}{ccc|c} 0 & 0 & 0 & 0 \\ 0 & 1 & 0 & 13 \\ 1 & 0 & 0 & 27 \end{array}\right]$

20. $\left[\begin{array}{ccc|c} 1 & 0 & 1 & -9 \\ 0 & 1 & 0 & 14 \\ 0 & 0 & 1 & 2.1 \end{array}\right]$

21. $\left[\begin{array}{ccc|c} 1 & 1 & 2 & 0 \\ 0 & 0 & 0 & 1 \\ 0 & 0 & 0 & 0 \end{array}\right]$

22. $\left[\begin{array}{cc|c} 1 & 0 & 5 \\ 0 & 1 & 9 \\ 0 & 0 & 0 \end{array}\right]$

23. $\left[\begin{array}{cccc|c} 1 & 0 & 1 & 0 & -2 \\ 0 & 1 & 0 & 0 & 4 \\ 0 & 0 & 0 & 1 & 0 \end{array}\right]$

24. $\left[\begin{array}{cccc|c} 1 & 0 & 0 & 1 & 3 \\ 0 & 1 & 1 & 0 & -2 \\ 0 & 0 & 0 & 1 & 5 \end{array}\right]$

In Exercises 25–32, rewrite the system of equations as an augmented matrix, then simplify the matrix to reduced row echelon form. Identify the solution(s) to the system of equations.

25. $\begin{aligned} 2.1x - y &= -8 \\ 3.4x + y &= 8 \end{aligned}$

26. $\begin{aligned} 5.2x - 1.3y &= 12 \\ -10.4x + 2.6y &= 24 \end{aligned}$

27. $\begin{aligned} 4x - 8y &= -16 \\ x + y &= 5 \end{aligned}$

28. $\begin{aligned} 2.9x - 8.1y &= 4 \\ 3.7x + 16.2y &= 1.5 \end{aligned}$

29. $\begin{aligned} x + y + z &= 6 \\ 2x - y + z &= 3 \\ 3x + 2z &= 9 \end{aligned}$

30. $\begin{aligned} 6x - 2y &= 6 \\ 4x + 2y &= 14 \\ x + y &= 5 \end{aligned}$

31. $\begin{aligned} x - y + 3z &= 5 \\ 2x + 4y + z &= -1 \end{aligned}$

32. $\begin{aligned} 3x - 2y + z &= 11 \\ x + 2y - 3z &= -1 \end{aligned}$

In Exercises 33–36, use technology to write the matrix in reduced row echelon form.

33. $\left[\begin{array}{ccc|c} 8.1 & 4.3 & 6.2 & 3.1 \\ 2.7 & -2.8 & 7.5 & 7.2 \\ 8.2 & -5.3 & -1.1 & 5.9 \end{array}\right]$

34. $\left[\begin{array}{ccc|c} 2.6 & 3.0 & 0.2 & 9.6 \\ 6.1 & 4.0 & 6.5 & 0.8 \\ 9.1 & -5.0 & -0.8 & 9.9 \end{array}\right]$

35. $\left[\begin{array}{ccc|c} 6.2 & 5.3 & -0.2 & 3.3 \\ 1.7 & -2.6 & 2.5 & 2.2 \\ 4.5 & 7.9 & -2.7 & 1.1 \end{array}\right]$

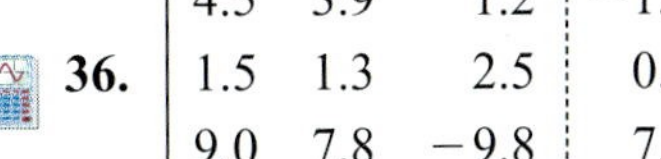

36. $\left[\begin{array}{ccc|c} 4.5 & 3.9 & 1.2 & -1.7 \\ 1.5 & 1.3 & 2.5 & 0.3 \\ 9.0 & 7.8 & -9.8 & 7.9 \end{array}\right]$

Section 2.3

For Exercises 37–40, do the following:

(a) Find the linear function that best models each data set as a function of time.

(b) Determine if and when the output of the two models will be equal.

(c) Interpret the results in part (b). Do they make sense in their real-world context?

(d) Explain how the results of your analysis could be used to improve business or society.

37. **Births to Unmarried Women**

Births to Unmarried Women in the UK and Iceland

Years Since 1980 (t)	United Kingdom (percent of all births) (K)	Iceland (percent of all births) (I)
0	12	40
18	38	64

Source: *Statistical Abstract of the United States, 2001,* Table 1331, p. 836.

38.  **Fruit and Vegetable Production**

Worldwide Produce Production

Years Since 1990 (t)	Fruit (millions of metric tons) (F)	Vegetables (millions of metric tons) (V)
0	352.2	461.4
2	379.0	478.6
4	391.0	532.1
6	427.3	591.2
8	432.8	626.0
10	459.2	670.1

Source: *Statistical Abstract of the United States, 2001,* Table 1357, p. 849.

 **39.**  **Basketball Game Attendance**

NCAA Basketball Game Attendance

Years Since 1985 (t)	Men's Games (thousands) (M)	Women's Games (thousands) (W)
0	26,584	2,072
5	28,741	2,777
10	28,548	4,962
14	29,025	8,698

Source: *Statistical Abstract of the United States, 2001,* Table 1241, p. 759.

 **40.**   **Education and Crime**

Years Since 1980 (t)	Private College Students (thousands) (S)	Adults in Prison, in Jail, on Probation, or on Parole (thousands) (A)
0	2,640	1,840
2	2,730	2,193
4	2,765	2,689
6	2,790	3,239
8	2,894	3,714
10	2,974	4,348
12	3,103	4,763
14	3,145	5,141
16	3,247	5,483
18	3,373	6,126

Source: *Statistical Abstract of the United States, 2001,* Tables 205, 335, pp. 133, 202.

In Exercises 41–43, set up the system of linear equations and solve it using matrices.

41. **Company Sales** The following table shows the sales revenue for the Starbucks Corporation from 1993 to 2002.

Starbucks Corporation Sales

Years Since 9/93	Sales Revenue (millions of dollars)
0	163.5
1	284.9
2	465.2
3	696.5
4	966.9
5	1,308.7
6	1,680.1
7	2,169.2
8	2,649.0
9	3,288.9

Source: moneycentral.msn.com.

Use matrices and the sales data from 1994, 1998, and 2002 to find a quadratic model for the income from sales. Then use the model to forecast the 2004 sales.

42. **Music Market Size** According to the Recording Industry Association of America, the size of the music market fluctuated between 1997 and 2002 as shown in the following table.

Music Market Size

Years Since 1997 (t)	Dollar Volume (millions) (M)
0	\$12,236.80
1	\$13,723.50
2	\$14,584.50
3	\$14,323.00
4	\$13,740.89
5	\$12,614.21

Source: www.riaa.com.

Use matrices and the dollar volume data from 1997, 1998, and 2002 to find a quadratic model for the dollar volume of the music market. Then use the model to forecast the market size in 2003.

43. **First-Aid Kit Supplies** Safetymax.com sells first-aid and emergency preparedness supplies to businesses. A company that assembles first-aid kits for consumers purchases 3275 1″ × 3″ plastic adhesive bandages, 1300 alcohol wipes, and 185 tubes of antibiotic ointment from Safetymax.com.

The company assembles compact, standard, and deluxe first-aid kits for sale to consumers. A compact first-aid kit contains 25 plastic adhesive bandages, 10 alcohol wipes, and 1 tube of antibiotic ointment. A standard first-aid kit contains 40 plastic adhesive bandages, 15 alcohol wipes, and 2 tubes of antibiotic ointment. A deluxe first-aid kit contains 60 plastic adhesive bandages, 25 alcohol wipes, and 4 tubes of antibiotic ointment.

How many of each type of kit should the company assemble in order to use up all of the bandages, wipes, and antibiotics ordered?

Make It Real

PROJECT 2

What to do

1. Find two sets of data from an area of personal interest that appear to be increasing or decreasing at a nearly constant rate. Both data sets should have the same domain (input values).
2. Find the linear function that best models each data set.
3. Determine if and when the output of the two models will be equal. (Find the point of intersection of their graphs.)
4. Interpret the results from Step 3. Do they make sense in their real-world context?
5. Explain how the results of your analysis could be used to improve business or society.

Where to look for data

U.S. Census Bureau
www.census.gov
Look at data on U.S. residents ranging from Internet usage to family size.

NADA
www.nada.com
Check out current trade-in and retail values of used cars, trucks, and vans. Compare the retail value of a model to its trade-in value over time.

State Office of Financial Management
Go to your state's web site and search for the Office of Financial Management. This office typically publishes city and county population data. Compare the growth of two nearby cities over the past decade.

Chapter 3

Matrix Algebra and Applications

As states faced budget crises in the early 2000s, citizens rallied to let state legislatures know which programs they saw as a priority. In Washington State, voters overwhelmingly approved an initiative in 2000 granting annual cost-of-living increases to K through 12 teachers and community college employees. Determining a new pay scale resulting from an across-the-board percentage increase in salaries, such as that approved by Washington voters, is easy to do using matrix algebra.

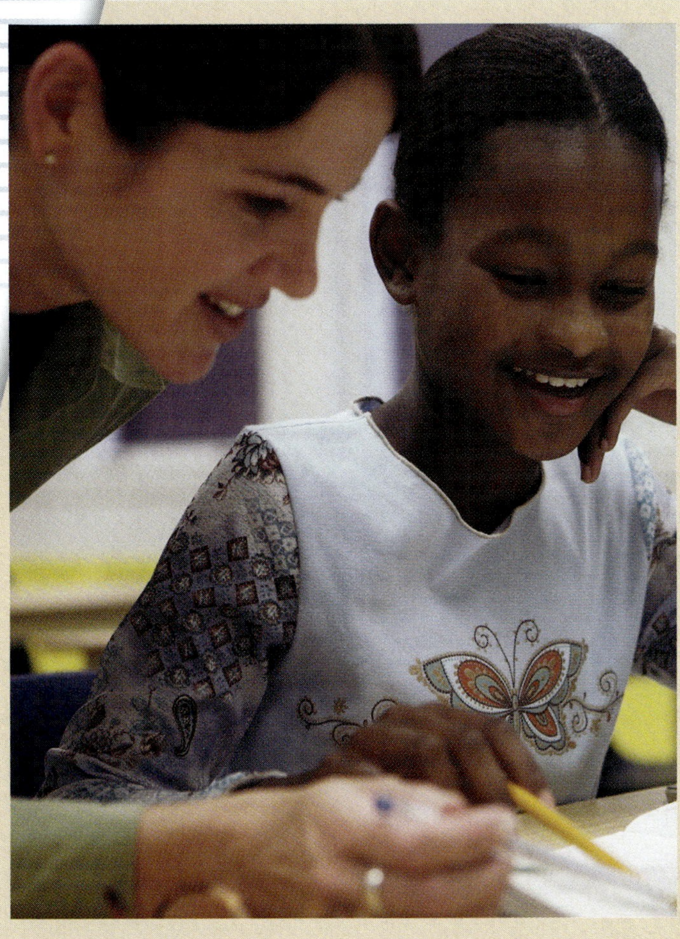

3.1 Matrix Addition and Scalar Multiplication

- Add and subtract matrices
- Do scalar multiplication of matrices
- Use matrices to analyze and interpret real-life data

GETTING STARTED Financial conflict is one of the leading causes of marital discord. To help alleviate future financial stress, many engaged couples take marriage preparation classes to help them confront marital issues, including finances, before they tie the knot.

In Chapter 2, we introduced augmented matrices as a tool for solving systems of linear equations. In this section, we will further develop the concept of matrices by demonstrating matrix addition and scalar multiplication. We will also demonstrate how to use these techniques to analyze real-life financial situations, such as a forthcoming marriage.

Matrix Addition

Two matrices, A and B, of the same dimension may be added together to form a new matrix $C = A + B$ by adding their corresponding entries. For example, if the 2×2 matrices $A = \begin{bmatrix} 1 & 2 \\ 3 & 4 \end{bmatrix}$ and $B = \begin{bmatrix} 4 & -2 \\ -3 & 1 \end{bmatrix}$ are added together, we get

$$\begin{aligned} C &= A + B \\ &= \begin{bmatrix} 1 & 2 \\ 3 & 4 \end{bmatrix} + \begin{bmatrix} 4 & -2 \\ -3 & 1 \end{bmatrix} \\ &= \begin{bmatrix} 1+4 & 2+(-2) \\ 3+(-3) & 4+1 \end{bmatrix} \qquad \text{Add corresponding entries} \\ &= \begin{bmatrix} 5 & 0 \\ 0 & 5 \end{bmatrix} \end{aligned}$$

MATRIX ADDITION

An $m \times n$ matrix A and an $m \times n$ matrix B may be added together to form a new $m \times n$ matrix, C. The value of the entry in the *ith* row and *jth* column of C is $c_{ij} = a_{ij} + b_{ij}$.

If A and B are not of the same dimension, matrix addition is undefined.

EXAMPLE 1

Calculating the Sum of Two Matrices

Calculate $C = A + B$ given $A = \begin{bmatrix} 2 & 0 \\ -1 & 6 \\ 4 & 5 \end{bmatrix}$ and $B = \begin{bmatrix} 9 & 2 \\ -7 & -6 \\ 8 & -1 \end{bmatrix}$.

SOLUTION

$$C = A + B$$

$$= \begin{bmatrix} 2 & 0 \\ -1 & 6 \\ 4 & 5 \end{bmatrix} + \begin{bmatrix} 9 & 2 \\ -7 & -6 \\ 8 & -1 \end{bmatrix}$$

$$= \begin{bmatrix} 2+9 & 0+2 \\ -1+(-7) & 6+(-6) \\ 4+8 & 5+(-1) \end{bmatrix}$$ Add corresponding entries

$$= \begin{bmatrix} 11 & 2 \\ -8 & 0 \\ 12 & 4 \end{bmatrix}$$

TECHNOLOGY TIP

Adding Two Matrices

1. Enter matrix A and matrix B into the calculator using the Matrix Editor.

```
MATRIX[B] 3 ×2
[ 9      2       ]
[ -7     -6      ]
[ 8      -1      ]

3,2=-1
```

2. Use the Matrix Names menu to place matrix A on the home screen. Press the [+] key. Then place matrix B on the home screen.

```
[A]+[B]
```

3. Press [ENTER] to display the sum of the two matrices.

```
[A]+[B]
              [[11 2]
               [-8 0]
               [12 4]]
```

Error Alert:
If you try to add two matrices of different dimensions, you will get a dimension mismatch error.

```
ERR:DIM MISMATCH
1:Quit
2:Goto
```

EXAMPLE 2 Using Matrix Addition to Solve Real-World Problems

Jamaal and Shina, an engaged couple, are discussing each of their financial obligations. Jamaal has the following debts:

$5000 student loan with Back Bay Credit Union
$2000 student loan with Aspen Bank
$7000 car loan with IntelliBank
$1500 credit card balance with IntelliBank
$800 credit card balance with Aspen Bank

Shina has the following debts:

$8000 student loan with Back Bay Credit Union
$4500 car loan with Back Bay Credit Union
$700 credit card balance with Aspen Bank
$1200 credit card balance with IntelliBank

Use matrix addition to determine their combined debts of each type at the various financial institutions.

SOLUTION Jamaal's debts (in thousands of dollars) may be represented as shown in Table 3.1.

TABLE 3.1

	Car Loan	Credit Card	Student Loan
Aspen Bank	0	0.8	2
Back Bay Credit Union	0	0	5
IntelliBank	7	1.5	0

The information may be further represented as a matrix whose rows and columns have the same meaning as the rows and columns of the table.

$$J = \begin{bmatrix} 0 & 0.8 & 2 \\ 0 & 0 & 5 \\ 7 & 1.5 & 0 \end{bmatrix}$$

Similarly, Shina's debts (in thousands of dollars) may be represented as shown in Table 3.2

TABLE 3.2

	Car Loan	Credit Card	Student Loan
Aspen Bank	0	0.7	0
Back Bay Credit Union	4.5	0	8
IntelliBank	0	1.2	0

and the matrix

$$S = \begin{bmatrix} 0 & 0.7 & 0 \\ 4.5 & 0 & 8 \\ 0 & 1.2 & 0 \end{bmatrix}$$

Their combined debt D is the sum of the two matrices.

$$\begin{aligned} D &= J + S \\ &= \begin{bmatrix} 0 & 0.8 & 2 \\ 0 & 0 & 5 \\ 7 & 1.5 & 0 \end{bmatrix} + \begin{bmatrix} 0 & 0.7 & 0 \\ 4.5 & 0 & 8 \\ 0 & 1.2 & 0 \end{bmatrix} \\ &= \begin{bmatrix} 0 & 1.5 & 2 \\ 4.5 & 0 & 13 \\ 7 & 2.7 & 0 \end{bmatrix} \end{aligned}$$

The result shows the types of debt owed to each institution (see Table 3.3).

TABLE 3.3

	Car Loan	Credit Card	Student Loan
Aspen Bank	0	1.5	2
Back Bay Credit Union	4.5	0	13
IntelliBank	7	2.7	0

To determine the total amount owed to each institution, the row entries should be summed. To determine the total amount of each type of debt, the column entries should be summed.

Using matrices to calculate the combined debts in Example 2 may have seemed trivial. Wouldn't it have been easier just to add the various quantities without introducing matrix notation? Perhaps. But what if we were looking at the merger of two companies with 40 creditors and 20 different types of debt? Would we want to calculate all 800 entries manually? Probably not. The notion of matrix addition, a feature embedded in most types of spreadsheet software, would greatly facilitate the calculation.

Scalar Multiplication

There are two types of multiplication that may be used with matrices: **scalar multiplication** and **matrix multiplication.** We will address scalar multiplication here and matrix multiplication in Section 3.2. The term **scalar** means *constant* or *number.* Scalar multiplication "scales" the entries of a matrix by making them larger or smaller by a given factor.

For example, if $A = \begin{bmatrix} 1 & 3 & 0 \\ -2 & 4 & 6 \\ 5 & 7 & 9 \end{bmatrix}$, then

$$\begin{aligned} 5A &= 5\begin{bmatrix} 1 & 3 & 0 \\ -2 & 4 & 6 \\ 5 & 7 & 9 \end{bmatrix} \\ &= \begin{bmatrix} 5 \cdot 1 & 5 \cdot 3 & 5 \cdot 0 \\ 5 \cdot (-2) & 5 \cdot 4 & 5 \cdot 6 \\ 5 \cdot 5 & 5 \cdot 7 & 5 \cdot 9 \end{bmatrix} && \text{Multiply each entry by the scalar 5} \\ &= \begin{bmatrix} 5 & 15 & 0 \\ -10 & 20 & 30 \\ 25 & 35 & 45 \end{bmatrix} \end{aligned}$$

In this example, the scalar 5 increased the magnitude of each entry by a factor of five. Any real number may be used as a scalar.

SCALAR MULTIPLICATION

An $m \times n$ matrix A and a real number k may be multiplied together to form a new $m \times n$ matrix, C. The value of the entry in the *ith* row and *jth* column of C is $c_{ij} = k \cdot a_{ij}$.

EXAMPLE 3

Multiplying by a Scalar

Calculate $C = -\frac{1}{2}A$ given $A = \begin{bmatrix} 2 & 0 \\ -4 & 6 \\ 1 & 8 \end{bmatrix}$.

SOLUTION

$$\begin{aligned} C &= -\frac{1}{2}A \\ &= -\frac{1}{2}\begin{bmatrix} 2 & 0 \\ -4 & 6 \\ 1 & 8 \end{bmatrix} \\ &= \begin{bmatrix} -\frac{1}{2} \cdot 2 & -\frac{1}{2} \cdot 0 \\ -\frac{1}{2} \cdot (-4) & -\frac{1}{2} \cdot 6 \\ -\frac{1}{2} \cdot 1 & -\frac{1}{2} \cdot 8 \end{bmatrix} && \text{Multiply each entry by the scalar } -\tfrac{1}{2} \\ &= \begin{bmatrix} -1 & 0 \\ 2 & -3 \\ -\frac{1}{2} & -4 \end{bmatrix} \end{aligned}$$

TECHNOLOGY TIP

Scalar Multiplication

1. Enter matrix A into the calculator using the Matrix Editor.

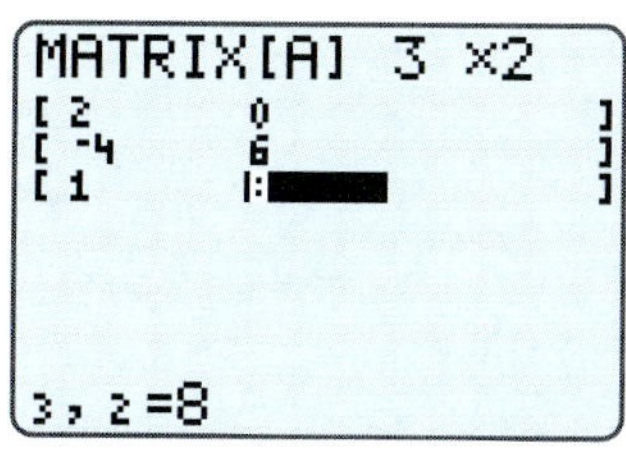

2. Type in the scalar on the home screen, press the [×] key, and then use the Matrix Names menu to place matrix A on the home screen.

```
-1/2*[A]
```

3. Press [ENTER] to display the product of the scalar and the matrix.

```
-1/2*[A]
      [[-1 0 ]
       [2  -3]
       [-.5 -4]]
```

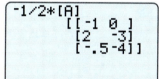

4. If you would like to convert decimal entries to fractions, press [MATH], select `1:►Frac`, then press [ENTER].

```
      [[-1   0 ]
       [2   -3]
       [-.5 -4]]
Ans►Frac
      [[-1   0 ]
       [2   -3]
       [-1/2 -4]]
```

EXAMPLE 4

Using Scalar Multiplication to Adjust a Pay Scale

In 2000, Washington State voters overwhelmingly approved an initiative to give state K through 12 and community college educators annual cost-of-living increases. (**Source:** I-732 Text.) Faculty at Green River Community College received a 3.7 percent salary increase for the 2001–2002 school year.

Each faculty member's base salary is calculated based on teaching experience and educational/professional development credits as shown in Table 3.4.

TABLE 3.4 2000–2001 Pay Scale

	Credits 240	Credits 300	Credits 360
Teaching Level 1	\$33,878	\$37,360	\$40,843
Teaching Level 2	\$35,695	\$39,177	\$42,660
Teaching Level 3	\$37,512	\$40,994	\$44,476
Teaching Level 4	\$39,329	\$42,811	\$46,293

Source: Green River Community College.

Determine the 2001–2002 pay scale for Green River Community College faculty.

SOLUTION We can solve the problem by writing the table as a matrix and multiplying by a scalar. Since salaries are to increase by 3.7 percent, the 2001–2002 salaries will be 103.7 percent of the 2000–2001 salaries.

$$\begin{aligned} 103.7\% &= \frac{103.7}{100} \\ &= 1.037 \end{aligned}$$

We will multiply the matrix by 1.037. This will increase each entry by 3.7 percent.

$$S = \begin{bmatrix} 33{,}878 & 37{,}360 & 40{,}843 \\ 35{,}695 & 39{,}177 & 42{,}660 \\ 37{,}512 & 40{,}994 & 44{,}476 \\ 39{,}329 & 42{,}811 & 46{,}293 \end{bmatrix}$$

$$\begin{aligned} 1.037S &= 1.037 \begin{bmatrix} 33{,}878 & 37{,}360 & 40{,}843 \\ 35{,}695 & 39{,}177 & 42{,}660 \\ 37{,}512 & 40{,}994 & 44{,}476 \\ 39{,}329 & 42{,}811 & 46{,}293 \end{bmatrix} \\ &\approx \begin{bmatrix} 35{,}131 & 38{,}742 & 42{,}354 \\ 37{,}016 & 40{,}627 & 44{,}238 \\ 38{,}900 & 42{,}511 & 46{,}122 \\ 40{,}784 & 44{,}395 & 48{,}006 \end{bmatrix} \end{aligned}$$

The new matrix represents the 2001–2002 pay scale for Green River Community College faculty, as shown in Table 3.5.

TABLE 3.5 2001–2002 Pay Scale

	Credits 240	Credits 300	Credits 360
Teaching Level 1	\$35,131	38,742	\$42,354
Teaching Level 2	\$37,016	\$40,627	\$44,238
Teaching Level 3	\$38,900	\$42,511	\$46,122
Teaching Level 4	\$40,784	\$44,395	\$48,006

Matrix Addition and Scalar Multiplication Properties

Let A, B, and C be $m \times n$ matrices and let c and k be real numbers. Let O be the $m \times n$ **zero matrix** (a matrix with entries of all zeros.) The following properties hold:

Matrix Addition Properties

1. Additive Associative $A + (B + C) = (A + B) + C$
2. Additive Commutative $A + B = B + A$
3. Additive Identity $A + O = O + A = A$
4. Additive Inverse $(-A) + A = A + (-A) = O$

Scalar Multiplication Properties

5. Distributive $c(A + B) = cA + cB$
6. Distributive $(c + k)A = cA + kA$
7. Multiplicative Associative $c(kA) = (ck)A$
8. Scalar Unit 1 $1 \cdot A = A$
9. Scalar Unit 0 $0 \cdot A = O$

You may have noticed that no matrix subtraction properties were listed. This is because we view matrix subtraction as a combination of scalar multiplication and matrix addition. The matrix expression $A - B$ is equivalent to $A + (-1)B$. As shown in Example 5, in practice we typically simplify the matrix expression $A - B$ by subtracting the entries of B from the corresponding entries of A.

EXAMPLE 5

Calculating the Difference of Two Matrices

Let $A = \begin{bmatrix} 4 & -1 & 2 \\ 3 & -5 & 0 \\ 9 & 11 & 20 \end{bmatrix}$ and $B = \begin{bmatrix} -4 & 7 & 3 \\ 10 & -9 & 10 \\ -2 & 4 & 8 \end{bmatrix}$. Show that $A - B$ is equivalent to $A + (-1)B$.

SOLUTION

$$A - B = \begin{bmatrix} 4 & -1 & 2 \\ 3 & -5 & 0 \\ 9 & 11 & 20 \end{bmatrix} - \begin{bmatrix} -4 & 7 & 3 \\ 10 & -9 & 10 \\ -2 & 4 & 8 \end{bmatrix}$$

$$= \begin{bmatrix} 4-(-4) & -1-7 & 2-3 \\ 3-10 & -5-(-9) & 0-10 \\ 9-(-2) & 11-4 & 20-8 \end{bmatrix}$$

Subtract the entries of B from the corresponding entries of A

$$= \begin{bmatrix} 8 & -8 & -1 \\ -7 & 4 & -10 \\ 11 & 7 & 12 \end{bmatrix}$$

$$A + (-1)B = \begin{bmatrix} 4 & -1 & 2 \\ 3 & -5 & 0 \\ 9 & 11 & 20 \end{bmatrix} + (-1)\begin{bmatrix} -4 & 7 & 3 \\ 10 & -9 & 10 \\ -2 & 4 & 8 \end{bmatrix}$$

$$= \begin{bmatrix} 4 & -1 & 2 \\ 3 & -5 & 0 \\ 9 & 11 & 20 \end{bmatrix} + \begin{bmatrix} (-1)(-4) & (-1)7 & (-1)3 \\ (-1)10 & (-1)(-9) & (-1)10 \\ (-1)(-2) & (-1)4 & (-1)8 \end{bmatrix}$$

Multiply the entries of B by the scalar -1

$$= \begin{bmatrix} 4 & -1 & 2 \\ 3 & -5 & 0 \\ 9 & 11 & 20 \end{bmatrix} + \begin{bmatrix} 4 & -7 & -3 \\ -10 & 9 & -10 \\ 2 & -4 & -8 \end{bmatrix}$$

$$= \begin{bmatrix} 8 & -8 & -1 \\ -7 & 4 & -10 \\ 11 & 7 & 12 \end{bmatrix}$$

Matrix expressions may combine one or more matrix operations. Example 6 includes matrix addition and scalar multiplication with three different matrices, including the zero matrix.

EXAMPLE 6

Solving a Matrix Algebra Problem Involving the Zero Matrix

Let $A = \begin{bmatrix} 1 & 0 & 4 \\ -1 & 2 & -5 \end{bmatrix}$, $B = \begin{bmatrix} 5 & 7 & 0 \\ 3 & -2 & -4 \end{bmatrix}$, and $O = \begin{bmatrix} 0 & 0 & 0 \\ 0 & 0 & 0 \end{bmatrix}$. Calculate $A - (2B + O)$.

SOLUTION

$$A - (2B + O) = \begin{bmatrix} 1 & 0 & 4 \\ -1 & 2 & -5 \end{bmatrix} - \left(2 \cdot \begin{bmatrix} 5 & 7 & 0 \\ 3 & -2 & -4 \end{bmatrix} + \begin{bmatrix} 0 & 0 & 0 \\ 0 & 0 & 0 \end{bmatrix}\right)$$

$$= \begin{bmatrix} 1 & 0 & 4 \\ -1 & 2 & -5 \end{bmatrix} + (-1)\left(\begin{bmatrix} 10 & 14 & 0 \\ 6 & -4 & -8 \end{bmatrix} + \begin{bmatrix} 0 & 0 & 0 \\ 0 & 0 & 0 \end{bmatrix}\right)$$

Multiply B by 2. Rewrite "$-$" as "$+$".

$$= \begin{bmatrix} 1 & 0 & 4 \\ -1 & 2 & -5 \end{bmatrix} + (-1)\begin{bmatrix} 10 & 14 & 0 \\ 6 & -4 & -8 \end{bmatrix}$$

$$= \begin{bmatrix} 1 & 0 & 4 \\ -1 & 2 & -5 \end{bmatrix} + \begin{bmatrix} -10 & -14 & 0 \\ -6 & 4 & 8 \end{bmatrix}$$

Multiply $2B$ by -1

$$= \begin{bmatrix} -9 & -14 & 4 \\ -7 & 6 & 3 \end{bmatrix}$$

3.1 Summary

In this section, you learned how to do matrix addition and scalar multiplication. You also saw how these techniques could be used to analyze real-life financial situations in life and in business.

3.1 Exercises

In Exercises 1–10, perform the indicated matrix operation, if possible, given the following matrices. Solve these problems without technology.

$$A = \begin{bmatrix} 1 & 6 \\ 3 & 4 \\ 5 & 2 \end{bmatrix}, \quad B = \begin{bmatrix} -5 & 0 \\ 7 & 8 \\ -9 & 1 \end{bmatrix},$$

$$C = \begin{bmatrix} 2 & 3 & 0 \\ -1 & 4 & 1 \\ 5 & -2 & 0 \end{bmatrix}, \quad D = \begin{bmatrix} 1 & -3 & 4 \\ 0 & 5 & 7 \\ 9 & 8 & 2 \end{bmatrix}$$

1. $A + B$ **2.** $A - B$ **3.** $B + C$

4. $C - D$ **5.** $D + C$ **6.** $2A$

7. $3B$ **8.** $2A + 3B$ **9.** $-2C + 2D$

10. $2A + 2C$

In Exercises 11–20, use technology to simplify the matrix expressions, given the following matrices.

$$A = \begin{bmatrix} 1.2 & 6.3 & 0.4 \\ -9.1 & 4.2 & 1.7 \\ 0.9 & -2.0 & 0.3 \end{bmatrix}$$

$$B = \begin{bmatrix} 1.4 & -0.3 & 0.4 \\ -2.8 & 5.5 & 7.1 \\ 9.2 & 8.6 & 2.0 \end{bmatrix}$$

11. $1.2A$ **12.** $-2.3B$

13. $1.2A - 2.3B$ **14.** $4.1A + 0.1B$

15. $-1.1A + 2.9B$ **16.** $2.9A + 0.1B$

17. $-8.7A + 8.7B$ **18.** $-A - 9.2B$

19. $7.8A + 9.9B$ **20.** $-1.7A - 2.1B$

In Exercises 21–36, use matrix addition and/or scalar multiplication to find the solution.

21. Personal Debt A couple is planning to get married. He has the following debts: a \$2700 consumer loan at Loan Shark Larry's, a \$26,500 car loan and an \$8200 credit card balance at Risky Bank, and a \$2700 consumer loan at Mastercraft Jewelers (for the ring). She has the following debts: a \$1200 car loan, a \$82,500 home mortgage, a \$200 credit card balance at Risky Bank, and a \$250 consumer loan at Mastercraft Jewelers. Determine their combined debts of each type at each of the various financial institutions.

22. Faculty Salaries For the 2002–2003 academic year, Green River Community College faculty received a 3.432 percent cost-of-living increase. (**Source:** Green River Community College.) The 2001–2002 pay scale is shown in the following table. Determine the 2002–2003 pay scale.

2001–2002 Pay Scale

	240 Credits (\$)	300 Credits (\$)	360 Credits (\$)
Level 1	35,131	38,742	42,354
Level 2	37,016	40,627	44,238
Level 3	38,900	42,511	46,122
Level 4	40,784	44,395	48,006

23. Faculty Salaries If the Green River Community College faculty in Exercise 22 get a 3.432 percent raise annually, determine the 2005–2006 pay scale.

24. Condiment Prices On July 20, 2002, Albertsons.com advertised the following items at the indicated prices.

	Albertsons	Hunts	Kraft
Ketchup, 24 oz	\$0.69	\$1.89	
Barbecue sauce, 18 oz		\$0.99	\$0.99
Mayonnaise, 32 oz	\$2.19		\$3.19

If the prices are affected only by inflation and the annual rate of inflation is 3 percent, determine the price of each of the items on July 20, 2006.

25. **Soda Prices** On July 20, 2002, Albertsons.com advertised the following items at the indicated prices.

	Albertsons	A&W	Henry Weinhards
Root beer, 6-pack	$1.59	$1.67	$4.50
Cream soda, 12-pack	$3.18	$3.34	$9.00
Club soda, 2 liters	$0.99		

If the prices are affected only by inflation and the annual rate of inflation is 3 percent, determine the price of each of the items on July 20, 2006.

26. **Auto Prices** The average trade-in value of a Volkswagen New Beetle and Volkswagen Golf in July 2002 is shown in the first table. The average retail value of the two vehicles is shown in the second table.

Average Trade-in Value

	Golf	New Beetle
2000 model	$11,000	$11,850
2001 model	$11,875	$13,175

Average Retail Value

	Golf	New Beetle
2000 model	$13,050	$14,000
2001 model	$14,025	$15,475

Source: www.nada.com.

Use matrices to create a table that shows the average dealer markup for each of the vehicles.

27. **Auto Prices** The average trade-in value of a Honda Civic and a Honda Accord in July 2002 is shown in the first table. The average retail value of the two vehicles is shown in the second table.

Average Trade-in Value

	Accord	Civic
2000 model	$14,800	$8,925
2001 model	$16,575	$9,850

Average Retail Value

	Accord	Civic
2000 model	$17,100	$10,800
2001 model	$18,975	$11,825

Source: www.nada.com.

Use matrices to create a table that shows the average dealer markup for each of the vehicles.

28. **Food Supply** Bed-and-breakfasts attract visitors with their intimate ambiance and their delicious cuisine. Many establishments serve wedding brunches in addition to providing lodging services.

A bed-and-breakfast hostess is planning an upcoming wedding brunch. She plans to make four dozen muffins (two dozen apple and two dozen blueberry) and six fruit crisps (three apple and three blueberry). The amount of flour, sugar, and fruit required to make a single fruit crisp is shown in the first table, and the amount required to make a dozen muffins is shown in the second table.

Fruit Crisp

	Flour	Sugar	Fruit
Apple	1/2 cup	3/4 cup	3 cups
Blueberry	1/2 cup	1 cup	4 cups

Muffins

	Flour	Sugar	Fruit
Apple	2 cups	1/4 cup	3/4 cup
Blueberry	2 cups	1/3 cup	1 cup

Use matrices to create a table that shows how much flour, sugar, and fruit will be required for the apple desserts and the blueberry desserts.

29. **Energy Usage** The amount of natural gas and coal energy produced and consumed in the United States is shown in the following tables. Use matrix operations to create a table that shows the difference between energy production and energy consumption.

Energy Production

Years Since 1960	Natural Gas (quadrillion BTUs)	Coal (quadrillion BTUs)
0	12.66	10.82
10	21.67	14.61
20	19.91	18.60
30	18.36	22.46
40	19.74	22.66

Energy Consumption

Years Since 1960	Natural Gas (quadrillion BTUs)	Coal (quadrillion BTUs)
0	12.39	9.84
10	21.80	12.27
20	20.39	15.42
30	19.30	19.25
40	23.33	22.41

Source: *Statistical Abstract of the United States, 2001*, Table 891, p. 569.

30. **Energy Policy** If you were a lobbyist for the natural gas industry, how would you use the results of Exercise 29 to persuade legislators to support further natural gas exploration?

31. **Energy Usage** The amount of nuclear electric power and coal energy produced and consumed in the United States is shown in the following tables. Use matrix operations to create a table that shows the difference between energy production and energy consumption.

Energy Production

Years Since 1960	Nuclear Electric Power (quadrillion BTUs)	Coal (quadrillion BTUs)
0	0.01	10.82
10	0.24	14.61
20	2.74	18.60
30	6.16	22.46
40	8.01	22.66

Energy Consumption

Years Since 1960	Nuclear Electric Power (quadrillion BTUs)	Coal (quadrillion BTUs)
0	0.01	9.84
10	0.24	12.27
20	2.74	15.42
30	6.16	19.25
40	8.01	22.41

Source: *Statistical Abstract of the United States, 2001*, Table 891, p. 569.

32. **Energy Policy** An environmentalist believes that the United States should produce only as much energy as it consumes. Based on the results of Exercise 31, which energy technology (coal or nuclear electric power) seems to best support the environmentalist's position? Defend your conclusion.

33. **Renewable Energy Consumption** The amount of renewable energy consumed in the United States in various years is shown in the following tables. Use matrices to construct a table showing the total amount of renewable energy consumed between the beginning of 1997 and the end of 1999.

1997 Renewable Energy Consumption

Renewable Energy Type	Energy Consumed (quadrillion BTUs)
Conventional hydroelectric power	3.94
Geothermal energy	0.33
Biomass	2.98
Solar energy	0.07
Wind energy	0.03

1998 Renewable Energy Consumption

Renewable Energy Type	Energy Consumed (Quadrillion BTUs)
Conventional hydroelectric power	3.55
Geothermal energy	0.34
Biomass	2.99
Solar energy	0.07
Wind energy	0.03

1999 Renewable Energy Consumption

Renewable Energy Type	Energy Consumed (quadrillion BTUs)
Conventional hydroelectric power	3.42
Geothermal energy	0.33
Biomass	3.51
Solar energy	0.08
Wind energy	0.04

Source: *Statistical Abstract of the United States, 2001*, Table 896, p. 572.

34. **Average Energy Consumption** Using the results from Exercise 33, use matrix operations to construct a table showing the average amount of energy consumed annually between the start of 1997 and the end of 1999.

35. **Energy Usage Trends** Using the results from Exercise 34 and the data tables from Exercise 33, construct a table showing the difference between the 1999 energy consumption and the average annual energy consumption. What conclusions can you draw from this table?

36. **Organized Physical Activity** The following tables show the percentage of high school students involved in physical education classes. Based on the information given, can you determine the percentage of ninth graders who exercise 20 or more minutes per class? Justify your answer.

Males Enrolled in a Physical Education Class

Grade	Enrolled in a P.E. Class (percent)	Exercised 20 Minutes or More per Class (percent)
9	60.7	82.1
10	82.3	84.4
11	65.3	79.4
12	44.6	82.0

Females Enrolled in a Physical Education Class

Grade	Enrolled in a P.E. Class (percent)	Exercised 20 Minutes or More per Class (percent)
9	51.5	69.6
10	75.6	72.5
11	56.6	70.2
12	36.8	68.0

Source: *Statistical Abstract of the United States, 2001*, Table 1246, p. 764.

Exercises 37–40 are intended to challenge your ability to apply matrix addition and scalar multiplication properties. For each exercise, $A = \begin{bmatrix} 2 & 15 & 0 \\ 6 & -4 & -9 \\ -8 & \frac{1}{2} & \frac{2}{3} \end{bmatrix}$ *and* $C = \begin{bmatrix} 12 & 1 & 4 \\ 3 & 0 & 5 \\ -2 & \frac{3}{2} & -\frac{1}{3} \end{bmatrix}$.

37. Solve the matrix equation for B.

$$5A - B = C$$

38. Solve the matrix equation for B.

$$-2A + 3B = C$$

39. Solve the matrix equation for B.

$$\frac{1}{2}A + \frac{2}{3}B = \frac{1}{6}C$$

40. Solve the matrix equation for B.

$$-100A - 200B = 330C$$

3.2 Matrix Multiplication and Inverses

- Do matrix multiplication
- Find the inverse of an invertible matrix with and without technology
- Use matrices to analyze and interpret real-life data

GETTING STARTED A theater charges \$8.50 per adult, \$5.50 per child, and \$6.50 per senior. What will be the total admission cost for a group of 12 adults, 16 children, and 4 seniors? This question may be answered using matrix multiplication.

In this section, we will introduce some special types of matrices, demonstrate how to do matrix multiplication, and show how to find the inverse of a 2×2 matrix. We will also look at real-life applications of matrix multiplication.

Special Types of Matrices

A $1 \times n$ **row matrix** is a matrix that consists of a single row and n columns. For example, $A = \begin{bmatrix} 2 & 5 & 3 \end{bmatrix}$ is a 1×3 row matrix and $B = \begin{bmatrix} 3 & 1 & 0 & 9 \end{bmatrix}$ is a 1×4 row matrix.

An $n \times 1$ **column matrix** is a matrix that consists of n rows and a single column. For example, $C = \begin{bmatrix} 2 \\ 1 \end{bmatrix}$ is a 2×1 column matrix and $D = \begin{bmatrix} -2 \\ 7 \\ 5 \end{bmatrix}$ is a 3×1 column matrix.

An $n \times n$ **square matrix** is a matrix with n rows and n columns. For example, $E = \begin{bmatrix} 0 & -1 & 4 \\ -5 & 5 & 7 \\ 9 & 4 & 1 \end{bmatrix}$ is a 3×3 square matrix and $F = \begin{bmatrix} 1 & 9 \\ 9 & 8 \end{bmatrix}$ is a 2×2 square matrix. We will use each of these types of matrices as we discuss matrix multiplication.

Matrix Multiplication

In the last section, we introduced the concept of scalar multiplication—the multiplication of a number and a matrix. There is another type of multiplication involving matrices—matrix multiplication. Since matrix multiplication is a somewhat strange process, we will introduce the concept with a brief example before giving a formal definition.

EXAMPLE 1

Using Matrix Multiplication to Calculate Theater Admission Cost

A theater charges \$8.50 per adult, \$5.50 per child, and \$6.50 per senior. What will be the total admission cost for a group of 12 adults, 16 children, and 4 seniors?

SOLUTION We know that the total cost is given by

$$\begin{aligned}\text{Total cost} &= \underset{\text{Adults}}{8.50(12)} + \underset{\text{Children}}{5.50(16)} + \underset{\text{Seniors}}{6.50(4)} \\ &= 102 + 88 + 26 \\ &= 216\end{aligned}$$

The total admission cost for the group is \$216.

The same result may be obtained by representing the individual admission cost as a row matrix and the number of guests as a column matrix.

$$C = \begin{bmatrix} 8.50 & 5.50 & 6.50 \end{bmatrix} \quad \text{Individual admission cost matrix}$$

$$N = \begin{bmatrix} 12 \\ 16 \\ 4 \end{bmatrix} \quad \text{Number of guests matrix}$$

The product of the 1×3 matrix C and the 3×1 matrix N is given by

$$CN = \begin{bmatrix} 8.50 & 5.50 & 6.50 \end{bmatrix} \begin{bmatrix} 12 \\ 16 \\ 4 \end{bmatrix}$$

From our initial solution, we know that the total cost is given by the expression

$$8.50(12) + 5.50(16) + 6.50(4)$$

How can we combine the elements of each matrix so that we end up with the desired expression? Observe that if we multiply the first entry in the row matrix by the first entry in the column matrix, we obtain 8.50(12), the first term of the expression. Similarly, multiplying the second entry in the row matrix by the second entry in the column matrix yields 5.50(16), the second term of the expression. The third term of the expression, 6.50(4), is obtained by multiplying the third term in the row matrix by the third term in the column matrix. The final result is obtained by summing the individual terms. Therefore,

$$\begin{aligned}CN &= \begin{bmatrix} 8.50 & 5.50 & 6.50 \end{bmatrix} \begin{bmatrix} 12 \\ 16 \\ 4 \end{bmatrix} \\ &= \begin{bmatrix} 8.50(12) + 5.50(16) + 6.50(4) \end{bmatrix} \\ &= \begin{bmatrix} 102 + 88 + 26 \end{bmatrix} \\ &= \begin{bmatrix} 216 \end{bmatrix}\end{aligned}$$

The total ticket cost is \$216.

How did we know the units of the 1×1 matrix? The units of C were dollars per person, since the ticket prices given were individual ticket prices. The units of N were people or persons. The units of the product of the matrices are the product of the units of the matrices, in the same order. Therefore,

$$\begin{aligned}\text{Units of } CN &= \left(\frac{\text{dollars}}{\text{person}}\right)(\text{persons}) \\ &= \left(\frac{\text{dollars}}{\cancel{\text{person}}}\right)(\cancel{\text{persons}}) \\ &= \text{dollars}\end{aligned}$$

In Example 1, we demonstrated how to multiply a row matrix and a column matrix. The process is summarized as follows.

THE PRODUCT OF A ROW MATRIX AND A COLUMN MATRIX

The product of a $1 \times n$ row matrix A and an $n \times 1$ column matrix B is the 1×1 square matrix given by

$$AB = \begin{bmatrix} a_1 & a_2 & \cdots & a_n \end{bmatrix} \begin{bmatrix} b_1 \\ b_2 \\ \vdots \\ b_n \end{bmatrix} = \begin{bmatrix} a_1b_1 + a_2b_2 + \cdots + a_nb_n \end{bmatrix}$$

EXAMPLE 2

Determining the Product of a Row Matrix and a Column Matrix

Calculate AB given $A = \begin{bmatrix} 1 & 2 \end{bmatrix}$ and $B = \begin{bmatrix} 3 \\ 4 \end{bmatrix}$.

SOLUTION

$$\begin{aligned} AB &= \begin{bmatrix} 1 & 2 \end{bmatrix} \begin{bmatrix} 3 \\ 4 \end{bmatrix} \\ &= \begin{bmatrix} 1(3) + 2(4) \end{bmatrix} \\ &= \begin{bmatrix} 3 + 8 \end{bmatrix} \\ &= \begin{bmatrix} 11 \end{bmatrix} \end{aligned}$$

The solution is the 1×1 matrix $[11]$. Note that this is a matrix, not the number 11.

EXAMPLE 3

Determining the Product of a Row Matrix and a Column Matrix

Calculate AB given $A = \begin{bmatrix} 2 & 0 & -1 & 8 \end{bmatrix}$ and $B = \begin{bmatrix} 5 \\ 7 \\ 4 \\ -2 \end{bmatrix}$.

SOLUTION

$$\begin{aligned} AB &= \begin{bmatrix} 2 & 0 & -1 & 8 \end{bmatrix} \begin{bmatrix} 5 \\ 7 \\ 4 \\ -2 \end{bmatrix} \\ &= \begin{bmatrix} 2(5) + 0(7) + (-1)(4) + 8(-2) \end{bmatrix} \\ &= \begin{bmatrix} 10 + 0 - 4 - 16 \end{bmatrix} \\ &= \begin{bmatrix} -10 \end{bmatrix} \end{aligned}$$

The method used to calculate the product of a $1 \times n$ row matrix and an $n \times 1$ column matrix may be used to calculate the entries of the product of any two matrices (provided the product exists).

MATRIX MULTIPLICATION

An $m \times n$ matrix A and an $n \times p$ matrix B may be multiplied together to form a new $m \times p$ matrix, C. The value of the entry in the ith row and the jth column of C is the product of the ith row of A and the jth column of B.

It is significant to note that matrix multiplication can be performed only if the number of columns of the first matrix is equal to the number of rows of the second matrix. Consider the $m \times n$ matrix A and the $n \times p$ matrix B shown here. We have boxed the ith row of A and the jth column of B.

$$C = AB$$

$$= \begin{bmatrix} a_{11} & a_{12} & \cdots & a_{1n} \\ a_{21} & a_{22} & \cdots & a_{2n} \\ \vdots & \vdots & \ddots & \vdots \\ \boxed{a_{i1}} & \boxed{a_{i2}} & \boxed{\cdots} & \boxed{a_{in}} \\ \vdots & \vdots & \ddots & \vdots \\ a_{m1} & a_{m2} & \cdots & a_{mn} \end{bmatrix} \cdot \begin{bmatrix} b_{11} & b_{12} & \cdots & \boxed{b_{1j}} & \cdots & b_{1p} \\ b_{21} & b_{22} & \cdots & \boxed{b_{2j}} & \cdots & b_{2p} \\ \vdots & \vdots & \ddots & \boxed{\vdots} & \ddots & \vdots \\ b_{n1} & b_{n2} & \cdots & \boxed{b_{nj}} & \cdots & b_{np} \end{bmatrix}$$

The entry c_{ij} of matrix C is determined by calculating

$$[c_{ij}] = [a_{i1} \quad a_{i2} \quad \cdots \quad a_{in}] \begin{bmatrix} b_{1j} \\ b_{2j} \\ \vdots \\ b_{nj} \end{bmatrix}$$

$$= [a_{i1}b_{1j} + a_{i2}b_{2j} + \cdots + a_{in}b_{nj}]$$

EXAMPLE 4

Determining the Product of Two Matrices

Find $C = AB$ given $A = \begin{bmatrix} 1 & 0 \\ 3 & 7 \\ 4 & -2 \end{bmatrix}$ and $B = \begin{bmatrix} 5 \\ 6 \end{bmatrix}$.

SOLUTION A is a 3×2 matrix and B is a 2×1 matrix. Since the "inside" dimensions match (both equal 2), the matrix $C = AB$ will take on the "outside" dimensions. That is, C will be a 3×1 matrix.

$$\begin{aligned} C &= AB \\ &= \begin{bmatrix} 1 & 0 \\ 3 & 7 \\ 4 & -2 \end{bmatrix} \begin{bmatrix} 5 \\ 6 \end{bmatrix} \\ &= \begin{bmatrix} 1(5) + 0(6) \\ 3(5) + 7(6) \\ 4(5) + (-2)(6) \end{bmatrix} && \begin{matrix} \text{Row 1 of } A \text{ times Column 1 of } B \\ \text{Row 2 of } A \text{ times Column 1 of } B \\ \text{Row 3 of } A \text{ times Column 1 of } B \end{matrix} \\ &= \begin{bmatrix} 5 + 0 \\ 15 + 42 \\ 20 - 12 \end{bmatrix} \\ &= \begin{bmatrix} 5 \\ 57 \\ 8 \end{bmatrix} \end{aligned}$$

Notice that the entries of C are the product of the rows of A and the column of B. For example, c_{21} is the product of the second row of A, $[3 \quad 7]$, and the first column of B, $\begin{bmatrix} 5 \\ 6 \end{bmatrix}$.

EXAMPLE 5

Determining the Product of Two Matrices

Find $C = DA$ given $D = \begin{bmatrix} 2 & 5 \\ 7 & 6 \end{bmatrix}$ and $A = \begin{bmatrix} 1 & 0 \\ 3 & 7 \\ 4 & -2 \end{bmatrix}$.

SOLUTION D is a 2×2 matrix and A is a 3×2 matrix. The matrix $C = DA$ cannot be computed, since the number of columns of D (two) does not equal the number of rows of A (three). Let's try to do the multiplication anyway, just to see what happens. To calculate the entry c_{21}, we need to multiply the second row of D by the first column of A.

$$\begin{aligned} [c_{21}] &= [7 \quad 6] \begin{bmatrix} 1 \\ 3 \\ 4 \end{bmatrix} \\ &= [7(1) + 6(3) + ?(4)] \end{aligned}$$

We can't perform the calculation. The conflict shown here will appear anytime you try to multiply two matrices whose rows and columns don't match up.

We will use the matrices from Example 5 in Example 6. However, this time we will reverse the order in which the matrices are to be multiplied.

EXAMPLE 6 Determining the Product of Two Matrices

Find $C = AD$ given $A = \begin{bmatrix} 1 & 0 \\ 3 & 7 \\ 4 & -2 \end{bmatrix}$ and $D = \begin{bmatrix} 2 & 5 \\ 7 & 6 \end{bmatrix}$.

SOLUTION A is a 3×2 matrix and D is a 2×2 matrix. The matrix $C = AD$ is a 3×2 matrix.

$$\begin{aligned} C &= AD \\ &= \begin{bmatrix} 1 & 0 \\ 3 & 7 \\ 4 & -2 \end{bmatrix} \begin{bmatrix} 2 & 5 \\ 7 & 6 \end{bmatrix} \\ &= \begin{bmatrix} 1(2) + 0(7) & 1(5) + 0(6) \\ 3(2) + 7(7) & 3(5) + 7(6) \\ 4(2) + (-2)(7) & 4(5) + (-2)(6) \end{bmatrix} \\ &= \begin{bmatrix} 2 & 5 \\ 55 & 57 \\ -6 & 8 \end{bmatrix} \end{aligned}$$

From Examples 5 and 6, we see that matrix multiplication is not commutative. That is, in general, $AB \neq BA$.

EXAMPLE 7 Using Technology to Determine the Product of Two Matrices

Find $C = AB$ given $A = \begin{bmatrix} 1 & 2 & 3 & 4 \\ 0 & 1 & 2 & 3 \\ 4 & 5 & 6 & 7 \\ 5 & 6 & 7 & 0 \end{bmatrix}$ and $B = \begin{bmatrix} -1.1 & 1.8 \\ 0.4 & 4.2 \\ -0.2 & 1.0 \\ -3.9 & 1.1 \end{bmatrix}$.

SOLUTION A is a 4×4 matrix and B is a 4×2 matrix, so AB will be a 4×2 matrix. Because of the size of the matrices and the complexity of the entries in matrix B, we will use technology to calculate the product, as demonstrated in the following Technology Tip. The result is

$$\begin{aligned} C &= AB \\ &= \begin{bmatrix} -16.5 & 17.6 \\ -11.7 & 9.5 \\ -30.9 & 41.9 \\ -4.5 & 41.2 \end{bmatrix} \end{aligned}$$

TECHNOLOGY TIP

Matrix Multiplication

1. Enter matrix A and matrix B into the calculator, using the Matrix Editor. Use the Matrix Names menu to place matrix A on the home screen. Then place matrix B on the home screen.

[A][B]

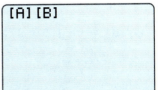

2. Press ENTER to display the product of the two matrices.

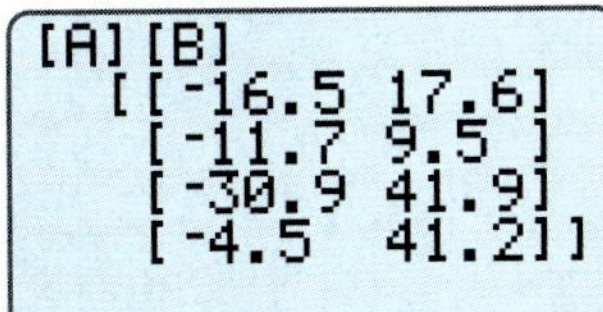

Error Alert:
If a dimension mismatch error occurs, double-check to make sure you have entered your matrices correctly. If you have, this error tells you that matrix multiplication is not possible because the number of columns of the first matrix is not the same as the number of rows of the second matrix.

Matrix Multiplication Properties

Let A, B, and C be matrices. Let I be an identity matrix (see page 135), and let O be a zero matrix. Given that the dimensions of the matrices allow each of the operations to be performed, the following properties hold:

1. Multiplicative Associative $A(BC) = (AB)C$
2. Multiplicative Identity $AI = IA = A$
3. Distributive $A(B + C) = AB + AC$
4. Distributive $(A + B)C = AC + BC$
5. Multiplication by a Zero Matrix $OA = AO = O$

Real-Life Applications of Matrix Multiplication

Up to this point, we have focused on the mathematical skill of matrix multiplication without demonstrating its usefulness in addressing real-world problems. In the next two examples, we will illustrate real-world uses of matrix multiplication.

EXAMPLE 8 Using Matrix Multiplication to Analyze Livestock Inventory Values

The U.S. Department of Agriculture monitors livestock inventories and their values. In a 1999 report on New York state hog and pig production, the USDA provided the data in Table 3.6.

TABLE 3.6 All Hogs and Pigs

Year	Number (thousands)	Value per Head (dollars)
1992	105	72.00
1993	90	77.00
1994	72	54.00
1995	66	71.00
1996	82	92.00
1997	79	81.00
1998	60	45.00

Source: www. nass.usda.gov.

From 1992 through 1998, what was the cumulative value of hogs and pigs in New York state?

SOLUTION To determine the annual value of hog and pig inventories in a particular year, we simply need to multiply the number of hogs and pigs by the value per head. To calculate the cumulative value of the hog and pig inventories, we need to add up the annual inventory values. These operations may be performed using matrix multiplication.

We'll first define a column matrix N. The rows of N represent the years 1992 through 1998, and the column of N represents the number of hogs and pigs (in thousands):

$$N = \begin{bmatrix} 105 \\ 90 \\ 72 \\ 66 \\ 82 \\ 79 \\ 60 \end{bmatrix}$$

We will next define a row matrix V. The row of V represents the value per head, and the columns of V represent the years 1992 through 1998.

$$V = \begin{bmatrix} 72 & 77 & 54 & 71 & 92 & 81 & 45 \end{bmatrix}$$

We now must determine how to multiply the matrices so that the results make sense. Let's calculate VN. The dimensions of V are 1 value per head (in dollars) $\times$ 7 years. The dimensions of N are 7 years $\times$ 1 number of hogs and pigs (in thousands). Since the columns of V and the rows of N represent the same thing, matrix multiplication is possible and meaningful. The dimensions of VN will be 1×1.

$$VN = \begin{bmatrix} 72 & 77 & 54 & 71 & 92 & 81 & 45 \end{bmatrix} \begin{bmatrix} 105 \\ 90 \\ 72 \\ 66 \\ 82 \\ 79 \\ 60 \end{bmatrix}$$

$$= [72(105) + 77(90) + 54(72) + 71(66) + 92(82) + 81(79) + 45(60)]$$

$$= [39{,}707]$$

What are the units of VN? We are multiplying value per head (in dollars) by number of head of hogs and pigs (in thousands).

$$\frac{\text{dollars}}{\text{pig or hog}} \cdot \text{thousand of pigs or hogs} = \frac{\text{dollars}}{\cancel{\text{pig or hog}}} \cdot \text{thousand } \cancel{\text{of pigs or hogs}}$$

$$= \text{thousand dollars}$$

The cumulative value of the hog and pig inventory in New York from 1992 through 1998 is \$39,707,000.

Let's now look at NV. The dimensions of N are 7 years $\times$ 1 number of hogs and pigs (in thousands). The dimensions of V are 1 value per head (in dollars) $\times$ 7 years. Since the number of columns of N is the same as the number of rows of V, matrix multiplication is possible. However, since the meaning of the column of N (number of hogs and pigs) is not the same as the meaning of the rows of V (value per head), NV is meaningless in a real-world context.

EXAMPLE 9

Using Matrix Multiplication to Find the Nutritional Content of a Mix

Breakfast cereal connoisseurs often enjoy mixing cereals to create a new breakfast taste. A connoisseur working at a bed-and-breakfast wants to report the nutritional content of various mixtures of Honey Nut Cheerios®, Rice Crunch-Ems!, and Corn Crunch-Ems! to his health-conscious guests. From the package labeling, he determines the nutritional content of each cereal and records it in Table 3.7.

TABLE 3.7

	Honey Nut Cheerios	Rice Crunch-Ems!	Corn Crunch-Ems!
Protein	3 grams/cup	1.6 grams/cup	2 grams/cup
Carbohydrates	24 grams/cup	20.8 grams/cup	27 grams/cup
Fat	1.5 grams/cup	0 grams/cup	0 grams/cup

Source: Health Valley Rice Crunch-Ems! and Corn Crunch-Ems! labels and General Mills Honey Nut Cheerios label.

His first mixture will contain 1 cup of Honey Nut Cheerios, 2 cups of Rice Crunch-Ems!, and 1 cup of Corn Crunch-Ems! His second mixture will contain 2 cups of Honey Nut Cheerios, 4 cups of Rice Crunch-Ems!, and 3 cups of Corn Crunch-Ems!. Determine the amount of protein, carbohydrates, and fat in a 1-cup serving of each mixture.

SOLUTION The nutrition content table may be represented by the matrix

$$N = \begin{array}{c} \begin{array}{ccc} \text{Honey} & \text{Rice} & \text{Corn} \end{array} \\ \begin{bmatrix} 3.0 & 1.6 & 2.0 \\ 24.0 & 20.8 & 27.0 \\ 1.5 & 0.0 & 0.0 \end{bmatrix} \end{array} \begin{array}{l} \\ \text{Protein} \\ \text{Carbs} \\ \text{Fat} \end{array}$$

The mixture ingredients may be represented by Table 3.8

TABLE 3.8

	Mixture 1	Mixture 2
Honey Nut Cheerios	1.0 cup	2.0 cups
Rice Crunch-Ems!	2.0 cups	4.0 cups
Corn Crunch-Ems!	1.0 cup	3.0 cups

and the corresponding matrix

$$M = \begin{array}{c} \begin{array}{cc} \text{Mix 1} & \text{Mix 2} \end{array} \\ \begin{bmatrix} 1 & 2 \\ 2 & 4 \\ 1 & 3 \end{bmatrix} \end{array} \begin{array}{l} \\ \text{Honey} \\ \text{Rice} \\ \text{Corn} \end{array}$$

Notice that the columns of N and the rows of M represent the same cereals (Honey Nut Cheerios, Rice Crunch-Ems!, and Corn Crunch-Ems!). The matrix NM will be a 3×2 matrix with rows representing protein, carbohydrates, and fat and columns representing Mixture 1 and Mixture 2.

$$NM = \begin{bmatrix} 3.0 & 1.6 & 2.0 \\ 24.0 & 20.8 & 27.0 \\ 1.5 & 0.0 & 0.0 \end{bmatrix} \begin{bmatrix} 1 & 2 \\ 2 & 4 \\ 1 & 3 \end{bmatrix}$$

$$= \begin{array}{c} \begin{array}{cc} \text{Mix 1} & \text{Mix 2} \end{array} \\ \begin{bmatrix} 8.2 & 18.4 \\ 92.6 & 212.2 \\ 1.5 & 3.0 \end{bmatrix} \end{array} \begin{array}{l} \\ \text{Protein} \\ \text{Carbs} \\ \text{Fat} \end{array}$$

What are the units of the entries in matrix NM? The units of N are grams/cup, and the units of M are cups.

UNITS

$$\frac{\text{grams}}{\text{cup}} \cdot \text{cup} = \frac{\text{grams}}{\cancel{\text{cup}}} \cdot \cancel{\text{cup}}$$

$$= \text{grams}$$

So the units of the product is grams.

Converting the matrix back to a table, we have Table 3.9.

TABLE 3.9

	Mixture 1	Mixture 2
Protein	8.2 grams	18.4 grams
Carbohydrates	92.6 grams	212.2 grams
Fat	1.5 grams	3.0 grams

The table shows the total amount of protein, carbohydrates, and fat in each mixture. However, the first mixture contains 4 cups of cereal, and the second contains 9 cups of cereal.

Dividing the terms in the first column by 4 cups and the entries in the second column by 9 cups, we get the results in Table 3.10.

TABLE 3.10

	Mixture 1	Mixture 2
Protein	2.05 grams/cup	2.04 grams/cup
Carbohydrates	23.15 grams/cup	23.58 grams/cup
Fat	0.375 gram/cup	0.33 gram/cup

Since the original data were accurate to one decimal place, we will round the table entries as shown in Table 3.11.

TABLE 3.11

	Mixture 1	Mixture 2
Protein	2.1 grams/cup	2.0 grams/cup
Carbohydrates	23.2 grams/cup	23.6 grams/cup
Fat	0.4 gram/cup	0.3 gram/cup

Table 3.11 shows the nutritional content of each of the mixtures.

Inverse Matrices

The **identity matrix** I_n is the $n \times n$ square matrix with 1s along the main diagonal of the matrix and 0s elsewhere. For example, $I_3 = \begin{bmatrix} 1 & 0 & 0 \\ 0 & 1 & 0 \\ 0 & 0 & 1 \end{bmatrix}$, $I_2 = \begin{bmatrix} 1 & 0 \\ 0 & 1 \end{bmatrix}$, and $I_1 = [1]$ are identity matrices. (The subscript on the I is often omitted, but the dimensions of I are typically implied by the context of the problem.) We are often interested in matrices that have the property that $AB = BA = I$. These types of matrices are often used to solve systems of linear equations. In general, $AB \neq BA$, since matrix multiplication is not commutative. However, if two $n \times n$ matrices are inverses of each other, then $AB = BA = I$.

INVERSE MATRICES

An $n \times n$ matrix A and an $n \times n$ matrix B are inverses of each other if and only if $AB = BA = I_n$. We say that $B = A^{-1}$ (read "A inverse").

A matrix with an inverse is said to be **invertible.** A matrix without an inverse is said to be **singular.**

EXAMPLE 10 Determining If a Matrix Is the Inverse of Another Matrix

Let $A = \begin{bmatrix} 3 & 5 \\ 1 & 2 \end{bmatrix}$ and $B = \begin{bmatrix} 2 & -5 \\ -1 & 3 \end{bmatrix}$. Determine if matrix B is the inverse of matrix A.

SOLUTION

$$\begin{aligned} AB &= \begin{bmatrix} 3 & 5 \\ 1 & 2 \end{bmatrix}\begin{bmatrix} 2 & -5 \\ -1 & 3 \end{bmatrix} \\ &= \begin{bmatrix} 6-5 & -15+15 \\ 2-2 & -5+6 \end{bmatrix} \\ &= \begin{bmatrix} 1 & 0 \\ 0 & 1 \end{bmatrix} \\ &= I_2 \end{aligned}$$

$$\begin{aligned} BA &= \begin{bmatrix} 2 & -5 \\ -1 & 3 \end{bmatrix}\begin{bmatrix} 3 & 5 \\ 1 & 2 \end{bmatrix} \\ &= \begin{bmatrix} 6-5 & 10-10 \\ -3+3 & -5+6 \end{bmatrix} \\ &= \begin{bmatrix} 1 & 0 \\ 0 & 1 \end{bmatrix} \\ &= I_2 \end{aligned}$$

Since $AB = BA = I_2$, $B = A^{-1}$.

In Example 10, the matrices looked remarkably similar to each other. In fact, for a 2×2 invertible matrix, it is fairly simple to calculate the inverse.

INVERSE OF A 2 × 2 MATRIX

The inverse of an invertible 2×2 matrix $A = \begin{bmatrix} a & b \\ c & d \end{bmatrix}$ is

$$A^{-1} = \frac{1}{ad - bc}\begin{bmatrix} d & -b \\ -c & a \end{bmatrix}$$

The quantity $ad - bc$ is called the **determinant** of the matrix A and is often written $\det(A)$. If $\det(A) = 0$, then $\frac{1}{ad - bc}$ is undefined and A is singular.

This method works only for 2×2 matrices. For larger matrices, finding the inverse requires a bit more effort. We will demonstrate how to find the inverse of a 3×3 matrix algebraically in Section 3.3.

EXAMPLE 11

Finding the Inverse of a 2 × 2 Matrix

Find the inverse of matrix $A = \begin{bmatrix} 4 & 2 \\ 5 & 3 \end{bmatrix}$, if it exists.

SOLUTION

$$\det(A) = 4(3) - 2(5)$$
$$\det(A) = 2$$

Since the determinant of A is not zero, the inverse of A exists.

$$A^{-1} = \frac{1}{2}\begin{bmatrix} 3 & -2 \\ -5 & 4 \end{bmatrix}$$
$$= \begin{bmatrix} 1.5 & -1 \\ -2.5 & 2 \end{bmatrix}$$

EXAMPLE 12

Using the Determinant to Determine If a Matrix Is Singular

Find the inverse of matrix $A = \begin{bmatrix} 2 & 6 \\ 1 & 3 \end{bmatrix}$, if it exists.

SOLUTION Since $\det(A) = 2(3) - 6(1) = 0$, A is singular. That is, A does not have an inverse.

Notice that Row 1 is equal to twice Row 2. Whenever one row of a matrix is a multiple of another row, the determinant of the matrix will be zero.

3.2 Summary

In this section, you learned about row, column, square, identity, and inverse matrices. You discovered matrix multiplication and learned techniques for finding the inverse of an invertible 2×2 matrix. You were also introduced to real-world applications of matrix multiplication.

3.2 Exercises

In Exercises 1–20, perform the indicated operation. As appropriate, use the matrices

$$A = \begin{bmatrix} 3 & 4 \\ -1 & -2 \end{bmatrix}, B = \begin{bmatrix} 2 & -4 \\ 1 & 2 \end{bmatrix},$$
$$C = \begin{bmatrix} 8 & 4 \\ 6 & 1 \\ 7 & 3 \end{bmatrix}, \text{ and } D = \begin{bmatrix} -2 & 0 & 3 \\ 0 & 4 & 3 \\ 1 & 7 & 0 \end{bmatrix}.$$

If the specified operation is undefined, so state.

1. $[3 \quad 9 \quad -2]\begin{bmatrix} 2 \\ 2 \\ 12 \end{bmatrix}$

2. $[-2 \quad 9 \quad 0.5]\begin{bmatrix} 1.5 \\ 4 \\ -6 \end{bmatrix}$

3. $A + B$

4. $A - B$

5. $4A$

6. $-2B$

7. $4A - 2B$

8. $C + D$

9. AB

10. BA

11. CD

12. DC

13. CA

14. AC

15. CB

16. A^{-1}

17. B^{-1}

18. $A^{-1}A$

19. AA^{-1}

20. $DC + C$

In Exercises 21–30, use the determinant formula to determine if the matrix is invertible or singular. If the matrix is invertible, find its inverse.

21. $A = \begin{bmatrix} 2 & 3 \\ 1 & 2 \end{bmatrix}$

22. $B = \begin{bmatrix} -2 & 4 \\ 1 & 2 \end{bmatrix}$

23. $C = \begin{bmatrix} 9 & 6 \\ 3 & 2 \end{bmatrix}$

24. $D = \begin{bmatrix} 0 & -1 \\ 4 & 4 \end{bmatrix}$

25. $E = \begin{bmatrix} -0.5 & -0.7 \\ 4.0 & 3.4 \end{bmatrix}$

26. $A = \begin{bmatrix} -1 & 1 \\ 1 & 1 \end{bmatrix}$

27. $B = \begin{bmatrix} 5 & 10 \\ 2 & 4 \end{bmatrix}$

28. $C = \begin{bmatrix} 0.1 & 0.6 \\ 2.0 & 1.2 \end{bmatrix}$

29. $D = \begin{bmatrix} 1.0 & -1.1 \\ 0.9 & 1.0 \end{bmatrix}$

30. $E = \begin{bmatrix} -0.9 & -0.9 \\ -0.9 & 0.9 \end{bmatrix}$

In Exercises 31–40, use matrix multiplication to find the answer to each question.

31. **Lumber Manufacturing Payroll** The number of employees working in the lumber and wood products manufacturing industry is shown in the following table.

Manufacturing Full-Time Employees: Lumber and Wood Products

Years Since 1995 (t)	Employees (E) (thousands)
0	772
1	782
2	794
3	816
4	843

Source: *Statistical Abstract of the United States, 2001,* Table 979, p. 622.

The average annual wage/salary of a full-time employee working in the lumber and wood products manufacturing industry is shown in the following table.

Manufacturing Salaries: Lumber and Wood Products

Years Since 1995 (t)	Average Annual Wage/Salary (S) (dollars)
0	25,110
1	26,148
2	27,382
3	28,278
4	29,040

Source: *Statistical Abstract of the United States, 2001,* Table 979, p. 622.

From 1995 through 1999, what was the total amount of money spent on employee wages and salaries?

32. **Rubber and Plastics Payroll** The number of employees working in the rubber and plastics manufacturing industry is shown in the following table.

Manufacturing Full-Time Employees: Rubber and Plastics Industry

Years Since 1995 (t)	Employees (N) (thousands)
0	963
1	965
2	984
3	998
4	994

Source: *Statistical Abstract of the United States, 2001*, Table 979, p. 622.

The average annual wage/salary of a full-time employee working in the rubber and plastics manufacturing industry is shown in the following table.

Manufacturing Rubber and Plastics Industry Employee Average Earnings

Years Since 1995 (t)	Earnings (E) (dollars)
0	29,867
1	30,898
2	32,237
3	33,574
4	34,508

Source: *Statistical Abstract of the United States, 2001*, Table 979, p. 622.

From 1995 through 1999, what was the total amount of money spent on employee earnings?

33. **Fruit Smoothie Nutritional Content** The Vita-Mix® Super 5000 is a blender-like kitchen appliance with a 2+-horsepower motor. The Vita-Mix blade tips move at up to 240 miles per hour, easily converting whole fruits into luscious smoothies or grinding wheat kernels into fine flour. (*Source:* www.vita-mix.com.)

The author uses his Vita-Mix regularly to make breakfast beverages and was curious about the nutritional content of two different types of whole fruit smoothies. The ingredients of the first smoothie are 1 large apple, 1 large orange, 1 large banana, and 1 cup of water. The ingredients of the second smoothie are 2 large oranges, 1 large banana, and 1 cup of water.

Assuming that each fruit yields a 1-cup serving, we have the following nutrition information. A large orange contains 53.2 milligrams (mg) of vitamin C (ascorbic acid), 40 mg of calcium, and 2.4 grams (g) of fiber. A large banana contains 9.1 mg of vitamin C, 6 mg of calcium, and 2.4 g of fiber. A large apple contains 5.7 mg of vitamin C, 7 mg of calcium, and 2.7 g of fiber. (*Source:* USDA.) How much vitamin C, calcium, and fiber are in each smoothie?

34. **Fruit Smoothie Nutritional Content** One cup of flaxseed contains 1.3 mg of vitamin C, 199 mg of calcium, and 27.9 g of fiber. If one tablespoon of flaxseed is added to each of the smoothies in Exercise 33, how much vitamin C, calcium, and fiber are in each smoothie? (*Hint:* There are 16 tablespoons in a cup.)

35. **Fruit Smoothie Nutritional Content** A tropical fruit smoothie is made from 2 cups of chopped mango, 1 cup of pineapple, 1 cup of coconut, and 2 cups of ice. A pina colada smoothie is made from 2 cups of pineapple, 1 cup of coconut, and 2 cups of ice.

A cup of fresh mango contains 10 mg of calcium, 27.7 mg of vitamin C, and 1.8 g of fiber. A cup of fresh pineapple contains 7 mg of calcium, 15.4 mg of vitamin C, and 1.2 g of fiber. A cup of raw coconut contains 14 mg of calcium, 3.3 mg of vitamin C, and 9 g of fiber. (*Source:* USDA.) How much vitamin C, calcium, and fiber are in each smoothie?

36. **Alcohol-Related Homicides** The following table shows the annual number of homicides resulting from an alcohol-related brawl.

Homicides Resulting from an Alcohol-Related Brawl

Years Since 1990 (t)	Homicides (H)
1	500
2	429
3	383
4	316
5	254
6	256
7	239
8	213
9	203
10	181

Source: Federal Bureau of Investigation.

Write the number of homicides as a 10×1 column matrix H. Let

$$A = [1\ \ 1\ \ 1\ \ 1\ \ 1\ \ 1\ \ 1\ \ 1\ \ 1\ \ 1]$$

Calculate and interpret the meaning of AH.

37. **Military Personnel** The table shows the number of military personnel in 1996 and 1998.

Year (t)	All Military Personnel (m) (thousands)	Army Personnel (a) (thousands)
1996	1,056	491
1998	1,004	484

Source: *Statistical Abstract of the United States, 2001,* Tables 499 and 500, pp. 328–329.

We can create a 2×2 matrix P with rows representing 1996 and 1998 and columns representing the number of military personnel.

$$\begin{array}{cc} & \begin{array}{cc} m & a \end{array} \\ P = & \begin{bmatrix} 1056 & 491 \\ 1004 & 484 \end{bmatrix} \begin{array}{l} \text{Year 1996} \\ \text{Year 1998} \end{array} \end{array}$$

Let $R = [-1\ \ \ 1]$ and $C = \begin{bmatrix} 1 \\ -1 \end{bmatrix}$. Calculate and interpret the meaning of RP and PC.

38. **Movie Theater Revenue** The following table shows the average movie theater ticket price and total theater attendance from 1995 to 1999.

Years Since 1995	Price (dollars per person)	Attendance (millions of people)
0	4.35	1,263
1	4.42	1,339
2	4.59	1,388
3	4.69	1,481
4	5.08	1,465

Source: *Statistical Abstract of the United States, 2001*, Table 1244, p. 761.

Use matrices to calculate the accumulated movie theater revenue from ticket sales from the start of 1995 through 1999.

39. **Company Profit** The following table shows the net sales and cost of goods sold for the Kellogg Company from 1999 through 2001.

Year	Net Sales (millions of dollars)	Cost of Goods Sold (millions of dollars)
1999	6,984.2	3,325.1
2000	6,954.7	3,327.0
2001	8,853.3	4,128.5

Source: Kellogg Company 2001 Annual Report, pp. 7, 27.

Use matrices to calculate the accumulated net sales, accumulated cost of goods sold, and accumulated profit for the time period 1999 through 2001.

40. **Company Profit** The following table shows the net revenue and cost of goods sold for the Coca-Cola Company from 1999 through 2001.

Year	Net Revenue (millions of dollars)	Cost of Goods Sold (millions of dollars)
1999	19,284	6,009
2000	19,889	6,204
2001	20,092	6,044

Source: Coca-Cola Company 2001 Annual Report, p. 57.

Use matrices to calculate the accumulated net revenue, accumulated cost of goods sold, and accumulated profit for the time period 1999 through 2001.

Exercises 41–45 are intended to challenge your understanding of matrix multiplication and matrix inverses.

41. Find three different 2×2 singular matrices.

42. For real numbers a and b, $(a + b)^2 = a^2 + 2ab + b^2$. For $n \times n$ matrices A and B, does $(A + B)^2 = A^2 + 2AB + B^2$? Explain.

43. A **diagonal matrix** A is a matrix with the property that $a_{ij} = 0$ when $i \neq j$. Show that if A is invertible, the inverse of the diagonal matrix

$$A = \begin{bmatrix} a & 0 & 0 \\ 0 & b & 0 \\ 0 & 0 & c \end{bmatrix} \text{ is } A^{-1} = \begin{bmatrix} \frac{1}{a} & 0 & 0 \\ 0 & \frac{1}{b} & 0 \\ 0 & 0 & \frac{1}{c} \end{bmatrix}.$$ Under

what conditions is A singular?

44. Find three 3×3 diagonal matrices that are their own inverses.

45. A **symmetric matrix** A is a matrix with the property that $a_{ij} = a_{ji}$. Show that if A is a 3×3 symmetric matrix, then A^2 is a 3×3 symmetric matrix. (*Hint:* $A^2 = A \cdot A$.)

3.3 Solving Matrix Equations

- Use matrix algebra to solve systems of linear equations
- Use matrices to analyze and interpret real-life data
- Find the inverse of an invertible matrix with and without technology

GETTING STARTED Many students seek to maintain or increase their GPA to be eligible for academic recognition, including scholarships, dean's lists, honor societies, and so on. A common question that students ask is, "If I get straight A's from now on, how high can I raise my GPA?" By using the techniques demonstrated in this section, you will be able to determine your highest possible cumulative GPA.

In this section, we will demonstrate how to find the inverse of a square matrix algebraically. We will also show how to use the inverse and matrix algebra to solve systems of equations. We will conclude by further investigating how matrix algebra may be used in business and consumer applications.

Finding the Inverse of a Matrix Algebraically

Recall that a matrix A is said to be invertible if there exists a matrix A^{-1} such that $AA^{-1} = A^{-1}A = I$. In Section 3.2, we introduced a quick method for finding the inverse of a 2×2 matrix; however, we did not explain how the method was derived. The inverse of an invertible matrix of any size may be determined algebraically by augmenting the matrix with the identity matrix and then row reducing the matrix. For example, if $A = \begin{bmatrix} 1 & 2 \\ 3 & 4 \end{bmatrix}$, we add on the identity matrix $I = \begin{bmatrix} 1 & 0 \\ 0 & 1 \end{bmatrix}$.

$$\left[\begin{array}{cc|cc} 1 & 2 & 1 & 0 \\ 3 & 4 & 0 & 1 \end{array}\right] \quad \text{Augment } A \text{ with } I$$

$$\left[\begin{array}{cc|cc} 1 & 2 & 1 & 0 \\ 0 & 2 & 3 & -1 \end{array}\right] \quad 3R_1 - R_2$$

$$\left[\begin{array}{cc|cc} 1 & 0 & -2 & 1 \\ 0 & 2 & 3 & -1 \end{array}\right] \quad R_1 - R_2$$

$$\left[\begin{array}{cc|cc} 1 & 0 & -2 & 1 \\ 0 & 1 & \frac{3}{2} & -\frac{1}{2} \end{array}\right] \quad \tfrac{1}{2}R_2$$

When the left-hand side of the augmented matrix is reduced to the identity matrix, the matrix on the right-hand side of the augmented matrix is A^{-1}. We may verify the accuracy of our result by multiplying A by A^{-1}:

$$\begin{aligned} AA^{-1} &= \begin{bmatrix} 1 & 2 \\ 3 & 4 \end{bmatrix}\begin{bmatrix} -2 & 1 \\ \frac{3}{2} & -\frac{1}{2} \end{bmatrix} \\ &= \begin{bmatrix} 1(-2) + 2\left(\frac{3}{2}\right) & 1(1) + 2\left(-\frac{1}{2}\right) \\ 3(-2) + 4\left(\frac{3}{2}\right) & 3(1) + 4\left(-\frac{1}{2}\right) \end{bmatrix} \\ &= \begin{bmatrix} -2 + 3 & 1 + (-1) \\ -6 + 6 & 3 + (-2) \end{bmatrix} \\ &= \begin{bmatrix} 1 & 0 \\ 0 & 1 \end{bmatrix} \\ &= I \end{aligned}$$

Multiplying A^{-1} by A will yield the same result.

We may determine the inverse matrix for all invertible 2×2 matrices by augmenting a generic matrix $A = \begin{bmatrix} a & b \\ c & d \end{bmatrix}$ with the identity matrix $I = \begin{bmatrix} 1 & 0 \\ 0 & 1 \end{bmatrix}$ and row reducing.

$$\left[\begin{array}{cc|cc} a & b & 1 & 0 \\ c & d & 0 & 1 \end{array}\right] = \left[\begin{array}{cc|cc} a & b & 1 & 0 \\ 0 & ad-bc & -c & a \end{array}\right] \quad -cR_1 + aR_2$$

$$= \left[\begin{array}{cc|cc} a & b & 1 & 0 \\ 0 & 1 & \dfrac{-c}{ad-bc} & \dfrac{a}{ad-bc} \end{array}\right] \quad \frac{1}{ad-bc}R_2$$

$$= \left[\begin{array}{cc|cc} a & 0 & 1 - b\left(\dfrac{-c}{ad-bc}\right) & -b\left(\dfrac{a}{ad-bc}\right) \\ 0 & 1 & \dfrac{-c}{ad-bc} & \dfrac{a}{ad-bc} \end{array}\right] \quad R_1 - bR_2$$

$$= \left[\begin{array}{cc|cc} a & 0 & 1 + \dfrac{bc}{ad-bc} & \dfrac{-ab}{ad-bc} \\ 0 & 1 & \dfrac{-c}{ad-bc} & \dfrac{a}{ad-bc} \end{array}\right] \quad \text{Simplify}$$

$$= \left[\begin{array}{cc|cc} 1 & 0 & \dfrac{1}{a}\left(1 + \dfrac{bc}{ad-bc}\right) & \dfrac{1}{a}\left(\dfrac{-ab}{ad-bc}\right) \\ 0 & 1 & \dfrac{-c}{ad-bc} & \dfrac{a}{ad-bc} \end{array}\right] \quad \frac{1}{a}R_1$$

Although the left-hand side of the augmented matrix is I, the right-hand side doesn't yet look like $\frac{1}{ad-bc}\begin{bmatrix} d & -b \\ -c & a \end{bmatrix}$. However, we can show that our matrix is equivalent to the general 2×2 inverse matrix with a few additional steps.

$$A^{-1} = \begin{bmatrix} \dfrac{1}{a}\left(1 + \dfrac{bc}{ad-bc}\right) & \dfrac{1}{a}\left(\dfrac{-ab}{ad-bc}\right) \\ \dfrac{-c}{ad-bc} & \dfrac{a}{ad-bc} \end{bmatrix}$$

$$= \begin{bmatrix} \dfrac{1}{a}\left(\dfrac{ad-bc}{ad-bc} + \dfrac{bc}{ad-bc}\right) & \dfrac{-ab}{a(ad-bc)} \\ \dfrac{-c}{ad-bc} & \dfrac{a}{ad-bc} \end{bmatrix}$$

$$= \frac{1}{ad-bc}\begin{bmatrix} \dfrac{ad-bc+bc}{a} & \dfrac{-ab}{a} \\ -c & a \end{bmatrix}$$

$$= \frac{1}{ad-bc}\begin{bmatrix} d & -b \\ -c & a \end{bmatrix}$$

The process of augmenting a matrix with the identity and row reducing may be used for any size square matrix.

Consider the 3×3 matrix $A = \begin{bmatrix} 2 & 3 & 4 \\ 3 & 4 & 5 \\ 4 & 0 & 6 \end{bmatrix}$. We will calculate the inverse of matrix A by creating the augmented matrix $[A \mid I]$. Then, using the row operations demonstrated in Section 2.2, we will reduce the left-hand side of the augmented matrix to reduced row echelon form. If the matrix is invertible, the reduced matrix will be $[I \,\vdots\, A^{-1}]$.

$$[A \,\vdots\, I] = \left[\begin{array}{ccc:ccc} 2 & 3 & 4 & 1 & 0 & 0 \\ 3 & 4 & 5 & 0 & 1 & 0 \\ 4 & 0 & 6 & 0 & 0 & 1 \end{array}\right]$$

We would like to have a leading 1 in column 1 to use to zero out the remaining entries in that column. Observe that $R_2 - R_1 \rightarrow R_1$ will give us a leading 1 in row 1.

$$\left[\begin{array}{ccc:ccc} 1 & 1 & 1 & -1 & 1 & 0 \\ 3 & 4 & 5 & 0 & 1 & 0 \\ 4 & 0 & 6 & 0 & 0 & 1 \end{array}\right] \quad R_2 - R_1$$

Note that we performed the row operation on all six entries of row 1. We will now zero out the entries a_{21} and a_{31} with the operations $3R_1 - R_2 \rightarrow R_2$ and $4R_1 - R_3 \rightarrow R_3$.

$$\left[\begin{array}{ccc:ccc} 1 & 1 & 1 & -1 & 1 & 0 \\ 0 & -1 & -2 & -3 & 2 & 0 \\ 0 & 4 & -2 & -4 & 4 & -1 \end{array}\right] \quad \begin{array}{l} \\ 3R_1 - R_2 \\ 4R_1 - R_3 \end{array}$$

Recall that we are trying to make the left-hand side look like the identity matrix. Next we will use $-R_2 \rightarrow R_2$ to get a 1 in a_{22}.

$$\left[\begin{array}{ccc:ccc} 1 & 1 & 1 & -1 & 1 & 0 \\ 0 & 1 & 2 & 3 & -2 & 0 \\ 0 & 4 & -2 & -4 & 4 & -1 \end{array}\right] \quad -R_2$$

We will now zero out a_{32} with the operation $4R_2 - R_3 \rightarrow R_3$.

$$\left[\begin{array}{ccc:ccc} 1 & 1 & 1 & -1 & 1 & 0 \\ 0 & 1 & 2 & 3 & -2 & 0 \\ 0 & 0 & 10 & 16 & -12 & 1 \end{array}\right] \quad 4R_2 - R_3$$

We can create a 1 in a_{33} by multiplying row 3 by 0.1; however, that will introduce decimals on the right-hand side of the matrix, which will make mental computations more difficult. Consequently, we will reduce the matrix as far as possible before creating decimal entries. We will next zero out a_{12} with the operation $R_1 - R_2 \rightarrow R_1$.

$$\left[\begin{array}{ccc:ccc} 1 & 0 & -1 & -4 & 3 & 0 \\ 0 & 1 & 2 & 3 & -2 & 0 \\ 0 & 0 & 10 & 16 & -12 & 1 \end{array}\right] \quad R_1 - R_2$$

Observe that the first two columns of the augmented matrix are identical to the first two columns of the identity matrix I_3. We cannot simplify the matrix any further without creating decimal entries. We will proceed by creating a 1 in a_{33} with the operation $0.1R_3 \rightarrow R_3$.

$$\left[\begin{array}{ccc:ccc} 1 & 0 & -1 & -4 & 3 & 0 \\ 0 & 1 & 2 & 3 & -2 & 0 \\ 0 & 0 & 1 & 1.6 & -1.2 & 0.1 \end{array}\right] \quad 0.1R_3$$

Finally, we will zero out a_{13} and a_{23} with the operations $R_1 + R_3 \rightarrow R_1$ and $R_2 - 2R_3 \rightarrow R_2$.

$$\left[\begin{array}{ccc:ccc} 1 & 0 & 0 & -2.4 & 1.8 & 0.1 \\ 0 & 1 & 0 & -0.2 & 0.4 & -0.2 \\ 0 & 0 & 1 & 1.6 & -1.2 & 0.1 \end{array}\right] \begin{array}{l} R_1 + R_3 \\ R_2 - 2R_3 \\ \end{array}$$

The matrix on the left is I_3, and the matrix on the right is A^{-1}. That is,

$$A^{-1} = \begin{bmatrix} -2.4 & 1.8 & 0.1 \\ -0.2 & 0.4 & -0.2 \\ 1.6 & -1.2 & 0.1 \end{bmatrix}$$

which may alternatively be written as

$$A^{-1} = 0.1\begin{bmatrix} -24 & 18 & 1 \\ -2 & 4 & -2 \\ 16 & -12 & 1 \end{bmatrix}$$

We can check our work by calculating $A^{-1}A$.

$$A^{-1}A = 0.1\begin{bmatrix} -24 & 18 & 1 \\ -2 & 4 & -2 \\ 16 & -12 & 1 \end{bmatrix}\begin{bmatrix} 2 & 3 & 4 \\ 3 & 4 & 5 \\ 4 & 0 & 6 \end{bmatrix}$$

$$= 0.1\begin{bmatrix} 10 & 0 & 0 \\ 0 & 10 & 0 \\ 0 & 0 & 10 \end{bmatrix} = \begin{bmatrix} 1 & 0 & 0 \\ 0 & 1 & 0 \\ 0 & 0 & 1 \end{bmatrix}$$

Since $A^{-1}A = I_3$, the matrices are inverses of each other.

The demonstrated method will work for any size square matrix. Let's recap the method.

HOW TO Finding the Inverse of a Matrix A

1. Augment the matrix A with the identity matrix I.
2. Use row operations to reduce the left-hand side of the augmented matrix to reduced row echelon form.
3. If the left-hand side of the reduced matrix is the identity matrix, the right-hand side is A^{-1}. If the left-hand side of the reduced matrix contains a row of zeros, the matrix A is singular.

EXAMPLE 1

Finding the Inverse of a 3 × 3 Matrix

Find the inverse of the matrix $A = \begin{bmatrix} 3 & 4 & 2 \\ -1 & 0 & -1 \\ 5 & 2 & 5 \end{bmatrix}$, if it exists.

SOLUTION

$$\left[\begin{array}{rrr|rrr} 3 & 4 & 2 & 1 & 0 & 0 \\ -1 & 0 & -1 & 0 & 1 & 0 \\ 5 & 2 & 5 & 0 & 0 & 1 \end{array}\right] \quad \text{Augment } A \text{ with } I$$

$$= \left[\begin{array}{rrr|rrr} -1 & 0 & -1 & 0 & 1 & 0 \\ 3 & 4 & 2 & 1 & 0 & 0 \\ 5 & 2 & 5 & 0 & 0 & 1 \end{array}\right] \quad \begin{array}{l} R_2 \\ R_1 \\ \end{array}$$

$$= \left[\begin{array}{rrr|rrr} 1 & 0 & 1 & 0 & -1 & 0 \\ 0 & 4 & -1 & 1 & 3 & 0 \\ 0 & 2 & 0 & 0 & 5 & 1 \end{array}\right] \quad \begin{array}{l} -R_1 \\ 3R_1 + R_2 \\ 5R_1 + R_3 \end{array}$$

$$= \left[\begin{array}{rrr|rrr} 1 & 0 & 1 & 0 & -1 & 0 \\ 0 & 0 & -1 & 1 & -7 & -2 \\ 0 & 2 & 0 & 0 & 5 & 1 \end{array}\right] \quad \begin{array}{l} \\ R_2 - 2R_3 \\ \end{array}$$

$$= \left[\begin{array}{rrr|rrr} 1 & 0 & 1 & 0 & -1 & 0 \\ 0 & 2 & 0 & 0 & 5 & 1 \\ 0 & 0 & -1 & 1 & -7 & -2 \end{array}\right] \quad \begin{array}{l} \\ R_3 \\ R_2 \end{array}$$

$$= \left[\begin{array}{rrr|rrr} 1 & 0 & 0 & 1 & -8 & -2 \\ 0 & 1 & 0 & 0 & 2.5 & 0.5 \\ 0 & 0 & 1 & -1 & 7 & 2 \end{array}\right] \quad \begin{array}{l} R_1 + R_3 \\ 0.5R_2 \\ -R_3 \end{array}$$

$$= [I \mid A^{-1}]$$

So $A^{-1} = \begin{bmatrix} 1 & -8 & -2 \\ 0 & 2.5 & 0.5 \\ -1 & 7 & 2 \end{bmatrix}$.

EXAMPLE 2

Determining If a 3 × 3 Matrix Is Singular

Find the inverse of the matrix $A = \begin{bmatrix} 5 & 2 & 6 \\ -2 & 2 & -4 \\ 1 & 6 & -2 \end{bmatrix}$, if it exists.

SOLUTION

$$\left[\begin{array}{rrr|rrr} 5 & 2 & 6 & 1 & 0 & 0 \\ -2 & 2 & -4 & 0 & 1 & 0 \\ 1 & 6 & -2 & 0 & 0 & 1 \end{array}\right] \quad \text{Augment } A \text{ with } I$$

$$= \left[\begin{array}{rrr|rrr} 1 & 6 & -2 & 0 & 0 & 1 \\ -2 & 2 & -4 & 0 & 1 & 0 \\ 5 & 2 & 6 & 1 & 0 & 0 \end{array}\right] \quad \begin{array}{l} R_3 \\ \\ R_1 \end{array}$$

$$= \left[\begin{array}{rrr|rrr} 1 & 6 & -2 & 0 & 0 & 1 \\ 0 & 14 & -8 & 0 & 1 & 2 \\ 0 & 28 & -16 & -1 & 0 & 5 \end{array}\right] \quad \begin{array}{l} \\ 2R_1 + R_2 \\ 5R_1 - R_3 \end{array}$$

$$= \left[\begin{array}{rrr|rrr} 1 & 6 & -2 & 0 & 0 & 1 \\ 0 & 14 & -8 & 0 & 1 & 2 \\ 0 & 0 & 0 & 1 & 2 & -1 \end{array}\right] \quad \begin{array}{l} \\ \\ 2R_2 - R_3 \end{array}$$

The left-hand side of the augmented matrix has a row of zeros. As a result, in reduced row echelon form, it will not be the identity matrix. Consequently, A is singular.

The TI-83 Plus calculator has a built-in function that allows you to easily calculate the inverse of an invertible matrix. This feature should be used primarily to check your work for accuracy in larger matrices (3×3 and above). For the Technology Tip, we will find the inverse of $A = \begin{bmatrix} 6 & 14 & 16 \\ 4 & 1 & 14 \\ 1 & 4 & 6 \end{bmatrix}$.

TECHNOLOGY TIP

Finding the Inverse of a Matrix

1. Enter matrix A into the calculator, using the Matrix Editor. Use the Matrix Names menu to place matrix A on the home screen. Then press the x^{-1} button.

```
[A]-1
```

2. Press ENTER to display the inverse of the matrix.

```
[A]-1
[[.25    .1   -.9...
 [.05    -.1  .1...
 [-.075  .05  .25...
```

3. To convert the entries to fractions, press MATH, then 1:▶Frac. Press ENTER.

```
[[.25    .1   -.9...
 [.05    -.1  .1...
 [-.075  .05  .25...
Ans▶Frac
[[1/4    1/10   -...
 [1/20   -1/10  1...
 [-3/40  1/20   1...
```

Error Alert:
If, when attempting to find the inverse of a matrix, a singular matrix error occurs, double-check to make sure that you have entered the matrix correctly. If you have, this error tells you that the matrix is singular (not invertible).

```
ERR:SINGULAR MAT
1:Quit
2:Goto
```

EXAMPLE 3 Using Technology to Find the Inverse of a Square Matrix

Use technology to find the inverse of the matrix $A = \begin{bmatrix} 6 & 14 & 16 \\ 4 & 1 & 14 \\ 1 & 4 & 6 \end{bmatrix}$.

SOLUTION

Using technology, we determine that $A^{-1} = \begin{bmatrix} 0.25 & 0.1 & -0.9 \\ 0.05 & -0.1 & 0.1 \\ -0.075 & 0.05 & 0.25 \end{bmatrix}$. The inverse matrix may alternatively be written as

$$A^{-1} = \begin{bmatrix} \frac{1}{4} & \frac{1}{10} & -\frac{9}{10} \\ \frac{1}{20} & -\frac{1}{10} & \frac{1}{10} \\ -\frac{3}{40} & \frac{1}{20} & \frac{1}{4} \end{bmatrix} \quad \text{or} \quad A^{-1} = \frac{1}{40}\begin{bmatrix} 10 & 4 & -36 \\ 2 & -4 & 4 \\ -3 & 2 & 10 \end{bmatrix}$$

Solving Systems of Linear Equations Using Matrix Equations

We will now return to the process of solving systems of linear equations, introduced in Chapter 2, with the added capability of matrix algebra. Consider the system of linear equations

$$\begin{aligned} x - 4y - z &= -5 \\ 3y + z &= 7 \\ 2x + y \quad &= 10 \end{aligned}$$

Let's define A to be the coefficient matrix of the system. That is, the entries of A are the coefficients of the variables of the system.

$$A = \begin{bmatrix} 1 & -4 & -1 \\ 0 & 3 & 1 \\ 2 & 1 & 0 \end{bmatrix}$$

Let's define a column matrix X to be the variable matrix of the system. That is, the entries of X are the variables of the system.

$$X = \begin{bmatrix} x \\ y \\ z \end{bmatrix}$$

Finally, let's define the column matrix B to be the constant matrix of the system. That is, the entries of B are the constants from the right-hand side of the equal sign of the system:

$$B = \begin{bmatrix} -5 \\ 7 \\ 10 \end{bmatrix}$$

Consider the matrix product AX.

$$AX = \begin{bmatrix} 1 & -4 & -1 \\ 0 & 3 & 1 \\ 2 & 1 & 0 \end{bmatrix} \begin{bmatrix} x \\ y \\ z \end{bmatrix}$$

$$= \begin{bmatrix} 1x - 4y - 1z \\ 0x + 3y + 1z \\ 2x + 1y + 0z \end{bmatrix}$$

But from the system of equations we know that

$$1x - 4y - 1z = -5$$
$$0x + 3y + 1z = 7$$
$$2x + 1y + 0z = 10$$

So

$$AX = \begin{bmatrix} -5 \\ 7 \\ 10 \end{bmatrix}$$
$$= B$$

Therefore, a system of linear equations may be represented by the matrix equation $AX = B$. The solution to the system of equations is given by the matrix X. Is there a way to solve the matrix equation for X? Let's left multiply both sides by the matrix A^{-1}. (Since matrix multiplication is not commutative, we use the terms *left multiply* and *right multiply* to designate on which side to place the matrix being inserted into the equation.)

$$AX = B$$
$$A^{-1}(AX) = A^{-1}B \qquad \text{Left multiply by } A^{-1}$$
$$(A^{-1}A)X = A^{-1}B \qquad \text{Associative Property}$$
$$IX = A^{-1}B \qquad A^{-1}A = I$$
$$X = A^{-1}B \qquad IX = X$$

So the product of the inverse of the coefficient matrix and the constant matrix is the solution matrix. Using the algebraic or technological methods previously introduced, it may be shown that the inverse of $A = \begin{bmatrix} 1 & -4 & -1 \\ 0 & 3 & 1 \\ 2 & 1 & 0 \end{bmatrix}$ is $A^{-1} = \begin{bmatrix} \frac{1}{3} & \frac{1}{3} & \frac{1}{3} \\ -\frac{2}{3} & -\frac{2}{3} & \frac{1}{3} \\ 2 & 3 & -1 \end{bmatrix}$. Thus, the solution to the matrix equation

$$\begin{bmatrix} 1 & -4 & -1 \\ 0 & 3 & 1 \\ 2 & 1 & 0 \end{bmatrix}\begin{bmatrix} x \\ y \\ z \end{bmatrix} = \begin{bmatrix} -5 \\ 7 \\ 10 \end{bmatrix} \text{ is the product of } A^{-1} \text{ and } B.$$

$$\begin{aligned} X &= A^{-1}B \\ &= \begin{bmatrix} \frac{1}{3} & \frac{1}{3} & \frac{1}{3} \\ -\frac{2}{3} & -\frac{2}{3} & \frac{1}{3} \\ 2 & 3 & -1 \end{bmatrix}\begin{bmatrix} -5 \\ 7 \\ 10 \end{bmatrix} \\ &= \begin{bmatrix} 4 \\ 2 \\ 1 \end{bmatrix} \end{aligned}$$

So $x = 4$, $y = 2$, and $z = 1$.

This method of solving systems of linear equations is extremely efficient and works well for large systems of linear equations with unique solutions. Unfortunately, if a system has multiple solutions, this method cannot be used because the coefficient matrix will not be invertible.

EXAMPLE 4 Solving a System of Equations Using Matrix Algebra

Write the system of equations as a matrix equation and solve.

$$\begin{aligned} x + y + z &= 7 \\ 3x - y + z &= 21 \\ -x + 2y + 2z &= 2 \end{aligned}$$

SOLUTION The system of equations is equivalent to the matrix equation

$$\begin{bmatrix} 1 & 1 & 1 \\ 3 & -1 & 1 \\ -1 & 2 & 2 \end{bmatrix}\begin{bmatrix} x \\ y \\ z \end{bmatrix} = \begin{bmatrix} 7 \\ 21 \\ 2 \end{bmatrix}$$

The solution to the system is given by

$$\begin{bmatrix} x \\ y \\ z \end{bmatrix} = \begin{bmatrix} 1 & 1 & 1 \\ 3 & -1 & 1 \\ -1 & 2 & 2 \end{bmatrix}^{-1}\begin{bmatrix} 7 \\ 21 \\ 2 \end{bmatrix}$$

Using technology, we determine that

$$\begin{bmatrix} 1 & 1 & 1 \\ 3 & -1 & 1 \\ -1 & 2 & 2 \end{bmatrix}^{-1} = \begin{bmatrix} \frac{2}{3} & 0 & -\frac{1}{3} \\ \frac{7}{6} & -\frac{1}{2} & -\frac{1}{3} \\ -\frac{5}{6} & \frac{1}{2} & \frac{2}{3} \end{bmatrix}$$

Therefore,

$$\begin{bmatrix} x \\ y \\ z \end{bmatrix} = \begin{bmatrix} 1 & 1 & 1 \\ 3 & -1 & 1 \\ -1 & 2 & 2 \end{bmatrix}^{-1} \begin{bmatrix} 7 \\ 21 \\ 2 \end{bmatrix}$$

$$= \begin{bmatrix} \frac{2}{3} & 0 & -\frac{1}{3} \\ \frac{7}{6} & -\frac{1}{2} & -\frac{1}{3} \\ -\frac{5}{6} & \frac{1}{2} & \frac{2}{3} \end{bmatrix} \begin{bmatrix} 7 \\ 21 \\ 2 \end{bmatrix}$$

$$= \begin{bmatrix} \frac{14}{3} + 0 + \left(-\frac{2}{3}\right) \\ \frac{49}{6} + \left(-\frac{21}{2}\right) + \left(-\frac{2}{3}\right) \\ -\frac{35}{6} + \frac{21}{2} + \frac{4}{3} \end{bmatrix}$$

$$= \begin{bmatrix} \frac{12}{3} \\ \frac{49}{6} + \left(-\frac{63}{6}\right) + \left(-\frac{4}{6}\right) \\ -\frac{35}{6} + \frac{63}{6} + \frac{8}{6} \end{bmatrix}$$

$$= \begin{bmatrix} 4 \\ -\frac{18}{6} \\ \frac{36}{6} \end{bmatrix}$$

$$= \begin{bmatrix} 4 \\ -3 \\ 6 \end{bmatrix}$$

So $x = 4$, $y = -3$, and $z = 6$ is the solution to the system of equations.

Analyzing Real-Life Data with Matrices

Throughout this chapter, we have focused on developing your matrix algebra skills. With a well-developed set of matrix algebra skills, you will be able to set up matrix equations with relative ease. Since the entries in the matrices in many real-life situations are decimal numbers, we will often use technology to determine the desired solution.

EXAMPLE 5

Using Matrix Algebra to Make Informed Investment Decisions

Financial advisors often counsel their clients to diversify their investments into a variety of accounts with different levels of performance and risk. The average annual return (over the 10-year period prior to June 30, 2002) of two mutual funds offered by Harbor Fund is shown in Table 3.12.

TABLE 3.12

	Average Annual Return
Capital Appreciation Fund	12.69%
Bond Fund	7.97%

Source: Harbor Fund account statement.

High-performance accounts typically have greater volatility than lower-performance accounts. For example, although the Capital Appreciation Fund has a higher average annual return over the 10-year period, it earned −23.42 percent in a recent year. The Bond Fund has a lower average annual return over the 10-year period, but it earned 10.86 percent in the same year that the Capital Appreciation Fund suffered the 23.42 percent loss.

An investor has \$1,000 to invest in the two accounts. Assuming that the accounts will earn the returns specified in the table over the next year, how much should she invest in each account if she wants to earn 8 percent, 10 percent, or 12 percent?

SOLUTION Let x be the amount invested in the Capital Appreciation Fund and y be the amount invested in the Bond Fund. Since the sum of the individual investments is \$1000, we have

$$x + y = 1000$$

The annual return on each account is the product of the rate of return and the amount of money invested in the account. For the Capital Appreciation Fund, the annual return is given by $0.1269x$. For the Bond Fund, the annual return is given by $0.0797y$. The combined return on the two accounts is given by the expression $0.1269x + 0.0797y$. If we let r be the desired rate of return on the \$1000 investment, then $1000r$ is the dollar amount of the return. Since these two expressions must be equal, we have

$$0.1269x + 0.0797y = 1000r$$

For the 8 percent return, $r = 0.08$. Therefore,

$$\begin{aligned} 0.1269x + 0.0797y &= 1000(0.08) \\ 0.1269x + 0.0797y &= 80 \end{aligned}$$

Consequently, we have the system of equations

$$\begin{aligned} x + y &= 1000 && \text{Total investment} \\ 0.1269x + 0.0797y &= 80 && \text{Total return on investment} \end{aligned}$$

and the corresponding matrix equation

$$\overset{A}{\begin{bmatrix} 1 & 1 \\ 0.1269 & 0.0797 \end{bmatrix}} \overset{X}{\begin{bmatrix} x \\ y \end{bmatrix}} = \overset{B}{\begin{bmatrix} 1000 \\ 80 \end{bmatrix}}$$

Similarly, for the 10 percent return, we have the system of equations

$$\begin{aligned} x + y &= 1000 && \text{Total investment} \\ 0.1269x + 0.0797y &= 100 && \text{Total return on investment} \end{aligned}$$

and the corresponding matrix equation

$$\overset{A}{\begin{bmatrix} 1 & 1 \\ 0.1269 & 0.0797 \end{bmatrix}} \overset{X}{\begin{bmatrix} x \\ y \end{bmatrix}} = \overset{B}{\begin{bmatrix} 1000 \\ 100 \end{bmatrix}}$$

Likewise, for the 12 percent return, we have the system of equations

$$\begin{aligned} x + y &= 1000 && \text{Total investment} \\ 0.1269x + 0.0797y &= 120 && \text{Total return on investment} \end{aligned}$$

and the corresponding matrix equation

$$\overset{A}{\begin{bmatrix} 1 & 1 \\ 0.1269 & 0.0797 \end{bmatrix}} \overset{X}{\begin{bmatrix} x \\ y \end{bmatrix}} = \overset{B}{\begin{bmatrix} 1000 \\ 120 \end{bmatrix}}$$

Notice that although the constant matrices of the three matrix equations differ, each of the equations has the same coefficient matrix A. We know that the solution to the matrix equation $AX = B$ is $X = A^{-1}B$. The inverse of A is given by

$$A^{-1} \approx \begin{bmatrix} -1.689 & 21.19 \\ 2.689 & -21.19 \end{bmatrix}$$

We will use the inverse matrix shown on our calculator, instead of the rounded matrix entries shown here, to increase the accuracy of our results.

Therefore, for the 8 percent return, we have

$$\begin{aligned} X &= A^{-1}B \\ &= \begin{bmatrix} -1.689 & 21.19 \\ 2.689 & -21.19 \end{bmatrix} \begin{bmatrix} 1000 \\ 80 \end{bmatrix} \\ &= \begin{bmatrix} 6.36 \\ 993.64 \end{bmatrix} \end{aligned}$$

She should invest \$6.36 in the Capital Appreciation Fund and \$993.64 in the Bond Fund in order to earn an 8 percent return.

For the 10 percent return, we have

$$\begin{aligned} X &= A^{-1}B \\ &= \begin{bmatrix} -1.689 & 21.19 \\ 2.689 & -21.19 \end{bmatrix} \begin{bmatrix} 1000 \\ 100 \end{bmatrix} \\ &= \begin{bmatrix} 430.08 \\ 569.92 \end{bmatrix} \end{aligned}$$

She should invest \$430.08 in the Capital Appreciation Fund and \$569.92 in the Bond Fund in order to earn a 10 percent return.

For the 12 percent return, we have

$$\begin{aligned} X &= \begin{bmatrix} -1.689 & 21.19 \\ 2.689 & -21.19 \end{bmatrix} \begin{bmatrix} 1000 \\ 120 \end{bmatrix} \\ &= \begin{bmatrix} 853.81 \\ 146.19 \end{bmatrix} \end{aligned}$$

She should invest \$853.81 in the Capital Appreciation Fund and \$146.19 in the Bond Fund in order to earn a 12 percent return.

Recall that we *assumed* that she would earn the 10-year average annual return on all of her investments. Since none of the returns are guaranteed, which blend of investments she decides to choose will depend upon her risk tolerance.

In Example 5, we could have solved each matrix equation individually without using the inverse; however, since each equation had the same coefficient matrix, we greatly reduced our computations by using the inverse of the coefficient matrix to help find the solution.

EXAMPLE 6 Using Matrix Algebra to Forecast a Grade Point Average

Students at Green River Community College must earn 90 credits to obtain an Associate of Arts degree. Three friends discover that they all have the same grade point average (GPA), even though they have earned different numbers of credits. They each have a 3.42 GPA, and they hope to raise their GPAs to 3.70 by the time they have earned 90 credits. The first student has earned 45 credits; the second student, 65 credits; and the third student, 75 credits. Is it possible for all of the students to increase their cumulative GPAs to 3.7 by the time they obtain 90 credits?

SOLUTION Let x be the number of credits a student has earned and y be the number of credits remaining. We know that for each student,

$$x + y = 90$$

In order to have a 3.7 cumulative GPA, each student must earn $3.7(90) = 333$ grade points. Each student has earned $3.42x$ grade points from the courses already taken. Assuming that 4.00 is the highest grade a student can earn in any course, the maximum number of grade points that can be earned on the remaining credits is $4.00y$. Combining these two expressions, we have

$$3.42x + 4.00y = 333$$

Setting up and solving the matrix equation for the system of equations, we get

$$\begin{aligned} x + y &= 90 && \text{Total number of credits} \\ 3.42x + 4.00y &= 333 && \text{Total number of grade points} \end{aligned}$$

$$\begin{bmatrix} 1 & 1 \\ 3.42 & 4.00 \end{bmatrix}\begin{bmatrix} x \\ y \end{bmatrix} = \begin{bmatrix} 90 \\ 333 \end{bmatrix}$$

$$\begin{bmatrix} x \\ y \end{bmatrix} = \begin{bmatrix} 1 & 1 \\ 3.42 & 4.00 \end{bmatrix}^{-1}\begin{bmatrix} 90 \\ 333 \end{bmatrix}$$

$$\begin{bmatrix} x \\ y \end{bmatrix} \approx \begin{bmatrix} 6.90 & -1.72 \\ -5.90 & 1.72 \end{bmatrix}\begin{bmatrix} 90 \\ 333 \end{bmatrix}$$

$$\begin{bmatrix} x \\ y \end{bmatrix} \approx \begin{bmatrix} 46.55 \\ 43.45 \end{bmatrix}$$

(Although we rounded A^{-1} when we wrote it down, we used the more accurate calculator result in our computations to come up with $X = \begin{bmatrix} 46.55 \\ 43.45 \end{bmatrix}$.)

If a student has earned fewer than 46.55 credits, it is possible for her to raise her GPA to 3.7 by the time she has earned 90 credits. If she has earned more than 46.55 credits, it is mathematically impossible for her to raise her GPA to 3.7 by the time she has earned 90 credits. In the case of the three friends, only the student with 45 credits could conceivably raise her GPA to 3.7. To do so, she must earn a 4.0 grade on the majority of her remaining coursework.

EXAMPLE 7

Using Matrix Algebra to Calculate Management Bonuses

A small company rewards its upper management by offering annual bonuses. Each executive receives a percentage of the profits that remain after the bonuses of all the executives have been deducted from the company's profits. The CEO receives 4 percent; the CFO, 4 percent; and the vice president, 2 percent. What will be the bonus amount of each executive if the company's annual profit is $200,000? If it is $300,000? If it is $500,000?

SOLUTION Let x be the CEO's bonus, y be the CFO's bonus, z be the vice president's bonus, and P be the annual profit of the company. We know that the amount of profit remaining after the executives' bonuses are deducted is $P - (x + y + z)$. Since each executive receives a different percentage of this quantity, we end up with three different equations.

$$x = 0.04[P - (x + y + z)] \quad \text{CEO receives 4.0\%}$$

$$y = 0.04[P - (x + y + z)] \quad \text{CFO receives 4.0\%}$$

$$z = 0.02[P - (x + y + z)] \quad \text{Vice president receives 2.0\%}$$

Each equation must be simplified before we can write the matrix equation.

$$\begin{aligned} x &= 0.04[P - (x + y + z)] \\ x &= 0.04P - 0.04x - 0.04y - 0.04z \\ 1.04x + 0.04y + 0.04z &= 0.04P \\ \frac{1.04}{0.04}x + y + z &= P \\ 26x + y + z &= P \end{aligned}$$

$$\begin{aligned} y &= 0.04[P - (x + y + z)] \\ y &= 0.04P - 0.04x - 0.04y - 0.04z \\ 0.04x + 1.04y + 0.04z &= 0.04P \\ x + \frac{1.04}{0.04}y + z &= P \\ x + 26y + z &= P \end{aligned}$$

$$\begin{aligned} z &= 0.02[P - (x + y + z)] \\ z &= 0.02P - 0.02x - 0.02y - 0.02z \\ 0.02x + 0.02y + 1.02z &= 0.02P \\ x + y + \frac{1.02}{0.02}z &= P \\ x + y + 51z &= P \end{aligned}$$

Combining the simplified equations yields the system of equations

$$\begin{aligned} 26x + y + z &= P \\ x + 26y + z &= P \\ x + y + 51z &= P \end{aligned}$$

and the corresponding matrix equation

$$\overset{A}{\begin{bmatrix} 26 & 1 & 1 \\ 1 & 26 & 1 \\ 1 & 1 & 51 \end{bmatrix}} \overset{X}{\begin{bmatrix} x \\ y \\ z \end{bmatrix}} = \overset{B}{\begin{bmatrix} P \\ P \\ P \end{bmatrix}}$$

Using technology, we determine that

$$A^{-1} = \begin{bmatrix} 0.03855 & -0.001455 & -0.0007273 \\ -0.001455 & 0.03855 & -0.0007273 \\ -0.0007273 & -0.0007273 & 0.01964 \end{bmatrix}$$

To minimize round-off error, we store this matrix on our calculator before proceeding further.

The solution to the matrix equation is

$$\begin{aligned} X &= A^{-1}B \\ &= \begin{bmatrix} 0.03855 & -0.001455 & -0.0007273 \\ -0.001455 & 0.03855 & -0.0007273 \\ -0.0007273 & -0.0007273 & 0.01964 \end{bmatrix} \begin{bmatrix} P \\ P \\ P \end{bmatrix} \\ &= P\begin{bmatrix} 0.03855 & -0.001455 & -0.0007273 \\ -0.001455 & 0.03855 & -0.0007273 \\ -0.0007273 & -0.0007273 & 0.01964 \end{bmatrix} \begin{bmatrix} 1 \\ 1 \\ 1 \end{bmatrix} \qquad \text{Since } \begin{bmatrix} P \\ P \\ P \end{bmatrix} = P\begin{bmatrix} 1 \\ 1 \\ 1 \end{bmatrix} \\ &= P\begin{bmatrix} 0.03636 \\ 0.03636 \\ 0.01818 \end{bmatrix} = P\begin{bmatrix} \frac{2}{55} \\ \frac{2}{55} \\ \frac{1}{55} \end{bmatrix} = \begin{bmatrix} \frac{2}{55}P \\ \frac{2}{55}P \\ \frac{1}{55}P \end{bmatrix} \end{aligned}$$

The CEO and CFO each receive $\frac{2}{55}$ of the annual profits. The vice president receives $\frac{1}{55}$ of the annual profits. Table 3.13 shows each bonus amount for the identified profit levels.

TABLE 3.13

Profit (P)	$\frac{2}{55}P$	$\frac{1}{55}P$
\$200,000	\$7,272.73	\$3,636.36
\$300,000	\$10,909.09	\$5,454.55
\$500,000	\$18,181.82	\$9,090.91

3.3 Summary

In this section, you learned how to find the inverse of a matrix algebraically. You also used the inverse matrix to solve systems of equations, and you saw how matrix algebra may be used in both business and consumer applications.

3.3 Exercises

In Exercises 1–10, find the inverse of the matrix A algebraically. If A is singular, so state.

1. $A = \begin{bmatrix} 2 & 5 & 3 \\ 1 & 4 & 2 \\ 0 & 2 & 4 \end{bmatrix}$ **2.** $A = \begin{bmatrix} 4 & 2 & 1 \\ 5 & 3 & 2 \\ 0 & 2 & 3 \end{bmatrix}$

3. $A = \begin{bmatrix} 4 & 0 & 2 \\ 2 & 1 & 2 \\ 3 & 2 & 4 \end{bmatrix}$ **4.** $A = \begin{bmatrix} 3 & 5 & 1 \\ 0 & 3 & 1 \\ 2 & 9 & 3 \end{bmatrix}$

5. $A = \begin{bmatrix} 6 & 1 & 0 \\ 5 & 4 & 8 \\ 0 & 1 & 2 \end{bmatrix}$ **6.** $A = \begin{bmatrix} -1 & 0 & 0 \\ -1 & 1 & 1 \\ 0 & 1 & 0 \end{bmatrix}$

7. $A = \begin{bmatrix} 1 & 1 & 0 \\ 0 & 1 & 1 \\ 1 & 0 & -1 \end{bmatrix}$ **8.** $A = \begin{bmatrix} 2 & 3 & 3 \\ 8 & 2 & 1 \\ 7 & 1 & 0 \end{bmatrix}$

9. $A = \begin{bmatrix} 2 & 4 & 6 \\ 3 & 6 & 9 \\ 1 & 7 & 5 \end{bmatrix}$ **10.** $A = \begin{bmatrix} 2 & 0 & 0 \\ 8 & 1 & 0 \\ 4 & 1 & 1 \end{bmatrix}$

In Exercises 11–20, write the system of equations as a matrix equation, $AX = B$, and solve the equation, either algebraically or by using technology.

11. $\begin{aligned} x + y + z &= 6 \\ 2x - y + z &= 3 \\ 4x - 2y + 3z &= 9 \end{aligned}$

12. $\begin{aligned} 5x + y + z &= 1 \\ 4x - 2y + 5z &= -2 \\ x - 7y + 6z &= -7 \end{aligned}$

13. $\begin{aligned} 3x + 2y + 3z &= 6 \\ 2x - 5y + z &= -11 \\ 4x + 2y + 3z &= 3 \end{aligned}$

14. $\begin{aligned} x + y + z &= 1 \\ 2x - y + z &= 4 \\ 4x - 2y + 3z &= 9 \end{aligned}$

15. $\begin{aligned} x + y + z &= 3 \\ 2x - y + z &= 13 \\ 4x - 2y + 3z &= 28 \end{aligned}$

16. $\begin{aligned} 3x - y + 2z &= 17 \\ 4x + 3y + 5z &= 12 \\ 6x + 8y + 4z &= 4 \end{aligned}$

17. $\begin{aligned} 3x - y + 2z &= 12 \\ 4x + 3y + 5z &= 45 \\ 6x + 8y + 4z &= 64 \end{aligned}$

18. $\begin{aligned} 2x + 5y + 3z &= 10 \\ x + 4y + 2z &= 7 \\ 0x + 2y + 4z &= 6 \end{aligned}$

19. $\begin{aligned} 2x + 5y + 3z &= 0 \\ x + 4y + 2z &= -1 \\ 0x + 2y + 4z &= 2 \end{aligned}$

20. $\begin{aligned} x + y + z &= 1 \\ 2x - y + z &= 1 \\ 4x - 2y + 3z &= 3 \end{aligned}$

In Exercises 21–30, use technology to find the inverse of the matrix, if it exists. If the matrix is singular, so state.

21. $A = \begin{bmatrix} 1 & 1 & 2 \\ 3 & 5 & 8 \\ 13 & 21 & 34 \end{bmatrix}$

22. $B = \begin{bmatrix} 1.0 & 5.2 & -2.8 \\ 0 & 4.0 & 9.4 \\ 0 & 0 & 0.5 \end{bmatrix}$

23. $C = \begin{bmatrix} 3.4 & 1.2 & 0.5 \\ 1.8 & 4.2 & 0 \\ 0.3 & 0 & 0 \end{bmatrix}$

24. $D = \begin{bmatrix} 1 & 1 & 0 \\ 0 & 1 & 1 \\ 1 & 0 & 1 \end{bmatrix}$

25. $E = \begin{bmatrix} 1 & 2 & 3 & 4 \\ 2 & 3 & 4 & 5 \\ 3 & 0 & 5 & 6 \\ 4 & 0 & 0 & 7 \end{bmatrix}$

26. $A = \begin{bmatrix} 3 & 2 & 1 \\ 2 & 1 & 3 \\ 1 & 3 & 2 \end{bmatrix}$

27. $B = \begin{bmatrix} 2.0 & 9.2 & -2.9 \\ 6.1 & 4.7 & 1.4 \\ 0.0 & 2.0 & 3.5 \end{bmatrix}$

28. $D = \begin{bmatrix} 1.1 & 2.2 & 3.3 \\ -1.1 & 2.2 & 1.1 \\ 0.0 & 4.4 & 1.1 \end{bmatrix}$

29. $D = \begin{bmatrix} 1 & 1 & 0 & 2 \\ 0 & 1 & 1 & 2 \\ 1 & 0 & 1 & 2 \\ 2 & 1 & 0 & 3 \end{bmatrix}$

30. $E = \begin{bmatrix} 1 & 7 & 8 & 0 \\ 7 & 2 & 6 & 9 \\ 8 & 6 & 3 & 5 \\ 0 & 9 & 5 & 4 \end{bmatrix}$

In Exercises 31–36, determine the solution by setting up and solving the matrix equation.

31. Nutritional Content A nut distributor wants to determine the nutritional content of various mixtures of pecans (oil-roasted, salted), cashews (dry-roasted, salted), and almonds (honey-roasted, unblanched). Her supplier has provided the following nutrition information:

	Almonds	Cashews	Pecans
Protein	26.2 grams/cup	21.0 grams/cup	10.1 grams/cup
Carbohydrates	40.2 grams/cup	44.8 grams/cup	14.3 grams/cup
Fat	71.9 grams/cup	63.5 grams/cup	82.8 grams/cup

Source: www.Nutri-facts.com.

Her first mixture, Protein Blend, contains 6 cups of almonds, 3 cups of cashews, and 1 cup of pecans. Her second mixture, Low-Fat Mix, contains 3 cups of almonds, 6 cups of cashews, and 1 cup of pecans. Her final mixture, Low-Carb Mix, contains 3 cups of almonds, 1 cup of cashews, and 6 cups of pecans. Determine the amount of protein, carbohydrates, and fat in a 1-cup serving of each of the mixtures.

32. Floral Costs A florist purchases her flowers from an online flower wholesaler. White daisies are \$3.38 per bunch, football mums are \$7.40 per bunch, super blue purple statice is \$4.25 per bunch, and misty blue limonium is \$5.25 per bunch. (**Source:** www.FlowerSales.com.) From these flowers, she will make three types of bouquets.

	Type 1	Type 2	Type 3
Daisies	1 bunch	1 bunch	2 bunches
Mums	1 bunch	1 bunch	None
Statice	1/2 bunch	None	1/2 bunch
Limonium	None	1/2 bunch	None

What is her flower cost for each type of bouquet? How much should she charge for each bouquet if her markup is 50 percent of her flower cost?

33. Return on Investments The average annual return (over the 10-year period prior to June 30, 2002) of two mutual funds offered by Harbor Fund is shown in the table.

	Average Annual Return
Money Market	4.50%
Large-Cap Value	11.01%

Source: Harbor Fund.

Suppose you have \$2,000 to invest in these two accounts. Assuming that the accounts will earn the returns specified in the table over the next year, how much should you invest in each account if you want to earn 6 percent, 8 percent, or 10 percent?

34. Nutritional Content A nut distributor wants to determine the nutritional content of various mixtures of pecans (oil-roasted, salted), cashews (dry-roasted, salted), and almonds (honey-roasted, unblanched). Her supplier has provided the following nutrition information.

	Almonds	Cashews	Pecans
Protein	26.2 grams/cup	21.0 grams/cup	10.1 grams/cup
Sugars	20.5 grams/cup	40.7 grams/cup	3.9 grams/cup
Fiber	19.7 grams/cup	4.1 grams/cup	10.4 grams/cup

Source: www.Nutri-facts.com.

Her first mixture, Protein Blend, contains 6 cups of almonds, 3 cups of cashews, and 1 cup of pecans. Her second mixture, Low-Sugar Mix, contains 2 cups of almonds, 1 cup of cashews, and 7 cups of pecans. Her final mixture, High-Fiber Mix, contains 5 cups of almonds, 1 cup of cashews, and 4 cups of pecans. Determine the amount of protein, sugar, and fiber in each 1-cup serving of the mixtures.

35. Grade Point Average A student at Green River Community College must earn 90 credits to obtain an Associate of Arts degree. Three students with a 2.9 existing GPA hope to increase their cumulative GPA to 3.5. The first student has earned 30 credits; the second, 45 credits; and the third, 55 credits. Is it possible for all of the students to increase their cumulative GPA to a 3.5 GPA by the time they obtain 90 credits?

36. **Executive Bonus Plan** A small company rewards its upper management by offering annual bonuses. Each executive receives a percentage of the profits that remain after the bonuses of all of the executives have been deducted from the company's profits. The CEO receives 5 percent; the CFO, 4 percent; and the vice president, 3 percent. What will be the bonus amount of each executive if the company's annual profit is \$500,000? \$800,000? \$1,000,000?

Exercises 37–46 are intended to challenge your understanding of the process of finding matrix inverses algebraically.

37. Find the inverse of the matrix

$$A = \begin{bmatrix} 1 & 0 & 0 & a \\ 0 & 1 & 1 & 0 \\ 0 & 1 & 0 & 1 \\ 0 & 0 & 1 & 1 \end{bmatrix}.$$

38. Under what conditions is the matrix

$$A = \begin{bmatrix} 1 & 0 & a \\ 0 & 1 & 0 \\ a & 0 & a \end{bmatrix} \text{ singular?}$$

39. Determine if the following statement is true or false: The matrix $A = \begin{bmatrix} 1 & a & 1 \\ 0 & a & 0 \\ 1 & -a & 1 \end{bmatrix}$ is singular for any value of a.

40. Determine if the following statement is true or false: The matrix $A = \begin{bmatrix} 1 & a & b \\ 0 & 1 & c \\ 0 & 0 & 1 \end{bmatrix}$ is invertible for all values of a, b, and c.

41. Determine if the following statement is true or false: The matrix $A = \begin{bmatrix} a & 1 & b \\ 0 & 1 & c \\ 0 & 0 & 1 \end{bmatrix}$ is invertible for all values of a, b, and c.

42. Show that if A is invertible, then A^2 is invertible. [*Hint:* $(AB)^{-1} = B^{-1}A^{-1}$.]

43. Show that if A and B are invertible and AB is defined, then AB is invertible. [*Hint:* $(AB)^{-1} = B^{-1}A^{-1}$.]

44. Find three different 2×2 matrices A with the property that $A^2 = I$.

45. Find two different 3×3 matrices that are their own inverses. That is, $A \cdot A = I$.

46. Given $A^2 = I$, write A^{-1} in terms of A.

3.4 Leontief Input-Output Models

- Do basic input-output analysis using Leontief input-output models

GETTING STARTED Economists typically divide the economies they analyze into industrial sectors. For example, a primitive economy may consist of energy, manufacturing, and agriculture sectors. All of the sectors are interrelated. The energy industry must produce enough energy to meet its own needs as well as the energy needs of the manufacturing and agriculture sectors. Similarly, the manufacturing industry must produce enough manufactured goods to meet its own needs as well as the manufactured goods needs of the energy and agriculture industries.

The economist Wassily Leontief earned the Nobel Prize in Economics in 1973, in part, for his innovative approach to analyzing economies. Leontief studied the interrelationship among 500 different sectors in the U.S. economy and laid the groundwork for the field of mathematics referred to as **input-output analysis.** In this section, we will demonstrate how to do basic input-output analysis.

A primitive economy consists of two sectors: energy and manufacturing. A unit of a product is interpreted to be \$1 worth of that product. To produce 1 unit of energy requires 0.1 unit of energy and 0.3 unit of manufactured products. To produce 1 unit of manufactured products requires 0.3 unit of energy and 0.6 unit of manufactured products. These are the economy's *internal demands*. The economy also demands a surplus of energy and manufactured products. This demand is referred to as an *external demand.* A \$27,000 surplus of energy and an \$81,000 surplus of manufactured goods are required. How many units of energy and manufacturing does the economy need to produce in order to meet its internal and external demands?

Let x be the number of units of energy produced and y be the number of units of manufactured products produced. To produce \$1 of energy costs \$0.10 in energy and \$0.30 in manufactured products. That is, $0.1x + 0.3y$ is the input required to meet the internal demand for energy. Similarly, to produce \$1 of manufactured products costs \$0.30 of energy and \$0.60 of manufactured goods. That is, $0.3x + 0.6y$ is the input required to meet the internal demand for manufactured goods. In addition to meeting the internal demands, the energy and manufacturing sectors must meet the external demands for their products. We have

$$\text{Total output} = \text{internal demand} + \text{external demand}$$

$$x = (0.1x + 0.3y) + 27{,}000 \quad \text{Energy sector}$$

$$y = (0.3x + 0.6y) + 81{,}000 \quad \text{Manufacturing sector}$$

This system may be alternatively written as a matrix equation.

$$\begin{bmatrix} x \\ y \end{bmatrix} = \begin{bmatrix} 0.1 & 0.3 \\ 0.3 & 0.6 \end{bmatrix}\begin{bmatrix} x \\ y \end{bmatrix} + \begin{bmatrix} 27{,}000 \\ 81{,}000 \end{bmatrix}$$

$$\text{Total output} = \text{internal demand} + \text{external demand}$$

The matrix $X = \begin{bmatrix} x \\ y \end{bmatrix}$ is called the **output matrix** (or **production matrix**), and the matrix $D = \begin{bmatrix} 27{,}000 \\ 81{,}000 \end{bmatrix}$ is called the **external demand matrix** (or **final demand matrix**). The matrix $T = \begin{bmatrix} 0.1 & 0.3 \\ 0.3 & 0.6 \end{bmatrix}$ is called the **technology matrix.** Each entry of the technology matrix has a specific meaning.

$T =$

Input from energy sector required to produce \$1 of energy	Input from energy sector required to produce \$1 of manufactured goods
Input from manufacturing sector required to produce \$1 of energy	Input from manufacturing sector required to produce \$1 of manufactured goods

Using the matrix names $X, D,$ and T, we can write the matrix equation as $X = TX + D$ and solve for X as shown.

$$\begin{bmatrix} x \\ y \end{bmatrix} = \begin{bmatrix} 0.1 & 0.3 \\ 0.3 & 0.6 \end{bmatrix}\begin{bmatrix} x \\ y \end{bmatrix} + \begin{bmatrix} 27{,}000 \\ 81{,}000 \end{bmatrix}$$

$$X = TX + D$$

$$IX = TX + D$$ Recall that I is the identity matrix, $I = \begin{bmatrix} 1 & 0 \\ 0 & 1 \end{bmatrix}$

$$IX - TX = D$$ Subtract TX from both sides

$$(I - T)X = D$$ Factor out X

$$(I - T)^{-1}(I - T)X = (I - T)^{-1}D$$ Left multiply both sides by the inverse of $I - T$

$$IX = (I - T)^{-1}D$$ The product of a matrix and its inverse is I

$$X = (I - T)^{-1}D$$ The product of a matrix X and matrix I is X

Note that the general solution $X = (I - T)^{-1}D$ is not dependent upon the specific values of $X, T,$ and D used in this problem. Consequently, the solution to all problems of the form $X = TX + D$ will be $X = (I - T)^{-1}D$, provided that $I - T$ is invertible. Returning to our specific problem, we have

$$\begin{aligned} I - T &= \begin{bmatrix} 1 & 0 \\ 0 & 1 \end{bmatrix} - \begin{bmatrix} 0.1 & 0.3 \\ 0.3 & 0.6 \end{bmatrix} \\ &= \begin{bmatrix} 0.9 & -0.3 \\ -0.3 & 0.4 \end{bmatrix} \end{aligned}$$

Recall that the inverse of a 2×2 matrix $A = \begin{bmatrix} a & b \\ c & d \end{bmatrix}$ is

$$A^{-1} = \frac{1}{ad - bc}\begin{bmatrix} d & -b \\ -c & a \end{bmatrix}$$

Consequently,

$$\begin{aligned} (I - T)^{-1} &= \frac{1}{(0.9)(0.4) - (-0.3)(-0.3)}\begin{bmatrix} 0.4 & 0.3 \\ 0.3 & 0.9 \end{bmatrix} \\ &= \frac{1}{0.36 - 0.09}\begin{bmatrix} 0.4 & 0.3 \\ 0.3 & 0.9 \end{bmatrix} \\ &= \frac{1}{0.27}\begin{bmatrix} 0.4 & 0.3 \\ 0.3 & 0.9 \end{bmatrix} \\ &= \begin{bmatrix} \frac{40}{27} & \frac{10}{9} \\ \frac{10}{9} & \frac{10}{3} \end{bmatrix} \end{aligned}$$

Therefore,

$$X = (I - T)^{-1}D$$

$$X = \begin{bmatrix} \frac{40}{27} & \frac{10}{9} \\ \frac{10}{9} & \frac{10}{3} \end{bmatrix} \begin{bmatrix} 27{,}000 \\ 81{,}000 \end{bmatrix}$$

$$\begin{bmatrix} x \\ y \end{bmatrix} = \begin{bmatrix} 130{,}000 \\ 300{,}000 \end{bmatrix}$$

The economy needs to produce \$130,000 worth of energy and \$300,000 worth of manufactured goods in order to meet its internal and external demands. We can check our work by verifying that $X = TX + D$.

$$X = TX + D$$

$$\begin{bmatrix} 130{,}000 \\ 300{,}000 \end{bmatrix} \stackrel{?}{=} \begin{bmatrix} 0.1 & 0.3 \\ 0.3 & 0.6 \end{bmatrix} \begin{bmatrix} 130{,}000 \\ 300{,}000 \end{bmatrix} + \begin{bmatrix} 27{,}000 \\ 81{,}000 \end{bmatrix}$$

$$\begin{bmatrix} 130{,}000 \\ 300{,}000 \end{bmatrix} \stackrel{?}{=} \begin{bmatrix} 103{,}000 \\ 219{,}000 \end{bmatrix} + \begin{bmatrix} 27{,}000 \\ 81{,}000 \end{bmatrix}$$

$$\begin{bmatrix} 130{,}000 \\ 300{,}000 \end{bmatrix} = \begin{bmatrix} 130{,}000 \\ 300{,}000 \end{bmatrix}$$

The result checks out, so we are confident that we have solved the problem correctly.

Although we developed the theory of solving input-output problems using an economy containing exactly two sectors, the same problem-solving approach works for economies with any number of sectors.

INPUT-OUTPUT PROBLEM SOLUTION

An economy with a technology matrix T and an external demand matrix D has an output matrix X. The matrix X satisfies the input-output equation

$$X = TX + D$$

The input-output equation has the solution

$$X = (I - T)^{-1}D$$

provided that $I - T$ is invertible.

The entry t_{ij} of the technology matrix T represents the input from sector i required to produce 1 unit of output in sector j.

Within the U.S. government, the Bureau of Economic Analysis (www.bea.gov) is responsible for conducting analyses of the economy. The bureau publishes annual input-output tables for various sectors of the economy. Many of the examples and exercises throughout this section are based on data from the 2002 input-output tables published by the bureau. We will typically focus on four or fewer sectors of the economy in a given problem. This approach will not give a complete picture of the economy because it assumes that the output of the industries in a given problem is dependent only upon the sectors given in the problem. However, it will give us an idea of how various sectors of the economy are interrelated.

EXAMPLE 1

Using Input-Output Analysis to Determine the Relationship Between Two Industries

The *motion picture and sound recording* (movie) industry and the *broadcasting and telecommunications* (telecom) industry are interrelated. Based on data from 2002, it requires roughly \$0.27 from the movie industry and \$0.01 from the telecom industry to produce \$1 of movie industry products. Similarly, it requires \$0.03 from the movie industry and \$0.24 from the telecom industry to produce \$1 of telecom products. If the external demand for movie products is \$42 billion and the external demand for telecom products is \$381 billion, what output from each sector is required to meet the demand for movie and telecom products?

SOLUTION Let x be the output of the movie industry and y be the output of the telecom industry, both in billions of dollars. The total output of the sectors is $X = \begin{bmatrix} x \\ y \end{bmatrix}$. The technology matrix T is given by

$$T = \begin{array}{c} \text{Movie output} \quad \text{Telecom output} \\ \begin{bmatrix} 0.27 & 0.03 \\ 0.01 & 0.24 \end{bmatrix} \end{array} \quad \begin{array}{l} \text{Movie input} \\ \text{Telecom input} \end{array}$$

The external demand matrix is $D = \begin{bmatrix} 42 \\ 381 \end{bmatrix}$. From our earlier discussion, we know that $X = (I - T)^{-1}D$.

$$\begin{aligned} X &= (I - T)^{-1}D \\ &= \left(\begin{bmatrix} 1 & 0 \\ 0 & 1 \end{bmatrix} - \begin{bmatrix} 0.27 & 0.03 \\ 0.01 & 0.24 \end{bmatrix} \right)^{-1} \begin{bmatrix} 42 \\ 381 \end{bmatrix} \\ &= \begin{bmatrix} 0.73 & -0.03 \\ -0.01 & 0.76 \end{bmatrix}^{-1} \begin{bmatrix} 42 \\ 381 \end{bmatrix} \\ &\approx \begin{bmatrix} 1.37 & 0.054 \\ 0.018 & 1.32 \end{bmatrix} \begin{bmatrix} 42 \\ 381 \end{bmatrix} \quad \text{We calculated the inverse using technology} \\ &\approx \begin{bmatrix} 78 \\ 502 \end{bmatrix} \end{aligned}$$

We estimate that \$78 billion of movie products and \$502 billion of telecom products need to be produced to meet the demand.

EXAMPLE 2

Using Input-Output Analysis to Determine the Relationship Between Three Industries

The *wood products* (wood), *paper products* (paper), and *printing and related support activities* (print) industries are interrelated. The following technology matrix quantifies the relationship between the industries:

$$T = \begin{array}{c} \text{Wood output} \quad \text{Paper output} \quad \text{Print output} \\ \begin{bmatrix} 0.216 & 0.023 & 0.000 \\ 0.004 & 0.219 & 0.241 \\ 0.000 & 0.001 & 0.060 \end{bmatrix} \end{array} \quad \begin{array}{l} \text{Wood input} \\ \text{Paper input} \\ \text{Print input} \end{array}$$

Explain the meaning of the entries t_{11}, t_{23}, and t_{31}. Then determine how much each industry needs to produce in order to meet internal and external demands, given that the external demand for wood products, paper products, and print products are \$64 billion, \$96 billion, and \$64 billion, respectively.

SOLUTION t_{11} is the entry in the first row and first column of the technology matrix. It represents the amount of wood products input that is required to produce \$1 of wood products. It requires nearly \$0.22 (\$0.216) of wood products to produce \$1 of wood products.

t_{23} is the entry in the second row and third column of the technology matrix. It represents the amount of paper products input that is required to produce \$1 of print products. It requires roughly \$0.24 (\$0.241) of paper products to produce \$1 of print products.

t_{31} is the entry in the third row and first column of the technology matrix. It represents the amount of print products input that is required to produce \$1 of wood products. Since the entry is equal to 0, no print products are required to produce \$1 of wood products.

To determine the total demand for wood, paper, and print products, we calculate $X = (I - T)^{-1}D$.

$$
\begin{aligned}
X &= (I - T)^{-1}D \\
&= \left(\begin{bmatrix} 1 & 0 & 0 \\ 0 & 1 & 0 \\ 0 & 0 & 1 \end{bmatrix} - \begin{bmatrix} 0.216 & 0.023 & 0.000 \\ 0.004 & 0.219 & 0.241 \\ 0.000 & 0.001 & 0.060 \end{bmatrix} \right)^{-1} \begin{bmatrix} 64 \\ 96 \\ 64 \end{bmatrix} \\
&= \begin{bmatrix} 0.784 & -0.023 & 0.000 \\ -0.004 & 0.781 & -0.241 \\ 0.000 & -0.001 & 0.940 \end{bmatrix}^{-1} \begin{bmatrix} 64 \\ 96 \\ 64 \end{bmatrix} \\
&\approx \begin{bmatrix} 1.276 & 0.038 & 0.010 \\ 0.007 & 1.281 & 0.328 \\ 0.000 & 0.001 & 1.064 \end{bmatrix} \begin{bmatrix} 64 \\ 96 \\ 64 \end{bmatrix} \\
&\approx \begin{bmatrix} 86 \\ 144 \\ 68 \end{bmatrix}
\end{aligned}
$$

In order to satisfy internal and external demands, \$86 billion of wood products, \$144 billion of paper products, and \$68 billion of print products need to be produced.

3.4 Summary

In this section, you learned how to do basic input-output analysis. These skills allow you to analyze the interrelationships among sectors of the economy.

3.4 Exercises

In Exercises 1–4, interpret the meaning of the indicated entries in the technology matrix.

1. Entries t_{11}, t_{12}, t_{21}, and t_{22}.

$$T = \begin{bmatrix} 0.2 & 0.5 \\ 0.4 & 0.8 \end{bmatrix} \begin{matrix} \text{Sector A} \\ \text{Sector B} \end{matrix}$$

2. Entries t_{11}, t_{12}, t_{21}, and t_{22}.

$$T = \begin{bmatrix} 0.27 & 0.03 \\ 0.01 & 0.24 \end{bmatrix} \begin{matrix} \text{Movie} \\ \text{Telecom} \end{matrix}$$

3. Entries t_{12}, t_{21}, t_{23},and t_{31}.

$$T = \begin{bmatrix} 0.14 & 0.01 & 0.00 \\ 0.00 & 0.10 & 0.00 \\ 0.02 & 0.02 & 0.01 \end{bmatrix} \begin{matrix} \text{Oil} \\ \text{Mining} \\ \text{Support} \end{matrix}$$

4. Entries t_{13}, t_{22}, t_{32},and t_{33}.

$$T = \begin{bmatrix} 0.01 & 0.01 & 0.00 \\ 0.00 & 0.00 & 0.02 \\ 0.03 & 0.00 & 0.06 \end{bmatrix} \begin{matrix} \text{Computer} \\ \text{Management} \\ \text{Support} \end{matrix}$$

In Exercises 5–10, use the given technology and external demand matrices to find the answers to the questions. The demand matrices are in millions of dollars.

$$T = \begin{bmatrix} 0.2 & 0.1 \\ 0.6 & 0.3 \end{bmatrix} \begin{matrix} \text{Agriculture} \\ \text{Manufacturing} \end{matrix}$$

$$D_1 = \begin{bmatrix} 210 \\ 300 \end{bmatrix} \qquad D_2 = \begin{bmatrix} 240 \\ 360 \end{bmatrix}$$

5. How much agriculture and manufacturing input is required to produce \$1 of manufacturing?

6. How much agriculture and manufacturing input is required to produce \$1 of agriculture?

7. Calculate $I - T$ and $(I - T)^{-1}$.

8. How much agriculture and manufacturing input is required to meet the internal demands of each industry and the external demands given in D_1 ?

9. How much agriculture and manufacturing input is required to meet the internal demands of each industry and the external demands given in D_2 ?

10. Given that the external demand for agriculture and manufacturing products is given by $D = \begin{bmatrix} d_1 \\ d_2 \end{bmatrix}$, how much agriculture and manufacturing input is required to meet the internal demands of each industry and the external demands given in D ?

Exercises 11–20 are derived from economic data for 2002 published by the U.S. Bureau of Economic Analysis.

11. **Industry Analysis: Two Industries** The *textile mills* and *textile product mills* (textile) industry and the *apparel and leather and allied products* (apparel) industry are interrelated. Based on data from 2002, it requires roughly \$0.23 from the textile industry and \$0.01 from the apparel industry to produce \$1 of textile products. Similarly, it requires \$0.20 from the textile industry and \$0.10 from the apparel industry to produce \$1 of apparel products. If the external demand for textile products is \$43 billion and the external demand for apparel products is \$48 billion, what output from each sector is required to meet the demand for textile and apparel products?

12. **Industry Analysis: Two Industries** The *fabricated metal products* (metal) industry and the *machinery* (machine) industry are interrelated. Based on data from 2002, it requires roughly \$0.11 from the metal industry and \$0.01 from the machine industry to produce \$1 of metal products. Similarly, it requires \$0.10 from the metal industry and \$0.09 from the machine industry to produce \$1 of machine products. If the external demand for metal products is \$185 billion and the external demand for machine products is \$210 billion, what output from each sector is required to meet the demand for metal and machine products?

13. **Industry Analysis: Two Industries** The *motor vehicles, bodies and trailers, and parts* (vehicle) industry and the *other transportation equipment* (transportation) industry are interrelated. Based on data from 2002, it requires roughly \$0.31 from the vehicle industry and \$0.00 from the transportation industry to produce \$1 of vehicle products. Similarly, it requires \$0.01 from the vehicle industry and \$0.18 from the transportation industry to produce \$1 of transportation products. If the external demand for vehicle products is \$296 billion and the external demand for transportation products is \$134 billion, what output from each sector is required to meet the demand for vehicle and transportation products?

 **14.**  **Industry Analysis: Two Industries** The *computer and electronic products* (computer) industry and the *electrical equipment, appliances, and components* (equipment) industry are interrelated. Based on data from 2002, it requires roughly \$0.22 from the computer industry and \$0.02 from the equipment industry to produce \$1 of computer products. Similarly, it requires \$0.03 from the computer industry and \$0.06 from the equipment industry to produce \$1 of equipment products. If the external demand for computer products is \$284 billion and the external demand for equipment products is \$82 billion, what output from each sector is required to meet the demand for computer and equipment products?

 **15.**  **Industry Analysis: Three Industries** The *primary metals* (metals), *fabricated metal products* (products), and *machinery* (machinery) industries are interrelated. Based on data from 2002, it requires \$0.26 of metals, \$0.02 of products, and \$0.01 of machinery to produce \$1 of metals. Similarly, it requires \$0.17 of metals, \$0.11 of products, and \$0.01 of machinery to produce \$1 of products. Likewise, it requires \$0.09 of metals, \$0.10 of products, and \$0.09 of machinery to produce \$1 of machinery. If the external demand for metals is \$37 billion, the external demand for products is \$181 billion, and the external demand for machinery is \$208 billion, what output from each sector is required to meet the demand for metals, products, and machinery?

 **16.**  **Industry Analysis: Three Industries** The *oil and gas extraction* (oil), *mining, except oil and gas* (mining), and *support activities for mining* (support) industries are interrelated. Based on data from 2002, it requires \$0.14 of oil, \$0.00 of mining, and \$0.02 of support to produce \$1 of oil. Similarly, it requires \$0.01 of oil, \$0.10 of mining, and \$0.02 of support to produce \$1 of mining. Likewise, it requires \$0.00 of oil, \$0.00 of mining, and \$0.01 of support to produce \$1 of support. If the external demand for oil is \$87 billion, the external demand for mining is \$43 billion, and the external demand for support is \$26 billion, what output from each sector is required to meet the demand for oil, mining, and support?

 **17.**  **Industry Analysis: Three Industries** The *computer systems design and related services* (computer), *management of companies and enterprises* (management), and *administrative and support services* (support) industries are interrelated. Based on data from 2002, \$0.01 of computer, \$0.00 of management, and \$0.03 of support is required to produce \$1 of computer. To produce \$1 of management requires \$0.01 of computer, \$0.00 of management, and \$0.00 of support. To produce \$1 of support requires \$0.00 of computer, \$0.02 of management, and \$0.06 of support. If the external demand for computer products is \$139 billion, the external demand for management is \$276 billion, and the external demand for support is \$401 billion, what output from each sector is required to meet the demand?

 **18.** **Industry Analysis: Four Industries** The *petroleum and coal products* (petrol), *chemical products* (chem), *plastics and rubber products* (plastic), and *nonmetallic mineral products* (mineral) industries are interrelated as shown in the given technology matrix.

$$T = \begin{bmatrix} 0.09 & 0.02 & 0.00 & 0.00 \\ 0.01 & 0.21 & 0.23 & 0.03 \\ 0.00 & 0.02 & 0.07 & 001 \\ 0.00 & 0.00 & 0.01 & 0.10 \end{bmatrix} \begin{matrix} \text{Petrol} \\ \text{Chem} \\ \text{Plastic} \\ \text{Mineral} \end{matrix}$$

The estimated external demand for each of these products in 2002 is given in the table.

Industry	External Demand
Petrol	\$168 billion
Chem	\$268 billion
Plastic	\$149 billion
Mineral	\$74 billion

What production level in each industry is required in order to satisfy internal and external demands?

 **19.** **Industry Analysis: Four Industries** The *Federal Reserve banks, credit intermediation, and related activities* (banks), *securities, commodity contracts, and investments* (securities), *insurance carriers and related activities* (insurance), and *funds, trusts, and other financial vehicles* (funds) industries are interrelated as shown in the given technology matrix.

$$T = \begin{bmatrix} 0.08 & 0.03 & 0.01 & 0.03 \\ 0.05 & 0.12 & 0.01 & 0.54 \\ 0.00 & 0.00 & 0.38 & 0.02 \\ 0.00 & 0.00 & 0.01 & 0.00 \end{bmatrix} \begin{matrix} \text{Banks} \\ \text{Securities} \\ \text{Insurance} \\ \text{Funds} \end{matrix}$$

The estimated external demand for each of these products in 2002 is given in the table.

Industry	External Demand
Banks	\$453 billion
Securities	\$129 billion
Insurance	\$281 billion
Funds	\$67 billion

What production levels are required to meet product demand?

 **20.**  **Industry Analysis: Four Industries** The *publishing (includes software)* (publish), *motion picture and sound recording* (movie), *broadcasting and telecommunications* (telecom), and *information and data processing services* (data) industries are interrelated as shown in the given technology matrix.

$$T = \begin{bmatrix} 0.09 & 0.00 & 0.00 & 0.00 \\ 0.00 & 0.27 & 0.03 & 0.00 \\ 0.03 & 0.01 & 0.24 & 0.06 \\ 0.02 & 0.00 & 0.00 & 0.02 \end{bmatrix} \begin{matrix} \text{Publish} \\ \text{Movie} \\ \text{Telecom} \\ \text{Data} \end{matrix}$$

The estimated external demand for each of these products in 2002 is given in the table.

Industry	External Demand
Publish	\$144 billion
Movie	\$42 billion
Telecom	\$370 billion
Data	\$88 billion

What production levels are required to meet internal and external demands?

Exercises 21–26 are intended to challenge your understanding of input-output tables.

21. A classmate presents you with the following "technology" matrix.

$$T = \begin{bmatrix} 0.6 & 0.3 \\ 0.5 & 0.9 \end{bmatrix}$$

What causes you to immediately question his results?

22. The following input-output problem has no solution. Explain why.

$$x = 0.5x + 0.4y + 2000$$
$$y = 0.5x + 0.6y + 3000$$

23. The relationship between three sectors of an economy are summarized in the given technology matrix.

$$T = \begin{bmatrix} 0.2 & 0.0 & 0.0 \\ 0.0 & 0.3 & 0.4 \\ 0.0 & 0.2 & 0.1 \end{bmatrix} \begin{matrix} \text{Sector A} \\ \text{Sector B} \\ \text{Sector C} \end{matrix}$$

Assume that the output of each sector is dependent only upon input from the indicated sectors. If production in Sector A is halted because of an employee strike, how will that affect Sectors B and C?

24. Referring to the technology matrix in Exercise 23, explain how Sectors A and B will be affected if production in Sector C is halted.

25. Let T be any technology matrix. For a functioning economy, between what values must the sum of the entries in a given column of T range?

26. Let T be any technology matrix. What numerical restrictions, if any, are placed upon the entries in T?

Chapter 3 Review Exercises

Section 3.1 *In Exercises 1–5, perform the indicated matrix operation, if possible, given the following matrices. Solve these problems without technology.*

$$A = \begin{bmatrix} 2 & 3 \\ -3 & -4 \\ 0 & 7 \end{bmatrix}, \ B = \begin{bmatrix} 5 & 6 \\ 3 & -4 \\ 9 & 2 \end{bmatrix}$$

1. $A + B$

2. $A - B$

3. $2A$

4. $3B$

5. $2A + 3B$

In Exercises 6–10, use technology to simplify the matrix expressions, given

$$A = \begin{bmatrix} 1.2 & 6.1 & -0.4 \\ -9.1 & 4.2 & 1.7 \\ 0.9 & 3.0 & 3.3 \end{bmatrix} \quad \text{and}$$

$$B = \begin{bmatrix} 3.4 & -0.3 & -0.4 \\ -3.8 & 5.6 & 7.2 \\ 2.2 & 2.6 & 2.0 \end{bmatrix}$$

 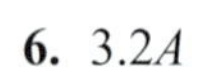 **6.** $3.2A$  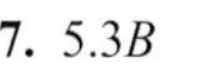 **7.** $5.3B$ 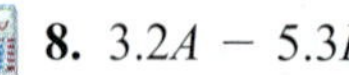 **8.** $3.2A - 5.3B$

  **9.** $4.1A + 0.1B$ **10.** $-1.1A + 2.9B$

11.  **Auto Prices** The average *trade-in value* of a Toyota Celica and a Toyota MR2 Spyder in July 2002 is shown in the first table. The average *retail value* of the two vehicles is shown in the second table.

Average Trade-In

	Celica	MR2 Spyder
2000 Model	\$13,025	\$18,750
2001 Model	\$14,850	\$20,100

Average Retail

	Celica	MR2 Spyder
2000 Model	\$15,300	\$21,300
2001 Model	\$17,250	\$22,725

Source: www.nada.com.

Use matrices to create a table that shows the average dealer markup for each of the vehicles.

Section 3.2

In Exercises 12–20, perform the indicated operation. As appropriate, use the matrices

$$A = \begin{bmatrix} 5 & 8 \\ 1 & 2 \end{bmatrix}, B = \begin{bmatrix} 4 & 5 \\ 2 & 2 \end{bmatrix},$$

$$C = \begin{bmatrix} 2 & 1 \\ 5 & 4 \\ 8 & 7 \end{bmatrix}, \text{and } D = \begin{bmatrix} -1 & 0 & 4 \\ -2 & 2 & -2 \\ -6 & 4 & 1 \end{bmatrix}.$$

If the specified operation is undefined, so state.

12. AB **13.** BA **14.** CD

15. DC **16.** CA **17.** AC

18. A^{-1} **19.** B^{-1} **20.** D^{-1}

In Exercises 21–22, use the determinant formula to determine if the matrix is invertible or singular.

21. $A = \begin{bmatrix} 9 & 6 \\ -3 & 2 \end{bmatrix}$ **22.** $B = \begin{bmatrix} -8 & -4 \\ 5 & 3 \end{bmatrix}$

In Exercises 23–24, use technology to find the inverse of the matrix, if it exists. If the matrix is singular, so state.

23. $B = \begin{bmatrix} 2.0 & 6.2 & -0.8 \\ 0 & 3.2 & 5.4 \\ 0 & 0 & -0.5 \end{bmatrix}$

 **24.** $B = \begin{bmatrix} 2.0 & 6.2 & -0.8 \\ 4.2 & 3.2 & 5.4 \\ 6.2 & 9.4 & 4.6 \end{bmatrix}$

Section 3.3

In Exercises 25–27, find the inverse of the matrix A algebraically. If A is singular, so state.

25. $A = \begin{bmatrix} 6 & -2 & 1 \\ 3 & -1 & 1 \\ 0 & 2 & 4 \end{bmatrix}$ **26.** $A = \begin{bmatrix} 9 & 7 & 8 \\ 3 & 3 & 3 \\ 6 & 10 & 11 \end{bmatrix}$

27. $A = \begin{bmatrix} 9 & 8 & 7 \\ 6 & 5 & 4 \\ -2 & 0 & 1 \end{bmatrix}$

In Exercises 28–30, write the system of equations as a matrix equation, $AX = B$, and solve.

28.
$$\begin{aligned} x + y + z &= 2 \\ 2x - y + z &= -6 \\ 4x - 2y + 3z &= -17 \end{aligned}$$

29.
$$\begin{aligned} x + y + z &= 14 \\ x - y + z &= 26 \\ x - y - z &= 2 \end{aligned}$$

30.
$$\begin{aligned} 3x + 2y + 3z &= 0 \\ 2x - 5y + z &= 0 \\ 4x + 2y + 3z &= 0 \end{aligned}$$

In Exercises 31–33, determine the solution by setting up and solving the matrix equation.

31.  **Return on Investment** The average annual return (over the 10-year period prior to June 30, 2002) of two mutual funds offered by Harbor Fund is shown in the following table.

	Average Annual Return
Growth Fund	5.84%
Capital Appreciation Fund	12.69%

Source: Harbor Fund account statement.

Suppose you have $2,000 to invest in these two accounts. Assuming that the accounts will earn the returns specified in the table over the next year, how much should you invest in each account if you want to earn 7 percent, 9 percent, or 11 percent?

32. **Grade Point Average** A student at Green River Community College must earn 90 credits to obtain an Associate of Arts degree. A student with a 3.1 existing GPA hopes to increase her cumulative GPA to 3.4. If she has 36 credits now and anticipates that she will be able to earn a 3.7 GPA on her remaining coursework, is it possible for her to increase her cumulative GPA to 3.4 by the time she obtains 90 credits?

33. **Floral Costs** A florist purchases her flowers from an online flower wholesaler. White daisies are $3.38 per bunch, football mums are $7.40 per bunch, and super blue purple statice is $4.25 per bunch. (**Source:** www.FlowerSales.com.) From these flowers she will make three types of jumbo bouquets:

	Type 1	Type 2	Type 3
Daisies	2 bunches	3 bunches	2 bunches
Mums	1 bunch	2 bunches	2 bunches
Statice	1 bunch	None	1 bunch

What is her flower cost for each type of bouquet? How much should she charge for each bouquet if her markup is 50 percent of her flower cost?

Exercises 3.4

Exercises 34–35 are derived from economic data for 2002 published by the U.S. Bureau of Economic Analysis.

34. **Industry Analysis: Three Industries** The *farms* (farm), *food and beverage and tobacco products* (food), and *food services and drinking places* (dining) industries are interrelated. Based on data from 2002, it requires $0.14 of farm products, $0.09 of food products, and $0.00 of dining products to produce $1 of farm products. Similarly, it requires $0.19 of farm products, $0.13 of food products, and $0.00 of dining products to produce $1 of food products. Likewise, it requires $0.01 of farm products, $0.18 of food products, and $0.01 of dining products to produce $1 of dining products. If the external demand for farm products is $53 billion, the external demand for food products is $422 billion, and the external demand for dining products is $379 billion, what output from each sector is required to meet the demand for farm, food, and dining products?

35. **Industry Analysis: Two Industries** The *machinery* (machine) and *motor vehicles, bodies and trailers, and parts* (auto) industries are interrelated. Based on data from 2002, it requires roughly $0.09 from the machine industry and $0.02 from the auto industry to produce $1 of machine products. Similarly, it requires $0.02 from the machine industry and $0.31 from the auto industry to produce $1 of auto products. If the external demand for machine products is $202 billion and the external demand for auto products is $293 billion, what output from each sector is required to meet the demand for machine and auto products?

Make It Real

PROJECT 3

What to do

1. Find out your current cumulative grade point average and your total number of graded credits.
2. Determine how many credits are required for your degree program.
3. Find two or three scholarships that require a minimum GPA.
4. If your cumulative GPA is below the required minimum for these scholarships, complete Step 5. Otherwise, complete Step 6.
5. Set up and solve a system of equations to determine how many credits with an "A" grade (4.0) a student with your cumulative GPA must earn in order to meet the GPA requirement. To do this, let x be the number of credits already earned and y be the number of credits left to be earned. Proceed to Step 7.
6. Set up and solve a system of equations to determine how many credits with a "C" grade (2.0) a student with your cumulative GPA can earn and still meet the GPA requirement. To do this, let x be the number of credits already earned and y be the number of credits left to be earned.
7. Suppose that one of your friends has the same GPA as you but has earned 12 credits fewer than you. Based upon your previous calculations, determine if it is mathematically possible for both you and your friend to meet the grade point average requirement for each of the scholarships by the time you have earned the required number of credits.

Chapter 4

Linear Programming

Businesses seek to maximize their profits while operating under budget, supply, labor, and space constraints. Determining which combination of variables will result in the maximum profit may be done through the use of linear programming. Although many factors affect a business's profitability, linear programming can help a business owner determine the "ideal" conditions for business success.

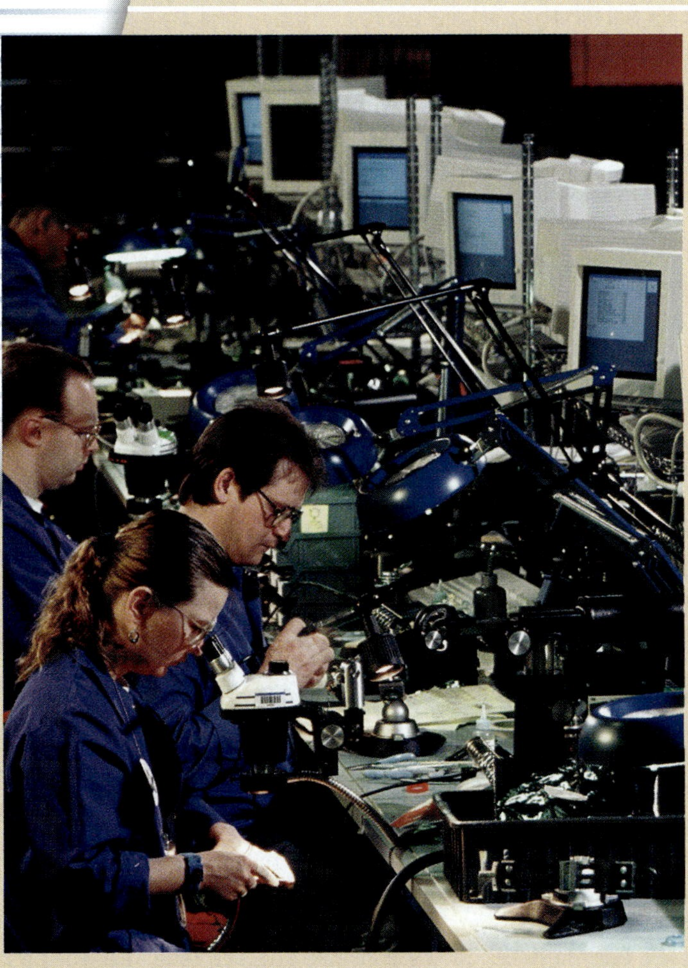

4.1 Graphing Linear Inequalities

- Graph linear inequalities
- Set up and solve systems of linear inequalities

GETTING STARTED Many students work multiple part-time jobs to finance their education. Often the jobs pay different wages and offer varying hours. Suppose that a student earns \$10.50 per hour delivering pizza and \$8.00 per hour working in a campus computer lab. If the student has only 30 hours per week to work and must earn \$252 during that period, how many hours must he spend at each job in order to meet his earnings goal? In this section, we will explain how linear inequalities may be used to answer this question. We will demonstrate how to graph linear inequalities and show that the solution region of a system of linear inequalities is the intersection of the graphs of the individual inequalities.

Linear Inequalities

In many real-life applications, we are interested in a range of possible solutions instead of a single solution. For example, when you prepare to buy a house, a lender will calculate the maximum amount of money it is willing to lend you; however, the lender doesn't require you to borrow the maximum amount. You may borrow any amount of money up to the maximum. Recall that in mathematics, we use inequalities to represent the range of possible solutions that meet the given criteria.

INEQUALITY NOTATION

$x \leq y$ is the set of all values of x less than or equal to y.

$x \geq y$ is the set of all values of x greater than or equal to y.

$x < y$ is the set of all values of x less than but not equal to y.

$x > y$ is the set of all values of x greater than but not equal to y.

The inequalities $x < y$ and $x > y$ are called **strict inequalities** because the two variables cannot ever be equal. Although strict inequalities have many useful applications, we will focus on the nonstrict inequalities in this chapter.

An easy way to keep track of the meaning of an inequality is to remember that the inequality sign always points toward the smaller number. Consider these everyday examples of inequalities:

You must be at least 16 years old to get a driver's license. $(16 \leq a)$ or $(a \geq 16)$.

You must be at least 21 years old to legally buy alcohol. $(21 \leq a)$ or $(a \geq 21)$.

The maximum fine for littering is \$200. $(200 \geq f)$ or $(f \leq 200)$.

Your carry-on bag must be no more than 22 inches long. $(22 \geq l)$ or $(l \leq 22)$.

A linear inequality looks like a linear equation with an inequality sign in the place of the equal sign. Recall that linear inequalities may be manipulated algebraically in the same way as linear equations, with one major exception: When we multiply or divide both sides of an inequality by a negative number, we must reverse the direction of the inequality sign. For example, if we multiply both sides of $-3x + 2y \leq 10$ by -1, we get $3x - 2y \geq -10$, not $3x - 2y \leq -10$.

Graphing Linear Inequalities

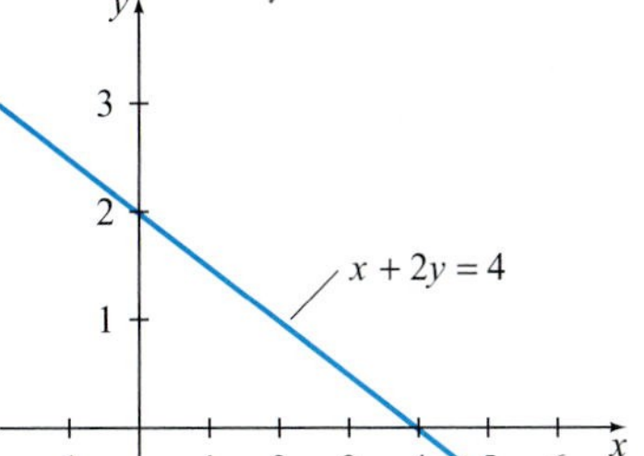

FIGURE 4.1

The graph of a linear inequality is a region bordered by a line called a **boundary line.** The **solution region** of a linear inequality is the set of all points (including the boundary line) that satisfy the inequality.

Consider the inequality $x + 2y \geq 4$ (see Figure 4.1). The boundary line of the solution region is $x + 2y = 4$, since the points satisfying this linear equation are on the border of the solution region.

We need to find all points (x, y) that satisfy the inequality. We know that all points on the line satisfy the inequality. Which points off the line satisfy the inequality? Let's pick a few points off the line (see Table 4.1) and test them to see if they satisfy the inequality. In order to satisfy the inequality, $x + 2y$ must be at least 4.

TABLE 4.1

x	y	$x + 2y$	In Solution Region?
-1	1	1	No
0	1	2	No
1	3	7	Yes
2	2	6	Yes
3	0	3	No
5	2	9	Yes
6	1	8	Yes

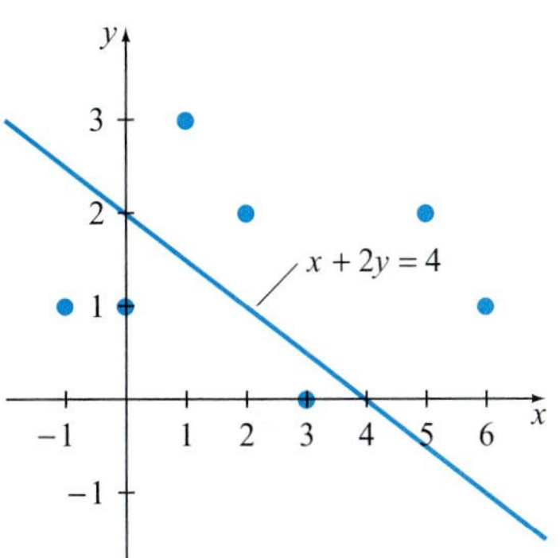

FIGURE 4.2

Graphically speaking, what do the four points in the solution region have in common (see Figure 4.2)?

They are all on the same side of the boundary line! In fact, all points on or above this boundary line satisfy the inequality. We represent this notion by shading the region above the boundary line, as shown in Figure 4.3.

Although we checked multiple points in this problem, we need to check only one point off the boundary line in order to determine which region to shade.

The linear inequality graphing process is summarized as follows.

FIGURE 4.3

HOW TO Linear Inequality Graphing Technique

To graph the solution region of the linear inequality $ax + by \leq c$ (or $ax + by \geq c$), do the following:

1. Graph the boundary line $ax + by = c$.
2. Select a point on one side of the line. [If the line doesn't pass through the origin, (0, 0) is an excellent choice for easy computations.]
3. Substitute the point into the linear inequality and simplify. If the simplified statement is true, the selected point and all other points on the same side of the line are in the solution region. If the simplified statement is false, all points on the opposite side of the line are in the solution region.
4. Shade the solution region.

EXAMPLE 1 Graphing the Solution Region of a Linear Inequality

Graph the solution region of the linear inequality $2x + y \leq 4$.

SOLUTION As shown in Chapter 1, the x-intercept is easily found by dividing the constant term by the coefficient on the x term.

$$x = \frac{4}{2} = 2$$

The point $(2, 0)$ is the x-intercept.

The y-intercept is found by dividing the constant term by the coefficient on the y term.

$$y = \frac{4}{1} = 4$$

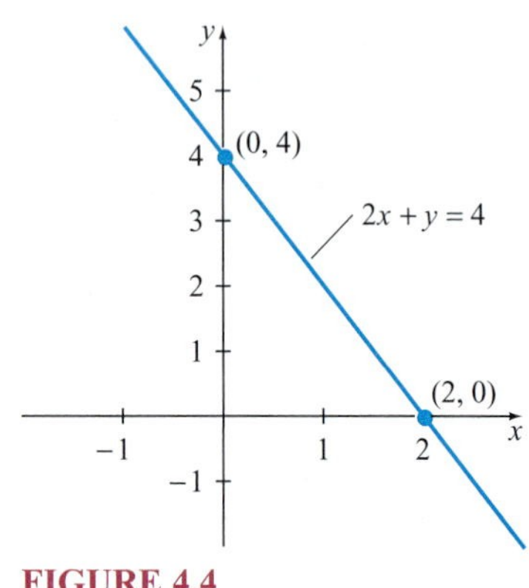

FIGURE 4.4

The point $(0, 4)$ is the y-intercept. We graph the x- and y-intercepts and then draw the line through the intercepts, as shown in Figure 4.4.

Next, we will pick the point (0, 0) to plug into the inequality.

$$2(0) + 0 \leq 4$$
$$0 \leq 4$$

The statement is true, so all points on the same side of the line as the origin are in the solution region. We shade the solution region (see Figure 4.5).

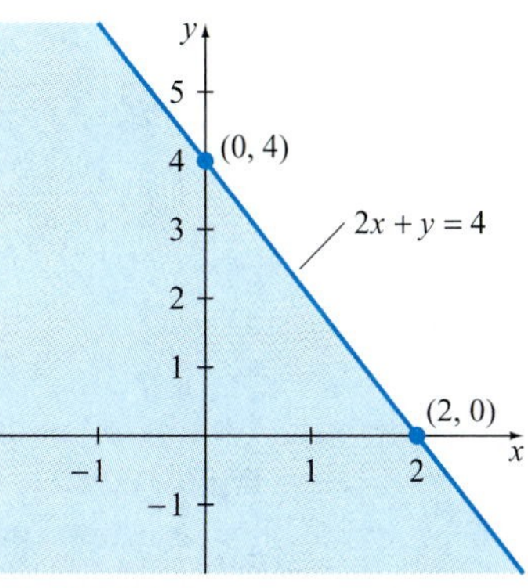

FIGURE 4.5

If you choose to convert lines from standard to slope-intercept form before graphing them, the following properties will help you to quickly identify the solution region without having to check a point.

SOLUTION REGION OF A LINEAR INEQUALITY

The solution region of a linear inequality $y \geq mx + b$ contains the line $y = mx + b$ and the shaded region **above** the line.

The solution region of a linear inequality $y \leq mx + b$ contains the line $y = mx + b$ and the shaded region **below** the line.

Graphing Systems of Linear Inequalities

Just as we can graph systems of linear equations, we can graph systems of linear inequalities. The solution region of a system of linear inequalities is the intersection of the solution regions of the individual inequalities. When we graph a solution region by hand, we will typically place arrows on the boundary lines to indicate which side of the lines satisfies the given inequality. Once all of the linear inequality graphs have been drawn, we will shade the region that has arrows pointing into the interior of the region from all sides.

EXAMPLE 2

Graphing the Solution Region of a System of Linear Inequalities

Graph the solution region of the system of linear inequalities.

$$3x + 2y \leq 5$$
$$x \geq 0$$
$$y \geq 0$$

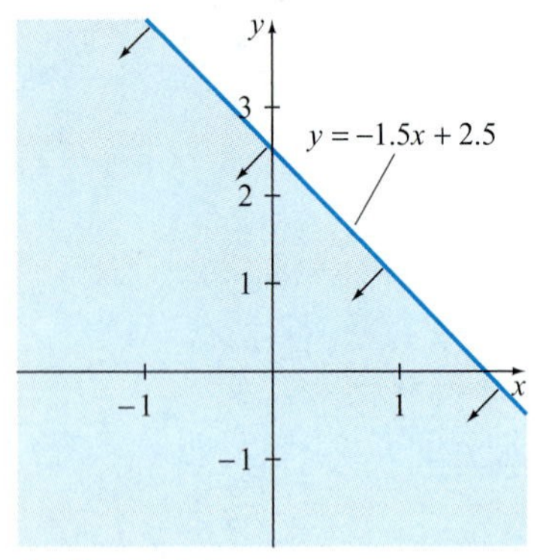

FIGURE 4.6

SOLUTION We first rewrite the linear inequality as a linear inequality in slope-intercept form.

$$3x + 2y \leq 5$$

$$2y \leq -3x + 5 \qquad \text{Subtract } 3x \text{ from both sides}$$

$$y \leq -\frac{3}{2}x + \frac{5}{2} \qquad \text{Divide both sides by 2}$$

$$y \leq -1.5x + 2.5 \qquad \text{Write as a decimal (optional)}$$

The boundary line is a line with slope -1.5 and y-intercept $(0, 2.5)$. Since y is less than or equal to the expression $-1.5x + 2.5$, we will shade the region below the line, as shown in Figure 4.6.

The next two inequalities ($x \geq 0$, $y \geq 0$) limit the solution region to positive values of x and y. The line $x = 0$ is the y axis. The line $y = 0$ is the x axis. Therefore, the solution region of the system of inequalities is the triangular region to the right of the line $x = 0$, above the line $y = 0$, and below the line $y = -1.5x + 2.5$ (see Figure 4.7).

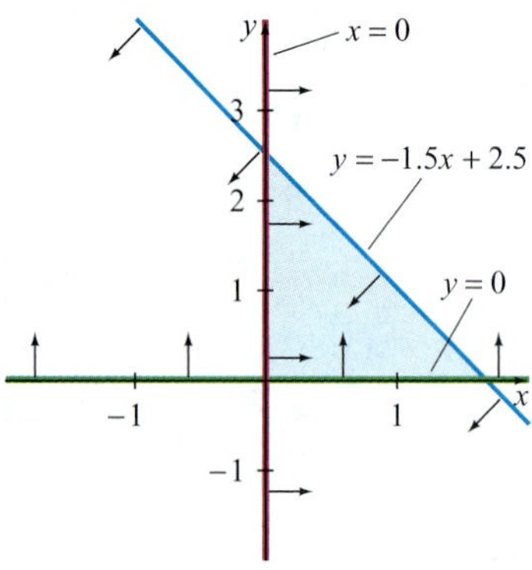

FIGURE 4.7

If it is possible to draw a circle around the solution region, the solution region is **bounded.** If no circle can be drawn that will enclose the entire solution region, the solution region is **unbounded.** The solution region in Example 2 was bounded. The solution region in Example 3 will be unbounded.

EXAMPLE 3

Graphing the Solution Region of a System of Linear Inequalities

Graph the solution region of the system of linear inequalities.

$$4x + y \geq 4$$

$$-x + y \geq 1$$

SOLUTION The x-intercept of the boundary line $4x + y = 4$ is $(1, 0)$, and the y-intercept is $(0, 4)$. Plugging in the point $(0, 0)$, we get

$$4(0) + (0) \geq 4$$

$$0 \geq 4$$

Since the statement is false, we graph $4x + y = 4$ and place arrows on the side of the line not containing the origin, as shown in Figure 4.8.

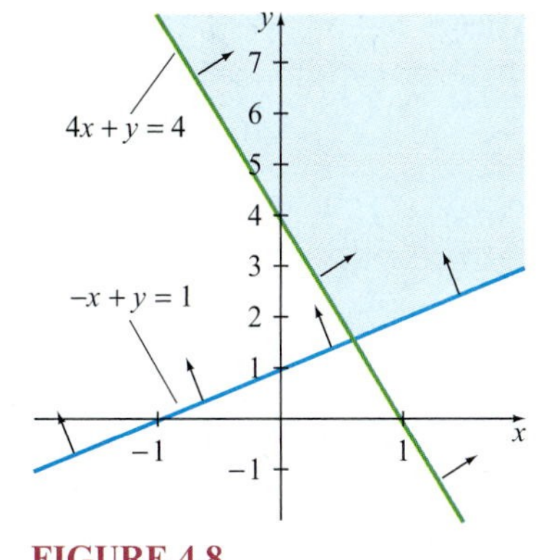

FIGURE 4.8

The x-intercept of the boundary line $-x + y = 1$ is $(-1, 0)$, and the y-intercept is $(0, 1)$. Plugging in the point $(0, 0)$, we get

$$-(0) + (0) \geq 1$$
$$0 \geq 1$$

Since the statement is false, we graph $-x + y = 1$ and place arrows on the side of the line not containing the origin, as shown in Figure 4.8.

The solution region is not bounded above the line $y = x + 1$ or above the line $y = -4x + 4$. Consequently, the shaded solution region is unbounded.

EXAMPLE 4

Graphing a System of Linear Inequalities with an Empty Solution Region

Graph the solution region of the system of linear inequalities.

$$-2x + 2y \geq 6$$
$$-x + y \leq 1$$

SOLUTION We will graph the boundary lines by first rewriting them in slope-intercept form.

Solving the first inequality for y, we get $y \geq x + 3$ and draw arrows pointing to the region above the line $y = x + 3$. Solving the second inequality for y, we get $y \leq x + 1$ and draw arrows pointing to the region below the line $y = x + 1$ (see Figure 4.9).

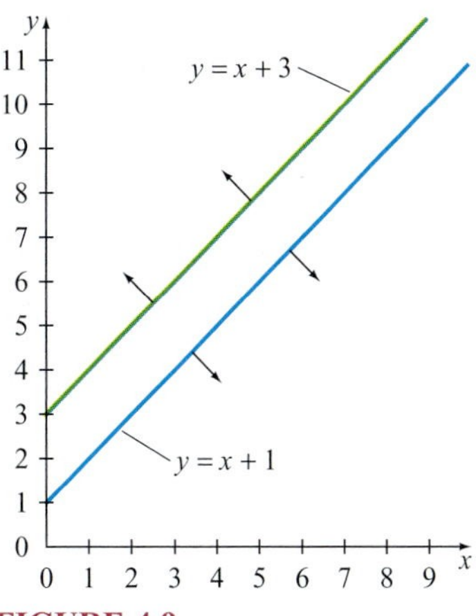

FIGURE 4.9

Because the lines have the same slope ($m = 1$), they are parallel. Consequently, the lines will never intersect. As seen in the graph, the two regions also will never intersect; they will always be separated by the region between the two lines. Therefore, this system of linear inequalities does not have a solution. That is, there is no ordered pair (x, y) that can satisfy both inequalities simultaneously.

The corners of a solution region are called **corner points.** To find the coordinates of each corner point, we solve the system of equations formed by the two intersecting boundary lines that create the corner.

EXAMPLE 5

Finding the Corner Points of a Solution Region

Graph the solution region of the system of linear inequalities and determine the coordinates of each corner point.

$$\begin{aligned} 2x + y &\leq 6 \\ -x + y &\geq 0 \\ x &\geq 0 \end{aligned}$$

SOLUTION We will graph the boundary line $2x + y = 6$ using its x- and y-intercepts. The x-intercept is $(3, 0)$, since $\frac{6}{2} = 3$. The function has y-intercept $(0, 6)$, since $\frac{6}{1} = 6$. Plugging in the point $(0, 0)$, we get

$$\begin{aligned} 2(0) + (0) &\leq 6 \\ 0 &\leq 6 \end{aligned}$$

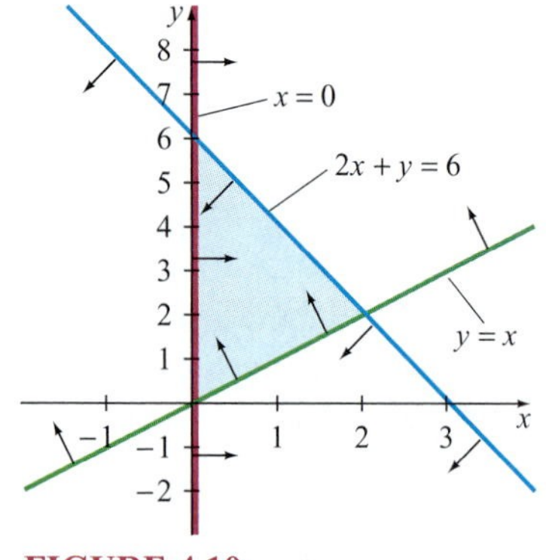

FIGURE 4.10

The statement is true, so we shade the side of the line containing the point $(0, 0)$. Solving the second inequality for y, we get $y \geq x$, and we shade the region above the line $y = x$. The third inequality restricts x to all positive values, so we shade the region to the right of the y axis.

The triangular region shown in Figure 4.10 is the intersection of the three regions and is the solution region of the system of linear inequalities. From the graph, it appears that the solution region has corner points at or near $(0, 0)$, $(0, 6)$, and $(2, 2)$. We will verify these results algebraically.

The coordinates of the first corner point may be found by solving the system of equations formed by the boundary lines that make up the corner.

$$\begin{aligned} -x + y &= 0 \\ x \quad &= 0 \end{aligned}$$

Adding the first equation to the second equation yields $y = 0$. Since the second equation tells us that $x = 0$, the coordinates of the first corner point are $(0, 0)$.

The coordinates of the second corner point may be found by solving the system of equations formed by the boundary lines that make up the corner.

$$\begin{aligned} 2x + y &= 6 \\ x \quad &= 0 \end{aligned}$$

The second equation tells us that $x = 0$. Substituting this value of x into the first equation yields

$$\begin{aligned} 2x + y &= 6 \\ 2(0) + y &= 6 \qquad \text{Substitute } x = 0 \\ y &= 6 \end{aligned}$$

The coordinates of the second corner point are $(0, 6)$.

The coordinates of the third corner point may be found by solving the system of equations formed by the boundary lines that make up the corner.

$$2x + y = 6$$
$$-x + y = 0$$

The second equation may be rewritten as $y = x$. Substituting this result into the first equation yields

$$2x + y = 6$$
$$2x + (x) = 6 \qquad \text{Since } y = x$$
$$3x = 6$$
$$x = 2$$

Since $x = y$, the coordinates of the corner point are $(2, 2)$.

It may have seemed superfluous to calculate the coordinates of the corner points algebraically in Example 5 when the coordinates were readily apparent from the graph of the solution region. Despite the apparent redundancy of the procedure, it is a necessary step. Example 6 illustrates the hazards of relying solely upon a graph for the coordinates of the corner points.

EXAMPLE 6

Finding the Corner Points of a Solution Region

Graph the solution region for the system of inequalities and determine the coordinates of the corner points.

$$x + y \leq 5$$
$$-5x + 5y \leq 6$$
$$y \geq 2$$

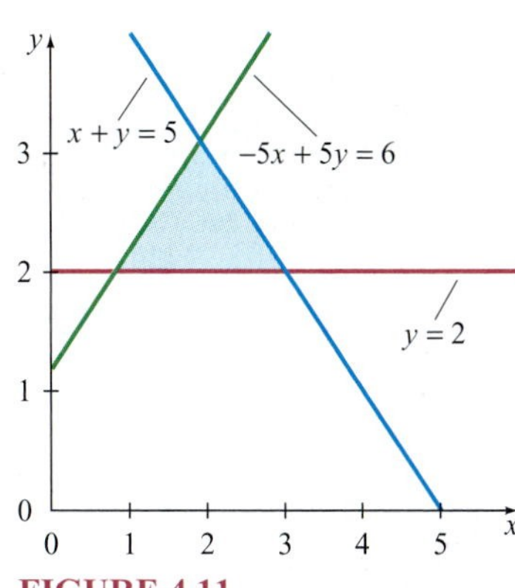

FIGURE 4.11

SOLUTION The graph of the solution region is shown in Figure 4.11.

From the graph, it appears that the corner points of the region are at or near $(2, 3)$, $(3, 2)$, and $(0.75, 2)$.

The coordinates of the first corner point may be found by solving the system of equations formed by the boundary lines that make up the corner.

$$x + y = 5$$
$$-5x + 5y = 6$$

Adding five times the first equation to the second equation yields

$$0x + 10y = 31$$
$$y = 3.1$$

Substituting this result back into the first equation yields

$$x + y = 5$$
$$x + (3.1) = 5$$
$$x = 1.9$$

The coordinates of the first corner point are $(1.9, 3.1)$. [From the graph, it looked as if the corner point was $(2, 3)$.]

The coordinates of the second corner point may be found by solving the system of equations formed by the boundary lines that make up the corner.

$$x + y = 5$$
$$y = 2$$

Since the second equation tells us that $y = 2$, we may substitute this value into the first equation.

$$x + y = 5$$
$$x + (2) = 5 \qquad \text{Since } y = 2$$
$$x = 3$$

The coordinates of the second corner point are $(3, 2)$. Unlike the first corner point, this result agrees with our graphical conclusion.

The coordinates of the third corner point may be found by solving the system of equations formed by the boundary lines that make up the corner.

$$-5x + 5y = 6$$
$$y = 2$$

Since the second equation tells us that $y = 2$, we may substitute this value into the first equation.

$$-5x + 5y = 6$$
$$-5x + 5(2) = 6 \qquad \text{Since } y = 2$$
$$-5x + 10 = 6$$
$$-5x = -4$$
$$x = 0.8$$

The coordinates of the third corner point are $(0.8, 2)$. [From the graph, it looked as if the corner point was $(0.75, 2)$.]

Using Technology to Graph Linear Inequalities

Many graphing calculators can draw the graphs of linear inequalities, as detailed in the following Technology Tip. Often, however, it is quicker to draw the graphs by hand.

TECHNOLOGY TIP

Graphing a System of Linear Inequalities

1. Enter the linear equations associated with each inequality by using the [Y=] editor. (We will use the system $y \leq -3x + 6$ and $y \geq 2x + 4$ for this example.)

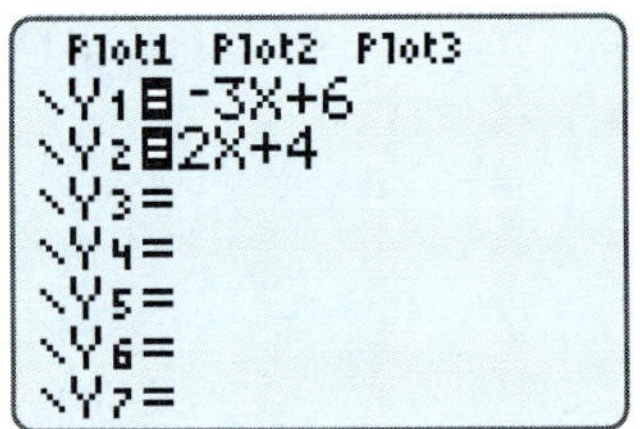

(Continued)

2. Move the cursor to the \ to the left of Y1 and press ENTER repeatedly. This will cycle through several graphing options. We want to shade the region below the line, so we will pick the lower triangular option.

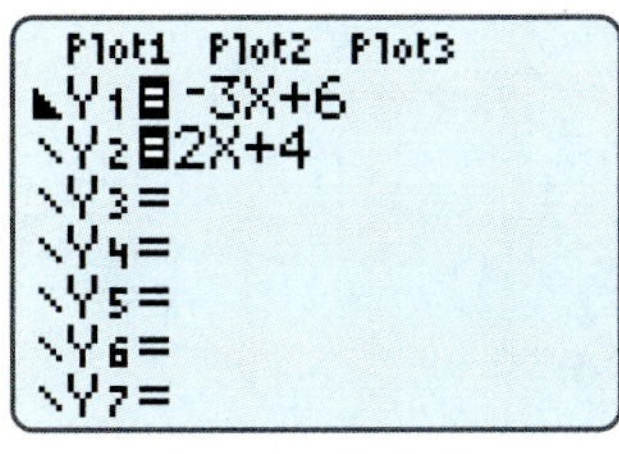

3. Move the cursor to the \ to the left of Y2 and press ENTER repeatedly. We want to shade the region above the line, so we will pick the upper triangular option.

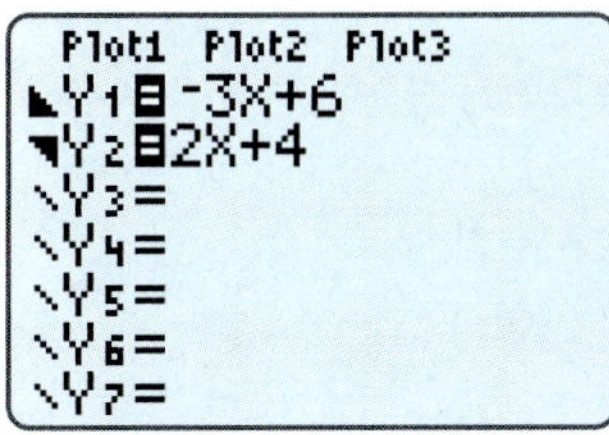

4. Press GRAPH to draw each of the shaded regions. The region with the crisscross pattern is the solution region.

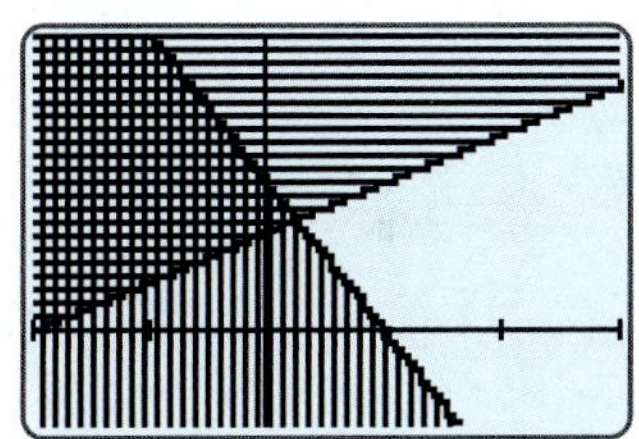

Linear Inequality Applications

Many real-life problems are subject to multiple constraints, such as budget, staffing, resources, and so on. Yet even when subjected to these constraints, the problems often have multiple solutions. As will be shown in Example 7, the linear inequality graphing techniques introduced earlier in this section may often be used to find solutions to real-life problems.

EXAMPLE 7

Using a Linear System of Inequalities to Find the Ideal Work Schedule

A student earns \$8.00 per hour working in a campus computer lab and \$10.50 per hour delivering pizza. If he has only 30 hours per week to work and must earn at least \$252 during that period, how many hours can he spend at each job in order to earn at least \$252?

SOLUTION Let c be the number of hours the student works in the computer lab and p be the number of hours he works delivering pizza. He can work at most 30 hours. This is represented by the inequality

$$c + p \leq 30 \qquad \text{The maximum number of work hours is 30}$$

The amount he earns working in the lab is $8.00c$, and the amount of money he earns delivering pizza is $10.50p$. His total income must be at least \$252. That is,

$$8c + 10.5p \geq 252 \qquad \text{The minimum amount of income is \$252}$$

Solving the inequalities for p in terms of c, we get the following system of inequalities and its associated graph (Figure 4.12). (We add the restrictions $p \geq 0$ and $c \geq 0$, since it doesn't make sense to work a negative number of hours at either job.)

$$p \leq -c + 30$$

$$p \geq -\frac{16}{21}c + 24 \qquad \text{(In decimal form, } p \geq -0.7619c + 24 \text{ approximately)}$$

$$p \geq 0, c \geq 0$$

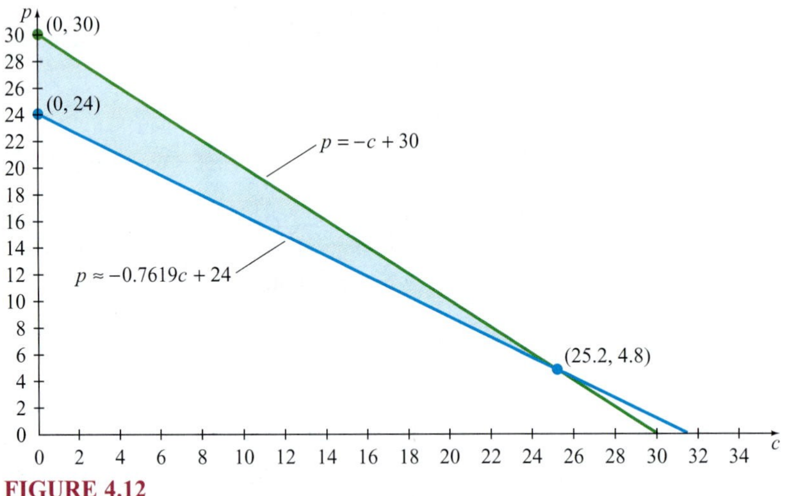

FIGURE 4.12

Every point of the solution region represents a combination of hours at the two jobs that will result in earnings of at least \$252.

The corner points of the solution region are (0, 24), (0, 30), and (25.2, 4.8). (The last point is found by calculating the intersection of the two boundary lines.) Let's calculate the student's weekly earnings at the corner points and some of the other points in the region (see Table 4.2).

TABLE 4.2

	Lab Hours	Pizza Hours	Weekly Earnings
Corner point	0	24	\$252.00
Corner point	0	30	\$315.00
Corner point	25.2	4.8	\$252.00
Interior point	5	25	\$302.50
Interior point	10	20	\$290.00
Interior point	15	14	\$267.00

The weekly earnings vary; however, in every case the number of work hours is less than or equal to 30 hours and the earnings are greater than or equal to \$252.

4.1 Summary

In this section, you learned how to graph linear inequalities. You discovered that the solution to a system of linear inequalities is the intersection of the graphs of the solution regions of the individual inequalities.

4.1 Exercises

In Exercises 1–10, graph the solution region of the linear inequality. Then use the graph to determine if point P is in the solution region.

1. $2x + y \le 6$; $P = (2, 4)$

2. $4x + y \le 0$; $P = (1, 1)$

3. $x + 5y \le 10$; $P = (0, 0)$

4. $5x + 6y \le 30$; $P = (0, 5)$

5. $-2x + 4y \ge -2$; $P = (1, 2)$

6. $x - y \le 10$; $P = (5, -5)$

7. $5x - 4y \le 0$; $P = (1, 0)$

8. $-3x - 3y \le 9$; $P = (2, -1)$

9. $2x - y \ge 8$; $P = (-3, 2)$

10. $7x - 6y \ge 12$; $P = (0, -1)$

In Exercises 11–25, graph the solution region of the system of linear equations. If there is no solution, explain why.

11. $-4x + y \ge 2$
$-2x + y \ge 1$
$x \le 0$

12. $-5x + y \ge 0$
$2x + y \le 4$
$y \ge 0$

13. $-2x + 6y \le 8$
$4x - 12y \le -6$

14. $10x - y \ge 12$
$9x - 2y \ge 2$

15. $3x - 2y \le 4$
$11x - 20y \ge 2$

16. $9x - 6y \le 0$
$4x + 5y \le 23$

17. $x - y \le -5$
$9x + y \le 25$

18. $2x + 5y \le 2$
$3x - 5y \le 3$

19. $6x + 2y \le 10$
$-x - 2y \ge -5$
$x \ge 0$
$y \ge 0$

20. $x - y \ge 3$
$6x + 7y \le 44$
$6x - 7y \le 16$

21. $2x - 4y \ge 16$
$9x + y \le -4$
$-3x + 6y \le -24$

22. $2x - 2y \ge 0$
$3x + y \le 4$
$5x - y \ge 5$

23. $8x - y \ge 3$
$x + 2y \le 11$
$9x + y \le 14$

24. $-4x + y \ge 2$
$-2x + y \ge 1$
$y \ge 1$
$x \le 1$

25. $-5x + y \ge 0$
$2x + y \le 4$
$y \le 1$
$x \le 1$

In Exercises 26–30, set up the system of linear inequalities that can be used to solve the problem. Then graph the system of equations and solve the problem.

26. **Nutritional Content** A 32-gram serving of Skippy® Creamy Peanut Butter contains 150 milligrams of sodium and 17 grams of fat. A 56-gram serving of Bumble Bee® Chunk Light Tuna in Water contains 250 milligrams of sodium and 0.5 gram of fat. (**Source:** Product labeling.) Some health professionals advise that a person on a 2500-calorie diet should consume no more than 2400 mg of sodium and 80 grams of fat. Graph the region showing all possible serving combinations of peanut butter and tuna that a person could eat and still meet the dietary guidelines.

27. **Nutritional Content** A Nature Valley® Strawberry Yogurt Chewy Granola Bar contains 130 milligrams of sodium and 3.5 grams of fat. A Nature's Choice® Multigrain Strawberry Cereal Bar contains 65 milligrams of sodium and 1.5 grams of fat. (**Source:** Product labeling.) Some health professionals advise that a person on a 2500-calorie diet should consume no more than 2400 mg of sodium and 80 grams of fat. Graph the region showing all possible serving combinations of granola bars and cereal bars that a person could eat and still meet the dietary guidelines.

28. **Student Wages** A student earns \$15.00 per hour designing web pages and \$9.00 per hour supervising a campus tutoring center. She has at most 30 hours per week to work, and she needs to earn at least \$300. Graph the region showing all possible work-hour allocations that meet her time and income requirements.

29. **Wages** A salaried employee earns \$900 per week managing a copy center. He is required to work a minimum of 35 hours but no more than 45 hours weekly. As a side business, he earns \$25 per hour designing brochures for local business clients. In order to maintain his standard of living, he must earn \$1100 per week. In order to maintain his quality of life, he limits his workload to 50 hours per week. Given that he has no control over the number of hours he has to work managing the copy center, will he be able to consistently meet his workload and income goals? Explain.

30. **Commodity Prices** Today's Market Prices (www.todaymarket.com) is a daily fruit and vegetable wholesale market price service. Produce retailers who subscribe to the service can use the wholesale prices to aid them in setting retail prices for the fruits and vegetables they sell.

A 25-pound carton of peaches holds 60 medium peaches or 70 small peaches. In August 2002, the wholesale price for local peaches in Los Angeles was \$9.00 per carton for medium peaches and \$10.00 per carton for small peaches. (Source: Today's Market Prices.) A fruit vendor has budgeted up to \$100 to spend on peaches. He estimates that weekly demand for peaches is at least 420 peaches but no more than 630 peaches. He wants to buy enough peaches to meet the minimum estimated demand but no more than the maximum estimated demand. Graph the region showing which small- and medium-size peach carton combinations meet his demand and budget restrictions.

Exercises 31–40 are intended to challenge your understanding of the graphs of linear inequalities.

31. Graph the solution region of the system of linear inequalities and identify the coordinates of the corner points.

$$\begin{aligned} 2x + 3y &\leq 6 \\ -2x + 4y &\geq 4 \\ -5x + y &\leq 15 \\ x &\leq 5 \\ y &\geq 2 \end{aligned}$$

32. Graph the solution region of the system of linear inequalities and identify the coordinates of the corner points.

$$\begin{aligned} -2x + y &\leq 4 \\ 7x + 2y &\geq 8 \\ x &\leq 0 \end{aligned}$$

33. Graph the solution region of the system of linear inequalities and identify the coordinates of the corner points.

$$\begin{aligned} -x + y &\leq 0 \\ -x - y &\geq -4 \\ y &\geq 2 \end{aligned}$$

34. Write a system of inequalities whose solution region has the corner points $(0, 0)$, $(1, 3)$, $(3, 5)$, and $(2, 1)$.

35. Write a system of inequalities whose solution region has the corner points $(1, 1)$, $(1, 3)$, $(5, 3)$, and $(2, 1)$.

36. Write a system of inequalities whose *unbounded* solution region has the corner points $(0, 5)$, $(2, 1)$, and $(5, 0)$.

37. Write a system of inequalities whose *unbounded* solution region has the corner points $(0, 5)$, $(4, 4)$, and $(5, 0)$.

38. A student concludes that the corner points of a solution region defined by a system of linear inequalities are $(0, 0)$, $(1, 1)$, $(0, 2)$, and $(2, 2)$. After looking at the graph of the region, the instructor immediately concludes that the student is incorrect. How did the instructor know?

39. Is it possible to have a bounded solution region with exactly one corner point? If so, give a system of inequalities whose solution region is bounded and has exactly one corner point.

40. Is it possible to have an *unbounded* solution region with exactly one corner point? If so, give a system of inequalities whose solution region is unbounded and has exactly one corner point.

4.2 Solving Linear Programming Problems Graphically

- Determine the feasible region of a linear programming problem
- Solve linear and integer programming problems in two variables

GETTING STARTED Many products, such as printer ink, are sold to business customers at a discount if large quantities are ordered. Profitable businesses want to minimize their supply costs yet have sufficient ink on hand to fulfill their printing requirements. How much ink should they order? The process of acquiring, producing, and distributing supplies can often be made more efficient by setting up and solving systems of linear inequalities.

In this section, we show how a mathematical method called **linear programming** can help businesses determine the most cost-effective way to manage their resources. We will demonstrate how linear programming is used to optimize an objective function subject to a set of linear constraints. We will also reveal how to find the whole-number solution of an integer programming problem. We will begin our discussion with the following set of definitions.

LINEAR PROGRAMMING PROBLEM

A linear equation $z = ax + by$, called an **objective function,** may be maximized or minimized subject to a set of linear **constraints** of the form

$$cx + dy \leq f \quad \text{or} \quad cx + dy \geq f$$

where x and y are variables (called **decision variables**) and a, b, c, d and f are real numbers. A problem consisting of an objective function and a set of linear constraints is called a **linear programming problem.** The values of x and y that optimize (maximize or minimize) the value of the objective function are called the **optimal solution.** A linear programming problem with the additional constraint that x and y are integers is called an **integer programming problem.**

In Section 4.1, we gave the example of a student earning \$10.50 an hour delivering pizza and \$8.00 an hour working in a campus computer lab. He had only 30 hours per week to work, and he had to earn at least \$252 during that period. We let p be the number of hours he spent delivering pizza and c be the number of hours he spent working in the computer lab. We had the constraints

$$c + p \leq 30 \qquad \text{He can work at most 30 hours}$$

$$8c + 10.5p \geq 252 \qquad \text{He must earn at least \$252}$$

$$p \geq 0, c \geq 0 \qquad \text{He must work a nonnegative number of hours at each job}$$

which we rewrote as

$$p \leq -c + 30$$

$$p \geq -\frac{16}{21}c + 24$$

$$p \geq 0, c \geq 0$$

From this scenario, we can set up linear programming problems to address each of the following questions:

1. What is the largest amount of money he can earn?

Objective function: Maximize $z = 8c + 10.5p$ Total amount earned

$$\text{Subject to} \begin{cases} p \leq -c + 30 \\ p \geq -\frac{16}{21}c + 24 \\ p \geq 0, c \geq 0 \end{cases}$$

2. What is the least number of hours he can work?

Objective function: Minimize $z = c + p$ Total hours worked

$$\text{Subject to} \begin{cases} p \leq -c + 30 \\ p \geq -\frac{16}{21}c + 24 \\ p \geq 0, c \geq 0 \end{cases}$$

3. What is the maximum number of hours he can work in the computer lab?

Objective function: Maximize $z = c$ Total computer lab hours

$$\text{Subject to} \begin{cases} p \leq -c + 30 \\ p \geq -\frac{16}{21}c + 24 \\ p \geq 0, c \geq 0 \end{cases}$$

Although each of the three objective functions has the same constraints, the optimal solution to each linear programming problem will differ based upon the objective function. However, all solutions will lie within the solution region of the system of constraints. In the context of linear programming, we call the solution region of the system of constraints the **feasible region** and the points within the region **feasible points.** Which of all the feasible points will optimize each of the objective functions? Testing all of the points in the feasible region would be an impossible task! Fortunately, we don't have to. The Fundamental Theorem of Linear Programming limits the number of points we have to test.

FUNDAMENTAL THEOREM OF LINEAR PROGRAMMING

1. If the solution to a linear programming problem exists, it will occur at a corner point.
2. If two adjacent corner points are optimal solutions, then all points on the line segment between them are also optimal solutions.
3. Linear programming problems with bounded feasible regions will always have optimal solutions.
4. Linear programming problems with unbounded feasible regions may or may not have optimal solutions.

Recall that the graph of the feasible region for the computer lab and pizza delivery problem had the corner points $(0, 24)$, $(0, 30)$, and $(25.2, 4.8)$, as shown in Figure 4.13.

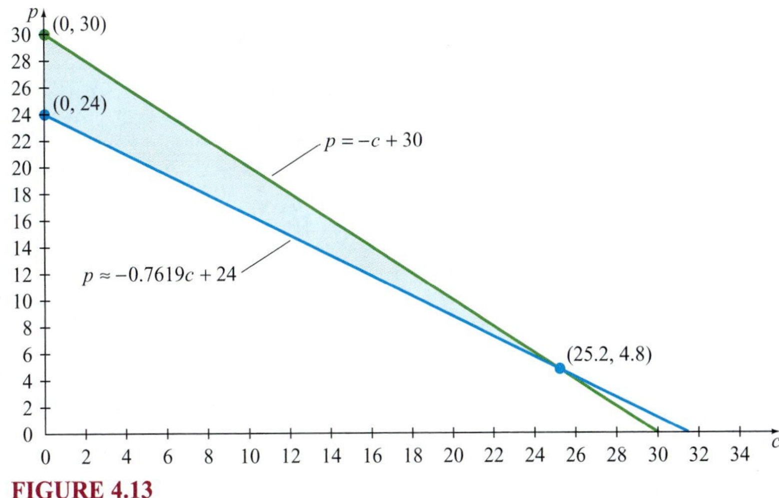

FIGURE 4.13

The solution to each one of the linear programming problems will occur at one of these points.

1. What is the largest amount of money he can earn?

$$\text{Maximize} \quad z = 8c + 10.5p.$$

We evaluate the objective function at each corner point (see Table 4.3).

TABLE 4.3

Corner Point		Objective Function
Computer Lab Hours (c)	Pizza Delivery Hours (p)	Total Earnings (in dollars) ($z = 8c + 10.5p$)
0	24	252
0	30	315
25.2	4.8	252

The maximum value of z occurs at corner point (0, 30). The optimal solution is $c = 0$ and $p = 30$. To maximize his earnings, he should work 0 hours in the lab and 30 hours delivering pizza.

2. What is the least number of hours he can work?

$$\text{Minimize} \quad z = c + p.$$

We evaluate the objective function at each corner point (see Table 4.4).

TABLE 4.4

Corner Point		Objective Function
Computer Lab Hours (c)	Pizza Delivery Hours (p)	Total Work Hours ($z = c + p$)
0	24	24
0	30	30
25.2	4.8	30

The minimum value of z occurs at corner point (0, 24). The optimal solution is $c = 0$ and $p = 24$. To minimize his work hours, he should work 0 hours in the lab and 24 hours delivering pizza.

3. What is the maximum number of hours he can work in the computer lab?

$$\text{Maximize} \quad z = c.$$

We evaluate the objective function at each corner point (see Table 4.5).

TABLE 4.5

Corner Point		Objective Function
Computer Lab Hours (c)	Pizza Delivery Hours (p)	Lab Work Hours ($z = c$)
0	24	0
0	30	0
25.2	4.8	25.2

The maximum value of z occurs at corner point (25.2, 4.8). The optimal solution is $c = 25.2$ and $p = 4.8$. The maximum amount of time he can work in the computer lab is 25.2 hours. He will still have to work 4.8 hours delivering pizza to reach his earnings goal.

Why do we have to check only the corner points of the feasible region? We'll address this question by considering a "family" of objective functions. Suppose we are asked to maximize the objective function $z = 3x + y$ subject to the following constraints:

$$\begin{aligned} 4x + y &\leq 12 \\ 2x + y &\leq 8 \\ x &\geq 0 \\ y &\geq 0 \end{aligned}$$

The graph of the feasible region and the coordinates of the corner points are shown in Figure 4.14.

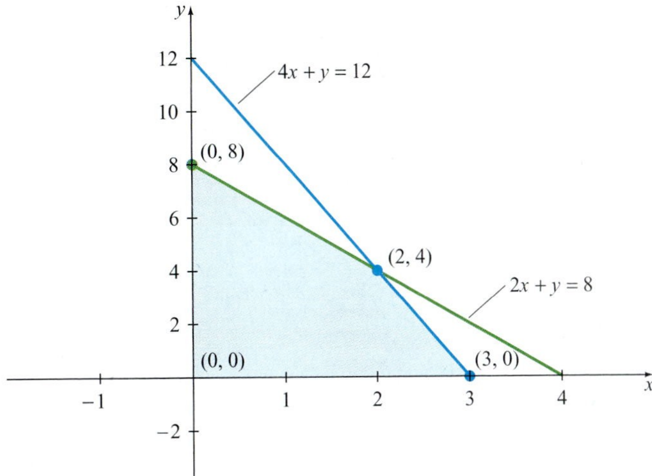

FIGURE 4.14

If we set the variable z in the objective function $z = 3x + y$ to a fixed value c, then the graph of the line $3x + y = c$ will pass through all points (x, y) that satisfy the equation $3x + y = c$. If we repeat this for different values of c, we will end up with a "family" of objective function lines. The lines will be parallel with different y-intercepts. (In Figure 4.15, we set z equal to the following constant values: 2, 4, 6, 8, 10, 12.)

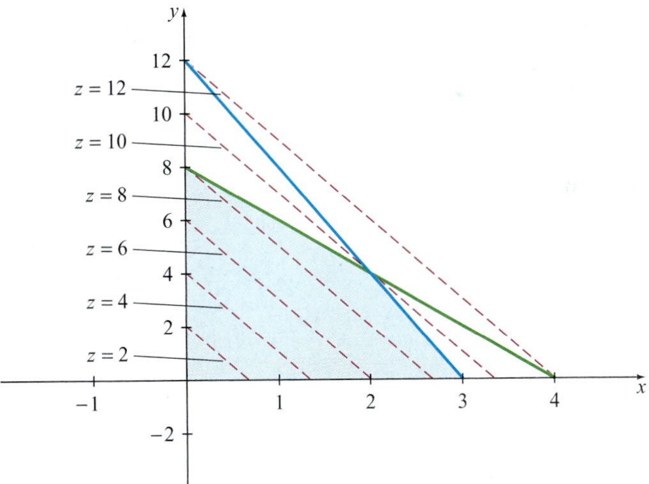

FIGURE 4.15

Recall that we want to maximize the objective function subject to the constraints. Observe that although $z = 12$ is the largest value of z shown in the figure, the line $3x + y = 12$ does not intersect the feasible region. Consequently, no feasible solution lies on the line $3x + y = 12$.

The first line above the feasible region that intersects the feasible region is the line $3x + y = 10$. This line intersects the feasible region at a single point: the corner point $(2, 4)$. At this corner point, $z = 10$. All other objective function lines that cross the feasible region take on values of z less than 10. Therefore, the objective function $z = 3x + y$ has an optimal solution at $(2, 4)$. At this point, the objective function takes on its maximum value: $z = 10$. This value of the objective function is referred to as the **optimal value** for the linear programming problem. Regardless of the objective function, the maximum (or minimum) value of the objective function will occur at a corner point of the feasible region. An argument similar to that given here can be made for any objective function and any feasible region.

The process of solving linear programming problems graphically is summarized in the following box.

HOW TO Graphical Method for Solving Linear Programming (LP) Problems

1. Graph the feasible region determined by the constraints.
2. Find the corner points of the feasible region.
3. Find the value of the objective function at each of the corner points.
4. If the feasible region is bounded, the maximum or minimum value of the objective function will occur at one of the corner points.
5. If the feasible region is an unbounded region in the first quadrant and the coefficients of the objective function are positive, then the objective function has a minimum value at a corner point. The objective function will not have a maximum value.

EXAMPLE 1 Solving a Linear Programming Problem Graphically

Solve the linear programming problem:

$$\text{Maximize} \quad z = 6x + 2y$$

$$\text{Subject to} \quad \begin{cases} -3x + y \geq 2 \\ x + y \leq 10 \\ x \geq 0 \\ y \geq 0 \end{cases}$$

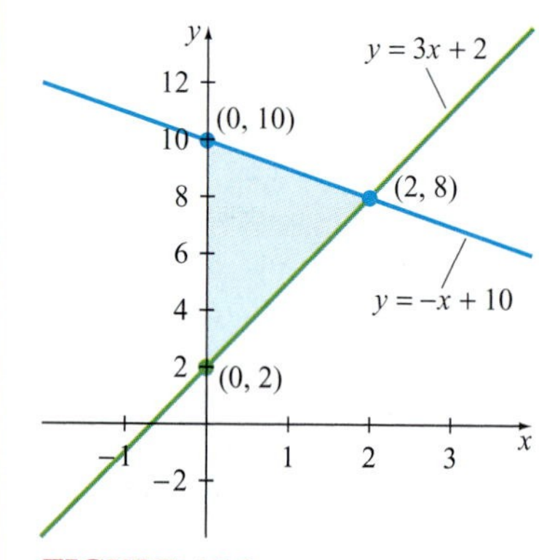

FIGURE 4.16

SOLUTION We begin by solving each inequality for y.

$$-3x + y \geq 2 \qquad\qquad x + y \leq 10$$

$$y \geq 3x + 2 \qquad\qquad y \leq -x + 10$$

Graphing the feasible region yields the graph shown in Figure 4.16.

The corner points occur where the boundary lines intersect. The corner points of the feasible region are (0, 2), (0, 10), and (2, 8). Substituting each of these points into the objective function, $z = 6x + 2y$, we get the results in Table 4.6.

TABLE 4.6

Corner Point		Objective Function
x	y	$z = 6x + 2y$
0	2	4
0	10	20
2	8	28

The objective function $z = 6x + 2y$ is maximized when $x = 2$ and $y = 8$. At that point, $z = 28$.

EXAMPLE 2

Solving an LP Problem with an Unbounded Feasible Region

Solve the linear programming problem:

$$\text{Minimize} \quad z = 2x + 5y$$

$$\text{Subject to} \quad \begin{cases} 4x + y \geq 4 \\ -x + y \geq 1 \\ x \geq 0 \\ y \geq 0 \end{cases}$$

SOLUTION The objective function is

$$z = 2x + 5y$$

The constraints are

$$4x + y \geq 4 \qquad x \geq 0$$
$$-x + y \geq 1 \qquad y \geq 0$$

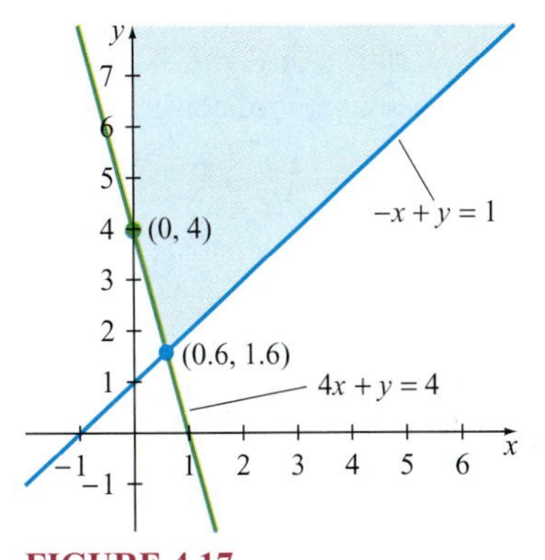

FIGURE 4.17

We draw each boundary line and shade the feasible region, as shown in Figure 4.17.

The region is unbounded and has two corner points. Since the coefficients of the objective function are both positive, the objective function will have a minimum. The first corner point of the feasible region is (0, 4), the y-intercept of one of the constraints. We determine the coordinates of the second corner point by finding the intersection point of the boundary lines, $y = x + 1$ and $y = -4x + 4$.

$$-4x + 4 = x + 1 \qquad \text{Solve the system of equations by the substitution method}$$
$$-5x = -3$$
$$x = 0.6$$
$$y = (0.6) + 1 \qquad \text{Substitute } x = 0.6 \text{ into } y = x + 1$$
$$y = 1.6$$

The second corner point is (0.6, 1.6). Substituting each of the points into the objective function, $z = 2x + 5y$, we obtain the values in Table 4.7.

TABLE 4.7

Corner Point		Objective Function
x	y	$z = 2x + 5y$
0	4	20
0.6	1.6	9.2

The objective function has a minimum at (0.6, 1.6). At that point, $z = 9.2$.

In Example 2, we had an unbounded feasible region. We were able to find a solution because we were asked to minimize the objective function $z = 2x + 5y$. If we had been asked to maximize the objective function, the problem would have had no solution. For a linear programming problem with an unbounded feasible region and an objective function with positive coefficients, no matter what "optimal solution" we pick, we will always be able to find a point (x, y) in the feasible region that yields a greater "optimal value."

EXAMPLE 3

Solving a Linear Programming Problem Graphically

Solve the linear programming problem:

$$\text{Minimize} \quad z = -5x + 3y$$

$$\text{Subject to} \quad \begin{cases} 6x + y \geq 6 \\ -2x + y \geq 1 \\ x \leq 2 \\ y \geq 1 \\ x \geq 0 \end{cases}$$

SOLUTION After graphing each of the five boundary lines, we will use arrows to show which side of the boundary line will be shaded, as shown in Figure 4.18. This technique is especially helpful when working with a large number of constraints.

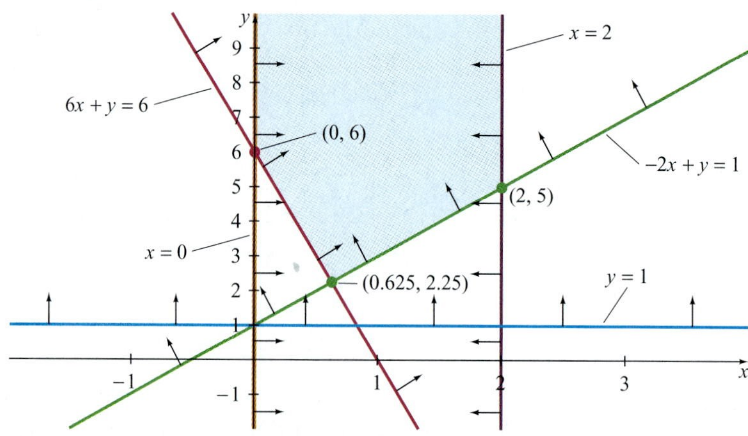

FIGURE 4.18

We shade the region that has boundary lines with all arrows pointing to the interior of the region. Notice that the line $y = 1$ does not form a boundary line of the feasible region. This is okay so long as the feasible region is on the appropriate side of $y = 1$. Since the arrows on the line $y = 1$ point toward the side that contains the feasible region, the constraint $y \geq 1$ is satisfied.

The first corner point $(0, 6)$ is easily determined, since it is the y-intercept of the boundary line $6x + y = 6$. The second corner point occurs at the intersection of $6x + y = 6$ and $-2x + y = 1$. We must solve the system of equations

$$6x + y = 6$$
$$-2x + y = 1$$

Subtracting the second equation from the first equation yields

$$8x = 5$$
$$x = 0.625$$

To determine the value of y, we substitute the x value back into the equation $-2x + y = 1$.

$$-2x + y = 1$$
$$-2(0.625) + y = 1 \qquad \text{Since } x = 0.625$$
$$-1.25 + y = 1$$
$$y = 2.25$$

The second corner point is $(0.625, 2.25)$.

The third point occurs at the intersection of $x = 2$ and $-2x + y = 1$. The x coordinate of the corner point is $x = 2$. To determine the y coordinate, we substitute this result into $-2x + y = 1$.

$$-2x + y = 1$$
$$-2(2) + y = 1 \qquad \text{Since } x = 2$$
$$-4 + y = 1$$
$$y = 5$$

The third corner point is $(2, 5)$.

With the corner points identified, we are ready to evaluate the objective function $z = -5x + 3y$ at each corner point, as shown in Table 4.8.

TABLE 4.8

Corner Point		Objective Function
x	y	$z = -5x + 3y$
0	6	18
0.625	2.25	3.625
2	5	5

Since we are looking for the minimum value of the objective function, the optimal solution is $(0.625, 2.25)$ and the optimal value is 3.625.

Real-Life Applications

As shown in Examples 4 and 5, many real-life problems can be analyzed using the techniques of linear programming.

EXAMPLE 4 Using Linear Programming to Do Investment Analysis

Table 4.9 shows the average annual rate of return on two TIAA-CREF investment accounts over a 10-year period.

TABLE 4.9

As of 6/30/04	
CREF Variable Annuity Accounts	10-Year Average
Bond Market	7.15%
Social Choice	10.31%

Source: www.tiaa-cref.com.

An investor wants to invest at least \$3000 in the Bond Market and Social Choice accounts. He assumes that he will be able to get a return equal to the 10-year average, and he wants the total return on his investment to be at least 9 percent. He assigns each share in the Bond Market account a risk rating of 2 and each share in the Social Choice account a risk rating of 4. The approximate share price at the end of June 2004 was \$74 per share for the Bond Market account and \$102 per share for the Social Choice account. He will use these prices in his analysis. How many shares of each account should he buy in order to minimize his overall risk? [Note that fractions of shares may be purchased. Also, to make computations easier, round each percentage to the nearest whole-number percent (i.e., 10.31 percent = 10 percent).]

SOLUTION Let x be the number of shares in the Bond Market account and y be the number of shares in the Social Choice account. The rounded rate of return for the Bond Market account is 7 percent, and that for the Social Choice account is 10 percent. The share price for the Bond Market account is \$74, and that for the Social Choice account is \$102. Each Bond Market share has a risk rating of 2, and each Social Choice share has a risk rating of 4. We want to minimize the overall risk. That is, we want to minimize $z = 2x + 4y$.

The amount of money invested in each account is equal to the share price times the number of shares. The amount invested is $74x$ for the Bond Market account and $102y$ for the Social Choice account. The total amount invested is given by the equation

$$74x + 102y \geq 3000$$ The total amount invested is at least \$3000

The dollar amount of the return on the investment is equal to the product of the rate and the amount invested. The return on the Bond Market account is $(0.07)(74x) = 5.18x$, and the return on the Social Choice account is $(0.10)(102y) = 10.20y$. Since we want to earn at least 9 percent on the total amount of money invested $(74x + 102y)$, the dollar amount of the minimum combined return is $(0.09)(74x + 102y) = 6.66x + 9.18y$. The combined return is given by the equation

$$5.18x + 10.20y \geq 6.66x + 9.18y$$ The combined return is at least 9 percent of the amount invested

$$-1.48x + 1.02y \geq 0$$

Combining the objective function and each of the constraints yields the following linear programming problem:

Minimize $z = 2x + 4y$

Subject to
$$\begin{cases} 74x + 102y \geq 3000 & \text{The total amount invested is at least \$3000} \\ -1.48x + 1.02y \geq 0 & \text{The combined return is at least 9 percent of the amount invested.} \\ x \geq 0 \\ y \geq 0 \end{cases}$$

The graph of the feasible region is shown in Figure 4.19.

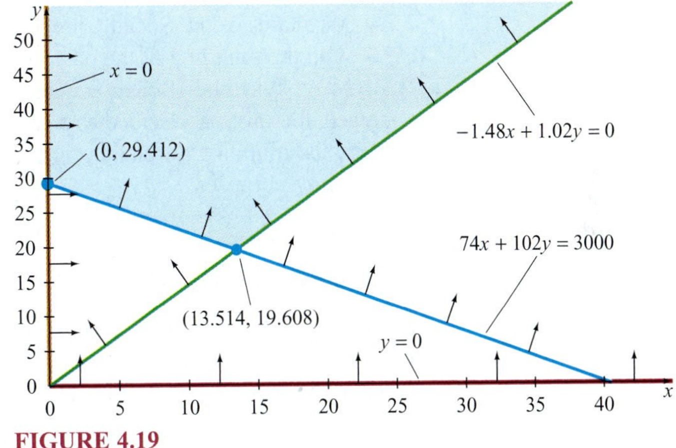

FIGURE 4.19

The first corner point is the y-intercept of the line $74x + 102y = 3000$. Since $\frac{3000}{102} \approx 29.412$, the corner point is $(0, 29.412)$.

The second corner point is the point of intersection of $74x + 102y = 3000$ and $-1.48x + 1.02y = 0$. We must solve the system of equations

$$\begin{aligned} 74x + 102y &= 3000 \\ -1.48x + 1.02y &= 0 \end{aligned}$$

We will find the solution using matrices.

$$\left[\begin{array}{cc|c} 74 & 102 & 3000 \\ -1.48 & 1.02 & 0 \end{array}\right]$$

$$\left[\begin{array}{cc|c} 74 & 102 & 3000 \\ 222 & 0 & 3000 \end{array}\right] \quad R_1 - 100R_2$$

$$\left[\begin{array}{cc|c} 0 & 306 & 6000 \\ 222 & 0 & 3000 \end{array}\right] \quad 3R_1 - R_2$$

$$\left[\begin{array}{cc|c} 0 & 1 & 19.608 \\ 1 & 0 & 13.514 \end{array}\right] \quad \begin{array}{l} 1/306\, R_1 \\ 1/222\, R_2 \end{array}$$

$$\left[\begin{array}{cc|c} 1 & 0 & 13.514 \\ 0 & 1 & 19.608 \end{array}\right] \quad \begin{array}{l} R_2 \\ R_1 \end{array}$$

The second corner point is $(13.514, 19.608)$.

We evaluate the objective function $z = 2x + 4y$ at each corner point, as shown in Table 4.10.

TABLE 4.10

Corner Point		Objective Function
x	y	$z = 2x + 4y$
0	29.412	117.648
13.514	19.608	105.460

The optimal solution is (13.514, 19.608), and the optimal value is 105.460.

The risk is minimized when 13.514 shares of the Bond Market account and 19.608 shares of the Social Choice account are purchased.

Did the rounding of the percentages and share prices affect the solution? Yes. However, since the 10-year average rate of return on each account was not guaranteed, the investor decided that rounding the percentages to whole-number percents and rounding the share prices to whole-dollar amounts was good enough for modeling purposes.

Integer Programming Problems

In many real-applications, we have the additional constraint that the objective function input values must be whole numbers. Problems of this type are called **integer programming problems.**

INTEGER PROGRAMMING PROBLEM

A linear programming problem with the additional constraint that the decision variables are integers is called an **integer programming problem.**

Integer programming problems containing many input variables and constraints can be very difficult to solve. Even in the two-decision-variable case, solving these types of problems requires several extra steps. Although we won't require you to solve any of these types of problems in the exercises, we will work an example here to illustrate the errors that may arise when the optimal solution of a linear programming problem is rounded to whole-number values.

EXAMPLE 5

Using Linear Programming to Make Business Decisions

Marlborough Printer Supplies sells generic replacement ink cartridges for a variety of printers through its eBay store. Shipping is free when 10 or more cartridges are ordered. In 2002, a single black ink cartridge for the Epson Color Stylus 660 printer cost \$2.50, and a three-pack of black ink cartridges cost \$6. (**Source:** eBay online store.) The owner of a small business needs to purchase at least 20 ink cartridges and wants to minimize her overall cost. How many single cartridges and how many three-packs should she buy?

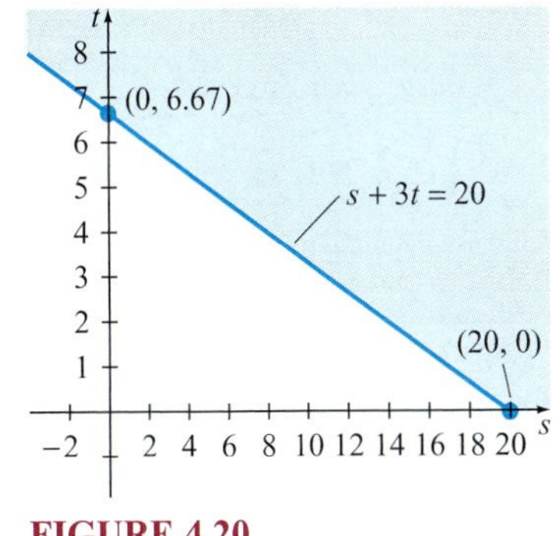

FIGURE 4.20

SOLUTION Let s be the number of single cartridges and t be the number of three-packs. Since 10 or more cartridges will be ordered, shipping will be free. Consequently, the equation of the objective function (the cost function) is

$$C = 2.5s + 6t$$

We have the constraints

$s + 3t \geq 20$	A minimum of 20 cartridges are ordered
$s \geq 0, t \geq 0$	The number of each type ordered is nonnegative

The feasible region is unbounded and has corner points $\left(0, 6\frac{2}{3}\right)$ and $(20, 0)$, as shown in Figure 4.20.

Evaluating the objective function $C = 2.5s + 6t$ at the two corner points yields the data in Table 4.11.

TABLE 4.11

Corner Point		Objective Function
Single (s)	Three-Pack (t)	$C = 2.5s + 6t$
20	0	50
0	$6\frac{2}{3}$	40

Since the smallest value of the objective function is 40, the optimal solution is $\left(0, 6\frac{2}{3}\right)$ and the optimal value is 40. However, since we can't order a fraction of a three-pack, we must find the whole-number solution. Our natural tendency might be to round the optimal solution to $(0, 7)$. However, doing so also alters the optimal value, as shown in Table 4.12.

TABLE 4.12

Rounded Optimal Solution		Objective Function
Single (s)	Three-Pack (t)	$C = 2.5s + 6t$
0	7	42

FIGURE 4.21

Using whole-number values for s and t, is it possible to further reduce the cost? We will investigate this question by adding additional constraints. We create Subproblem 1 by adding the constraint $t \leq 6$ and Subproblem 2 by adding the constraint $t \geq 7$, as shown in Figure 4.21. (These are the whole-number values on either side of $6\frac{2}{3}$.)

These new constraints split the feasible region into two separate regions. We will solve Subproblem 1 first. The unbounded feasible region of Subproblem 1 has corner points (20, 0) and (2, 6).

TABLE 4.13

Corner Point of Subproblem 1		Objective Function
Single (s)	Three-Pack (t)	$C = 2.5s + 6t$
20	0	50
2	6	41

The optimal solution for Subproblem 1 is (2, 6), as shown in Table 4.13. Since this is a whole-number solution, it makes sense in the context of the problem. When two single cartridges and six three-packs are purchased, the total ink cost is \$41.

The feasible region of Subproblem 2 has the corner point (0, 7) (see Table 4.14).

TABLE 4.14

Corner Point of Subproblem 2		Objective Function
Single (s)	Three-Pack (t)	$C = 2.5s + 6t$
0	7	42

The optimal solution for Subproblem 2 is (0, 7). Since this is also a whole-number solution, it makes sense in the context of the problem. When no single cartridges and seven three-packs are purchased, the total ink cost is \$42.

The optimal whole-number solution for the entire linear programming problem will be the subproblem solution that yields the smallest value of the objective function. The whole-number solution for Subproblem 1, (2, 6), had optimal value \$41. The whole-number solution for Subproblem 2, (0, 7), had optimal value \$42. Comparing the results of Subproblems 1 and 2, we conclude that the optimal whole-number solution of the entire problem is (2, 6). When two single cartridges and six three-packs are ordered, the overall cost is minimized. (It is important to note that if the business owner spent the extra dollar and ordered no single cartridges and seven three-packs, she would get 21 cartridges instead of 20 cartridges. She may decide that the extra cartridge is worth the extra dollar.)

As shown in Example 5, we must be aware that rounding an optimal solution to whole-number values does not guarantee that we have found the optimal whole-number solution.

4.2 Summary

In this section, you learned how a mathematical method called linear programming can help businesses determine the most cost-effective way to manage their resources. You used linear programming to optimize an objective function subject to a set of linear constraints.

4.2 Exercises

In Exercises 1–20, find the optimal solution and optimal value of the linear programming problem. If a solution does not exist, explain why.

1. Minimize $z = 3x + 7y$

Subject to $\begin{cases} 4x + y \geq 4 \\ -x + y \geq 1 \\ x \geq 0 \\ y \geq 0 \end{cases}$

2. Minimize $z = 6x + 2y$

Subject to $\begin{cases} 6x + y \geq 16 \\ -2x + y \geq 0 \\ x \geq 0 \\ y \geq 0 \end{cases}$

3. Minimize $z = 9x + y$

Subject to $\begin{cases} 6x + y \geq 16 \\ -2x + y \geq 0 \\ x \geq 0 \\ y \geq 0 \end{cases}$

4. Maximize $z = 9x + y$

Subject to $\begin{cases} 6x + y \leq 16 \\ -2x + y \leq 0 \\ x \geq 0 \\ y \geq 0 \end{cases}$

5. Maximize $z = x + 10y$

Subject to $\begin{cases} 6x + y \leq 16 \\ -2x + y \leq 0 \\ x \geq 0 \\ y \geq 0 \end{cases}$

6. Minimize $z = 2x - 5y$

Subject to $\begin{cases} 4x + y \leq 12 \\ -6x + 2y \leq 24 \\ x \geq 0 \\ y \geq 0 \end{cases}$

7. Maximize $z = 2x - 5y$

Subject to $\begin{cases} 4x + y \leq 12 \\ -6x + 2y \leq 24 \\ x \geq 0 \\ y \geq 0 \end{cases}$

8. Minimize $z = x - y$

Subject to $\begin{cases} -4x + y \geq 8 \\ -3x + y \leq 6 \\ x \geq 0 \\ y \geq 0 \end{cases}$

9. Maximize $z = -2x - y$

Subject to $\begin{cases} -4x + y \geq 8 \\ -3x + y \leq 6 \\ x \leq 4 \\ x \geq 0 \\ y \geq 0 \end{cases}$

10. Minimize $z = 5x - y$

Subject to $\begin{cases} -3x + y \geq 9 \\ -2x + y \leq 6 \\ x \leq 3 \\ x \geq 0 \\ y \geq 0 \end{cases}$

11. Minimize $z = -2x + 7y$

Subject to $\begin{cases} -x + 4y \geq 4 \\ -x + y \geq 1 \\ x \geq 0 \\ y \geq 0 \end{cases}$

12. Minimize $z = 3x + 5y$

Subject to $\begin{cases} 6x + y \geq 21 \\ -2x + y \geq 1 \\ x \geq 3 \\ y \geq 0 \end{cases}$

13. Minimize $z = 11x + 9y$

Subject to $\begin{cases} 6x + y \geq 16 \\ -2x + y \geq 0 \\ x \geq 0 \\ y \geq 2 \end{cases}$

14. Maximize $z = 11x + 9y$

Subject to $\begin{cases} 6x + y \geq 16 \\ -2x + y \geq 0 \\ x \geq 0 \\ y \geq 2 \end{cases}$

15. Maximize $z = x + 10y$

Subject to $\begin{cases} 6x + y \le 29 \\ -2x + y \le -3 \\ y \le 5 \\ x \ge 0 \\ y \ge 0 \end{cases}$

16. Minimize $z = 20x - y$

Subject to $\begin{cases} 4x + y \le 20 \\ -6x + 2y \le 40 \\ x \ge 0 \\ y \ge 0 \end{cases}$

17. Maximize $z = 20x - y$

Subject to $\begin{cases} 4x + y \le 20 \\ -6x + 2y \le 40 \\ x \ge 0 \\ y \ge 0 \end{cases}$

18. Minimize $z = x - y$

Subject to $\begin{cases} 2x + y \ge 0 \\ 3x - y \le 0 \\ x \ge 1 \\ y \ge 1 \end{cases}$

19. Maximize $z = -2x + y$

Subject to $\begin{cases} -4x + 3y \ge 0 \\ -3x + 4y \ge -1 \\ x \le 4 \\ x \ge 0 \\ y \ge 2 \end{cases}$

20. Minimize $z = x + y$

Subject to $\begin{cases} 10x + 2y \ge 8 \\ -20x - 4y \le -16 \\ x \le 4 \\ x \ge 0 \\ y \ge 0 \end{cases}$

For Exercises 21–32, identify the objective function and constraints of the linear programming problem. Then solve the problem and interpret the real-world meaning of the results.

21. **Minimum Commodity Cost** Today's Market Prices (www.todaymarket.com) is a daily fruit and vegetable wholesale market price service. Produce retailers who subscribe to the service can use wholesale prices to aid them in setting retail prices for the fruits and vegetables they sell.

A 25-pound carton of peaches holds 60 medium peaches or 70 small peaches. In August 2002, the wholesale price for local peaches in Los Angeles was \$9.00 per carton for medium peaches and \$10.00 per carton for small peaches. (**Source:** Today's Market Prices.) A fruit vendor sells the medium peaches for \$0.50 each and the small peaches for \$0.45 each. He estimates that weekly demand for peaches is at least 420 peaches but no more than 630 peaches. He wants to buy enough peaches to meet the minimum estimated demand, but no more than the maximum estimated demand. How many boxes of each size of peaches should he buy if he wants to minimize his wholesale cost?

22. **Painkiller Costs** An online drugstore sells Tylenol Extra Strength in a variety of bottle sizes. The 250-caplet bottle costs $\approx$ \$15, and the 150-caplet bottle costs $\approx$ \$12. (**Source:** www.drugstore.com.) A family wants to order a supply of at least 750 caplets. How many 150-caplet bottles and how many 250-caplet bottles should the family order if it wants to minimize costs?

23. **Investment Choices** The following table shows the average annual rate of return on a variety of TIAA-CREF investment accounts over a 10-year period.

As of 6/30/04	
CREF Variable Annuity Accounts	10-Year Average
Bond Market	7.15%
Equity Index	11.34%
Global Equities	7.24%
Growth	8.56%
Money Market	4.22%
Social Choice	10.31%
Stock	9.97%

Source: www.tiaa-cref.com.

An investor wants to invest at least \$2000 in the Stock and Growth accounts. He assumes that he will be able to get a return equal to the 10-year average, and he wants the total return on his investment to be at least 9 percent. He assigns each share in the Stock account a risk rating of 6 and each share in the Growth account a risk rating of 7. The approximate share price at the end of June 2004 was $174 per share for the Stock account and $55 per share for the Growth account. He will use these prices in his analysis. How many shares of each account should he buy in order to minimize his overall risk? [Note that fractions of shares may be purchased. Also, to make computations easier, round each percentage to the nearest whole-number percent (i.e., for 9.97 percent, use 10 percent).]

24. **Investment Choices** An investor wants to invest at least $5000 in the Global Equities and Equity Index accounts shown in Exercise 23. She assumes that she will be able to get a return equal to the 10-year average, and she wants the total return on her investment to be at least 10 percent. She assigns each share in the Global Equities account a risk rating of 6 and each share in the Equity Index account a risk rating of 5. The approximate share price at the end of June 2004 was $70 per share for the Global Equities account and $72 per share for the Equity Index account. She will use these prices in her analysis. How many shares of each account should she buy in order to minimize her overall risk? [Note that fractions of shares may be purchased. Also, to make computations easier, round each percentage to the nearest whole-number percent (i.e., for 11.34 percent, use 11 percent).]

25. **Pet Nutrition: Food Cost** PETsMART.com sold the following varieties of dog food in June 2003:

Nature's Recipe Venison Meal & Rice Canine, 20 percent protein, $21.99 per 20-pound bag

PETsMART Premier Oven Baked Lamb Recipe, 25 percent protein, $22.99 per 30-pound bag.

(**Source:** www.petsmart.com.)

A dog breeder wants to make at least 300 pounds of a mix containing at most 22 percent protein. How many bags of each dog food variety should the breeder buy in order to minimize cost? (*Hint:* Note that each bag is a different weight.)

26. **First Aid Kit Supplies** Safetymax.com sells first aid supplies to businesses. A company that assembles first aid kits for consumers purchases 3500 1″ × 3″ plastic adhesive bandages and 1800 alcohol wipes from Safetymax.com.

The company assembles standard and deluxe first aid kits for sale to consumers. A **standard** first aid kit contains 40 plastic adhesive bandages and 20 alcohol wipes. A **deluxe** first aid kit contains 50 plastic adhesive bandages and 28 alcohol wipes.

The company makes a profit of $3 from each standard first aid kit sold and $4 from each deluxe first aid kit sold. Assuming that every kit produced will be sold, how many of each type of kit should the company assemble in order to maximize profit?

27. **Furniture Production** In June 2004, an online furniture retailer offered the following items at the indicated prices:

Teak Double Rocker, $745
Teak Tennis Bench, $124
(**Source:** www.outdoordecor.com.)

Suppose that the number of hours required to produce each item is as shown in the following table.

	Cut	Finish	Package
Rocker	4	7	1
Bench	1	3	1

The company has a maximum of 360 labor hours available in the Cutting Department, a maximum of 730 labor hours available in the Finishing Department, and a maximum of 150 labor hours available in the Packaging Department. Suppose that the company makes a profit of $314 from the sale of each rocker and $57 from the sale of each bench. Assuming that all items produced are sold, how many rockers and how many benches should the company produce in order to maximize profit?

28. **Furniture Production** In June 2004, an online furniture retailer offered the following items at the indicated prices:

Avalon Teak Armchair, $378
Teak Tennis Bench, $124
(**Source:** www.outdoordecor.com.)

Suppose that the number of hours required to produce each item is as shown in the following table.

	Cut	Finish	Package
Armchair	2	4	1
Bench	1	3	1

The company has a maximum of 200 labor hours available in the Cutting Department, a maximum of 480 labor hours available in the Finishing Department, and a maximum of 150 labor hours available in the Packaging Department. Suppose the company makes a profit of $181 from the sale of each armchair and $57 from the sale of each bench. Assuming that all items produced are sold, how many armchairs and how many benches should the company produce in order to maximize profit?

29. **Transportation Costs** A high school PTA in southern Florida is planning an overnight trip to Orlando, Florida, for its graduating class. A Plus Transportation, a local charter transportation company, offers the following rates (as of December 2003):

Vehicle Capacity	Overnight Charter Rate (for 2 Days of Service)
29	$1000
49	$1800

Source: www.buscharter.net.

The school anticipates that 135 students will go on the trip. Each 29-passenger vehicle requires two chaperones, and each 49-passenger vehicle requires four chaperones. (The chaperones will be traveling with the students.) At most 16 chaperones are available to go on the trip. How many of each type of vehicle should the PTA charter in order to minimize transportation costs?

30. **Transportation Costs** A high school PTA in southern Florida is planning an overnight trip to Orlando, Florida, for its graduating class. A Plus Transportation, a local charter transportation company, offers the following rates (as of December 2003).

Vehicle Capacity	Overnight Charter Rate (for 2 Days of Service)
10	$625
57	$1995

Source: www.buscharter.net.

The school anticipates that 153 students will go on the trip. Each 10-passenger vehicle requires one chaperone, and each 57-passenger vehicle requires six chaperones. (The chaperones will be traveling with the students.) At most 24 chaperones are available to go on the trip. How many of each type of vehicle should the PTA charter in order to minimize transportation costs?

31. **Television Advertising** For the week of July 5-July 11, 2004, a national media research company estimated that 14,834,000 viewers watched *CSI* and 10,557,000 viewers watched *Law and Order*. (**Source:** www.nielsenmedia.com.)

The amount of money a network can charge for advertising is based in part on the size of the viewing audience. Suppose that a 30-second commercial running on *CSI* costs $3100 per spot and a 30-second commercial running on *Law and Order* costs $2500 per spot. A beverage company is willing to spend up to $87,000 for commercials run during episodes of the two programs. The company requires at least 10 spots to be run on each program. How many spots on each program should be purchased in order to maximize the number of viewers?

32. **Television Advertising** For the week of July 5-July 11, 2004, a national media research company estimated that 14,834,000 viewers watched *CSI* and 10,557,000 viewers watched *Law and Order*. (**Source:** www.nielsenmedia.com.)The amount of money a network can charge for advertising is based in part on the size of the viewing audience. Suppose that a 30-second commercial running on *CSI* costs $3100 per spot and a 30-second commercial running on *Law and Order* costs $2500 per spot. An athletic gear company is willing to spend up to $68,500 for commercials run during episodes of the two programs. The company requires at least 10 spots to be run on *CSI* and at least 15 spots on *Law and Order*. How many spots on each program should be purchased in order to maximize the number of viewers?

For Exercises 33–35, use the following data for publicly traded recreational vehicle companies. The information was accurate as of July 16, 2004.

Company	Share Price (dollars)	Earnings/ Share (dollars)	Dividend/ Share (dollars)
Harley-Davidson, Inc. (HDI)	62.70	2.56	0.40
Polaris Industries Inc. (PII)	50.50	2.50	0.92
Winnebago Industries Inc. (WGO)	33.42	1.80	0.20

Source: moneycentral.msn.com.

33. **Investment Choices** An investor has up to \$4000 to invest in Harley-Davidson, Inc., and Polaris Industries, Inc. The investor wants to earn at least \$60 in dividends while maximizing total earnings. How many shares of each company's stock should the investor buy? (Assume that portions of shares may be purchased.)

34. **Investment Choices** An investor has up to \$10,000 to invest in Harley-Davidson, Inc., and Winnebago, Inc. The investor wants to earn at least \$50 in dividends while maximizing total earnings. How many shares of each company's stock should the investor buy? (Assume that portions of shares may be purchased.)

35. **Investment Choices** An investor has up to \$17,000 to invest in Polaris Industries, Inc., and Winnebago, Inc. The investor wants to have total earnings of at least \$900 while maximizing total dividends. How many shares of each company's stock should the investor buy? (Assume that portions of shares may be purchased.)

Exercises 36–40 are intended to challenge your understanding of linear programming.

36. Consider the following linear programming problem.

Maximize $P = 3x + 4y$

Subject to $\begin{cases} x + y \le 2 \\ 3x + 5y \le 9 \\ x \ge 0, y \ge 0 \end{cases}$

Can a whole-number solution to a linear programming problem be obtained by simply rounding the noninteger solution to whole-number values? Explain.

37. Is it possible for a corner point (a, b) to simultaneously minimize and maximize an objective function? If yes, give an example.

38. Give an example of a linear programming problem that does not have a solution.

39. Is it possible for the feasible region of a linear programming problem not to have any corner points? If yes, give an example.

40. Given the following linear programming problem, what is the maximum possible number of corner points the feasible region could have?

Maximize $P = x + y$

Subject to $\begin{cases} -ax + by \le c \\ -dx + fy \le g \\ x \ge 0, y \ge 0 \end{cases}$

Assume a, b, c, d, f, and g are positive.

4.3 Solving Standard Maximization Problems with the Simplex Method

- Apply the simplex method to solve multivariable standard maximization problems

GETTING STARTED Suppose you are a wholesale fruit buyer for a grocery store. You have been given the assignment to purchase five varieties of apples, two varieties of peaches, and three varieties of pears. The grocery store has known budget and space constraints. You are asked to maximize the number of pieces of fruit purchased. This linear programming problem has 10 decision variables and cannot be solved graphically. Although the graphical method works well for linear programming problems with two decision variables, it does not work for problems containing more than two decision variables. In this section, we will introduce an alternative method that may be used to solve linear programming problems with any number of decision variables.

In 1947, George B. Dantzig developed the **simplex method** to solve linear programming problems. His method has been used to solve linear programming problems with hundreds of decision variables and hundreds of constraints. We will use the method to solve **standard maximization problems.**

STANDARD MAXIMIZATION PROBLEM

A **standard maximization problem** is a linear programming problem with an objective function that is to be maximized. The objective function is of the form

$$P = ax + by + cz + \cdots$$

where $a, b, c, \ldots$ are real numbers and $x, y, z, \ldots$ are decision variables.

The decision variables are constrained to nonnegative values. Additional constraints are of the form

$$Ax + By + Cz + \cdots \leq M$$

where $A, B, C, \ldots$ are real numbers and M is nonnegative.

Observe that substituting $x = 0$, $y = 0$, $z = 0, \ldots$ into each constraint inequality $Ax + By + Cz + \cdots \leq M$ yields $A(0) + B(0) + C(0) + \cdots \leq M$, which simplifies to $0 \leq M$. Since the inequality $0 \leq M$ is valid for all nonnegative values of M, the origin $(0, 0, \ldots, 0)$ is contained in the solution region of each constraint and, consequently, in the feasible region of the standard maximization problem. Furthermore, if at least one of the constraints $Ax + By + Cz + \cdots \leq M$ has all nonnegative coefficients, we are guaranteed that the feasible region is bounded. For the vast majority of our real-life applications, the constraints will have nonnegative coefficients.

Graphically speaking, the simplex method starts at the origin and moves from corner point to corner point of the feasible region, each time increasing the value of the objective function until it attains its maximum value. Remarkably, the method doesn't require that all corner points be tested, a fact that is appreciated by those solving linear programming problems with hundreds of corner points. Although the simplex method may be used for linear programming problems with any number of decision variables, we will restrict our use to problems with four variables or less. Larger problems are typically solved using computers.

We will introduce the simplex method with a two-variable example before formally listing the steps of the method.

Suppose we are asked to use the simplex method to maximize $P = 6x + 2y$ subject to the constraints

$$\begin{aligned} 2x + y &\leq 10 \\ x + y &\leq 8 \\ x &\geq 0 \\ y &\geq 0 \end{aligned}$$

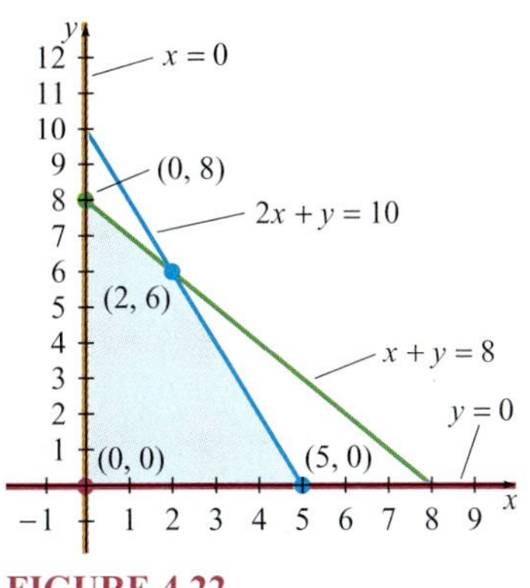

FIGURE 4.22

Since the decision variables are nonnegative and additional constraints are of the form $Ax + By + Cz + \cdots \leq M$ with M nonnegative, this linear programming problem is a standard maximization problem. Consequently, we will be able to maximize the objective function using the simplex method. Although this is not part of the simplex method, we will draw the graph of the feasible region to give you a frame of reference as we work through the problem. The feasible region has corner points (0, 0), (0, 8), (2, 6), and (5, 0) and is shown in Figure 4.22.

The value of the objective function, $P = 6x + 2y$, varies depending upon the corner point. Substituting each corner point (x, y) into $P = 6x + 2y$ and solving for P yields Table 4.15.

TABLE 4.15

Corner Point		Objective Function Value
x	y	$P = 6x + 2y$
0	0	0
0	8	16
2	6	24
5	0	30

We see that the objective function $P = 6x + 2y$ attains a maximum value of 30 at the corner point $(5, 0)$. Using the simplex method to solve the standard maximization problem will yield the same result.

Recall that we are to solve the following standard maximization problem:

$$\text{Maximize} \quad P = 6x + 2y$$

$$\text{Subject to} \quad \begin{cases} 2x + y \leq 10 \\ x + y \leq 8 \\ x \geq 0 \\ y \geq 0 \end{cases}$$

The first step of the simplex method requires us to convert each of the constraint inequalities into an equation. Since $2x + y \leq 10$, there is some value $s \geq 0$ such that $2x + y + s = 10$. The variable s is called a **slack variable** because it "takes up the slack." That is, s adds in whatever value is necessary to make the left-hand side of the equation equal 10. The value of s will vary depending upon the value of x and y (see Table 4.16).

TABLE 4.16

Corner Point		Slack Variable Value
x	y	$s = 10 - 2x - y$
0	0	10
0	8	2
2	6	0
5	0	0

For the constraint inequality $x + y \leq 8$, we observe that there is some value t such that $x + y + t = 8$. Again, the value of t will vary depending upon the value of x and y (see Table 4.17).

TABLE 4.17

Corner Point		Slack Variable Value
x	y	$t = 8 - x - y$
0	0	8
0	8	0
2	6	0
5	0	3

We rewrite the objective function equation $P = 6x + 2y$ by moving all terms to the left-hand side of the equation, which yields the equation $-6x - 2y + P = 0$. The system of equations representing the linear programming problem created by combining the constraint equations and the objective function equation is

$$\begin{aligned} 2x + y + s &= 10 && \text{Constraint 1} \\ x + y + t &= 8 && \text{Constraint 2} \\ -6x - 2y + P &= 0 && \text{Objective function} \end{aligned}$$

This is a system of three equations and five unknowns (x, y, s, t, P). Since the number of unknowns exceeds the number of equations, there will be more than one solution to the system, if a solution exists.

We can represent the system of equations using matrix notation. The augmented matrix is called the **initial simplex tableau.**

$$\begin{array}{c} \\ \text{Constraint 1} \\ \text{Constraint 2} \\ \text{Objective function} \end{array} \quad \begin{array}{c} \begin{array}{ccccc} x & y & s & t & P \end{array} \\ \left[\begin{array}{ccccc|c} 2 & 1 & 1 & 0 & 0 & 10 \\ 1 & 1 & 0 & 1 & 0 & 8 \\ \hline -6 & -2 & 0 & 0 & 1 & 0 \end{array}\right] \end{array}$$

The vertical and horizontal lines in the tableau are used to separate the bottom row and rightmost column from the rest of the matrix. This separation will become important when we calculate test ratios (as will be explained later).

The variables of the columns of the simplex tableau that contain exactly one nonzero entry are called **active (basic)** variables. The variables of the columns that contain more than one nonzero entry are called **inactive (nonbasic)** variables. For the initial tableau, the variables s, t, and P are the active variables, since their corresponding columns in the tableau contain exactly one nonzero value. The variables x and y are the inactive variables in the initial simplex tableau, since their corresponding columns contain more than one nonzero value.

$$\begin{array}{c} \begin{matrix} x & \; y & s & t & P \end{matrix} \\ \left[\begin{array}{rrrrr|r} 2 & 1 & 1 & 0 & 0 & 10 \\ 1 & 1 & 0 & 1 & 0 & 8 \\ \hline -6 & -2 & 0 & 0 & 1 & 0 \end{array}\right] \end{array} \quad \begin{array}{l} \text{Constraint 1} \\ \text{Constraint 2} \\ \text{Objective function} \end{array}$$

The basic feasible solution of the tableau always corresponds with the origin. The solution is obtained by substituting $x = 0$ and $y = 0$ into the equations that generated the tableau.

$$\begin{aligned} 2(0) + (0) + s &= 10 && \text{Constraint 1} \\ (0) + (0) + t &= 8 && \text{Constraint 2} \\ -6(0) - 2(0) + P &= 0 && \text{Objective function} \end{aligned}$$

Simplifying yields

$$\begin{aligned} s &= 10 && \text{Constraint 1} \\ t &= 8 && \text{Constraint 2} \\ P &= 0 && \text{Objective function} \end{aligned}$$

These results are summarized in Table 4.18.

TABLE 4.18

Corner Point		Objective Function Value	Slack Variable Value	
x	y	$P = 6x + 2y$	s	t
0	0	0	10	8

The objective function $P = 6x + 2y$ has a value of 0 at the corner point $(0, 0)$.

An alternative and quicker approach to finding the solution may be obtained from the initial tableau itself by crossing out each inactive variable column and reading the result from the tableau. Crossing out an inactive variable column is equivalent to setting the column variable to zero. For the initial simplex tableau, we have

$$\begin{array}{c} \begin{matrix} x & \; y & s & t & P \end{matrix} \\ \left[\begin{array}{rrrrr|r} \not{2} & \not{1} & 1 & 0 & 0 & 10 \\ \not{1} & \not{1} & 0 & 1 & 0 & 8 \\ \hline -\not{6} & -\not{2} & 0 & 0 & 1 & 0 \end{array}\right] \end{array} \quad \begin{array}{l} \text{Constraint 1} \\ \text{Constraint 2} \\ \text{Objective function} \end{array}$$

The first row of the resultant matrix is equivalent to

$$\begin{aligned} 1s + 0t + 0P &= 10 \\ s &= 10 \end{aligned}$$

The second row is equivalent to

$$0s + 1t + 0P = 8$$
$$t = 8$$

The bottom row is equivalent to

$$0s + 0t + 1P = 0$$
$$P = 0$$

These results are the same as those shown in Table 4.18. Clearly, the corner point $(0, 0)$ does not maximize the value of the objective function.

How can we make P as large as possible? The coefficients on the x and y terms of the objective function $P = 6x + 2y$ give us a clue. Since the coefficient on the x term is 6, each 1-unit increase in x will result in a 6-unit increase in P. Similarly, since the coefficient on the y term is 2, each 1-unit increase in y will result in a 2-unit increase in P. Consequently, our initial strategy is to make the x term as big as possible, since it contributes the most to the value of P. The pivoting process detailed next will achieve this objective.

We will now use row operations to convert the initial simplex tableau into a new tableau. We will first select a **pivot column.** Then, using row operations, we will make the pivot column look like one of the columns of the 3×3 identity matrix. To choose the pivot column, we find the negative number in the bottom row that is furthest away from zero. The corresponding column is the pivot column. In this tableau, the pivot column will be the x column, since -6 is further away from 0 than -2.

$$\begin{array}{c} \begin{matrix} x & y & s & t & P & \end{matrix} \\ \left[\begin{array}{ccccc|c} 2 & 1 & 1 & 0 & 0 & 10 \\ 1 & 1 & 0 & 1 & 0 & 8 \\ \hline -6 & -2 & 0 & 0 & 1 & 0 \end{array}\right] \\ \uparrow \text{ Pivot column} \end{array}$$

Why do we pick the column with the negative number that has the largest magnitude? Recall that we are trying to maximize the objective function $P = 6x + 2y$. As was pointed out earlier, the value of x contributes more to P than the value of y because of the differences in their coefficients. Picking the column with the negative number of the largest magnitude guarantees that we are making the variable that contributes the most to the value of P as large as possible.

The **pivot** is the nonzero entry in the pivot column that will remain after we zero out the other column entries. It must be a positive number. If the pivot column contains more than one positive number, we will select the pivot by using a **test ratio.** To calculate the test ratio, we divide the rightmost entry of each row (excluding the bottom row) by the positive entry in the pivot column of that row.

$$\begin{array}{c} \begin{matrix} x & y & s & t & P & \end{matrix} \\ \left[\begin{array}{ccccc|c} 2 & 1 & 1 & 0 & 0 & 10 \\ 1 & 1 & 0 & 1 & 0 & 8 \\ \hline -6 & -2 & 0 & 0 & 1 & 0 \end{array}\right] \begin{array}{l} 10/2 = 5 \leftarrow \text{Test ratio for first row} \\ 8/1 = 8 \leftarrow \text{Test ratio for second row} \\ \end{array} \\ \uparrow \text{ Pivot column} \end{array}$$

Since the x column is the pivot column, these test ratios correspond geometrically with the x-intercepts of the constraints (see Figure 4.23).

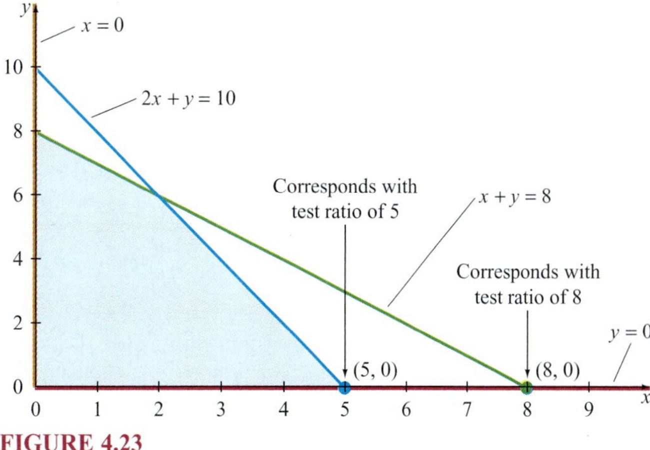

FIGURE 4.23

We select as the pivot the pivot-column entry that has the smallest test ratio. That is, we pick the x-intercept on the graph that is closest to the origin to be the pivot. This ensures that we remain in the feasible region. In this case, the first row of the tableau had the smallest test ratio. Consequently, the pivot will be the 2 in the first row and first column.

Pivot

Pivot row →

$$\begin{array}{c} \begin{matrix} x & y & s & t & P & \end{matrix} \\ \left[\begin{array}{ccccc|c} \boxed{2} & 1 & 1 & 0 & 0 & 10 \\ 1 & 1 & 0 & 1 & 0 & 8 \\ \hline -6 & -2 & 0 & 0 & 1 & 0 \end{array}\right] \end{array} \quad \begin{array}{l} 10/2 = 5 \leftarrow \text{Test ratio for first row} \\ 8/1 = 8 \leftarrow \text{Test ratio for second row} \end{array}$$

Pivot column

Why do we calculate test ratios only for positive column entries? When we calculate test ratios for the x column, we are actually determining the values of the x-intercepts of the constraints. If the column entry is a negative number, the x-intercept of the constraint will be to the left of the origin. However, we are constrained to nonnegative values for both x and y. Therefore, determining the value of the x-intercepts is necessary only for positive values. (If the column entry is zero, the corresponding line is horizontal and doesn't have an x-intercept.)

Recall that the pivot for this tableau is the 2 in the x column and the first row.

$$\begin{array}{c} \begin{matrix} x & y & s & t & P & \end{matrix} \\ \left[\begin{array}{ccccc|c} \boxed{2} & 1 & 1 & 0 & 0 & 10 \\ 1 & 1 & 0 & 1 & 0 & 8 \\ \hline -6 & -2 & 0 & 0 & 1 & 0 \end{array}\right] \end{array}$$

We will use the pivot to zero out the remaining entries in the column. However, to ensure that all active variables remain nonnegative, row operations must be of the form

$$aR_c \pm bR_p \rightarrow R_c \text{ with } a > 0 \text{ and } b > 0$$

where R_c is the row to be changed and R_p is the row containing the pivot. Using the row operations

$$2R_2 - R_1 \rightarrow R_2$$

and

$$R_3 + 3R_1 \rightarrow R_3$$

we get the new tableau

$$\begin{array}{c} \\ \\ \\ \end{array}\begin{array}{ccccc|c} x & y & s & t & P & \\ \boxed{2} & 1 & 1 & 0 & 0 & 10 \\ 0 & 1 & -1 & 2 & 0 & 6 \\ \hline 0 & 1 & 3 & 0 & 1 & 30 \end{array} \qquad \begin{array}{l} \\ \\ 2R_2 - R_1 \\ R_3 + {}_3R_1 \end{array}$$

Columns x and t almost look like columns of the identity matrix; however, their nonzero entry is a 2 instead of a 1. Using the row operations

$$\frac{1}{2}R_1 \rightarrow R_1$$

and

$$\frac{1}{2}R_2 \rightarrow R_2$$

we get

$$\begin{array}{ccccc|c} x & y & s & t & P & \\ \boxed{1} & \frac{1}{2} & \frac{1}{2} & 0 & 0 & 5 \\ 0 & \frac{1}{2} & -\frac{1}{2} & 1 & 0 & 3 \\ \hline 0 & 1 & 3 & 0 & 1 & 30 \end{array} \qquad \begin{array}{l} \\ 1/2\, R_1 \\ 1/2\, R_2 \\ \\ \end{array}$$

The three columns that look like columns from the 3×3 identity matrix are x, t, and P. Each of these columns has exactly one nonzero entry, so their corresponding variables are the active variables. The remaining columns, y and s, have more than one nonzero entry, so their corresponding variables are the inactive variables. The system of equations represented by the tableau is

$$x + \frac{1}{2}y + \frac{1}{2}s = 5$$

$$\frac{1}{2}y - \frac{1}{2}s + t = 3$$

$$1y + 3s + P = 30$$

Setting the inactive variables y and s to zero yields the system of equations

$$x + \frac{1}{2}(0) + \frac{1}{2}(0) = 5$$

$$\frac{1}{2}(0) - \frac{1}{2}(0) + t = 3$$

$$1(0) + 3(0) + P = 30$$

which simplifies to $x = 5$, $t = 3$, and $P = 30$. The inactive variables y and s are both zero. As noted earlier, a quick way to see the values of the active variables is to cross out the columns of the inactive variables and read the resultant values of the active variables directly from the tableau.

$$\begin{array}{cccccc} x & y & s & t & P & \\ \left[\begin{array}{ccccc|c} 1 & \frac{1}{2} & \frac{1}{2} & 0 & 0 & 5 \\ 0 & \frac{1}{2} & -\frac{1}{2} & 1 & 0 & 3 \\ \hline 0 & 1 & 3 & 0 & 1 & 30 \end{array}\right] \end{array}$$

Reading from the tableau, we have

$$x = 5$$
$$t = 3$$
$$P = 30$$

The inactive variables y and s are both equal to zero. Since $x = 5$ and $y = 0$, this solution corresponds with the corner point (5, 0) of the feasible region. At (5, 0), $P = 30$.

If there were negative entries in the bottom row of the tableau, we would select a new pivot column and repeat the process. However, since the bottom row contains only nonnegative entries, we are done. The objective function is maximized at (5, 0). This corroborates our earlier conclusion, which was based on Table 4.15.

The simplex method is not intuitive and takes some effort to learn. We will summarize the steps of the process and do several more examples to help you master the method.

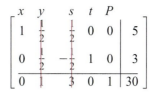

HOW TO The Simplex Method

1. Verify that the linear programming problem is a standard maximization problem.
2. Add slack variables to convert the constraints into linear equations.
3. Rewrite the objective function in the form $ax + by + \cdots + P = 0$, where $a, b, \ldots$ are real numbers and $x, y, \ldots$ are decision variables. (Make sure the coefficient on P is positive one.)
4. Set up the initial tableau from the system of equations generated in Steps 2 and 3. Be sure to put the objective function equation from Step 3 in the bottom row of the tableau.
5. Select the pivot column by identifying the negative entry in the bottom row with the largest magnitude.
6. Select the pivot by calculating test ratios for the positive entries in the pivot column and choosing the entry with the smallest test ratio. The test ratio of a row is calculated by dividing the last entry of the row by the entry in the pivot column of the row.
7. Use the pivot to zero out the remaining entries in the pivot column. All row operations must be of the form $aR_c \pm bR_P \rightarrow R_c$ with $a > 0$ and $b > 0$, where R_c is the row to be changed and R_p is the pivot row. As needed, multiply a row by a positive number to convert the pivot to a 1.
8. If the bottom row contains all nonnegative entries, cross out the columns of the inactive variables and read the solution from the tableau. If the bottom row contains negative entries, return to Step 5 and repeat the process for the new tableau.

EXAMPLE 1 Using the Simplex Method to Solve a Standard Maximization Problem

Maximize $P = 4x + 3y$ subject to

$$3x + 2y \leq 12$$
$$x + y \leq 5$$
$$x \geq 0, y \geq 0$$

SOLUTION We confirm that the problem is a standard maximization problem. Converting the linear programming problem into a system of linear equations, we get

$$\begin{aligned} 3x + 2y + s &= 12 \\ x + y + t &= 5 \\ -4x - 3y + P &= 0 \end{aligned}$$

and the initial simplex tableau

$$\begin{array}{c} \begin{matrix} x & \quad y & s & t & P & \quad \end{matrix} \\ \left[\begin{array}{ccccc|c} 3 & 2 & 1 & 0 & 0 & 12 \\ 1 & 1 & 0 & 1 & 0 & 5 \\ \hline -4 & -3 & 0 & 0 & 1 & 0 \end{array}\right] \end{array}$$

Since -4 is the negative number in the bottom row that has the largest magnitude, we select the x column as the pivot column. Because the entries in the first and second rows of the pivot column are positive, we must calculate the test ratios.

$$\begin{array}{c} \begin{matrix} x & \quad y & s & t & P & \quad \end{matrix} \\ \left[\begin{array}{ccccc|c} 3 & 2 & 1 & 0 & 0 & 12 \\ 1 & 1 & 0 & 1 & 0 & 5 \\ \hline -4 & -3 & 0 & 0 & 1 & 0 \end{array}\right] \end{array} \quad \begin{array}{l} 12/3 = 4 \\ 5/1 = 5 \\ \end{array}$$

↑ Pivot column

Since the x column is the pivot column, the test ratios give us the x-intercepts of the constraint equations, as shown in Figure 4.24.

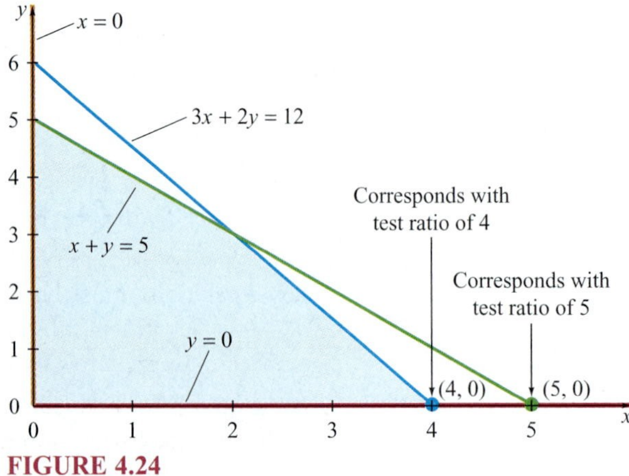

FIGURE 4.24

By picking the column entry with the smallest test ratio, we guarantee that our computations will yield a result that is in the feasible region. The pivot entry is the 3 in the first row and the x column. We zero out the remaining entries in the pivot column with the indicated operations. The resultant tableau is

$$\begin{array}{ccccc|c} x & y & s & t & P & \\ \boxed{3} & 2 & 1 & 0 & 0 & 12 \\ 0 & 1 & -1 & 3 & 0 & 3 \\ \hline 0 & -1 & 4 & 0 & 3 & 48 \end{array} \qquad \begin{array}{l} \\ \\ 3R_2 - R_1 \\ 3R_3 + 4R_1 \end{array}$$

Further simplifying the tableau yields

$$\begin{array}{ccccc|c} x & y & s & t & P & \\ 1 & \frac{2}{3} & \frac{1}{3} & 0 & 0 & 4 \\ 0 & \frac{1}{3} & -\frac{1}{3} & 1 & 0 & 1 \\ \hline 0 & -\frac{1}{3} & \frac{4}{3} & 0 & 1 & 16 \end{array} \qquad \begin{array}{l} \\ \frac{1}{3}R_1 \\ \frac{1}{3}R_2 \\ \frac{1}{3}R_3 \end{array}$$

We cross out the columns of the inactive variables and read the result:

$$\begin{array}{ccccc|c} x & y & s & t & P & \\ 1 & \frac{2}{3} & \frac{1}{3} & 0 & 0 & 4 \\ 0 & \frac{1}{3} & -\frac{1}{3} & 1 & 0 & 1 \\ \hline 0 & -\frac{1}{3} & \frac{4}{3} & 0 & 1 & 16 \end{array}$$

$x = 4$, $t = 1$, and $P = 16$. The inactive variables y and s are all zero. The corner point (4, 0) yields an objective function value of $P = 16$.

Observe that the fractional entries in the inactive variable columns of the tableau were eliminated when we set the inactive variables to zero. To avoid spending the time calculating fractional entries that will be eliminated anyway, an alternative approach is to cross out the inactive variable columns before converting any active variable column entries to a 1. We then read the resultant equations from the tableau and solve.

$$\begin{array}{ccccc|c} x & y & s & t & P & \\ 3 & 2 & 1 & 0 & 0 & 12 \\ 0 & 1 & -1 & 3 & 0 & 3 \\ \hline 0 & -1 & 4 & 0 & 3 & 48 \end{array}$$

$$3x = 12 \qquad 3t = 3 \qquad 3P = 48$$

$$x = 4 \qquad t = 1 \qquad P = 16$$

Observe that this approach yields the same result and requires fewer computations.

The bottom row of the tableau is equivalent to the equation

$$-y + 4s + 3P = 48$$

$$3P = y - 4s + 48$$

$$P = \frac{1}{3}y - \frac{4}{3}s + 16$$

Observe that since the coefficient on the y term is positive, increasing the value of y while leaving the value of s unchanged will yield a value of P greater than 16. Consequently, the tableau

$$\begin{array}{c} \begin{matrix} x & y & s & t & P \end{matrix} \\ \left[\begin{array}{ccccc|c} 3 & 2 & 1 & 0 & 0 & 12 \\ 0 & 1 & -1 & 3 & 0 & 3 \\ \hline 0 & -1 & 4 & 0 & 3 & 48 \end{array}\right] \end{array}$$

is **not** the final simplex tableau. The negative value in the bottom row indicates that the value of P may be increased.

We will repeat the simplex method for this new tableau.

$$\begin{array}{c} \begin{matrix} x & y & s & t & P \end{matrix} \\ \left[\begin{array}{ccccc|c} 3 & 2 & 1 & 0 & 0 & 12 \\ 0 & 1 & -1 & 3 & 0 & 3 \\ \hline 0 & -1 & 4 & 0 & 3 & 48 \end{array}\right] \end{array}$$

New pivot column

The y column is the pivot column, since it is the column with a negative entry in the bottom row. Because the entries in the first and second rows of the pivot column are positive, we must calculate the test ratios.

$$\begin{array}{c} \begin{matrix} x & y & s & t & P \end{matrix} \\ \left[\begin{array}{ccccc|c} 3 & 2 & 1 & 0 & 0 & 12 \\ 0 & \boxed{1} & -1 & 3 & 0 & 3 \\ \hline 0 & -1 & 4 & 0 & 3 & 48 \end{array}\right] \end{array} \qquad \begin{array}{l} 12/2 = 6 \\ 3/1 = 3 \\ \end{array}$$

The 1 in the second row is the pivot. We simplify the tableau using the indicated row operations.

$$\begin{array}{c} \begin{matrix} x & y & s & t & P \end{matrix} \\ \left[\begin{array}{ccccc|c} 3 & 0 & 3 & -6 & 0 & 6 \\ 0 & 1 & -1 & 3 & 0 & 3 \\ \hline 0 & 0 & 3 & 3 & 3 & 51 \end{array}\right] \end{array} \qquad \begin{array}{l} R_1 - 2R_2 \\ \\ R_3 + R_2 \end{array}$$

$$3x = 6 \qquad y = 3 \qquad 3P = 51$$

$$x = 2 \qquad\qquad P = 17$$

Because the bottom row of the tableau is entirely nonnegative, we do not need to repeat the pivoting process. The corner point (2, 3) yields a maximum objective function value of $P = 17$.

Observe that each new simplex tableau corresponds with a different corner point. The initial simplex tableau corresponded with (0, 0); the second tableau, with (4, 0); and the final tableau, with (2, 3), as shown in Figure 4.25.

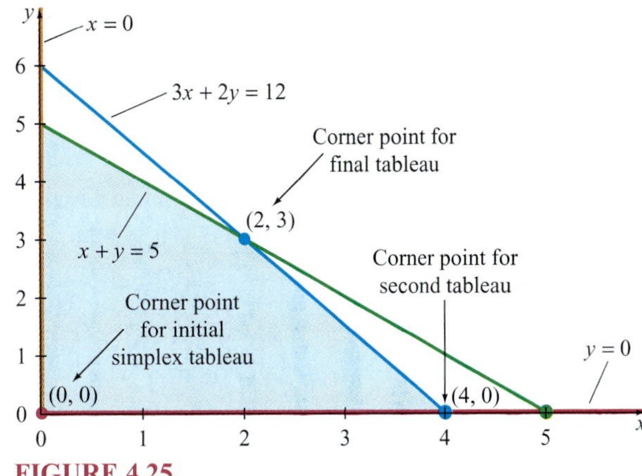

FIGURE 4.25

Graphically speaking, the simplex method moves from a corner point to an adjacent corner point, each time increasing the value of the objective function.

We can verify the accuracy of our result by calculating the value of the objective function at each corner point (see Table 4.19).

TABLE 4.19

Corner Points		Objective Function
x	y	$P = 4x + 3y$
0	0	0
4	0	16
2	3	17
0	5	15

EXAMPLE 2 Using the Simplex Method to Make Business Decisions

A store can buy at most 60 packages of three different brands of cheese. The first brand costs \$2 per package, the second brand costs \$3 per package, and the third brand costs \$4 per package. The store can spend up to \$120 on cheese. The first brand resells for \$5 per package, the second brand resells for \$6 per package, and the third brand resells for \$8 per package. How many packages of each brand of cheese should the store buy in order to maximize profit? (*Note:* Profit per item = revenue per item − cost per item.)

SOLUTION Let x be the number of packages of the first brand, y be the number of packages of the second brand, and z be the number of packages of the third brand. The total revenue from cheese sales is $R = 5x + 6y + 8z$. The total cost of the cheese is $C = 2x + 3y + 4z$. Since profit is the difference between revenue and cost,

$$\begin{aligned} P &= R - C \\ &= (5x + 6y + 8z) - (2x + 3y + 4z) \\ &= 3x + 3y + 4z \end{aligned}$$

Since we want to maximize profit, we must maximize $P = 3x + 3y + 4z$ subject to

$$x + y + z \leq 60 \quad \text{There are at most 60 packages}$$
$$2x + 3y + 4z \leq 120 \quad \text{At most, \$120 is spent on cheese}$$
$$x \geq 0, y \geq 0, z \geq 0 \quad \text{Nonnegative number of each type of cheese is purchased}$$

The problem is a standard maximization problem. Converting the linear programming problem into a system of linear equations, we get

$$x + y + z + s = 60 \quad \text{Package quantity constraint}$$
$$2x + 3y + 4z + t = 120 \quad \text{Budget constraint}$$
$$-3x - 3y - 4z + P = 0 \quad \text{Objective function}$$

and the initial simplex tableau

$$\begin{array}{cccccc|c} x & y & z & s & t & P & \\ 1 & 1 & 1 & 1 & 0 & 0 & 60 \\ 2 & 3 & 4 & 0 & 1 & 0 & 120 \\ \hline -3 & -3 & -4 & 0 & 0 & 1 & 0 \end{array}$$

Pivot column (z)

Since -4 is the negative number in the bottom row that has the largest magnitude, the z column is the pivot column. Because the entries in the first and second rows of the pivot column are positive, we must calculate the test ratios.

$$\begin{array}{cccccc|c} x & y & z & s & t & P & \\ 1 & 1 & 1 & 1 & 0 & 0 & 60 \\ 2 & 3 & \boxed{4} & 0 & 1 & 0 & 120 \\ \hline -3 & -3 & -4 & 0 & 0 & 1 & 0 \end{array} \quad \begin{array}{l} 60/1 = 60 \\ 120/4 = 30 \\ \end{array}$$

The pivot entry is 4, since the second row had the smallest test ratio. We zero out the remaining entries in the pivot column with the indicated row operations. The resultant tableau is

$$\begin{array}{cccccc|c} x & y & z & s & t & P & \\ 2 & 1 & 0 & 4 & -1 & 0 & 120 \\ 2 & 3 & 4 & 0 & 1 & 0 & 120 \\ \hline -1 & 0 & 0 & 0 & 1 & 1 & 120 \end{array} \quad \begin{array}{l} 4R_1 - R_2 \\ \\ R_3 + R_2 \end{array}$$

We could convert the nonzero entries in the active columns to a 1 by multiplying Rows 1 and 2 by $\frac{1}{4}$; however, that would result in messy fractional entries in the tableau. We will hold off on the reduction until we have obtained the final simplex tableau. Since there is still a negative value in the bottom row, we must again select a pivot column and entry. The pivot column x has two positive entries, so we must compute the test ratios.

$$\begin{array}{cccccc|c} x & y & z & s & t & P & \\ \boxed{2} & 1 & 0 & 4 & -1 & 0 & 120 \\ 2 & 3 & 4 & 0 & 1 & 0 & 120 \\ \hline -1 & 0 & 0 & 0 & 1 & 1 & 120 \end{array} \quad \begin{array}{l} 120/2 = 60 \\ 120/2 = 60 \\ \end{array}$$

Pivot column (x)

Since both entries have the same test ratio, we may pick either entry to be the pivot. We select the entry in Row 1 as the pivot and zero out the remaining entries in the pivot column with the indicated row operations. The resultant tableau is

$$\begin{array}{c} \\ \\ \\ \end{array}\begin{array}{cccccc|c} x & y & z & s & t & P & \\ \hline 2 & 1 & 0 & 4 & -1 & 0 & 120 \\ 0 & 2 & 4 & -4 & 2 & 0 & 0 \\ \hline 0 & 1 & 0 & 4 & 1 & 2 & 360 \end{array} \quad \begin{array}{l} \\ \\ R_2 - R_1 \\ 2R_3 + R_1 \end{array}$$

Since the bottom row of the tableau is entirely nonnegative, it is unnecessary to repeat the pivot selection process. Crossing out the inactive variable columns and reading the result yields

$$\begin{array}{cccccc|c} x & y & z & s & t & P & \\ \hline 2 & 1 & 0 & 4 & -1 & 0 & 120 \\ 0 & 2 & 4 & -4 & 2 & 0 & 0 \\ \hline 0 & 1 & 0 & 4 & 1 & 2 & 360 \end{array}$$

$$2x = 120 \qquad 4z = 0 \qquad 2P = 360$$

$$x = 60 \qquad z = 0 \qquad P = 180$$

The corner point (x, y, z) that maximizes the objective function is $(60, 0, 0)$. It is interesting to note that although z is an active variable, $z = 0$. Active variables may equal zero, but inactive variables must equal zero.

In the context of the scenario, the store should order 60 packages of the first brand to earn the maximum profit of \$180. Even though the profit per package for the third brand (\$4 per package) was higher than the profit per package for the first brand (\$3 per package), the higher wholesale cost of the third brand (\$4 versus \$2) limited the number of third-brand items available for sale. Ironically, the lowest-priced brand of cheese resulted in the greatest profit.

In Examples 1 and 2, there was a unique solution to the standard maximization problem; however, on occasion you'll encounter a standard maximization problem with infinitely many solutions. Example 3 is such a case.

EXAMPLE 3

Using the Simplex Method to Make Business Decisions

A store can purchase three brands of binders. The first brand, A, costs \$2 per binder. The second brand, B, costs \$1 per binder. The third brand, C, costs \$2 per binder. The store can spend up to \$100 and purchase up to 50 binders. It resells binder A for \$5 per binder, binder B for \$3 per binder, and binder C for \$5 per binder. Recall that profit per binder is revenue per binder minus cost per binder. How many of each type of binder should the store buy if it wants to maximize its profit?

SOLUTION Let x be the number of binders of the first brand, y be the number of binders of the second brand, and z be the number of binders of the third brand. The total revenue from binder sales is $R = 5x + 3y + 5z$. The total cost of the binders is $C = 2x + 1y + 2z$. Since profit is the difference between revenue and cost,

$$\begin{aligned} P &= R - C \\ &= (5x + 3y + 5z) - (2x + 1y + 2z) \\ &= 3x + 2y + 3z \end{aligned}$$

Since we want to maximize profit, we must maximize $P = 3x + 2y + 3z$ subject to

$x + y + z \leq 50$	At most 50 binders are purchased
$2x + y + 2z \leq 100$	At most \$100 is spent on binders
$x \geq 0, y \geq 0, z \geq 0$	A nonnegative number of each brand of binder is purchased

The problem is a standard maximization problem. Converting the linear programming problem into a system of linear equations, we get

$x + y + z + s = 50$	Binder quantity constraint
$2x + y + 2z + t = 100$	Budget constraint
$-3x - 2y - 3z + P = 0$	Objective function

and the initial simplex tableau

$$\begin{array}{c} \begin{matrix} x & y & z & s & t & P & \end{matrix} \\ \left[\begin{array}{cccccc|c} 1 & 1 & 1 & 1 & 0 & 0 & 50 \\ 2 & 1 & 2 & 0 & 1 & 0 & 100 \\ \hline -3 & -2 & -3 & 0 & 0 & 1 & 0 \end{array}\right] \end{array}$$

Pivot column

Since -3 is the negative number in the bottom row that has the largest magnitude, we may select either column x or column z as the pivot column. We select column z. Because the entries in the first and second rows of the pivot column are positive, we must calculate the test ratios.

$$\begin{array}{c} \begin{matrix} x & y & z & s & t & P & \end{matrix} \\ \left[\begin{array}{cccccc|c} 1 & 1 & \boxed{1} & 1 & 0 & 0 & 50 \\ 2 & 1 & 2 & 0 & 1 & 0 & 100 \\ \hline -3 & -2 & -3 & 0 & 0 & 1 & 0 \end{array}\right] \end{array} \quad \begin{array}{l} 50/1 = 50 \\ 100/2 = 50 \\ \end{array}$$

Since the entries have the same test ratio, we may select either one as the pivot. We select the entry in the first row. We zero out the remaining entries in the pivot column with the indicated row operations and obtain the tableau

$$\begin{array}{c} \begin{matrix} x & y & z & s & t & P & \end{matrix} \\ \left[\begin{array}{cccccc|c} 1 & 1 & 1 & 1 & 0 & 0 & 50 \\ 0 & -1 & 0 & -2 & 1 & 0 & 0 \\ \hline 0 & 1 & 0 & 3 & 0 & 1 & 150 \end{array}\right] \end{array} \quad \begin{array}{l} \\ R_2 - 2R_1 \\ R_3 + 3R_1 \end{array}$$

Crossing out the inactive variable columns and reading the resultant equations, we get

$$\begin{array}{c} \begin{matrix} x & y & z & s & t & P & \end{matrix} \\ \left[\begin{array}{cccccc|c} 1 & 1 & 1 & 1 & 0 & 0 & 50 \\ 0 & -1 & 0 & -2 & 1 & 0 & 0 \\ \hline 0 & 1 & 0 & 3 & 0 & 1 & 150 \end{array}\right] \end{array}$$

$$x + z = 50 \qquad t = 0 \qquad P = 150$$

$$z = -x + 50$$

There are infinitely many solutions (x, y, z) that maximize the objective function. Each solution is of the form $(n, 0, -n + 50)$ with $0 \leq n \leq 50$.

In the context of the real-world scenario, the store may purchase any combination of 50 binders from brands A and C and still maximize profits. This is because both brands have the same wholesale cost and the same retail price. As a result, the cost, revenue, and profit for each item are the same.

When a standard maximization problem has more than two constraints, additional slack variables will be needed. This will result in additional columns in the initial simplex tableau. We typically use the letters u, v, and so on for slack variables in addition to the letters s and t. Even with additional rows or columns, the simplex method is effective in finding the optimal solution to any standard maximization problem.

EXAMPLE 4

Using the Simplex Method to Make Business Decisions

Clothing retailers typically mark up the price of apparel by 100 percent. That is, the retail price is typically 200 percent of the wholesale cost. WholesaleFashion.com claims to be the number one Internet business-to-business source for fashion retailers of women's apparel and shoes. It sells directly to businesses instead of to consumers. In January 2003, WholesaleFashion.com offered a 3/4 Sleeves Stretch Top for \$9.50, a Long Sleeve Turtle Neck Top for \$12.50, a Heart Neck Tank Top for \$9.25, and an American Flag Tank Top for \$9.75. **(Source:** www.wholesalefashion.com.)

A women's apparel store has up to \$1000 to spend on the four items. It has enough rack space for up to 60 tops. It anticipates the demand for tank tops to be greater than the demand for the other tops, so it wants to order at least twice as many tank tops as sleeved tops. The store intends to mark up all of the items by 100 percent. Assuming that all of the items that the store orders will sell, how many of each item should it order if it wants to maximize revenue?

SOLUTION Let x be the number of 3/4 Sleeves Stretch Tops, y be the number of Long Sleeve Turtle Neck Tops, z be the number of Heart Neck Tank Tops, and w be the number of American Flag Tank Tops. The number of sleeved tops is $x + y$, and the number of tank tops is $z + w$. Since the number of tank tops is to be at least twice the number of sleeved tops, we have

$$z + w \geq 2(x + y)$$

$$z + w \geq 2x + 2y$$

$$-2x - 2y + z + w \geq 0$$

$$2x + 2y - z - w \leq 0 \quad \text{Dividing by } -1 \text{ reverses the inequality sign}$$

We are asked to maximize the revenue from sales

$$R = 19x + 25y + 18.50z + 19.50w$$

subject to the constraints

$$x + y + z + w \leq 60 \quad \text{At most 60 tops are purchased}$$

$$2x + 2y - z - w \leq 0 \quad \text{Twice as many tank tops are purchased}$$

$$9.5x + 12.5y + 9.25z + 9.75w \leq 1000 \quad \text{At most \$1000 is spent on tops}$$

$$x \geq 0, y \geq 0, z \geq 0, w \geq 0 \quad \text{A nonnegative number of each type is purchased}$$

Rewriting the objective function and adding the slack variables to the constraint inequalities yields the system of equations

$$\begin{aligned}
x + y + z + w + s &= 60\\
2x + 2y - z - w + t &= 0\\
9.5x + 12.5y + 9.25z + 9.75w + u &= 1000\\
-19x - 25y - 18.50z - 19.50w + R &= 0
\end{aligned}$$

and the associated initial simplex tableau

$$\begin{array}{cccccccc|c}
x & y & z & w & s & t & u & R & \\
1 & 1 & 1 & 1 & 1 & 0 & 0 & 0 & 60\\
2 & 2 & -1 & -1 & 0 & 1 & 0 & 0 & 0\\
9.5 & 12.5 & 9.25 & 9.75 & 0 & 0 & 1 & 0 & 1000\\
\hline
-19 & -25 & -18.5 & -19.5 & 0 & 0 & 0 & 1 & 0
\end{array}$$

Pivot column (the y column)

The y column is the pivot column, since -25 is the negative entry in the bottom row with the greatest magnitude. Since the first three entries in the y column are positive, we must calculate three test ratios.

$$\begin{array}{cccccccc|c}
x & y & z & w & s & t & u & R & \\
1 & 1 & 1 & 1 & 1 & 0 & 0 & 0 & 60\\
2 & \boxed{2} & -1 & -1 & 0 & 1 & 0 & 0 & 0\\
9.5 & 12.5 & 9.25 & 9.75 & 0 & 0 & 1 & 0 & 1000\\
\hline
-19 & -25 & -18.5 & -19.5 & 0 & 0 & 0 & 1 & 0
\end{array}$$

$60/1 = 60$

$0/2 = 0$

$1000/12.5 = 80$

The pivot entry is the 2 in the second column and second row, since the second row has the smallest test ratio. We zero out the remaining entries in the second column with the indicated operations.

Because of the numerical complexity of the entries in the tableau, we will use the Technology Tips following this example to do the row operations. The resultant tableau is

$$\begin{array}{cccccccc|c}
x & y & z & w & s & t & u & R & \\
0 & 0 & 3 & 3 & 2 & -1 & 0 & 0 & 120\\
2 & 2 & -1 & -1 & 0 & 1 & 0 & 0 & 0\\
-6 & 0 & 31 & 32 & 0 & -12.5 & 2 & 0 & 2000\\
\hline
12 & 0 & -62 & -64 & 0 & 25 & 0 & 2 & 0
\end{array}$$

$2R_1 - R_2$

$2R_3 - 12.5R_2$

$2R_4 + 25R_2$

New pivot column (the w column)

The w column is the new pivot column, since -64 is the negative entry in the bottom row that has the largest magnitude. Since the only positive entries in the w column are in the first and third rows, we need to calculate only two test ratios.

$$\begin{array}{cccccccc|c}
x & y & z & w & s & t & u & R & \\
0 & 0 & 3 & \boxed{3} & 2 & -1 & 0 & 0 & 120\\
2 & 2 & -1 & -1 & 0 & 1 & 0 & 0 & 0\\
-6 & 0 & 31 & 32 & 0 & -12.5 & 2 & 0 & 2000\\
\hline
12 & 0 & -62 & -64 & 0 & 25 & 0 & 2 & 0
\end{array}$$

$120/3 = 40$

$2000/32 = 62.5$

Since the entry in the second row of the pivot column was negative, we did not calculate a test ratio for the second row. The pivot entry is the 3 in the w column and first row. We will zero out the remaining entries in the w column with the indicated operations. The resultant tableau is

$$\begin{array}{c} \\ \\ \\ \\ \end{array}\begin{bmatrix} x & y & z & w & s & t & u & R & \\ 0 & 0 & 3 & 3 & 2 & -1 & 0 & 0 & 120 \\ 6 & 6 & 0 & 0 & 2 & 2 & 0 & 0 & 120 \\ -18 & 0 & -3 & 0 & -64 & -5.5 & 6 & 0 & 2160 \\ \hline 36 & 0 & 6 & 0 & 128 & 11 & 0 & 6 & 7680 \end{bmatrix}\begin{array}{l} \\ \\ 3R_2 + R_1 \\ 3R_3 - 32R_1 \\ 3R_4 + 64R_1 \end{array}$$

Since all entries in the bottom row are nonnegative, this is the final simplex tableau. Crossing out the inactive variable columns and reading the resultant equations, we get

$$\begin{bmatrix} x & y & z & w & s & t & u & R & \\ 0 & 0 & 3 & 3 & 2 & -1 & 0 & 0 & 120 \\ 6 & 6 & 0 & 0 & 2 & 2 & 0 & 0 & 120 \\ -18 & 0 & -3 & 0 & -64 & -5.5 & 6 & 0 & 2160 \\ \hline 36 & 0 & 6 & 0 & 128 & 11 & 0 & 6 & 7680 \end{bmatrix}$$

$$x = 0 \qquad 6y = 120 \qquad z = 0 \qquad 3w = 120$$
$$y = 20 \qquad w = 40$$
$$s = 0 \qquad t = 0 \qquad 6u = 2160 \qquad 6R = 7680$$
$$u = 360 \qquad R = 1280$$

The maximum revenue that can be attained is \$1280. This will be achieved when 20 Long Sleeve Turtle Neck Tops and 40 American Flag Tank Tops are purchased. Recall that u was the slack variable associated with the equation for the total cost of the tops. Since $u = 360$, the amount spent by the store is \$360 less than the maximum amount allowed. That is, \$640 is spent on the tops. The additional money remains unspent. It was the space constraint (a maximum of 60 tops), not the budget constraint (a maximum of \$1000), that limited how many tops could be ordered.

For large tableaus or tableaus with noninteger entries, it is often helpful to do row operations on the calculator. The next four Technology Tips detail how to do this on the TI-83 Plus.

TECHNOLOGY TIP

Interchanging Two Rows

1. Enter a matrix or simplex tableau using the Matrix Editor. (Press 2nd x^{-1} to access the Matrix Editor.)

```
[A]
[[1  2  1 0 0 6...
 [2  1  0 1 0 4...
 [-5 -3 0 0 1 0...
```

(Continued)

2. Close the Matrix Editor to store the matrix, then reopen the Matrix Editor and move the cursor to the MATH menu.

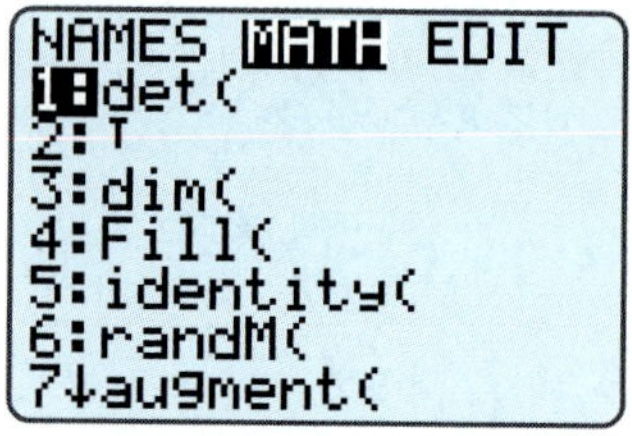

3. Scroll to C:rowSwap(and press ENTER. This operation is used to interchange one row with another row.

4. Type in the matrix name from the Matrix Menu, the first row number, and the second row number. For example, rowSwap([A],1,2)is equivalent to $R_1 \leftrightarrow R_2$.

```
rowSwap([A],1,2)
```

5. Press ENTER to display the new matrix or tableau.

```
rowSwap([A],1,2)
[[2  1  0 1 0 4…
 [1  2  1 0 0 6…
 [-5 -3 0 0 1 0…
```

TECHNOLOGY TIP

Adding One Row to Another Row

1. Enter a matrix or simplex tableau using the Matrix Editor. (Press 2nd x^{-1} to access the Matrix Editor.)

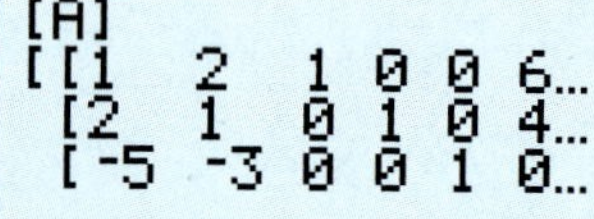

(Continued)

2. Close the Matrix Editor to store the matrix, then reopen the Matrix Editor and move the cursor to the `MATH` menu.

3. Scroll to `D:row+(` and press ENTER. This operation is used to add one row to another row.

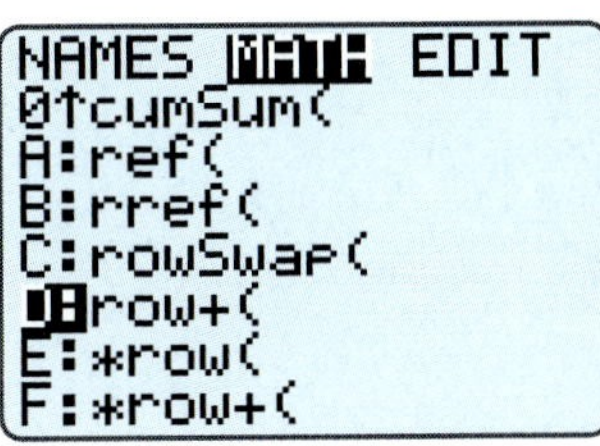

4. Type in the matrix name from the Matrix Menu, the first row number, and the second row number. The result will be placed in the last row listed. For example, `row+([A],2,3)` is equivalent to $R_2 + R_3 \rightarrow R_3$.

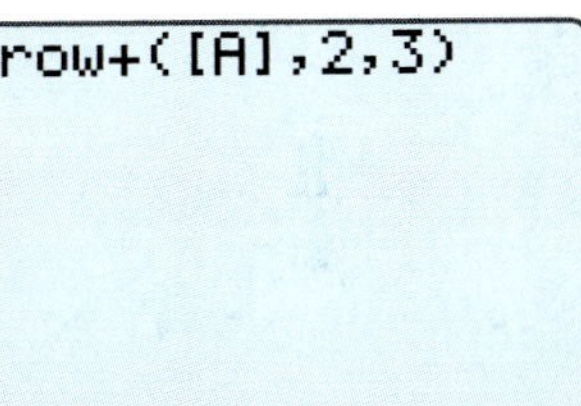

5. Press ENTER to display the new matrix or tableau.

```
row+([A],2,3)
[[1  2  1 0 0 6…
 [2  1  0 1 0 4…
 [-3 -2 0 1 1 4…
```

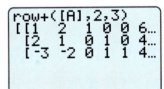

TECHNOLOGY TIP

Multiplying a Row by a Nonzero Constant

1. Enter a matrix or simplex tableau using the Matrix Editor. (Press 2nd x^{-1} to access the Matrix Editor.)

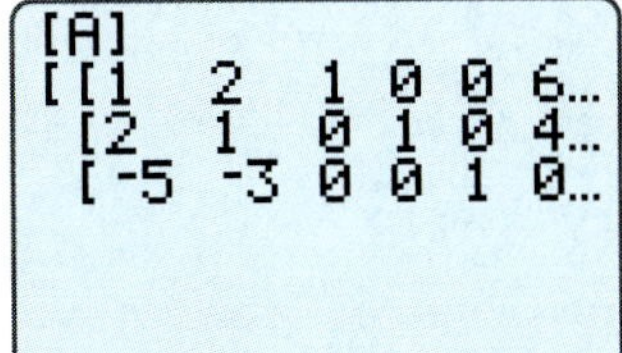

(Continued)

2. Close the Matrix Editor to store the matrix, then reopen the Matrix Editor and move the cursor to the `MATH` menu.

```
NAMES MATH EDIT
1:det(
2:T
3:dim(
4:Fill(
5:identity(
6:randM(
7↓augment(
```

3. Scroll to `E:*row(` and press ENTER. This operation is used to multiply a row by a nonzero constant.

```
NAMES MATH EDIT
0↑cumSum(
A:ref(
B:rref(
C:rowSwap(
D:row+(
E:*row(
F:*row+(
```

4. Type in the multiplier value, the matrix name from the Matrix Menu, and the row to be multiplied. The result will be placed in the last row listed. For example, `*row(-2,[A],1)` means $-2R_1 \rightarrow R_1$.

```
*row(-2,[A],1)
```

5. Press ENTER to display the new matrix or tableau.

```
*row(-2,[A],1)
[[-2 -4 -2 0 0 ...
 [2  1  0  1 0 ...
 [-5 -3 0  0 1 ...
```

TECHNOLOGY TIP

Adding a Multiple of One Row to Another Row

1. Enter a matrix or simplex tableau using the Matrix Editor. (Press 2nd x^{-1} to access the Matrix Editor.)

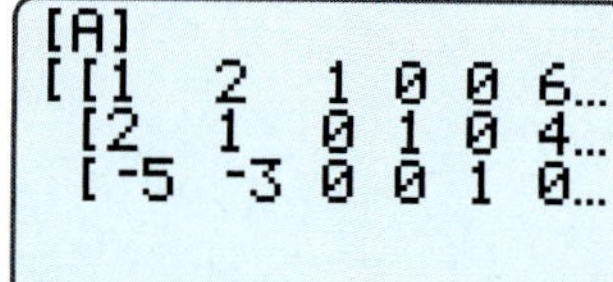

(Continued)

2. Close the Matrix Editor to store the matrix, then reopen the Matrix Editor and move the cursor to the MATH menu.

3. Scroll to `F:*row+(` and press ENTER. This operation is used to add a multiple of one row to another row.

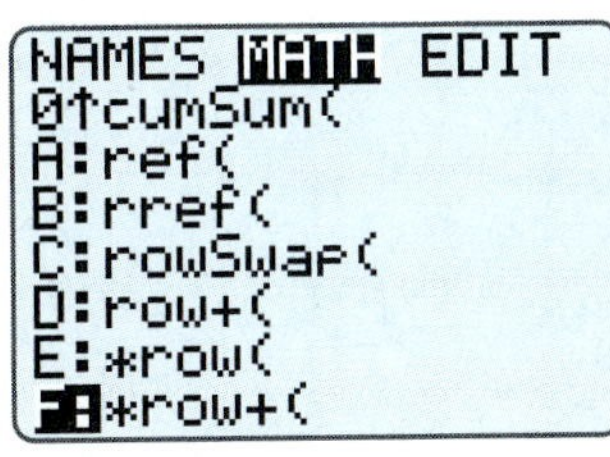

4. Type in the multiplier value, the matrix name, the row to be multiplied, and the row to be added. The result will be placed in the last row listed. For example, `*row+(4,[A],2,3)` means $4R_2 + R_3 \rightarrow R_3$.

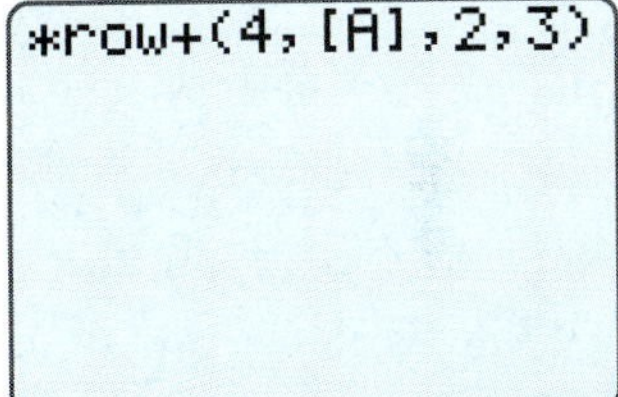

5. Press ENTER to display the new matrix or tableau.

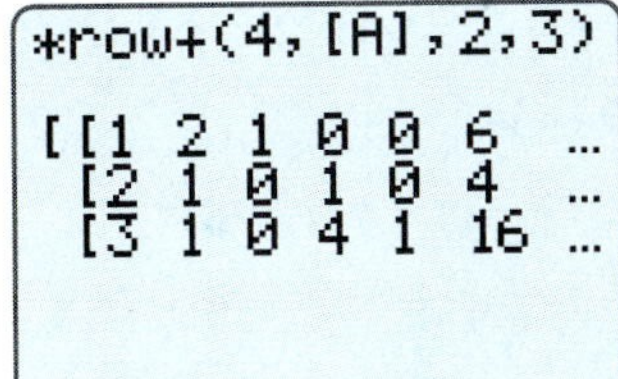

Recall that simplex method row operations must be of the form $aR_c \pm bR_p \rightarrow R_c$, with both a and b positive. Alternatively, we may write this expression as $\pm bR_p + aR_c \rightarrow R_c$. The latter form is more consistent with the calculator functionality and may make using the calculator easier.

EXAMPLE 5

Using the Simplex Method to Make Business Decisions

In January 2003, WholesaleFashion.com offered a Long Halter Dress for \$25.75, a Long Open Back Dress for \$21.75, a pair of Thai Silk Pump shoes for \$17.75, and a pair of Glitter Evening High Heels for \$17.50.
(**Source:** www.wholesalefashion.com.)

A women's apparel store has up to \$1000 to spend on the four items. It will mark up the dresses 100 percent but mark up the shoes only 50 percent. The store anticipates that it will sell 80 percent of each type of dress at full price and sell the remaining 20 percent of each type of dress at its cost. It expects to sell 90 percent of each type of shoes at full price and the remaining 10 percent of each shoe type at its cost. It expects to sell a pair of shoes with each dress it sells. (The store will market a dress and shoes as a package.) Additionally, it may sell some shoes separately. How many of each item should the store order to maximize its profit?

SOLUTION Let x be the number of Long Halter Dresses, y be the number of Long Open Back Dresses, z be the number of pairs of Thai Silk Pump shoes, and w be the number of pairs of Glitter Evening High Heels.

The revenue generated by selling 80 percent of each type of dress at full retail price is given by

$$\begin{aligned} R &= 51.50(0.8x) + 43.50(0.8y) \\ &= 41.20x + 34.80y \end{aligned}$$

The revenue generated by selling the remaining 20 percent of each type of dress at the wholesale cost is given by

$$\begin{aligned} R &= 25.75(0.2x) + 21.75(0.2y) \\ &= 5.15x + 4.35y \end{aligned}$$

The combined revenue from dress sales is

$$\begin{aligned} \text{Dress revenue} &= (41.20x + 34.80y) + (5.15x + 4.35y) \\ &= 46.35x + 39.15y \end{aligned}$$

The retail price of each shoe variety is 150 percent of the wholesale price.

$$\begin{aligned} \text{Thai Silk Pumps} &= 1.5(17.75) \\ &= 26.625 \\ &\approx 26.63 && \text{Rounded to the nearest cent} \end{aligned}$$

$$\begin{aligned} \text{Glitter Evening High Heels} &= 1.5(17.50) \\ &= 26.25 \end{aligned}$$

The revenue generated by selling 90 percent of each shoe variety at full retail price is given by

$$\begin{aligned} R &= 26.63(0.9z) + 26.25(0.9w) \\ &\approx 23.97z + 23.63w && \text{Rounded to the nearest cent} \end{aligned}$$

The revenue generated by selling 10 percent of each shoe variety at wholesale cost is

$$\begin{aligned} R &= 17.75(0.1z) + 17.50(0.1w) \\ &\approx 1.78z + 1.75w && \text{Rounded to the nearest cent} \end{aligned}$$

The combined revenue from shoe sales is

$$\begin{aligned} \text{Shoe revenue} &= (23.97z + 23.63w) + (1.78z + 1.75w) \\ &\approx 25.75z + 25.38w \end{aligned}$$

The combined revenue from dresses and shoes is the sum of the dress revenue and the shoe revenue.

$$\begin{aligned} \text{Combined revenue} &= \text{dress revenue} + \text{shoe revenue} \\ &\approx (46.35x + 39.15y) + (25.75z + 25.38w) \\ &\approx 46.35x + 39.15y + 25.75z + 25.38w \end{aligned}$$

The total cost of the dresses and shoes is

$$C = 25.75x + 21.75y + 17.75z + 17.5w$$

Since profit is the difference in revenue and cost,

$$\begin{aligned} P &= R - C \\ &\approx (46.35x + 39.15y + 25.75z + 25.38w) \\ &\quad - (25.75x + 21.75y + 17.75z + 17.50w) \\ &\approx 20.60x + 17.40y + 8.00z + 7.88w \end{aligned}$$

We need to order at least as many pairs of shoes as dresses, so

$$\text{Number of dresses} \leq \text{number of pairs of shoes}$$
$$(x + y) \leq (z + w)$$
$$x + y - z - w \leq 0$$

We are asked to maximize the profit $P = 20.6x + 17.4y + 8.00z + 7.88w$ subject to the constraints

$$\begin{aligned} x + y - z - w &\leq 0 && \text{Quantity constraint} \\ 25.75x + 21.75y + 17.75z + 17.5w &\leq 1000 && \text{Budget constraint} \\ x \geq 0, y \geq 0, z \geq 0, w &\geq 0 \end{aligned}$$

Rewriting the objective function and adding the slack variables to the constraint inequalities yields the following system of equations:

$$\begin{aligned} x + y - z - w + s &= 0 \\ 25.75x + 21.75y + 17.75z + 17.5w + t &= 1000 \\ -20.60x - 17.40y - 8.00z - 7.88w + P &= 0 \end{aligned}$$

and the associated initial simplex tableau

$$\begin{array}{c} \\ \\ \end{array}\left[\begin{array}{ccccccc|c} x & y & z & w & s & t & P & \\ 1 & 1 & -1 & -1 & 1 & 0 & 0 & 0 \\ 25.75 & 21.75 & 17.75 & 17.50 & 0 & 1 & 0 & 1000 \\ \hline -20.60 & -17.40 & -8.00 & -7.88 & 0 & 0 & 1 & 0 \end{array}\right]$$

Pivot column (x)

The x column is the pivot column, since -20.60 is the negative entry in the bottom row that has the largest magnitude. We must calculate test ratios for the first two rows.

$$\left[\begin{array}{ccccccc|c} x & y & z & w & s & t & P & \\ \boxed{1} & 1 & -1 & -1 & 1 & 0 & 0 & 0 \\ 25.75 & 21.75 & 17.75 & 17.50 & 0 & 1 & 0 & 1000 \\ \hline -20.60 & -17.40 & -8.00 & -7.88 & 0 & 0 & 1 & 0 \end{array}\right] \quad \begin{array}{l} \\ 0/1 = 0 \\ 1000/25.75 \approx 38.83 \\ \\ \end{array}$$

The pivot entry is the 1 in the x column and the first row. We will zero out the remaining entries in the first column with the indicated operations. Because of the numerical complexity of the entries in the tableau, we will use the Technology Tips described in this section to do the row operations.

The resulting tableau is

$$\begin{array}{ccccccc|c} x & y & z & w & s & t & P & \\ 1 & 1 & -1 & -1 & 1 & 0 & 0 & 0 \\ 0 & -4 & \boxed{43.5} & 43.25 & -25.75 & 1 & 0 & 1000 \\ \hline 0 & 3.2 & -28.60 & -28.48 & 20.60 & 0 & 1 & 0 \end{array} \quad \begin{array}{l} \\ \\ -25.75R_1 + R_2 \\ 20.60\,R_1 + R_3 \end{array}$$

New pivot column (z)

(Note: We wrote the row operations in the alternative form $\pm bR_p + aR_c \to R_c$.)

The z column is the new pivot column, since -28.60 is the negative entry in the bottom row that has the largest magnitude. Since there is only one positive entry in the z column above the bottom row, that entry is the pivot. The pivot entry is the 43.5 in the z column and the second row. We will zero out the remaining entries in the z column with the following row operations written in the alternative form:

$$R_2 + 43.5R_1 \to R_1$$
$$28.6R_2 + 43.5R_3 \to R_3$$

To do this on the calculator requires a sequence of operations. To perform the operation $R_2 + 43.5R_1 \to R_1$, we do the following to the tableau that is on the screen of our calculator:

Step 1: *row(43.5, *ans*, 1)	Multiplies R_1 by 43.5
Step 2: ENTER	Returns the modified matrix
Step 3: row + (*ans*, 2, 1)	Adds the first two rows of the modified matrix
Step 4: ENTER	Completes the operation

To perform the operation $28.6R_2 + 43.5R_3 \to R_3$, we do the following to the tableau that is on the screen of our calculator:

Step 1: *row(43.5, *ans*, 3)	Multiplies R_3 by 43.5
Step 2: ENTER	Returns the modified matrix
Step 3: *row + (28.6, *ans*, 2, 3)	Multiplies R_2 by 28.6 and adds it to the modified R_3
Step 4: ENTER	Completes the operation

The resultant tableau is

$$\begin{array}{ccccccc|c} x & y & z & w & s & t & P & \\ 43.5 & 39.5 & 0 & -0.25 & 17.75 & 1 & 0 & 1000 \\ 0 & -4 & 43.5 & \boxed{43.25} & -25.75 & 1 & 0 & 1000 \\ \hline 0 & 24.8 & 0 & -1.93 & 159.65 & 28.60 & 43.5 & 28600 \end{array}$$

New pivot column (w)

The bottom row contains a negative entry in the w column, so we must repeat the pivoting process. Since the only positive entry in the pivot column is 43.25, it is the new pivot. The indicated row operations (in alternative form) will zero out the additional terms in the pivot column.

$$0.25R_2 + 43.25R_1 \to R_1$$
$$1.93R_2 + 43.25R_3 \to R_3$$

To perform the operation $0.25R_2 + 43.25R_1 \rightarrow R_1$, we do the following to the tableau that is on the screen of our calculator:

Step 1: *row(43.25, *ans*, 1)	Multiplies R_1 by 43.25
Step 2: ENTER	Returns the modified matrix
Step 3: *row + (0.25, *ans*, 2, 1)	Multiplies R_2 by 0.25 and adds it to the modifed R_1
Step 4: ENTER	Completes the operation

To perform the operation $1.93R_2 + 43.25R_3 \rightarrow R_3$, we do the following to the tableau that is on the screen of our calculator:

Step 1: *row(43.25, *ans*, 3)	Multiplies R_3 by 43.25
Step 2: ENTER	Returns the modified matrix
Step 3: *row + (1.93, *ans*, 2, 3)	Multiplies R_2 by 1.93 and adds it to the modified R_3
Step 4: ENTER	Completes the operation

The resultant simplex tableau is

$$\begin{array}{ccccccc|c} x & y & z & w & s & t & P & \\ 1881.38 & 1707.38 & 10.88 & 0 & 761.25 & 43.5 & 0 & 43500 \\ 0 & -4 & 43.5 & 43.25 & -25.75 & 1 & 0 & 1000 \\ \hline 0 & 1064.88 & 83.96 & 0 & 6855.17 & 1238.88 & 1881.38 & 1238880 \end{array}$$

Crossing out the inactive variable columns and reading the resultant equations, we get

$$\begin{array}{ccccccc|c} x & y & z & w & s & t & P & \\ 1881.38 & 1707.38 & 10.88 & 0 & 761.25 & 43.5 & 0 & 43500 \\ 0 & -4 & 43.5 & 43.25 & -25.75 & 1 & 0 & 1000 \\ \hline 0 & 1064.88 & 83.96 & 0 & 6855.17 & 1238.88 & 1881.38 & 1238880 \end{array}$$

Reading from the tableau, we see that

$$1881.38x = 43500 \qquad y = 0 \qquad z = 0 \qquad 43.25w = 1000$$
$$x = 23.12 \qquad\qquad w = 23.12$$
$$s = 0 \qquad t = 0 \qquad 1881.38P = 1238880$$
$$P = 658.50$$

Since it doesn't make sense to talk about a fraction of a dress or pair of shoes, we will round the values to the nearest whole number. Rounding to whole numbers doesn't guarantee the maximum whole number solution. However, if the solution isn't the maximum whole number solution, it will be close to the optimal solution. We estimate that the maximum profit will be achieved when 23 Long Halter Dresses and 23 pairs of Glitter Evening High Heels are purchased. Since $t = 0$, the amount spent by the store appears to be exactly \$1000. (Actually, the cost is \$994.75, since we rounded the number of dresses and shoes.) The maximum profit is expected to be \$658.50. (Since we rounded down the number of shoes and dresses, the exact profit is \$655.04.)

Even though Example 5 required some rounding to make sense of the results, the numerical analysis provided some valuable input to help us make a sound business decision.

4.3 Summary

In this section, you learned the simplex method for solving standard maximization problems. You learned that this is the method of choice in solving problems with more than two decision variables.

4.3 Exercises

In Exercises 1–10, determine whether the problem is a standard maximization problem. If it isn't, explain why.

1. Maximize $P = 9x + 8y$

Subject to $\begin{cases} 2x + y \le 10 \\ -x + y \le 1 \\ x \ge 0 \\ y \ge 0 \end{cases}$

2. Maximize $P = -2x - 5y$

Subject to $\begin{cases} 4x + 9y \le 10 \\ -11x + y \le -21 \\ x \ge 0 \\ y \ge 0 \end{cases}$

3. Maximize $P = 6xy$

Subject to $\begin{cases} 6x + 7y \le 13 \\ -8x - 4y \le 12 \\ x \ge 0 \\ y \ge 0 \end{cases}$

4. Maximize $P = 4x - 2y + z$

Subject to $\begin{cases} 2x + y + 4z \le 24 \\ -2x + 3y + 3z \le 150 \\ x \ge 0 \\ y \ge 0 \\ z \ge 0 \end{cases}$

5. Maximize $P = -1.2x - 2.8y + 4.3z$

Subject to $\begin{cases} 3.2x + 1.5y + 7.4z \le 249.8 \\ -2.7x + 3.4y + 3.9z \le 190.1 \end{cases}$

6. Maximize $P = 4x + 4y + 9z$

Subject to $\begin{cases} -6x - y + 4z \le -24 \\ -9x - 1.5y + 6z \le 36 \\ x \ge 0 \\ y \ge 0 \\ z \ge 0 \end{cases}$

7. Maximize $P = x + 9y + 8z$

Subject to $\begin{cases} -6x - y + 4z \le -24 \\ -9x - 1.5y + 6z \le 36 \\ x \ge 0 \\ y \ge 0 \\ z \ge 0 \end{cases}$

8. Maximize $P = 4x + 4y + 9z$

Subject to $\begin{cases} -3x - 2y + 4z \ge -22 \\ 5x - 2.5y + 6z \le 36 \\ x \ge 0 \\ y \ge 0 \\ z \ge 0 \end{cases}$

9. Maximize $P = -8x + 2y - z$

Subject to $\begin{cases} z \le 4 \\ x - y \le 3 \\ 2x - z \le 5 \\ x + y + z \le 0 \end{cases}$

10. Maximize $P = 5x + 7z$

Subject to $\begin{cases} 4x + 2y \le 16 \\ 4x - 2y \le 3 \\ x \ge 0 \\ y \ge 0 \end{cases}$

In Exercises 11–20, solve the standard maximization problems by using the simplex method. Check your answer by graphing the feasible region and calculating the value of the objective function at each of the corner points.

11. Maximize $P = 9x + 8y$

Subject to $\begin{cases} 2x + y \le 10 \\ -x + y \le 1 \\ x \ge 0 \\ y \ge 0 \end{cases}$

12. Maximize $P = -2x + 4y$

Subject to $\begin{cases} 2x + y \le 6 \\ x + y \le 4 \\ x \ge 0 \\ y \ge 0 \end{cases}$

13. Maximize $P = 5x - 2y$

Subject to $\begin{cases} -3x + y \le 0 \\ 3x + y \le 12 \\ x \ge 0 \\ y \ge 0 \end{cases}$

14. Maximize $P = 7x + 10y$

Subject to $\begin{cases} x + y \le 4 \\ 5x + y \le 12 \\ x \ge 0 \\ y \ge 0 \end{cases}$

15. Maximize $P = 4x + 4y$

Subject to $\begin{cases} 4x + 3y \le 20 \\ -x + 2y \le 6 \\ x \ge 0 \\ y \ge 0 \end{cases}$

16. Maximize $P = 5x - 2y$

Subject to $\begin{cases} 4x + 5y \le 28 \\ -3x + 2y \le 2 \\ x \ge 0 \\ y \ge 0 \end{cases}$

17. Maximize $P = 6x + 5y$

Subject to $\begin{cases} 3x + 2y \le 12 \\ x + y \le 5 \\ x \ge 0 \\ y \ge 0 \end{cases}$

18. Maximize $P = 6x + 2y$

Subject to $\begin{cases} x + y \le 10 \\ 2x + y \le 16 \\ -2x + y \le 0 \\ x \ge 0 \\ y \ge 0 \end{cases}$

19. Maximize $P = 7x - 5y$

Subject to $\begin{cases} 6x - 5y \le 30 \\ 2x + 2y \le 10 \\ -5x + 3y \le 7 \\ x \ge 0 \\ y \ge 0 \end{cases}$

20. Maximize $P = x + y$

Subject to $\begin{cases} 5x + y \le 20 \\ -x + 2y \le 29 \\ 10x + y \le 30 \\ x \ge 0 \\ y \ge 0 \end{cases}$

In Exercises 21–30, solve the standard maximization problems by using the simplex method.

21. Maximize $P = 5x + 6y + 6z$

Subject to $\begin{cases} 2x + 3y + 2z \le 120 \\ -x + y + z \le 60 \\ x \ge 0 \\ y \ge 0 \\ z \ge 0 \end{cases}$

22. Maximize $P = 4x + 6y + 5z$

Subject to $\begin{cases} 2x + 3y + 3z \le 210 \\ x + y + z \le 100 \\ x \ge 0 \\ y \ge 0 \\ z \ge 0 \end{cases}$

23. Maximize $P = 10x + 6y + 12z$

Subject to $\begin{cases} 8x + 5y + 9z \le 360 \\ x + y + z \le 50 \\ x \ge 0 \\ y \ge 0 \\ z \ge 0 \end{cases}$

24. Maximize $P = x - 2y + 4z$

Subject to $\begin{cases} 6x - 5y + 10z \le 300 \\ 2x + 5y + 2z \le 500 \\ x \ge 0 \\ y \ge 0 \\ z \ge 0 \end{cases}$

25. Maximize $P = 5x - 2y + 4z$

Subject to $\begin{cases} 8x + 5y + 9z \le 360 \\ x + y + z \le 50 \\ x \ge 0 \\ y \ge 0 \\ z \ge 0 \end{cases}$

26. Maximize $P = 4x + 4y - 10z$

Subject to $\begin{cases} 3x + 3y + 6z \le 42 \\ 2x + y + z \le 8 \\ 3x + 2y + 5z \le 23 \\ x \ge 0 \\ y \ge 0 \\ z \ge 0 \end{cases}$

27. Maximize $P = -x - y + 10z$

Subject to $\begin{cases} 2x + 2y + 2z \leq 14 \\ 4x + 2y + 2z \leq 16 \\ 3x + 2y + 5z \leq 23 \\ x \geq 0 \\ y \geq 0 \\ z \geq 0 \end{cases}$

28. Maximize $P = x + 2y + 3z$

Subject to $\begin{cases} x + z \leq 20 \\ y + 2z \leq 30 \\ x + 2y + 3z \leq 60 \\ x \geq 0 \\ y \geq 0 \\ z \geq 0 \end{cases}$

29. Maximize $P = x + y + z$

Subject to $\begin{cases} x + z \leq 0 \\ y + 2z \leq 5 \\ x + 2y + 3z \leq 8 \\ x \geq 0 \\ y \geq 0 \\ z \geq 0 \end{cases}$

30. Maximize $P = x + 2y + 3z$

Subject to $\begin{cases} x - z \leq 20 \\ y + z \leq 30 \\ x - 2y + z \leq 40 \\ x \geq 0 \\ y \geq 0 \\ z \geq 0 \end{cases}$

In Exercises 31–36, set up and solve the standard maximization problem using the simplex method.

31. **Resource Allocation: Beverages** In the mood for a new beverage recipe? iDrink.com provides you with recipes for alcoholic and nonalcoholic drinks based on the ingredients you have on hand.

The Cranberry Cooler calls for 2 ounces of lemon-lime soda and 4 ounces of cranberry juice in addition to other ingredients. Nancy's Party Punch calls for 32 ounces of cranberry juice and no lemon-lime soda. Jimmy Wallbanger calls for 6 ounces of lemon-lime soda and no cranberry juice. (**Source:** www.idrink.com.) The Cranberry Cooler recipe yields 1 serving, the Nancy's Party Punch recipe yields 9 servings, and the Jimmy Wallbanger recipe yields 1 serving.

If you have 2 quarts (64 ounces) of lemon-lime soda and 1 gallon (128 ounces) of cranberry juice, at most how many drink servings can you make?

32. **Resource Allocation: Sandwiches** A plain hamburger requires one ground beef patty and a bun. A cheeseburger requires one ground beef patty, one slice of cheese, and a bun. A double cheeseburger requires two ground beef patties, two slices of cheese, and a bun.

Frozen hamburger patties are typically sold in packs of 12; hamburger buns, in packs of 8; and cheese slices, in packs of 24.

A family is in charge of providing burgers for a neighborhood block party. The family members have purchased 13 packs of buns, 11 packs of hamburger patties, and 3 packs of cheese slices. How many of each type of sandwich should the family prepare if its members want to maximize the number of burgers with cheese?

33. **Resource Allocation: Sandwiches** Repeat Exercise 32, except maximize the number of hamburger patties used.

34. **Resource Allocation: Sandwiches** Repeat Exercise 32. except maximize the number of double cheeseburgers.

35. **Battery Sales** AAA Alkaline Discount Batteries sells low-cost batteries to consumers. In August 2002, the company offered AA batteries at the following prices: 50-pack for \$10.00, 100-pack for \$18.00, and 600-pack for \$96.00. (**Source:** www.aaaalkalinediscountbatteries.com.)

An electronics store owner wants to buy at most 1000 batteries and spend at most \$175. She expects that she'll be able to resell all of the batteries she orders for \$4.00 per 4-pack. How many packs (50-packs, 100-packs, or 600-packs) should she order if she wants to maximize her revenue? What is her maximum revenue?

36. **Battery Sales** Repeat Exercise 35, except maximize profit. Assume that her only cost is the cost of the batteries. What is her maximum profit?

Exercises 37–38 are intended to expand your understanding of the simplex method.

37. Given the standard maximization problem

Maximize $z = 4x + 4y$

Subject to $\begin{cases} -x + y \leq 4 \\ x + 2y \leq 14 \\ 5x + 2y \leq 30 \\ x \geq 0, y \geq 0 \end{cases}$

do the following:

(a) Graph the feasible region and find the coordinates of the corner points.

(b) Solve the standard maximization problem using the simplex method.

(c) Find the feasible solution associated with each simplex tableau and label the point on the graph that is associated with each feasible solution.

(d) Explain what the simplex method does in terms of the graph of the feasible region.

38. Repeat Exercise 37; however, this time pick a different column to be the first pivot column. Compare and contrast your results with those of Exercise 37.

4.4 Solving Standard Minimization Problems with the Dual

- Find the solution to a standard minimization problem by solving the dual problem with the simplex method

GETTING STARTED A food distributor has two production facilities: one in Portland, Oregon, and the other in Spokane, Washington. Subject to production limitations and market demands, what shipment plan will minimize the distributor's shipment costs? Questions such as this may often be answered by solving a standard minimization problem.

In this section, we will introduce standard minimization problems and show how finding the maximum solution for a dual problem leads us to the minimum solution of the minimization problem.

STANDARD MINIMIZATION PROBLEM

A **standard minimization problem** is a linear programming problem with an objective function that is to be minimized. The objective function is of the form

$$P = ax + by + cz + \cdots$$

where $a, b, c, \ldots$ are real numbers and $x, y, z, \ldots$ are decision variables.

The decision variables are constrained to nonnegative values. Additional constraints are of the form

$$Ax + By + Cz + \cdots \geq M$$

where $A, B, C, \ldots$ are real numbers and M is nonnegative.

We are told that M is nonnegative. That is, $M \geq 0$. Is the origin in the feasible region? Substituting the origin $x = 0,\ y = 0,\ z = 0, \ldots$ into each constraint inequality $Ax + By + Cz + \cdots \geq M$ yields the inequality $A(0) + B(0) + C(0) + \cdots \geq M$ which simplifies to $0 \geq M$. Since M is nonnegative, the inequality $0 \geq M$ is satisfied if and only if $M = 0$. That is, the origin is in the feasible region if and only if $M = 0$ for every constraint. If $M \neq 0$ for one or more constraints, the origin is not in the feasible region.

EXAMPLE 1

Determining If a Linear Programming Problem Is a Standard Minimization Problem

Determine if the linear programming problem is a standard minimization problem.

$$\text{Minimize} \quad P = 2x + 3y$$

$$\text{Subject to} \begin{cases} 3x + y \geq 6 \\ 2x - y \geq 4 \\ \quad x \geq 0, y \geq 0 \end{cases}$$

SOLUTION The objective function is to be minimized, the constraints are of the form $Ax + By + Cz + \cdots \geq M$ with M nonnegative, and the decision variables are nonnegative. Therefore, the problem is a standard minimization problem.

EXAMPLE 2

Determining If a Linear Programming Problem Is a Standard Minimization Problem

Determine if the linear programming problem is a standard minimization problem.

$$\text{Minimize} \quad P = 4x + 2y$$

$$\text{Subject to} \begin{cases} -3x + 7y \leq -9 \\ 7x + 5y \geq 4 \\ \quad x \geq 0, y \geq 0 \end{cases}$$

SOLUTION At first glance, the problem doesn't look like a standard minimization problem. The first constraint has a less than or equal sign, and the constant term is negative. However, if we multiply the first constraint by -1, it becomes $3x - 7y \geq 9$. The problem may then be rewritten as

$$\text{Minimize} \quad P = 4x + 2y$$

$$\text{Subject to} \begin{cases} 3x - 7y \geq 9 \\ 7x + 5y \geq 4 \\ \quad x \geq 0, y \geq 0 \end{cases}$$

Therefore, the linear programming problem is a standard minimization problem.

When setting up a linear programming problem, if we make sure that the constant term of each constraint is nonnegative, then we will be able to readily see if the problem is a standard maximization or minimization problem.

The Dual

For a standard minimization problem whose objective function has nonnegative coefficients, we may construct a standard maximization problem called the **dual problem.** By solving the dual problem, we can find the solution to the standard minimization problem. Consider the standard minimization problem,

$$\text{Minimize} \quad P = 6x + 5y$$

$$\text{Subject to} \begin{cases} x + y \geq 2 \\ 2x + y \geq 3 \\ \quad x \geq 0, y \geq 0 \end{cases}$$

We first construct a matrix for the problem as shown.

$$A = \left[\begin{array}{cc|c} 1 & 1 & 2 \\ 2 & 1 & 3 \\ \hline 6 & 5 & 1 \end{array}\right] \qquad \begin{array}{l} x + y \geq 2 \\ 2x + y \geq 3 \\ 6x + 5y = P \end{array}$$

The constraints are placed in the first two rows of the matrix, and the coefficients of the objective function are placed in the last row of the matrix. The matrix for the dual problem is found by transposing the matrix. The **transpose** of a matrix A, which is written A^T, is created by switching the rows and columns of A. In this case, the first row of A is $[1 \quad 1 \quad 2]$, so the first column of A^T is $\begin{bmatrix} 1 \\ 1 \\ 2 \end{bmatrix}$. Repeating the process for the additional columns of A^T yields $A^T = \left[\begin{array}{cc|c} 1 & 2 & 6 \\ 1 & 1 & 5 \\ \hline 2 & 3 & 1 \end{array}\right]$. We construct the dual problem from A^T. We first extract the constraints and the objective function from the matrix.

$$A = \left[\begin{array}{cc|c} 1 & 2 & 6 \\ 1 & 1 & 5 \\ \hline 2 & 3 & 1 \end{array}\right] \qquad \begin{array}{l} x + 2y \leq 6 \\ x + y \leq 5 \\ 2x + 3y = P \end{array}$$

The dual problem is

$$\text{Maximize} \quad P = 2x + 3y$$

$$\text{Subject to} \begin{cases} x + 2y \leq 6 \\ x + y \leq 5 \\ x \geq 0, y \geq 0 \end{cases}$$

How are the optimal values of the standard minimization problem and its dual problem related? Let's look at the graphs of the feasible regions (Figures 4.26 and 4.27).

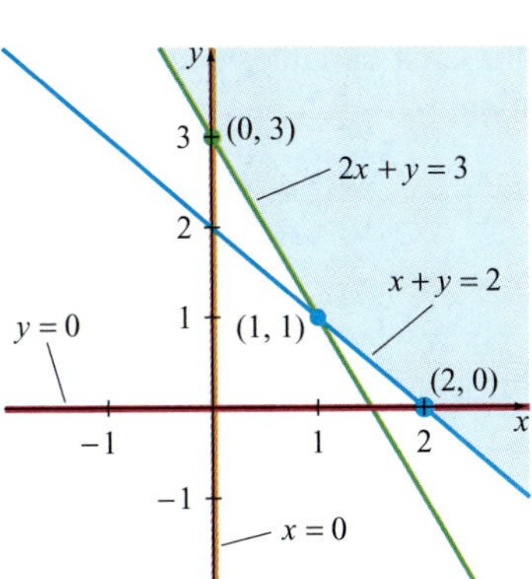

FIGURE 4.26 Standard Minimization Problem
Corner points:
(0, 3), (1, 1), (2, 0)

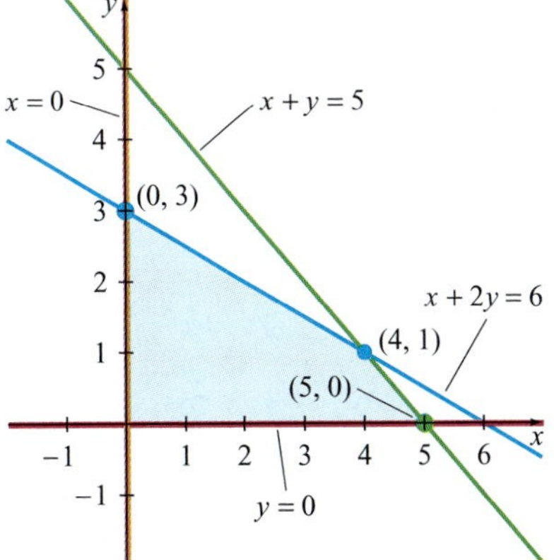

FIGURE 4.27 Standard Maximization Problem
Corner Points:
(0, 0), (0, 3), (4, 1), (5, 0)

For each problem, we calculate the value of the objective function at each corner point (see Tables 4.20 and 4.21).

TABLE 4.20 Standard Minimization Problem

x	y	$P = 6x + 5y$
0	3	15
1	1	**11**
2	0	12

TABLE 4.21 Standard Maximization Problem

x	y	$P = 2x + 3y$
0	0	0
0	3	9
4	1	**11**
5	0	10

Observe that the minimum value of the *minimization* problem objective function is 11. Similarly, the maximum value of the *maximization* problem objective function is 11. This leads us to the Fundamental Principle of Duality.

FUNDAMENTAL PRINCIPLE OF DUALITY

A standard minimization problem has a solution if and only if its dual problem has a solution. If a solution exists, the optimal value of the standard minimization problem and the optimal value of the dual problem are equal.

Although we already know the optimal value of the dual problem, we will demonstrate how using the simplex method to find the maximum value of the dual problem helps us determine the minimum value of the standard minimization problem. Recall that the dual problem is

$$\text{Maximize} \quad P = 2x + 3y$$

$$\text{Subject to} \begin{cases} x + 2y \le 6 \\ x + y \le 5 \\ x \ge 0, y \ge 0 \end{cases}$$

Adding in the slack variables and rewriting the objective function yield the system of equations

$$\begin{aligned} x + 2y + s \quad\quad &= 6 \\ x + y \quad + t \quad &= 5 \\ -2x - 3y \quad\quad + P &= 0 \end{aligned}$$

and the corresponding initial simplex tableau

$$\begin{array}{c} \begin{matrix} x & y & s & t & P & \end{matrix} \\ \left[\begin{array}{rrrrr|r} 1 & 2 & 1 & 0 & 0 & 6 \\ 1 & 1 & 0 & 1 & 0 & 5 \\ \hline -2 & -3 & 0 & 0 & 1 & 0 \end{array}\right] \end{array}$$

The y column is the pivot column, since it contains the negative entry in the bottom row with the largest magnitude. We calculate the test ratios and identify the pivot.

$$\begin{array}{c} \\ \\ \\ \\ \end{array}\begin{matrix} x & y & s & t & P & \\ \end{matrix}$$

$$\left[\begin{array}{rrrrr|r} 1 & \boxed{2} & 1 & 0 & 0 & 6 \\ 1 & 1 & 0 & 1 & 0 & 5 \\ \hline -2 & -3 & 0 & 0 & 1 & 0 \end{array}\right] \quad \begin{array}{l} 6/2 = 3 \\ 5/1 = 5 \\ \\ \end{array}$$

The 2 in the first row of the y column is the pivot, since the first row has the smallest test ratio. We zero out the remaining terms in the y column using the indicated operations.

$$\begin{matrix} x & y & s & t & P & \end{matrix}$$

$$\left[\begin{array}{rrrrr|r} 1 & 2 & 1 & 0 & 0 & 6 \\ 1 & 0 & -1 & 2 & 0 & 4 \\ \hline -1 & 0 & 3 & 0 & 2 & 18 \end{array}\right] \quad \begin{array}{l} \\ 2R_2 - R_1 \\ 2R_3 + 3R_1 \end{array}$$

The x column is the new pivot column, since it is the only column with a negative entry in the bottom row. We calculate the test ratios and identify the pivot.

$$\begin{matrix} x & y & s & t & P & \end{matrix}$$

$$\left[\begin{array}{rrrrr|r} 1 & 2 & 1 & 0 & 0 & 6 \\ \boxed{1} & 0 & -1 & 2 & 0 & 4 \\ \hline -1 & 0 & 3 & 0 & 2 & 18 \end{array}\right] \quad \begin{array}{l} 6/1 = 6 \\ 4/1 = 4 \\ \\ \end{array}$$

The 1 in the second row of the x column is the pivot, since the second row has the smallest test ratio. We zero out the remaining entries in the x column with the indicated operations.

$$\begin{matrix} x & y & s & t & P & \end{matrix}$$

$$\left[\begin{array}{rrrrr|r} 0 & 2 & 2 & -2 & 0 & 2 \\ 1 & 0 & -1 & 2 & 0 & 4 \\ \hline 0 & 0 & 2 & 2 & 2 & 22 \end{array}\right] \quad \begin{array}{l} R_1 - R_2 \\ \\ R_3 + R_2 \end{array}$$

We must perform one additional step to make the nonzero term in each active column equal 1.

$$\begin{matrix} x & y & s & t & P & \end{matrix}$$

$$\left[\begin{array}{rrrrr|r} 0 & 1 & 1 & -1 & 0 & 1 \\ 1 & 0 & -1 & 2 & 0 & 4 \\ \hline 0 & 0 & 1 & 1 & 1 & 11 \end{array}\right] \quad \begin{array}{l} 1/2\ R_1 \\ \\ 1/2\ R_3 \end{array}$$

Reading from the tableau, we have

$$x = 4 \qquad y = 1 \qquad P = 11$$

The corner point $(4, 1)$ yields the maximum value of the objective function of the dual problem $P = 2x + 3y$.

Ironically, the optimal solution of the standard minimization problem is also nested in the final simplex tableau. With all active variable columns containing only 1s and 0s, the values in the bottom row of the s and t columns correspond with the x and y values, respectively, of the optimal solution of the standard minimization problem. That is, $x = 1$ and $y = 1$. The optimal value is the same as that of the dual problem: $P = 11$. We can verify that the corner point $(1, 1)$ of the standard minimization problem yields an objective function value of 11 by substituting the point into the objective function of the standard minimization problem $P = 6x + 5y$.

$$\begin{aligned} P &= 6x + 5y \\ 11 &\stackrel{?}{=} 6(1) + 5(1) \\ 11 &= 11 \end{aligned}$$

The relationship between the standard minimization problem and its dual problem is somewhat remarkable, and the problem-solving method is straightforward.

HOW TO Using Duals to Solve Standard Minimization Problems

1. Find the dual standard maximization problem.
2. Use the simplex method to solve the maximization problem.
3. The maximum value of the objective function of the dual problem is the minimum value of the objective function of the minimization problem.
4. The optimum solution of the minimization problem is given in the entries of the bottom row of the final tableau corresponding with the columns of the slack variables (as long as the entry in the P column equals 1).

EXAMPLE 3 Solving a Standard Minimization Problem

Solve the standard minimization problem.

$$\text{Minimize} \quad P = 4x + 5y + 6z$$

$$\text{Subject to} \begin{cases} x + y + z \geq 3 \\ 2x - z \geq 0 \\ x \geq 0, y \geq 0, z \geq 0 \end{cases}$$

SOLUTION We begin by constructing a matrix from the constraints and objective function.

$$A = \left[\begin{array}{ccc|c} \overset{x}{1} & \overset{y}{1} & \overset{z}{1} & 3 \\ 2 & 0 & -1 & 0 \\ \hline 4 & 5 & 6 & 1 \end{array}\right] \qquad \begin{aligned} x + y + z &\geq 3 \\ 2x + 0y - z &\geq 0 \\ 4x + 5y + 6z &= P \end{aligned}$$

Next we find A^T and extract the objective function and constraints. Note that the constraints are of the form $ax + by \leq M$ with M nonnegative.

$$A^T = \left[\begin{array}{cc|c} 1 & 2 & 4 \\ 1 & 0 & 5 \\ 1 & -1 & 6 \\ \hline 3 & 0 & 1 \end{array}\right] \qquad \begin{aligned} x + 2y &\leq 4 \\ x + 0y &\leq 5 \\ x - y &\leq 6 \\ 3x + 0y &= P \end{aligned}$$

The dual problem is

$$\text{Maximize} \quad P = 3x$$

$$\text{Subject to} \quad \begin{cases} x + 2y \le 4 \\ x \le 5 \\ x - y \le 6 \\ x \ge 0, y \ge 0, z \ge 0 \end{cases}$$

The dual is a standard maximization problem. We solve the problem using the simplex method. We begin by adding slack variables to the constraints.

$$\begin{aligned} x + 2y + s &= 4 \\ x + t &= 5 \\ x - y + u &= 6 \\ x \ge 0, y \ge 0, z &\ge 0 \end{aligned}$$

Rewriting the objective function yields $-3x + P = 0$. The initial simplex tableau is

$$\begin{array}{cccccc|c} x & y & s & t & u & P & \\ \boxed{1} & 2 & 1 & 0 & 0 & 0 & 4 \\ 1 & 0 & 0 & 1 & 0 & 0 & 5 \\ 1 & -1 & 0 & 0 & 1 & 0 & 6 \\ \hline -3 & 0 & 0 & 0 & 0 & 1 & 0 \end{array}$$

The x column is the pivot column, since it is the only column with a negative entry in the bottom row. Mentally calculating the test ratios, we see that the entry in the first row and the x column is the pivot.

The final tableau is obtained by using the indicated row operations.

$$\begin{array}{cccccc|cl} x & y & s & t & u & P & & \\ 1 & 2 & 1 & 0 & 0 & 0 & 4 & \\ 0 & -2 & -1 & 1 & 0 & 0 & 1 & R_2 - R_1 \\ 0 & -3 & -1 & 0 & 1 & 0 & 2 & R_3 - R_1 \\ \hline 0 & 6 & 3 & 0 & 0 & 1 & 12 & R_4 + 3R_1 \end{array}$$

Reading from the tableau, we see that the objective function of the *maximization* problem has a maximum value of 12. This occurs when $x = 4$ and $y = 0$. According to the fundamental principle of duality, the minimum value of the objective function of the *minimization* problem is also 12. The solution to the *minimization* problem may be found by looking at the slack variable columns in the bottom row of the final tableau. The s, t, and u columns of the final simplex tableau are the x, y, and z values of the solution to the minimization problem. That is, $(3, 0, 0)$ yields an objective function value of 12. This can be verified by substituting $x = 3, y = 0$, and $z = 0$ into $P = 4x + 5y + 6z$.

$$\begin{aligned} P &= 4(3) + 5(0) + 6(0) \\ &= 12 \end{aligned}$$

It is important to note that a standard minimization problem and its dual problem may have different numbers of decision variables. In Example 3, the standard minimization problem had three decision variables, x, y, and z, whereas the dual problem had only two decision variables, x and y.

EXAMPLE 4 Solving a Standard Minimization Problem

Solve the standard minimization problem.

$$\text{Minimize} \quad P = 2x + y + 2z$$

$$\text{Subject to} \begin{cases} x + 2y + z \geq 6 \\ \quad 3y + 2z \geq 12 \\ \qquad x \geq 0, y \geq 0, z \geq 0 \end{cases}$$

SOLUTION We have

$$A = \left[\begin{array}{ccc|c} 1 & 2 & 1 & 6 \\ 0 & 3 & 2 & 12 \\ \hline 2 & 1 & 2 & 1 \end{array}\right] \qquad \begin{array}{r} x + 2y + z \geq 6 \\ 3y + 2z \geq 12 \\ 2x + y + 2z = P \end{array}$$

Therefore,

$$A^T = \left[\begin{array}{cc|c} 1 & 0 & 2 \\ 2 & 3 & 1 \\ 1 & 2 & 2 \\ \hline 6 & 12 & 1 \end{array}\right] \quad \text{and} \quad \begin{array}{r} x \leq 2 \\ 2x + 3y \leq 1 \\ x + 2y \leq 2 \\ 6x + 12y = P \end{array}$$

The dual problem is

$$\text{Maximize} \quad P = 6x + 12y$$

$$\text{Subject to} \begin{cases} x \leq 2 \\ 2x + 3y \leq 1 \\ x + 2y \leq 2 \\ x \geq 0, \ y \geq 0 \end{cases}$$

Adding the slack variables and rewriting the objective function yields

$$\begin{aligned} x + s &= 2 \\ 2x + 3y + t &= 1 \\ x + 2y + u &= 2 \\ -6x - 12y + P &= 0 \end{aligned}$$

The initial simplex tableau is

$$\begin{array}{c} \begin{array}{cccccc|c} x & y & s & t & u & P & \end{array} \\ \left[\begin{array}{cccccc|c} 1 & 0 & 1 & 0 & 0 & 0 & 2 \\ 2 & \boxed{3} & 0 & 1 & 0 & 0 & 1 \\ 1 & 2 & 0 & 0 & 1 & 0 & 2 \\ \hline -6 & -12 & 0 & 0 & 0 & 1 & 0 \end{array}\right] \end{array}$$

The y column is the pivot, since it contains the negative entry in the bottom row with the largest magnitude. The pivot is the 3 in the second row and the y column because the second row's test ratio of $\frac{1}{3}$ is smaller than the third row's test ratio of $\frac{2}{2}$. Using the indicated row operations yields

$$\begin{array}{cccccc|c} x & y & s & t & u & P & \\ 1 & 0 & 1 & 0 & 0 & 0 & 2 \\ 2 & 3 & 0 & 1 & 0 & 0 & 1 \\ -1 & 0 & 0 & -2 & 3 & 0 & 4 \\ \hline 2 & 0 & 0 & 4 & 0 & 1 & 4 \end{array} \quad \begin{array}{l} \\ \\ \\ 3R_3 - 2R_2 \\ R_4 + 4R_2 \end{array}$$

To obtain the final simplex tableau, we must ensure that the nonzero entry of each active variable column is equal to 1.

$$\begin{array}{cccccc|c} x & y & s & t & u & P & \\ 1 & 0 & 1 & 0 & 0 & 0 & 2 \\ \frac{2}{3} & 1 & 0 & \frac{1}{3} & 0 & 0 & \frac{1}{3} \\ -\frac{1}{3} & 0 & 0 & -\frac{2}{3} & 1 & 0 & \frac{4}{3} \\ \hline 2 & 0 & 0 & 4 & 0 & 1 & 4 \end{array} \quad \begin{array}{l} \\ \\ \frac{1}{3}R_2 \\ \frac{1}{3}R_3 \\ \end{array}$$

The solution to the dual problem is $x = 0$ and $y = \frac{1}{3}$. The optimal value is 4. From the bottom row of the final tableau, we read the solution of the minimization problem from the slack variable columns: $x = 0$, $y = 4$, and $z = 0$. The minimal value of the objective function is 4. This can be verified by substituting the solution into the objective function equation of the standard minimization problem.

$$\begin{aligned} P &= 2x + y + 2z \\ &= 2(0) + (4) + 2(0) \\ &= 4 \end{aligned}$$

Substituting the point $(0, 4, 0)$ back into the constraints of the standard minimization problem yields

$$\begin{aligned} x + 2y + z &\geq 6 & \qquad 3y + 2z &\geq 12 \\ (0) + 2(4) + (0) &\geq 6 & \qquad 3(4) + 2(0) &\geq 12 \\ 8 &\geq 6 & \qquad 12 &\geq 12 \end{aligned}$$

We are further convinced that we have found the optimal solution, since the solution satisfies both constraints (as it should). Also, $3y + 2z$ was made as small as the constraint would allow.

EXAMPLE 5

Using a Standard Minimization Problem to Minimize Shipping Costs

A northwest region canned food distributor has two production facilities: one in Spokane, Washington, and one in Portland, Oregon. For two supermarkets, the Portland facility produces a minimum of 700 cases of canned food per week and the Spokane facility produces a minimum of 300 cases per week. A Seattle-area supermarket requires at least 600 cases per week, and a supermarket in Ellensburg, Washington, requires at least 400 cases per week. Shipping costs (based on a rate of \$0.36 per mile) vary based on the point of origin and destination as shown in Table 4.22.

TABLE 4.22

		Destination	
		Seattle	Ellensburg
Origin	Portland	$0.62 per case	$0.80 per case
	Spokane	$1.01 per case	$0.63 per case

What shipping schedule will minimize the food distributor's shipping costs?

SOLUTION

Let x be the number of cases shipped from Portland to Seattle.
Let y be the number of cases shipped from Spokane to Seattle.
Let z be the number of cases shipped from Portland to Ellensburg.
Let w be the number of cases shipped from Spokane to Ellensburg.

We need to minimize the shipment cost function, $C = 0.62x + 1.01y + 0.80z + 0.63w$. We have the following constraints:

$x + z \geq 700$ The Portland facility produces at least 700 cases

$y + w \geq 300$ The Spokane facility produces at least 300 cases

$x + y \geq 600$ The Seattle-area supermarket needs at least 600 cases

$z + w \geq 400$ The Ellensburg supermarket needs at least 400 cases

$x \geq 0, y \geq 0, z \geq 0, w \geq 0$ The number of cases shipped is nonnegative

This is a standard minimization problem. We will solve it by using the dual and the simplex method. We have

$$A = \left[\begin{array}{cccc|c} 1 & 0 & 1 & 0 & 700 \\ 0 & 1 & 0 & 1 & 300 \\ 1 & 1 & 0 & 0 & 600 \\ 0 & 0 & 1 & 1 & 400 \\ \hline 0.62 & 1.01 & 0.80 & 0.63 & 1 \end{array}\right]$$

Thus

$$A^T = \left[\begin{array}{cccc|c} 1 & 0 & 1 & 0 & 0.62 \\ 0 & 1 & 1 & 0 & 1.01 \\ 1 & 0 & 0 & 1 & 0.80 \\ 0 & 1 & 0 & 1 & 0.63 \\ \hline 700 & 300 & 600 & 400 & 1 \end{array}\right]$$

The dual problem is

$$\text{Maximize} \quad P = 700x + 300y + 600z + 400w$$

$$\text{Subject to} \begin{cases} x + z \leq 0.62 \\ y + z \leq 1.01 \\ x + w \leq 0.80 \\ y + w \leq 0.63 \\ \quad x \geq 0, y \geq 0, z \geq 0, w \geq 0 \end{cases}$$

The initial tableau is given by

x	y	z	w	s	t	u	v	P	
[1]	0	1	0	1	0	0	0	0	0.62
0	1	1	0	0	1	0	0	0	1.01
1	0	0	1	0	0	1	0	0	0.80
0	1	0	1	0	0	0	1	0	0.63
−700	−300	−600	−400	0	0	0	0	1	0

The pivot is the 1 in the first row of the x column of the tableau. The second tableau is generated with the indicated row operations. The 1 in the third row of the w column is the new pivot.

x	y	z	w	s	t	u	v	P		
1	0	1	0	1	0	0	0	0	0.62	
0	1	1	0	0	1	0	0	0	1.01	
0	0	−1	[1]	−1	0	1	0	0	0.18	$R_3 - R_1$
0	1	0	1	0	0	0	1	0	0.63	
0	−300	100	−400	700	0	0	0	1	434	$R_5 + 700R_1$

The third tableau is generated with the indicated row operations. We may choose either the y column or the z column to be the new pivot column. We pick the 1 in the fourth row of the y column to be the new pivot.

x	y	z	w	s	t	u	v	P		
1	0	1	0	1	0	0	0	0	0.62	
0	1	1	0	0	1	0	0	0	1.01	
0	0	−1	1	−1	0	1	0	0	0.18	
0	[1]	1	0	1	0	−1	1	0	0.45	$R_4 - R_3$
0	−300	−300	0	300	0	400	0	1	506	$R_5 + 400R_3$

The final simplex tableau is generated with the indicated row operations.

x	y	z	w	s	t	u	v	P		
1	0	1	0	1	0	0	0	0	0.62	
0	0	0	0	−1	1	1	−1	0	0.56	$R_2 - R_4$
0	0	−1	1	−1	0	1	0	0	0.18	
0	1	1	0	1	0	−1	1	0	0.45	
0	0	0	0	600	0	100	300	1	641	$R_5 + 300R_4$

The optimal solution of the dual problem is $x = 0.62$, $y = 0.45$, $z = 0$, and $w = 0.18$ with optimal value $P = 641$. We read the optimal solution for the minimization problem from the bottom row of the final tableau. When $x = 600$, $y = 0$, $z = 100$, and $w = 300$, the optimal value of 641 is obtained. In the context of the problem, the total shipping cost is minimized when 600 cases are shipped from Portland to Seattle, no cases are shipped from Spokane to Seattle, 100 cases are shipped from Portland to Ellensburg, and 300 cases are shipped from Spokane to Ellensburg. The total shipping cost is \$641. (*Note:* This result assumes that the minimum number of cases were produced at each facility.)

EXAMPLE 6 Using a Standard Minimization Problem to Minimize Training Costs

Many companies with large government contracts use a competitive bidding process to hire subcontractors to do much of the work, A company has a contract to create at least 1000 hours of training and plans to hire three subcontractors (Trainum, Teachum, and Schoolum) to help with the work. Each subcontractor requires a contract for at least 200 hours of training development. Trainum charges \$250 per hour of training development, Teachum charges \$300 per hour of training development, and Schoolum charges \$250 per hour of training development. How many hours should the company allocate to each subcontractor in order to minimize costs?

SOLUTION

Let x be the number of hours allocated to Trainum.
Let y be the number hours allocated to Teachum.
Let z be the number of hours allocated to Schoolum.

We need to minimize the training development cost function, $C = 250x + 300y + 250z$, subject to the following constraints:

$$x + y + z \geq 1000 \quad \text{At least 1000 training hours are produced}$$
$$x \geq 200 \quad \text{Trainum is allocated at least 200 hours}$$
$$y \geq 200 \quad \text{Teachum is allocated at least 200 hours}$$
$$z \geq 200 \quad \text{Schoolum is allocated at least 200 hours}$$

This is a standard minimization problem. We will solve it by using the dual and the simplex method.

$$A = \left[\begin{array}{ccc|c} 1 & 1 & 1 & 1000 \\ 1 & 0 & 0 & 200 \\ 0 & 1 & 0 & 200 \\ 0 & 0 & 1 & 200 \\ \hline 250 & 300 & 250 & 1 \end{array}\right] \quad \text{and} \quad A^T = \left[\begin{array}{cccc|c} 1 & 1 & 0 & 0 & 250 \\ 1 & 0 & 1 & 0 & 300 \\ 1 & 0 & 0 & 1 & 250 \\ \hline 1000 & 200 & 200 & 200 & 1 \end{array}\right]$$

The dual problem is

$$\text{Maximize} \quad P = 1000x + 200y + 200z + 200w$$

$$\text{Subject to} \quad \begin{cases} x + y \leq 250 \\ x + z \leq 300 \\ x + w \leq 250 \\ x \geq 0, y \geq 0, z \geq 0, w \geq 0 \end{cases}$$

The initial simplex tableau is

$$\begin{array}{c} \begin{array}{cccccccc} x & y & z & w & s & t & u & P \end{array} \\ \left[\begin{array}{cccccccc|c} \boxed{1} & 1 & 0 & 0 & 1 & 0 & 0 & 0 & 250 \\ 1 & 0 & 1 & 0 & 0 & 1 & 0 & 0 & 300 \\ 1 & 0 & 0 & 1 & 0 & 0 & 1 & 0 & 250 \\ \hline -1000 & -200 & -200 & -200 & 0 & 0 & 0 & 1 & 0 \end{array}\right] \end{array}$$

The pivot column is the x column, since -1000 is the negative entry in the bottom row with the largest magnitude. The test ratios of the first and third rows are both equal to 250. Since this is smaller than the test ratio for the second row

(300), either the first or the third entry in the x column may be selected as the pivot. We pick the first entry. Using the indicated row operations, we obtain the second tableau.

$$\begin{array}{c} \\ \\ \\ \\ \\ \end{array}\begin{matrix} x & y & z & w & s & t & u & P & \\ \end{matrix}$$

$$\left[\begin{array}{cccccccc|c} 1 & 1 & 0 & 0 & 1 & 0 & 0 & 0 & 250 \\ 0 & -1 & 1 & 0 & -1 & 1 & 0 & 0 & 50 \\ 0 & -1 & 0 & \boxed{1} & -1 & 0 & 1 & 0 & 0 \\ \hline 0 & 800 & -200 & -200 & 1000 & 0 & 0 & 1 & 250000 \end{array}\right] \begin{array}{l} \\ R_2 - R_1 \\ R_3 - R_1 \\ R_4 + 1000R_1 \end{array}$$

Since the magnitude of the negative entries in the bottom row is the same, we may pick either the z column or the w column to be the pivot column. We pick the w column. The pivot is the 1 in the third row. Using the indicated row operation, we get the third tableau. The new pivot is the entry in the second row of the z column.

$$\begin{matrix} x & y & z & w & s & t & u & P \end{matrix}$$

$$\left[\begin{array}{cccccccc|c} 1 & 1 & 0 & 0 & 1 & 0 & 0 & 0 & 250 \\ 0 & -1 & \boxed{1} & 0 & -1 & 1 & 0 & 0 & 50 \\ 0 & -1 & 0 & 1 & -1 & 0 & 1 & 0 & 0 \\ \hline 0 & 600 & -200 & 0 & 800 & 0 & 200 & 1 & 250000 \end{array}\right] \begin{array}{l} \\ \\ \\ R_4 + 200R_3 \end{array}$$

Using the indicated row operation, we obtain the final tableau.

$$\begin{matrix} x & y & z & w & s & t & u & P \end{matrix}$$

$$\left[\begin{array}{cccccccc|c} 1 & 1 & 0 & 0 & 1 & 0 & 0 & 0 & 250 \\ 0 & -1 & 1 & 0 & -1 & 1 & 0 & 0 & 50 \\ 0 & -1 & 0 & 1 & -1 & 0 & 1 & 0 & 0 \\ \hline 0 & 400 & 0 & 0 & 600 & 200 & 200 & 1 & 260000 \end{array}\right] \begin{array}{l} \\ \\ \\ R_4 + 200R_2 \end{array}$$

The optimal solution for the dual problem is $x = 250$, $y = 0$, $z = 50$, and $w = 0$. The optimal value is 260,000.

An optimal solution for the minimization problem is $x = 600$, $y = 200$, and $z = 200$ with optimal value 260,000. In the context of the problem, the training development costs are minimized if Trainum develops 600 hours of training, Teachum develops 200 hours of training, and Schoolum develops 200 hours of training. The minimum training costs are $260,000. Since the hourly training development costs for Trainum and Schoolum are the same, another optimal solution is $(200, 200, 600)$. In fact, any point of the form $(t, 200, 800 - t)$ for $0 \le t \le 800$ will yield an optimal solution.

As shown in Example 6, a standard minimization problem may have more than one optimal solution. Each solution, however, will generate the same optimal value. The method of using the dual guarantees that we will find *an* optimal solution, if there is one.

EXAMPLE 7

Using a Standard Minimization Problem to Minimize Food Costs

PETsMART.com, an online retailer, allows pet owners to quickly compare nutrition and pricing for different brands of pet food sold on its web site. A June 2003 query revealed the data given in Table 4.23.

TABLE 4.23

Brand	Protein (percent)	Fiber (percent)	Price (dollars per 20-lb bag)
Eukanuba Adult Maintenance Formula	25	5	18.99
Nutro Natural Choice Plus	27	3	17.99
Science Diet Active Formula Canine Maintenance	26.5	3.5	17.99

Source: www.petsmart.com.

A dog breeder wants to create at least 300 pounds of a dog food mix that is at least 26 percent protein and 4 percent fiber while minimizing dog food cost. How many 20-pound bags of each type of dog food should the breeder buy?

SOLUTION

Let x be the number of bags of the Eukanuba brand.
Let y be the number of bags of the Nutro brand.
Let z be the number of bags of the Science Diet brand.

To make computations easier, we will convert the percentages per bag to pounds per bag by multiplying the percentage per bag by 20 pounds and dividing by 100, as shown in Table 4.24.

TABLE 4.24

Brand	Protein (pounds)	Fiber (pounds)
Eukanuba Adult Maintenance Formula	5.0	1.0
Nutro Natural Choice Plus	5.4	0.6
Science Diet Active Formula Canine Maintenance	5.3	0.7

The total weight of the mix is $20x + 20y + 20z$ pounds.
The total amount of protein in the mix is $5.0x + 5.4y + 5.3z$ pounds. We want the mix to be at least 26 percent protein. Thus we require that

$$0.26(20x + 20y + 20z) \leq 5.0x + 5.4y + 5.3z$$
$$5.2x + 5.2y + 5.2z \leq 5.0x + 5.4y + 5.3z$$
$$0.2x - 0.2y - 0.1z \leq 0$$
$$-0.2x + 0.2y + 0.1z \geq 0$$

The total amount of fiber in the mix is $1.0x + 0.6y + 0.7z$ pounds. We want the mix to be at least 4 percent fiber. Thus we require that

$$\begin{aligned} 0.04(20x + 20y + 20z) &\le 1.0x + 0.6y + 0.7z \\ 0.8x + 0.8y + 0.8z &\le 1.0x + 0.6y + 0.7z \\ -0.2x + 0.2y + 0.1z &\le 0 \\ 0.2x - 0.2y - 0.1z &\ge 0 \end{aligned}$$

For computational ease, we round the prices to the nearest dollar. We must minimize $C = 19x + 18y + 18z$ subject to the following constraints:

$$\begin{aligned} 20x + 20y + 20z &\ge 300 && \text{At least 300 pounds must be purchased} \\ -0.2x + 0.2y + 0.1z &\ge 0 && \text{The mix is at least 26 percent protein} \\ 0.2x - 0.2y - 0.1z &\ge 0 && \text{The mix is at least 4 percent fiber} \\ x \ge 0,\ y \ge 0,\ z &\ge 0 && \text{A nonnegative number of bags are purchased} \end{aligned}$$

We have

$$A = \left[\begin{array}{ccc|c} 20 & 20 & 20 & 300 \\ -0.2 & 0.2 & 0.1 & 0 \\ 0.2 & -0.2 & -0.1 & 0 \\ \hline 19 & 18 & 18 & 1 \end{array}\right] \quad \text{and} \quad A^T = \left[\begin{array}{ccc|c} 20 & -0.2 & 0.2 & 19 \\ 20 & 0.2 & -0.2 & 18 \\ 20 & 0.1 & -0.1 & 18 \\ \hline 300 & 0 & 0 & 1 \end{array}\right]$$

The dual problem is

$$\text{Maximize} \quad P = 300x + 0y + 0z$$

$$\text{Subject to} \quad \begin{cases} 20x - 0.2y + 0.2z \le 19 \\ 20x + 0.2y - 0.2z \le 18 \\ 20x + 0.1y - 0.1z \le 18 \\ x \ge 0, y \ge 0, z \ge 0 \end{cases}$$

with the corresponding initial tableau

$$\begin{array}{c} \quad x \quad\quad y \quad\quad z \quad s \quad t \quad u \quad P \\ \left[\begin{array}{ccccccc|c} 20 & -0.2 & 0.2 & 1 & 0 & 0 & 0 & 19 \\ \boxed{20} & 0.2 & -0.2 & 0 & 1 & 0 & 0 & 18 \\ 20 & 0.1 & -0.1 & 0 & 0 & 1 & 0 & 18 \\ \hline -300 & 0 & 0 & 0 & 0 & 0 & 1 & 0 \end{array}\right] \end{array}$$

The x column is the pivot column. Either the second or the third entry in the x column may be used as a pivot. Using the entry in the second row of the x column as a pivot and the indicated row operations, we obtain the second tableau. The entry in the third row of the z column is the new pivot, since it has the smallest test ratio.

$$\begin{array}{c} \quad x \quad\quad y \quad\quad z \quad s \quad t \quad u \quad P \\ \left[\begin{array}{ccccccc|c} 0 & -0.4 & 0.4 & 1 & -1 & 0 & 0 & 1 \\ 20 & 0.2 & -0.2 & 0 & 1 & 0 & 0 & 18 \\ 0 & -0.1 & \boxed{0.1} & 0 & -1 & 1 & 0 & 0 \\ \hline 0 & 3 & -3 & 0 & 15 & 0 & 1 & 270 \end{array}\right] \begin{array}{l} R_1 - R_2 \\ \\ R_3 - R_2 \\ R_4 + 15R_2 \end{array} \end{array}$$

The indicated row operations give the next tableau. The new pivot is the 3 in the first row of the t column.

$$\begin{array}{ccccccc|c} x & y & z & s & t & u & P & \\ 0 & 0 & 0 & 1 & \boxed{3} & -4 & 0 & 1 \\ 20 & 0 & 0 & 0 & -1 & 2 & 0 & 18 \\ 0 & -0.1 & 0.1 & 0 & -1 & 1 & 0 & 0 \\ \hline 0 & 0 & 0 & 0 & -15 & 30 & 1 & 270 \end{array} \quad \begin{array}{l} R_1 - 4R_3 \\ R_2 + 2R_3 \\ \\ R_4 + 30R_3 \end{array}$$

The indicated row operations give the next tableau.

$$\begin{array}{ccccccc|c} x & y & z & s & t & u & P & \\ 0 & 0 & 0 & 1 & 3 & -4 & 0 & 1 \\ 60 & 0 & 0 & 1 & 0 & 2 & 0 & 55 \\ 0 & -0.3 & 0.3 & 1 & 0 & -1 & 0 & 1 \\ \hline 0 & 0 & 0 & 5 & 0 & 10 & 1 & 275 \end{array} \quad \begin{array}{l} \\ 3R_2 + R_1 \\ 3R_3 + R_1 \\ R_4 + 5R_1 \end{array}$$

The final simplex tableau is obtained by converting the nonzero entry of each column with a single nonzero entry to a 1 by using the indicated row operations.

$$\begin{array}{ccccccc|c} x & y & z & s & t & u & P & \\ 0 & 0 & 0 & \frac{1}{3} & 1 & -\frac{4}{3} & 0 & \frac{1}{3} \\ 1 & 0 & 0 & \frac{1}{60} & 0 & \frac{1}{30} & 0 & \frac{11}{12} \\ 0 & -1 & 1 & \frac{10}{3} & 0 & -\frac{10}{3} & 0 & \frac{10}{3} \\ \hline 0 & 0 & 0 & 5 & 0 & 10 & 1 & 275 \end{array} \quad \begin{array}{l} \frac{1}{3}R_1 \\ \frac{1}{60}R_2 \\ \frac{1}{0.3}R_3 \\ \end{array}$$

Setting the inactive variable columns s and u to 0 yields

$$x = \frac{11}{12} \qquad -y + z = \frac{10}{3} \qquad t = \frac{1}{3} \qquad P = 275$$

$$z = y + \frac{10}{3}$$

Any solution of the form $\left(\frac{11}{12}, r, r + \frac{10}{3}\right)$ for nonnegative r maximizes the objective function of the dual problem. The optimal value is 275.

The minimization problem has the solution $x = 5$, $y = 0$, and $z = 10$. When 5 bags of the Eukanuba brand and 10 bags of the Science Diet brand are purchased, the total cost is \$275. This is the minimum cost for a mix that meets the breeder's nutrition requirements.

4.4 Summary

In this section, you learned how to solve standard minimization problems using the notion of the dual problem. You discovered that finding the maximum solution of a dual problem leads us to the minimum solution of the minimization problem.

4.4 Exercises

In Exercises 1–10, determine if the problem is a standard minimization problem. If it isn't, explain why.

1. Minimize $P = 9x + 8y$

Subject to $\begin{cases} 2x + y \geq 10 \\ -x + y \geq 1 \\ x \geq 0 \\ y \geq 0 \end{cases}$

2. Minimize $P = -2x - 5y$

Subject to $\begin{cases} 4x + 9y \geq 10 \\ -11x + y \geq -21 \\ x \geq 0 \\ y \geq 0 \end{cases}$

3. Minimize $P = 6xy$

Subject to $\begin{cases} 6x + 7y \geq 13 \\ -8x - 4y \geq 12 \\ x \geq 0 \\ y \geq 0 \end{cases}$

4. Minimize $P = 4x - 2y + z$

Subject to $\begin{cases} 2x + y + 4z \geq 24 \\ -2x + 3y + 3z \geq 150 \\ x \geq 0 \\ y \geq 0 \\ z \geq 0 \end{cases}$

5. Minimize $P = -1.2x - 2.8y + 4.3z$

Subject to $\begin{cases} 3.2x + 1.5y + 7.4z \geq 249.8 \\ -2.7x + 3.4y + 3.9z \geq 190.1 \end{cases}$

6. Minimize $P = 4x + 4y + 9z$

Subject to $\begin{cases} -6x - y + 4z \leq -24 \\ -9x - 1.5y + 6z \geq 36 \\ x \geq 0 \\ y \geq 0 \\ z \geq 0 \end{cases}$

7. Minimize $P = x + 9y + 8z$

Subject to $\begin{cases} -6x - y + 4z \geq -24 \\ -9x - 1.5y + 6z \geq 36 \\ x \geq 0 \\ y \geq 0 \\ z \geq 0 \end{cases}$

8. Minimize $P = 4x + 4y + 9z$

Subject to $\begin{cases} -3x - 2y + 4z \leq -22 \\ 5x - 2.5y + 6z \geq 36 \\ x \geq 0 \\ y \geq 0 \\ z \geq 0 \end{cases}$

9. Minimize $P = -8x + 2y - z$

Subject to $\begin{cases} z \geq 4 \\ x - y \geq 3 \\ 2x - z \geq 5 \\ x + y + z \leq 0 \end{cases}$

10. Minimize $P = 5x + 7z$

Subject to $\begin{cases} 4x + 2y \geq 16 \\ 4x - 2y \geq 3 \\ x \geq 0 \\ y \geq 0 \end{cases}$

In Exercises 11–20, find the transpose of the given matrix.

11. $A = \begin{bmatrix} 2 & 2 & 1 & 4 \\ 5 & 9 & 7 & 3 \\ 8 & 1 & 0 & 1 \end{bmatrix}$

12. $A = \begin{bmatrix} 1 & 1 & 1 & 4 \\ 1 & 1 & 1 & 5 \\ 4 & 5 & 1 & 1 \end{bmatrix}$

13. $A = \begin{bmatrix} 1 & 2 & 4 \\ 6 & 8 & 1 \end{bmatrix}$

14. $A = \begin{bmatrix} 8 & 2 & 1 \\ 2 & 5 & 7 \\ 1 & 7 & 1 \end{bmatrix}$

15. $A = \begin{bmatrix} 9 & 3 & -1 & 4 \\ 7 & 2 & -2 & 1 \end{bmatrix}$

16. $A = \begin{bmatrix} 1 & -2 & 1 & 4 & 11 \\ -5 & 4 & -7 & 3 & 9 \\ 3 & 7 & 9 & 3 & 1 \end{bmatrix}$

17. $A = \begin{bmatrix} 9 & 2 & 18 \\ -3 & -2 & 6 \\ 4 & 0 & 0 \\ 5 & 6 & 1 \end{bmatrix}$

18. $A = \begin{bmatrix} 6 & 5 & 4 & 3 & 2 \\ 5 & 6 & 7 & 8 & 9 \\ 4 & 5 & 6 & 5 & 1 \end{bmatrix}$

19. $A = \begin{bmatrix} 0 & 2 & 1 \\ -5 & 4 & 7 \\ 8 & 1 & 0 \\ -1 & 2 & 1 \\ 4 & 2 & 1 \end{bmatrix}$

20. $A = \begin{bmatrix} 2 & 4 & 6 & 5 & 1 \end{bmatrix}$

In Exercises 21–30, do the following:

(i) Write the dual problem for the given standard minimization problem.
(ii) Solve the dual problem using the simplex method.
(iii) Use the final simplex tableau of the dual problem to solve the standard minimization problem.
(iv) Check your answer by graphing the feasible region of the standard minimization problem and calculating the value of the objective function at each of the corner points.

21. Minimize $P = x + 2y$

Subject to $\begin{cases} 3x + y \geq 6 \\ 2x + 3y \geq 11 \\ x \geq 0, y \geq 0 \end{cases}$

22. Minimize $P = 4x + 2y$

Subject to $\begin{cases} 5x + 7y \geq 19 \\ -2x + 3y \geq 5 \\ x \geq 0, y \geq 0 \end{cases}$

23. Minimize $P = 9x + 6y$

Subject to $\begin{cases} 5x + 2y \geq 30 \\ 10x - 5y \geq 15 \\ x \geq 0, y \geq 0 \end{cases}$

24. Minimize $P = 4x + 4y$

Subject to $\begin{cases} 3x + 4y \geq 18 \\ 5x + 3y \geq 19 \\ x \geq 0, y \geq 0 \end{cases}$

25. Minimize $P = 5x + 20y$

Subject to $\begin{cases} -5x + 2y \geq 10 \\ 5x + y \geq 25 \\ x \geq 0, y \geq 0 \end{cases}$

26. Minimize $P = 2x + 5y$

Subject to $\begin{cases} x + y \geq 9 \\ 2x - 2y \geq 6 \\ x \geq 0, y \geq 0 \end{cases}$

27. Minimize $P = 3x + 2y$

Subject to $\begin{cases} x + y \geq 5 \\ 3x + 2y \geq 11 \\ 4x - y \geq 0 \\ x \geq 0, y \geq 0 \end{cases}$

28. Minimize $P = 9x + 7y$

Subject to $\begin{cases} 5x + 3y \geq 8 \\ 3x + 5y \geq 8 \\ 8x - 8y \geq 0 \\ x \geq 0, y \geq 0 \end{cases}$

29. Minimize $P = 6x + 5y$

Subject to $\begin{cases} x + y \geq 4 \\ x - y \geq 2 \\ 2x - y \geq 6 \\ x \geq 0, y \geq 0 \end{cases}$

30. Minimize $P = 5x + 7y$

Subject to $\begin{cases} x + y \geq 4 \\ x - y \geq 2 \\ 2x - y \geq 6 \\ x \geq 0, y \geq 0 \end{cases}$

In Exercises 31–40, set up and solve the standard minimization problem.

31. **Pet Nutrition: Food Cost**
PETsMART.com sold the following varieties of dog food in June 2003. The price shown is for an 8-pound bag.

Pro Plan Adult Chicken & Rice Formula, 25 percent protein, 3 percent fiber, $7.99
Pro Plan Adult Lamb & Rice Formula, 28 percent protein, 3 percent fiber, $7.99
Pro Plan Adult Turkey & Barley Formula, 26 percent protein, 3 percent fiber, $8.49
(**Source:** www.petsmart.com.)

A dog breeder wants to make at least 120 pounds of a mix containing at least 27 percent protein and at least 3 percent fiber. How many 8-pound bags of each dog food variety should the breeder buy in order to minimize cost? (Round prices up to the nearest dollar.)

32. **Pet Nutrition: Food Cost** PETsMART.com sold the following varieties of dog food in June 2003.

Authority Chicken Adult Formula, 32 percent protein, 3 percent fiber, \$19.99 per 33-pound bag

Bil-Jac Select Dog Food, 27 percent protein, 4 percent fiber, \$18.99 per 18-pound bag

Iams Minichuncks, 26 percent protein, 4 percent fiber, \$8.99 per 8-pound bag

(**Source:** www.petsmart.com.)

A dog kennel wants to make at least 2178 pounds of a mix containing at least 29 percent protein and at least 3.5 percent fiber. How many bags of each dog food variety should the kennel buy in order to minimize cost? (Round prices up to the nearest dollar.)

33. **Pet Nutrition: Food Cost** PETsMART.com sold the following varieties of dog food in June 2003.

Nature's Recipe Venison Meal & Rice Canine, 20 percent protein, 10 percent fat, \$21.99 per 20-pound bag

Nutro Max Natural Dog Food, 27 percent protein, 16 percent fat, \$12.99 per 17.5-pound bag

PETsMART Premier Oven Baked Lamb Recipe, 25 percent protein, 14 percent fat, \$22.99 per 30-pound bag

(**Source:** www.petsmart.com.)

A dog breeder wants to make at least 175 pounds of a mix containing at least 25 percent protein and at least 14 percent fat. How many bags of each dog food variety should the breeder buy in order to minimize cost? (Round prices *up* to the nearest dollar.)

34. **Pet Nutrition: Fat Content** PETsMART.com sold the following varieties of dog food in June 2003.

Nature's Recipe Venison Meal & Rice Canine, 20 percent protein, 10 percent fat, \$21.99 per 20-pound bag

Nutro Max Natural Dog Food, 27 percent protein, 16 percent fat, \$12.99 per 17.5-pound bag

PETsMART Premier Oven Baked Lamb Recipe, 25 percent protein, 14 percent fat, \$22.99 per 30-pound bag

(**Source:** www.petsmart.com.)

A dog breeder wants to make at least 210 pounds of a mix containing at least 25 percent protein. How many bags of each dog food variety should the breeder buy in order to minimize fat content?

35. **Food Distribution Cost** Wal-Mart Stores, Inc., has food distribution centers in Monroe, Georgia, and Shelbyville, Tennessee, and Wal-Mart Supercenters in Birmingham, Alabama, and Scottsboro, Alabama. (**Source:** www.walmart.com.) Suppose that the Monroe distribution center must ship at least 600 cases of peanut butter weekly and the Shelbyville distribution center must ship at least 400 cases of peanut butter weekly.* If the Birmingham store requires at least 700 cases of peanut butter weekly and the Scottsboro store requires at least 300 cases of peanut butter weekly, what shipment plan will minimize the distribution cost?

Estimated Distribution Cost per Case

	Monroe	Shelbyville
Birmingham	\$0.68	\$0.57
Scottsboro	\$0.65	\$0.32

*Distribution costs and amounts are hypothetical.

36. **Food Distribution Cost** Wal-Mart Stores, Inc., has food distribution centers in Monroe, Georgia, and Shelbyville, Tennessee, and Wal-Mart Supercenters in Birmingham, Alabama, and Calhoun, Georgia. (**Source:** www.walmart.com.) Suppose that the Monroe distribution center must ship at least 600 cases of pickles weekly and the Shelbyville distribution center must ship at least 700 cases of pickles weekly.* If the Birmingham store requires at least 1000 cases of pickles weekly and the Calhoun store requires at least 300 cases of pickles weekly, what shipment plan will minimize the distribution cost?

Estimated Distribution Cost per Case

	Monroe	Shelbyville
Birmingham	\$0.68	\$0.57
Calhoun	\$0.39	\$0.50

*Distribution costs and amounts are hypothetical.

37. **Food Distribution Cost** Wal-Mart Stores, Inc., has food distribution centers in Monroe, Georgia, and Shelbyville, Tennessee, and Wal-Mart Supercenters in Scottsboro, Alabama, and Calhoun, Georgia. (**Source:** www.walmart.com.) Suppose that the Monroe distribution center must ship at least 400 cases of potato chips weekly and the Shelbyville distribution center must ship at least 200 cases of potato chips weekly.* If the Scottsboro store requires at least 300 cases of potato chips weekly and the Calhoun store requires at least 300 cases of potato chips weekly, what shipment plan will minimize the distribution cost?

Estimated Distribution Cost per Case

	Monroe	Shelbyville
Calhoun	\$0.39	\$0.50
Scottsboro	\$0.65	\$0.32

*Distribution costs and amounts are hypothetical.

38. **Food and Entertainment** It costs \$35 for a family to dine out and \$10 for the family to eat at home. Eating out has a fun rating of 10 points, while eating at home has a fun rating of 2 points.

It costs \$32 for a family to play a game of miniature golf and \$30 to watch a movie. Miniature golf has a fun rating of 8 points, and watching a movie has a fun rating of 6 points.

The family must eat at least 21 meals weekly and must go out for food or miniature golf at least five times weekly.The family wants to earn at least 82 fun points per week. How many times a week should the family participate in each activity in order to minimize food and entertainment costs?

39. **Marital Harmony** It costs a couple 3 hours and \$200 to go clothes shopping together. She gives shopping a fun rating of 10 points, while he gives it a fun rating of 1 point (a total of 11 fun points for the couple).

It costs a couple 4 hours and \$250 to go to a major league baseball playoff game. She gives the game a fun rating of 4 points, while he gives it a fun rating of 10 points (a total of 14 fun points for the couple).

The couple wants to spend at least 10 hours together (shopping and watching baseball) while earning at least 36 fun points. She insists that they go shopping at least twice. How shall they spend their time if they want to minimize their financial costs?

40. **Marital Discord** It costs a couple 3 hours to go clothes shopping together. He gives shopping with her a fun rating of 6 points while she gives it a fun rating of 1 point (a total of 7 fun points for the couple).

It costs a couple 4 hours to go to a major league baseball game. She gives going to a game with him a fun rating of 7 points while he gives it a fun rating of 2 points (a total of 9 fun points for the couple).

The couple wants to earn at least 30 fun points. She refuses to go shopping with him unless he goes with her to at least one game. How shall they spend their time if they want to minimize their time together? (Things are not going well in the relationship.)

4.5 Solving General Linear Programming Problems with the Simplex Method

- Solve general linear programming problems with the simplex method

GETTING STARTED An investor plans to invest at most \$3000 in the three publicly traded recreational vehicle companies shown in Table 4.25.

TABLE 4.25

Company	Share Price (dollars)	Dividend/Share (dollars)
Harley- Davidson, Inc. (HDI)	62.70	0.40
Polaris Industries Inc. (PII)	50.50	0.92
Winnebago Industries Inc. (WGO)	33.42	0.20

Source: moneycentral.msn.com. (Accurate as of July 16, 2004.)

He wants to earn at least \$50 in dividends while maximizing the number of shares purchased. How many shares of each company should he purchase?

At first glance, this looks like a standard maximization problem. However, a closer analysis reveals that it is not. While the investment constraint is of the form $Ax + By + Cz + \cdots \leq M$, the dividend constraint is of the form $Ax + By + Cz + \cdots \geq M$ instead of $Ax + By + Cz + \cdots \leq M$. This is a **general linear programming problem** or a linear programming problem with *mixed constraints*.

In this section, we will show how to solve general linear programming problems. We will demonstrate how minimization problems may be solved by maximizing the negative of the objective function. We will return to the recreational vehicle investment problem in Example 3.

Recall that the feasible region of any standard maximization problem always includes the origin. Graphically speaking, the simplex method starts at the origin and moves from corner point to adjacent corner point, each time increasing the value of the objective function until the maximum value is reached. The simplex method works with standard maximization problems because we are guaranteed that the origin is in the feasible region. But what if the origin is outside of the feasible region? The feasible region of a linear programming problem with mixed constraints often does not contain the origin.

A linear programming problem with mixed constraints has constraints in two or more of the following forms: $Ax + By + Cz + \cdots \leq M$, $Ax + By + Cz + \cdots \geq M$, or $Ax + By + Cz + \cdots = M$. In all cases, $M \geq 0$. Consider the following linear programming problem with mixed constraints.

$$\text{Maximize} \quad P = 3x + 2y$$

$$\text{Subject to} \begin{cases} -x + 2y \leq 4 \\ 2x - y \leq 4 \\ x + y \geq 5 \\ x \geq 0, y \geq 0 \end{cases}$$

Since the constraint $x + y \geq 5$ contains a $\geq$ sign instead of a $\leq$ sign, this is not a standard maximization problem. It is a problem with mixed constraints. The graph of the feasible region is shown in Figure 4.28.

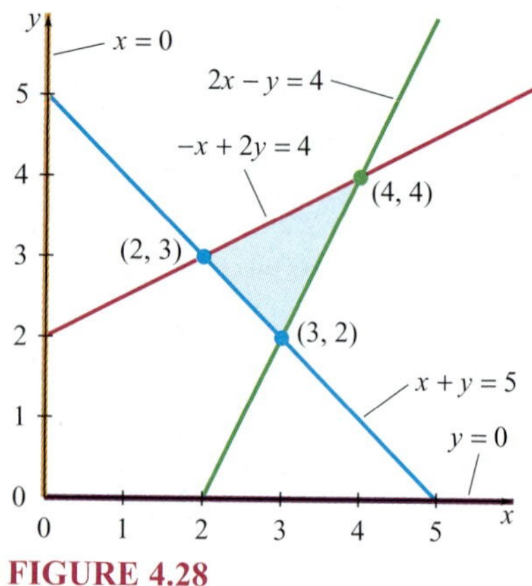

FIGURE 4.28

Notice that the origin is not contained within the feasible region. Our strategy for solving linear programming problems with mixed constraints is a two-stage process.

Stage 1: Get inside the feasible region.
Stage 2: Solve the problem with the simplex method.

Since our problem contains only two decision variables, we have the luxury of observing how the process works graphically. This is not the case in problems involving three or more decision variables.

We set up the problem for the simplex method as usual by adding in slack variables to constraints of the form $Ax + By + Cz + \cdots \leq M$.

$$-x + 2y + s = 4$$
$$2x - y + t = 4$$

Can we add in slack variables to constraints of the form $Ax + By + Cz + \cdots \geq M$? Let's see. We have the constraint $x + y \geq 5$. Recall that slack variables add in what is necessary to make the inequality an equality. By definition, slack variables must be nonnegative. What must we do to the inequality $x + y \geq 5$ to make it an equality? Adding a slack variable to the inequality will increase instead of decrease the value of the left-hand side. Since the left-hand side of the inequality is greater than or equal to 5, we must subtract some nonnegative value u in order to make $x + y$ equal to 5. That is,

$$x + y - u = 5$$

Since this variable takes away the surplus, it is referred to as a **surplus variable.** We have

$$\begin{aligned} -x + 2y + s &= 4 \\ 2x - y + t &= 4 \\ x + y - u &= 5 \\ -3x - 2y + P &= 0 \end{aligned}$$

and the corresponding initial tableau

$$\begin{array}{c} \\ \end{array}\begin{bmatrix} \begin{array}{cccccc|c} x & y & s & t & u & P & \\ -1 & 2 & 1 & 0 & 0 & 0 & 4 \\ 2 & -1 & 0 & 1 & 0 & 0 & 4 \\ 1 & 1 & 0 & 0 & -1 & 0 & 5 \\ \hline -3 & -2 & 0 & 0 & 0 & 1 & 0 \end{array} \end{bmatrix}$$

Setting the x and y variables equal to zero yields the solution to this tableau.

$$x = 0 \qquad y = 0 \qquad s = 4 \qquad t = 4 \qquad -u = 5 \qquad P = 0$$
$$u = -5$$

Recall that all decision, slack, and surplus variables are required to be nonnegative. Since $u < 0$, this solution is not in the feasible region. This may be easily seen graphically (Figure 4.29). The origin is not in the feasible region.

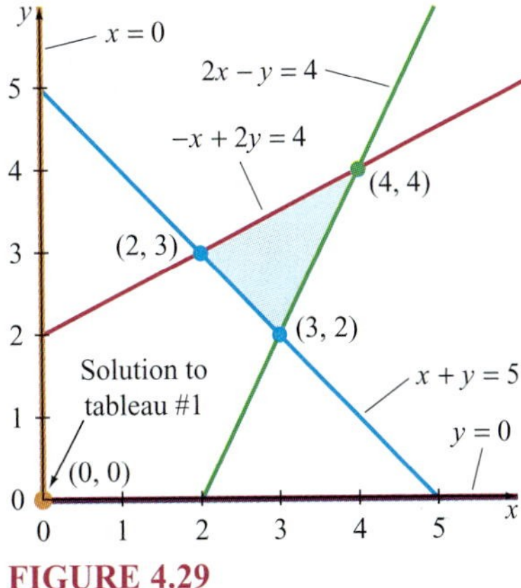

FIGURE 4.29

In the initial tableau, columns s, t, u, and P were the active variable columns. Since the active variable u was negative, we want to modify the tableau in such a way as to make u an inactive variable. We place a * on the left of the tableau to highlight the rows above the bottom row that resulted in negative variable values. In this case, the third row was the only one with a negative variable value.

$$\begin{array}{c} \\ \\ \\ * \\ \\ \end{array}\begin{bmatrix} \begin{array}{cccccc|c} x & y & s & t & u & P & \\ -1 & 2 & 1 & 0 & 0 & 0 & 4 \\ 2 & -1 & 0 & 1 & 0 & 0 & 4 \\ 1 & 1 & 0 & 0 & -1 & 0 & 5 \\ \hline -3 & -2 & 0 & 0 & 0 & 1 & 0 \end{array} \end{bmatrix}$$

Since all surplus variables in the initial tableau are negative, each row of the initial tableau containing surplus variables will be starred. We select the column with the largest positive entry in the starred row as the pivot column. Since the positive values in the starred row are equal, we may select either the x column or the y column to be the pivot column. We select the x column. To determine the pivot, we calculate the test ratios and select the row with the smallest test ratio. This will ensure that all nonnegative variables will remain nonnegative.

$$\begin{array}{c} \\ \\ \\ * \\ \\ \end{array}\begin{array}{c} \begin{matrix} x & y & s & t & u & P & \end{matrix} \\ \left[\begin{array}{cccccc|c} -1 & 2 & 1 & 0 & 0 & 0 & 4 \\ \boxed{2} & -1 & 0 & 1 & 0 & 0 & 4 \\ 1 & 1 & 0 & 0 & -1 & 0 & 5 \\ \hline -3 & -2 & 0 & 0 & 0 & 1 & 0 \end{array}\right] \end{array} \quad \begin{array}{l} \\ 4/2 = 2 \\ 5/1 = 5 \\ \\ \end{array}$$

Pivot column

The 2 in the second row of the x column is the pivot. We modify the initial tableau with the indicated operations and star any rows corresponding with negative active variables.

$$\begin{array}{c} \\ \\ \\ * \\ \\ \end{array}\begin{array}{c} \begin{matrix} x & y & s & t & u & P & \end{matrix} \\ \left[\begin{array}{cccccc|c} 0 & 3 & 2 & 1 & 0 & 0 & 12 \\ \boxed{2} & -1 & 0 & 1 & 0 & 0 & 4 \\ 0 & 3 & 0 & -1 & -2 & 0 & 6 \\ \hline 0 & -7 & 0 & 3 & 0 & 2 & 12 \end{array}\right] \end{array} \quad \begin{array}{l} \\ 2R_1 + R_2 \\ \\ 2R_3 - R_2 \\ 2R_4 + 3R_2 \end{array}$$

$$\begin{array}{cccccc} 2x = 4 & y = 0 & 2s = 12 & t = 0 & -2u = 6 & 2P = 12 \\ x = 2 & & s = 6 & & u = -3 & P = 6 \end{array}$$

We starred the third row, since the value of u is negative. Although u is still negative, it is less negative than it was before. We will continue the process until u is nonnegative. However, before continuing, let's examine what is happening graphically (Figure 4.30).

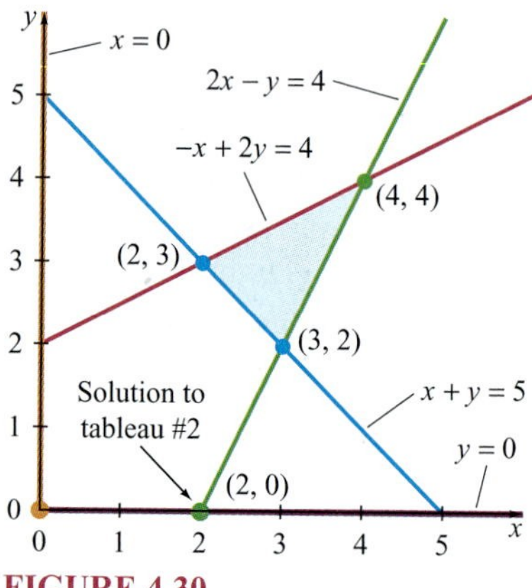

FIGURE 4.30

We have moved from the origin to the x-intercept of one of the constraints. However, we still are outside of the feasible region. This is due to the fact that the surplus variable u is still negative. We will pivot again and further increase the value of u.

The largest value in the starred row is the 3 in the y column. The y column is the new pivot column.

$$\begin{array}{c} \\ \\ * \\ \\ \end{array}\begin{array}{c} \begin{matrix} x & y & s & t & u & P & \end{matrix} \\ \left[\begin{array}{cccccc|c} 0 & 3 & 2 & 1 & 0 & 0 & 12 \\ 2 & -1 & 0 & 1 & 0 & 0 & 4 \\ 0 & \boxed{3} & 0 & -1 & -2 & 0 & 6 \\ \hline 0 & -7 & 0 & 3 & 0 & 2 & 12 \end{array}\right] \end{array} \quad \begin{array}{l} 12/3 = 4 \\ \\ 6/3 = 2 \\ \\ \end{array}$$

Pivot column

The pivot is the 3 in the third row of the y column, since that row has the smallest test ratio. We zero out the remaining terms in the column with the indicated operations and read the solution from the tableau.

$$\begin{array}{c} \begin{matrix} x & y & s & t & u & P & \end{matrix} \\ \left[\begin{array}{cccccc|c} 0 & 0 & 2 & 2 & 2 & 0 & 6 \\ 6 & 0 & 0 & 2 & -2 & 0 & 18 \\ 0 & 3 & 0 & -1 & -2 & 0 & 6 \\ \hline 0 & 0 & 0 & 2 & -14 & 6 & 78 \end{array}\right] \end{array} \quad \begin{array}{l} R_1 - R_3 \\ 3R_2 + R_3 \\ \\ 3R_4 + 7R_3 \end{array}$$

$$\begin{array}{cccccc} 6x = 18 & 3y = 6 & 2s = 6 & t = 0 & u = 0 & 6P = 78 \\ x = 3 & y = 2 & s = 3 & & & P = 13 \end{array}$$

Since none of the variables are negative, this solution is in the feasible region (Figure 4.31). Since we are now in the feasible region, this completes Stage 1 of the process.

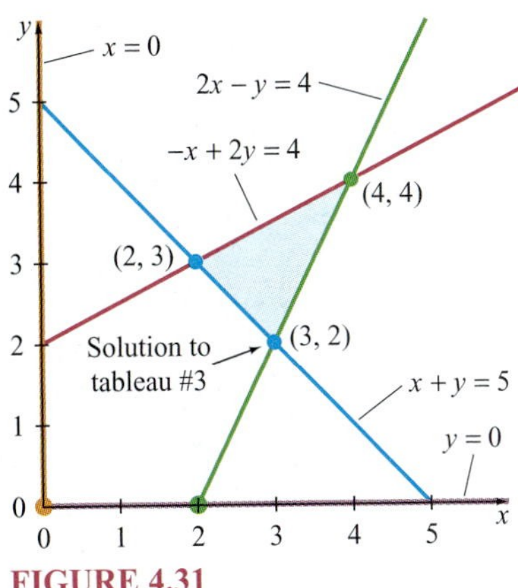

FIGURE 4.31

Stage 2 allows us to apply the simplex method to find the optimal solution. The u column is the pivot column. The 2 in the first row of that column is the pivot, since it is the only nonnegative value in the column.

$$\begin{array}{c} \begin{matrix} x & y & s & t & u & P & \end{matrix} \\ \left[\begin{array}{cccccc|c} 0 & 0 & 2 & 2 & \boxed{2} & 0 & 6 \\ 6 & 0 & 0 & 2 & -2 & 0 & 18 \\ 0 & 3 & 0 & -1 & -2 & 0 & 6 \\ \hline 0 & 0 & 0 & 2 & -14 & 6 & 78 \end{array}\right] \end{array}$$

Pivot column

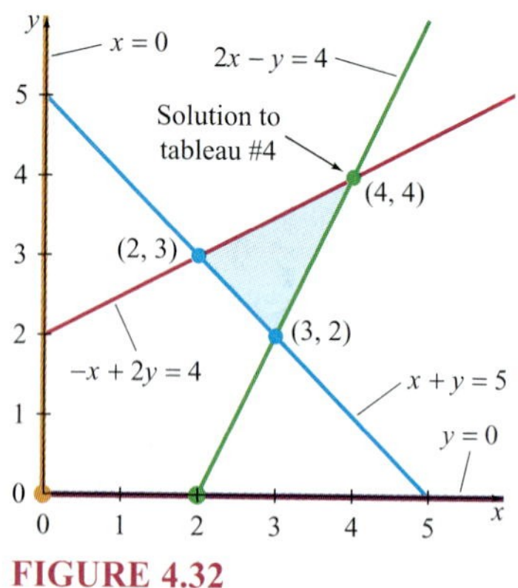

FIGURE 4.32

We zero out the remaining terms in the u column using the indicated operations.

$$\begin{array}{c} \begin{matrix} x & y & s & t & u & P & \end{matrix} \\ \left[\begin{array}{cccccc|c} 0 & 0 & 2 & 2 & 2 & 0 & 6 \\ 6 & 0 & 2 & 4 & 0 & 0 & 24 \\ 0 & 3 & 2 & 1 & 0 & 0 & 12 \\ \hline 0 & 0 & 14 & 16 & 0 & 6 & 120 \end{array}\right] \end{array} \begin{array}{l} \\ \\ R_2 + R_1 \\ R_3 + R_1 \\ R_4 + 7R_1 \end{array}$$

$$\begin{array}{cccccc} 6x = 24 & 3y = 12 & s = 0 & t = 0 & 2u = 6 & 6P = 120 \\ x = 4 & y = 4 & & & u = 3 & P = 20 \end{array}$$

Since the bottom row of the tableau does not contain any negative values, this is the final simplex tableau. The solution that maximizes the objective function $P = 3x + 2y$ is $x = 4$ and $y = 4$ (Figure 4.32). The maximum value of the objective function is $P = 20$.

The following box details the steps used to solve the general linear programming problem.

HOW TO Solving General Linear Programming Problems

Stage 1: Get inside the feasible region

1. Star all rows that correspond with a negative value of a decision, slack, or surplus variable.
2. Identify the largest positive entry in the starred row. The corresponding column is the pivot column.
3. Calculate the test ratios for the positive entries above the bottom row of the pivot column.
4. Pick the pivot column entry with the smallest test ratio as the pivot.
5. Row reduce the tableau using operations of the form $aR_c \pm bR_p \to R_c$ with a and b positive. (Recall that R_c is the row we want to change and R_p is the pivot row.)
6. From the tableau, calculate the value of all decision, slack, and surplus variables. If any of the variables are negative, repeat Steps 1 through 5 for the new tableau. Otherwise, go to Stage 2.

Stage 2: Solve the maximization problem with the simplex method.

EXAMPLE 1 Solving a Linear Programming Problem with No Optimal Solution

Graph the feasible region associated with the given linear programming problem. Then solve the problem using the two-stage method described previously, indicating on the graph the solution that corresponds with each tableau.

$$\text{Maximize} \quad P = 3x + 10y$$

$$\text{Subject to} \begin{cases} 4x - y \geq 11 \\ x + 2y \geq 5 \\ -x + y \leq 1 \\ x \geq 0, y \geq 0 \end{cases}$$

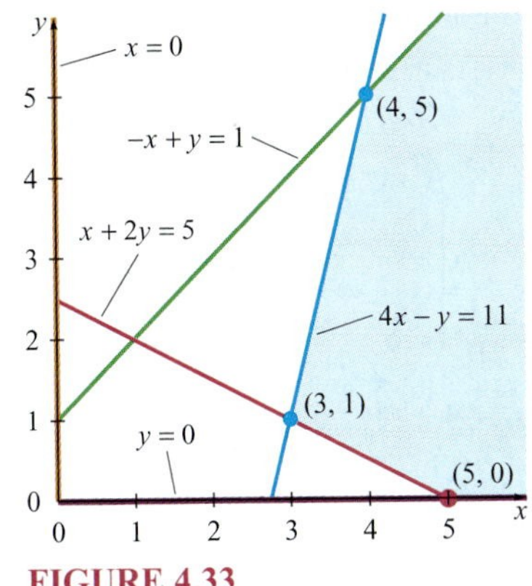

FIGURE 4.33

SOLUTION The graph of the feasible region is shown in Figure 4.33.

We can see from the graph that the feasible region is unbounded. Since the objective function has positive coefficients and the region is unbounded, the objective function will not have a maximum value. Nevertheless, we will proceed with the two-stage method to discover what indicators in the tableau let us know that there is no objective function maximum.

We add the slack and surplus variables and rewrite the objective function. We have

$$4x - y - s = 11$$
$$x + 2y - t = 5$$
$$-x + y + u = 1$$
$$-3x - 10y + P = 0$$

and the corresponding initial tableau

$$\begin{array}{c} \\ \\ \\ \\ \end{array}\begin{array}{cccccc|c} x & y & s & t & u & P & \\ \hline 4 & -1 & -1 & 0 & 0 & 0 & 11 \\ 1 & 2 & 0 & -1 & 0 & 0 & 5 \\ -1 & 1 & 0 & 0 & 1 & 0 & 1 \\ \hline -3 & -10 & 0 & 0 & 0 & 1 & 0 \end{array}$$

Reading from the tableau, we have

$$x = 0 \qquad y = 0 \qquad -s = 11 \qquad -t = 5 \qquad u = 1 \qquad P = 0$$
$$s = -11 \qquad t = -5$$

Since both s and t are negative, we will star the corresponding rows.

$$\begin{array}{c} \\ * \\ * \\ \\ \\ \end{array}\begin{array}{cccccc|c} x & y & s & t & u & P & \\ \hline 4 & -1 & -1 & 0 & 0 & 0 & 11 \\ 1 & 2 & 0 & -1 & 0 & 0 & 5 \\ -1 & 1 & 0 & 0 & 1 & 0 & 1 \\ \hline -3 & -10 & 0 & 0 & 0 & 1 & 0 \end{array}$$

We may choose either starred row to work with. We pick the row corresponding with the variable t (second row). The largest value in the labeled columns of the second row is the 2 in the y column. Consequently, the y column is the pivot column. We calculate the test ratios and locate the pivot.

$$\begin{array}{c} \\ * \\ * \\ \\ \\ \end{array}\begin{array}{cccccc|c} x & y & s & t & u & P & \\ \hline 4 & -1 & -1 & 0 & 0 & 0 & 11 \\ 1 & 2 & 0 & -1 & 0 & 0 & 5 \\ -1 & \boxed{1} & 0 & 0 & 1 & 0 & 1 \\ \hline -3 & -10 & 0 & 0 & 0 & 1 & 0 \end{array} \quad \begin{array}{l} \\ \\ 5/2 = 2.5 \\ 1/1 = 1 \\ \\ \end{array}$$

Pivot column (↑ y column)

We zero out the remaining terms in the pivot column using the indicated operations and star the rows that correspond with negative variable values.

$$\begin{array}{c} \\ * \\ * \\ \\ \\ \end{array}\begin{array}{cccccc|c} x & y & s & t & u & P & \\ 3 & 0 & -1 & 0 & 1 & 0 & 12 \\ 3 & 0 & 0 & -1 & -2 & 0 & 3 \\ -1 & 1 & 0 & 0 & 1 & 0 & 1 \\ \hline -13 & 0 & 0 & 0 & 10 & 1 & 10 \end{array} \quad \begin{array}{l} \\ R_1 + R_3 \\ R_2 - 2R_3 \\ \\ R_4 + 10R_3 \end{array}$$

$$x = 0 \quad y = 1 \quad -s = 12 \quad -t = 3 \quad u = 0 \quad P = 10$$
$$s = -12 \quad t = -3$$

We observe that although t remains negative, it has become less negative. We will again select the largest value in a labeled column of the second row. The 3 in the x column is the largest value. Consequently, the x column is the pivot column.

$$\begin{array}{c} \\ * \\ * \\ \\ \\ \end{array}\begin{array}{cccccc|c} x & y & s & t & u & P & \\ 3 & 0 & -1 & 0 & 1 & 0 & 12 \\ \boxed{3} & 0 & 0 & -1 & -2 & 0 & 3 \\ -1 & 1 & 0 & 0 & 1 & 0 & 1 \\ \hline -13 & 0 & 0 & 0 & 10 & 1 & 10 \end{array} \quad \begin{array}{l} \\ 12/3 = 4 \\ 3/3 = 1 \\ \\ \\ \end{array}$$

Pivot column

The 3 in the second row of the x column is the pivot, since the second row has the smallest test ratio. We zero out the remaining entries in the x column using the indicated operations. We then star the row corresponding to the negative variable.

$$\begin{array}{c} \\ * \\ \\ \\ \\ \end{array}\begin{array}{cccccc|c} x & y & s & t & u & P & \\ 0 & 0 & -1 & 1 & 3 & 0 & 9 \\ 3 & 0 & 0 & -1 & -2 & 0 & 3 \\ 0 & 3 & 0 & -1 & 1 & 0 & 6 \\ \hline 0 & 0 & 0 & -13 & 4 & 3 & 69 \end{array} \quad \begin{array}{l} \\ R_1 - R_2 \\ \\ 3R_3 + R_2 \\ 3R_4 + 13R_2 \end{array}$$

$$3x = 3 \quad 3y = 6 \quad -s = 9 \quad t = 0 \quad u = 0 \quad 3P = 69$$
$$x = 1 \quad y = 2 \quad s = -9 \quad P = 23$$

Observe that the surplus variable t is now nonnegative. We need to continue the process to make s nonnegative. The largest value in the labeled columns of the starred row is the 3 in the u column. Consequently, the u column is the pivot column.

$$\begin{array}{c} \\ * \\ \\ \\ \\ \end{array}\begin{array}{cccccc|c} x & y & s & t & u & P & \\ 0 & 0 & -1 & 1 & \boxed{3} & 0 & 9 \\ 3 & 0 & 0 & -1 & -2 & 0 & 3 \\ 0 & 3 & 0 & -1 & 1 & 0 & 6 \\ \hline 0 & 0 & 0 & -13 & 4 & 3 & 69 \end{array} \quad \begin{array}{l} \\ 9/3 = 3 \\ \\ 6/1 = 6 \\ \\ \end{array}$$

Pivot column

Using the 3 in the first row of the u column as a pivot, we zero out the remaining column entries using the indicated operations.

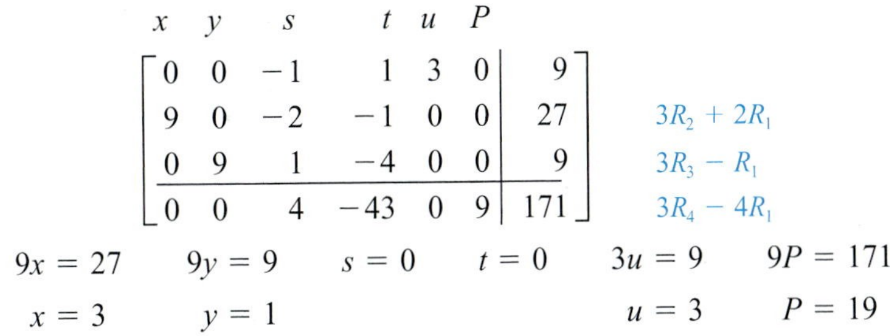

$$\begin{array}{c} \\ \\ \\ \\ \\ \end{array}\begin{array}{cccccc|c} x & y & s & t & u & P & \\ \hline 0 & 0 & -1 & 1 & 3 & 0 & 9 \\ 9 & 0 & -2 & -1 & 0 & 0 & 27 \\ 0 & 9 & 1 & -4 & 0 & 0 & 9 \\ \hline 0 & 0 & 4 & -43 & 0 & 9 & 171 \end{array} \quad \begin{array}{l} \\ \\ 3R_2 + 2R_1 \\ 3R_3 - R_1 \\ 3R_4 - 4R_1 \end{array}$$

$$\begin{array}{cccccc} 9x = 27 & 9y = 9 & s = 0 & t = 0 & 3u = 9 & 9P = 171 \\ x = 3 & y = 1 & & & u = 3 & P = 19 \end{array}$$

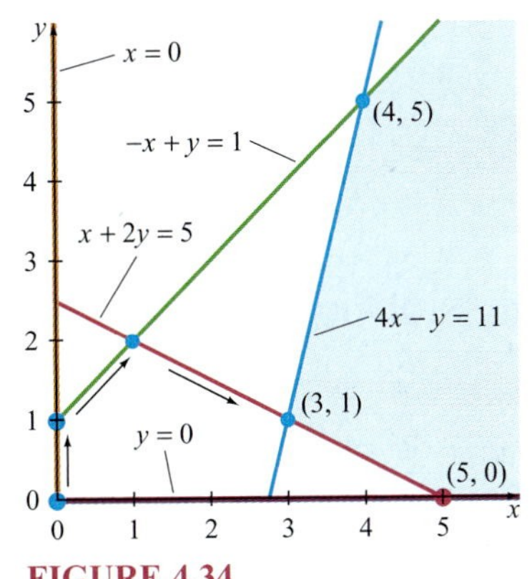

FIGURE 4.34

Since all decision, slack, and surplus variables are nonnegative, we are in the feasible region. We can now proceed to Stage 2 of the process: applying the simplex method. Before going on, let's observe what has happened graphically (see Figure 4.34).

The blue dots on the graph indicate the solutions to the various tableaus. We started at $(0, 0)$, proceeded to $(0, 1)$, then to $(1, 2)$, and finally to $(3, 1)$.

Returning to the tableau, we apply the simplex method. The only negative entry in the bottom row is in the t column, so the t column is our pivot column. The only positive entry in the t column is the 1 in the first row, so that is our pivot.

$$\begin{array}{cccccc|c} x & y & s & t & u & P & \\ \hline 0 & 0 & -1 & \boxed{1} & 3 & 0 & 9 \\ 9 & 0 & -2 & -1 & 0 & 0 & 27 \\ 0 & 9 & 1 & -4 & 0 & 0 & 9 \\ \hline 0 & 0 & 4 & -43 & 0 & 9 & 171 \end{array}$$

Pivot column (the t column)

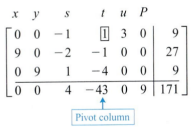

We zero out the remaining entries in the t column with the indicated operations.

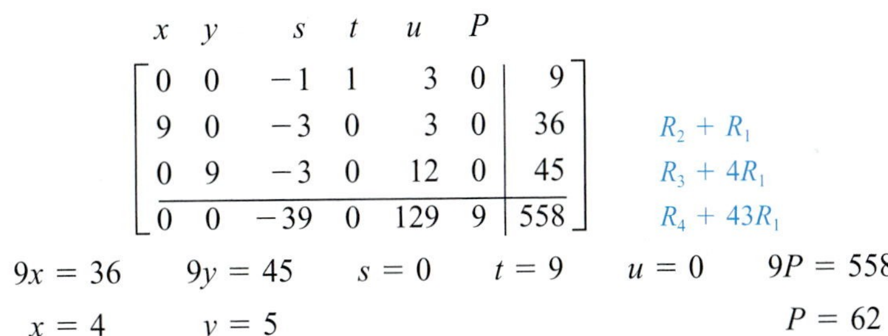

$$\begin{array}{cccccc|c} x & y & s & t & u & P & \\ \hline 0 & 0 & -1 & 1 & 3 & 0 & 9 \\ 9 & 0 & -3 & 0 & 3 & 0 & 36 \\ 0 & 9 & -3 & 0 & 12 & 0 & 45 \\ \hline 0 & 0 & -39 & 0 & 129 & 9 & 558 \end{array} \quad \begin{array}{l} \\ \\ R_2 + R_1 \\ R_3 + 4R_1 \\ R_4 + 43R_1 \end{array}$$

$$\begin{array}{cccccc} 9x = 36 & 9y = 45 & s = 0 & t = 9 & u = 0 & 9P = 558 \\ x = 4 & y = 5 & & & & P = 62 \end{array}$$

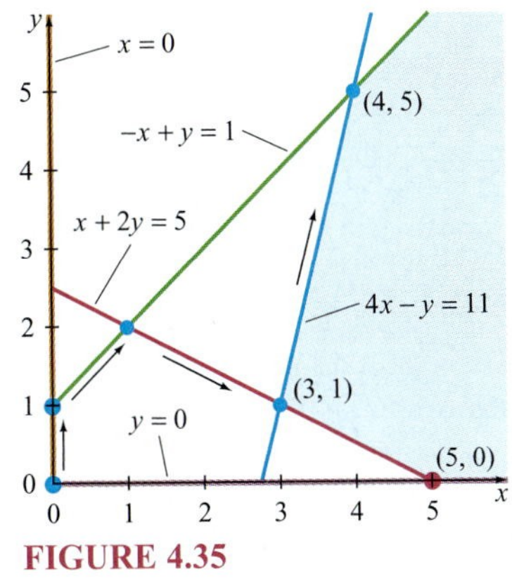

FIGURE 4.35

The s column is the new pivot column. However, we are unable to select a pivot because every entry in the s column is negative. This signifies that the objective function has no optimal solution. Figure 4.35 shows what has happened graphically.

We increased the value of the objective function by moving from the corner point $(3, 1)$ to the corner point $(4, 5)$. Out of all of the corner points, this is the corner point that yields the largest value of the objective function. However, since the region is unbounded, we can continue to increase the value of the objective function indefinitely.

LINEAR PROGRAMMING PROBLEMS WITH NO OPTIMAL SOLUTIONS

If none of the entries in the pivot column of a simplex tableau are positive, the corresponding feasible region is unbounded and the objective function has no optimal solution.

EXAMPLE 2

Solving a General Linear Programming Problem

Solve the general linear programming problem:

$$\text{Maximize} \quad P = 4x + 3y + 2z$$
$$\text{Subject to} \begin{cases} -x - y + z \geq 1 \\ -2x + y + z \geq 3 \\ -x + y + z \leq 3 \\ x \geq 0, y \geq 0, z \geq 0 \end{cases}$$

SOLUTION We add the slack and surplus variables and rewrite the objective function.

$$-x - y + z - s = 1$$
$$-2x + y + z - t = 3$$
$$-x + y + z + u = 3$$
$$-4x - 3y - 2z + P = 0$$

The corresponding initial tableau is shown. We calculate the basic solution and star the rows with negative variable values.

$$\begin{array}{r} \\ * \\ * \\ \\ \\ \end{array}\begin{array}{c} \begin{array}{ccccccc} x & y & z & s & t & u & P \end{array} \\ \left[\begin{array}{rrrrrrr|r} -1 & -1 & 1 & -1 & 0 & 0 & 0 & 1 \\ -2 & 1 & 1 & 0 & -1 & 0 & 0 & 3 \\ -1 & 1 & 1 & 0 & 0 & 1 & 0 & 3 \\ \hline -4 & -3 & -2 & 0 & 0 & 0 & 1 & 0 \end{array}\right] \end{array}$$

$x = 0 \quad y = 0 \quad z = 0 \quad -s = 1 \quad -t = 3 \quad u = 3 \quad P = 0$

$s = -1 \quad t = -3$

We may work with either starred row. We choose the second row. (This row corresponds to $t = -3$.) The largest positive entries in the labeled columns of the second row are the 1 in the y column and the 1 in the z column. We may pick either column as the pivot column. We pick the y column. We calculate the test ratios and locate the pivot. The test ratios are equal, so we may pick either the second or the third row as the pivot row. We choose the second row.

$$\begin{array}{c} \begin{array}{ccccccc} x & y & z & s & t & u & P \end{array} \\ \left[\begin{array}{rrrrrrr|r} -1 & -1 & 1 & -1 & 0 & 0 & 0 & 1 \\ -2 & \boxed{1} & 1 & 0 & -1 & 0 & 0 & 3 \\ -1 & 1 & 1 & 0 & 0 & 1 & 0 & 3 \\ \hline -4 & -3 & -2 & 0 & 0 & 0 & 1 & 0 \end{array}\right] \end{array} \begin{array}{l} \\ \\ 3/1 = 3 \\ 3/1 = 3 \\ \\ \end{array}$$

Pivot column (↑ under the y column)

We zero out the y column using the indicated operations and star the row corresponding with a negative variable value.

$$\begin{array}{c} \\ * \\ \\ \\ \\ \end{array}\begin{array}{c} \begin{array}{ccccccc|c} x & y & z & s & t & u & P & \end{array} \\ \left[\begin{array}{ccccccc|c} -3 & 0 & 2 & -1 & -1 & 0 & 0 & 4 \\ -2 & 1 & 1 & 0 & -1 & 0 & 0 & 3 \\ 1 & 0 & 0 & 0 & 1 & 1 & 0 & 0 \\ \hline -10 & 0 & 1 & 0 & -3 & 0 & 1 & 9 \end{array}\right] \end{array} \quad \begin{array}{l} \\ R_1 + R_2 \\ \\ R_3 - R_2 \\ R_4 + 3R_2 \end{array}$$

$$x = 0 \quad y = 3 \quad z = 0 \quad -s = 4 \quad t = 0 \quad u = 0 \quad P = 9$$
$$s = -4$$

The 2 in the z column is the largest positive value in the starred row. Consequently, the z column is the pivot column. We calculate the test ratios and locate the pivot.

$$\begin{array}{ccccccc|c} x & y & z & s & t & u & P & \end{array}$$
$$\left[\begin{array}{ccccccc|c} -3 & 0 & \boxed{2} & -1 & -1 & 0 & 0 & 4 \\ -2 & 1 & 1 & 0 & -1 & 0 & 0 & 3 \\ 1 & 0 & 0 & 0 & 1 & 1 & 0 & 0 \\ \hline -10 & 0 & 1 & 0 & -3 & 0 & 1 & 9 \end{array}\right] \quad \begin{array}{l} 4/2 = 2 \\ 3/1 = 3 \\ \\ \\ \end{array}$$

Pivot column (z)

We zero out the z column with the indicated operations.

$$\begin{array}{ccccccc|c} x & y & z & s & t & u & P & \end{array}$$
$$\left[\begin{array}{ccccccc|c} -3 & 0 & 2 & -1 & -1 & 0 & 0 & 4 \\ -1 & 2 & 0 & 1 & -1 & 0 & 0 & 2 \\ 1 & 0 & 0 & 0 & 1 & 1 & 0 & 0 \\ \hline -17 & 0 & 0 & 1 & -5 & 0 & 2 & 14 \end{array}\right] \quad \begin{array}{l} \\ 2R_2 - R_1 \\ \\ 2R_4 - R_1 \end{array}$$

$$x = 0 \quad 2y = 2 \quad 2z = 4 \quad s = 0 \quad t = 0 \quad u = 0 \quad 2P = 14$$
$$y = 1 \quad z = 2 \quad P = 7$$

Since all of the decision, slack, and surplus variables are positive, we are in the feasible region. We may now move to Stage 2 and apply the simplex method. Since the x column contains the negative value in the bottom row with the largest magnitude, it is the pivot column. Since there is only one positive entry in the x column, it is the pivot.

$$\begin{array}{ccccccc|c} x & y & z & s & t & u & P & \end{array}$$
$$\left[\begin{array}{ccccccc|c} -3 & 0 & 2 & -1 & -1 & 0 & 0 & 4 \\ -1 & 2 & 0 & 1 & -1 & 0 & 0 & 2 \\ \boxed{1} & 0 & 0 & 0 & 1 & 1 & 0 & 0 \\ \hline -17 & 0 & 0 & 1 & -5 & 0 & 2 & 14 \end{array}\right]$$

Pivot column (x)

We zero out the remaining terms in the x column using the indicated operations.

$$\begin{array}{c} \begin{matrix} x & y & z & s & t & u & P & \end{matrix} \\ \left[\begin{array}{ccccccc|c} 0 & 0 & 2 & -1 & 2 & 3 & 0 & 4 \\ 0 & 2 & 0 & 1 & 0 & 1 & 0 & 2 \\ 1 & 0 & 0 & 0 & 1 & 1 & 0 & 0 \\ \hline 0 & 0 & 0 & 1 & 12 & 17 & 2 & 14 \end{array}\right] \end{array} \begin{array}{l} R_1 + 3R_3 \\ R_2 + R_3 \\ \\ R_4 + 17R_3 \end{array}$$

$$x = 0 \quad 2y = 2 \quad 2z = 4 \quad s = 0 \quad t = 0 \quad u = 0 \quad 2P = 14$$
$$y = 1 \quad z = 2 \quad P = 7$$

Since all of the entries in the bottom row of the tableau are nonnegative, this solution is the optimal solution. When $x = 0$, $y = 1$, and $z = 2$, the objective function $P = 4x + 3y + 2z$ attains a maximum value of 7.

Figure 4.36 shows the graph of the constraints and the optimal solution. Each constraint is a plane. The planes intersect at (0, 1, 2).

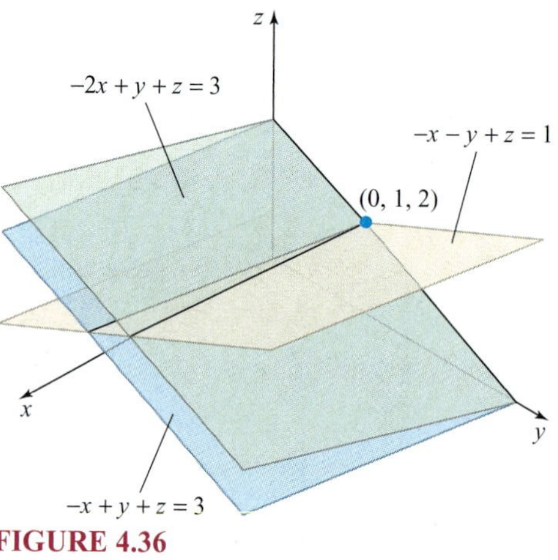

FIGURE 4.36

EXAMPLE 3

Using Linear Programming to Make Investment Decisions

An investor plans to invest at most \$3000 in the three publicly traded recreational vehicle companies shown in Table 4.26. (Share prices are rounded to the nearest dollar, and dividends per share are rounded to the nearest dime.)

TABLE 4.26

Company	Share Price (dollars)	Dividends/Share (dollars)
Harley-Davidson, Inc. (HDI)	63	0.40
Polaris Industries Inc. (PII)	51	0.90
Winnebago Industries Inc. (WGO)	33	0.20

Source: moneycentral.msn.com. (Accurate as of July 16, 2004.)

He wants to earn at least \$50 in dividends while maximizing the number of shares purchased. How many shares of each company should he purchase?

SOLUTION Let x be the number of shares of Harley-Davidson, y be the number of shares of Polaris Industries, and z be the number of shares of Winnebago Industries. We must maximize $P = x + y + z$ subject to

$$63x + 51y + 33z \leq 3000 \quad \text{The total amount invested is at most \$3000}$$

$$0.4x + 0.9y + 0.2z \geq 50 \quad \text{Total dividends are at least \$50}$$

$$x \geq 0, y \geq 0, z \geq 0$$

We add in the slack and surplus variables and rewrite the objective function.

$$63x + 51y + 33z + s = 3000$$
$$0.4x + 0.9y + 0.2z - t = 50$$
$$-x - y - z + P = 0$$

We then write the initial tableau and star the row corresponding with a negative variable value.

$$\begin{array}{c} \\ \\ * \\ \\ \end{array}\begin{array}{c} \begin{array}{cccccc} x & y & z & s & t & P \end{array} \\ \left[\begin{array}{cccccc|c} 63 & 51 & 33 & 1 & 0 & 0 & 3000 \\ 0.4 & 0.9 & 0.2 & 0 & -1 & 0 & 50 \\ \hline -1 & -1 & -1 & 0 & 0 & 1 & 0 \end{array}\right] \end{array}$$

$x = 0 \quad y = 0 \quad z = 0 \quad s = 3000 \quad -t = 50 \quad P = 0$

$t = -50$

The largest entry in the starred row is the 0.9 in the y column; hence, the y column is the pivot column. We calculate the test ratios and locate the pivot.

$$\begin{array}{c} \begin{array}{cccccc} x & y & z & s & t & P \end{array} \\ \left[\begin{array}{cccccc|c} 63 & 51 & 33 & 1 & 0 & 0 & 3000 \\ 0.4 & \boxed{0.9} & 0.2 & 0 & -1 & 0 & 50 \\ \hline -1 & -1 & -1 & 0 & 0 & 1 & 0 \end{array}\right] \end{array} \quad \begin{array}{l} 3000/51 \approx 58.82 \\ 50/0.9 \approx 55.56 \\ \\ \end{array}$$

We zero out the y column with the indicated operations and star the row corresponding with a negative variable value.

$$\begin{array}{c} \begin{matrix} x & y & z & s & t & P & \end{matrix} \\ \left[\begin{array}{cccccc|c} 36.3 & 0 & 19.5 & 0.9 & 51 & 0 & 150 \\ 0.4 & 0.9 & 0.2 & 0 & -1 & 0 & 50 \\ \hline -0.5 & 0 & -0.7 & 0 & -1 & 0.9 & 50 \end{array}\right] \begin{array}{l} 0.9R_1 - 51R_2 \\ \\ 0.9R_3 + R_2 \end{array} \end{array}$$

$x = 0 \quad 0.9y = 50 \quad z = 0 \quad 0.9s = 150 \quad t = 0 \quad 0.9P = 50$

$y \approx 55.56 \quad s \approx 166.67 \quad P \approx 55.56$

Since all of the decision, slack, and surplus variables are nonnegative, we are in the feasible region and may move to Stage 2.

The negative entry in the bottom row of a labeled column with the largest magnitude is the -1 in the t column. Consequently, the t column is the pivot column. Since 51 is the only positive entry in the t column, it is the pivot. We zero out the t column with the indicated operations.

$$\begin{array}{c} \begin{matrix} x & y & z & s & t & P & \end{matrix} \\ \left[\begin{array}{cccccc|c} 36.3 & 0 & 19.5 & 0.9 & \boxed{51} & 0 & 150 \\ 56.7 & 45.9 & 29.7 & 0.9 & 0 & 0 & 2700 \\ \hline 10.8 & 0 & -16.2 & 0.9 & 0 & 45.9 & 2700 \end{array}\right] \begin{array}{l} \\ 51R_2 + R_1 \\ 51R_3 + R_1 \end{array} \end{array}$$

$x = 0 \quad 45.9y = 2700 \quad z = 0 \quad s = 0 \quad 51t = 150 \quad 45.9P = 2700$

$y \approx 58.82 \quad t \approx 2.94 \quad P \approx 58.82$

The z column is the new pivot column, since it contains a negative entry in the bottom row. We calculate the test ratios and identify the pivot.

$$\begin{array}{c} \begin{matrix} x & y & z & s & t & P & \end{matrix} \\ \left[\begin{array}{cccccc|c} 36.3 & 0 & \boxed{19.5} & 0.9 & 51 & 0 & 150 \\ 56.7 & 45.9 & 29.7 & 0.9 & 0 & 0 & 2700 \\ \hline 10.8 & 0 & -16.2 & 0.9 & 0 & 45.9 & 2700 \end{array}\right] \begin{array}{l} 150/19.5 \approx 7.69 \\ 2700/29.7 \approx 90.91 \\ \\ \end{array} \end{array}$$

We zero out the z column with the indicated operations.

$$\begin{array}{c} \begin{matrix} x & y & z & s & t & P & \end{matrix} \\ \left[\begin{array}{cccccc|c} 36.3 & 0 & 19.5 & 0.9 & 51 & 0 & 150 \\ 27.54 & 895.05 & 0 & -9.18 & -1514.7 & 0 & 48195 \\ \hline 798.66 & 0 & 0 & 32.13 & 826.2 & 895.05 & 55080 \end{array}\right] \begin{array}{l} \\ 19.5R_2 - 29.7R_1 \\ 19.5R_3 + 16.2R_1 \end{array} \end{array}$$

$x = 0 \quad 895.05y = 48195 \quad 19.5z = 150 \quad s = 0 \quad t = 0 \quad 895.05P = 55080$

$y \approx 53.85 \quad z \approx 7.69 \quad P \approx 61.54$

The investor should purchase 53.85 shares of Polaris Industries and 7.69 shares of Winnebago Industries in order to maximize the number of shares while simultaneously earning dividends of at least \$50.

Minimization Problems

Examples 1, 2, and 3 showed how to *maximize* an objective function of a linear programming problem with mixed constraints. What if we want to *minimize* the objective function? Fortunately, with one minor modification to the objective function, the same procedure works. We begin by observing the relationship between a function f and the function $-f = -1 \cdot f$ shown in Figure 4.37.

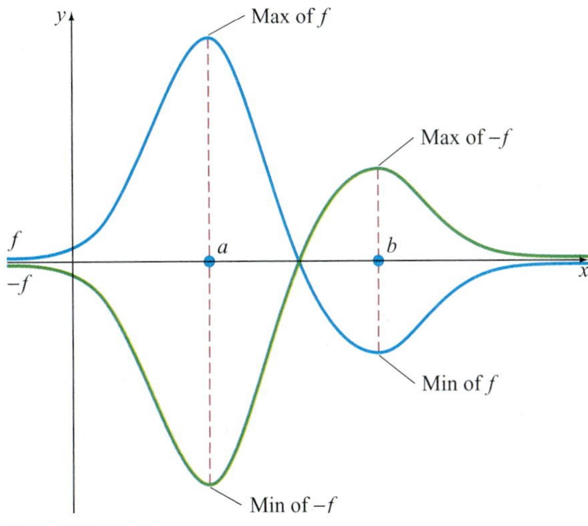

FIGURE 4.37

Observe that the maximum of f and the minimum of $-f$ both occur at $x = a$. Similarly, the minimum of f and the maximum of $-f$ both occur at $x = b$. This relationship holds true for all functions. Consequently, if we want to *minimize* an objective function P, we need only *maximize* the function $-P$.

EXAMPLE 4

Solving a General Linear Programming Problem

Solve the general linear programming problem:

$$\text{Minimize} \quad P = 3x + 5y + 2z$$

$$\text{Subject to} \begin{cases} x + 2y - z \le 0 \\ -2x + 4y + z \ge 3 \\ -x + y + z \le 3 \\ x \ge 0, y \ge 0, z \ge 0 \end{cases}$$

SOLUTION We first introduce a new function $C = -P$.

$$\begin{aligned} C &= -P \\ &= -(3x + 5y + 2z) \\ &= -3x - 5y - 2z \end{aligned}$$

The solution that maximizes C will minimize P. Therefore, the linear programming problem may be rewritten as

$$\text{Maximize} \quad C = -3x - 5y - 2z$$

$$\text{Subject to} \begin{cases} x + 2y - z \le 0 \\ -2x + 4y + z \ge 3 \\ -x + y + z \le 3 \\ x \ge 0, y \ge 0, z \ge 0 \end{cases}$$

We add in the slack and surplus variables and rewrite the objective function. The system of equations is

$$\begin{aligned} x + 2y - z + s &= 0 \\ -2x + 4y + z - t &= 3 \\ -x + y + z + u &= 3 \\ 3x + 5y + 2z + C &= 0 \end{aligned}$$

and has the corresponding tableau

$$\begin{array}{c} \begin{matrix} x & y & z & s & t & u & C & \end{matrix} \\ \left[\begin{array}{rrrrrrr|r} 1 & 2 & -1 & 1 & 0 & 0 & 0 & 0 \\ -2 & 4 & 1 & 0 & -1 & 0 & 0 & 3 \\ -1 & 1 & 1 & 0 & 0 & 1 & 0 & 3 \\ \hline 3 & 5 & 2 & 0 & 0 & 0 & 1 & 0 \end{array}\right] \end{array}$$

$x = 0 \quad y = 0 \quad z = 0 \quad s = 0 \quad -t = 3 \quad u = 3 \quad C = 0$

$t = -3$

Since the surplus variable t is negative, we star the corresponding row. The y column is the pivot column because it contains the largest positive entry in the starred row. Calculating the test ratios, we select the pivot.

$$\begin{array}{c} \begin{matrix} x & y & z & s & t & u & C & \end{matrix} \\ \begin{matrix} \\ * \\ \\ \\ \end{matrix} \left[\begin{array}{rrrrrrr|r} 1 & \boxed{2} & -1 & 1 & 0 & 0 & 0 & 0 \\ -2 & 4 & 1 & 0 & -1 & 0 & 0 & 3 \\ -1 & 1 & 1 & 0 & 0 & 1 & 0 & 3 \\ \hline 3 & 5 & 2 & 0 & 0 & 0 & 1 & 0 \end{array}\right] \begin{matrix} 0/2 = 0 \\ 3/4 = 0.75 \\ 3/1 = 3 \\ \\ \end{matrix} \end{array}$$

We zero out the remaining terms in the y column using the indicated operations and star the row corresponding with the negative variable value.

$$\begin{array}{c} \begin{matrix} x & y & z & s & t & u & C & \end{matrix} \\ \begin{matrix} \\ * \\ \\ \\ \end{matrix} \left[\begin{array}{rrrrrrr|r} 1 & 2 & -1 & 1 & 0 & 0 & 0 & 0 \\ -4 & 0 & 3 & -2 & -1 & 0 & 0 & 3 \\ -3 & 0 & 3 & -1 & 0 & 2 & 0 & 6 \\ \hline 1 & 0 & 9 & -5 & 0 & 0 & 2 & 0 \end{array}\right] \begin{matrix} \\ R_2 - 2R_1 \\ 2R_3 - R_1 \\ 2R_4 - 5R_1 \end{matrix} \end{array}$$

$x = 0 \quad 2y = 0 \quad z = 0 \quad s = 0 \quad -t = 3 \quad 2u = 6 \quad 2C = 0$

$y = 0 \quad t = -3 \quad u = 3 \quad C = 0$

The largest positive value in the starred row is the 3 in the z column. Consequently, the z column is the pivot column. We calculate the test ratios and select the pivot.

$$\begin{array}{c} \begin{matrix} x & y & z & s & t & u & C & \end{matrix} \\ \left[\begin{array}{rrrrrrr|r} 1 & 2 & -1 & 1 & 0 & 0 & 0 & 0 \\ -4 & 0 & \boxed{3} & -2 & -1 & 0 & 0 & 3 \\ -3 & 0 & 3 & -1 & 0 & 2 & 0 & 6 \\ \hline 1 & 0 & 9 & -5 & 0 & 0 & 2 & 0 \end{array}\right] \begin{matrix} \\ 3/3 = 1 \\ 6/3 = 2 \\ \\ \end{matrix} \end{array}$$

We zero out the remaining terms in the z column with the indicated operations.

$$\begin{array}{c} \begin{matrix} x & y & z & s & t & u & C \end{matrix} \\ \left[\begin{array}{ccccccc|c} -1 & 6 & 0 & 1 & -1 & 0 & 0 & 3 \\ -4 & 0 & 3 & -2 & -1 & 0 & 0 & 3 \\ 1 & 0 & 0 & 1 & 1 & 2 & 0 & 3 \\ \hline 13 & 0 & 0 & 1 & 3 & 0 & 2 & -9 \end{array}\right] \end{array} \quad \begin{array}{l} 3R_1 + R_2 \\ \\ R_3 - R_2 \\ R_4 - 3R_2 \end{array}$$

$$x = 0 \quad 6y = 3 \quad 3z = 3 \quad s = 0 \quad t = 0 \quad 2u = 3 \quad 2C = -9$$

$$y = \frac{1}{2} \quad z = 1 \quad u = \frac{3}{2} \quad C = -\frac{9}{2}$$

Since all decision, slack, and surplus variables are nonnegative, we are now in the feasible region. (There is no restriction on the sign of the objective function. In this case, C is negative.) We may now move to Stage 2 of the process. However, since no entry in the bottom row of a labeled column is negative, this is the final tableau. The objective function $C = -3x - 5y - 2z$ is maximized at $\left(0, \frac{1}{2}, 1\right)$. Consequently, the objective function $P = 3x + 5y + 2z$ is minimized at $\left(0, \frac{1}{2}, 1\right)$. The maximum value of C is $-\frac{9}{2}$. Since $P = -C$, the minimum value of P is $\frac{9}{2}$.

So far, the general linear programming problems we have demonstrated have included only linear inequalities as constraints. However, some linear programming problems have linear equations as constraints. Any linear equation may be written as a system of linear inequalities. For example, consider the linear equation $2x + 3y = 5$. This equation is equivalent to the following system of linear inequalities:

$$2x + 3y \leq 5$$
$$2x + 3y \geq 5$$

When we are given a linear equation as a constraint, we will rewrite it as a system of linear inequalities, as demonstrated in Example 5.

EXAMPLE 5

Using Linear Programming to Make Investment Decisions

An investor has \$5000 to invest in three mutual funds: Bond, Index, and Growth. The Bond fund is expected to earn 7 percent; the Index fund, 11 percent; and the Growth fund, 9 percent. The investor wants to earn an annual return of at least 10 percent while minimizing the amount invested in the Index fund. How much should she invest in each account?

SOLUTION Let x be the amount invested in the Bond fund, y be the amount invested in the Index fund, and z be the amount invested in the Growth fund. We must solve the following linear programming problem.

Minimize $P = y$

$$\text{Subject to } \begin{cases} x + y + z = 5000 & \text{\$5000 is invested} \\ 0.07x + 0.11y + 0.09z \geq 0.1(5000) & \text{The return is at least 10\% of \$5000} \\ x \geq 0, y \geq 0, z \geq 0 \end{cases}$$

We rewrite the first constraint as two inequalities and simplify the second constraint. We then create a new objective function $C = -P$. Since $P = y$, $C = -y$. The new linear programming problem is

$$\text{Maximize} \quad C = -y$$

$$\text{Subject to} \begin{cases} x + y + z \leq 5000 \\ x + y + z \geq 5000 \\ 0.07x + 0.11y + 0.09z \geq 5 \\ x \geq 0, y \geq 0, z \geq 0 \end{cases}$$

We add the surplus and slack variables and rewrite the objective function.

$$x + y + z + s = 5000$$
$$x + y + z - t = 5000$$
$$0.07x + 0.11y + 0.09z - u = 500$$
$$y + C = 0$$

We write the initial tableau and star the appropriate rows.

$$\begin{array}{c} \\ \\ * \\ * \\ \\ \end{array} \begin{array}{ccccccc|c} x & y & z & s & t & u & C & \\ 1 & 1 & 1 & 1 & 0 & 0 & 0 & 5000 \\ 1 & 1 & 1 & 0 & -1 & 0 & 0 & 5000 \\ 0.07 & 0.11 & 0.09 & 0 & 0 & -1 & 0 & 500 \\ \hline 0 & 1 & 0 & 0 & 0 & 0 & 1 & 0 \end{array}$$

$x = 0 \quad y = 0 \quad z = 0 \quad s = 5000 \quad -t = 5000 \quad -u = 500 \quad C = 0$

$t = -5000 \quad u = -500$

Although we may use either starred row, we pick the second row of the tableau because the entries are simpler. We may pick the x, y, or z column as the pivot column. We pick the x column. We calculate the test ratios and identify the pivot.

$$\begin{array}{ccccccc|c} x & y & z & s & t & u & C & \\ \boxed{1} & 1 & 1 & 1 & 0 & 0 & 0 & 5000 \\ 1 & 1 & 1 & 0 & -1 & 0 & 0 & 5000 \\ 0.07 & 0.11 & 0.09 & 0 & 0 & -1 & 0 & 500 \\ \hline 0 & 1 & 0 & 0 & 0 & 0 & 1 & 0 \end{array} \quad \begin{array}{l} \\ 5000/1 = 5000 \\ 5000/1 = 5000 \\ 500/0.07 \approx 7143 \\ \\ \end{array}$$

We may pick either the first or the second term in the x column to be the pivot. We pick the first term and zero out the remaining entries with the indicated operations. We then star the appropriate row.

$$\begin{array}{c} \\ \\ \\ * \\ \\ \end{array} \begin{array}{ccccccc|c} x & y & z & s & t & u & C & \\ 1 & 1 & 1 & 1 & 0 & 0 & 0 & 5000 \\ 0 & 0 & 0 & -1 & -1 & 0 & 0 & 0 \\ 0 & 0.04 & 0.02 & -0.07 & 0 & -1 & 0 & 150 \\ \hline 0 & 1 & 0 & 0 & 0 & 0 & 1 & 0 \end{array} \quad \begin{array}{l} \\ \\ R_2 - R_1 \\ R_3 - 0.07R_1 \\ \\ \end{array}$$

$x = 5000 \quad y = 0 \quad z = 0 \quad s = 0 \quad -t = 0 \quad -u = 150 \quad C = 0$

$t = 0 \quad u = -150$

The y column is the pivot column, since the largest positive entry in the starred row is the 0.04 in the y column. We calculate the test ratios and locate the pivot.

$$
\begin{array}{c}
\\ \\ \\ * \\ \\
\end{array}
\begin{array}{c}
\begin{array}{ccccccc|c} x & y & z & s & t & u & C & \end{array} \\
\left[\begin{array}{ccccccc|c}
1 & 1 & 1 & 1 & 0 & 0 & 0 & 5000 \\
0 & 0 & 0 & -1 & -1 & 0 & 0 & 0 \\
0 & \boxed{0.04} & 0.02 & -0.07 & 0 & -1 & 0 & 150 \\
\hline
0 & 1 & 0 & 0 & 0 & 0 & 1 & 0
\end{array}\right]
\end{array}
\quad
\begin{array}{l}
5000/1 = 5000 \\
\\
150/0.04 = 3750 \\
\\
\end{array}
$$

We zero out the y column with the indicated operations.

$$
\begin{array}{c}
\begin{array}{ccccccc|c} x & y & z & s & t & u & C & \end{array} \\
\left[\begin{array}{ccccccc|c}
0.04 & 0 & 0.02 & 0.11 & 0 & 1 & 0 & 50 \\
0 & 0 & 0 & -1 & -1 & 0 & 0 & 0 \\
0 & 0.04 & 0.02 & -0.07 & 0 & -1 & 0 & 150 \\
\hline
0 & 0 & -0.02 & 0.07 & 0 & 1 & 0.04 & -150
\end{array}\right]
\end{array}
\quad
\begin{array}{l}
0.04R_1 - R_3 \\
\\
\\
0.04R_4 - R_3
\end{array}
$$

$$
\begin{array}{lllllll}
0.04x = 50 & 0.04y = 150 & z = 0 & s = 0 & -t = 0 & u = 0 & 0.04C = -150 \\
x = 1250 & y = 3750 & & & t = 0 & & C = -3750
\end{array}
$$

Since all decision, slack, and surplus variables are nonnegative, we are in the feasible region and may proceed to Stage 2.

Since the only negative entry in the bottom row of a labeled column is the -0.02 in the z column, the z column is the pivot column. We calculate the test ratios and identify the pivot.

$$
\begin{array}{c}
\begin{array}{ccccccc|c} x & y & z & s & t & u & C & \end{array} \\
\left[\begin{array}{ccccccc|c}
0.04 & 0 & \boxed{0.02} & 0.11 & 0 & 1 & 0 & 50 \\
0 & 0 & 0 & -1 & -1 & 0 & 0 & 0 \\
0 & 0.04 & 0.02 & -0.07 & 0 & -1 & 0 & 150 \\
\hline
0 & 0 & -0.02 & 0.07 & 0 & 1 & 0.04 & -150
\end{array}\right]
\end{array}
\quad
\begin{array}{l}
50/0.02 = 2500 \\
\\
150/0.04 = 3750 \\
\\
\end{array}
$$

We zero out the remaining terms in the pivot column with the indicated operations.

$$
\begin{array}{c}
\begin{array}{ccccccc|c} x & y & z & s & t & u & C & \end{array} \\
\left[\begin{array}{ccccccc|c}
0.04 & 0 & 0.02 & 0.11 & 0 & 1 & 0 & 50 \\
0 & 0 & 0 & -1 & -1 & 0 & 0 & 0 \\
-0.04 & 0.04 & 0 & -0.18 & 0 & -2 & 0 & 100 \\
\hline
0.04 & 0 & 0 & 0.18 & 0 & 2 & 0.04 & -100
\end{array}\right]
\end{array}
\quad
\begin{array}{l}
\\
\\
R_3 - R_1 \\
R_4 + R_1
\end{array}
$$

$$
\begin{array}{lllllll}
x = 0 & 0.04y = 100 & 0.02z = 50 & s = 0 & -t = 0 & u = 0 & 0.04C = -100 \\
 & y = 2500 & z = 2500 & & t = 0 & & C = -2500
\end{array}
$$

Since all entries in the bottom row are positive, this is the final tableau. The objective function C reaches its maximum value when $x = 0$, $y = 2500$, and $z = 2500$. The maximum value is $C = -2500$. The objective function $P = -C$ attains its minimum value of 2500 at the same point. In the context of the problem, a 10 percent return is earned when \$0 is invested in the Bond fund, \$2500 is invested in the Index fund, and \$2500 is invested in the Growth fund. This solution minimizes the amount of money invested in the Index fund.

4.5 Summary

In this section, you learned how to solve general linear programming problems. You discovered that minimization problems may be solved by maximizing the negative of the objective function.

4.5 Exercises

In Exercises 1–5, rewrite the general linear programming problem as a system of equations with slack and surplus variables and a rewritten objective function.

1. Maximize $P = 4x + 2y + 3z$

$$\text{Subject to } \begin{cases} 2x + y + z \leq 10 \\ x + z \geq 3 \\ x - 2z \geq 1 \\ x \geq 0, y \geq 0, z \geq 0 \end{cases}$$

2. Maximize $P = 2x + 10y + z$

$$\text{Subject to } \begin{cases} -x + 3y \leq 2 \\ 3x + z \geq 4 \\ 5x - 4z \geq 1 \\ x \geq 0, y \geq 0, z \geq 0 \end{cases}$$

3. Maximize $P = 3x - y + 9z$

$$\text{Subject to } \begin{cases} -12x + 8y + 2z \leq 15 \\ 8x + 5y - 2z \leq 10 \\ 3x - 2y + 6z \geq 5 \\ x \geq 0, y \geq 0, z \geq 0 \end{cases}$$

4. Maximize $P = x - z$

$$\text{Subject to } \begin{cases} -x + y + z \leq 12 \\ x + 3y - 5z \geq 5 \\ 2x + y - 2z \leq 1 \\ x \geq 0, y \geq 0, z \geq 0 \end{cases}$$

5. Maximize $P = 10x + 9y + z$

$$\text{Subject to } \begin{cases} 4x - 3y - 2z \leq 12 \\ -2x + 4y + 6z \leq 12 \\ 3x - 2y + 4z \geq 12 \\ x \geq 0, y \geq 0, z \geq 0 \end{cases}$$

In Exercises 6–10, star the rows of the tableau that correspond with negative variable values.

6.

$$\begin{array}{cccccc|c} x & y & s & t & u & P & \\ -1 & 2 & 1 & 0 & 0 & 0 & 4 \\ 2 & -3 & 0 & 1 & 0 & 0 & 6 \\ 4 & 1 & 0 & 0 & -1 & 0 & 5 \\ \hline -6 & -2 & 0 & 0 & 0 & 1 & 0 \end{array}$$

7.

$$\begin{array}{cccccc|c} x & y & s & t & u & P & \\ 9 & 4 & -1 & 0 & 0 & 0 & 10 \\ -2 & 5 & 0 & 1 & 0 & 0 & 8 \\ 1 & 1 & 0 & 0 & -1 & 0 & 12 \\ \hline 5 & 4 & 0 & 0 & 0 & 1 & 0 \end{array}$$

8.

$$\begin{array}{cccccc|c} x & y & s & t & u & P & \\ -1 & 2 & -1 & 0 & 0 & 0 & 4 \\ 2 & 0 & -2 & 1 & 0 & 0 & 6 \\ 4 & 0 & 1 & 0 & -5 & 0 & 5 \\ \hline -6 & 0 & 3 & 0 & 0 & 4 & 24 \end{array}$$

9.

$$\begin{array}{cccccc|c} x & y & s & t & u & P & \\ 0 & 6 & -2 & 0 & 0 & 0 & 16 \\ 5 & 1 & 0 & 1 & 0 & 0 & 1 \\ 0 & 3 & 0 & 4 & -4 & 0 & 6 \\ \hline 0 & 10 & 0 & 5 & 0 & 3 & 9 \end{array}$$

10.

$$\begin{array}{cccccc|c} x & y & s & t & u & P & \\ 0 & 3 & 0 & 0 & 3 & 0 & 15 \\ 0 & 0 & 1 & 1 & 2 & 0 & 1 \\ 2 & 0 & 0 & -2 & -4 & 0 & 4 \\ \hline 0 & 0 & 0 & 6 & 3 & 1 & 22 \end{array}$$

In Exercises 11–20, graph the feasible region of the general linear programming problem. Then solve the problem by using the methods demonstrated in the section. For each tableau, indicate on the graph of the feasible region the point that corresponds with the solution of the tableau.

11. Maximize $P = 2x - 3y$

$$\text{Subject to} \begin{cases} 2x + y \leq 10 \\ x + y \geq 3 \\ 3x - y \geq 1 \\ x \geq 0, y \geq 0 \end{cases}$$

12. Maximize $P = 4x + y$

$$\text{Subject to} \begin{cases} 2x + y \leq 10 \\ 2x + 3y \geq 14 \\ -2x + y \leq 2 \\ x \geq 0, y \geq 0 \end{cases}$$

13. Maximize $P = 5x - 8y$

$$\text{Subject to} \begin{cases} 2x + y \leq 13 \\ 9x - 3y \leq 6 \\ -x + 2y \geq 6 \\ x \geq 0, y \geq 0 \end{cases}$$

14. Maximize $P = 6x + 3y$

$$\text{Subject to} \begin{cases} -x + y \geq 0 \\ 2x + y \geq 9 \\ x + 2y \geq 6 \\ x \geq 0, y \geq 0 \end{cases}$$

15. Maximize $P = -3x + 4y$

$$\text{Subject to} \begin{cases} 3x + 12y \geq 12 \\ 2x + y \leq 8 \\ x + y \leq 6 \\ x \geq 0, y \geq 0 \end{cases}$$

16. Minimize $P = 3x + 5y$

$$\text{Subject to} \begin{cases} -5x + 2y \leq 2 \\ -x + y \leq 4 \\ -7x + 4y \geq 4 \\ x \geq 0, y \geq 0 \end{cases}$$

17. Minimize $P = 7x + 4y$

$$\text{Subject to} \begin{cases} 3x + y \geq 10 \\ x + y \geq 6 \\ 3x + 2y \leq 17 \\ x \geq 0, y \geq 0 \end{cases}$$

18. Minimize $P = -3x + 8y$

$$\text{Subject to} \begin{cases} y \leq 6 \\ -2x + y \geq 0 \\ 2x + y \geq 8 \\ x \geq 0, y \geq 0 \end{cases}$$

19. Minimize $P = -x + y$

$$\text{Subject to} \begin{cases} x + y \geq 1 \\ 2x - 3y \geq 2 \\ x + 4y \geq 4 \\ x \geq 0, y \geq 0 \end{cases}$$

20. Maximize $P = 4x + 2y$

$$\text{Subject to} \begin{cases} 5x + 2y \geq 17 \\ x - 2y \leq 1 \\ x + y \geq 7 \\ x \geq 0, y \geq 0 \end{cases}$$

In Exercises 21–35, solve the general linear programming problem. If there is no solution, so state.

21. Maximize $P = 3x + 2y + z$

$$\text{Subject to} \begin{cases} x + y \leq 10 \\ x + z \leq 5 \\ y + z \geq 4 \\ x \geq 0, y \geq 0, z \geq 0 \end{cases}$$

22. Maximize $P = 5x + 6y + 2z$

$$\text{Subject to} \begin{cases} x + 2y \leq 10 \\ 2x + z \leq 8 \\ y + 2z \geq 6 \\ x \geq 0, y \geq 0, z \geq 0 \end{cases}$$

23. Maximize $P = x - y + z$

$$\text{Subject to} \begin{cases} 4x + 3y + 2z \geq 12 \\ 2x + 2y + z \leq 8 \\ x + y + 4z \geq 4 \\ x \geq 0, y \geq 0, z \geq 0 \end{cases}$$

24. Maximize $P = 6x + y$

$$\text{Subject to} \begin{cases} 5x + 2y + 4z \leq 23 \\ -x + y + z \geq 4 \\ x - y + z \geq 4 \\ x \geq 0, y \geq 0, z \geq 0 \end{cases}$$

25. Maximize $P = 2z$

$$\text{Subject to} \begin{cases} x + 2y + 3z \geq 12 \\ 3x + y + 2z \leq 12 \\ 2x + 3y + z \leq 12 \\ x \geq 0, y \geq 0, z \geq 0 \end{cases}$$

26. Maximize $P = 2x + y + z$

Subject to $\begin{cases} -x + y + z \le 5 \\ x - y + z \le 7 \\ -x - y + z \ge 3 \\ x \ge 0, y \ge 0, z \ge 0 \end{cases}$

27. Maximize $P = 6x - 2y + 6z$

Subject to $\begin{cases} -x - y + z \le 1 \\ x + y + z \le 7 \\ 2x - 3y + z \ge 12 \\ x \ge 0, y \ge 0, z \ge 0 \end{cases}$

28. Minimize $P = 4y + 9z$

Subject to $\begin{cases} -2x - 3y + z \ge 0 \\ 4x + y + z \ge 10 \\ -4x + 2y + z \ge 3 \\ x \ge 0, y \ge 0, z \ge 0 \end{cases}$

29. Minimize $P = -x - y + 4z$

Subject to $\begin{cases} -2x + 3y + z \ge 4 \\ 2x + 4y + z \le 4 \\ -4x + y + z \le 12 \\ x \ge 0, y \ge 0, z \ge 0 \end{cases}$

30. Minimize $P = 5x + 4y + 3z$

Subject to $\begin{cases} 2x - 5y + z \ge 4 \\ 2x + 4y + z \le 13 \\ z \ge 3 \\ x \ge 0, y \ge 0, z \ge 0 \end{cases}$

31. Minimize $P = 9x - 3y - 3z$

Subject to $\begin{cases} -3x - 4y + z \le 4 \\ 2x - 5y + z \le 14 \\ 3x - 2y + z \ge 8 \\ x \ge 0, y \ge 0, z \ge 0 \end{cases}$

32. Minimize $P = x - y + 3z$

Subject to $\begin{cases} 2x + 2y - z \le 4 \\ 5x - 4y + z \ge 16 \\ x - 4y + z \ge 5 \\ x \ge 0, y \ge 0, z \ge 0 \end{cases}$

33. Minimize $P = 6x - 2y + 4z$

Subject to $\begin{cases} 8x + y - z \le 8 \\ x - y + z \ge 1 \\ x - y + z \le 10 \\ x \ge 0, y \ge 0, z \ge 0 \end{cases}$

34. Minimize $P = x + 2y + 3z$

Subject to $\begin{cases} x + y + z \ge 6 \\ x - y + z \ge 2 \\ x + z \le 4 \\ x \ge 0, y \ge 0, z \ge 0 \end{cases}$

35. Maximize $P = 4x + 3y + 2z$

Subject to $\begin{cases} x + y + z \ge 8 \\ -2x + z \ge 1 \\ 3y + 5z \le 27 \\ x \ge 0, y \ge 0, z \ge 0 \end{cases}$

In Exercises 36–45, use the techniques demonstrated in the section to set up and solve each problem.

36. **Commodity Prices** Today's Market Prices (www.todaymarket.com) is a daily fruit and vegetable wholesale market price service. Produce retailers who subscribe to the service can use wholesale prices to aid them in setting retail prices for the fruits and vegetables they sell.

A 25-pound carton of peaches holds 60 medium peaches or 70 small peaches. In August 2002, the wholesale price for local peaches in Los Angeles was \$9.00 per carton for medium peaches and \$10.00 per carton for small peaches. (**Source:** Today's Market Prices.) A fruit vendor has budgeted up to \$100 to spend on peaches. He estimates that weekly demand for peaches is no more than 660 peaches. He wants to buy at least four boxes of each size of peach. Subject to these constraints, how many boxes of each size of peach should he buy in order to maximize the number of peaches available for sale?

37. **Battery Sales** AAA Alkaline Discount Batteries sells low-cost batteries to consumers. In August 2002, the firm offered AA batteries at the following prices: 50-pack for \$10.00, 100-pack for \$18.00, and 600-pack for \$96.00. (**Source:** www.aaaalkalinediscountbatteries.com.)

An electronics store owner wants to buy at least 900 batteries and spend at most \$150. She expects that she'll be able to resell all of the batteries she orders for \$1.50 each. How many packs (50-packs, 100-packs, or 600-packs) should she order if she wants to maximize her revenue? What is her maximum revenue? (*Hint:* At most one 600-pack may be purchased without exceeding the \$150 limit.)

38. **Resource Allocation: Food** An ice cream parlor wants to make three different types of ice cream: vanilla, strawberry, and peach-cherry. The parlor has 120 cups of cream, 48 eggs, and 32 cups of sugar on hand. The vanilla ice cream recipe calls for 4 cups of cream, 1 egg, and 0.75 cup of sugar. The strawberry ice cream recipe calls for 2 cups of cream, 2 eggs, and 0.75 cup of sugar. The peach-cherry ice cream recipe calls for 4 cups of cream, 1 egg, and 1.25 cups of sugar. Each recipe yields 1.5 quarts of ice cream. The parlor needs at least 18 quarts of vanilla and at least 6 quarts of each of the other varieties. The parlor wants to maximize the amount of ice cream produced. How many batches of each variety of ice cream does the parlor need to produce?

39. **Resource Allocation: Food** A delicatessen makes three types of pudding: rice, tapioca, and vanilla. The deli has 108 cups of milk, 150 cups of sugar, and 84 eggs on hand. The rice pudding recipe requires 12 cups of milk, 1.5 cups of sugar, and 9 eggs and yields 24 servings. The tapioca pudding recipe requires 12 cups of milk, 1.5 cups of sugar, and 9 eggs and yields 18 servings. The vanilla pudding recipe requires 6 cups of milk, 1.5 cups of sugar, and 6 eggs and yields 12 servings. The deli requires that the sum of the number of batches of tapioca pudding and the number of batches of rice pudding be at least two. How many batches of each recipe must the deli produce in order to maximize the number of servings?

40. **Advertising** A large company advertises through magazines, radio, and television. For every $10,000 spent on magazine advertising, the company estimates that it reaches 150,000 people. For every $10,000 spent on radio advertising, the company estimates that it reaches 250,000 people. For every $10,000 spent on television advertising, the company estimates that it reaches 300,000 people. The company has at most $2.5 million to spend on advertising. It requires that at least twice as much be spent on radio as on television and that the amount spent on magazines be at least $150,000 more than the amount spent on radio. How much should the company spend on each type of advertising in order to maximize the number of people reached?

41. **Nutrition** The nutritional content of canned beans varies based on the type of bean and the manufacturer. A 1/2-cup serving of Fred Meyer Pinto Beans contains 7 grams of fiber, 1 gram of sugar, and 6 grams of protein. A 1/2-cup serving of Fred Meyer Kidney Beans contains 11 grams of fiber, 1 gram of sugar, and 8 grams of protein. A 1/2-cup serving of Trader Joe's Cuban Style Black Beans contains 2 grams of fiber, 1 gram of sugar, and 6 grams of protein. (**Source:** Package labeling.) A bean dish is to be made using the three bean varieties. The dish can contain at most 3 cups of beans but must include at least 39 grams of fiber and 42 grams of protein. How much of each type of bean should be included in the dish in order to minimize the amount of kidney beans used? (Fractions of cups may be used.)

42. **Nutrition** The nutritional content of canned vegetables varies based on the type of vegetable and the manufacturer. A 1/2-cup serving of S&W Cut Blue Lake Green Beans contains 2 grams of fiber, 2 grams of sugar, and 20 calories. A 1/2-cup serving of Safeway Golden Sweet Whole Kernel Corn contains 2 grams of fiber, 6 grams of sugar, and 80 calories. A 1/2-cup serving of Pot O'Gold Sliced Carrots contains 2 grams of fiber, 4 grams of sugar, and 30 calories. (**Source:** Package labeling.) A dish is to be made using the three types of vegetables. The dish can contain at most 5 cups of vegetables but must include at least 18 grams of fiber and at least 32 grams of sugar (for flavor). How much of each type of vegetable should be included in the dish in order to minimize the amount of calories in the dish?

43. **Resource Allocation: Schools** A city has two elementary schools. The first school has a maximum enrollment of 500 students, and the second school has a maximum enrollment of 420 students. The city is divided into two regions: North and South. There are at least 400 students in the North region and at least 430 students in the South region. The annual transportation cost varies by region as shown in the table.

Transportation Costs

	School 1 (cost per student)	School 2 (cost per student)
North	120	180
South	100	150

Based on these constraints, what is the minimum possible transportation cost, and under what conditions does it occur?

44. Resource Allocation: Schools A city has two high schools. The first school has a maximum enrollment of 900 students, and the second school has a maximum enrollment of 550 students. The city is divided into two regions: Inner City and Suburbs. There are at least 500 students in the Inner City region and at least 800 students in the Suburbs region. The annual transportation cost varies by region as shown in the table.

Transportation Costs

	School 1 (cost per student)	School 2 (cost per student)
Inner City	220	260
Suburbs	200	280

Based on these constraints, what is the minimum possible transportation cost, and under what conditions does it occur?

45. Transportation Costs A furniture company has warehouses in Phoenix, Arizona, and Las Vegas, Nevada. The company has customers in Kingman, Arizona, and Flagstaff, Arizona. Its Phoenix warehouse has 500 desks in stock, and its Las Vegas warehouse has 200 desks in stock. Its Kingman customer needs at least 250 desks, and its Flagstaff customer needs at least 400 desks. Based on a rate of \$0.01 per mile per item, the cost of delivery per item is shown in the table.

	Phoenix	Las Vegas
Kingman	\$1.85	\$1.05
Flagstaff	\$1.45	\$2.51

Subject to these constraints, under what conditions are the company's delivery costs minimized?

Exercises 46–50 are intended to challenge your understanding of general linear programming problems.

46. Maximize $P = x + y + z + w$

Subject to $\begin{cases} x + y + z \leq 9 \\ y + z + w \leq 8 \\ x + z + w \geq 7 \\ x + y + w \geq 6 \\ x \geq 0, y \geq 0, z \geq 0, w \geq 0 \end{cases}$

47. Minimize $P = x + y + z + w$

Subject to $\begin{cases} x + y + z \leq 6 \\ y + z + w \leq 8 \\ x + z + w \geq 9 \\ x + y + w \geq 7 \\ x \geq 0, y \geq 0, z \geq 0, w \geq 0 \end{cases}$

48. Create a two-variable linear programming problem with three or more mixed constraints with the property that the feasible region consists of exactly one point. Provide evidence that the feasible region contains exactly one point.

49. Create a three-variable linear programming problem with three or more mixed constraints with the property that the feasible region is empty. Provide evidence that the feasible region is empty.

50. Explain what happens graphically in Stage 1 of the general linear programming problem-solving process.

Chapter 4 Review Exercises

Section 4.1 *In Exercises 1–4, graph the solution region of the linear inequality. Then use the graph to determine if point P is a solution.*

1. $4x - 2y \leq 6; P = (2, 9)$

2. $3x + 5y \leq 0; P = (1, 7)$

3. $9x - 8y \leq 12; P = (8, 9)$

4. $7x + 6y \leq 42; P = (3, 4)$

In Exercises 5–8, graph the solution region of the system of linear equations. If there is no solution, explain why.

5. $-4x + 3y \geq 2$
$-3x + 2y \geq 1$

6. $-10x + y \geq 0$
$2x + y \leq 4$

7. $2x + 4y \leq 8$
$6x - 2y \leq -6$

8. $x + y \leq 8$
$-x + y \leq 0$
$4x - 2y \leq 8$

In Exercise 9, set up and graphically solve the system of linear inequalities.

9. Wages A salaried employee earns \$800 per week managing a copy center. She is required to work a minimum of 40 hours but no more than 50 hours weekly. As a side business, she earns \$30 per hour designing brochures for local business clients. In order to maintain her standard of living, she must earn \$1000 per week. In order to maintain her quality of life, she limits her workload to 50 hours per week. Given that she has no control over the number of hours she will have to work managing the copy center, will she be able to consistently meet her workload and income goals? Explain.

Section 4.2

In Exercises 10–13, find the optimal solution to the linear programming problem, if it exists. If a solution does not exist, explain why.

10. Minimize $z = 5x - 7y$

Subject to $\begin{cases} 4x + y \geq 4 \\ -x + y \geq 1 \\ x \geq 0 \\ y \geq 0 \end{cases}$

11. Minimize $z = 9x + 4y$

Subject to $\begin{cases} 6x + y \geq 16 \\ -2x + y \geq 0 \\ x \geq 0 \\ y \geq 0 \end{cases}$

12. Maximize $z = 6x + 10y$

Subject to $\begin{cases} 6x + y \leq 16 \\ -2x + y \leq 0 \\ x \geq 0 \\ y \geq 0 \end{cases}$

13. Maximize $z = 5x - y$

Subject to $\begin{cases} 6x + 2y \leq 16 \\ -3x - y \leq 10 \\ x \geq 0 \\ y \geq 0 \end{cases}$

For Exercise 14, identify the objective function and constraints of the linear programming problem. Then solve the problem and interpret the real-world meaning of your results.

14. Family Food Storage A family wants to purchase at least 75 pounds of beans. A #10 can of pinto beans weighs 5.0 pounds and costs \$2.75. A #10 can of white beans weighs 5.3 pounds and costs \$2.88. (**Source:** Kent Washington Cannery.) The family wants to buy at least 25 pounds of pinto beans and at least 53 pounds of white beans. It has budgeted \$50 and, because of limited storage space, wants to minimize the number of cans purchased. How many cans of each type of beans should the family purchase?

Section 4.3

In Exercises 15–16, determine if the problem is a standard maximization problem. If it isn't, explain why.

15. Maximize $P = -2x + 8y$

Subject to $\begin{cases} 2x + y \leq 1 \\ -x + -y \leq -20 \end{cases}$

16. Maximize $P = -9x + 8y$

Subject to $\begin{cases} 42x + 19y \leq 10 \\ -11x + 19y \leq 21 \\ x \geq 0 \\ y \geq 0 \end{cases}$

In Exercises 17–18, solve the standard maximization problems by using the simplex method. Check your answer by graphing the feasible region and calculating the value of the objective function at each of the corner points.

17. Maximize $P = -2x + 10y$

Subject to $\begin{cases} 2x + y \leq 10 \\ -x + y \leq 1 \\ x \geq 0 \\ y \geq 0 \end{cases}$

18. Maximize $P = 5x - 2y$

Subject to $\begin{cases} 2x + y \le 6 \\ 2x + 2y \le 8 \\ x \ge 0 \\ y \ge 0 \end{cases}$

In Exercises 19–20, solve the standard maximization problems by using the simplex method.

19. Maximize $P = 6x + 4y + 5z$

Subject to $\begin{cases} 2x + 3y + 2z \le 120 \\ x + y + z \le 60 \\ x \ge 0 \\ y \ge 0 \\ z \ge 0 \end{cases}$

20. Maximize $P = 4x - y + z$

Subject to $\begin{cases} 2x + 3y + 3z \le 210 \\ x + y + z \le 100 \\ x \ge 0 \\ y \ge 0 \\ z \ge 0 \end{cases}$

In Exercises 21–22, set up and solve the standard maximization problem using the simplex method.

21. **Battery Sales** AAA Alkaline Discount Batteries sells low-cost batteries to consumers. In August 2002, the company offered AAA batteries at the following prices: 100-pack for \$18.00, 600-pack for \$96.00, and 1200-pack for \$180.00.
(**Source:** www.aaaalkalinediscountbatteries.com.)

An electronics store owner wants to buy at most 1800 batteries and spend at most \$300. She expects that she'll be able to resell all of the batteries she orders for \$1.00 each. How many packs (100-packs, 600-packs, or 1200-packs) should she order if she wants to maximize her revenue? What is her maximum revenue?

22. **Battery Sales** Repeat Exercise 21 except maximize profit. Assume that her only cost is the cost of the batteries. What is her maximum profit?

Section 4.4

In Exercises 23–25, determine if the problem is a standard minimization problem. If it isn't, explain why.

23. Minimize $P = 11x + 8y + 2z$

Subject to $\begin{cases} 2x + y + z \ge 10 \\ -x + y - z \ge 1 \\ x \ge 0 \\ y \ge 0 \\ z \ge 0 \end{cases}$

24. Maximize $P = -2x - 5y + 7z$

Subject to $\begin{cases} x + 2y + z \ge 11 \\ -11x + y \le -21 \\ x \ge 0 \\ y \ge 0 \\ z \ge 0 \end{cases}$

25. Maximize $P = 4x - 2y$

Subject to $\begin{cases} 6x + 7y \ge 13 \\ -8x - 4y \ge -12 \\ x \ge 0 \\ y \ge 0 \end{cases}$

In Exercises 26–28, find the transpose of the given matrix.

26. $A = \begin{bmatrix} 7 & -2 & 0 & 4 \\ 5 & 3 & 7 & 6 \\ -1 & 1 & 2 & 1 \end{bmatrix}$

27. $A = \begin{bmatrix} 12 & 10 & 8 \\ 14 & 12 & 10 \\ 16 & 14 & 12 \end{bmatrix}$

28. $A = \begin{bmatrix} 1 & 0 & 1 & 0 & 2 \\ 0 & 1 & 0 & 1 & 9 \\ 1 & 1 & 1 & 1 & 11 \end{bmatrix}$

In Exercises 29–32, do the following:

(i) Write the dual problem for the given standard minimization problem.

(ii) Solve the dual problem using the simplex method.

(iii) Use the final simplex tableau of the dual problem to solve the standard minimization problem.

(iv) Check your answer by graphing the feasible region of the standard minimization problem and calculating the value of the objective function at each of the corner points.

29. Minimize $P = 4x + 2y$

Subject to $\begin{cases} 4x + 5y \geq 20 \\ 2x + 3y \geq 11 \\ x \geq 0, y \geq 0 \end{cases}$

30. Minimize $P = x + 6y$

Subject to $\begin{cases} 2x + 9y \geq 19 \\ -x + 10y \geq 5 \\ x \geq 0, y \geq 0 \end{cases}$

31. Minimize $P = 2x + 3y$

Subject to $\begin{cases} 5x + 2y \geq 26 \\ 10x - 5y \geq 25 \\ x \geq 0, y \geq 0 \end{cases}$

32. Minimize $P = 3x + 5y$

Subject to $\begin{cases} 9x + 4y \geq 36 \\ 3x + 2y \geq 18 \\ x \geq 0, y \geq 0 \end{cases}$

In Exercises 33–34, set up and solve the standard minimization problem.

33. **Pet Nutrition: Food Cost** PETsMART.com sold the following varieties of dog food in June 2003.

Authority Chicken Adult Formula, 32 percent protein, 3 percent fiber, \$19.99 per 33-pound bag

Bil-Jac Select Dog Food, 27 percent protein, 4 percent fiber, \$18.99 per 18-pound bag

Iams Minichuncks, 26 percent protein, 4 percent fiber, \$8.99 per 8-pound bag

(**Source:** www.petsmart.com.)

A dog kennel wants to make at least 330 pounds of a mix containing at least 30 percent protein and at least 3 percent fiber. How many bags of each dog food variety should the kennel buy in order to minimize cost? (To make computations easier, round the price of each bag to the nearest dollar.)

34. **Pet Nutrition: Fat Content** PETsMART.com sold the following varieties of dog food in June 2003.

Nature's Recipe Venison Meal & Rice Canine, 20 percent protein, 10 percent fat, \$21.99 per 20-pound bag

Nutro Max Natural Dog Food, 27 percent protein, 16 percent fat, \$12.99 per 17.5-pound bag

PETsMART Premier Oven Baked Lamb Recipe, 25 percent protein, 14 percent fat, \$22.99 per 30-pound bag

(**Source:** www.petsmart.com.)

A dog breeder wants to make at least 300 pounds of a mix containing at least 24 percent protein. How many bags of each dog food variety should the breeder buy in order to minimize fat content?

Section 4.5

In Exercises 35–36, rewrite the general linear programming problem as a system of equations with slack and surplus variables and a rewritten objective function.

35. Maximize $P = 2x + 3y + z$

Subject to $\begin{cases} 2x + 4y + z \leq 10 \\ x + 2z \geq 3 \\ 3x - 2z \geq 1 \\ x \geq 0, y \geq 0, z \geq 0 \end{cases}$

36. Maximize $P = x - 10y + 6z$

Subject to $\begin{cases} -2x + 2y + 3z \leq 2 \\ 5x + 2y + 3z \geq 4 \\ 7x - 3y + 2z \geq 1 \\ x \geq 0, y \geq 0, z \geq 0 \end{cases}$

In Exercises 37–38, solve the general linear programming problem. If there is no solution, so state.

37. Maximize $P = 4x + y + 10z$

Subject to $\begin{cases} x + 3y + 2z \leq 5 \\ 2x + y + 2x \geq 4 \\ 2y + 3z \leq 6 \\ x \geq 0, y \geq 0, z \geq 0 \end{cases}$

38. Minimize $P = 4x + y + 10z$

Subject to $\begin{cases} x + 3y + 2z \leq 5 \\ 2x + y + 2x \geq 4 \\ 2y + 3z \leq 6 \\ x \geq 0, y \geq 0, z \geq 0 \end{cases}$

Make It Real

PROJECT 4

What to do

1. Find the wholesale and retail prices of three items of personal interest.
2. Estimate the number of cubic feet each item (or box of items) occupies.
3. Find a place in your home or workplace where you could store the items. Calculate the number of cubic feet in the storage area.
4. Assume that you may spend up to 100 times the retail price of the most expensive item in purchasing a combination of the three items.
5. Assume that you will be able to sell at retail price all of the items you order at wholesale. Determine how many of the items you should order if you want to maximize your profit subject to the spending and storage area constraints.

Where to look for prices

Wholesale Prices

Fruits: www.todaymarket.com
Vegetables: www.todaymarket.com
Flowers: www.flowersales.com
Computer parts: www.tcwo.com
Batteries: www.aaaalkalinediscountbatteries.com
Various items: Costco, Sam's Club, etc.
Search the Internet at www.yahoo.com and type in the key word "wholesale".

Retail Prices

Local grocery stores, clothing stores, etc.
Advertisements from newspapers
Search the Internet and type in a key word for your product of interest.

Chapter 5

Nonlinear Models

Mathematical functions are commonly used to model real-world data. Although every model has its limitations, models are often used to forecast expected results. Selecting which mathematical model to use is relatively easy once you become familiar with the basic types of mathematical functions. Remarkably, these basic functions may be used to effectively model many real-world data sets. For example, based on data from 1993 to 2002, the sales revenue from Starbucks stores may be effectively modeled by a quadratic function.

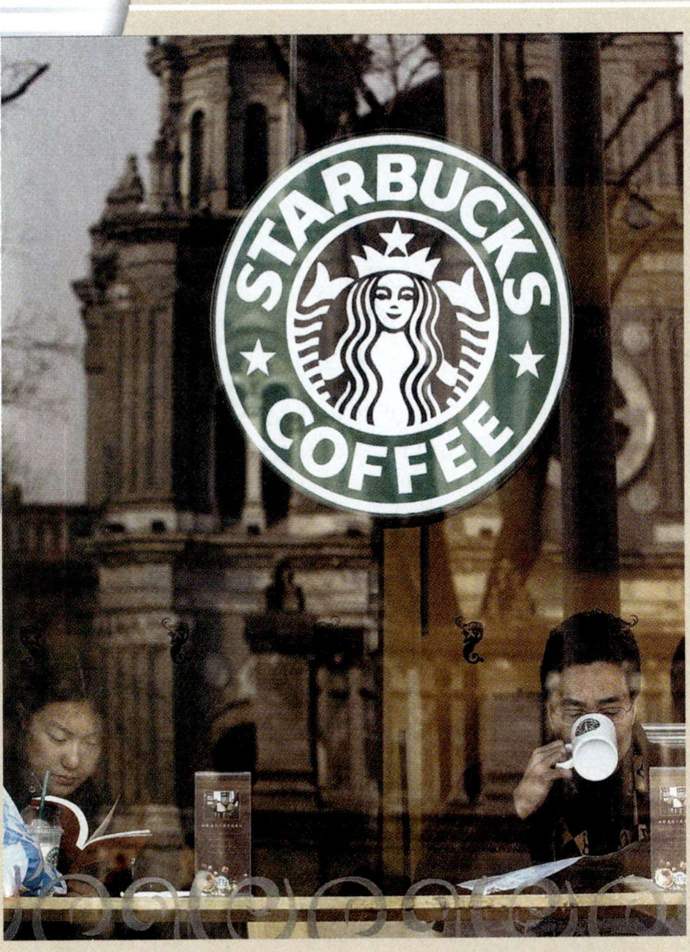

5.1 Quadratic Function Models
- Use nonlinear functions to model real-life phenomena
- Interpret the meaning of mathematical models in their real-world context
- Model real-life data with quadratic functions

5.2 Higher-Order Polynomial Function Models
- Use nonlinear functions to model real-life phenomena
- Interpret the meaning of mathematical models in their real-world context
- Model real-life data with higher-order polynomial functions

5.3 Exponential Function Models
- Use nonlinear functions to model real-life phenomena
- Interpret the meaning of mathematical models in their real-world context
- Graph exponential functions
- Find the equation of an exponential function from a table
- Model real-life data with exponential functions

5.4 Logarithmic Function Models
- Use nonlinear functions to model real-life phenomena
- Interpret the meaning of mathematical models in their real-world context
- Graph logarithmic functions
- Model real-life data with logarithmic functions
- Apply the rules of logarithms to simplify logarithmic expressions
- Solve logarithmic equations

5.5 Choosing a Mathematical Model
- Use critical thinking skills in selecting a mathematical model

5.1 Quadratic Function Models

- Use nonlinear functions to model real-life phenomena
- Interpret the meaning of mathematical models in their real-world context
- Model real-life data with quadratic functions

GETTING STARTED Many college-bound high school students enroll in Advanced Placement Program (AP®) courses to earn college credit while still in high school. The College Board® offers 35 courses in 19 disciplines ranging from Studio Art: 3D Design to Calculus AB. (**Source:** The College Board®.) The number of students taking the Calculus AB test has increased dramatically since 1969, as shown in Figure 5.1.

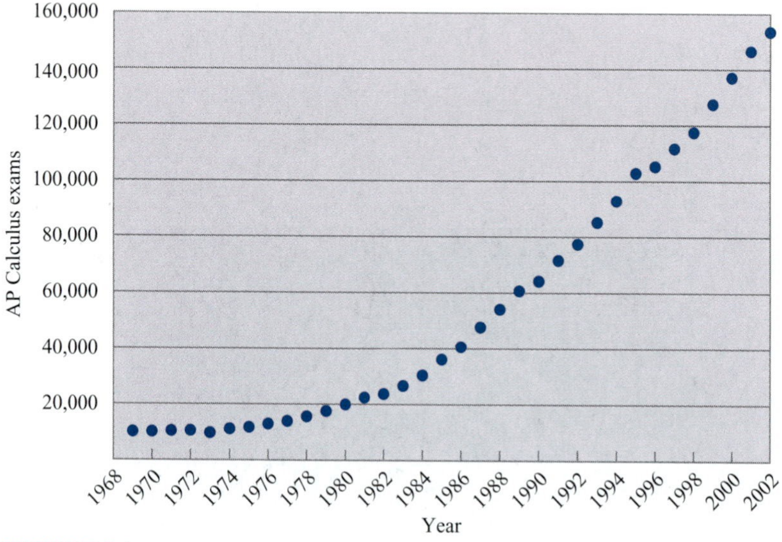

FIGURE 5.1

How many students will take the test in 2005? Nobody knows; however, using a mathematical model, we can predict what may happen.

In this section, we will discuss the identifying features of quadratic functions. We will then demonstrate how to use algebraic methods and quadratic regression to model data sets whose graphs open upward or downward over their domain.

A polynomial is a function that is the sum of terms of the form ax^n, where a is a real number and n is a nonnegative integer. For example, each of the following functions is a polynomial.

$$f(x) = 6x^4 - 3x^3 + 5x^2 - 2x + 9$$
$$g(x) = 4x^2 + 2x$$
$$h(x) = -1.3x + 9.7$$
$$j(x) = x^2 - 2x + 6$$

May a constant term like 9 be written in the form ax^n? Yes! Since $x^0 = 1$ for nonzero x, $9x^0 = 9$. Consequently, a constant function $s(x) = 9$ is also a polynomial.

The **degree of a polynomial** is the value of its largest exponent. For example, the degree of the polynomial $f(x) = 6x^4 - 3x^3 + 5x^2 - 2x + 9$ is 4, since 4 is the largest exponent. Since the equation of any line may be written as $y = ax^1 + b$, lines are polynomials of degree 1. We've worked extensively with first-degree polynomials (lines) in the preceding chapters. In this section, we are interested in polynomials of degree 2. Polynomials of degree 2 are called **quadratic functions.**

QUADRATIC FUNCTION

A polynomial function of the form $f(x) = ax^2 + bx + c$ with $a \neq 0$ is called a **quadratic function.** The graph of a quadratic function is a **parabola.**

When a parabola opens upward, we say that it is "concave up." When the parabola opens downward, we say that it is "concave down." The steepness of the sides of the parabola and its concavity are controlled by the value of a, the coefficient on the x^2 term in its equation. If $a > 0$, the graph is concave up. If $a < 0$, the graph is concave down. As the magnitude of a increases, the steepness of the graph increases. (For $a > 0$, the magnitude of a is a. For $a < 0$, the magnitude of a is $-a$.) Consider the graphs in Figures 5.2 and 5.3. These graphs have the same values for b and c ($b = -2$ and $c = 3$) but differing values of a.

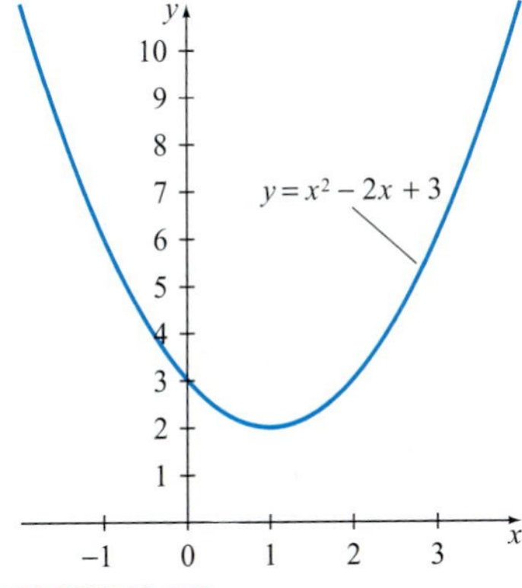

FIGURE 5.2

In Figure 5.2, since a is the coefficient on the x^2 term, $a = 1$. The magnitude of a is 1. Since $a > 0$, the graph is concave up.

In Figure 5.3, since a is the coefficient on the x^2 term, $a = 2$. The magnitude of a is 2. Increasing the magnitude of a increased the graph's steepness.

Consider the graphs in Figures 5.4 and 5.5. These graphs have the same values for b and c ($b = 4$ and $c = 0$) but differing values of a.

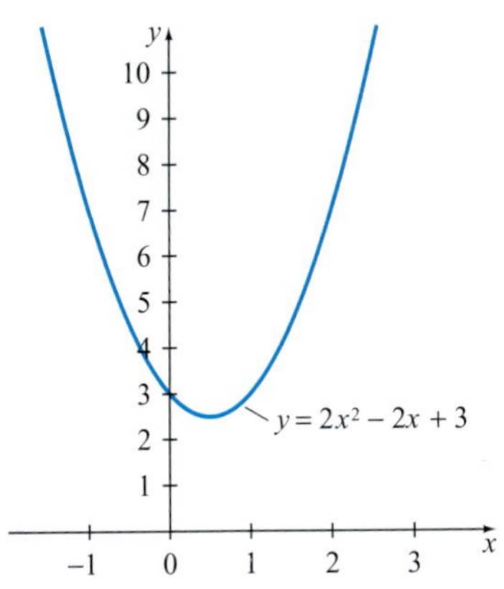

FIGURE 5.3

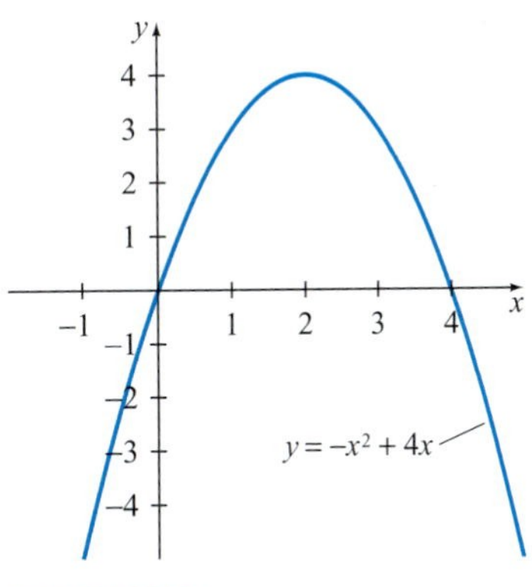

FIGURE 5.4

In Figure 5.4, since a is the coefficient on the x^2 term, $a = -1$. The magnitude of a is 1. Since $a < 0$, the graph is concave down.

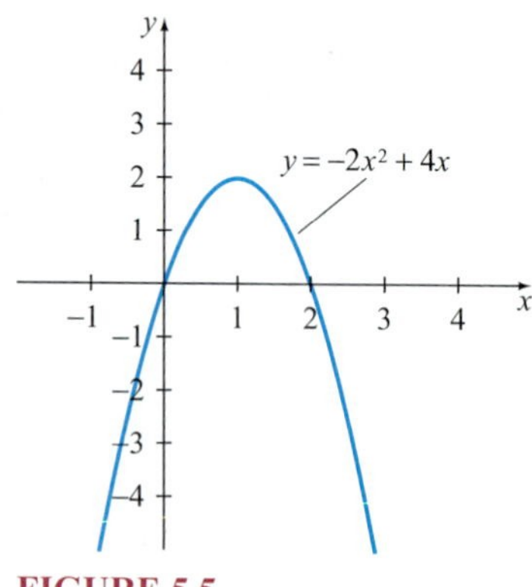

FIGURE 5.5

In Figure 5.5, since a is the coefficient on the x^2 term, $a = -2$. The magnitude of a is 2. Increasing the magnitude of a from 1 to 2 increased the graph's steepness.

The coefficient b in the equation affects the horizontal and vertical placement of the parabola. The constant term c in the equation indicates that the point $(0, c)$ is the y-intercept of the graph.

Recall that the graph of a function is said to be increasing if the value of y gets bigger as the value of x increases. Similarly, the graph of a function is said to be decreasing if the value of y gets smaller as the value of x increases. The vertex of a parabola is the point on the graph of a quadratic function where the curve changes from decreasing to increasing (or vice versa). The minimum y value of a concave up parabola occurs at the vertex (see Figure 5.6a). Similarly, the maximum y value of a concave down parabola occurs at the vertex (see Figure 5.6b).

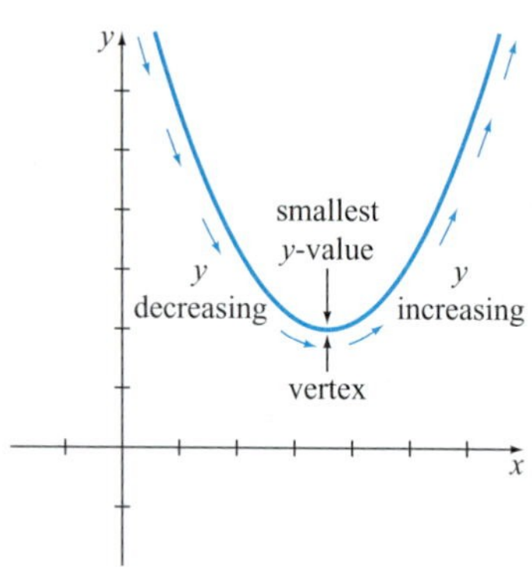

FIGURE 5.6a Concave Up Parabola

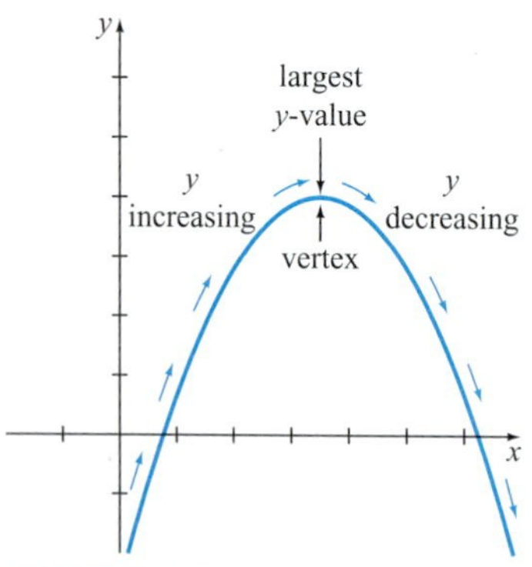

FIGURE 5.6b Concave Down Parabola

For parabolas with x-intercepts, the x coordinate of the vertex always lies halfway between the x-intercepts. Recall that as a result of the quadratic formula, we know that the x coordinate of the vertex is $x = \frac{-b}{2a}$. The y coordinate of the vertex is obtained by evaluating the quadratic function at this x value. For example, for the quadratic function $y = x^2 - 2x + 3$, we know that $a = 1$ and $b = -2$. The x coordinate of the vertex is

$$\begin{aligned} x &= \frac{-b}{2a} \\ &= \frac{-(-2)}{2(1)} \\ &= \frac{2}{2} \\ &= 1 \end{aligned}$$

Evaluating the function at $x = 1$ yields

$$\begin{aligned} y &= x^2 - 2x + 3 \\ &= (1)^2 - 2(1) + 3 \\ &= 1 - 2 + 3 \\ &= 2 \end{aligned}$$

Therefore, the vertex of the function is $(1, 2)$.

Parabolas are symmetrical. That is, if we draw a vertical line through the vertex of the parabola, the portion of the graph on the left of the line is the mirror image of the portion of the graph on the right of the line. The line is referred to as the **axis of symmetry** (see Figure 5.7).

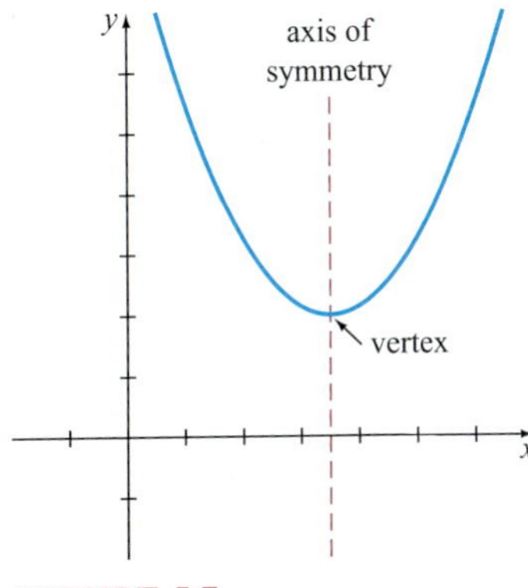

FIGURE 5.7

Since the axis of symmetry is a vertical line passing through the vertex, the equation of the axis of symmetry is $x = \frac{-b}{2a}$.

EXAMPLE 1

Describing the Graph of a Quadratic Function from Its Equation

Determine the concavity, the y-intercept, and the vertex of the quadratic function $y = 3x^2 + 6x - 1$.

SOLUTION We have $a = 3$, $b = 6$, and $c = -1$. Since $a > 0$, the parabola is concave up. Since $c = -1$, the y-intercept is $(0, -1)$. The x coordinate of the vertex is given by

$$\begin{aligned} x &= \frac{-b}{2a} \\ &= \frac{-6}{2(3)} \\ &= \frac{-6}{6} \\ &= -1 \end{aligned}$$

The y coordinate of the vertex is obtained by evaluating the function at $x = -1$.

$$\begin{aligned} y &= 3x^2 + 6x - 1 \\ &= 3(-1)^2 + 6(-1) - 1 \\ &= 3 - 6 - 1 \\ &= -4 \end{aligned}$$

The vertex of the parabola is $(-1, -4)$.

EXAMPLE 2 Determining the Equation of a Parabola from Its Graph

Determine the equation of the parabola shown in Figure 5.8.

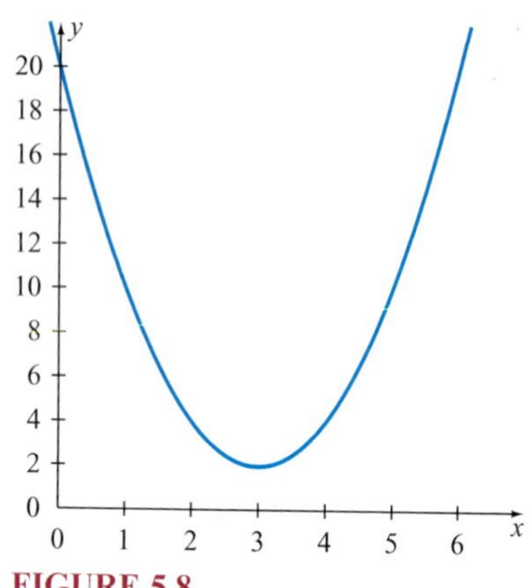

FIGURE 5.8

SOLUTION The y-intercept is $(0, 20)$, so $c = 20$. The vertex is $(3, 2)$. Since the x coordinate of the vertex is $\frac{-b}{2a}$, we know that

$$\frac{-b}{2a} = 3$$
$$-b = 6a$$
$$b = -6a$$

Since $f(x) = ax^2 + bx + c$, we have

$$f(x) = ax^2 + (-6a)x + (20) \qquad \text{Since } b = -6a$$
$$= ax^2 - 6ax + 20$$

The vertex is $(3, 2)$, so $f(3) = 2$. Therefore,

$$f(x) = ax^2 - 6ax + 20$$
$$2 = a(3)^2 - 6a(3) + 20 \qquad \text{Substitute in } (3, 2)$$
$$2 = 9a - 18a + 20$$
$$-18 = -9a$$
$$a = 2$$

Since $b = -6a$, $b = -12$. The equation of the parabola is $f(x) = 2x^2 - 12x + 20$.

We can check our work by substituting in a different point from the graph. The parabola passes through the point $(4, 4)$. Consequently, this point should satisfy the equation $f(x) = 2x^2 - 12x + 20$.

$$f(x) = 2x^2 - 12x + 20$$
$$4 \stackrel{?}{=} 2(4)^2 - 12(4) + 20 \qquad \text{Substitute in } (4, 4)$$
$$4 \stackrel{?}{=} 2(16) - 48 + 20$$
$$4 \stackrel{?}{=} 32 - 48 + 20$$
$$4 = 4$$

Recognizing the relationship between the graph of a parabola and its corresponding quadratic function provides a quick way to evaluate the accuracy of a quadratic model. Quadratic models may be determined algebraically or by using quadratic regression. We will demonstrate both methods in the next several examples.

Let's return to the AP Calculus exam data introduced at the beginning of the section. At first glance, the Calculus AB exam data don't look at all like a parabola; however, we do observe that the data appear to be increasing at an ever-increasing rate. (It took six years [1980 to 1986] for the number of exams to increase from 20,000 to 40,000, but it took only three years [1986 to 1989] for the number of exams to increase from 40,000 to 60,000.) Plotting the quadratic equation $f(t) = 157.8486t^2 - 622{,}378t + 613{,}500{,}479$ together with the data set, we observe that, in fact, a quadratic equation fits the data very well (Figure 5.9). (**Source:** The College Board®.) This equation is found by performing quadratic regression on the data, a process that we will detail in the forthcoming Technology Tip.

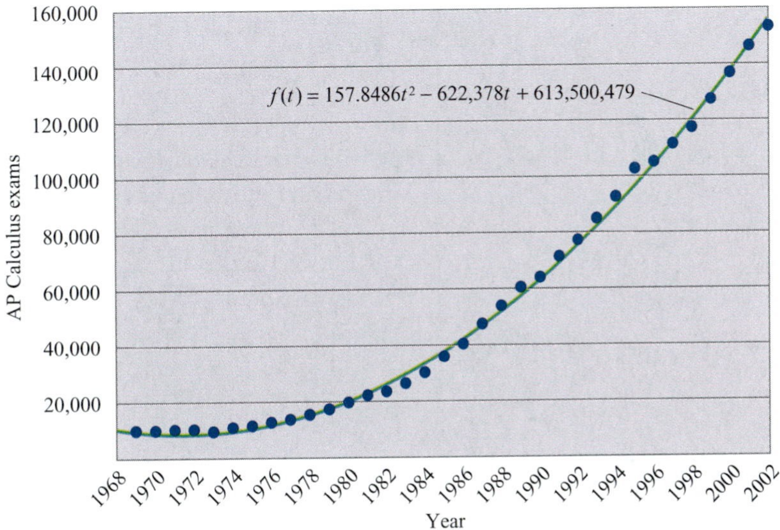

FIGURE 5.9

Observe that the coefficients of the variables in the quadratic equation are very large. We can come up with an equation with smaller coefficients by **aligning the data.** We'll let $t = 0$ in 1969, $t = 1$ in 1970, and so on. Doing quadratic regression on the aligned data yields the equation $f(t) = 157.85t^2 - 770.64t + 10{,}268$ with the coefficient of determination $r^2 = 0.9979$. Recall that the coefficient of determination is a measure of how well the model fits the data. The closer r^2 is to 1, the better the model fits the data. Although both of the models fit the data, the second model will make computations easier because of the smaller coefficients.

Using the model, we predict how many tests will be administered in 2005. In 2005, $t = 36$.

$$\begin{aligned} f(36) &= 157.85(36)^2 - 770.64(36) + 10{,}268 \\ &= 187{,}099 \end{aligned}$$

We estimate that 187,099 AP Calculus AB exams will be administered in 2005.

A quadratic function model for a data set may be generated by using the quadratic regression feature on a graphing calculator, as demonstrated in the following Technology Tip. However, the fact that the calculator can create a quadratic model does not guarantee that the model will be a good fit for the data.

TECHNOLOGY TIP

Quadratic Regression

1. Enter the data using the Statistics Menu List Editor. (Refer to Section 1.3 if you've forgotten how to do this.)

L1	L2	L3 3
0	10280	------
1	10273	
2	10592	
3	10611	
4	9871	
5	11213	
6	11804	

L3(1) =

2. Bring up the Statistics Menu Calculate feature by pressing STAT and using the blue arrows to move to the CALC menu. Then select item 5:QuadReg and press ENTER.

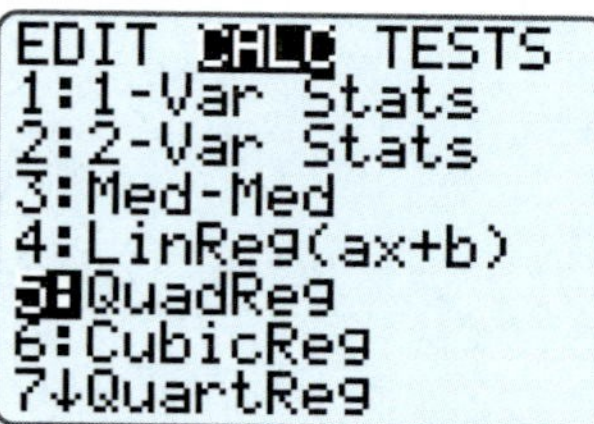

3. If you want to automatically paste the regression equation into the Y= editor, press the key sequence VARS Y-Vars; 1:Function; 1:Y1 and press ENTER. Otherwise press ENTER.

```
QuadReg
 y=ax²+bx+c
 a=157.8485883
 b=-770.6397775
 c=10268.35154
 R²=.9979069591
```

Although the quadratic model fits the AP Calculus AB exam data very well from 1969 to 2002, we must be cautious in using it to predict future behavior. (Predicting the output value for an input value outside of the interval of the input data is called **extrapolation.**) For this data set, we may feel reasonably comfortable with an estimate two or three years beyond the last data point; however, we would doubt the accuracy of the model 100 years beyond 2002. For example, in 2102 ($t = 133$), the estimated number of exams is 2,702,587. This figure is more than 17 times greater than the maximum number of exams that have ever been administered!

We also need to look at the population of students who could possibly take the exam. Between 1970 and 1999, the number of twelfth graders (those who typically take the AP Calculus exam) fluctuated from a high of 3,026,000 in 1977 to a low of 2,381,000 in 1990. (**Source:** *Statistical Abstract of the United States, 2001*, Table 232, p. 149.) If the number of twelfth graders continues to fluctuate between these two values, at some point our model estimate for the exams would exceed the number of people who could conceivably take the exam. This illustrates the necessity to verify that a model makes sense in its real-world context.

It is often possible to model a data set with more than one mathematical model. When selecting a model, we should consider the following:

1. The graphical fit of the model to the data
2. The correlation coefficient (r) or the coefficient of determination (r^2)
3. The known behavior of the thing being modeled

Recall that the closer the correlation coefficient is to 1 or -1, the better the model fits the data. Similarly, the closer the coefficient of determination is to 1, the better the model fits the data.

EXAMPLE 3 Using Quadratic Regression to Forecast Prescription Drug Sales

Retail prescription drug sales in the United States increased from 1995 to 2000 as shown in Table 5.1.

TABLE 5.1

Years Since 1995 (t)	Retail Sales (billions of dollars) [$S(t)$]
0	68.6
2	89.1
3	103.0
4	121.7
5	140.7

Source: *Statistical Abstract of the United States, 2001*, Table 127, p. 94.

Model the data using a quadratic function. Then use the model to predict retail prescription drug sales in 1996 and 2001.

SOLUTION We observe from the scatter plot that the data appear concave up everywhere (Figure 5.10).

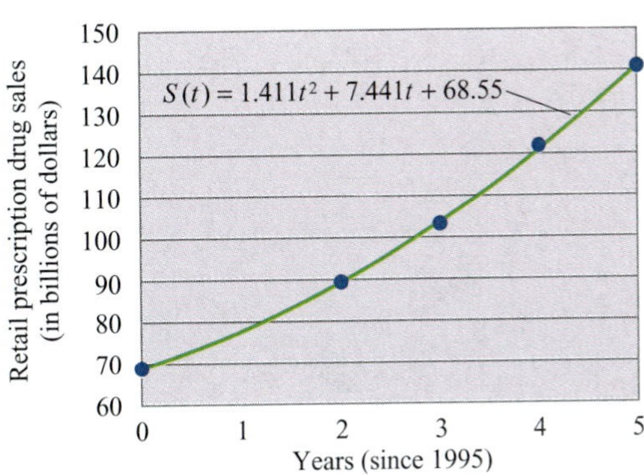

FIGURE 5.10

A quadratic function may fit the data well. We use quadratic regression to find the quadratic model that best fits the data.

Based on data from 1995 to 2000, retail prescription drug sales in the United States may be modeled by

$$S(t) = 1.411t^2 + 7.441t + 68.55 \text{ billion dollars}$$

where t is the number of years since 1995.

The coefficient of determination ($r^2 = 0.9997$) is extremely close to 1. The graph also appears to "touch" each data point. The model appears to fit the data well.

In 1996, $t = 1$, and in 2001, $t = 6$. Evaluating the function at each t value, we get

$$\begin{aligned} S(1) &= 1.411(1)^2 + 7.441(1) + 68.55 \\ &= 77.402 \\ S(6) &= 1.411(6)^2 + 7.441(6) + 68.55 \\ &= 163.992 \end{aligned}$$

Since the original data were accurate to only one decimal place, we will round our solutions to one decimal place as well. We estimate that prescription drug sales were \$77.4 billion in 1996 and \$164.0 billion in 2001.

You may ask, "Is there a way to find a quadratic model without using quadratic regression?" There is. The model may not be the model of best fit, but it may still model the data effectively. In Example 4, we repeat Example 3 using an algebraic method to find a quadratic model.

EXAMPLE 4 Using Algebraic Methods to Model Prescription Drug Sales

Retail prescription drug sales in the United States increased from 1995 to 2000 as shown in Table 5.2.

TABLE 5.2

Years Since 1995 (t)	Retail Sales (billions of dollars) [$S(t)$]
0	68.6
2	89.1
3	103.0
4	121.7
5	140.7

Source: *Statistical Abstract of the United States, 2001*, Table 127, p. 94.

Model the data using a quadratic function. Then use the model to predict retail prescription drug sales in 1996 and 2001.

SOLUTION Given any three data points, we can find a quadratic function that passes through the points, provided that the points define a nonlinear function. We will pick the points $(0, 68.6)$, $(3, 103.0)$, and $(5, 140.7)$ from the table. A quadratic function is of the form $S(t) = at^2 + bt + c$. Each of the points must satisfy this equation.

$$68.6 = a(0)^2 + b(0) + c \qquad \text{Substitute } t = 0, S(t) = 68.6$$
$$c = 68.6$$

$$103.0 = a(3)^2 + b(3) + c \qquad \text{Substitute } t = 3, S(t) = 103.0$$
$$103.0 = 9a + 3b + c$$

$$140.7 = a(5)^2 + b(5) + c \qquad \text{Substitute } t = 5, S(t) = 140.7$$
$$140.7 = 25a + 5b + c$$

Since we know $c = 68.6$, we can simplify the last two equations.

$$103.0 = 9a + 3b + 68.6$$
$$34.4 = 9a + 3b$$

$$140.7 = 25a + 5b + 68.6$$
$$72.1 = 25a + 5b$$

We can now find the values of a and b by solving the system of equations.

$$9a + 3b = 34.4$$
$$25a + 5b = 72.1$$

$$\left[\begin{array}{cc|c} 9 & 3 & 34.4 \\ 25 & 5 & 72.1 \end{array}\right]$$

$$\left[\begin{array}{cc|c} 9 & 3 & 34.4 \\ -30 & 0 & -44.3 \end{array}\right] \qquad 5R_1 - 3R_2$$

$$\left[\begin{array}{cc|c} 0 & 30 & 211.1 \\ -30 & 0 & -44.3 \end{array}\right] \qquad 10R_1 + 3R_2$$

$$-30a = -44.3 \qquad 30b = 211.1$$
$$a = 1.477 \qquad b = 7.037$$

A quadratic model for the data is

$$S(t) = 1.477t^2 + 7.037t + 68.6$$

Graphing this model with the data shows that it fits the data relatively well (Figure 5.11).

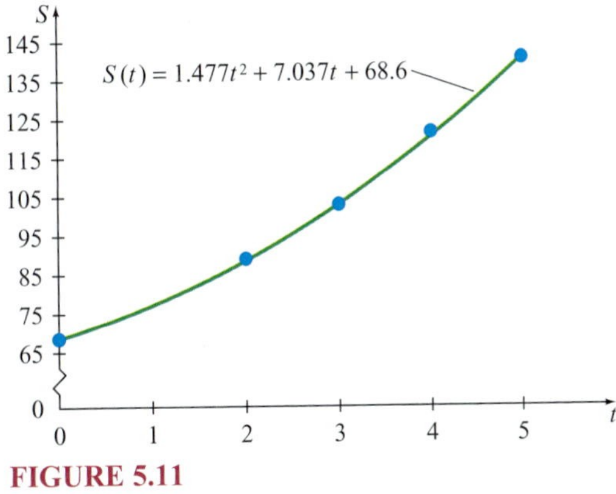

FIGURE 5.11

In 1996, $t = 1$, and in 2001, $t = 6$. Evaluating the function at each t value, we get

$$S(1) = 1.477(1)^2 + 7.037(1) + 68.6 \approx 77.1 \text{ billion} \qquad S(6) = 1.477(6)^2 + 7.037(6) + 68.6 \approx 164.0 \text{ billion}$$

These estimates are close to the estimates from Example 3. In Example 3, we estimated that prescription drug sales were \$77.1 billion in 1996 and \$164.0 billion in 2001.

EXAMPLE 5 Using Quadratic Regression to Forecast Nursing Home Care Costs

Because of their need for constant medical care, many elderly Americans are placed in nursing homes by their families. The amount of money spent on nursing home care has risen substantially since 1960, as shown in Table 5.3.

TABLE 5.3

Years Since 1960 (t)	Nursing Home Care (billions of dollars) [$N(t)$]
0	1
10	4
20	18
30	53
40	96

Source: *Statistical Abstract of the United States, 2001*, Table 119, p. 91.

The U.S. Centers for Medicare and Medicaid Services predict that by 2010 the cost of nursing home care will reach \$183 billion.

Model the data with a quadratic function and calculate the cost of nursing home care in 2010. Compare your result to the U.S. Centers for Medicare & Medicaid Services estimate.

SOLUTION Using quadratic regression, we determine that the quadratic model of best fit is

$$N(t) = 0.07214t^2 - 0.4957t + 1.029 \text{ billion dollars}$$

where t is the number of years since 1960.

Since the coefficient of determination ($r^2 = 0.9987$) is extremely close to 1, we anticipate that the model fits the data well. Graphing the data and the model together yields Figure 5.12.

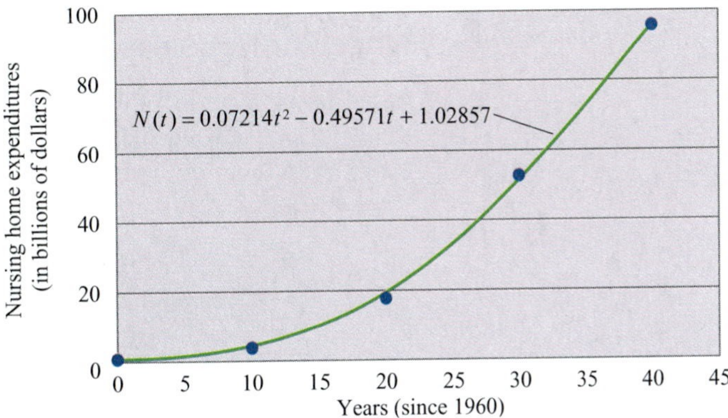

FIGURE 5.12

The graph passes near each data point. The model appears to fit the data fairly well. Zooming in, however, we see that the graph decreases from 1 billion to 0 billion between 1960 and 1963 (Figure 5.13). (We rounded to whole numbers, since the original data were in whole numbers.)

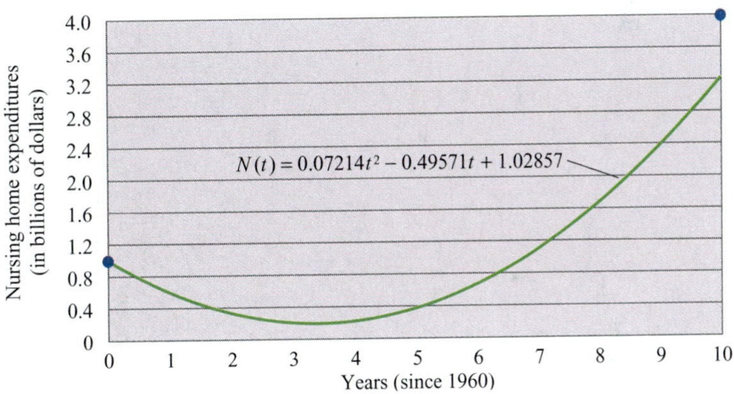

FIGURE 5.13

In reality, nursing home expenditures have increased every year since 1960. Nevertheless, despite this limitation in our model, we will use it to predict nursing home expenditures in 2010. In 2010, $t = 50$.

$$\begin{aligned} N(50) &= 0.721(50)^2 - 0.4957(50) + 1.0286 \\ &= 156.60 \end{aligned}$$

We estimate that in 2010, \$157 billion will be spent on nursing home care. Our model estimate is substantially less than the \$183 billion that the U.S. Centers for Medicare & Medicaid Services estimate. Their estimate probably anticipated the health care needs of the aging population of baby boomers, while ours did not.

Models projecting health care costs are useful for legislators, insurance companies, and consumers. By considering future costs, people can prepare for the future and avert financial crises.

EXAMPLE 6 Using Algebraic Methods to Find a Quadratic Model for Net Sales

Based on the data in Table 5.4, find a quadratic model for the net sales of the Kellogg Company algebraically.

TABLE 5.4

Years Since 1999 (t)	Kellogg Company Net Sales (millions of dollars) $R(t)$
0	6,984.2
1	6,954.7
2	8,853.3

Source: Kellogg Company 2001 Annual Report, pp. 7, 27.

SOLUTION Since we are given the y-intercept, $(0, 6984.2)$, we know that $c = 6984.2$. We have

$$\begin{aligned} R(t) &= at^2 + bt + c \\ &= at^2 + bt + 6984.2 \end{aligned}$$

$$\begin{aligned} R(1) &= a(1)^2 + b(1) + 6984.2 \\ 6954.7 &= a + b + 6984.2 \\ -29.5 &= a + b \end{aligned}$$

$$\begin{aligned} R(2) &= a(2)^2 + b(2) + 6984.2 \\ 8853.3 &= 4a + 2b + 6984.2 \\ 1869.1 &= 4a + 2b \end{aligned}$$

We must solve the system of equations

$$\begin{aligned} a + b &= -29.5 \\ 4a + 2b &= 1869.1 \end{aligned}$$

We will solve the system using the substitution method. Solving the first equation for a yields $a = -b - 29.5$. Substituting this result into the second equation $4a + 2b = 1869.1$ yields

$$\begin{aligned} 4(-b - 29.5) + 2b &= 1869.1 && \text{Since } a = -b - 29.5 \\ -4b - 118 + 2b &= 1869.1 \\ -2b &= 1987.1 \\ b &= -993.55 \end{aligned}$$

Since $a = -b - 29.5$,

$$\begin{aligned} a &= -(-993.55) - 29.5 \\ &= 964.05 \end{aligned}$$

The quadratic function that models the revenue of the Kellogg Company is $R(t) = 964.05t^2 - 993.55t + 6984.2$, where t is the number of years since the end of 1999 and $R(t)$ is the revenue from sales in millions of dollars.

In each of the preceding examples, the quadratic model fit the data well. This is not always the case, as demonstrated in Example 7.

EXAMPLE 7 Determining When a Quadratic Model Should Not Be Used

The per capita consumption of ready-to-eat and ready-to-cook breakfast cereal is shown in Table 5.5.

TABLE 5.5

Years Since 1980 (t)	Cereal Consumption (pounds) [$C(t)$]	Years Since 1980 (t)	Cereal Consumption (pounds) [$C(t)$]
0	12	10	15.4
1	12	11	16.1
2	11.9	12	16.6
3	12.2	13	17.3
4	12.5	14	17.4
5	12.8	15	17.1
6	13.1	16	16.6
7	13.3	17	16.3
8	14.2	18	15.6
9	14.9	19	15.5

Source: *Statistical Abstract of the United States, 2001*, Table 202, p. 129.

Explain why you do or do not believe that a quadratic function will model the data set well.

SOLUTION We first draw the scatter plot of the data set (Figure 5.14).

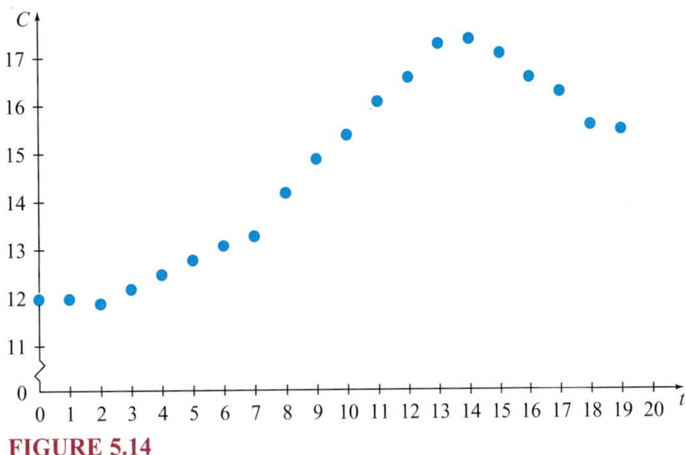

FIGURE 5.14

(*Note:* We have adjusted the viewing window so that we may better analyze the data.

Recall that a parabola is either concave up everywhere or concave down everywhere. This scatter plot appears to be concave up between $t = 1$ and $t = 10$, concave down between $t = 10$ and $t = 17$, and concave up between $t = 17$ and $t = 19$. The fact that the scatter plot changes concavity causes us to doubt that a quadratic model will fit the data well. In short, the scatter plot doesn't look like a parabola or a portion of a parabola.

The quadratic model that best fits the data is $C(t) = -0.0179t^2 + 0.637t + 10.8$ and is shown in Figure 5.15.

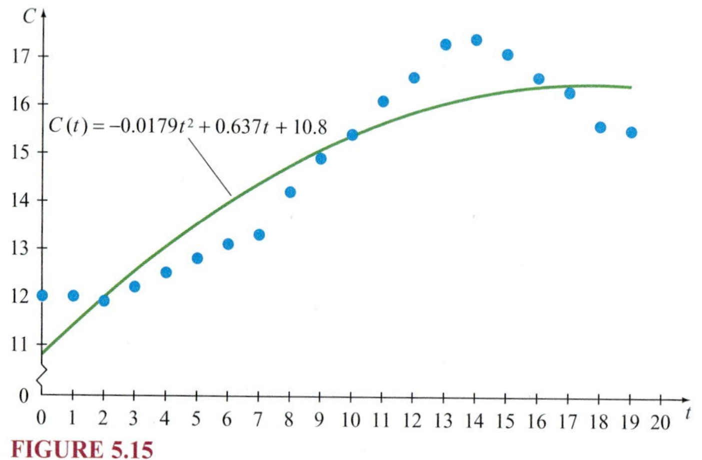

FIGURE 5.15

The coefficient of determination ($r^2 = 0.853$) is not as close to 1 as in our previous examples. Additionally, we can see visually that the model does not fit the data well.

Once a model has been created, we can often use it to make educated guesses about what may happen in the future. Forecasting the future is a key function for many businesses.

EXAMPLE 8

Using Quadratic Regression to Forecast Fuel Consumption

As vans, pickups, and SUVs have increased in popularity, the total fuel consumption of these types of vehicles has also increased (Table 5.6).

TABLE 5.6 Motor Fuel Consumption of Vans, Pickups, and SUVs

Years Since 1980 (t)	Fuel Consumption (billions of gallons) (F)
0	23.8
5	27.4
10	35.6
15	45.6
19	52.8

Source: *Statistical Abstract of the United States, 2001*, Table 1105, p. 691.

Model the fuel consumption with a quadratic function and forecast the year in which fuel consumption will reach 81.0 billion gallons.

SOLUTION Using quadratic regression, we determine the model to be

$$F(t) = 0.0407t^2 + 0.809t + 23.3 \text{ billion gallons}$$

where t is the number of years since the end of 1980. We want to know at what value of t does $F(t) = 81.0$. This problem may be solved algebraically or graphically using technology. We will solve the problem twice (once with each method) and allow you to use the method of your choice in the exercises.

ALGEBRAIC SOLUTION

$$81.0 = 0.0407t^2 + 0.809t + 23.3$$
$$0 = 0.0407t^2 + 0.809t - 57.7$$

This is a quadratic function with $a = 0.0407$, $b = 0.809$, and $c = -57.7$.

Recall that the solution to a quadratic equation of the form $at^2 + bt + c = 0$ is given by the Quadratic Formula,

$$t = \frac{-b \pm \sqrt{b^2 - 4ac}}{2a}$$

Substituting our values of a, b, and c into the Quadratic Formula yields

$$\begin{aligned} t &= \frac{-0.809 \pm \sqrt{(-0.809)^2 - 4(0.0407)(-57.7)}}{2(0.0407)} \\ &= \frac{-0.809 \pm \sqrt{10.0}}{0.0814} \\ &= \frac{-0.809 \pm 3.17}{0.0814} \end{aligned}$$

In the context of the problem, we know that t must be nonnegative. Consequently, we will calculate only the nonnegative solution.

$$t = \frac{-0.809 + 3.17}{0.0814}$$
$$= \frac{2.36}{0.0814}$$
$$= 29.0$$

We anticipate that at the end of 2009 (29 years after the end of 1980), the fuel consumption will have reached 81.0 billion gallons.

GRAPHICAL SOLUTION

We graph the function $F(t) = 0.0407t^2 + 0.809t + 23.3$ and the horizontal line $y = 81.0$ simultaneously (Figure 5.16). $F(t) = 81.0$ at the point at which these two functions intersect.

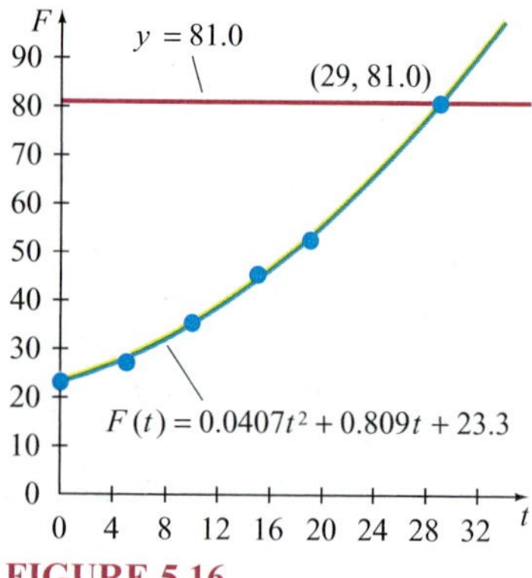

FIGURE 5.16

Using the intersect feature on the graphing calculator, we determine that the point of intersection is (29, 81.0). (The technique for finding a point of intersection on a graphing calculator is detailed in Section 2.1.) The interpretation of the solution is the same as that given in the discussion of the algebraic method.

5.1 Summary

In this section, you learned how to use quadratic regression to model a data set. You also discovered the importance of analyzing a mathematical model before using it to calculate unknown values.

5.1 Exercises

In Exercises 1–5, determine the concavity, y-intercept, and vertex of the quadratic equation.

1. $y = x^2 - 2x + 1$
2. $f(x) = -2x^2 + 4$
3. $g(x) = 3x^2 + 3x$
4. $h(t) = -1.2t^2 + 2.4t + 4.5$
5. $f(t) = 2.8t^2 - 1.4t + 2.1$

In Exercises 6–10, determine the equation of the parabola from the graph.

6.

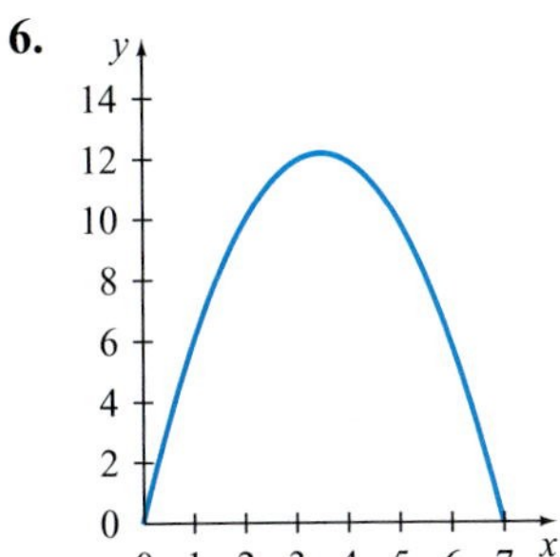

7.

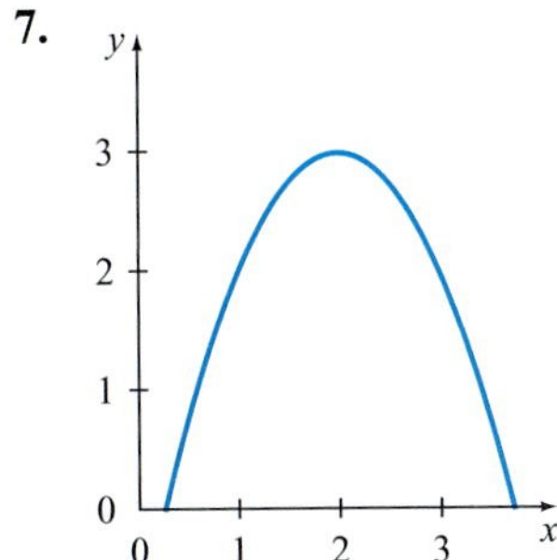

8.

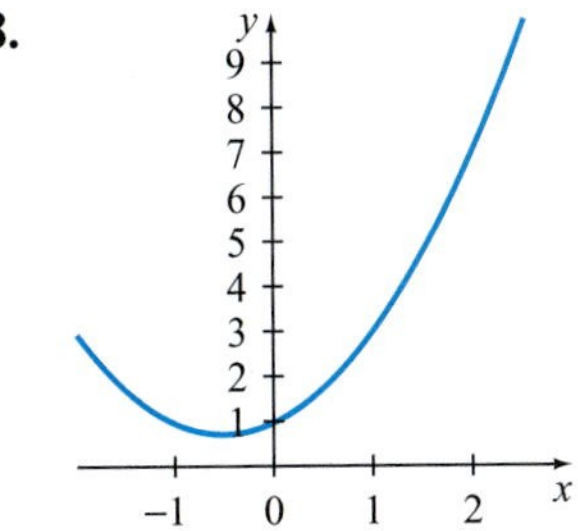

9.

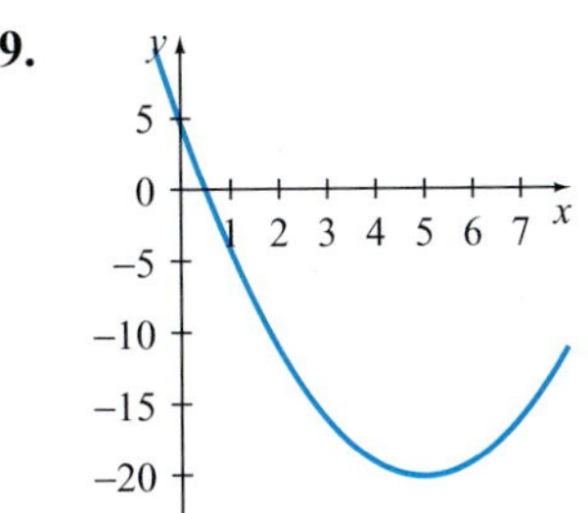

10.

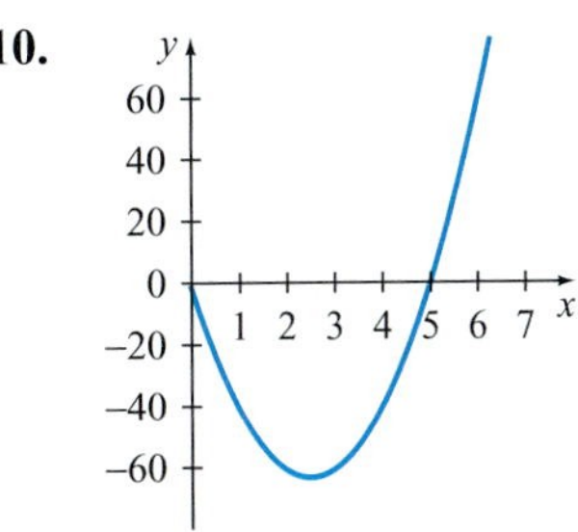

In Exercises 11–15, do the following:

(a) Find a quadratic model for the data algebraically.
(b) Graph the model together with the scatter plot.
(c) Explain why you do or do not believe that the function is a good model.

11. 

NCAA Basketball Game Attendance

Years Since 1985 (t)	Women's Games (thousands) (W)
0	2,072
5	2,777
10	4,962
14	8,698

Source: *Statistical Abstract of the United States, 2001*, Table 1241, p. 759.

12.  **Ice Cream Cost**

Prepackaged Ice Cream Cost

Years Since 1990 (t)	Ice Cream Cost (dollars per 1/2 gallon) (C)
0	2.54
3	2.59
4	2.62
5	2.68
6	2.94
7	3.02
8	3.30
9	3.40
10	3.66

Source: *Statistical Abstract of the United States, 2001*, Table 706, p. 468.

13.  **DVD Player Shipments**

DVD Player Shipments in North America

Years Since 1997 (t)	Shipments (thousands) (D)
1	1,080
2	4,100
3	9,700
4	16,700

Source: DVD Entertainment Group.

14. **Health Care Spending**

Health Care Spending by Insurance Companies

Years Since 1960 (t)	Spending (billions of dollars) (I)
0	6
10	16
20	68
30	233
40	438

Source: *Statistical Abstract of the United States, 2001*, Table 119, p. 91.

15.  **VHS Tape and CD-R Sales**

VHS Tape Versus CD-R Disk Sales, 2000-2002

Millions of Blank VHS Tapes Sold Worldwide (v)	Millions of CD-R Disks Sold Worldwide (C)
1,409	2,730
1,322	3,730
1,195	4,225

Source: International Recording Media Association.

In Exercises 16–50, do the following:

(a) Draw a scatter plot for the data.

(b) Explain why a quadratic function might or might not fit the data.

If it appears that a quadratic function may fit the data:

(c) Use quadratic regression to find the quadratic function that best fits the data.

(d) Graph the model together with the scatter plot.

(e) Explain why you do or do not believe that the function is a good model.

(f) Answer any additional questions that may be given.

 16.  **College Freshmen**

Male College Freshmen

Years Since 1970 (t)	Percentage of Freshmen That Are Male (F)
0	52.1
10	48.8
15	48.9
20	46.9
25	45.6
27	45.5
28	45.5
29	45.3
30	45.2

According to the model, in what year were there an equal number of male and female freshmen?

 17. 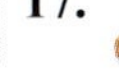 **Personal Income** Because of inflation, the buying power of a dollar decreases over time. As a result, your personal income may increase while your buying power decreases. In the following table, personal income per capita is listed using constant (1996) dollars. That is, the actual income figures are adjusted for inflation.

Personal Income per Capita, Connecticut, Ranked #1 in 2000

Years Since 1980 (t)	Income (dollars) (I)
0	22,530
10	31,223
20	37,854

Source: *Statistical Abstract of the United States, 2001*, Table 652, p. 426.

 18. **Personal Income** In the following table, personal income per capita is listed using constant (1996) dollars. That is, the actual income figures are adjusted for inflation.

Personal Income per Capita, Mississippi, ranked #50 in 2000

Years Since 1980 (t)	Income (dollars) (I)
0	12,817
10	15,373
20	19,554

Source: *Statistical Abstract of the United States, 2001*, Table 652, p. 426.

What is the vertex of the quadratic model, and what does it mean in the context of the problem?

 19.  **Game Sales**

U.S. Video Game Industry Sales

Years Since 1997 (t)	Sales (billions of dollars) (S)
0	5.1
1	6.2
2	6.9
3	6.6
4	9.4

Source: www.npdfunworld.com.

 **20.**   **Milk Consumption**

Per Capita Milk Beverage Consumption

Years Since 1980 (t)	Consumption (gallons) (M)
0	27.6
5	26.7
10	25.7
15	24.3
17	24.0
18	23.7
19	23.6

Source: *Statistical Abstract of the United States, 2001*, Table 202, p. 129.

 **21.**  **Conventional Mortgages**

Percentage of New Privately Owned One-Family Houses Financed with a Conventional Mortgage

Years Since 1970 (t)	Conventional Mortgages (percent) (M)
0	47
10	55
20	62
25	74
29	79

Source: *Statistical Abstract of the United States, 2001*, Table 938, p. 597.

According to the model, in what year were 50 percent of the new privately owned homes financed with a conventional mortgage?

 **22.**  **Home Square Footage**

Percentage of New One-Family Houses Under 1,200 Square Feet

Years Since 1970 (t)	Percentage of New Houses (percent) (H)
0	36
10	21
20	11
25	10
29	7

Source: *Statistical Abstract of the United States, 2001*, Table 938, p. 597.

What is the concavity of the graph of the model, and what does it mean in the context of the problem?

 **23.**  **Home Square Footage**

New One-Family Homes: Average Number of Square Feet

Years Since 1970 (t)	Houses (square feet) (H)
0	1,500
10	1,740
20	2,080
25	2,095
29	2,225

Source: *Statistical Abstract of the United States, 2001*, Table 938, p. 597.

 24. **Air Conditioning in Homes**

New One-Family Homes with Central Air Conditioning

Years Since 1970 (t)	Percentage of New Houses (percent) (H)
0	34
10	63
20	76
25	80
29	84

Source: *Statistical Abstract of the United States, 2001*, Table 938, p. 597.

 25. **Homes with Garages**

New One-Family Homes with a Garage

Years Since 1970 (t)	Percentage of New Houses (percent) (H)
0	58
10	69
20	82
25	84
29	87

Source: *Statistical Abstract of the United States, 2001*, Table 938, p. 597.

In what year were 90 percent of new one-family homes expected to have a garage?

 26.  **Home Sales Price**

Median Sales Price of a New One-Family House in the Southern United States

Years Since 1980 (t)	Price (thousands of dollars) (P)
0	59.6
5	75.0
10	99.0
15	124.5
20	148.0

Source: *Statistical Abstract of the United States, 2001,* Table 940, p. 598.

 27.  **Home Sales Price**

Median Sales Price of a New One-Family House in the Northeastern United States

Years Since 1980 (t)	Price (thousands of dollars) (P)
0	69.5
5	103.3
10	159.0
15	180.0
20	227.4

Source: *Statistical Abstract of the United States, 2001,* Table 940, p. 598.

 **28.**  **Home Sales Price**

Median Sales Price of a New One-Family House in the Western United States

Years Since 1980 (t)	Price (thousands of dollars) (P)
0	72.3
5	92.6
10	147.5
15	141.4
20	196.4

Source: *Statistical Abstract of the United States, 2001,* Table 940, p. 598.

 **29.**  **Vehicle Leasing**

Percentage of Households Leasing Vehicles

Years Since 1989 (t)	Percentage of Households (percent) (P)
0	2.5
3	2.9
6	4.5
9	6.4

Source: *Statistical Abstract of the United States, 2001*, Table 1086, p. 680.

In what year did the percentage of households leasing vehicles reach 5 percent?

 **30.**  **Fuel Consumption**

Motor Fuel Consumption of Vans, Trucks, and SUVs

Years Since 1980 (t)	Fuel Consumption (billions of gallons) (F)
0	23.8
5	27.4
10	35.6
15	45.6
19	52.8

Source: *Statistical Abstract of the United States, 2001*, Table 1105, p. 691.

 **31.**  **Coffee Sales**

Starbucks Corporation Sales

Years Since 09/93 (t)	Income from Sales (millions of dollars) (S)
0	163.5
1	284.9
2	465.2
3	696.5
4	966.9
5	1,308.7
6	1,680.1
7	2,169.2
8	2,649.0
9	3,288.9

Source: moneycentral.msn.com.

According to the model, when did Starbucks Corporation sales reach $5000 million?

 **32.** Organic Cropland

Certified Organic Cropland

Years Since 1992 (t)	Acres (thousands) (A)
0	403
1	465
2	557
3	639
4	850

Source: *Statistical Abstract of the United States, 2001*, Table 805, p. 526.

 **33.**  Poultry Pricing

Average Retail Price of Fresh Whole Chicken

Years Since 1985 (t)	Price (dollars per pound) (P)
0	0.78
5	0.86
6	0.86
7	0.88
8	0.91
9	0.90
10	0.94
11	1.00
12	1.00
13	1.06
14	1.05
15	1.08

Source: *Statistical Abstract of the United States, 2001*, Table 706, p. 468.

 **34.** Prison Rate

Federal and State Prison Rate

Years Since 1980 (t)	Rate (prisoners per 100,000 people) (R)
0	139
2	171
4	188
6	217
8	247
10	297
12	332
14	389
16	427
18	461

Source: *Statistical Abstract of the United States, 2001*, Table 332, p. 200.

 **35.**  College Enrollment

Private College Enrollment

Years Since 1980 (t)	Students (in thousands) (S)
0	2,640
2	2,730
4	2,765
6	2,790
8	2,894
10	2,974
12	3,103
14	3,145
16	3,247
18	3,373

Source: *Statistical Abstract of the United States, 2001*, Table 205, p. 133.

 **36.**  **Military Personnel**

Active-Duty Military Personnel

Years Since 1990 (t)	Personnel (thousands) (P)
1	1,263
2	1,214
3	1,171
4	1,131
5	1,085
6	1,056
7	1,045
8	1,004
9	1,003

Source: *Statistical Abstract of the United States, 2001*, Table 499, p. 328.

 **37.**  **Personal Income**

Per Capita Personal Income: California

Years Since 1993 (t)	Personal Income (dollars) (P)
0	22,833
1	23,348
2	24,339
3	25,373
4	26,521
5	28,240
6	29,772
7	32,149

Source: Bureau of Economic Analysis (www.bea.gov).

 **38.** **Personal Income**

Per Capita Personal Income: Colorado

Years Since 1993 (t)	Personal Income (dollars) (P)
0	22,196
1	23,055
2	24,289
3	25,514
4	27,067
5	28,764
6	30,206
7	32,434

Source: Bureau of Economic Analysis (www.bea.gov).

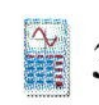 **39.**  **Personal Income**

Per Capita Personal Income: Louisiana

Years Since 1993 (t)	Personal Income (dollars) (P)
0	17,587
1	18,602
2	19,314
3	19,978
4	20,874
5	21,948
6	22,274
7	23,090

Source: Bureau of Economic Analysis (www.bea.gov).

 **40.**  **Personal Income**

Per Capita Personal Income: Minnesota

Years Since 1993 (t)	Personal Income (dollars) (P)
0	21,903
1	23,241
2	24,295
3	25,904
4	27,086
5	29,092
6	30,105
7	31,935

Source: Bureau of Economic Analysis (www.bea.gov).

 **41.**  **Pharmaceutical Company Profit**

Johnson & Johnson Pharmaceutical Operating Profit

Years Since 1997 (t)	Operating Profit (millions of dollars) (P)
0	2,332
1	3,114
2	3,735
3	4,394
4	4,928

Source: Johnson & Johnson Annual Report, 2001, p. 8.

In what year were Johnson & Johnson profits expected to reach $6000 million?

42. **Pharmaceutical Income**

Johnson & Johnson Net Income

Years Since 1990 (t)	Net Income (millions of dollars) (I)
0	1,195
1	1,441
2	1,572
3	1,786
4	1,998
5	2,418
6	2,958
7	3,385
8	3,798
9	4,348
10	4,998
11	5,899

Source: Johnson & Johnson Annual Report, 2001, p. 18.

In what year was Johnson & Johnson net income projected to reach $7000 million?

 43.  **Company Payroll**

Payroll of Private-Employer Firms*

Years Since 1990 (t)	Payroll (billions of dollars) (F)
0	2,104
1	2,145
2	2,272
3	2,363
4	2,488
5	2,666
6	2,849
7	3,048
8	3,309

*Firms are an aggregation of all establishments owned by a parent company.

Source: *Statistical Abstract of the United States, 2001*, Table 726, p. 486.

 44.  **Public-School Teachers**

Projected Number of Public-School Teachers

Years Since 2000 (t)	Public-School Teachers (thousands) (G)
0	2,850
1	2,865
2	2,877
3	2,891
4	2,905
5	2,914
6	2,919
7	2,927
8	2,932
9	2,937
10	2,940

Source: *Statistical Abstract of the United States, 2001*, Table 207, p. 134.

According to the model, in what year is the number of public-school teachers expected to reach 3 million?

 45.  **Private-School Teachers**

Projected Number of Private-School Teachers

Years Since 2000 (t)	Private-School Teachers (thousands) (G)
0	402
1	403
2	404
3	405
4	407
5	408
6	409
7	410
8	411
9	411
10	412

Source: *Statistical Abstract of the United States, 2001*, Table 207, p. 134.

 **46.**  **Apparel Production Wages**

Apparel Production Workers' Wages

Years Since 1980 (t)	Wage of Men's and Boys' Furnishings Production Workers (dollars per hour) (M)
0	4.23
10	6.06
15	7.19
16	7.40
17	7.72
18	7.97
19	8.27
20	8.54

Source: *Statistical Abstract of the United States, 2001*, Table 609, p. 394.

 **47.**  **Apparel Production Wages**

Apparel Production Workers' Wages

Years Since 1980 (t)	Wage of Women's and Misses' Outerwear Production Workers (dollars per hour) (W)
0	4.61
10	6.26
15	7.27
16	7.49
17	7.84
18	8.15
19	8.41
20	8.40

Source: *Statistical Abstract of the United States, 2001*, Table 609, p. 394.

48. **McDonalds Restaurant Sales**

McDonalds Franchised Sales

Years Since 1990 (t)	Franchised Sales (millions of dollars) (S)
1	12,959
2	14,474
3	15,756
4	17,146
5	19,123
6	19,969
7	20,863
8	22,330
9	23,830
10	24,463
11	24,838

Source: www.mcdonalds.com.

 **49.**  **Advertising Expenditures**

Advertising Expenditures: Billboards

Years Since 1990 (t)	Advertising Expenditures (millions of dollars) (A)
0	1,084
1	1,077
2	1,030
3	1,090
4	1,167
5	1,263
6	1,339
7	1,455
8	1,576
9	1,725
10	1,870

Source: *Statistical Abstract of the United States, 2001*, Table 1272, p. 777.

In what year were billboard advertising expenditures projected to reach \$2 billion?

 **50.**  **Advertising Expenditures**

Advertising Expenditures: Direct Mail

Years Since 1990 (t)	Advertising Expenditures (millions of dollars) (A)
0	23,370
1	24,460
2	25,392
3	27,266
4	29,638
5	32,866
6	34,509
7	36,890
8	39,620
9	41,403
10	44,715

Source: *Statistical Abstract of the United States, 2001*, Table 1272, p. 777.

In what year were direct mail advertising expenditures expected to reach \$50 billion?

Exercises 51–55 are intended to challenge your understanding of quadratic functions.

51. Find two different quadratic functions that pass through the points $(0, 0)$ and $(4, 0)$.

52. A classmate claims that any group of three points defines a unique quadratic function. Is your classmate correct? Explain.

53. Is there a single quadratic function that passes through each of the following points: $(1, 3)$, $(2, 5)$, $(4, 9)$? Explain.

54. What is the graphical significance of the point with x coordinate $x = \frac{-b + \sqrt{b^2 - 4ac}}{2a}$ on the graph of $y = ax^2 + bx + c$?

55. If $ax^2 + bx + c > 3$ for all values of x, what do we know about a and the graph of $y = ax^2 + bx + c$?

5.2 Higher-Order Polynomial Function Models

- Use nonlinear functions to model real-life phenomena
- Interpret the meaning of mathematical models in their real-world context
- Model real-life data with higher-order polynomial functions

GETTING STARTED Since 1975, the average wage of sailors working on the East Coast of the United States has trailed that of sailors working on the West Coast (Table 5.7). At one point, West Coast sailors made nearly $1\frac{1}{2}$ times the wage of East Coast sailors.

TABLE 5.7 Typical Basic Monthly Wage for Able-Bodied Seamen in Addition to Room and Board

Years Since 1980 (t)	East Coast Wage (dollars) (E)	West Coast Wage (dollars) (W)
0	967	1,414
5	1,419	2,029
10	1,505	2,218
13	1,721	2,438
14	1,790	2,536
15	1,918	2,637
16	2,014	2,769
17	2,094	2,879
18	2,178	2,994
19	2,265	3,114
20	2,453	3,114

Source: *Statistical Abstract of the United States, 2001*, Table 1072, p. 674.

Salaries of both groups of sailors have increased substantially since 1980, although pay increases have been somewhat irregular. Will East Coast sailors ever make as much as West Coast sailors?

In this section, we will use *cubic* and *quartic* polynomials to model data. To keep computations simple, we use these models only if we're dissatisfied with the fit of simpler models.

Cubic Functions

A constant function is a polynomial of degree 0. A nonconstant linear function is a polynomial of degree 1. A quadratic function is a polynomial of degree 2. A **cubic function** is a polynomial of degree 3 and has the form $f(x) = ax^3 + bx^2 + cx + d$. The graph of a cubic function has exactly one **inflection point** (a point where the graph of the function changes concavity). The term **concavity** refers to the curvature of the graph. Recall that when the graph curves upward, we say that it is "concave up," and when the graph curves downward, we say that it is "concave down." The following rhyme is helpful in remembering the meaning of the terms.

Concave up is like a cup.
Concave down is like a frown.

The graphs in Figure 5.17 are concave up.

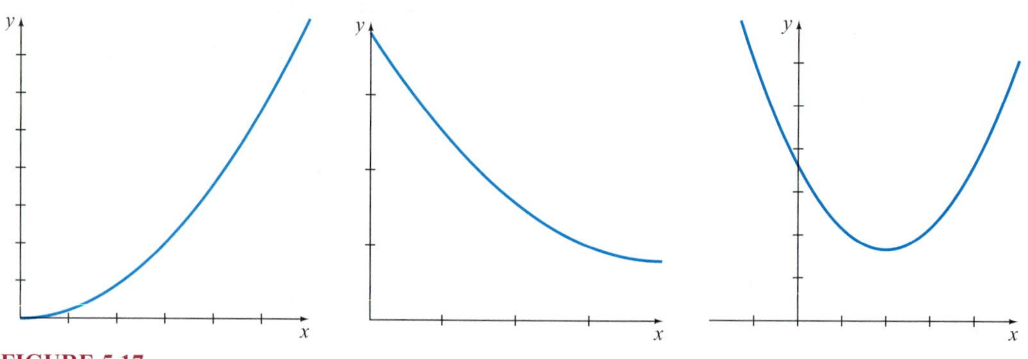

FIGURE 5.17

The graphs in Figure 5.18 are concave down.

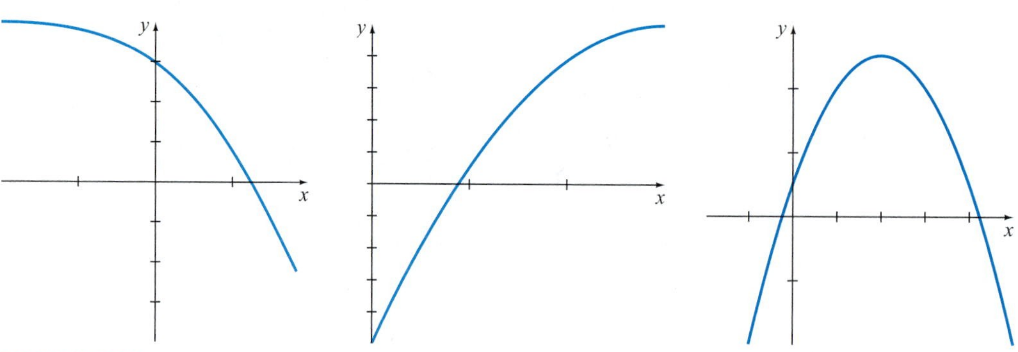

FIGURE 5.18

Cubic function graphs consist of two pieces: a concave up piece and a concave down piece. If $a > 0$, the graph of the cubic function is first concave down and then concave up. The graph of the cubic function will take on one of the basic shapes shown in Figure 5.19.

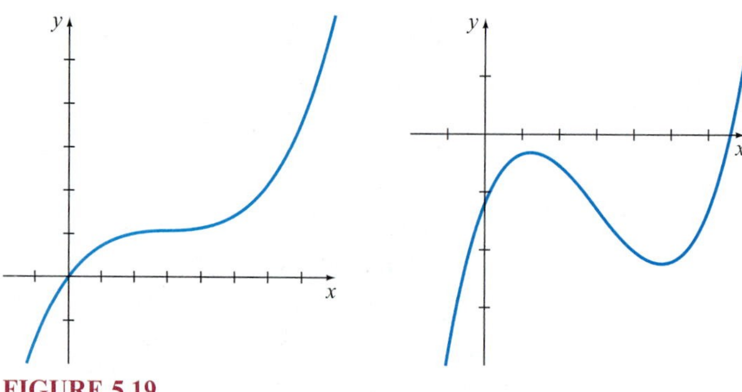

FIGURE 5.19

The size of the "bends" in the graph will vary depending upon the values of b and c. When a is positive, the graph of a cubic function will be increasing (rising from left to right) for sufficiently large values of x.

If $a < 0$, the graph of the cubic function is first concave up and then concave down. The graph of the cubic function will take on one of the shapes shown in Figure 5.20.

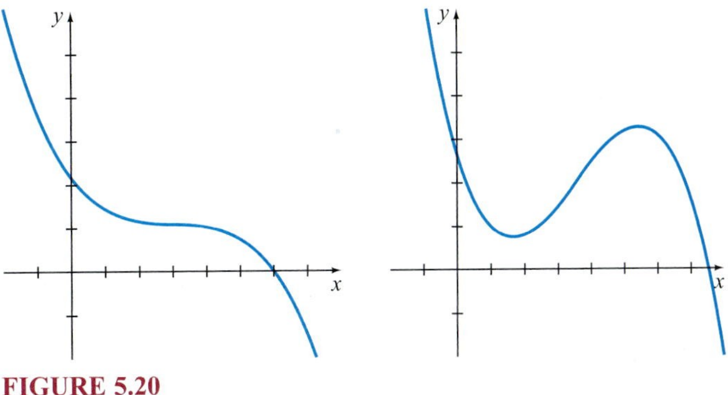

FIGURE 5.20

Again, the size of the "bends" in the graph will vary depending upon the values of b and c. When a is negative, the graph of a cubic function will be decreasing (falling from left to right) for sufficiently large values of x.

Returning to the sailors' wages, we can draw some conclusions by looking at a scatter plot of the data. Since a portion of the scatter plot appears linear, we first attempt to model the data with a linear function (Figure 5.21).

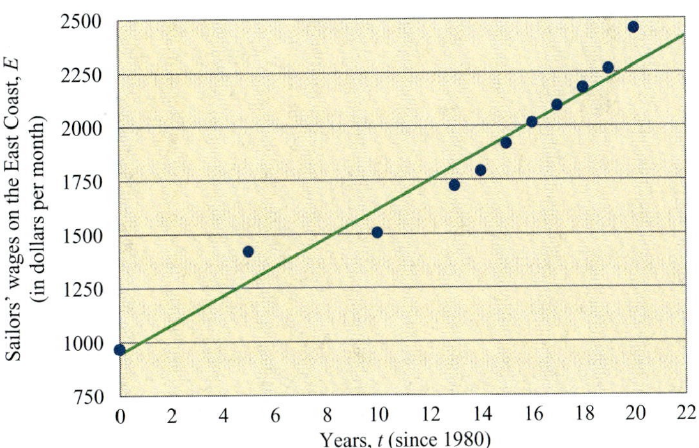

FIGURE 5.21

The linear model does not fit the data as well as we would like. We also note that from 1980 to 1990, the graph appears to be concave down, and from 1990 to 2000, the graph appears to be slightly concave up. We expect that a cubic model with $a > 0$ may fit the data well. Using the cubic regression feature on our calculator, we determine that the cubic function that best fits the data is

$$E(t) = 0.2792t^3 - 7.037t^2 + 102.6t + 983.6 \text{ dollars per month}$$

where t is the number of years since 1980. The model graph is shown in Figure 5.22.

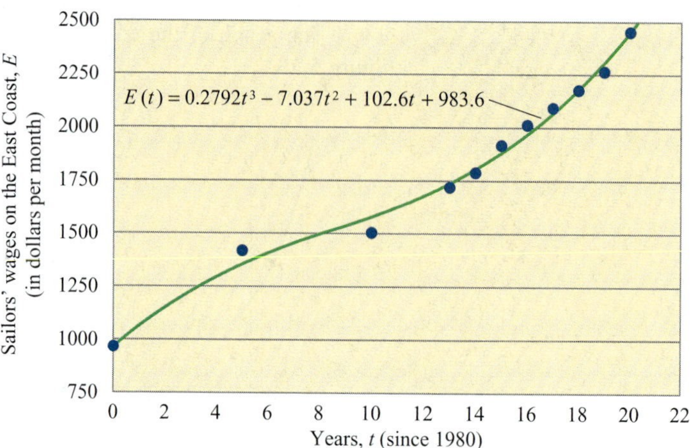

FIGURE 5.22

The model fits the data fairly well, especially after 1993. The coefficient of determination of $r^2 = 0.9905$ further increases our confidence in the model. Let's now look at the scatter plot of the data from the West Coast (Figure 5.23).

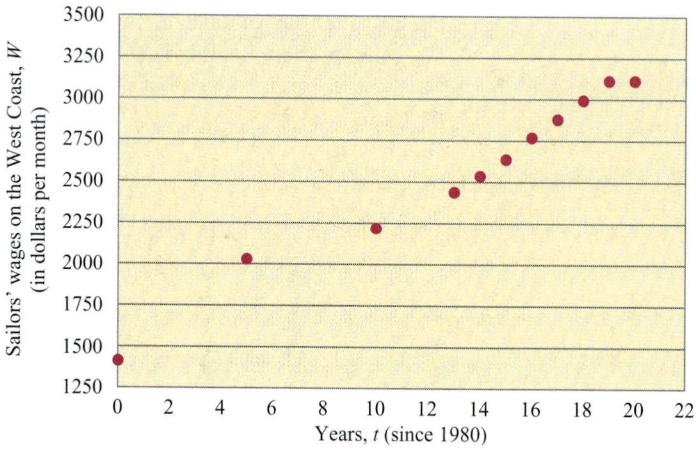

FIGURE 5.23

Again, the graph appears to be concave down from 1980 to 1990. From 1990 to 1999, the graph appears to be concave up. From 1999 to 2000, the wage did not change. Using the cubic regression feature on our calculator, we determine that the cubic equation that best fits the data is

$$W(t) = 0.2546t^3 - 7.277t^2 + 132.2t + 1437 \text{ dollars per month}$$

where t is the number of years since 1980. The model graph is shown in Figure 5.24.

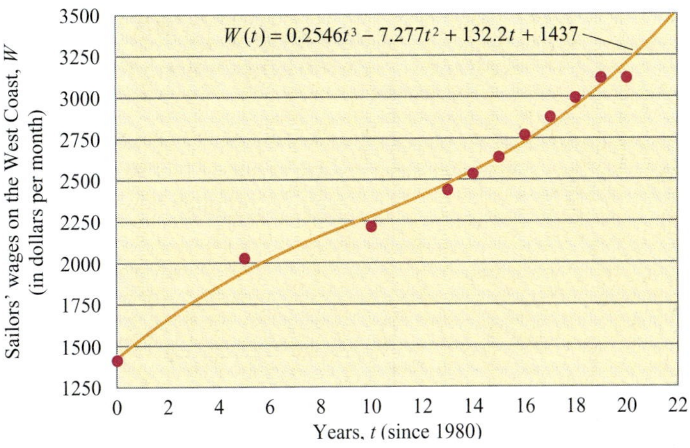

FIGURE 5.24

Will the two models ever be equal? In 2017, the graphs intersect with a monthly wage of nearly \$9100 (Figure 5.25)! Since 2017 is 17 years beyond the last point in our data set, we are somewhat skeptical of the result. Nevertheless, we can conclude that the wage gap isn't likely to be eliminated in the next couple of years.

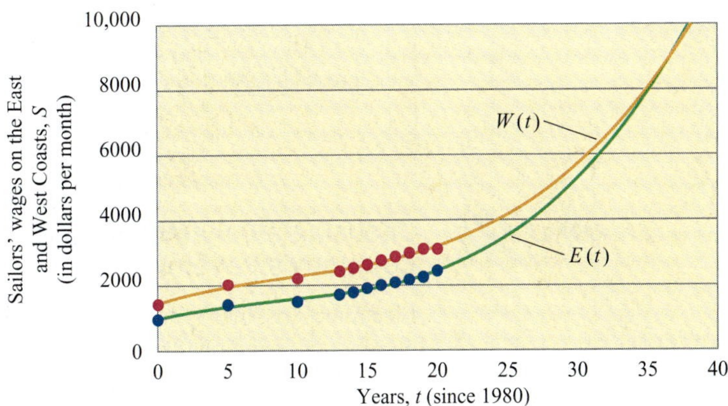

FIGURE 5.25

The following Technology Tip details how to use your TI-83 Plus calculator to do cubic and quartic regression.

TECHNOLOGY TIP

Cubic and Quartic Regression

1. Enter the data using the Statistics Menu List Editor. (Refer to Section 1.3 if you've forgotten how to do this.)

```
L1     L2     L3     1
0      612    ------
5      967
10     1419
15     1505
18     1721
19     1790
20     1918
L1(1) = 0
```

2. Bring up the Statistics Menu Calculate feature by pressing STAT and using the blue arrows to move to the `CALC` menu. Then select item `6:CubicReg` or item `7:QuadReg` and press ENTER.

```
EDIT CALC TESTS
1:1-Var Stats
2:2-Var Stats
3:Med-Med
4:LinReg(ax+b)
5:QuadReg
6:CubicReg
7↓QuartReg
```

3. If you want to automatically paste the regression equation into `Y1` in the Y= editor, press the key sequence VARS `Y-VARS`; `1:Function`; `1:Y1` and press ENTER. Otherwise, press ENTER.

```
CubicReg
 y=ax³+bx²+cx+d
 a=.1643275034
 b=-5.657940597
 c=112.3153143
 d=594.9549798
 R²=.9927303408
```

EXAMPLE 1

Choosing a Polynomial Function to Model a Company's Revenue

Electronic Arts, Inc., is one of the premier producers of interactive electronic games playable on game platforms such as the Sony PlayStation 2®, Microsoft Xbox®, Nintendo GameCube™, and Sony PlayStation. Despite the sluggish economy in the early 2000s, Electronic Arts showed an overall strong growth in revenue (Table 5.8).

TABLE 5.8

Years Since End of Fiscal Year 1998 (t)	Net Revenues (millions of dollars) [$R(t)$]
0	908.852
1	1221.863
2	1420.011
3	1322.273
4	1724.675
5	2482.244

Source: Electronic Arts, Inc., Annual Reports, March 31, 2002 and 2004.

Find a polynomial model for the annual revenue of Electronic Arts, Inc., and forecast in what year the revenue will reach \$3 billion.

SOLUTION From the scatter plot, it appears that a cubic function will best fit the data (Figure 5.26). Using cubic regression, we determine that the cubic model that best fits the data is

$$R(t) = 43.87t^3 - 274.8t^2 + 594.7t + 900.0 \text{ million dollars}$$

where t is the number of years since the end of fiscal year 1998.

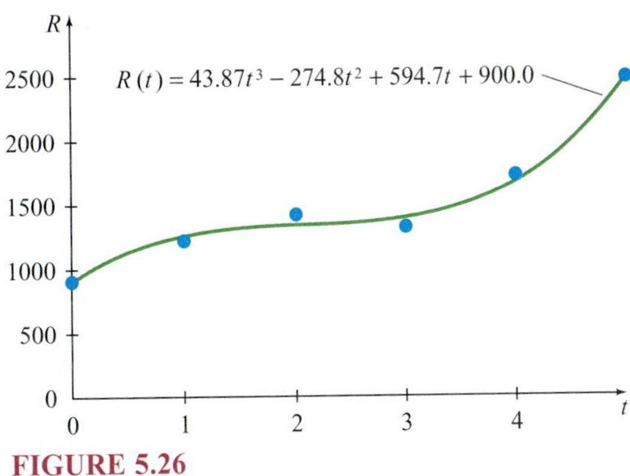

FIGURE 5.26

Since \$3 billion is the same as \$3000 million, we must determine when $R(t) = 3000$. To do this, we simultaneously graph $R(t) = 43.87t^3 - 274.8t^2 + 594.7t + 900.0$ and $y = 3000$. Then, using the intersection feature on our graphing calculator, we determine when the two functions will be equal (Figure 5.27).

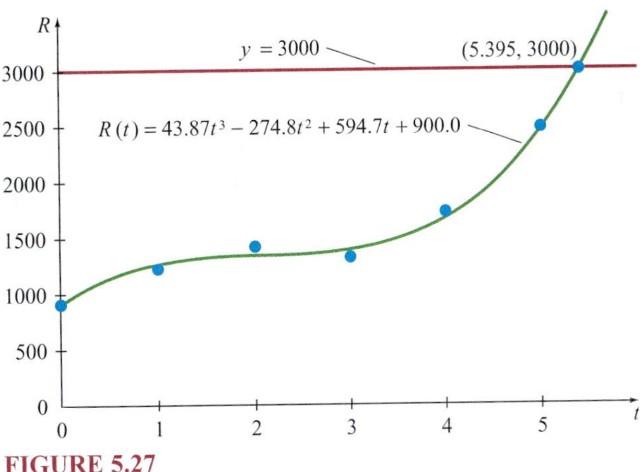

FIGURE 5.27

According to the model, 5.395 years after the end of fiscal year 1998, the annual revenue is expected to reach \$3 billion. To add meaning to the result, we convert 0.395 years into months.

$$0.395 \text{ ~~years~~} \cdot \frac{12 \text{ months}}{1 \text{ ~~year~~}} \approx 5 \text{ months}$$

During the one-year period prior to the end of the fifth month of fiscal year 2004, we anticipate that \$3 billion in revenues will be earned. (*Note:* Since $t = 5$ is the end of fiscal year 2003, $t = 5.395$ is in fiscal year 2004.)

Quartic Functions

A fourth-degree polynomial is a function of the form $f(x) = ax^4 + bx^3 + cx^2 + dx + e$ and is called a **quartic** function. A quartic function may have zero or two inflection points. A quartic graph with $a > 0$ will have one of the basic shapes shown in Figure 5.28.

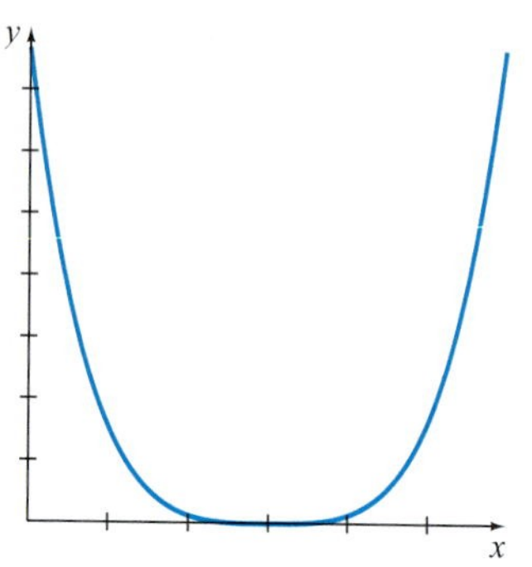

Concave up everywhere
No inflection points

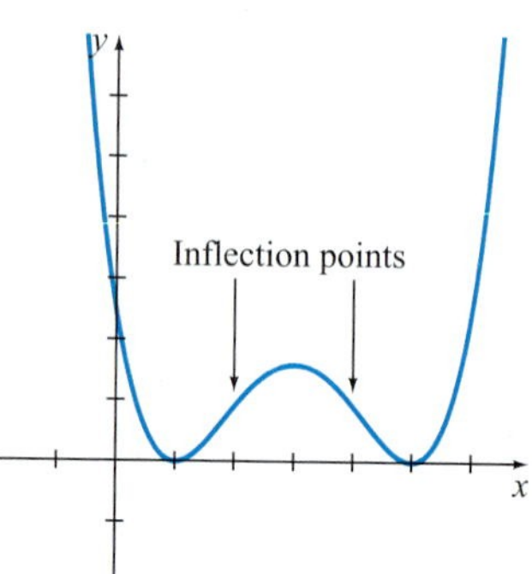

Concave up, then down, then up
Two inflection points

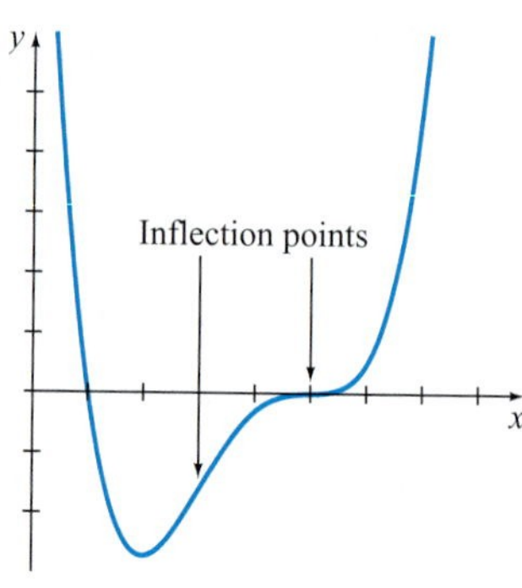

Concave up, then down, then up
Two inflection points

FIGURE 5.28

Notice that each graph opens upward. The size of the "bends" in the graph will vary depending upon the values of b, c, and d. When a is positive, the graph of a quartic function will be increasing (rising from left to right) for sufficiently large values of x.

A quartic graph with $a < 0$ will have one of the basic shapes shown in Figure 5.29.

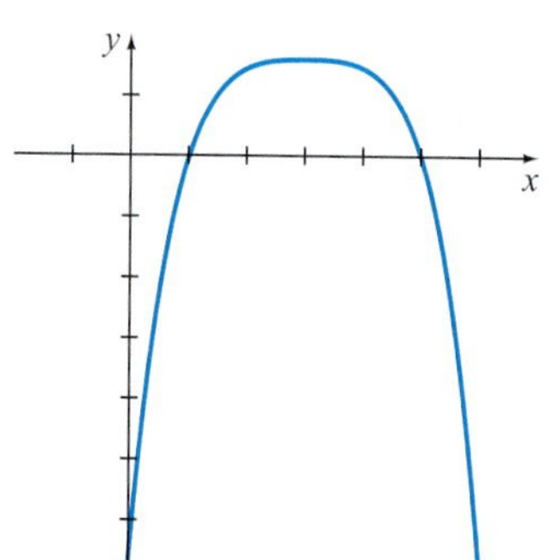

Concave down everywhere
No inflection points

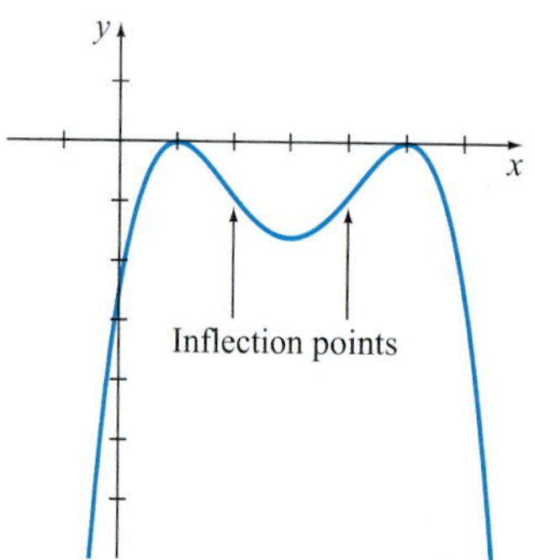

Concave down, then up, then down
Two inflection points

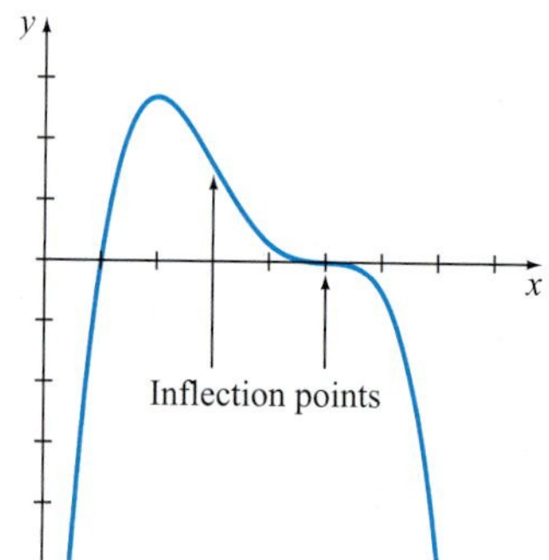

Concave down, then up, then down
Two inflection points

FIGURE 5.29

Notice that each graph opens downward. Again, the size of the "bends" in the graph will vary depending upon the values of b, c, and d. When a is negative, the graph of a quartic function will be decreasing (falling from left to right) for sufficiently large values of x.

EXAMPLE 2

Choosing a Nonlinear Function to Model Interest Rates

The average fixed interest rate for a conventional mortgage on a new home fluctuated between 1985 and 1995 as shown in Table 5.9.

TABLE 5.9 Average Fixed Interest Rates on New Home Conventional Mortgages

Years Since 1985 (t)	Percentage Rate (percent) (R)
0	11.90
2	9.50
4	10.20
6	9.32
8	7.27
10	7.95

Source: *Statistical Abstract of the United States, 2001*, Table 1185, p. 733.

Model the data with a nonlinear function. Do you think the model will be a good indicator of future interest rates? Explain.

SOLUTION

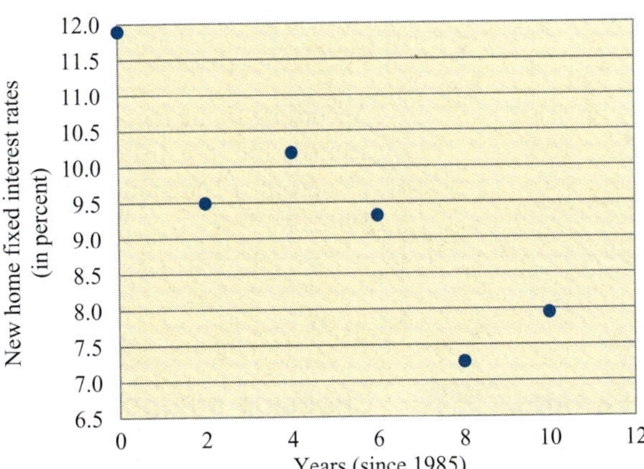

FIGURE 5.30

As shown in Figure 5.30, the graph appears to be concave up between 1985 and 1988, concave down from 1988 to 1992, and concave up from 1992 to 1995. A quartic model may work well (Figure 5.31).

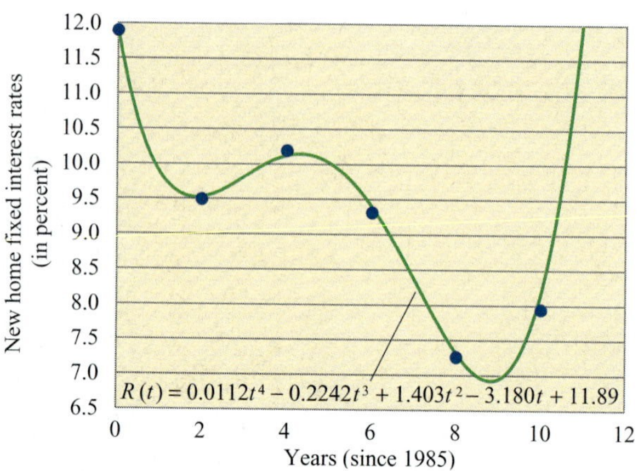

FIGURE 5.31

Using the quartic regression feature on our calculator, we determine that fixed interest rates for mortgages on new homes between 1985 and 1995 may be modeled by the equation

$$R(t) = 0.0112t^4 - 0.2242t^3 + 1.403t^2 - 3.1796t + 11.89 \text{ percent}$$

where t is the number of years since the end of 1985.

The model shows that interest rates will rise rapidly every year after 1995. Let's calculate the predicted rate at the end of 1997 ($t = 12$).

$$\begin{aligned} R(12) &= 0.0112(12)^4 - 0.2242(12)^3 + 1.403(12)^2 - 3.1796(12) + 11.89 \\ &= 20.59 \end{aligned}$$

According to the model, interest rates reached 20.59 percent in 1997. This figure seems a bit unrealistic. (Between 1970 and 2000, interest rates remained below 15 percent.) The model may be a better indicator for intermediate years, such as 1992 or 1994.

EXAMPLE 3 Choosing a Function to Model the Number of Mutual Funds

The number of mutual funds available to investors has increased dramatically since 1970, as shown in Table 5.10.

TABLE 5.10 Number of Mutual Funds

Years Since 1970 (t)	Mutual Funds (M)
0	361
2	410
4	431
6	452
8	505
10	564
12	857
14	1,241
16	1,835
18	2,708
20	3,081
22	3,826
24	5,330
26	6,254
28	7,314
30	8,171

Source: *Statistical Abstract of the United States, 2001*, Table 1214, p. 744.

Find a function that models the data and estimate the number of mutual funds in 2005.

SOLUTION We could model the data with a quadratic function (Figure 5.32); however, even though the correlation coefficient is close to 1, the quadratic model doesn't fit the data very well between 1970 and 1980.

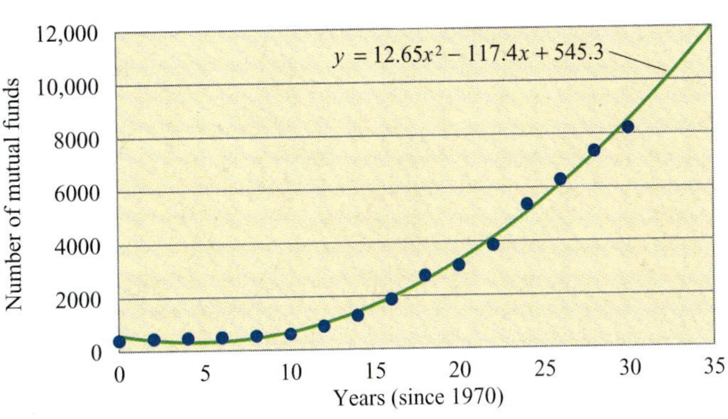

FIGURE 5.32

A quartic model will give a better fit (Figure 5.33).

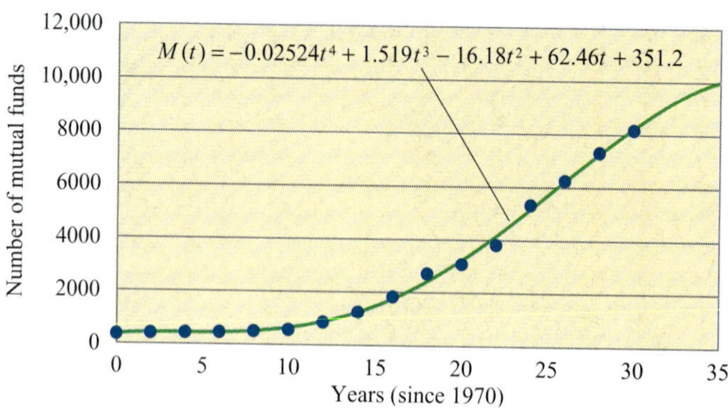

FIGURE 5.33

The number of mutual funds is modeled by the quartic equation

$$M(t) = -0.02524t^4 + 1.519t^3 - 16.18t^2 + 62.46t + 351.2 \text{ mutual funds}$$

where t is the number of years since 1970. For the year 2005, $t=35$. To estimate the number of mutual funds in 2005, we evaluate $M(t)$ at $t=35$.

$$\begin{aligned} M(35) &= -0.02524(35)^4 + 1.519(35)^3 - 16.18(35)^2 + 62.46(35) + 351.2 \\ &= 9968 \end{aligned}$$

According to the model, in 2005 ($t=35$), there will be 9968 mutual funds. Given the historically rapid growth in the number of mutual funds, this estimate seems reasonable.

5.2 Summary

In this section, you learned how to use cubic and quartic polynomials to model data. You also discovered the importance of being cautious in using model predictions.

5.2 Exercises

In Exercises 1–20, find the quadratic, cubic, or quartic model that best fits the data. If multiple models appear to fit the data well, use the simplest model to answer the questions.

1. Public Ninth- to Twelfth-Grade Enrollment

Years Since 1984 (t)	Students (in thousands) (S)
0	12,304
1	12,388
2	12,333
3	12,076
4	11,687
5	11,390
6	11,338
7	11,541
8	11,735
9	11,961
10	12,213
11	12,500
12	12,847
13	13,054
14	13,193
15	13,369

Source: *Statistical Abstract of the United States, 2001,* Table 232, p. 149.

According to your model, how many students were enrolled in high school (ninth to twelfth grades) in 2000?

2. Student-to-Teacher Ratio at Private Elementary and Secondary Schools

Years Since 1960 (t)	Students per Teacher (R)
0	30.7
5	28.3
10	23.0
11	22.6
12	21.6
13	21.2
14	20.4
15	19.6
20	17.7
25	16.2
26	15.7
27	15.5
28	15.2
29	14.2
30	14.7
31	14.7
32	14.6
33	14.8
34	14.9
35	14.9
36	14.9
37	15.1
38	15.2

Source: *Statistical Abstract of the United States, 2001,* Table 235, p. 150.

According to your model, what will be the student-to-teacher ratio in 2006? Does this seem reasonable? Explain.

 3.  **Public-School Teachers**

Public Elementary- and Secondary-School Teachers

Years Since 1980 (t)	Teachers (thousands) (R)
0	2,211
1	2,192
2	2,158
3	2,134
4	2,144
5	2,175
6	2,215
7	2,249
8	2,282
9	2,324
10	2,362
11	2,409
12	2,429
13	2,466
14	2,512
15	2,568
16	2,607
17	2,673
18	2,745
19	2,813
20	2,886

Source: *Statistical Abstract of the United States, 2001,* Table 237, p. 151.

According to the model, in what year is the number of teachers expected to reach 3 million?

 4.  **Principals' Salaries**

High School Principals' Salaries

Years Since 1980 (t)	Salary (dollars) (S)
0	29,207
1	32,231
2	34,776
3	37,602
4	39,334
5	42,094
6	44,986
7	47,896
8	50,512
9	52,987
10	55,722
11	59,106
12	61,768
13	63,054
14	64,993
15	66,596
16	69,277
17	72,410
18	74,380
19	76,768
20	79,839

Source: *Statistical Abstract of the United States, 2001,* Table 238, p. 152.

According to the model, when will the average high school principal's salary reach \$100,000?

 5.  **School Internet Access**

Percentage of Public-School Classrooms with Internet Access

Years Since 1994 (t)	Percentage of Classrooms (percent) (C)
0	3
1	8
2	14
3	27
4	51
5	64
6	77

Source: *Statistical Abstract of the United States, 2001,* Table 243, p. 155.

According to the model, what percentage of classrooms had Internet access in 2001? In 2002? Do the model estimates seem reasonable? Explain.

 6.  **Banks**

Number of Different Banks (Not Bank Branches)

Years Since 1984 (t)	Banks (B)
0	17,900
1	18,033
2	17,876
3	17,325
4	16,562
5	15,829
6	15,192
7	14,517
8	13,891
9	13,261
10	12,641
11	12,002
12	11,478
13	10,923
14	10,463
15	10,221
16	9,908

Source: *Statistical Abstract of the United States, 2001,* Table 1173, p. 728.

According to the model, when will the number of banks drop below 9000? What do you think is causing the number of different banking companies to decrease?

 7.  **Debit Cards**

Number of Debit Cards

Years Since 1990 (t)	Cards (millions) (C)
0	164
5	201
6	205
7	211
8	217
9	228

Source: *Statistical Abstract of the United States, 2001,* Table 1189, p. 734.

A consulting firm has projected that the number of debit cards will reach 270 million by 2005. Does this projection agree with your model?

 8.  **NASDAQ Volume**

NASDAQ Average Daily Volume

Years Since 1980 (t)	Shares (millions) (S)
0	27
2	33
4	60
6	114
8	123
10	132
12	191
14	295
16	544
18	802
20	1,757

Source: *Statistical Abstract of the United States, 2001,* Table 1205, p. 741.

According to your model, when will the NASDAQ market have an average daily volume of 2 billion shares? Visit www.marketdata.nasdaq.com to see if your model estimate is accurate.

 9.  **Stock Trading**

New York Stock Exchange Shares Traded

Years Since 1980 (t)	Shares (millions) (S)
0	11,562
2	16,669
4	23,309
6	36,009
8	41,118
10	39,946
12	51,826
14	74,003
16	105,477
18	171,188
20	265,499

Source: *Statistical Abstract of the United States, 2001,* Table 1207, p. 742.

According to the model, how many shares will be traded in 2005? Does this seem reasonable? Explain.

 **10.**  **Listed Stocks**

Companies with Stocks Listed on the NYSE

Years Since 1980 (t)	Companies (C)
0	1,570
2	1,526
4	1,543
6	1,575
8	1,681
10	1,774
12	2,088
14	2,570
16	2,907
18	3,114
20	2,862

Source: *Statistical Abstract of the United States, 2001,* Table 1207, p. 742.

According to the model, in which years is the number of companies listed on the New York Stock Exchange expected to exceed 2700 companies?

 **11.**  **Computer Retailers' Wages**

Average Annual Wage per Worker for Computer and Equipment Retailers

Years Since 1992 (t)	Wages (dollars) (W)
1	32,200
2	30,500
3	32,100
4	33,800
5	35,000
6	37,300

Source: *Statistical Abstract of the United States, 2001,* Table 1123, p. 703.

In what year do you expect the average wage to exceed $40,000?

 **12.**  **Computer Wholesalers' Wages**

Average Annual Wage per Worker for Computer and Equipment Wholesalers

Years Since 1992 (t)	Wages (dollars) (W)
1	52,500
2	52,900
3	52,900
4	54,300
5	56,700
6	62,200

Source: *Statistical Abstract of the United States, 2001,* Table 1123, p. 703.

Calculate the average wage in 2000. Does this seem reasonable? Explain.

 **13.**  **Software Developers' Wages**

Average Wage per Worker at Prepackaged Software Development Firms

Years Since 1992 (t)	Wages (dollars) (W)
0	57,000
1	54,500
2	57,000
3	63,700
4	70,100
5	79,200
6	94,100

Source: *Statistical Abstract of the United States, 2001,* Table 1123, p. 703.

In what year will the wage reach $120,000?

 14.   **Software Retailers' Wages**

Average Wage per Worker at Prepackaged Software Retailers

Years Since 1992 (t)	Wages (dollars) (W)
0	32,200
1	30,500
2	32,100
3	33,800
4	35,000
5	37,300
6	40,400

Source: *Statistical Abstract of the United States, 2001*, Table 1123, p. 703.

What do you expect the average worker's wage was in 1999?

 15.  **Software Wholesalers' Wages**

Average Wage per Worker at Prepackaged Software Wholesalers

Years Since 1992 (t)	Wages (dollars) (W)
0	52,500
1	52,900
2	52,900
3	54,300
4	56,700
5	62,200
6	69,700

Source: *Statistical Abstract of the United States, 2001*, Table 1123, p. 703.

Compare the wholesalers' average wage model with the retailers' average wage model from Exercise 14. According to the models, will wages at retailers ever catch up with wages at wholesalers?

 16. **Radio Broadcasters' Wages**

Average Annual Wage per Worker in Radio Broadcasting

Years Since 1992 (t)	Annual Wage (dollars) (W)
0	23,500
1	24,300
2	26,000
3	27,200
4	29,300
5	31,300
6	34,200

Source: *Statistical Abstract of the United States, 2001*, Table 1123, p. 703.

According to the model, what was the average wage of a radio broadcasting employee in 2006? Does this estimate seem reasonable? Explain.

 17.  **Television Broadcasters' Wages**

Average Annual Wage per Worker in Television Broadcasting

Years Since 1992 (t)	Annual Wage (dollars) (W)
0	41,400
1	42,200
2	43,700
3	47,200
4	51,100
5	51,000
6	54,600

Source: *Statistical Abstract of the United States, 2001*, Table 1123, p. 703.

According to the model, when will the average annual wage for television broadcasting employees exceed $60,000? Does this estimate seem reasonable? Explain.

 **18.** Internet Usage

Projected Internet Usage, Hours per Person per Year (Based on 1995–1999 Data)

Years Since 1995 (t)	Usage per Person (hours per year) (H)
0	5
1	10
2	34
3	61
4	99
5	135
6	162
7	187
8	208
9	228

Source: *Statistical Abstract of the United States, 2001*, Table 1125, p. 704.

What do you project the Internet usage will be in 2006? Do you believe your projection is a good estimate? Explain.

  **19.**  Homes with Cable

Percentage of TV Homes with Cable

Years Since 1970 (t)	TV Homes with Cable (percent) (C)
0	6.7
5	12.6
10	19.9
15	42.8
16	45.6
17	47.7
18	49.4
19	52.8
20	56.4
21	58.9
22	60.2
23	61.4
24	62.4
25	63.4
26	65.3
27	66.5
28	67.2
29	67.5

(Continued)

(Continued)

Source: *Statistical Abstract of the United States, 2001*, Table 1126, p. 705.

According to the model, what percentage of TV homes had cable in 2002? Does this seem reasonable? Explain.

20. Homes with VCRs

Percentage of TV Homes with a VCR

Years Since 1980 (t)	Homes with a VCR (percent) (C)
0	1.1
5	20.8
6	36.0
7	48.7
8	58.0
9	64.6
10	68.6
11	71.9
12	75.0
13	77.1
14	79.0
15	81.0
16	82.2
17	84.2
18	84.6
19	84.6

Source: *Statistical Abstract of the United States, 2001*, Table 1126, p. 705.

Using the model from Exercise 19, determine in what year the percentage of homes with VCRs and the percentage of homes with cable was the same. (*Hint:* The meaning of t in the two exercises is different.)

Exercises 21–25 are intended to challenge your understanding of higher-order polynomial functions.

21. A "bend" in a graph occurs when the graph changes from increasing to decreasing or from decreasing to increasing. A quadratic function has exactly one bend. A cubic function has zero or two bends. A quartic function has one or three bends. In general, how many bends may the graph of a polynomial of degree n have?

22. Can the number of x-intercepts of a graph of a polynomial function ever exceed the degree of the polynomial? Explain.

23. A function of the form $f(x) = x^n$ is called a **power function.** If n is a positive even integer, what is the relationship between $f(x)$ and $f(-x)$?

24. Find the equation of the cubic function that passes through the points $(-1, 0)$, $(0, 0)$, $(1, 0)$, and $(3, 24)$.

25. Find the equation of the quartic function that passes through the points $(-1, 0)$, $(0, 0)$, $(1, 0)$, $(2, 0)$, and $(3, 24)$.

5.3 Exponential Function Models

- Use nonlinear functions to model real-life phenomena
- Interpret the meaning of mathematical models in their real-world context
- Graph exponential functions
- Find the equation of an exponential function from a table
- Model real-life data with exponential functions

GETTING STARTED Michael Jordan is arguably the most renowned athlete of all time. With millions of dollars in endorsements in addition to his multimillion-dollar salary, he is one of the wealthiest athletes in history. Since his entry into the NBA in 1984, the average salary for NBA players has increased *exponentially*.

In this section, we will demonstrate how exponential functions can be used to model rapidly increasing data sets such as the average salary for NBA players. We will develop exponential models from tables of data and verbal descriptions. In addition, we will show what an exponential function graph looks like.

The average annual salary of an NBA player increased from \$170 thousand in 1980 to \$2.6 million in 1998, as shown in Table 5.11.

If we plot the NBA salary data together with the graph of the function $S(t) = 161.4(1.169)^t$, we see that the graph of S fits the data fairly well (Figure 5.34).

TABLE 5.11 Average NBA Athlete's Salary

Years Since 1980 (t)	Annual Salary (thousands of dollars) (S)
0	170
5	325
10	750
15	1,900
16	2,000
17	2,200
18	2,600

Source: *Statistical Abstract of the United States, 2001,* Table 1324, p. 829.

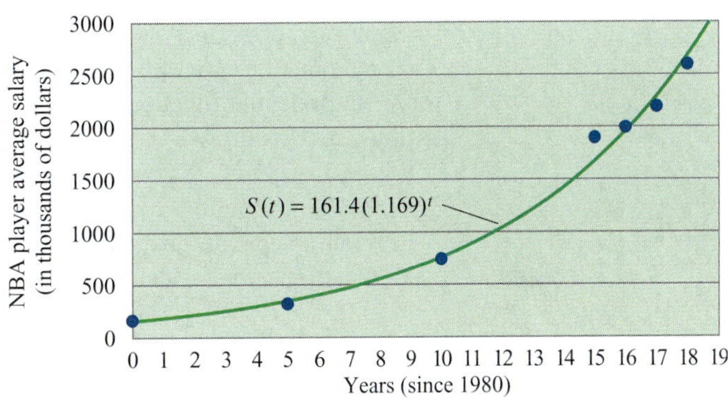

FIGURE 5.34

Notice that the independent variable, t, appears as an exponent. The function S is called an **exponential function.**

EXPONENTIAL FUNCTION

If a and b are real numbers with $a \neq 0$, $b > 0$, and $b \neq 1$, then the function

$$y = ab^x$$

is called an **exponential function.** The value b is called the **base** of the exponential function.

Why must b be positive? Consider the function $y = (-1)^x$. For integer values, y oscillates between -1 and 1 (Table 5.12). However, the function is undefined for numerous noninteger values of x (Table 5.13).

TABLE 5.12

x	y
-2	1
-1	-1
0	1
1	-1
2	1

TABLE 5.13

x	y
-0.5	Undefined
-0.3	Undefined
0.1	Undefined
1.4	-1
1.5	Undefined

On the other hand, if $b > 0$, then the value of b^x is defined for *all* integer and noninteger values of x.

Why don't we allow b to be 1? If $b = 1$, then

$$y = ab^x$$

$$y = a(1)^x$$

$$y = a \qquad \text{Since } (1)^x = 1 \text{ for all } x$$

The graph of the function $y = a$ is a horizontal line and does not exhibit the same graphical behavior as all other functions of the form $y = ab^x$ with $b > 0$. By eliminating the case of $b = 1$, we are able to talk about a family of like functions.

Exponential functions are used frequently to model growth and decay situations, such as growth in population, depreciation of a vehicle, or growth in a retirement account. When the growth or decay of a quantity is modeled by an exponential function $y = ab^x$, the independent variable is frequently time. The beginning value of the quantity at time $x = 0$ is referred to as the **initial value** of the function and, as demonstrated below, is equal to a.

$$y = ab^0$$

$$= a(1) \qquad \text{Since } b^0 = 1 \text{ for all nonzero real numbers } b$$

$$= a$$

Graphically speaking, the constant a in $y = ab^x$ corresponds with the y-intercept of the exponential graph. In most real-life applications, a will be positive. The initial value of the NBA player salary function $S(t) = 161.4(1.169)^t$ is 161.4. That is, the model estimates that the average NBA player's salary was about \$161 thousand in the year $t = 0$ (1980).

The base b of $y = ab^x$ is often referred to as the **growth factor** of the function. Increasing the value of x by one unit increases y by a factor of b. The growth factor of the NBA player salary function $S(t) = 161.4(1.169)^t$ is 1.169, since $b = 1.169$. That is, the average NBA player's salary is increasing by a factor of 1.169 annually. Next year's salary is forecast to be 1.169 times this year's salary.

Recall that the x-intercept of a function occurs when $y = 0$. For what values of x does $ab^x = 0$? Let's consider the exponential function $y = 2^x$. We'll generate a table of values and plot a few points to get an idea about what is happening graphically (Figure 5.35).

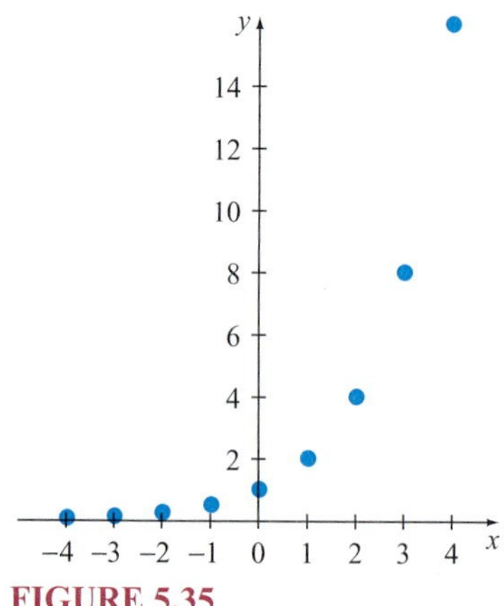

FIGURE 5.35

x	$y = 2^x$
-4	$\frac{1}{16}$
-3	$\frac{1}{8}$
-2	$\frac{1}{4}$
-1	$\frac{1}{2}$
0	1
1	2
2	4
3	8
4	16

Observe that as x increases, y also increases. Also, observe that y is positive for all values of x. That is, *there are no x-intercepts*. This remains true even if we pick a negative number with a larger magnitude, say $x = -33$.

$$2^{-33} = \frac{1}{2^{33}}$$

$$2^{-33} = \frac{1}{8{,}589{,}934{,}592}$$

We call the line $y = 0$ a **horizontal asymptote** of the function $y = 2^x$. For sufficiently small values of x, the graph of $y = 2^x$ approaches the graph of the line $y = 0$. All exponential functions of the form $y = ab^x$ have a horizontal asymptote at $y = 0$.

Exponential Function Graphs

Exponential function graphs will take on one of the four basic shapes specified in Table 5.14. Each graph has a horizontal asymptote at $y = 0$ and a y-intercept at $(0, a)$.

TABLE 5.14 Exponential Function Graphs: $y = ab^x$

Value of a	Value of b	Concavity of Graph	Increasing/ Decreasing	Sample Graph
$a > 0$	$b > 1$	Concave up	Increasing	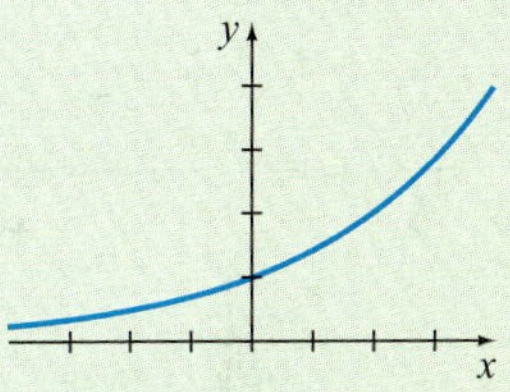
$a > 0$	$0 < b < 1$	Concave up	Decreasing	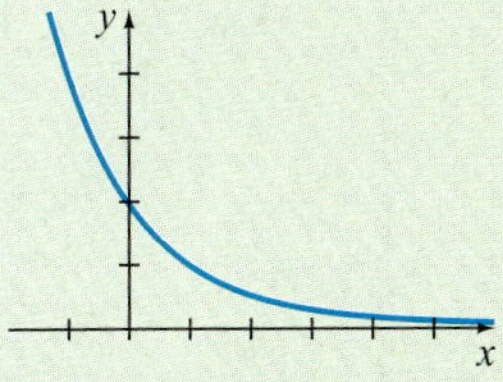
$a < 0$	$b > 1$	Concave down	Decreasing	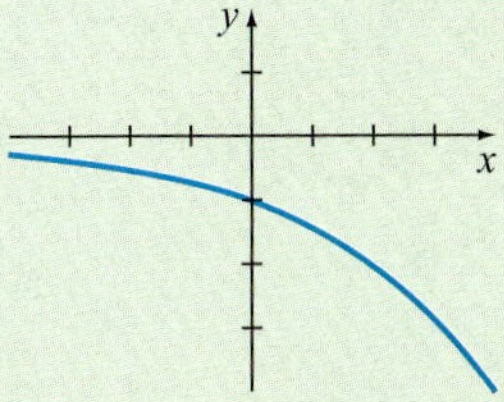
$a < 0$	$0 < b < 1$	Concave down	Increasing	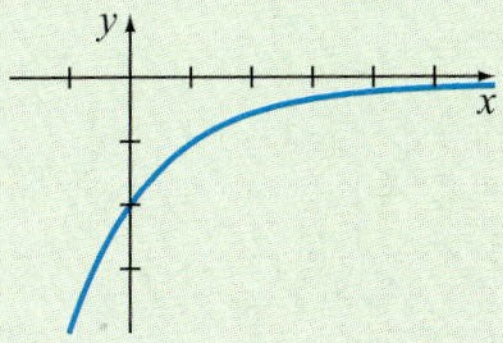

EXAMPLE 1

Comparing Exponential Graphs

Compare and contrast the graphs of $f(x) = 3(2)^x$ and $g(x) = 4(0.5)^x$. (You may sketch the graphs by hand by plotting several points, or you may use technology to graph the functions.)

SOLUTION Both f and g are exponential functions. In the function equation of f, $a = 3$ and $b = 2$. The graph of f is concave up, since $a > 0$, and is increasing, since $b > 1$. The graph of f has a y-intercept at (0, 3).

In the function equation of g, $a = 4$ and $b = 0.5$. The graph of g is concave up, since $a > 0$, and is decreasing, since $b < 1$. The graph of g has a y-intercept at (0, 4). The graphs of both functions have a horizontal asymptote at $y = 0$ (Figure 5.36).

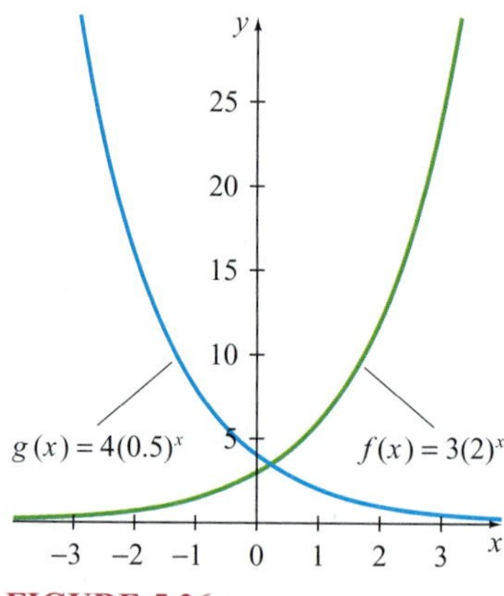

FIGURE 5.36

Properties of Exponents

As seen earlier in this section, it is often necessary to apply the properties of exponents when working with exponential functions. Since these properties are typically covered in depth in an algebra course, we will only summarize them here.

PROPERTIES OF EXPONENTS

If b, m, and n are real numbers with $b > 0$, the following properties hold.

Property	Example
1. $b^{-n} = \dfrac{1}{b^n}$	1. $2^{-3} = \dfrac{1}{2^3}$
2. $b^m \cdot b^n = b^{m+n}$	2. $3^2 \cdot 3^4 = 3^6$
3. $\dfrac{b^m}{b^n} = b^{m-n}$	3. $\dfrac{5^6}{5^4} = 5^2$
4. $b^{mn} = (b^m)^n = (b^n)^m$	4. $6^{2 \cdot 3} = (6^2)^3 = (6^3)^2$

Finding an Exponential Function from a Table

We can easily determine if a table of values models an exponential function by calculating the ratio of consecutive outputs for evenly spaced inputs. The ratio will be constant if the table of values models an exponential function. Consider Table 5.15.

TABLE 5.15

x	y
0	6
2	24
4	96
6	384

The domain values are equally spaced (two units apart). Calculating the ratios of the consecutive range values, we get

$$\frac{24}{6} = 4 \qquad \frac{96}{24} = 4 \qquad \frac{384}{96} = 4$$

In each case, the ratio was 4. Therefore, the table of values models an exponential function. But what is the equation of the function?

We can find the equation of the function algebraically. We know that an exponential function must be of the form $y = ab^x$. Substituting the point (0, 6) into the equation and solving, we get

$$y = ab^x$$
$$6 = ab^0 \quad \text{Substitute } x = 0 \text{ and } y = 6$$
$$6 = a$$

Since $a = 6$, $y = 6b^x$. Substituting the point (2, 24) into the equation $y = 6b^x$, we get

$$y = 6b^x$$
$$24 = 6b^2 \quad \text{Substitute } x = 2 \text{ and } y = 24$$
$$4 = b^2$$
$$b = 2 \quad b \neq -2 \text{ since the base of an exponential function must be positive}$$

Therefore, the exponential function that models the table data is $y = 6(2)^x$.

EXAMPLE 2

Finding an Exponential Equation from a Table

Find the equation of the exponential function modeled by Table 5.16.

TABLE 5.16

x	y
1	21
2	63
3	189
4	567

SOLUTION We know that an exponential function must be of the form $y = ab^x$. Substituting the point (2, 63) into the equation, we get

$$y = ab^x$$
$$63 = ab^2 \quad \text{Substitute } x = 2 \text{ and } y = 63$$

Substituting the point (1, 21) into the equation $y = ab^x$, we get

$$y = ab^x$$
$$21 = ab^1 \quad \text{Substitute } x = 1 \text{ and } y = 21$$

We can eliminate the a variable by dividing the first equation by the second equation. (*Note:* What we are really doing is dividing both sides of the first equation by the same nonzero quantity, expressed in two different forms.)

$$\frac{63}{21} = \frac{ab^2}{ab^1}$$

$$3 = \frac{a}{a} \cdot b^{2-1} \qquad \text{Since } \frac{b^m}{b^n} = b^{m-n}$$

$$3 = 1 \cdot b$$

$$b = 3$$

We may then substitute the value of b into either equation to find a.

$$21 = ab^1$$

$$21 = a \cdot 3^1 \qquad \text{Substitute } b = 3$$

$$\frac{21}{3} = a$$

$$a = 7$$

The exponential function that models the table data is $y = 7(3)^x$.

When using the method of dividing one equation by the other, computations will tend to be easier if we divide the equation with the largest exponent by the equation with the smallest exponent.

The technique shown in Example 2 may be generalized for all exponential functions. Using the result from the generalized solution will allow us to determine the exponential equation more quickly.

Consider an exponential function $y = ab^x$ whose graph goes through the points (x_1, y_1) and (x_2, y_2). We have

$$y = ab^x$$

$$y_1 = ab^{x_1} \qquad \text{Substitute } x = x_1 \text{ and } y = y_1$$

and

$$y = ab^x$$

$$y_2 = ab^{x_2} \qquad \text{Substitute } x = x_2 \text{ and } y = y_2$$

Dividing the second equation by the first equation yields a quick way to calculate b.

$$\frac{y_2}{y_1} = \frac{ab^{x_2}}{ab^{x_1}}$$

$$\frac{y_2}{y_1} = \frac{a}{a} b^{x_2 - x_1}$$

$$\frac{y_2}{y_1} = b^{x_2 - x_1}$$

$$\left(\frac{y_2}{y_1}\right)^{\frac{1}{x_2 - x_1}} = \left(b^{x_2 - x_1}\right)^{\frac{1}{x_2 - x_1}}$$

$$\left(\frac{y_2}{y_1}\right)^{\frac{1}{x_2 - x_1}} = b^{\frac{x_2 - x_1}{x_2 - x_1}}$$

$$b = \left(\frac{y_2}{y_1}\right)^{\frac{1}{x_2 - x_1}}$$

Using the points (1, 21) and (3, 189) from Example 2, we get

$$b = \left(\frac{y_2}{y_1}\right)^{\frac{1}{x_2 - x_1}}$$

$$b = \left(\frac{189}{21}\right)^{\frac{1}{3-1}} \quad \text{Substitute } x_1 = 0, x_2 = 2, y_1 = 21, \text{ and } y_2 = 189$$

$$= 9^{1/2}$$

$$= \sqrt{9} \quad \text{Recall that for } b \geq 0, b^{1/2} = \sqrt{b}$$

$$= 3$$

This method is especially useful when checking to see if a data table with unequally spaced inputs is an exponential function. If it is an exponential function, the value of b will be constant regardless of which two points we substitute into the formula.

Using Exponential Regression to Model Data

Data sets with near-constant ratios of change may be modeled using the exponential regression feature on our graphing calculator. It is often helpful to do a scatter plot of the data first to see if the graph looks like an exponential function.

EXAMPLE 3

Using Exponential Regression to Model the Population of Akron, Ohio

The population of Akron, Ohio, is shown in Table 5.17. Model the population with an exponential function and forecast the population of Akron in 2010.

TABLE 5.17

Years Since 1970 (x)	Population (thousands) (y)
0	275
10	237
20	223
30	217

Source: *Statistical Abstract of the United States, 2001,* Table 34, p. 34.

SOLUTION We first draw a scatter plot so that we can visually predict whether an exponential function will fit the data well (Figure 5.37). (Note that because we have zoomed in on the data, the "x axis" is given by $y = 210$ instead of $y = 0$.)

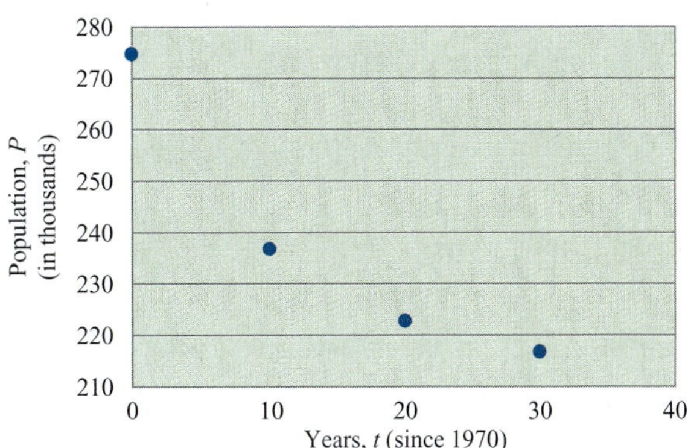

FIGURE 5.37

Since the scatter plot is decreasing and concave up, an exponential function may fit the data well. Using techniques given in the Technology Tip following this example, we can calculate an exponential model for the data set (Figure 5.38).

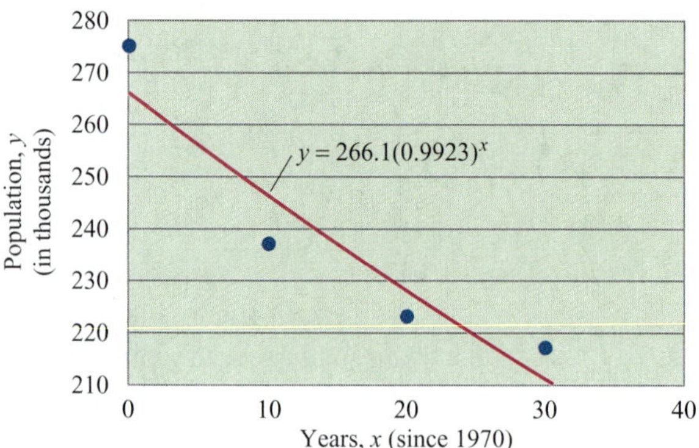

FIGURE 5.38

TABLE 5.18

Year (x)	Population −210 (thousands) (y)
0	65
10	27
20	13
30	7

The model $y = 266.1(0.9923)^x$ is a terrible fit for the data. What happened? Recall that an exponential function of the form $y = ab^x$ has a horizontal asymptote at $y = 0$. From the scatter plot of the data, it appears that the population is approaching a constant value of $y = 210$ instead of $y = 0$. (This value is not precise. We are *guessing* what the horizontal asymptote would be based on the scatter plot of the data set.) If we subtract 210 from each of the y values, we get a new data set with a horizontal asymptote of $y = 0$. This is referred to as an **aligned data set** (Table 5.18).

Using exponential regression to find a model for the aligned data set yields $y = 60.81(0.9285)^x$, a function that fits the aligned data well (Figure 5.39).

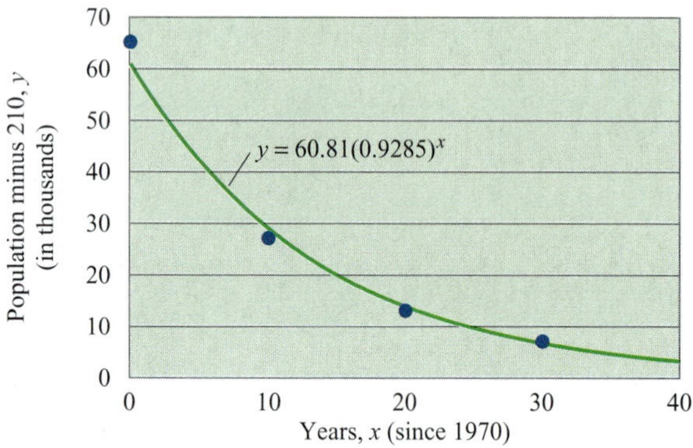

FIGURE 5.39

Subtracting 210 from each of the y values of the original data set had the graphical implication of moving the data down 210 units. To return the aligned data model to the position of the original data, we need to move the model up 210 units. To do this, we add back the 210.

$$P(x) = 60.81(0.9285)^x + 210 \quad \text{Add 210 to shift the data graph upward}$$

The model created from the aligned data fits those data much better than the unaligned data model (Figure 5.40).

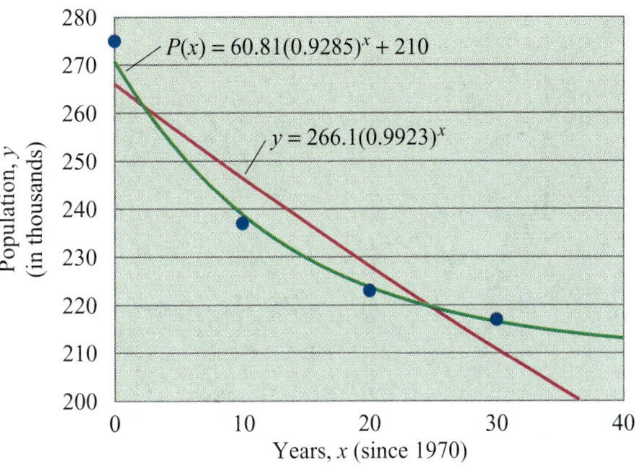

FIGURE 5.40

To estimate the population of Akron, Ohio, in 2010, we evaluate $P(x) = 60.81(0.9285)^x + 210$ at $x = 40$.

$$P(40) = 60.81(0.9285)^{40} + 210$$
$$\approx 213$$

We forecast that the population of Akron, Ohio, in 2010 will be 213,000.

The methods of aligning data and of performing exponential regression are detailed in the following Technology Tips.

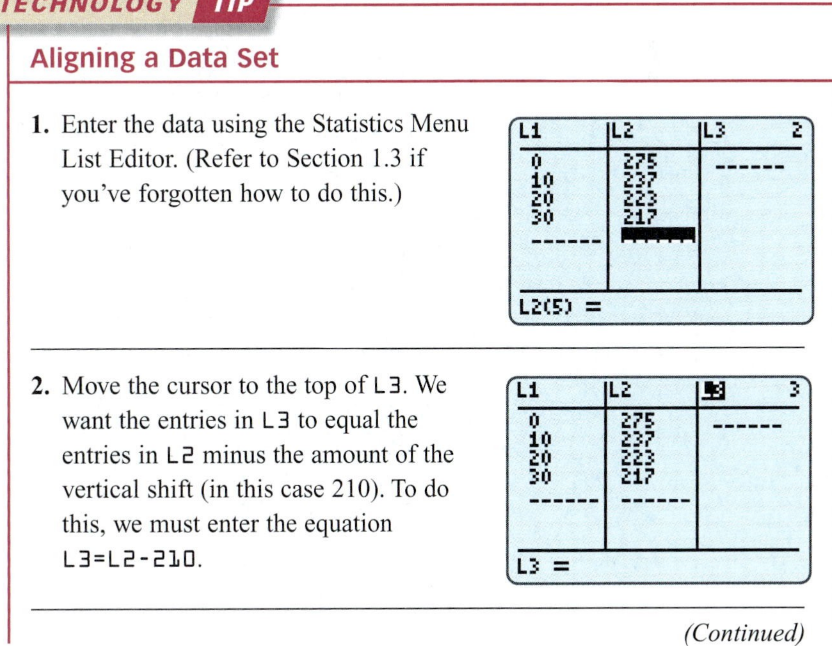

TECHNOLOGY TIP

Aligning a Data Set

1. Enter the data using the Statistics Menu List Editor. (Refer to Section 1.3 if you've forgotten how to do this.)

2. Move the cursor to the top of L3. We want the entries in L3 to equal the entries in L2 minus the amount of the vertical shift (in this case 210). To do this, we must enter the equation L3=L2-210.

(Continued)

3. Press [2nd] then [2] to place L2 on the equation line at the bottom of the viewing window. Then press [−] and [2][1][0] to subtract 210.

```
L1     L2     L3     3
0      275    ------
10     237
20     223
30     217
------ ------
L3 =L2-210
```

4. Press [ENTER] to display the list of aligned values in L3.

```
L1     L2     L3     3
0      275    65
10     237    27
20     223    13
30     217    7
------ ------
L3(5) =
```

TECHNOLOGY TIP

Exponential Regression

1. Enter the data using the Statistics Menu List Editor. (Refer to Section 1.3 if you've forgotten how to do this.)

```
L1     L2     L3     3
0      275    65
10     237    27
20     223    13
30     217    7
------ ------
L3(5) =
```

2. Bring up the Statistics Menu Calculate feature by pressing [STAT] and using the blue arrows to move to the CALC menu. Then select item 0:ExpReg and press [ENTER].

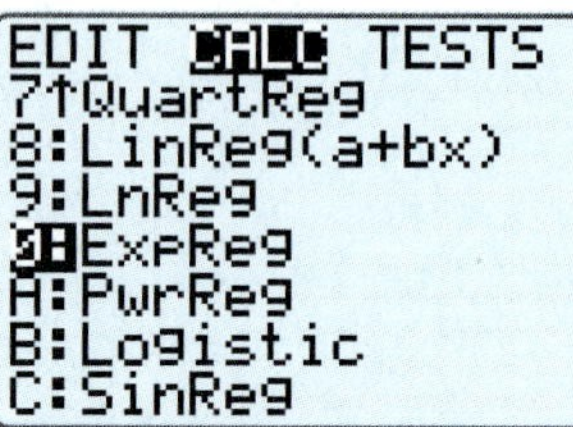

3. If the data to be evaluated are in L1 and L2, press [ENTER]. Otherwise, go to step 4. (If you want to automatically paste the regression equation into the [Y=] editor, press the key sequence [VARS] Y-VARS; 1:Function; 1:Y1 before pressing [ENTER].)

(Continued)

4. If the data to be evaluated are in L1 and L3, press the key sequence [2nd] [1] [,] [2nd] [3] to place the entries L1 and L3 on the home screen.

```
ExpReg L1, L3
```

5. Press [ENTER] to display the exponential model. (If you want to automatically paste the regression equation into the [Y=] editor, press the key sequence [VARS] Y-VARS; 1:Function; 1:Y1 before pressing [ENTER].)

```
ExpReg
 y=a*b^x
 a=60.8078408
 b=.9285201575
 r²=.9938920452
 r=-.9969413449
```

Finding an Exponential Function from a Verbal Description

Much of the information we encounter in the media is given to us in terms of percentages. Consider these typical news headlines. "Gasoline Prices Flare Up 9 Percent" and "Home Values Increase by 6 Percent." Anything that increases or decreases at a constant percentage rate may be modeled with an exponential function. The value of the growth factor b is $1 + r$, where r is the decimal form of the percentage (i.e., 5 percent = 0.05). Note that if $r < 0$, then $b < 1$. A negative rate of growth is referred to as **depreciation,** while a positive rate of growth is referred to as **appreciation.**

ANNUAL GROWTH RATE

Let $y = ab^t$ model the amount of a quantity at time t years. The annual growth rate r of the quantity y is given by $r = b - 1$. Similarly, the annual growth factor is $b = r + 1$.

For example, the author ordered an appraisal of his home in August 2004. In her report, the appraiser asserted that home values in the area were increasing at a rate of 5 percent per year. In other words, $r = 0.05$. The corresponding growth factor is $b = 1.05$. To forecast the value of his home in August 2005, the author multiplied the appraised value by the growth factor.

EXAMPLE 4

Using an Exponential Model to Forecast a Car's Value

New cars typically lose 50 percent of their value in the first three years. Many cars lose more than 25 percent of their value in the first year alone. (**Source:** Runzheimer International.) Depreciating approximately 13.4 percent annually, the Saturn

Sedan 2 holds its value better than most vehicles. Estimate the value of a \$12,810 Saturn Sedan 2 three years after it is purchased.

SOLUTION Since the car is depreciating at a constant percentage rate, an exponential function may be used to model the data. The initial value of the car is \$12,810, so $a = 12{,}810$. The car is losing value at 13.4 percent per year, so $r = -0.134$. Since $b = 1 + r$, the exponential function is

$$\begin{aligned} V(t) &= 12{,}810(b)^t \\ &= 12{,}810[1 + (-0.134)]^t \quad \text{Since } b = 1 + (-0.134) \\ &= 12{,}810(0.866)^t \text{ dollars} \end{aligned}$$

where t is the age of the car in years.

We want to know the value of the car after three years, so $t = 3$.

$$\begin{aligned} V(3) &= 12{,}810(0.866)^3 \\ &= 12{,}810(0.649) \\ &= 8320 \end{aligned}$$

We estimate that the car will be valued at \$8320 when it is three years old.

EXAMPLE 5 Using an Exponential Function to Model Inflation

Inflation (rising prices) causes money to lose its buying power. In the United States, inflation hovers around 3 percent annually.

(a) If a candy bar costs \$0.59 today, what will it cost 10 years from now?
(b) When will a candy bar cost \$1.00?

SOLUTION Since we're assuming a constant percentage increase in the price of the candy bar, an exponential model should be used. The initial value is \$0.59, so $a = 0.59$. The price of the candy bar is increasing by 3 percent annually, so $r = 0.03$. The cost of the candy bar is given by

$$\begin{aligned} C(t) &= 0.59(1 + 0.03)^t \text{ dollars} \\ &= 0.59(1.03)^t \text{ dollars} \end{aligned}$$

where t is the number of years from today.

(a) Evaluating the function at $t = 10$, we get

$$\begin{aligned} C(10) &= 0.59(1.03)^{10} \\ &= 0.59(1.344) \\ &= 0.79 \end{aligned}$$

We estimate that the candy bar will cost \$0.79 ten years from now.

(b) We want to know when $C(t) = 1$.

$$\begin{aligned} 1 &= 0.59(1.03)^t \\ \frac{1}{0.59} &= 1.03^t \\ 1.695 &= 1.03^t \end{aligned}$$

At this point, we are stuck. We haven't yet developed the mathematical machinery (logarithms) to get the variable out of the exponent. We can, however, attack the problem graphically by breaking the equation $1.695 = 1.03^t$ into two separate functions: $y_1 = 1.695$ and $y_2 = 1.03^t$. By graphing these functions simultaneously, we can determine the point of intersection of the two functions using our graphing calculator.

Using the graphing calculator, we generate the graph in Figure 5.41 and determine the point of intersection.

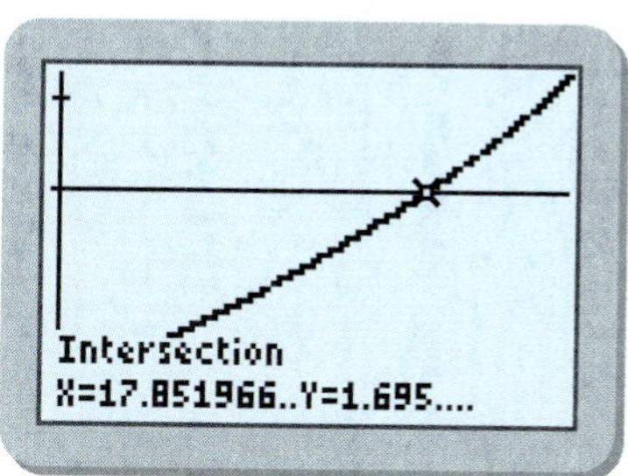

FIGURE 5.41

The functions intersect when $t = 17.85$. In about 18 years, we estimate that the cost of a candy bar will be \$1.00.

We also could have used the guess-and-check method to approximate the result by evaluating $C(t) = 0.59(1.03)^t$ at various values of t. Since

$$C(17) = 0.59(1.03)^{17} = 0.9752 \quad \text{and} \quad C(18) = 0.59(1.03)^{18} = 1.004$$

we determine that the price reaches \$1.00 between the seventeenth and eighteenth years.

5.3 Summary

In this section, you learned how exponential functions can be used to model rapidly increasing (or decreasing) data sets. You graphed exponential functions and developed exponential models from tables of data and verbal descriptions.

5.3 Exercises

In Exercises 1–10, do the following:

(a) Determine if the graph of the function is increasing or decreasing.

(b) Determine if the graph of the function is concave up or concave down.

(c) Identify the coordinate of the y-intercept.

(d) Graph the function to verify your conclusions.

1. $y = 4(0.25)^x$

2. $y = -2(0.5)^x$

3. $y = 0.5(2)^x$

4. $y = 6(0.1)^x$

5. $y = 0.4(5)^x$

6. $y = -0.1(0.2)^x$

7. $y = -1.2(2.3)^x$

8. $y = 5(0.4)^x$

9. $y = 3(0.9)^x$

10. $y = -5(3)^x$

In Exercises 11–20, use algebraic methods to find the equation of the exponential function that fits the data in the table.

11.

x	y
0	2
1	6
2	18
3	54

12.

x	y
0	3
1	12
2	48
3	192

13.

x	y
1	10
2	20
3	40
4	80

14.

x	y
1	10
2	50
3	250
4	1,250

15.

x	y
2	16
4	4
6	1
8	0.25

16.

x	y
2	100
4	1
6	0.01
8	0.0001

17.

x	y
1	1
2	4
4	64
5	256

18.

x	y
2	4
5	32
9	512
11	2,048

19.

x	y
0	256
5	8
7	2
10	0.25

20.

x	y
0	10,000
2	900
5	24.3
7	2.187

In Exercises 21–25, use exponential regression to model the data in the table. Use the model to predict the value of the function when $t = 25$, and interpret the real-world meaning of the result.

The Consumer Price Index is used to measure the increase in prices over time. In each of the following tables, the index is assumed to have the value 100 in the year 1984.

21. **Dental Prices**

Price of Dental Services

Years Since 1980 (t)	Price Index (I)
0	78.9
5	114.2
10	155.8
15	206.8
20	258.5

Source: *Statistical Abstract of the United States, 2001,* Table 694, p. 455.

 **22.** **Wine Prices**

Price of Wine Consumed at Home

Years Since 1980 (t)	Price Index (I)
0	89.5
5	100.2
10	114.4
15	133.6
20	151.6

Source: *Statistical Abstract of the United States, 2001,* Table 694, p. 455.

  **23.**  **Television Prices**

Price of a Television Set

Years Since 1980 (t)	Price Index (I)
0	104.6
5	88.7
10	74.6
15	68.1
20	49.9

Source: *Statistical Abstract of the United States, 2001,* Table 694, p. 455.

 **24.**  **Alcohol Prices**

Price of Distilled Spirits Consumed at Home

Years Since 1980 (t)	Price Index (I)
0	89.8
5	105.3
10	125.7
15	145.7
20	162.3

Source: *Statistical Abstract of the United States, 2001,* Table 694, p. 455.

 **25.**  **Entertainment Prices**

Price of Admission to Entertainment Venues

Years Since 1980 (t)	Price Index (I)
0	83.8
5	112.8
10	151.2
15	181.5
20	230.5

Source: *Statistical Abstract of the United States, 2001,* Table 694, p. 455.

In Exercises 26–30, find the exponential function that fits the verbal description.

26. **Salaries** Instructors' salaries are \$45,000 per year and are expected to increase by 3.5 percent annually. What will instructors' salaries be five years from now?

27. **Savings Account** A savings account balance is currently \$235 and is earning 2.32 percent per year. When will the balance reach \$250?

28. **Television Price** The cost of a 27-inch flat-screen television in 2002 was \$599.99. (**Source:** www.bestbuy.com.) Television prices are expected to decrease by 16 percent per year. How much is the flat-screen television expected to cost in 2007?

29. **Concert Admission** Reserved seating at a Dave Matthews Band concert at the Gorge Amphitheater in George, Washington, cost \$59.90 in September 2004. (**Source:** www.ticketmaster.com.) If admission fees for concerts are expected to increase by 28 percent per year, what is the expected price of a Dave Matthews Band ticket in September 2006?

30. **Tuition** Tuition is currently \$2024 per year and is increasing by 12 percent annually. In how many years from now will tuition reach \$3567?

In Exercises 31–35, find the solution by solving the equation graphically.

 **31.** $(1.2)^x = 5$

 **32.** $(0.9)^x = 0.5$

 **33.** $(1.05)^x = 2$

 **34.** $(1.12)^x = 2$

 **35.** $(0.3)^x = 0.09$

Exercises 36–40 are intended to challenge your understanding of exponential functions.

36. **Retirement Investments** An investor has $3000 to invest in two different investment accounts: CREF Social Choice and TIAA Retirement Annuity. Based on data from January 1, 1993, through December 31, 2002, the average annual return on the CREF Social Choice account was 8.67 percent, and the average annual return on the TIAA Retirement Annuity was 6.93 percent. (**Source:** TIAA-CREF.) The investor plans to invest twice as much in the Retirement Annuity as in the Social Choice account. Assuming that the future rates of return will be the same as the past rates of return, do the following:

(a) Find the equation of the function for the predicted value of the $3000 investment *t* years from now.

(b) What is the predicted value of the investment 20 years from now?

37. **Retirement Investments** Repeat Exercise 36, except this time invest twice as much in the Social Choice account as in the Retirement Annuity. What factors should the investor consider before changing the amount invested in each of the accounts?

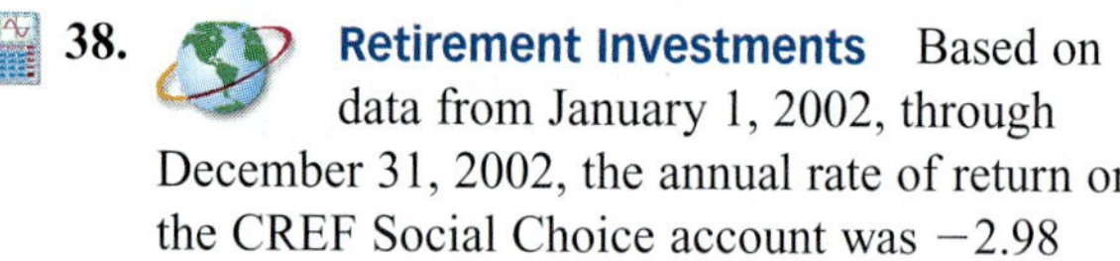

38. **Retirement Investments** Based on data from January 1, 2002, through December 31, 2002, the annual rate of return on the CREF Social Choice account was −2.98 percent, substantially lower than the 10-year average return of 8.67 percent. Over the same period, the CREF Stock account earned −20.73 percent, substantially lower than its 10-year average return of 7.69 percent. (**Source:** TIAA-CREF.)

An investor invested $1000 in each of the two accounts at the start of 2002.

(a) What was the value of the investment in each account at the end of 2002?

(b) Assuming that in each year following 2002, the investments earned an annual return equal to the 10-year average, predict how long it will take for the combined value of the accounts to reach $2000.

 **39.**  **Retirement Investments** Based on data from January 1, 2002, through December 31, 2002, the annual rate of return of the CREF Growth account was −30.06 percent, substantially lower than the five-year average return of −5.48 percent. Over the same period, the CREF Inflation-linked Bond account earned 16.32 percent, substantially higher than its five-year average return of 8.33 percent. (**Source:** TIAA-CREF.)

An investor invested $1000 in each of the two accounts at the start of 2002.

(a) What was the value of the investment in each account at the end of 2002?

(b) Assuming that in each year following 2002 the investments earned an annual return equal to the five-year average, predict how long it will take for the combined $2000 investment in the accounts to double.

 **40.** **Investments** How long will it take for a $1000 investment earning 10 percent annually to equal the value of a $2000 investment earning 6 percent annually?

5.4 Logarithmic Function Models

- Use nonlinear functions to model real-life phenomena
- Interpret the meaning of mathematical models in their real-world context
- Graph logarithmic functions
- Model real-life data with logarithmic functions
- Apply the rules of logarithms to simplify logarithmic expressions
- Solve logarithmic equations

GETTING STARTED In Section 5.3, we showed that the NBA players' average salary could be modeled by

$$S(t) = 161.4(1.169)^t \text{ thousand dollars}$$

where t is the number of years since 1980. We could use the model to estimate the average salary in any year t. But what if we wanted to algebraically determine the year in which the salary would be $1,000,000?

In this section, we will demonstrate how logarithmic functions can be used to model data sets that are increasing at a decreasing rate. We will show how to graph logarithmic functions and will illustrate graphically the inverse relationship between logarithmic and exponential functions. We will explain how the rules of logarithms can be used to simplify logarithmic expressions and solve logarithmic equations. We will also show how to develop logarithmic models from tables of data, such as NBA players' salaries, by using logarithmic regression.

We want to determine the year algebraically in which the NBA players' average salary reached $1 million. We can estimate the year by looking at the data in Table 5.19.

TABLE 5.19 NBA Players' Average Salary

Salary (thousands of dollars) (s)	Years Since 1980 (t)
170	0
325	5
750	10
1,900	15
2,000	16
2,200	17
2,600	18

Source: *Statistical Abstract of the United States, 2001*, Table 1324, p. 829.

Notice that *salary* is the independent variable and *years* is the dependent variable. The average salary was $750,000 in 1990 and $1,900,000 in 1995, so we know that $1,000,000 was reached sometime between 1990 and 1995.

Returning to the exponential model for the NBA players' average salary, we have

$$s(t) = 161.4(1.169)^t \text{ thousand dollars}$$

where t is the number of years since 1980. Recall that \$1,000,000 is equal to \$1000 thousand. We must determine when $s(t) = 1000$. In other words, we must solve the equation

$$1000 = 161.4(1.169)^t \quad \text{Since the units of the salary are thousands of dollars}$$

$$\frac{1000}{161.4} = \frac{161.4(1.169)^t}{161.4}$$

$$6.196 = 1.169^t$$

If we want to solve this equation algebraically, at this point we are stuck. We must come up with some way to get the t out of the exponent. Observe that t is the exponent that we place on 1.169 in order to get 6.196. We need to create a symbol to represent the phrase "is the exponent we place on." The chosen symbol, as bizarre as it may seem, is **log**.

$$\underbrace{t \text{ is}}_{t =} \quad \underbrace{\text{the exponent we place on}}_{\log} \quad \underbrace{1.169}_{1.169} \quad \underbrace{\text{in order to get } 6.196}_{6.196}$$

Compressing these terms into a single equation yields

$$t = \log_{1.169} 6.196$$

Using the calculator LOG button and some log rules that will be introduced later, we determine that $t \approx 11.7$ years. That is, between 1991 $(t = 11)$ and 1992 $(t = 12)$, the NBA players' average salary reached \$1 million.

Logarithmic Functions

The notation **log** is short for **logarithm.** It is helpful to think of **log** as representing the phrase "the exponent we place on." The equation $y = \log_b(x)$, which is read "y equals log base b of x," means "y is the exponent we place on b in order to get x." For example, $y = \log_2(8)$ means "y is the exponent we place on 2 in order to get 8." That is, $2^y = 8$. Since $2^3 = 8$, we conclude that $y = 3$.

EXAMPLE 1

Calculating a Logarithm

Find the value of y given $y = \log_3(9)$.

SOLUTION $y = \log_3(9)$ means "y is the exponent we place on 3 in order to get 9." That is, $3^y = 9$. Since $3^2 = 9$, we conclude that $y = 2$.

LOGARITHMIC FUNCTION

Let b and x be real numbers with $b > 0$ and $b \neq 1$ and $x > 0$. The function

$$y = \log_b(x)$$

is called a **logarithmic function.** The value b is called the **base** of the logarithmic function. We read the expression $\log_b(x)$ as "log base b of x."

The use of parentheses around the x in $y = \log_b(x)$ is optional. That is, $y = \log_b(x)$ is equivalent to $y = \log_b x$. However, we will consistently use parentheses throughout this text to remind ourselves that the expression "$\log_b$" is meaningless by itself.

Although any positive number not equal to 1 may be used as the base of a logarithmic function, the most commonly used bases are 10 and e. The irrational number e is approximately equal to 2.71828. The number e, like π, has many wonderful uses. Although we will see some of the power of e in this chapter, we will witness its full strength when we move into calculus. The number e is generated by calculating $\left(1 + \frac{1}{n}\right)^n$ for infinitely large n. You might think that $\left(1 + \frac{1}{n}\right)^n$ will get infinitely big as n goes to infinity. Surprisingly, $\left(1 + \frac{1}{n}\right)^n \approx 2.718281828$ for large values of n (see Table 5.20).

TABLE 5.20

n	$\left(1 + \frac{1}{n}\right)^n$
1	2.000
10	2.594
100	2.705
1,000	2.717
10,000	2.718
100,000	2.718

The value of e, 2.718281828. . . , may be accessed on the TI-83 Plus calculator by pressing the key sequence [2nd] and [÷].

The expression $\log_{10}(x)$ is often written $\log(x)$ and is called the **common log.** We will assume that the base is 10 when no base is written.

The expression $\log_e(x)$ is typically written $\ln(x)$ and is called the **natural log.**

Graphs of Logarithmic Functions

Like polynomial and exponential function graphs, logarithmic function graphs share many common characteristics. Consider the logarithmic functions $f(x) = \log_2(x)$, $g(x) = \log_3(x)$, and $h(x) = \log_4(x)$ (Figure 5.42).

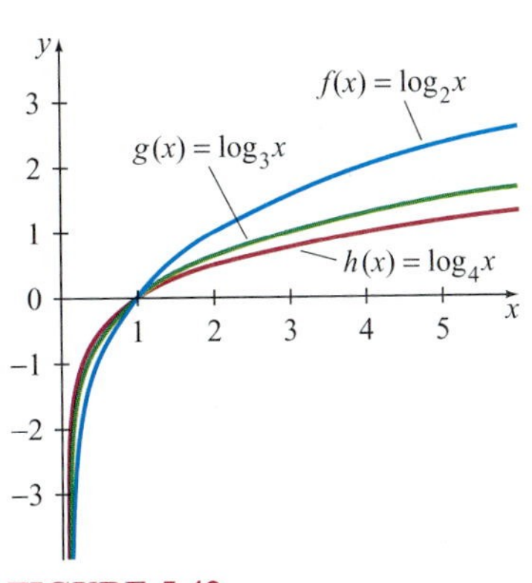

FIGURE 5.42

Each graph is concave down and increasing. Each graph passes through the point $(1, 0)$ and appears to have a vertical asymptote at $x = 0$ (the y axis). In fact, for all positive values of b not equal to 1, the graph of $y = \log_b(x)$ has a vertical asymptote at $x = 0$ and an x-intercept at $(1, 0)$. Logarithmic functions are undefined for $x < 0$. The shape of the graph is determined by b. The base b controls how rapidly the graph increases (or decreases). For each of the graphs in Figure 5.42, $b > 1$. What if $b < 1$? Consider the logarithmic functions $f(x) = \log_{0.3}(x)$, $g(x) = \log_{0.5}(x)$, and $h(x) = \log_{0.7}(x)$ (Figure 5.43).

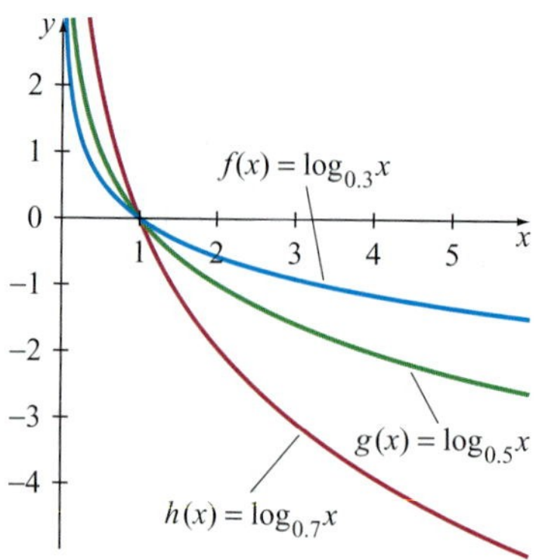

FIGURE 5.43

Each graph is concave up and decreasing. However, these graphs also have a vertical asymptote at $x = 0$ and pass through the point $(1,0)$. Our graphical observations are summarized in Table 5.21.

TABLE 5.21 Logarithmic Function Graphs: $y = \log_b(x)$

Value of b	Concavity	Increasing/Decreasing	Graph
$b > 1$	Concave down	Increasing	
$0 < b < 1$	Concave up	Decreasing	

EXAMPLE 2

Determining the Shape of a Logarithmic Function

Determine the concavity and increasing/decreasing behavior of $y = \log_2(x)$ and $y = \log_{0.5}(x)$. Then graph both functions to verify your results.

SOLUTION Since $2 > 1$, $y = \log_2(x)$ will be concave down and increasing. In order to graph the function by hand, we must generate a table of values for the function. You may find it helpful to write $y = \log_2(x)$ as $x = 2^y$, then select the values of y and calculate the values of x (Table 5.22). Plotting the points and connecting the dots yields the graph shown in Figure 5.44.

TABLE 5.22

$x = 2^y$	y
1	0
2	1
4	2
8	3
16	4

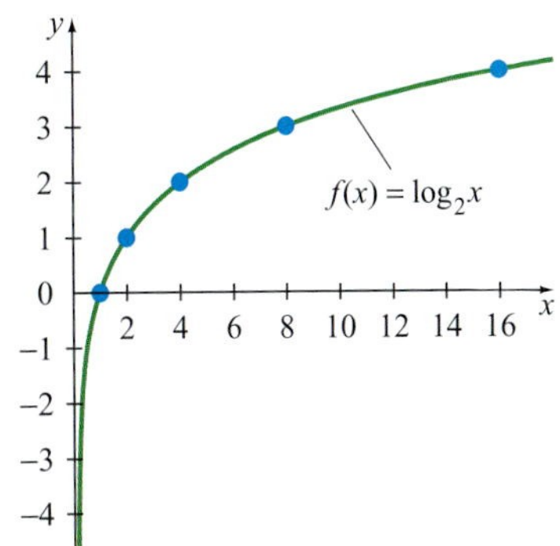

FIGURE 5.44

Since $0.5 < 1$, $y = \log_{0.5}(x)$ will be concave up and decreasing. You may find it helpful to write $y = \log_{0.5}(x)$ as $x = (0.5)^y$, then select the values of y and calculate the values of x (Table 5.23). Plotting the points and connecting the dots yields the graph shown in Figure 5.45.

TABLE 5.23

$x = (0.5)^y$	y
16	−4
8	−3
4	−2
2	−1
1	0

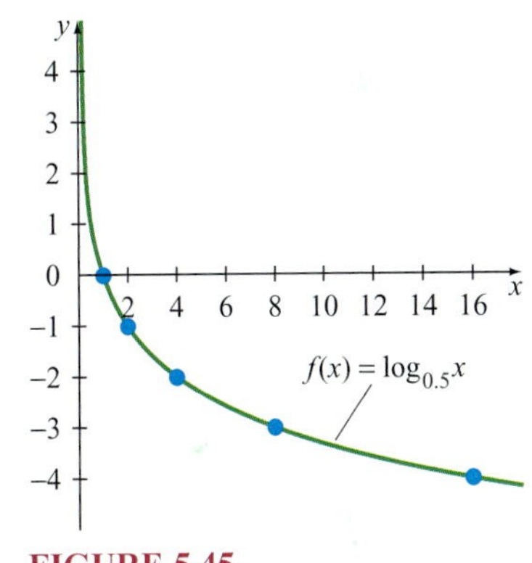

FIGURE 5.45

EXAMPLE 3

Comparing the Graphs of Logarithmic Functions

Graph $f(x) = \ln(x)$ and $g(x) = \log(x)$ simultaneously. Then determine where $f(x) < g(x)$ and where $g(x) > f(x)$.

SOLUTION

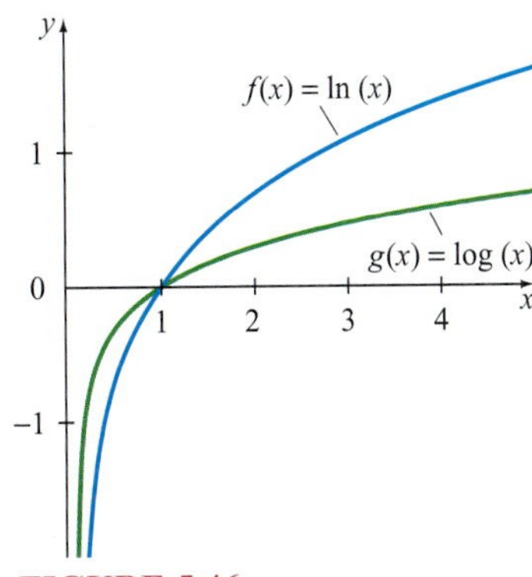

FIGURE 5.46

(To graph the function on the TI-83 Plus, use the LOG and LN keys as appropriate.) Based on the graphs in Figure 5.46, we conclude that for $x < 1$, $\ln(x) < \log(x)$. For $x > 1$, $\ln(x) > \log(x)$.

Relationship Between Logarithmic and Exponential Functions

Exponential and logarithmic functions are intimately related. Any logarithmic function may be rewritten as an exponential function (and vice versa).

LOGARITHMIC AND EXPONENTIAL FUNCTION RELATIONSHIP

For $b > 0$ with $b \neq 1$ and $x > 0$, the following statements are equivalent:

1. $y = \log_b(x)$

2. $b^y = x$

Consider tables of values for the functions $f(x) = 10^x$ and $g(x) = \log(x)$ (Table 5.24).

TABLE 5.24

x	$f(x) = 10^x$
−2	0.01
−1	0.1
0	1
1	10
2	100

x	$g(x) = \log(x)$
0.01	−2
0.1	−1
1	0
10	1
100	2

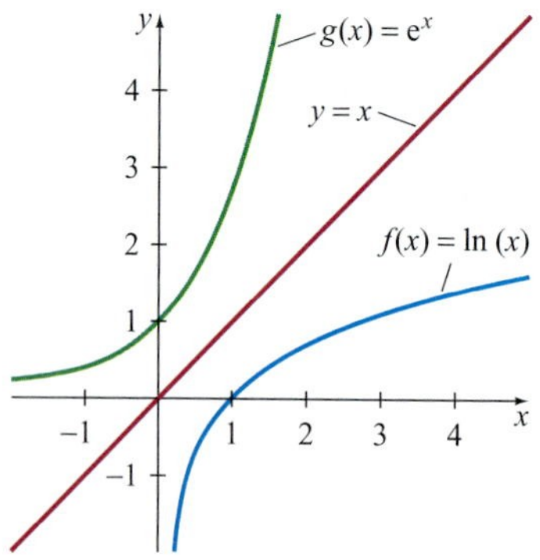

Notice that the *input* of the exponential function is the *output* of the logarithmic function. Similarly, the *output* of the exponential function is the *input* of the logarithmic function. A point (x_1, y_1) is on the graph of f if and only if the point (y_1, x_1) is on the graph of g. Functions with this property are called **inverse** functions. Exponential functions and logarithmic functions are inverse functions of each other. The inverse relationship of logarithmic and exponential functions may also be seen by graphing $y = \ln(x)$ and $y = e^x$ together with the graph of $y = x$ (Figure 5.47). The graph of $y = \ln(x)$ is the reflection of $y = e^x$ about the line $y = x$. That is, if we folded our paper along the line $y = x$, the graphs of the two functions would lie directly on top of each other.

FIGURE 5.47

When we solve a logarithmic function $y = \log_b(x)$ for x, we get the exponential function, $x = b^y$. This relationship between logarithmic and exponential functions will be critical to solving the logarithmic and exponential equations.

Rules of Logarithms

There are several rules that are used to manipulate logarithms. Each rule is important; however, we will use Rule 7 most extensively. It is so essential that we comically call it "the mother of all log rules."

RULES OF LOGARITHMS

If x, y, and n are real numbers with $x > 0$ and $y > 0$, then the following rules apply.

Common Logarithm	Natural Logarithm
1. $\log(1) = 0$	1. $\ln(1) = 0$
2. $\log(10) = 1$	2. $\ln(e) = 1$
3. $\log(10^x) = x$	3. $\ln(e^x) = x$
4. $10^{\log(x)} = x$	4. $e^{\ln(x)} = x$
5. $\log(xy) = \log(x) + \log(y)$	5. $\ln(xy) = \ln(x) + \ln(y)$
6. $\log\left(\frac{x}{y}\right) = \log(x) - \log(y)$	6. $\ln\left(\frac{x}{y}\right) = \ln(x) - \ln(y)$
7. $\log(x^p) = p\log(x)$	7. $\ln(x^p) = p\ln(x)$

In fact, Rules 1, 5, 6, and 7 apply for any base, not just 10 and e. Rules similar to Rules 2 through 4 may easily be formulated for any other base. In our applications, we will be using base 10 and base e predominantly. This is because the following change-of-base formula allows us to convert any logarithm into a logarithm with one of these two bases.

CHANGE-OF-BASE FORMULA

$$\log_b(a) = \frac{\log(a)}{\log(b)} = \frac{\ln(a)}{\ln(b)}$$

The change-of-base formula is easily derived.

$$\text{Let } y = \log_b(a)$$

$$b^y = a \qquad \text{Rewrite in exponential form}$$

$$\log(b^y) = \log(a) \qquad \text{Take the log of both sides}$$

$$y\log(b) = \log(a) \qquad \text{Apply log Rule 7}$$

$$y = \frac{\log(a)}{\log(b)} \qquad \text{Divide both sides by } \log(b)$$

$$\text{But } y = \log_b(a)$$

so

$$\log_b(a) = \frac{\log(a)}{\log(b)}$$

The ability to apply the log rules to manipulate expressions and equations is one of the critical skills required to be successful in solving exponential and logarithmic equations. To assist you in mastering those skills, we will do several "skill-and-drill" examples.

EXAMPLE 4 Using Log Rules to Simplify a Logarithmic Expression

Use the log rules to rewrite the logarithmic expression as a single term.

$$\log(3) - \log(27) + \log(12)$$

SOLUTION

$$\begin{aligned}\log(3) - \log(27) + \log(12) &= \log\left(\frac{3}{27}\right) + \log(12) && \text{Rule 6}\\ &= \log\left(\frac{3}{27} \cdot 12\right) && \text{Rule 5}\\ &= \log\left(\frac{36}{27}\right)\\ &= \log\left(\frac{4}{3}\right)\end{aligned}$$

EXAMPLE 5 Writing a Logarithmic Expression as a Single Logarithm

Rewrite the logarithmic expression as a single logarithm.

$$\log(x) + \log(x^3) - 4\log(x^2)$$

SOLUTION

$$\begin{aligned}&\log(x) + \log(x^3) - 4\log(x^2)\\ &\quad= \log(x) + 3\log(x) - 2 \cdot 4\log(x) && \text{Rule 7}\\ &\quad= 4\log(x) - 8\log(x) && \text{Combine like terms}\\ &\quad= -4\log(x) && \text{Combine like terms}\end{aligned}$$

The solution may also be written as

$$-\log(x^4),\ \log(x^{-4}),\ \text{or } \log\left(\frac{1}{x^4}\right)$$

When checking a solution with the answers in the back of the book, it is important to recognize that the correct solution may be written in a variety of forms.

EXAMPLE 6 Writing a Logarithmic Expression as a Single Logarithm

Rewrite the expression $2\log\left(\frac{2}{3x}\right) - \log(x^3)$ as a single logarithm.

SOLUTION

$$\begin{aligned}&2\log\left(\frac{2}{3x}\right) - \log(x^3)\\ &\quad= \log\left(\frac{2}{3x}\right)^2 - \log(x^3) && \text{Rule 7}\\ &\quad= \log\left[\frac{(2)^2}{(3x)^2}\right] + \log(x^3)^{-1} && \text{Rule 7, rules of exponents}\end{aligned}$$

$$= \log\left(\frac{4}{9x^2}\right) + \log(x^{-3}) \quad \text{Rules of exponents}$$

$$= \log\left(\frac{4}{9x^2} \cdot x^{-3}\right) \quad \text{Rule 5}$$

$$= \log\left(\frac{4}{9x^2} \cdot \frac{1}{x^3}\right) \quad \text{Rules of exponents}$$

$$= \log\left(\frac{4}{9x^5}\right) \quad \text{Rules of exponents and fractions}$$

EXAMPLE 7

Solving a Logarithmic Equation

Solve the equation $\log_3(x) = 4$ for x.

SOLUTION

$$\log_3(x) = 4$$

$$3^4 = x \quad \text{Rewrite in exponential form}$$

$$x = 81$$

EXAMPLE 8

Solving an Exponential Equation

Solve the equation $3^x = \frac{1}{12}$ for x.

SOLUTION

$$3^x = \frac{1}{12}$$

$$\ln(3^x) = \ln\left(\frac{1}{12}\right) \quad \text{Take the natural log of both sides}$$

$$x\ln(3) = \ln(1) - \ln(12) \quad \text{Rule 7 and Rule 6}$$

$$x\ln(3) = 0 - \ln(12) \quad \text{Rule 1}$$

$$x\ln(3) = -\ln(12)$$

$$x = -\frac{\ln(12)}{\ln(3)} \quad \text{Divided by } \ln(3)$$

$$x \approx -2.262$$

It is important to note that $-\frac{\ln(12)}{\ln(3)} \neq -\ln\left(\frac{12}{3}\right)$.

Logarithmic Models

Functions that are increasing at a decreasing rate may often be modeled by logarithmic functions. Using the logarithmic regression feature, **LnReg,** on our calculator returns a function of the form $y = a + b\ln(x)$. The a shifts the graph of $y = b\ln(x)$ vertically by $|a|$ units. The resultant graph passes through the point $(1, a)$ instead of $(1, 0)$.

EXAMPLE 9

Using Logarithmic Regression to Forecast the Population of Hawaii

TABLE 5.25 Projected Population of Hawaii

People (thousands) (p)	Years Since 1995 (T)
1,187	0
1,257	5
1,342	10
1,553	20
1,812	30

Source: www.census.gov.

Find the logarithmic model that best fits the data in Table 5.25. Then evaluate the function at $p = 1700$ and interpret your result.

SOLUTION Using the logarithmic regression feature on our calculator (as shown in the following Technology Tip), we get

$$y = -496.0 + 70.17\ln(x)$$

or, in terms of our variables,

$$T(p) = -496.0 + 70.17\ln(p)$$

Evaluating the function when $p = 1700$, we get

$$\begin{aligned} T(1700) &= -496.0 + 70.17\ln(1700) \\ &= -496.0 + 70.17(7.438) \\ &= -496.0 + 522.0 \\ &= 26.0 \end{aligned}$$

The population of Hawaii is projected to reach 1,700,000 in 2021 (26 years after 1995).

TECHNOLOGY TIP

Logarithmic Regression

1. Enter the data using the Statistics Menu List Editor. (Refer to Section 1.3 if you've forgotten how to do this.)

```
L1      L2      L3   3
1187    0
1257    5
1342    10
1553    20
1812    30
------  ------
L3(1) =
```

2. Bring up the Statistics Menu Calculate feature by pressing STAT and using the blue arrows to move to the CALC menu. Then select item 9:LnReg and press ENTER.

3. If you want to automatically paste the regression equation into the Y= editor, press the key sequence VARS Y-VARS; 1:Function; 1:Y1 and press ENTER. Otherwise, press ENTER.

```
LnReg
 y=a+blnx
 a=-496.0111246
 b=70.17413909
 r²=.9976542757
 r=-.9988264492
```

5.4 Summary

In this section, you learned how to graph logarithmic functions. You discovered the inverse relationship between logarithmic and exponential functions and practiced using rules of logarithms to simplify logarithmic expressions and solve logarithmic equations. You also developed logarithmic models from data tables by using logarithmic regression.

5.4 Exercises

In Exercises 1–5, solve the logarithmic equation for y without using a calculator.

1. $y = \log_5(25)$

2. $y = \log_3(81)$

3. $y = \log_2(64)$

4. $y = \log_2(4^{-1})$

5. $y = \log_3\left(\frac{1}{9}\right)$

In Exercises 6–10, determine the concavity and increasing/decreasing behavior of the graph of the function. Then graph the function. (Hint: You will have to use the change-of-base formula if you graph the function on your calculator. That is, $y = \log_b(x) \Rightarrow y = \frac{\ln(x)}{\ln(b)}$.)

6. $y = \log_2(x)$

7. $y = \log_4(x)$

8. $y = \log_{0.2}(x)$

9. $y = \log_{0.7}x$

10. $y = \log_{0.8}x$

In Exercises 11–20, use the inverse relationship between logarithmic and exponential functions to solve the equations for x. Simplify your answers.

11. $2^x = 64$

12. $5^x = 125$

13. $2^x = \frac{1}{2}$

14. $3^x = \frac{1}{9}$

15. $4^x = 64$

16. $\log_5(x) = 3$

17. $\log_4(x) = -2$

18. $\log_2(x) = 6$

19. $\log_3(x) = 5$

20. $\log_4(x) = \frac{1}{2}$

In Exercises 21–40, use the rules of logarithms to rewrite each expression as a single logarithmic expression.

21. $\log(2) + \log(8)$

22. $\log(2) - \log(8)$

23. $\log(2x) + \log(x^3)$

24. $\log(5x) + \log(20)$

25. $\log(j) + \log(a) + \log(m)$

26. $3\log(x^2) - 4\log(x^2) + \log(2x^3)$

27. $7\log(x^{-1}) - 4\log(3x^2) - \log(4x)$

28. $2\log(3x)^2 + \log(x)^{-12}$

29. $4\log\left(\frac{1}{x}\right) + 3\log(x)$

30. $3\log\left(\frac{2}{5x}\right) - \log(x^3)$

31. $\ln(x) - \ln\left(\frac{1}{x}\right)$

32. $2\ln(2x)^2 + \ln\left(\frac{1}{x^2}\right)$

33. $\ln(1) - \log(1) + \ln(x^2)$

34. $-2\ln\left(\frac{x}{2}\right) + \ln(2)$

35. $3\ln(3x^2) - \ln(x^6)$

36. $2\ln(3x) - \ln(3x^2) + 2\ln(3)$

37. $-\ln(4x^2)^3 + 2\ln(x^6) - 6\ln(x)$

38. $-\ln(3x)^2 - 3\ln(x^{-2}) - 5\ln(x)$

39. $4\ln(3x) + \ln(81x^2) - \ln(9x)$

40. $\ln(3x) - \ln(9x^2) + \ln(3) - \ln(x)$

In Exercises 41–45, use logarithmic regression to model the data in the table. Use the model to predict the value of the function when $i = 125$, and interpret the real-world meaning of the result.

The Consumer Price Index is used to measure the increase in prices over time. In each of the following tables, the index is assumed to have the value 100 in the year 1984.

 **41.**  **Dental Services**

Price of Dental Services

Price Index (i)	Years Since 1980 (t)
78.9	0
114.2	5
155.8	10
206.8	15
258.5	20

Source: *Statistical Abstract of the United States, 2001*, Table 694, p. 455.

 **42.**  **Price of Wine**

Price of Wine Consumed at Home

Price Index (i)	Years Since 1980 (t)
89.5	0
100.2	5
114.4	10
133.6	15
151.6	20

Source: *Statistical Abstract of the United States, 2001*, Table 694, p. 455.

 **43.**  **Price of a TV**

Price of a Television Set

Price Index (i)	Years Since 1980 (t)
104.6	0
88.7	5
74.6	10
68.1	15
49.9	20

Source: *Statistical Abstract of the United States, 2001*, Table 694, p. 455.

 **44.**  **Price of Alcohol**

Price of Distilled Spirits Consumed at Home

Price Index (i)	Years Since 1980 (t)
89.8	0
105.3	5
125.7	10
145.7	15
162.3	20

Source: *Statistical Abstract of the United States, 2001*, Table 694, p. 455.

 **45.**  **Entertainment Admission Price**

Price of Admission to Entertainment Venues

Price Index (i)	Years Since 1980 (t)
83.8	0
112.8	5
151.2	10
181.5	15
230.5	20

Source: *Statistical Abstract of the United States, 2001*, Table 694, p. 455.

Exercises 46–50 are intended to challenge your understanding of logarithmic and exponential functions.

 **46.**  **Homes with VCRs** Given the following table, perform each of the tasks identified as **(a)** through **(c)**.

Percentage of TV Homes with a VCR

Years Since 1984 (t)	Homes with a VCR (percent) (V)
1	20.8
2	36.0
3	48.7
4	58.0
5	64.6
6	68.6
7	71.9
8	75.0
9	77.1
10	79.0
11	81.0
12	82.2
13	84.2
14	84.6

Source: *Statistical Abstract of the United States, 2001*, Table 1126, p. 705.

(a) Use logarithmic regression to find the logarithmic function that best fits the data.

(b) Solve the equation for the input variable, t.

(c) Determine the year in which 90 percent of TV homes were expected to have VCRs. Do you think this prediction will be accurate? Explain.

47. The domain of the function $y = \log_b(x)$ is the set of all positive real numbers. Explain why the negative real numbers and zero are not in the domain of the function.

48. If $b > c > 1$, for what values of x is $\log_b(x) > \log_c(x)$?

49. Does the equation $e^x = \ln(x)$ have a solution? Explain.

50. In order for the equation $\log_b(x) = 2\log_c(x)$ to be true for all positive values of x, what must be the relationship between b and c?

5.5 Choosing a Mathematical Model

- Use critical thinking skills in selecting a mathematical model

GETTING STARTED Many data sets may be effectively fitted with more than one mathematical model. Determining which model to use requires critical thinking. In this section, we will discuss strategies and techniques for choosing a mathematical model. However, before delving into model selection strategies, we will *briefly* introduce one other type of mathematical model: the logistic function. More detailed coverage of logistic functions is given in Chapter 14 of the *Finite Math and Applied Calculus* version of this text.

Logistic Functions

A logistic function graph is an s-shaped curve that is bounded above and below by horizontal asymptotes (Figures 5.48 and 5.49). It may be either an increasing function or a decreasing function.

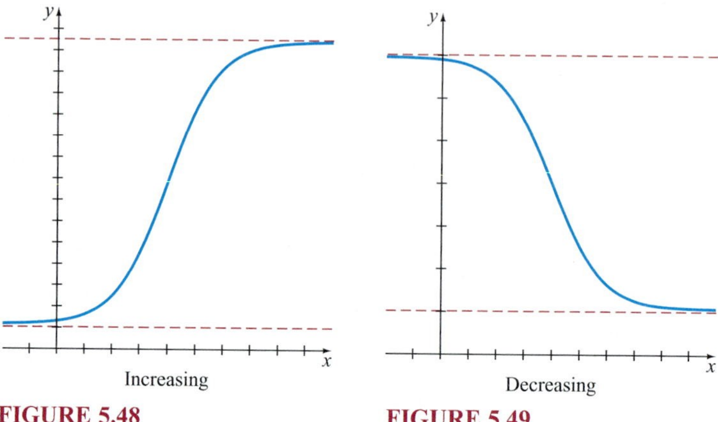

FIGURE 5.48

FIGURE 5.49

The equation of a logistic function is of the form

$$f(x) = \frac{c}{1 + ae^{-bx}} + k$$

where a, b, c, and k are all constants. Logistic functions exhibit slow growth, then rapid growth, followed by slow growth or slow decay, then rapid decay, followed by slow decay.

Selecting a Mathematical Model

Graphing calculators and computer spreadsheets are extremely useful tools for mathematical modeling. When you are given a table of data to model, use the following model selection strategies.

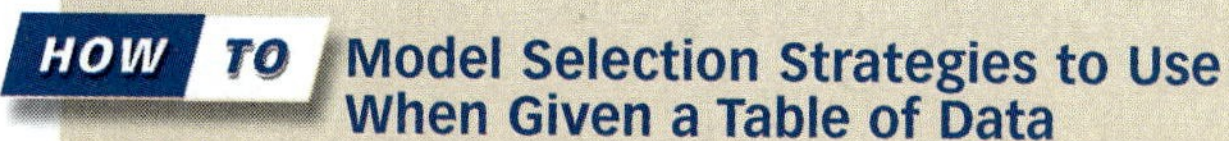

Model Selection Strategies to Use When Given a Table of Data

1. Draw a scatter plot.
2. Determine whether the scatter plot exhibits the behavior of the graph of one or more of the standard mathematical functions: linear, quadratic, cubic, quartic, exponential, logarithmic, or logistic.
3. Find a mathematical model for each function type selected in Step 2.
4. Use all available information to anticipate the expected behavior of the thing being modeled outside of the data set. Eliminate models that don't exhibit the expected behavior. (Sometimes it is convenient to switch the order of Steps 3 and 4.)
5. Choose the simplest model from among the models that meet your criteria.

In accomplishing Step 2, it is helpful to recognize key graphical features exhibited by the data set. Often we can eliminate one or more model types by noting the concavity of the scatter plot. Table 5.26 identifies key graphical features to look for.

TABLE 5.26

Function Type	Model Equation	Key Graphical Features
Linear	$y = mx + b$	Straight line
Quadratic	$y = ax^2 + bx + c$	Concave up everywhere or concave down everywhere
Cubic	$y = ax^3 + bx^2 + cx + d$	Changes concavity exactly once; no horizontal asymptotes
Quartic	$y = ax^4 + bx^3 + cx^2 + dx + f$	Changes concavity zero or two times; no horizontal asymptotes
Exponential	$y = ab^x$ with $a > 0$	Concave up; horizontal asymptote at $y = 0$
Logarithmic	$y = a + b\ln(x)$ with $b > 0$	Concave down; vertical asymptote at $x = 0$
Logistic	$y = \dfrac{c}{1 + ae^{-bx}} + k$	Changes concavity exactly once; horizontal asymptotes at $y = k$ and $y = c + k$

EXAMPLE 1

Choosing a Mathematical Model to Forecast a Room Rate

As shown in Table 5.27, the average room rate for a hotel/motel increased between 1990 and 1999. Find a mathematical model for the data and forecast the average hotel/motel room rate for 2004.

TABLE 5.27 Average Hotel/Motel Room Rate

Years Since 1990 (t)	Room Rate (dollars) (R)
0	57.96
1	58.08
2	58.91
3	60.53
4	62.86
5	66.65
6	70.93
7	75.31
8	78.62
9	81.33

Source: *Statistical Abstract of the United States, 2001,* Table 1266, p. 774.

SOLUTION We first draw a scatter plot of the data on our graphing calculator (see Figure 5.50).

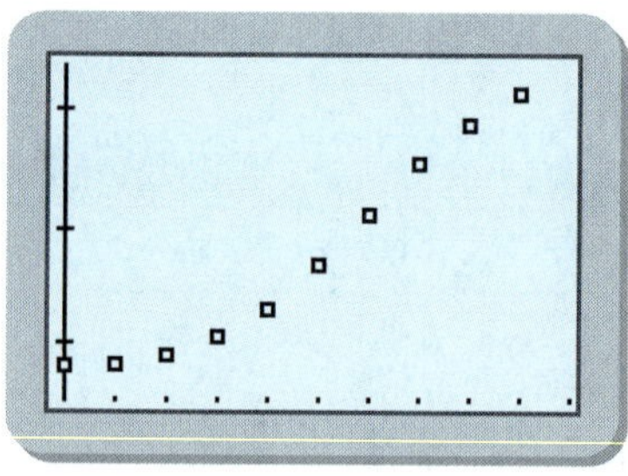

FIGURE 5.50

The graph is initially concave up, but it changes concavity near $t = 6$. Since the graph changes concavity exactly once, a cubic or a logistic model (Figure 5.51) may fit the data.

R as a function of t

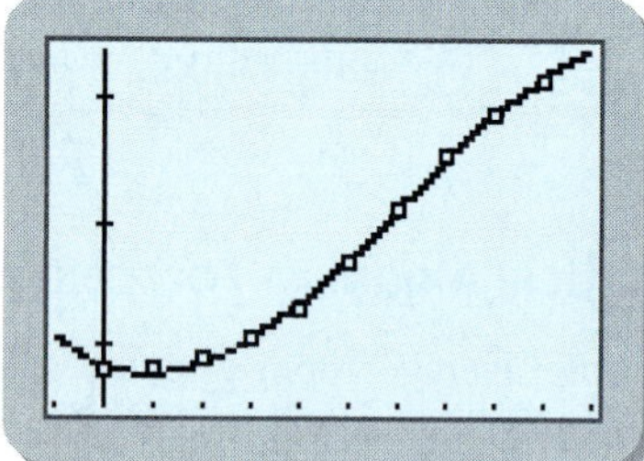

$$R(t) = -0.04815t^3 + 0.8868t^2 - 1.495t + 58.32$$

R as a function of t

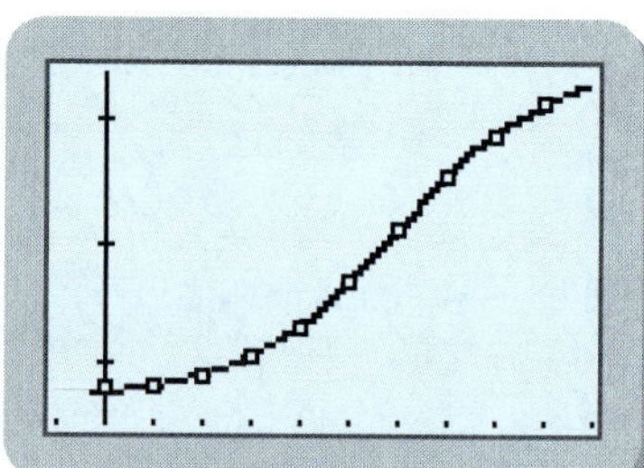

$$R(t) = \frac{27.80}{(1 + 47.23e^{-0.6425t})} + 57$$

FIGURE 5.51

Both the cubic model and the logistic model appear to fit the data extremely well. Which model will be best for forecasting R in 2004 ($t = 14$)? To answer this question, we extend the viewing rectangle for each graph to include the interval $[-1, 15]$ (see Figure 5.52).

R as a function of t

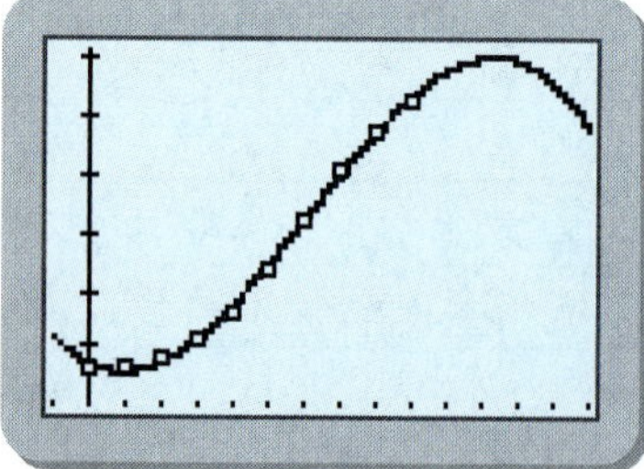

$$R(t) = -0.04815t^3 + 0.8868t^2 - 1.495t + 58.32$$

R as a function of t

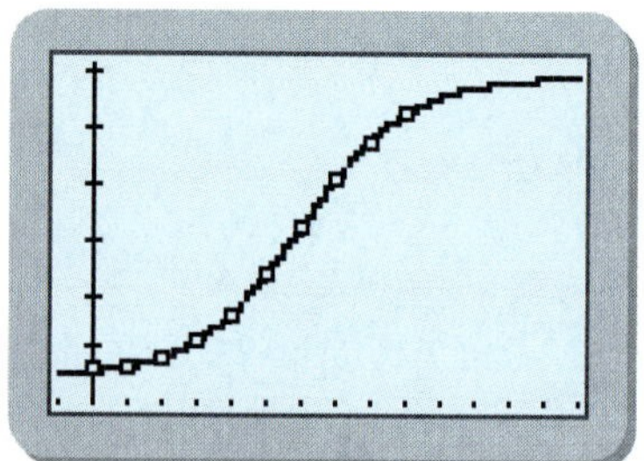

$$R(t) = \frac{27.80}{(1 + 47.23e^{-0.6425t})} + 57$$

FIGURE 5.52

The cubic model forecasts a decrease in the room rate by 2004 ($t = 14$). According to the cubic model, the rate will drop to $79.07. Since room rates have increased every year since 1990, it is unlikely that the room rate will drop at any point in the future.

The logistic model forecasts a leveling off of the room rate. According to the logistic model, the 2004 room rate will be $84.64. This forecast is likely to be more accurate than the cubic model projection.

It is important to note that there is a certain degree of uncertainty when picking a model to use from a group of models that seem to fit the data. It is possible that two different people may select different models as the "best" model. For this reason, it is important to always explain the reasoning behind the selection of a particular model.

EXAMPLE 2 Choosing a Mathematical Model to Forecast a Movie Ticket Price

As shown in Table 5.28, the average price of a movie ticket increased between 1975 and 1999. Find a mathematical model for the data and forecast the average price of a movie ticket in 2004.

TABLE 5.28

Years Since 1975 (t)	Movie Ticket Price (dollars) (P)
0	2.05
5	2.69
10	3.55
15	4.23
20	4.35
24	5.08

Source: *Statistical Abstract of the United States, 2001*, Table 1244, p. 761.

SOLUTION We first draw the scatter plot (Figure 5.53).

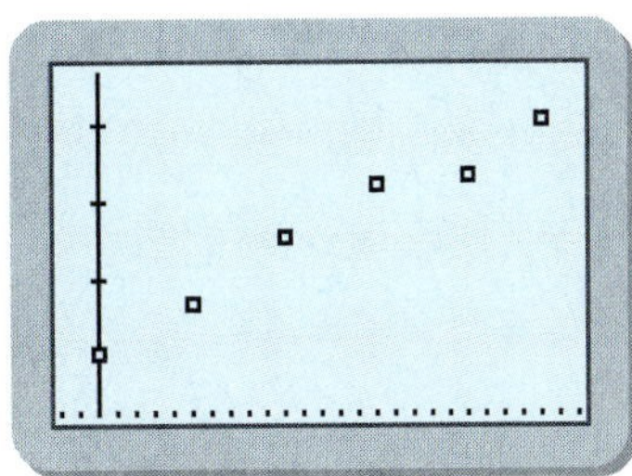

FIGURE 5.53

The first four data points appear to be somewhat linear, so our initial impression is that a linear model may fit the data well. However, the fifth data point is not aligned with the first four, so we know that the data set isn't perfectly linear. However, a linear model may still fit the data fairly well.

Alternatively, we may look at the scatter plot and conclude that the function is concave down on $[0, 15]$ and concave up on $[15, 24]$. Since the scatter plot appears to change concavity once and does not have any horizontal asymptotes, a cubic model may work well. Figure 5.54 shows the two models.

P as a function of t

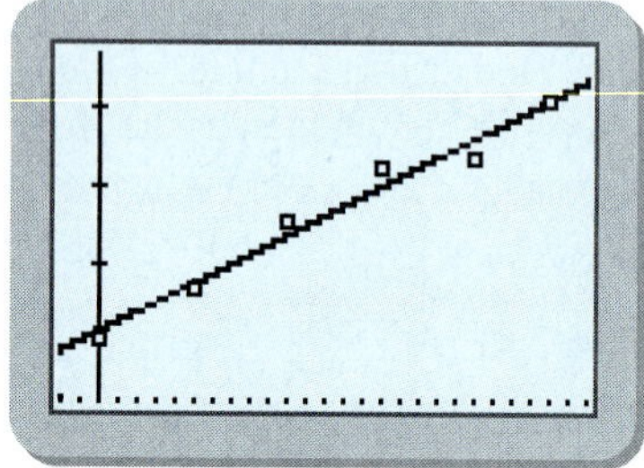

$P(t) = 0.1224t + 2.148$

P as a function of t

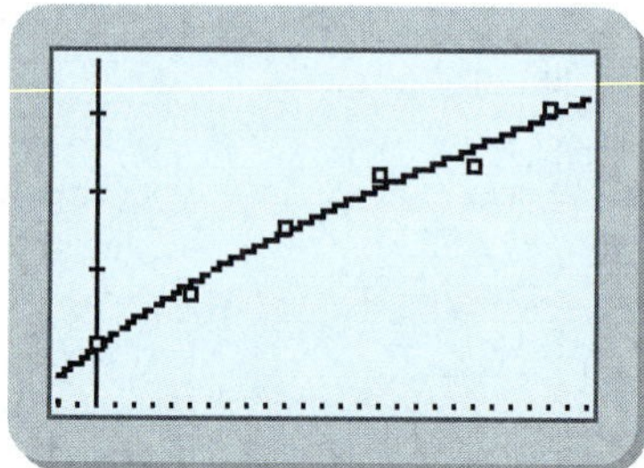

$P(t) = 0.00007426t^3 - 0.004278t^2 + 0.1846t + 2.001$

FIGURE 5.54

In 2004, $t = 29$. Evaluating each function at $t = 29$ yields the following:

$$\begin{aligned} P(29) &= 0.1224(29) + 2.148 \\ &\approx \$5.70 \end{aligned}$$

$$\begin{aligned} P(29) &= 0.00007426(29)^3 - 0.004278(29)^2 + 0.1846(29) + 2.001 \\ &\approx \$5.57 \end{aligned}$$

It is difficult to know which of these estimates of the average price of a movie ticket is more accurate. The models seem to fit the data equally well. Furthermore, we have no other information that leads us to believe that one model would be better than the other. Therefore, we pick the simpler model: $P(t) = 0.1224t + 2.148$.

Certain verbal descriptions often hint at the particular mathematical model to use. By watching for key phrases, we can narrow the model selection process. Table 5.29 presents some examples of typical phrases, their interpretation, and possible models.

TABLE 5.29 Finding a Mathematical Model from a Verbal Description

Phrase	Interpretation	Possible Model
Salaries are projected to increase by $800 per year for the next several years.	The graph of the salary function will have a constant slope, $m = 800$.	Linear $S(t) = 800t + b$
The company's revenue has decreased by 5 percent annually for the past six years.	Since the annual decrease is a *percentage* of the previous year's revenue, the dollar amount by which revenue decreases annually is decreasing. The revenue graph is decreasing and concave up.	Exponential $R(t) = a(1 - 0.05)^t = a(0.95)^t$
Product sales were initially slow when the product was introduced, but sales increased rapidly as the popularity of the product increased. Sales are continuing to increase, but not as quickly as before.	The graph of the sales function may have a horizontal asymptote at or near $y = 0$ and a horizontal asymptote slightly above the maximum projected sales amount.	Logistic
Enrollments have been dropping for years. Each year we lose more students than we did the year before.	The graph of the enrollment function is decreasing, since enrollments are dropping. It will also be concave down, since the rate at which students are dropping continues to increase in magnitude.	Quadratic
Company profits increased rapidly in the early 1990s but leveled off in the late 1990s. In the early 2000s, profits again increased rapidly.	The graph of the profit function increases rapidly, then levels off, then increases rapidly again.	Cubic

EXAMPLE 3

Choosing a Mathematical Model to Forecast Coca-Cola Production

In its 2001 Annual Report, the Coca-Cola Company reported the following:

> Our worldwide unit case volume increased 4 percent in 2001, on top of a 4 percent increase in 2000. The increase in unit case volume reflects consistent performance across certain key operations despite difficult global economic conditions. Our business system sold 17.8 billion unit cases in 2001.
>
> (**Source:** Coca-Cola Company 2001 Annual Report, p. 46.)

Find a mathematical model for the unit case volume of the Coca-Cola Company.

SOLUTION Since the unit case volume is increasing at a constant percentage rate (4 percent), an exponential model will fit the data well. Furthermore, the initial number of unit cases sold was 17.8 billion.

$$\begin{aligned} V(t) &= ab^t \text{ where } t \text{ is the number of years since the end of 2001} \\ &= 17.8(1 + 0.04)^t \text{ billion unit cases} \\ &= 17.8(1.04)^t \text{ billion unit cases} \end{aligned}$$

The mathematical model for the unit case volume is $V(t) = 17.8(1.04)^t$.

Sometimes a data set may not be effectively modeled by any of the aforementioned functions. In these cases, we look to see if we can model the data with a piecewise function. That is, we use one function to model a portion of the data and a different function to model another portion of the data. Even with this approach, there are some data sets (e.g., daily stock prices) that can rarely be modeled by one of the standard functions.

EXAMPLE 4

Finding a Piecewise Model for AIDS Deaths in the United States

From 1981 to 1995, the number of adult and adolescent AIDS deaths in the United States increased dramatically. However, from 1995 to 2001, the annual death rate plummeted, as shown in Table 5.30.

TABLE 5.30 Adult and Adolescent AIDS Deaths in the United States

Years Since 1981 (t)	Number of Deaths During Year (D)
0	122
1	453
2	1,481
3	3,474
4	6,877
5	12,016
6	16,194
7	20,922
8	27,680
9	31,436
10	36,708

(Continued)

(Continued)

Years Since 1981 (t)	Number of Deaths During Year (D)
11	41,424
12	45,187
13	50,071
14	50,876
15	37,646
16	21,630
17	18,028
18	16,648
19	14,433
20	8,963

Source: Centers for Disease Control and Prevention, "HIV/AIDS Surveillance Report," December 2001, p. 30.

Find the mathematical model that best models the data and forecast the number of adult and adolescent AIDS deaths in 2004.

SOLUTION We first draw a scatter plot of the data (Figure 5.55).

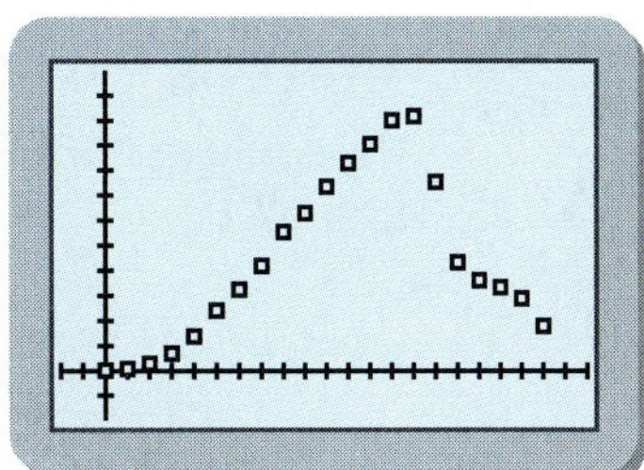

FIGURE 5.55

The data set appears to exhibit logistic behavior up until 1995. After 1995, the graph appears to exhibit cubic behavior. We will use a piecewise function to model the data set. We determine each of the model pieces by using logistic and cubic regression. (Although the data point associated with $t = 14$ was used in finding both pieces of the model, we must assign the domain value $t = 14$ to one piece or the other. We choose to assign $t = 14$ to the logistic piece of the model.)

$$P(t) = \begin{cases} \dfrac{53{,}955}{1 + 38.834e^{-0.45127t}} & 0 \le t \le 14 \\ -381.06t^3 + 20{,}770t^2 - 379{,}469t + 2{,}339{,}211 & t > 14 \end{cases}$$

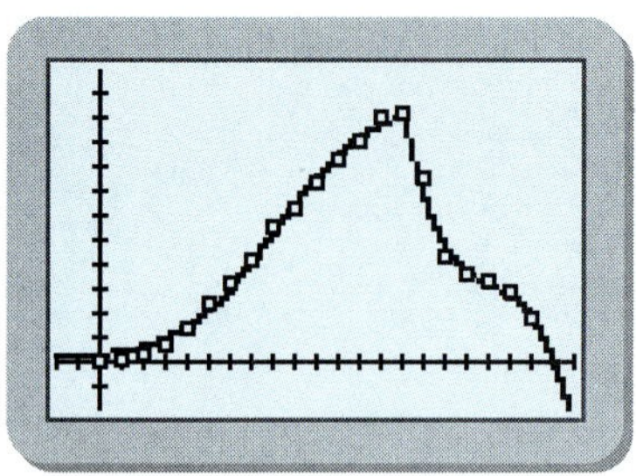

FIGURE 5.56

Our piecewise model (Figure 5.56) appears to fit the data set very well. We are asked to forecast the number of AIDS deaths in 2004 ($t = 23$). Since $23 > 14$, we will use the second function in the piecewise model.

$$P(23) = -381.06(23)^3 + 20{,}770(23)^2 - 379{,}469(23) + 2{,}339{,}211$$
$$= -37{,}603 \text{ adult and adolescent AIDS deaths}$$

It is impossible to have a negative number of deaths! Despite the fact that the model fit the data well, using the model to forecast the 2004 mortality rate yielded an unreasonable result. Returning to the data set, we estimate that the number of AIDS deaths in the years beyond 2001 will be somewhere between 0 and 8963 (the 2001 figure).

If a data set cannot be effectively modeled by one of the standard mathematical functions or a piecewise function, we may conclude that we don't know how to effectively model the data. In such cases, we may estimate a future result by identifying a range of seemingly reasonable values.

EXAMPLE 5

Analyzing Data Not Easily Modeled with a Common Function

The number of firearms detected during airport passenger screening is shown in Table 5.31.

TABLE 5.31

Years Since 1980 (t)	Firearms Detected (F)
0	1,914
5	2,913
10	2,549
11	1,644
12	2,608
13	2,798

Years Since 1980 (t)	Firearms Detected (F)
14	2,994
15	2,390
16	2,155
17	2,067
18	1,515
19	1,552

Source: *Statistical Abstract of the United States, 2001*, Table 1062, p. 669.

Estimate the number of firearms detected by airport screeners in 2002.

SOLUTION

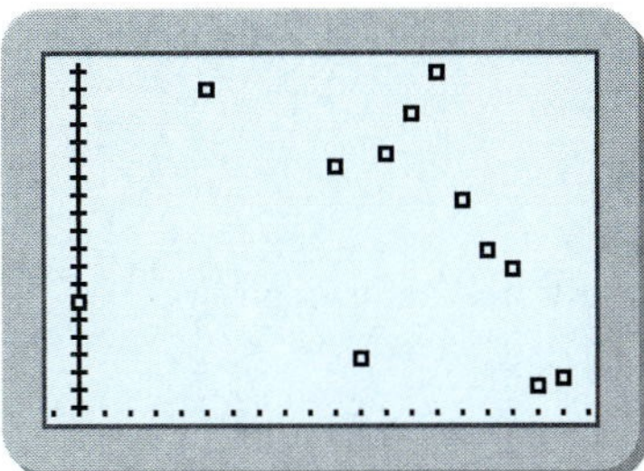

FIGURE 5.57

The scatter plot (Figure 5.57) does not resemble any of the standard mathematical functions. It also does not appear that a piecewise model will fit the data well. The number of firearms detected in a given year appears to be somewhat random, ranging from about 1500 firearms to roughly 3000 firearms.

After September 11, 2001, airline screening became much more thorough. The increased security may be having a deterrent effect. According to the Bureau of Transportation Statistics (www.bts.gov), there were 1071 firearms detected in 2001. As of March 2004, airline screening data for 2002 had not yet been published on the bureau's web site. We estimate that the number of firearms detected in 2002 will be in the 1000–2000 range. Our estimate is based in part on the additional data we discovered through research.

5.5 Summary

In this section, you learned strategies and techniques for selecting mathematical models. You also discovered that oftentimes more than one function may be used to model the same data set.

5.5 Exercises

In Exercises 1–10, find the equation of the mathematical model (if possible) that you believe will most accurately forecast the indicated result. Justify your conclusions.

1. **Gaming Hardware Sales**

Electronic Gaming Hardware Factory Sales

Years Since 1990 (t)	Sales (millions of dollars) (S)
0	975
1	1,275
2	1,575
3	1,650
4	1,575
5	1,500
6	1,600
7	1,650
8	1,980
9	2,250

Source: *Statistical Abstract of the United States, 2001*, Table 1005, p. 634.

Forecast gaming hardware sales for 2002.

2.  **Number of Farms**

Years Since 1978 (t)	Farms (thousands) (f)
0	1,015
4	987
9	965
14	946
19	932

Source: *Statistical Abstract of the United States, 2001*, Table 803, p. 523.

Forecast the number of farms in 2000.

3.  **Community College Education Costs**

Maricopa Community College District Tuition and Fees

Years Since 1997–98 (t)	Cost per Credit (dollars) (C)
0	37
1	38
2	40
3	41
4	43
5	46
6	51

Source: www.dist.maricopa.edu.

Forecast the cost per credit for students attending college in the district in 2005–06.

4.  **Breakfast Cereal Consumption**

Per Capita Consumption of Breakfast Cereals

Years Since 1980 (t)	Consumption (pounds) (C)
0	12
1	12
2	11.9
3	12.2
4	12.5
5	12.8
6	13.1
7	13.3
8	14.2
9	14.9
10	15.4
11	16.1

(Continued)

(Continued)

12	16.6
13	17.3
14	17.4
15	17.1
16	16.6
17	16.3
18	15.6
19	15.5

Source: *Statistical Abstract of the United States, 2001*, Table 202, p. 129.

Forecast the per capita consumption of breakfast cereal in 2001.

5.  **State University Enrollment**

Washington State Public University Enrollment

Years Since 1990 (t)	Students (S)
0	81,401
1	81,882
2	83,052
3	84,713
4	85,523
5	86,080
6	87,309
7	89,365
8	90,189
9	91,543
10	92,821

Source: Washington State Higher Education Coordinating Board, Higher Education Statistics, September 2001.

Forecast the Washington state public university enrollment in 2010.

6.  **Community College Education Costs**

Average Annual Undergraduate Tuition and Fees at Washington State Community Colleges

Years Since 1984-85 (t)	Tuition and Fees (dollars) (F)
0	581
1	699
2	699
3	759
4	780
5	822
6	867
7	945
8	999
9	1,125
10	1,296
11	1,350
12	1,401
13	1,458
14	1,515
15	1,584
16	1,641
17	1,743

Source: Washington State Higher Education Coordinating Board, Higher Education Statistics, September 2001.

Forecast the annual tuition and fees at a Washington state community college in 2005.

7.  **AIDS Incidence in Children**

Estimated Pediatric AIDS Incidence (United States only)

Years Since 1992 (t)	Number of Cases (C)
0	954
1	927
2	821
3	687
4	515
5	329
6	235
7	179
8	120
9	101

Source: Centers for Disease Control and Prevention, "HIV/AIDS Surveillance Report," December 2001, p. 36.

Forecast the number of pediatric AIDS cases in the United States in 2005.

8.  **Per Capita Personal Income**

Per Capita Personal Income-Florida

Years Since 1993 (t)	Personal Income (dollars) (P)
0	21,320
1	21,905
2	22,942
3	23,909
4	24,869
5	26,161
6	26,593
7	27,764

Source: Bureau of Economic Analysis (www.bea.gov).

Forecast the per capita personal income in Florida in 2004.

9. 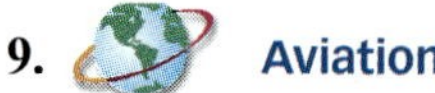 **Aviation**

Air Carrier Accidents

Year (t)	Accidents (A)
1992	18
1993	23
1994	23
1995	36
1996	37
1997	49
1998	50
1999	52
2000	54

Source: Statistical Abstract of the United States, 2001, Table 1063, p. 669.

Forecast the number of air carrier accidents in 2002.

10.  **Federal Funds for Elections**

Federal Funds for Presidential Election Campaigns

Years Since 1980 (t)	Federal Funds (millions of dollars) (F)
0	62.7
4	80.3
8	92.2
12	110.4
16	152.6

Source: *Statistical Abstract of the United States, 2001*, Table 409, p. 255.

Forecast the amount of federal funds spent in the 2004 presidential election.

In Exercises 11–20, find mathematical models for each of the verbal descriptions.

11. Candy Bar Prices Candy bars currently cost \$0.60 each. The price of a candy bar is expected to increase by 3 percent per year in the future.

12. **Housing Prices** On March 12, 2004, a builder priced a new home in Queen Creek, Arizona, at \$198.9K. The builder's sales representative told the author that the price for that home style would increase on March 16, March 30, and April 13 to \$205.9K, \$209.9K, and \$212.9K, respectively. (**Source:** Fulton Homes.)

Find a mathematical model for the price of the new home style.

13. **Calculator Prices** A calculator currently costs \$87. The price of the calculator is expected to increase by \$3 per year.

Find a mathematical model for the price of the calculator.

14. **Mortality Rates** There are presently 95 members of a high school graduating class who are still living. The number of surviving class members is decreasing at a rate of 4 percent per year.

Find a mathematical model for the number of surviving class members.

15. **Product Sales Growth** We are introducing a new product next year. We anticipate that sales will initially be slow but will increase rapidly once people become aware of our product. We anticipate that our monthly sales will start to level off in 18 months at about \$200,000. We predict that sales for the first two months will be \$12,000 and \$19,000, respectively.

Develop a mathematical model to forecast monthly product sales.

16. **Club Membership** We are concerned about the decreasing number of members of our business club. Two years ago, we had 200 members. Last year we had 165 members, and this year we have 110 members. If something doesn't change, we expect to lose even more members next year than we lost this year.

Develop a mathematical model for the club membership.

17. **Population Growth** The town of Queen Creek, Arizona, was founded in 1989. In 1990, there were 2667 people living in the town. The town grew rapidly in the 1990s, in large part because of new home construction in the area. There were 4316 people living in the town in 2000 and 4940 people in 2001. The Arizona Department of Commerce estimated the 2002 population of Queen Creek at 5555 people.

If a logistic model is used to model the population of Queen Creek, what is the maximum projected population of the city? (*Hint:* First align the data.)

The Arizona Department of Commerce estimated the 2003 population of Queen Creek at 7480. In light of this additional information, is there a different type of model that would have better predicted the 2003 population of Queen Creek? Explain.

18. **Housing Prices** On March 12, 2004, a builder priced a new home in Queen Creek, Arizona, at \$143.9K. The builder's sales representative told the author that the price for that home style would increase on March 16, March 30, and April 13 to \$148.9K, \$151.9K, and \$153.9K, respectively. (**Source:** Fulton Homes.)

Find a mathematical model for the price of the new home style.

19. **Federal Tax Rates** Federal income tax rates are dependent upon the amount of taxable income received. In 2003, federal income taxes were calculated as follows. For single filers, the first \$7000 earned was taxed at 10 percent. The next \$21,400 earned was taxed at 15 percent. The next \$40,400 earned was taxed at 25 percent.

For example, the tax of a single woman who earned \$25,000 would be calculated as follows:

10% tax on the first \$7,000

$\$7{,}000 \times 0.10 = \$700$

Amount to be taxed at a higher rate

$\$25{,}000 - \$7000 = \$18{,}000$

15% tax on the next \$18,000

$\$18{,}000 \times 0.15 = \$2{,}700$

The person's total tax is

$\$700 + \$2{,}700 = \$3{,}400$

Find a mathematical model for income tax as a function of taxable income for single filers.

20. **Federal Tax Rates** In 2003, federal income taxes for married individuals filing jointly were calculated as follows. The first \$14,000 was taxed at 10 percent. The next \$42,800 was taxed at 15 percent. The next \$57,850 was taxed at 25 percent.

The tax of a married couple who earned $65,000 would be calculated as follows:

10% tax on the first $14,000

$14,000 × 0.10 = $1,400

Amount to be taxed at a higher rate

$65,000 − $14,000 = $51,000

15% tax on the next $42,800

$42,800 × 0.15 = $6420

Amount to be taxed at a higher rate

$51,000 − $42,800 = $8,200

25% tax on next $8,200

$8,200 × 0.25 = $2,050

The couple's total tax is

$1,400 + $6,420 + $2,050 = $9,870

Find a mathematical model for income tax as a function of taxable income for couples filing jointly.

Exercises 21–25 are intended to challenge your understanding of mathematical modeling.

21. The graph of a mathematical model passes through all of the points of a data set. A student claims that the model is a perfect forecaster of future results. How would you respond?

22. A scatter plot is concave up and increasing on [0, 5], concave down and increasing on [5, 8], and decreasing at a constant rate on [8, 15]. Describe two different mathematical models that may fit the data set.

23. Describe how a business owner can benefit from mathematical modeling, despite the imprecision of a model's results.

24. Daily fluctuations in the stock market make the share price of a stock very difficult to model. What approach would you take if you wanted to model the long-term performance of a particular stock?

25. You are asked by your boss to model the data shown in the following scatter plot. How would you respond?

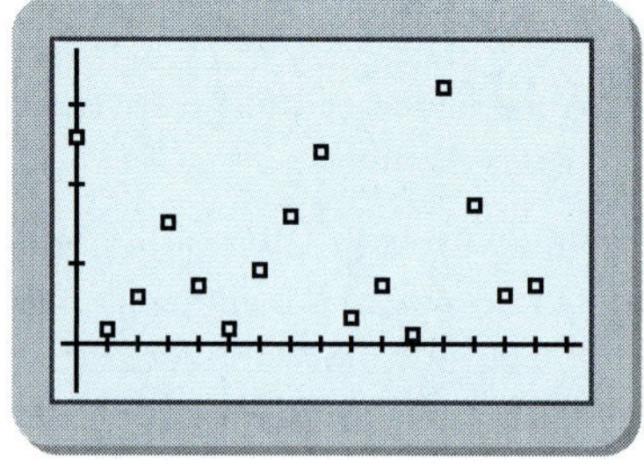

Chapter 5 Review Exercises

Section 5.1 *In Exercises 1–13, find the model that best fits the data. Use the model to answer the given questions.*

1.  **Advertising Expenditures: Magazines**

Years Since 1990 (t)	Advertising Expenditures (millions of dollars) (A)
0	6,803
1	6,524
2	7,000
3	7,357
4	7,916
5	8,580
6	9,010
7	9,821
8	10,518
9	11,433
10	12,348

Source: *Statistical Abstract of the United States, 2001,* Table 1272, p. 777.

According to the model, how much money was spent on magazine advertising in 2002?

2. 

Advertising Expenditures: Cable TV

Years Since 1990 (t)	Advertising Expenditures (millions of dollars) (A)
0	2,457
1	2,728
2	3,201
3	3,678
4	4,302
5	5,108
6	6,438
7	7,237
8	8,301
9	10,429
10	12,364

Source: *Statistical Abstract of the United States, 2001,* Table 1272, p. 777.

According to the model, how much money was spent on cable television advertising in 2002?

3. Using the models from Exercises 1 and 2, determine in what year cable television advertising expenditures are expected to exceed magazine advertising expenditures.

4.  **Advertising Expenditures: Radio**

Years Since 1990 (t)	Advertising Expenditures (millions of dollars) (A)
0	8,726
1	8,476
2	8,654
3	9,457
4	10,529
5	11,338
6	12,269
7	13,491
8	15,073
9	17,215
10	19,585

Source: *Statistical Abstract of the United States, 2001,* Table 1272, p. 777.

According to the model, when will radio advertising exceed $25 billion?

5.  **Advertising Expenditures: Yellow Pages**

Years Since 1990 (t)	Advertising Expenditures (millions of dollars) (A)
0	8,926
1	9,182
2	9,320
3	9,517
4	9,825
5	10,236
6	10,849
7	11,423
8	11,990
9	12,652
10	13,367

Source: *Statistical Abstract of the United States, 2001,* Table 1272, p. 777.

According to the model, when will Yellow Pages advertising exceed $15 billion?

Section 5.2

6.  **Federal Credit Unions**

Years Since 1975 (t)	Credit Unions (C)
0	12,737
5	12,440
10	10,125
15	8,511
20	7,329
25	6,336

Source: *Statistical Abstract of the United States, 2001,* Table 1184, p. 732.

According to the model, how many federal credit unions were there in 1999? Does this seem reasonable? Explain.

7.  **Advertising Expenditures: Newspapers**

Years Since 1990 (t)	Advertising Expenditures (millions of dollars) (A)
0	32,281
1	30,409
2	30,737
3	32,025
4	34,356
5	36,317
6	38,402
7	41,670
8	44,292
9	46,648
10	49,246

Source: *Statistical Abstract of the United States, 2001,* Table 1272, p. 777.

According to the model, how much money was spent on newspaper advertising in 2001?

8.  **Full-Service Restaurant Sales**

Years Since 1980 (t)	Sales (millions of dollars) (S)
0	39,307
2	46,443
4	54,815
6	61,474
8	69,356
10	77,811
12	83,561
14	91,457
16	100,830
18	117,774
20	134,461

Source: *Statistical Abstract of the United States, 2001,* Table 1268, p. 775.

According to the model, when will full-service restaurant sales exceed $150 billion?

9.  **Advertising Expenditures: Broadcast TV**

Years Since 1990 (t)	Advertising Expenditures (millions of dollars) (A)
0	26,616
1	25,461
2	27,249
3	28,020
4	31,133
5	32,720
6	36,046
7	36,893
8	39,173
9	40,011
10	44,438

Source: *Statistical Abstract of the United States, 2001,* Table 1272, p. 777.

According to the model, when will broadcast television advertising expenditures exceed $50 billion?

10.  **Manufacturing Full-Time Employees: Leather and Leather Products**

Years Since 1995 (t)	Employees (thousands) (N)
0	106
1	95
2	89
3	84
4	76

Source: *Statistical Abstract of the United States, 2001,* Table 979, p. 622.

According to the model, in what year will the number of full-time employees in the leather and leather products manufacturing industry drop below 65,000?

11. **Wages**

Average Hourly Earnings in Manufacturing Industries: Michigan

Years Since 1980 (t)	Average Earnings (dollars per hour) (E)
0	9.52
1	10.53
2	11.18
3	11.62
4	12.18
5	12.64
6	12.80
7	12.97
8	13.31
9	13.51
10	13.86
11	14.52
12	14.81
13	15.36
14	16.13
15	16.31
16	16.67
17	17.18
18	17.61
19	18.38
20	19.20

Source: *Statistical Abstract of the United States, 2001,* Table 978, p. 622.

According to the model, what will the average hourly wage in Michigan manufacturing industries be in 2003?

12.  **Wages**

Average Hourly Earnings in Manufacturing Industries: Florida

Years Since 1980 (t)	Average Earnings (dollars per hour) (E)
0	5.98
1	6.53
2	7.02
3	7.33
4	7.62
5	7.86
6	8.02
7	8.16
8	8.39
9	8.67
10	8.98
11	9.30
12	9.59
13	9.76
14	9.97
15	10.18
16	10.55
17	10.95
18	11.43
19	11.83
20	12.28

Source: *Statistical Abstract of the United States, 2001*, Table 978, p. 622.

According to the model, what will the average hourly wage in Florida manufacturing industries be in 2003?

13. Based on the wage data in Exercises 11–12, do you think it would be better to start up a manufacturing business in Florida or in Michigan? Justify your answer and explain what other issues might affect your decision.

Section 5.3

In Exercises 14–15, determine if the graph of the function is increasing or decreasing and if it is concave up or concave down. Identify the coordinates of the y-intercept. Then graph the function to verify your conclusions.

14. $y = 2(0.75)^x$

15. $y = -0.3(2.8)^x$

In Exercises 16–17, find the equation of the exponential function that fits the data in the table algebraically.

16.

x	y
0	3
1	15
2	75
3	375

17.

x	y
2	36
4	144
6	576
8	2,304

In Exercises 18–19, use exponential regression to model the data in the table. Use the model to predict the value of the function when $t = 10$, and interpret the real-world meaning of the result.

 **18.**  **Number of Subway Restaurants**

Years Since 1996 (t)	Restaurants (N)
0	12,516
1	13,066
2	13,600
3	14,162
4	14,662

Source: www.subway.com.

19.  **Number of McDonalds Restaurants**

Years Since 1997 (t)	Restaurants (N)
0	22,928
1	24,513
2	26,309
3	28,707
4	30,093

Source: www.mcdonalds.com.

In Exercises 20–21, find the exponential function that fits the verbal description. Calculate the value of the function five years from now and interpret its real-world meaning.

20. My car is depreciating at a rate of 17 percent per year. It is currently valued at $6000.

21. My monthly household expenses are increasing by 2.3 percent annually. It currently costs $4500 per month to maintain my household.

In Exercises 22–23, find the solution by graphically solving the equation.

22. $1.9(2.6)^x = 10$

23. $9.7(0.4)^x = 2$

Section 5.4

In Exercises 24–25, solve the logarithmic equation for y without using a calculator.

24. $y = \log_6(36)$

25. $y = \log_5(0.2)$

In Exercises 26–27, determine the concavity and increasing/decreasing behavior of the graph of the function. Then graph the function to verify your results. (Hint: You will have to use the change-of-base formula if you graph the function on your calculator. That is, $y = \log_b(x) \Rightarrow y = \frac{\ln(x)}{\ln(b)}$)

26. $y = \log_5(x)$

27. $y = \log_{0.4}(x)$

In Exercises 28–31, use the inverse relationship between logarithmic and exponential functions to solve the equations for x. Simplify your answers.

28. $\left(\frac{1}{2}\right)^x = 64$

29. $5^x = 625$

30. $\log_3(x) = 4$

31. $\log_5(x) = -2$

In Exercises 32–35, use the rules of logarithms to rewrite each expression as a single logarithm.

32. $\log(5) + \log(3)$

33. $2\log(4x) - \log(8)$

34. $3\log(2x)^2 - \log(2x^3)$

35. $-\log(3x)^2 + \log(3x^2)$

In Exercises 36–37, use logarithmic regression to model the data in the table. Use the model to predict the value of the function when t = 10, and interpret the real-world meaning of the result.

 **36.**  **McDonalds' Systemwide Sales**

Years Since 1997 (t)	Sales (millions of dollars) (S)
1	33,638
2	35,979
3	38,491
4	40,181
5	40,630

Source: www.mcdonalds.com.

 **37.**  **Non-Alcohol-Related Auto Accident Fatalities**

Years Since 1989 (t)	Fatalities (percentage) (F)
1	50.5
3	54.5
4	56.5
5	59.3
6	58.8
7	59.1
8	61.5
9	61.4
10	61.7

Source: *Statistical Abstract of the United States, 2001,* Table 1099, p. 688.

Does the model estimate at $t = 10$ agree with the raw data value?

Section 5.5

In Exercises 38–40, find the equation of the mathematical model (if possible) that you believe will most accurately forecast the future behavior of the thing being modeled. Justify your conclusions.

38.  **Projected Teacher Salaries**

Years Since 1990 (t)	Public School Teacher Average Annual Salary (thousands of dollars) (S)
0	31.4
1	33.1
2	34.1
3	35.0
4	35.7
5	36.7
6	37.7
7	38.5
8	39.5
9	40.6
10	41.7

Source: *Statistical Abstract of the United States, 2001,* Table 237, p. 151.

39.  **Cassette Tape Market Share**

Cassette Tape Sales

Years Since 1993 (t)	Percent of Music Market (percentage points) (P)
0	38.0
1	32.1
2	25.1
3	19.3
4	18.2
5	14.8
6	8.0
7	4.9
8	3.4
9	2.4

Source: Recording Industry Association of America.

40.  **Music Market Size**

Music Market Size

Years Since 1997 (t)	Dollar Volume (millions) (P)
0	$12,236.80
1	$13,723.50
2	$14,584.50
3	$14,323.00
4	$13,740.89
5	$12,614.21

Source: Recording Industry Association of America.

Make It Real

PROJECT 5

What to do

1. Visit the Bureau of Labor Statistics web site (www.bls.gov/cpi) and access the most recent Consumer Price Index news report.

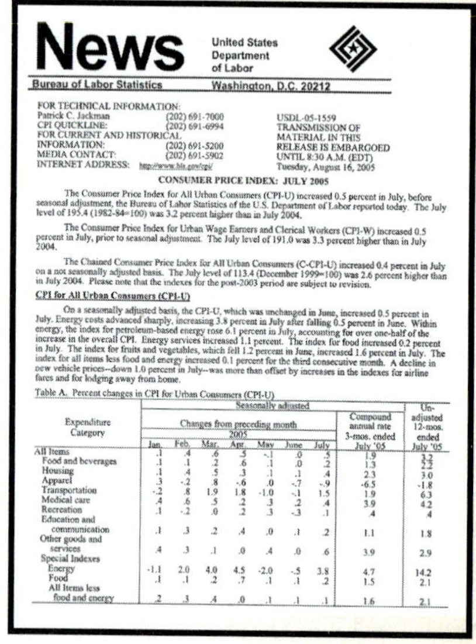

News United States Department of Labor

Bureau of Labor Statistics Washington, D.C. 20212

FOR TECHNICAL INFORMATION:
Patrick C. Jackman (202) 691-7000
CPI QUICKLINE: (202) 691-6994
FOR CURRENT AND HISTORICAL
INFORMATION: (202) 691-5200
MEDIA CONTACT: (202) 691-5902
INTERNET ADDRESS: http://www.bls.gov/cpi/

USDL-05-1559
TRANSMISSION OF MATERIAL IN THIS RELEASE IS EMBARGOED UNTIL 8:30 A.M. (EDT)
Tuesday, August 16, 2005

CONSUMER PRICE INDEX: JULY 2005

The Consumer Price Index for All Urban Consumers (CPI-U) increased 0.5 percent in July, before seasonal adjustment, the Bureau of Labor Statistics of the U.S. Department of Labor reported today. The July level of 195.4 (1982-84=100) was 3.2 percent higher than in July 2004.

The Consumer Price Index for Urban Wage Earners and Clerical Workers (CPI-W) increased 0.5 percent in July, prior to seasonal adjustment. The July level of 191.0 was 3.3 percent higher than in July 2004.

The Chained Consumer Price Index for All Urban Consumers (C-CPI-U) increased 0.4 percent in July on a not seasonally adjusted basis. The July level of 113.4 (December 1999=100) was 2.6 percent higher than in July 2004. Please note that the indexes for the post-2003 period are subject to revision.

CPI for All Urban Consumers (CPI-U)

On a seasonally adjusted basis, the CPI-U, which was unchanged in June, increased 0.5 percent in July. Energy costs advanced sharply, increasing 3.8 percent in July after falling 0.5 percent in June. Within energy, the index for petroleum-based energy rose 6.1 percent in July, accounting for over one-half of the increase in the overall CPI. Energy services increased 1.1 percent. The index for food increased 0.2 percent in July. The index for fruits and vegetables, which fell 1.2 percent in June, increased 1.6 percent in July. The index for all items less food and energy increased 0.1 percent for the third consecutive month. A decline in new vehicle prices--down 1.0 percent in July--was more than offset by increases in the indexes for airline fares and for lodging away from home.

Table A. Percent changes in CPI for Urban Consumers (CPI-U)

Expenditure Category	Seasonally adjusted								Unadjusted 12-mos. ended July '05
	Changes from preceding month 2005							Compound annual rate 3-mos. ended July '05	
	Jan.	Feb.	Mar.	Apr.	May	June	July		
All Items	.1	.4	.6	.5	-.1	.0	.5	1.9	3.2
Food and beverages	.1	.1	.2	.6	.1	.0	.2	1.3	2.2
Housing	.1	.4	.5	.3	.1	.1	.4	2.3	3.0
Apparel	.3	-.2	.8	-.6	.0	-.7	-.9	-6.5	-1.8
Transportation	-.2	.8	1.9	1.8	-1.0	-.1	1.5	1.9	6.3
Medical care	.4	.6	.5	.2	.3	.2	.4	3.9	4.2
Recreation	.1	-.2	.0	.2	.3	-.3	.1	.4	.4
Education and communication	.1	.3	.2	.4	.0	.1	.2	1.1	1.8
Other goods and services	.4	.3	.1	.0	.4	.0	.6	3.9	2.9
Special Indexes									
Energy	-1.1	2.0	4.0	4.5	-2.0	-.5	3.8	4.7	14.2
Food	.1	.1	.2	.7	.1	.1	.2	1.5	2.1
All Items less food and energy	.2	.3	.4	.0	.1	.1	.1	1.6	2.1

2. Select an expenditure category.
3. Record the annual percentage rate of change in that category. (This is called the compound annual rate in the report.)
4. Find the price of a product that belongs to the category you selected at a local retailer.
5. Using the Consumer Price Index information, estimate the price of the item five years from now.

Consumer Price Index explained

"The Consumer Price Index (CPI) is a measure of the average change in prices over time of goods and services purchased by households. The CPI for All Urban Consumers (CPI-U) and the Chained CPI for All Urban Consumers (C-CPI-U), which cover approximately 87 percent of the total population and include in addition to wage earners and clerical worker households, groups such as professional, managerial, and technical workers, the self-employed, short-term workers, the unemployed, and retirees and others not in the labor force." (**Source:** Bureau of Labor Statistics News, July 2002, USDL-020-480.)

Chapter 6

Mathematics of Finance

Effective retirement planning can help ensure peace and prosperity in one's golden years. One of the greatest benefits young workers have is time. By investing wisely early in their careers, young workers have more time for their investments to grow before retirement than their older counterparts have. By understanding compound interest and annuities, you can better plan for your retirement.

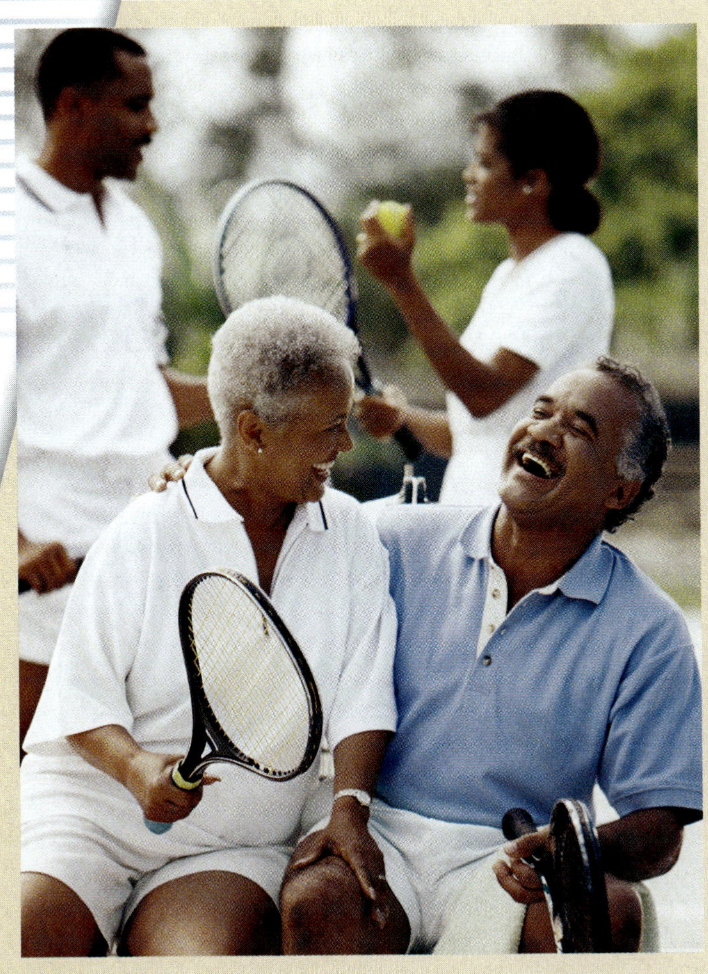

6.1 Solving Exponential Equations

- Solve exponential equations algebraically

GETTING STARTED The price of whole chicken between 1990 and 2000 may be modeled by

$$P(t) = 0.8443(1.025)^t \text{ dollars per pound}$$

where t is the number of years since the end of 1990. (**Source:** Modeled from *Statistical Abstract of the United States, 2001,* Table 707, p. 468.) A catering business is forecasting future food costs. It plans to increase the price of its Chicken Divan entrée when the cost of chicken reaches \$1.25 per pound. When should the catering business increase the price of the entrée? This question can be answered by using logarithms to solve an exponential equation.

In this section, we will refine our technique of solving exponential equations algebraically. When algebraic methods fail to isolate the variable, we will use graphical methods to estimate the solution.

Returning to the whole chicken price problem, we want to know when $P = 1.25$. That is, we must solve

$$1.25 = 0.8443(1.025)^t$$

$$1.481 = 1.025^t \quad \text{Divide both sides by 0.8443}$$

To get the t out of the exponent, we will take the natural log of both sides and apply Log Rule 7.

$$\ln(1.481) = \ln(1.025)^t \quad \text{Take the natural log of both sides}$$

$$\ln(1.481) = t\ln(1.025) \quad \text{Rule 7}$$

$$t = \frac{\ln(1.481)}{\ln(1.025)} \quad \text{Divide both sides by } \ln(1.025)$$

$$t \approx 15.9 \quad \text{Evaluate on calculator}$$

The price of whole chicken is expected to reach \$1.25 per pound by 2006 (16 years after 1990). The catering business should raise its prices by 2006.

The following process for solving exponential equations minimizes the number of log rules we have to use.

HOW TO **Method for Solving Exponential Equations of the Form $y = ab^x$**

To solve an exponential equation $y = ab^x$, do the following:

1. Divide both sides by a.

$$\frac{y}{a} = b^x$$

(Continued)

2. Take the log or natural log of both sides.

$$\log\left(\frac{y}{a}\right) = \log(b)^x \text{ or } \ln\left(\frac{y}{a}\right) = \ln(b)^x$$

3. Bring the x down in front of the log or natural log.

$$\log\left(\frac{y}{a}\right) = x\log(b) \text{ or } \ln\left(\frac{y}{a}\right) = x\ln(b)$$

4. Divide both sides by $\log(b)$ or $\ln(b)$, respectively.

$$x = \frac{\log\left(\frac{y}{a}\right)}{\log(b)} \text{ or } x = \frac{\ln\left(\frac{y}{a}\right)}{\ln(b)}$$

We will repeat this basic process in the next couple of examples to give you a feel for the method.

EXAMPLE 1

Determining the Doubling Time for Prescription Drug Spending

The amount of money spent on prescription drugs in the United States may be modeled by

$$D(t) = 2.333(1.099)^t \text{ billion dollars}$$

where t is the number of years since the end of 1960. (**Source:** Modeled from *Statistical Abstract of the United States, 2001*, Table 119, p. 91.) According to the model, how many years does it take for the amount of money spent on prescription drugs to double?

SOLUTION According to the model, \$2.333 billion was spent on prescription drugs in 1960. Double the initial amount is \$4.666 billion.

$$4.666 = 2.333(1.099)^t$$

$$\frac{4.666}{2.333} = (1.099)^t \quad \text{Divide both sides by 2.333}$$

$$2 = (1.099)^t \quad \text{Simplify}$$

$$\ln(2) = \ln(1.099)^t \quad \text{Take the natural log of both sides}$$

$$\ln(2) = t\ln(1.099) \quad \text{Rule 7}$$

$$t = \frac{\ln(2)}{\ln(1.099)} \quad \text{Divide both sides by } \ln(1.099)$$

$$t \approx 7.34 \quad \text{Evaluate on calculator}$$

According to the model, prescription drug spending doubles every seven years or so. So why do we care? Increases in drug costs are typically passed on to the consumer through higher premiums or higher co-payments.

EXAMPLE 2 Forecasting Tuition Costs with an Exponential Function

The cost of full-time resident *quarterly* tuition at Green River Community College may be modeled by

$$E(t) = 430.6(1.042)^t \text{ dollars}$$

where t is the number of years since the end of 1994. (**Source:** Modeled from Green River Community College data.)

When their daughter was born in 1994, a couple invested \$2000 in a college savings account earning 3 percent annually. They hope to use the money to pay for the first year of their daughter's college tuition, but, with rising tuition costs, they are uncertain whether their savings will cover the annual cost of the tuition. How long do they have before the annual cost of tuition (three quarters) will exceed the amount of their savings?

SOLUTION The value of the couple's investment can be modeled by

$$A(t) = 2000(1.03)^t \text{ dollars}$$

where t is the number of years since the end of 1994.

The annual cost of tuition is $3E$, since three quarters is considered a full academic year. We want to find out when $3E = A$.

$$3E = A$$

$$3[430.6(1.042)^t] = 2000(1.03)^t$$

$$1291.8(1.042)^t = 2000(1.03)^t \quad \text{Simplify}$$

$$\frac{(1.042)^t}{(1.03)^t} = \frac{2000}{1291.8} \quad \text{Divide both sides by } (1.03)^t \text{ and by } 1291.8$$

$$\left(\frac{1.042}{1.03}\right)^t = 1.548 \quad \text{Simplify}$$

$$(1.012)^t = 1.548 \quad \text{Simplify}$$

$$\ln(1.012)^t = \ln(1.548) \quad \text{Take the natural log of both sides}$$

$$t\ln(1.012) = \ln(1.548) \quad \text{Rule 7}$$

$$t = \frac{\ln(1.548)}{\ln(1.012)} \quad \text{Divide both sides by } \ln(1.012)$$

$$t \approx 37.74 \quad \text{Evaluate on calculator}$$

(*Note:* As in other examples and exercises in this text, we are keeping the actual values in our calculator even though we round the values when we write them down. By waiting to round until we have our final answer, we obtain a more accurate result.)

According to the model, the annual cost of tuition will exceed the balance of the savings account in 2032 (38 years after 1994). What will the annual cost of tuition be that year? Evaluating either function at $t = 37.74$, we find that the *annual* tuition is \$6102. That is, in 2032, we expect the *quarterly* tuition to be about \$2034! Fortunately, the couple's savings will be sufficient to cover a year of tuition provided that their daughter goes to college before she's 38 years old.

Example 2 was a bit more challenging because it required us to solve an exponential equation of the form $ab^t = cd^t$. An equation of this form can always be reduced as detailed in the following box.

HOW TO Method for Solving Exponential Equations of the Form $ab^x = cd^x$

To solve an exponential equation $ab^x = cd^x$, do the following:

1. Divide both sides by a.

$$b^x = \frac{cd^x}{a}$$

2. Divide both sides by d^x.

$$\frac{b^x}{d^x} = \frac{c}{a}$$

3. Rewrite $\frac{b^x}{d^x}$ as $\left(\frac{b}{d}\right)^x$.

$$\left(\frac{b}{d}\right)^x = \frac{c}{a}$$

4. Solve using the method for solving exponential equations of the form $y = b^x$.

$$\ln\left(\frac{b}{d}\right)^x = \ln\left(\frac{c}{a}\right)$$

$$x\ln\left(\frac{b}{d}\right) = \ln\left(\frac{c}{a}\right)$$

$$x = \frac{\ln\left(\frac{c}{a}\right)}{\ln\left(\frac{b}{d}\right)}$$

EXAMPLE 3

Solving a System of Equations Graphically

Health care expenditures by insurance companies and patients have risen sharply over the past four decades. Based on data from 1960 to 2000, patient out-of-pocket expenditures can be modeled by

$$P(t) = 13.16(1.075)^t \text{ billion dollars}$$

and insurance company expenditures can be modeled by

$$I(t) = 0.3593t^2 - 3.561t + 7.857 \text{ billion dollars}$$

where t is the number of years since the end of 1960. (**Source:** Modeled from *Statistical Abstract of the United States, 2001*, Table 119, p. 91.)

Patient spending initially exceeded insurance company spending. In what year, were insurance company and patient out-of-pocket spending the same?

SOLUTION We must find when $P(t) = I(t)$.

$$P(t) = I(t)$$
$$13.16(1.075)^t = 0.3593t^2 - 3.561t + 7.857$$

Observe that the left-hand side of the equation has a variable in the exponent. The only way we will be able to get the variable out of the exponent is by taking the log of both sides. However, if we take the log of both sides, the equation becomes

$$\log[13.16(1.075)^t] = \log(0.3593t^2 - 3.561t + 7.857)$$

leaving us with the variable trapped inside the logarithm on the right-hand side. Because of this dilemma, we are unable to solve the problem algebraically. However, we can use a graphing calculator to graph $P(t) = 13.16(1.075)^t$ and $I(t) = 0.3593t^2 - 3.561t + 7.857$ and find their points of intersection. Doing so yields Figure 6.1.

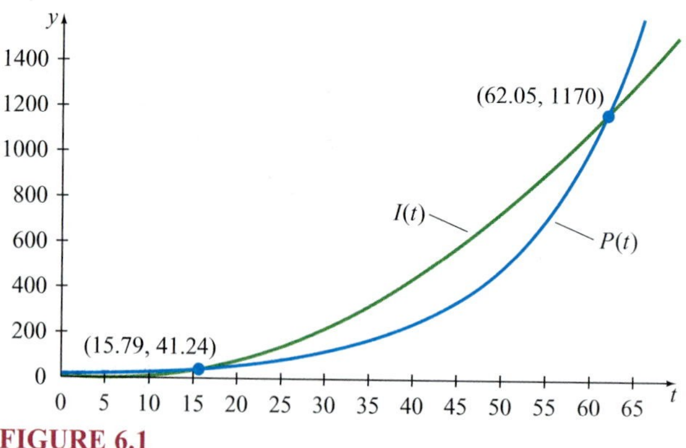

FIGURE 6.1

The graphs intersect at (15.79, 41.24) and (62.05, 1170). [The graphs also intersect at $(-1.089, 12.16)$; however, we are interested only in positive values of t.] According to the models, insurance spending first surpassed out-of-pocket expenditures in 1976 ($t = 16$). However, out-of pocket expenditures are expected to exceed insurance spending in 2022 ($t = 62$). Since our model was based on data from $t = 0$ to $t = 40$, we are a bit skeptical about the accuracy of the second projection because $t = 62$ is relatively far away from the data values used to create the model. On the other hand, we feel very comfortable with our first projection, since it is within the range of data values used to create the model. Referring back to the source data used to create the model (not included here), we discover that in 1977, out-of-pocket and insurance expenditures were both equal to \$45 billion. Our projection of \$41 billion in 1976 was off by about a year and about \$4 billion.

Predicting the output value for an input value outside of the interval of the input data is called **extrapolation.** Predicting the output value for an input value inside the interval of the input data is called **interpolation.** We must use extreme caution when considering the extrapolated values of a model. Typically, the further away the input value is from the domain of the data set, the less likely it is that the extrapolated value is accurate in modeling the real-life scenario.

6.1 Summary

In this section, you refined your ability to solve exponential equations algebraically. When algebraic methods failed to isolate the variable, you used graphical methods to estimate the solution.

6.1 Exercises

In Exercises 1–10, solve the exponential equation algebraically.

1. $2(3^x) = 162$

2. $3(2^x) = 768$

3. $1.1^x = 1.21$

4. $5(0.5)^x = 0.125$

5. $3(2^x) = 4^x$

6. $4(3^x) = 10(2^x)$

7. $2(1.09)^t = 3(2.11)^t$

8. $4(3.9)^t = 1.7$

9. $2(3.1)^t = 3.6$

10. $3(7.1)^t = 99$

In Exercises 11–20, solve the problems algebraically.

11. **Church Growth** The number of religious congregations of the Church of Jesus Christ of Latter-Day Saints in the United States may be modeled by

$$C(t) = 8679(1.023)^t \text{ congregations}$$

where t is the number of years since the end of 1988. (**Source:** Modeled from data compiled by Mark Davies, Brigham Young University.)

In what year is the number of congregations expected to reach 15,000?

12. **Video Game Sales** In the first quarter of 2001, U.S. video game industry sales totaled \$1.6 billion. Sales increased by 20 percent in the first quarter of 2002. (**Source:** www.npdfunworld.com.) If this pattern continues, video game industry sales in the first quarter, S, may be modeled by

$$S(t) = 1.6(1.2)^t \text{ billion dollars}$$

where t is the number of years since the end of 2001.

In what year are first-quarter sales expected to exceed \$3.2 billion?

13. **Land in Farms** The amount of land in U.S. farms may be modeled by

$$F(t) = 115.1(0.9352)^t + 900 \text{ million acres}$$

where t is the number of years since the end of 1978. (**Source:** Modeled from *Statistical Abstract of the United States, 2001*, Table 796, p. 523.)

According to the model, in what year will the amount of land in U.S. farms drop below 910 million acres?

14. **Milk Beverage Consumption** The annual per capita milk beverage consumption may be modeled by

$$M(t) = 27.76(0.9914)^t \text{ gallons}$$

where t is the number of years since the end of 1980. (**Source:** Modeled from *Statistical Abstract of the United States, 2001*, Table 202, p. 129.)

According to the model, in what year will the average person drink 20 gallons of milk?

15. **Bottled Water Consumption** The annual per capita bottled water consumption may be modeled by

$$W(t) = 2.593(1.106)^t \text{ gallons}$$

where t is the number of years since the end of 1980. (**Source:** Modeled from *Statistical Abstract of the United States, 2001*, Table 204, p. 130.)

According to the model, in what year will the average person drink 20 gallons of bottled water?

16. **Beverage Consumption** The annual per capita milk beverage consumption may be modeled by

$$M(t) = 27.76(0.9914)^t \text{ gallons}$$

and the annual per capita bottled water consumption may be modeled by

$$W(t) = 2.593(1.106)^t \text{ gallons}$$

where t is the number of years since the end of 1980. (**Source:** Modeled from *Statistical Abstract of the United States, 2001*, Tables 202, 204, pp. 129–130.)

A dairy farmer discovered a freshwater spring on his property and, recognizing the decreasing demand for milk and the rising demand for bottled water, is considering producing bottled water instead of milk. As a part of his business plan, he wants to know when the consumption of bottled water is projected to exceed the consumption of milk. According to the models, when will bottled water consumption surpass milk beverage consumption?

17. **Church Growth** Based on data from 1990 to 2001, the number of American adults who consider themselves Lutherans may be modeled by

$$L(t) = 9110(1.005)^t \text{ thousand adults}$$

and the number who consider themselves nondenominational Christians may be modeled by

$$C(t) = 8073(1.053)^t \text{ thousand adults}$$

where t is the number of years since the end of 1990. (**Source:** Modeled from American Religious Identification Survey.)

According to the models, in what year will the number of adults in each religious group be the same?

18. **Bread Prices** The price of white bread has been increasing exponentially since 1980. The consumer price index for white bread may be modeled by

$$I(t) = 86.56(1.043)^t \text{ index points}$$

where t is the number of years since the end of 1980. (**Source:** Modeled from *Statistical Abstract of the United States, 2001*, Table 694, p. 455.)

In 1984, the index was 100. When the index reaches 300, the price of bread will have tripled over its 1984 price. In what year is that expected to occur?

19. **Population Growth** The projected population of Ohio may be modeled by

$$O(t) = 11{,}160(1.003)^t \text{ thousand people}$$

and the projected population of Illinois may be modeled by

$$I(t) = 12{,}050(1.002)^t \text{ thousand people}$$

where t is the number of years since the end of 1995.

The number of representatives a state has in Congress is based in part on the state's population. According to the models, in what year will the population of the two states (and therefore their representation in Congress) be equal?

20. **Prescription Drug Expenditures** Based on data from 1960 to 2000, prescription drug expenditures in the United States may be modeled by

$$D(t) = 2.333(1.099)^t \text{ billion dollars}$$

and out-of-pocket health care expenditures by consumers may be modeled by

$$P(t) = 13.16(1.075)^t \text{ billion dollars}$$

where t is the number of years since the end of 1960. (**Source:** Modeled from *Statistical Abstract of the United States, 2001*, Table 119, p. 91.)

According to the models, in what year will out-of-pocket health care spending and prescription drug spending be equal?

In Exercises 21–25, solve the problem graphically.

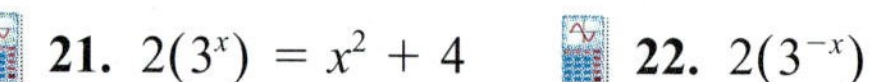

21. $2(3^x) = x^2 + 4$

22. $2(3^{-x}) = -x^2 + 2$

23. **Teacher Salaries** Based on data from 1990 to 2000, the average salary of public elementary and secondary school teachers may be modeled by

$$S(t) = 32.05(1.027)^t \text{ thousand dollars}$$

where t is the number of years since the end of 1990. (**Source:** Modeled from *Statistical Abstract of the United States, 2001*, Table 237, p. 151.)

Based on data from 1998 to 2000, the average salary of an assistant professor at a public university may be modeled by

$$P(x) = 1.7x + 42.4 \text{ thousand dollars}$$

where x is the number of years since the end of 1998. (**Source:** Modeled from *Statistical Abstract of the United States, 2001*, Table 280, p. 173.)

In what year are the salaries of public elementary and secondary school teachers and public university assistant professors expected to be equal? (*Hint:* Before graphing the functions, make sure they are in terms of the same variable, t or x.)

24.  **Beverage Consumption** The annual per capita milk beverage consumption may be modeled by

$$M(t) = -0.219t + 27.7 \text{ gallons}$$

and the annual per capita bottled water consumption may be modeled by

$$W(t) = 2.593(1.106)^t \text{ gallons}$$

where t is the number of years since the end of 1980. (**Source:** Modeled from *Statistical Abstract of the United States, 2001*, Tables 202, 204, pp. 129–130.)

In what year is the consumption of milk and bottled water expected to be the same?

Compare your results for this exercise to the solution of Exercise 16. Did the exponential and linear models for milk consumption yield the same result?

25. **Coca-Cola Company Sales** Based on data from 1999-2001, the number of unit cases of Coca-Cola brand beverages sold worldwide may be modeled by

$$S(t) = 16.5(1.04)^t \text{ billion unit cases}$$

and the net operating revenue of the Coca-Cola Company may be modeled by

$$R(t) = -0.2t^2 + 0.8t + 19.3 \text{ billion dollars}$$

where t is the number of years since the end of 1999. (**Source:** Modeled from Coca-Cola Company 2001 Annual Report, pp. 46, 57.)

When is the number of unit cases sold (in billions) expected to equal the number of dollars of net operating revenue (in billions)?

In Exercises 26–30, use algebraic or graphical methods to find the solution.

26. **Investments** Between August 1, 1952, and December 31, 2002, the CREF Stock Account earned an average annual return of 10.11 percent. (**Source:** TIAA-CREF.) If an investor invests \$500 in an account earning 10.11 percent annually, how long will it take for the investment to grow to \$1000?

27. **Investments** Between October 2, 1995, and December 31, 2002, the TIAA Real Estate Account earned an average annual return of 7.74 percent. (**Source:** TIAA-CREF.) If an investor invests \$500 in an account earning 7.74 percent annually, how long will it take for the investment to grow to \$1000?

28. **Investments** If \$500 is invested in an account earning 9.2 percent annually and \$1000 is invested in an account earning 5.3 percent annually, how long will it take for the values of the two accounts to be equal?

29. **Investments** An investor puts \$500 into an account earning 8 percent annually and \$1000 into an account losing 8 percent annually.

(a) How long will it take for the value of the accounts to be equal?

(b) When the value of the two accounts are equal, what is the combined value of the investments?

(c) Thirty years after the initial investment is made, what will be the combined value of the investments?

(d) If the investor had invested the initial \$1500 in an account paying an annual rate of 4.15 percent annually, what would the account value have been at the end of 30 years?

(e) Compare the results of parts (c) and (d). As an investor, how could you use this information?

30. **Investments** If \$500 is invested in an account earning 10 percent annually and \$500 is invested in an account losing 10 percent annually, will the combined account value remain at a constant \$1000? Explain.

Exercises 31–40 are intended to challenge your understanding of exponential equations.

31. **Investments** If \$100 is invested in an account earning 5 percent annually and \$100 is invested in an account losing 6 percent annually, will the combined value of the accounts increase over time? Explain.

32. Does the equation

$$3^x = -x^2 + 4x - 2$$

have a solution? Explain.

33. Solve the exponential equation algebraically.

$$3(2^x) = 3^{2x}$$

34. Solve the exponential equation algebraically.

$$5(2^{x+1}) = 2^{2x-1}$$

35. For what values of a will the equation

$$-x^2 = a(2^x)$$

have a solution?

36. For what values of b will the equation

$$x^2 = 2(b^x)$$

have a solution?

37. At least how many solutions will the given equation have?

$$a^{3x} = b^{2x};\ a > 0, b > 0$$

38. For what values of a and b will the given equation have an infinite number of solutions?

$$a^{3x} = b^{2x};\ a > 0, b > 0$$

39. For what values of a and b do the graphs of $g(x) = a^x$ and $f(x) = b^x$ intersect in more than one place? Explain.

40. Given that $ac \neq bd$, do the graphs of $g(x) = ca^x$ and $f(x) = db^x$ intersect in more than one place? Explain.

6.2 Simple and Compound Interest

- Solve logarithmic equations algebraically
- Use the simple and compound interest formulas to find the future value of an investment

GETTING STARTED Inflation diminishes the buying power of money over time. A common challenge faced by all investors is earning a high enough return on an investment to stay ahead of inflation. A commonly used, conservative investment is the certificate of deposit (CD). CDs are FDIC-insured up to $100,000 per person. They have a fixed maturity date, usually from three months to five years, and they typically pay higher interest than a savings account. A penalty is often charged for withdrawing funds before the maturity date.

Bankrate.com compiles advertised rates from financial institutions across the nation and allows consumers to search for the best rates locally or nationwide. In this section, we will demonstrate how to use the simple and compound interest formulas to compare investment options such as those found on Bankrate.com.

Simple Interest

An investment paying an annual rate of simple interest pays a fixed amount annually. This amount, called **interest,** is calculated as a percentage of the initial investment. The simple interest earned after t years on an investment of P dollars earning an interest rate r annually is $I = Prt$. The initial investment, P, is commonly called the **principal.** The value of the investment after t years is referred to as the **future value** of the investment.

SIMPLE INTEREST

The future value A of an initial investment P earning a simple interest rate r is given by

$$\begin{aligned} A &= P + Prt \\ &= P(1 + rt) \end{aligned}$$

where t is the number of years after the initial investment is made.

The rate, r, is the decimal form of the percentage rate. That is, 5 percent is written as 0.05. The amount of interest earned after t years is Prt.

EXAMPLE 1

Determining the Future Value of a Certificate of Deposit

On August 30, 2002, the one-year CD with the highest advertised rate nationally was a CD paying an annual rate of 3 percent simple interest annually, offered by ING DIRECT. (**Source:** www.bankrate.com.) What will be the value of a \$200 investment when it matures?

SOLUTION We have $P = 200$ dollars, $r = 0.03$, and $t = 1$ year (since it is a one-year CD). The future value of the investment is

$$\begin{aligned} A &= P + Prt \\ &= 200 + 200(0.03)(1) \\ &= 200 + 6 \\ &= 206 \end{aligned}$$

When the CD matures, it will be worth \$206.

EXAMPLE 2

Determining a Simple Interest Rate

How high a return would we have to earn if we wanted a \$200 simple-interest investment to be valued at \$300 after five years?

SOLUTION We have $A = 300$ dollars, $P = 200$ dollars, and $t = 5$ years.

$$\begin{aligned} A &= P + Prt \\ 300 &= 200 + 200(r)(5) \\ 100 &= 1000r \\ r &= 0.1 \end{aligned}$$

We would have to earn an annual rate of 10 percent simple interest in order for a \$200 investment to reach \$300 in five years.

Compound Interest

Simple interest investments earn interest only on the initial amount invested. On the other hand, compound interest investments earn interest on the initial amount invested and any previously earned interest. Consequently, simple interest investments grow linearly, while compound interest investments grow exponentially.

For example, consider a \$1000 investment into an account paying a 5 percent interest rate compounded quarterly. What will be the future value of the account after one year? As demonstrated in Chapter 5, an annual rate of 5 percent corresponds with an annual growth factor of 1.05 (recall that $b = 1 + r$). However, since interest is paid quarterly, not annually, we must determine the quarterly rate. Since there are four quarters in one year, the quarterly interest rate is calculated by dividing the annual interest rate by 4. The quarterly rate is $\frac{5 \text{ percent}}{4} = 1.25$ percent. The quarterly growth factor is 1.0125. So to determine the value of our investment after interest is paid at the end of the first quarter, we multiply our \$1000 investment by 1.0125.

$$1000(1.0125) = \$1012.50$$

Over the course of the first quarter, \$12.50 in interest was earned.

What will the value of the account be after interest is paid at the end of the *second* quarter? Since the rate and **compounding frequency** (the number of times interest is paid each year) remain unchanged, the quarterly growth factor is still 1.0125. Therefore, if we multiply the \$1012.50 account value by the growth factor, we will obtain the investment value after interest is paid at the end of the second quarter.

$$1012.50(1.0125) = \$1025.16$$

Over the course of the second quarter, \$12.66 in interest was earned. Why was this amount greater than the interest for the first quarter? Because the first-quarter interest earned interest itself during the second quarter.

What will the value of the account be after interest is paid at the end of the *third* quarter? Since the rate and compounding frequency remain unchanged, the quarterly growth factor is still 1.0125. Therefore, if we multiply the \$1025.16 account value by the growth factor, we will obtain the investment value after interest is paid at the end of the third quarter.

$$1025.16(1.0125) = \$1037.97$$

Over the course of the third quarter, \$12.81 in interest was earned.

What will the value of the account be after interest is paid at the end of the *fourth* quarter? Again, we multiply \$1037.97 by the growth factor of 1.0125. The investment value after interest is paid at the end of the fourth quarter is given by

$$1037.97(1.0125) = \$1050.95$$

Over the course of the fourth quarter, \$12.98 in interest was earned.

Observe that for each quarter, we calculated the future value of the account by multiplying by the growth factor one or more times.

$$\begin{aligned} 1000(1.0125)^1 &= 1012.50 \\ 1000(1.0125)^2 &= 1025.16 \\ 1000(1.0125)^3 &= 1037.97 \\ 1000(1.0125)^4 &= 1050.95 \end{aligned}$$

Observe that for each subsequent quarter, the exponent on the growth factor is increased by 1. If we want to know the account value at the end of t years, we must first convert the years to quarters. Since there are 4 quarters in a year, there are $4t$ quarters in t years.

Recall that the growth factor itself was determined by adding 1 to the annual percentage rate divided by the number of times interest was paid each year.

$$\left(1 + \frac{0.05}{4}\right) = 1.0125$$

Combining all of these observations, we can find an equation for calculating the future value A of the account after t years.

$$A = 1000\left(1 + \frac{0.05}{4}\right)^{4t}$$

This equation is the compound interest formula for the account earning 5 percent interest compounded quarterly with an initial investment of \$1000.

COMPOUND INTEREST

The future value A of an initial investment P earning a compound interest rate $100r$ percent is given by

$$A = P\left(1 + \frac{r}{n}\right)^{nt}$$

where n is the number of times interest is paid annually and t is the number of years after the initial investment is made. As with simple interest, the rate r is the decimal form of the percentage rate.

As noted earlier, the number of times interest is paid annually, n, is called the *compounding frequency*. In solving interest rate problems, we must convert the verbal description of the compounding frequency into a numeric value, as illustrated in Table 6.1.

TABLE 6.1

Interest Is Compounded	Compounding Frequency n
Annually	1
Semiannually	2
Quarterly	4
Monthly	12
Daily	365

EXAMPLE 3 Using the Compound Interest Formula to Forecast Investment Value

On August 30, 2002, the one-year CD with the second highest advertised rate nationally was a CD offered by DeepGreen Bank paying an annual rate of 2.95 percent interest compounded daily. A \$1000 minimum deposit was required. (**Source:** www.bankrate.com.) What will be the value of a \$1000 investment when it matures?

SOLUTION We have $P = 1000$ dollars, $r = 0.0295$, $t = 1$ year (since it is a one-year CD), and $n = 365$ (since interest is calculated daily). The future value of the investment is

$$\begin{aligned} A &= P\left(1 + \frac{r}{n}\right)^{nt} \\ &= 1000\left(1 + \frac{0.0295}{365}\right)^{365(1)} \\ &= 1000(1.0000808)^{365} \\ &= 1000(1.0299) \\ &= 1029.94 \end{aligned}$$

When the CD matures, it will be worth \$1029.94.

Observe that $1029.94 = 1000\left(1 + \frac{0.02994}{1}\right)^{1(1)}$. That is, earning 2.95 percent interest compounded daily for one year is equivalent to earning 2.994 percent interest compounded annually for one year.

The **annual percentage yield** of an account is the percentage rate at which interest would need to be paid if interest were calculated and paid only once a year. The annual percentage yield is the percentage that reflects the total interest to be received for a 365-day year based on an institution's compounding method. To protect consumers, Truth-in-Savings regulations require that the annual percentage yield on interest-bearing accounts be publicly disclosed.

ANNUAL PERCENTAGE YIELD

The annual percentage yield, APY, for an account earning $100r$ percent compounded n times per year is

$$APY = \left(1 + \frac{r}{n}\right)^n - 1$$

For a simple interest account,

$$APY = r$$

It is informative to look deeper into the origin of the annual percentage yield formulas. Observe that the compound interest formula

$$A = P\left(1 + \frac{r}{n}\right)^{nt}$$

may be written as

$$A = P\left[\left(1 + \frac{r}{n}\right)^n\right]^t$$

If r and n are known values, then $b = \left(1 + \frac{r}{n}\right)^n$ is a constant. The compound interest formula may then be rewritten as an exponential function:

$$A = P(b)^t$$

As discussed earlier, the annual rate of return of a quantity with growth factor b is equal to $b - 1$. In terms of investments, the annual percentage yield, APY, is given by

$$\begin{aligned} APY &= b - 1 \\ &= \left(1 + \frac{r}{n}\right)^n - 1 \qquad \text{Since } b = \left(1 + \frac{r}{n}\right)^n \end{aligned}$$

For *simple interest* investments, the annual percentage yield is equivalent to the simple interest rate.

EXAMPLE 4

Comparing Investments with Different Interest Payment Methods

On August 30, 2002, NetBank offered a five-year CD paying an annual rate of 4.74 percent compounded daily and ING DIRECT offered a five-year CD paying an annual rate of 4.80 percent simple interest. An investor has \$1000 to invest. Which CD will have a higher value when it matures?

SOLUTION

NetBank:

$$\begin{aligned} A &= P\left(1 + \frac{r}{n}\right)^{nt} \\ &= 1000\left(1 + \frac{0.0474}{365}\right)^{365(5)} \\ &= 1000(1.0001299)^{1825} \\ &= 1000(1.26742) \\ &= 1267.42 \end{aligned}$$

The NetBank CD will be worth \$1267.42.

ING DIRECT:

$$\begin{aligned} A &= P + Prt \\ &= 1000 + 1000(0.0480)(5) \\ &= 1000 + 240 \\ &= 1240 \end{aligned}$$

The ING DIRECT CD will be worth \$1240.00.

The NetBank CD will be worth \$27.42 more than the ING DIRECT CD. Surprisingly, the CD with the lower rate earned a greater return. The interest calculation method used made a substantial difference.

EXAMPLE 5

Determining the Annual Percentage Yield of an Investment

Returning to the investments introduced in Example 4, determine the annual percentage yield (*APY*) for each CD. Recall that NetBank offered a five-year CD paying an annual rate of 4.74 percent compounded daily and ING DIRECT offered a five-year CD paying an annual rate of 4.80 percent simple interest.

SOLUTION Since the annual percentage yield is the simple interest rate on a one-year investment, the *APY* for ING DIRECT is 4.80 percent. For NetBank,

$$\begin{aligned} APY &= \left(1 + \frac{r}{n}\right)^n - 1 \\ &= \left(1 + \frac{0.0474}{365}\right)^{365} - 1 \\ &= 1.0485 - 1 \\ &= 0.0485 \end{aligned}$$

The *APY* for NetBank is 4.85 percent.

When comparing savings options, it is essential to look at the *APY* in addition to the advertised rate. As we saw in Example 5, sometimes the investment with the higher annual rate has a lower *APY*.

Continuous Compound Interest

What happens to the *APY* as the compounding frequency increases? Will it increase dramatically if interest is calculated every hour ($n = 8760$), every minute ($n = 525{,}600$), or every second ($n = 31{,}536{,}000$)? Consider an account earning an annual rate of 6 percent interest (see Table 6.2).

TABLE 6.2

Compounding Frequency	Number of Times Compounded Each Year (n)	*APY* $\left(1 + \frac{0.06}{n}\right)^n - 1$
Annually	1	0.06000
Quarterly	4	0.06136
Monthly	12	0.06168
Daily	365	0.06183
Hourly	8,760	0.06184
Every minute	525,600	0.06184
Every second	31,536,000	0.06184

It appears that as the compounding frequency increases, the annual percentage yield approaches a constant value of 0.06184. Recall from Section 5.4 that the expression $\left(1 + \frac{1}{n}\right)^n$ was observed to approach the irrational number

$e = 2.718281828\ldots$ as n approached infinity. It is interesting to note that

$$\left(1 + \frac{0.06}{n}\right)^n \approx e^{0.06}$$
$$\approx 1.06184$$

In fact, as the compounding frequency increases, $\left(1 + \frac{r}{n}\right)^n \to e^r$ for any r. (The symbol $\to$ means "approaches" and is commonly used when one quantity is getting infinitely close to another quantity. When we use the phrase *infinitely close,* we mean that we can make the first value get as close to the second value as we would like.) This result is not immediately obvious; however, we can easily convince ourselves of its validity.

Recall that $\left(1 + \frac{1}{x}\right)^x \to e$ as $x \to \infty$. Let $x = \frac{n}{r}$. Observe that if r is any positive constant, then $x \to \infty$ as $n \to \infty$. Since $x = \frac{n}{r}$, $n = rx$. Let's return to the compound interest formula, $A = P\left(1 + \frac{r}{n}\right)^{nt}$. This may be rewritten as

$$A = P\left[\left(1 + \frac{r}{n}\right)^n\right]^t$$
$$= P\left[\left(1 + \frac{1}{x}\right)^{rx}\right]^t \qquad \text{Since } \frac{r}{n} = \frac{1}{x} \text{ and } n = rx$$
$$= P\left[\left(1 + \frac{1}{x}\right)^x\right]^{rt}$$

But $\left(1 + \frac{1}{x}\right)^x \to e$ as $x \to \infty$. So, as the compounding frequency n gets infinitely large, the compound interest formula becomes

$$A = Pe^{rt}$$

We represent the notion of infinitely large n by using the term **continuous compounding.**

CONTINUOUS COMPOUND INTEREST

The future value A of an initial investment P earning a continuous compound interest rate $100r$ percent is given by

$$A = Pe^{rt}$$

where t is the number of years after the initial investment is made. The annual percentage yield is $APY = e^r - 1$.

To understand how we determined the APY, observe that the continuous compound interest formula may be converted to the compound interest formula as shown here.

$$A = Pe^{rt}$$
$$= P(e^r)^t$$
$$= Pb^t \qquad \text{Let } b = e^r$$
$$= P(1 + APY)^t \qquad \text{Since } b = 1 + APY$$

Notice that $e^r = 1 + APY$ or, in other terms, $APY = e^r - 1$.

EXAMPLE 6

Determining the Effect of Increasing the Compounding Frequency

Verify that the annual percentage yield of an account earning a compound interest rate of 5.00 percent approaches $e^{0.05} - 1$ as the compounding frequency increases.

SOLUTION

TABLE 6.3

Compounding Frequency	n	$\left(1 + \frac{0.05}{n}\right)^n - 1$
Annually	1	0.05000
Monthly	12	0.05116
Daily	365	0.05127
Hourly	8,760	0.05127
Continuous	$n \to \infty$	$e^{0.05} - 1 = 0.05127$

EXAMPLE 7

Converting a Quarterly Rate to a Continuous Rate

An account earning an annual interest rate of 12.00 percent compounded quarterly earns the same as an account earning what continuous rate?

SOLUTION Let r be the rate (as a decimal) for the continuously compounding account. The compound interest formula for each account may be written in the form $A = Pb^t$. For the 12.00 percent account, we have

$$\begin{aligned} b &= \left(1 + \frac{0.12}{4}\right)^4 \\ &= (1 + 0.03)^4 \\ &= (1.03)^4 \\ &= 1.1255 \end{aligned}$$

For the continuously compounding account, we have $b = e^r$. Since $b = e^r$, $\ln(b) = r$.

$$\begin{aligned} r &= \ln(1.1255) \\ &= 0.1182 \end{aligned}$$

A 12.00 percent interest rate compounded quarterly yields the same as an 11.82 percent rate compounded continuously.

EXAMPLE 8

Comparing the Values of Different Types of Investments

How many years will it take for a \$300 investment earning 5 percent compounded continuously to be worth more than a \$500 investment earning an annual interest rate of 4 percent compounded quarterly?

SOLUTION

Continuous: $A = 300e^{0.05t}$

Quarterly: $A = 500\left(1 + \frac{0.04}{4}\right)^{4t}$

We want to know when the accounts will have the same balance, so we set them equal to each other and solve.

$$300e^{0.05t} = 500\left(1 + \frac{0.04}{4}\right)^{4t}$$

$$300e^{0.05t} = 500(1.01)^{4t}$$

$$0.6e^{0.05t} = (1.01)^{4t} \qquad \text{Divide both sides by 500}$$

$$0.6(1.0513)^t = (1.0406)^t \qquad \text{Evaluate the exponents}$$

$$0.6 = \left(\frac{1.0406}{1.0513}\right)^t \qquad \text{Divide both sides by } (1.0513)^t$$

$$0.6 = (0.9899)^t \qquad \text{Simplify}$$

$$\ln(0.6) = t\ln(0.9899) \qquad \text{Take the natural log of both sides}$$

$$t = \frac{\ln(0.6)}{\ln(0.9899)} \qquad \text{Divide both sides by } \ln(0.9899)$$

$$t \approx 50.09 \qquad \text{Evaluate}$$

It will take a little more than 50 years for the accounts to reach the same balance. At the end of the first quarter of the 51st year ($t = 50.25$), the account paying an annual rate of interest compounded quarterly will have a balance of \$3694.59 and the account paying an annual rate of continuously compounded interest will have a balance of \$3700.72. It is important to note that although the mathematical solution is 50.09, the balance of the account paying an annual rate of interest compounded quarterly increases only at the end of a quarter (i.e., 50.00, 50.25, 50.50, etc.). Therefore, the quarterly-interest formula gives an accurate value for the account only if the input is one of these quarterly values.

EXAMPLE 9

Using Compound Interest to Analyze Payday Loans

On February 17, 2003, mycashnow.com published the following loan terms:

Loan amount: \$500
Loan term: 7–14 days
Finance charge: \$93.10

If a consumer borrows the money for seven days, what is the annual percentage yield on the investment for the lender? (Assume that the finance charge is entirely interest.)

SOLUTION In this problem, the lender may be viewed as the investor and the consumer is the one paying interest. The lender's initial investment is \$500. At the end of the seven-day period, the lender gets back its \$500 investment plus \$93.10 in interest. We have

$$A = P\left(1 + \frac{r}{n}\right)^{nt}$$

$$593.10 = 500\left(1 + \frac{r}{n}\right)^{n\left(\frac{7}{365}\right)} \quad \text{Since 7 days is } \frac{7}{365} \text{ of a year}$$

$$593.10 = 500(b)^{\left(\frac{7}{365}\right)} \quad \text{Let } b \text{ equal } \left(1 + \frac{r}{n}\right)^n$$

$$1.1862 = b^{\frac{7}{365}} \quad \text{Divide both sides by 500}$$

$$(1.1862)^{\frac{365}{7}} = \left(b^{\frac{7}{365}}\right)^{\frac{365}{7}} \quad \text{Raise both sides to the } \frac{365}{7}$$

$$7358.78 = b \quad \text{Evaluate}$$

Since $APY = b - 1$, $APY = 7357.78$. Converting this to a percentage gives an annual percentage yield of 735,778 percent! To put this in perspective, a \$500 investment earning 735,778 percent annually would be worth \$3,679,390 after one year.

6.2 Summary

In this section, you learned how to use the simple and compound interest formulas to compare investment options. You also learned the importance of looking at the annual percentage yield in addition to the advertised rate.

6.2 Exercises

In Exercises 1–10, answer each of the questions using the simple and/or compound interest formulas. For each exercise, quoted rates are from Bankrate.com and are accurate as of August 2002.

1. **Certificates of Deposit** First NB of Baldwin County offers a five-year CD paying an annual rate of 4.5 percent simple interest with a minimum investment of \$500. How much will the CD be worth at maturity if the minimum amount is invested?

2. **Certificates of Deposit** State Bank of Texas offers a five-year CD paying an annual rate of 4.25 percent simple interest with a minimum investment of \$1000. How much will the CD be worth at maturity if the minimum amount is invested?

3. **Certificates of Deposit** New South Federal Savings offers a one-year CD paying an annual rate of 2.75 percent simple interest with a minimum investment of \$5000. How much will the CD be worth at maturity if the minimum amount is invested?

4. **Doubling Time** What simple interest rate would an investor need to earn in order to double his or her investment in seven years?

5. **Certificates of Deposit** Ascencia Bank offers a five-year CD paying an annual rate of 4.70 percent interest compounded monthly with a minimum investment of \$500. How much will the CD be worth at maturity if the minimum amount is invested?

6. **Certificates of Deposit** Countrywide Bank offers a five-year CD paying an

annual rate of 4.69 percent interest compounded monthly with a minimum investment of $1000. How much will the CD be worth at maturity if the minimum amount is invested?

7. **Certificates of Deposit** ING DIRECT offers a five-year CD paying an annual rate of 4.80 percent simple interest. Is this CD a better investment than the Ascencia Bank CD in Exercise 5? Explain.

8. **Certificates of Deposit** Resource Bank offers a three-month CD paying an annual rate of 2.59 percent compounded quarterly with a minimum investment of $10,000. What will the value of the CD be at maturity?

9. **Certificates of Deposit** Medford Savings Bank offers a three-month CD paying an annual rate of 2.00 percent simple interest. Everbank.com offers a three-month CD paying an annual rate of 1.98 percent compounded daily. Which account will be worth more at maturity?

10. **Certificates of Deposit** ING DIRECT, Ascencia Bank, and interState NetBank offer five-year CDs with a 4.80 percent *APY*. ING DIRECT pays simple interest, Ascencia Bank pays 4.70 percent compounded monthly, and interState NetBank pays 4.69 percent compounded daily. Which account will yield the highest return at maturity? Explain.

In Exercises 11–15, calculate the annual percentage yield for each investment.

11. A savings account paying an annual rate of 3.50 percent compounded quarterly.

12. A checking account paying an annual rate of 1.25 percent compounded daily.

13. A three-year certificate of deposit paying an annual rate of 3.43 percent compounded continuously.

14. A three-year certificate of deposit paying an annual rate of 2.45 percent simple interest.

15. A savings account paying an annual rate of 2.35 percent compounded continuously.

In Exercises 16–20, determine how many years it will take for the two accounts to reach the same balance.

16. A $1000 investment in an account paying an annual rate of 3.00 percent compounded monthly and an $1100 investment in an account paying an annual rate of 3.02 percent compounded annually.

17. A $200 investment in an account paying an annual rate of 5.00 percent compounded continuously and a $300 investment in an account paying an annual rate of 6.00 percent compounded semiannually (twice a year).

18. A $2000 investment in an account paying an annual rate of 4.00 percent compounded continuously and a $3000 investment in an account paying an annual rate of 5.00 percent compounded monthly.

19. A $1000 investment in an account paying an annual rate of 5.25 percent compounded continuously and an $800 investment in an account paying an annual rate of 6.25 percent compounded continuously.

20. A $2000 investment at a simple interest rate of 12.00 percent and a $1500 investment at an interest rate of 12.00 percent compounded monthly.

In Exercises 21–28, do the following. For continuous rates, determine the annual rate that has the same APY *as the continuous rate. For annual rates, determine the continuous rate that has the same* APY *as the annual rate.*

21. Continuous rate: 7.23 percent

22. Continuous rate: 2.98 percent

23. Annual rate: 3.29 percent

24. Annual rate: 4.75 percent

25. Continuous rate: 11.02 percent

26. Annual rate: 6.54 percent

27. Annual rate: 17.20 percent

28. Continuous rate: 1.00 percent

Exercises 29–32 deal with payday loans. Some consumers turn to payday loans for needed cash. The lender gives the consumers an advance on their forthcoming paycheck and, when the consumers get paid, they pay off the lender. Although many payday lenders don't consider the finance charges they assess to be "interest", we will treat the charges as interest.

29. **Payday Loans** On February 17, 2003, mycashnow.com published the following loan terms:

Loan amount: $100
Loan term: 7–14 days
Finance charge: $18.62

If the consumer borrows the money for seven days, what is the annual percentage rate on the loan?

30. **Payday Loans** On February 17, 2003, mycashnow.com published the following loan terms:

Loan amount: $500
Loan term: 7–14 days
Finance charge: $93.10

If the consumer borrows the money for 14 days, what is the annual percentage rate on the loan?

31. **Payday Loans** On February 17, 2003, mycashnow.com published the following loan terms:

Loan amount: $200
Loan term: 7–14 days
Finance charge: $37.24

If the consumer borrows the money for 14 days, what is the annual percentage rate on the loan?

32. **Payday Loans** On February 17, 2003, mycashnow.com published the following loan terms:

Loan amount: $300
Loan term: 7–14 days
Finance charge: $55.86

If the consumer borrows the money for 10 days, what is the annual percentage rate on the loan?

Exercises 33–36 deal with inflation. Inflation is defined to be the percentage rate at which prices are increasing. The Consumer Price Index is often used to measure inflation.

33. **White Bread Prices** The average cost of a loaf of bread in 1983 was assigned a Consumer Price Index value of 100. By 2000, the index number had increased to 199.1, indicating that the price had nearly doubled. (**Source:** *Statistical Abstract of the United States, 2001*, Table 694, p. 454.)

(a) On average, at what annual percentage rate was the cost of white bread increasing between 1983 and 2000?

(b) If the average price of a loaf of bread was $1.12 in 1993 and prices continue to increase at the same rate, what will be the cost of a loaf of bread in 2008?

34. **Tomato Prices** The average cost of a pound of tomatoes in 1980 was assigned a Consumer Price Index value of 81.9. By 2000, the index number had increased to 234.7, indicating that the price had nearly tripled. (**Source:** *Statistical Abstract of the United States, 2001*, Table 694, p. 454.)

(a) On average, at what annual percentage rate was the cost of tomatoes increasing between 1980 and 2000?

(b) In 1990, the average price of field-grown tomatoes was $0.86 per pound. (**Source:** *Statistical Abstract of the United States, 2001*, Table 706, p. 468.) If prices continue to increase at the same percentage rate, what will be the cost of a pound of tomatoes in 2010?

35. **Rental Costs** The average cost of renting a place to live was assigned a Consumer Price Index value of 103.0 in 1983. By 1997, the index number had increased to 186.4. (**Source:** *Statistical Abstract of the United States, 2001*, Table 694, p. 454.)

(a) On average, at what annual percentage rate was the cost of rent increasing between 1983 and 1997?

(b) If an apartment cost $220 a month to rent in 1983, in what year is it expected to cost $500 per month?

36. **Interstate Long Distance Costs** The average cost per minute to make an interstate long distance phone call was assigned a Consumer Price Index value of 101.5 in 1983. By 2000, the index number had fallen to 68.0. (**Source:** *Statistical Abstract of the United States, 2001*, Table 694, p. 454.)

(a) On average, at what annual percentage rate was the cost of interstate long distance calls decreasing between 1983 and 2000?

(b) What do you think may have been the cause of the price reduction?

Exercises 37–43 deal with savings bonds. The U.S. government issues savings bonds to consumers at half of their face value. For example, a $50 savings bond is issued to a consumer for $25. The bond is guaranteed to be worth its face value no later than its original maturity date.

37. **Savings Bonds** Series E/EE bonds issued on 12/01/81 have an original maturity date of 12/01/89. (**Source:** www.publicdebt.treas.gov.)

What is the minimum annual percentage yield earned by the bonds from the date of issue to the original maturity date?

38. **Savings Bonds** Series E/EE bonds issued on 1/1/93 have an original maturity date of 1/1/05.
(**Source:** www.publicdebt.treas.gov.)
What is the minimum annual percentage yield that will be earned by the bonds from the date of issue to the original maturity date?

39. **Savings Bonds** Series E/EE bonds issued on 1/1/96 have an original maturity date of 1/1/13.
(**Source:** www.publicdebt.treas.gov.)
What is the minimum annual percentage yield earned by the bonds from the date of issue to the original maturity date?

40. **Savings Bonds** A \$500 EE savings bond issued for \$250 in July 2000 was worth \$283.60 in July 2003.
(**Source:** www.publicdebt.treas.gov.)
What is the annual percentage yield on the bond?

41. **Savings Bonds** A \$50 EE savings bond issued for \$25 in January 2000 was worth \$28.68 in February 2003.
(**Source:** www.publicdebt.treas.gov.)
What is the annual percentage yield on the bond?

42. **Savings Bonds** A \$50 EE savings bond issued for \$25 in January 1980 was worth \$125.62 in February 2003.
(**Source:** www.publicdebt.treas.gov.)
What is the annual percentage yield on the bond?

43. **Savings Bonds** A \$50 EE savings bond issued for \$25 in January 1990 was worth \$52.88 in February 2003.
(**Source:** www.publicdebt.treas.gov.)
What is the annual percentage yield on the bond?

Exercises 44–48 are intended to challenge your understanding of simple and compound interest.

44. Explain the practical meaning of $\frac{r}{12}$ from the compound interest formula $A = P\left(1 + \frac{r}{12}\right)^{12t}$.

45. A \$1000 simple interest investment is earning twice the interest rate of a \$1000 investment earning interest compounded annually. If, after *two* years, the investments have the same value, what is the compound interest rate?

46. A \$1000 simple interest investment is earning twice the interest rate of a \$1000 investment earning interest compounded annually. If, after *three* years, the investments have the same value, what is the compound interest rate?

47. You decide to open up a payday loan company. You charge 20 percent of the loan amount for a one-month loan. Treating the finance charge as interest, what is your annual percentage yield on a \$500 loan?

48. Explain how an understanding of the compound interest and simple interest formulas can benefit you personally.

6.3 Future Value of an Increasing Annuity

- Use the future value of an increasing annuity formula to answer questions related to accounts with systematic investments, such as retirement savings plans

GETTING STARTED You can be a millionaire in 40 years. Some of you will become millionaires much sooner. Becoming a millionaire is not a rich man's game. With a little financial discipline and a basic knowledge of annuities, you can make your money work for you, rather than always working for money.

In this section, we will discuss how to use the future value of an increasing annuity formula to help prepare for a comfortable retirement. Using this formula, we can analyze a variety of financial scenarios with relative ease.

Financial instruments that require a series of payments of set size and frequency are called **annuities.** Suppose we want to have \$1,000,000 in our retirement account after 40 years of investing. We plan on making monthly contributions to an account earning 12 percent interest compounded monthly. How much should we invest each month? The monthly periodic interest rate is $i = \frac{0.12}{12} = 0.01$. Since we plan on making monthly contributions at the end of each month for 40 years, we will make a total of 480 contributions. Since interest is also paid at the end of each month, our first payment will earn interest for 479 months. (It didn't earn any interest the first month, since the contribution was made at the end of the month.) Our second payment will earn interest for 478 months. Our third payment will earn interest for 477 months, and so on. Our second to last payment will earn interest for one month, and our final payment, made at the end of the 480th month, will not earn any interest at all. If R is the amount of the monthly contribution, then the compound interest formula tells us that the future value of a monthly contribution after m months is

$$\begin{aligned} FV &= R(1 + 0.01)^m \\ &= R(1.01)^m \end{aligned}$$

The future value of our entire investment is the sum of the future values of each monthly contribution (see Table 6.4).

TABLE 6.4

Contribution Number	Contribution Amount	Future Value
1	R	$FV = R(1.01)^{479}$
2	R	$FV = R(1.01)^{478}$
3	R	$FV = R(1.01)^{477}$
...	...	...
479	R	$FV = R(1.01)^1$
480	R	$FV = R(1.01)^0$

We want the future value of our investment to be \$1,000,000.

$$\begin{aligned} 1{,}000{,}000 &= R(1.01)^{479} + R(1.01)^{478} + \cdots + R(1.01)^1 + R(1.01)^0 \\ 1{,}000{,}000 &= R[(1.01)^{479} + (1.01)^{478} + \cdots + (1.01)^1 + 1] \end{aligned}$$

Calculating the value of the right-hand side of the equation looks like an arduous task; however, we will be able to greatly simplify it with some clever algebraic manipulation. We will begin by multiplying the equation by 1.01. The resultant equation is

$$\begin{aligned} 1.01(1{,}000{,}000) &= 1.01R[(1.01)^{479} + (1.01)^{478} + \cdots + (1.01)^1 + 1] \\ 1{,}010{,}000 &= R[(1.01)^{480} + (1.01)^{479} + \cdots + (1.01)^2 + (1.01)^1] \end{aligned}$$

We will now subtract the original equation $1{,}000{,}000 = R[(1.01)^{479} + (1.01)^{478} + \cdots + (1.01)^1 + 1]$ from the new equation $1{,}010{,}000 = R[(1.01)^{480} + (1.01)^{479} + \cdots + (1.01)^2 + (1.01)^1]$.

$$\begin{aligned} 1{,}010{,}000 &= R[(1.01)^{480} + (1.01)^{479} + \cdots + (1.01)^2 + (1.01)^1] \\ -1{,}000{,}000 &= R[(1.01)^{479} + (1.01)^{478} + \cdots + (1.01)^1 + 1] \\ \hline 10{,}000 &= R[(1.01)^{480} \quad - 1] \end{aligned}$$

Solving for R, we get

$$\begin{aligned} R &= \frac{10{,}000}{(1.01)^{480} - 1} \\ &= \$85.00 \end{aligned}$$

You will be a millionaire in 40 years if you invest \$85.00 a month in an account earning an annual rate of 12 percent interest compounded monthly. Isn't that amazing? Most of us spend three to four times that much on car payments alone.

Generalizing the formula used in the million-dollar example, we have

$$R = \frac{FV(i)}{(1 + i)^m - 1}$$

where $m = nt$ is the number of payments to be made, $i = \frac{r}{n}$ is the periodic interest rate, FV is the future value of the annuity, and R is the payment amount. It is common to represent the payment amount R by PMT. Solving the formula for FV, we get the future value of an increasing annuity formula.

FUTURE VALUE OF AN INCREASING ANNUITY (WITH A ZERO PRESENT VALUE)

The future value FV of an increasing annuity with an initial balance of 0 dollars is given by

$$FV = PMT\,\frac{(1 + i)^m - 1}{i}$$

where $i = \frac{r}{n}$ is the periodic interest rate, $m = nt$ is the number of payments, and PMT is the payment amount.

If the annuity has a nonzero present value (a balance), the present value will grow in accordance with the compound interest formula. The future value of the entire investment will be the sum of the two quantities.

FUTURE VALUE OF AN INCREASING ANNUITY (WITH A NONZERO PRESENT VALUE)

The future value FV of an annuity with a present value PV is given by

$$FV = PV(1 + i)^m + PMT\,\frac{(1 + i)^m - 1}{i}$$

where $i = \frac{r}{n}$ is the periodic interest rate, $m = nt$ is the number of payments, and PMT is the payment amount.

The future value of an increasing annuity formulas are typically used with sinking funds. A **sinking fund** is an account established for the purpose of accumulating money to pay off future debts or obligations.

EXAMPLE 1

Forecasting Retirement Savings with an Increasing Annuity

Since its inception on August 1, 1952, the CREF Stock Account has performed very well. As of April 30, 2005, it had earned an average of 10.39 percent annually. (**Source:** TIAA-CREF.) A 35-year-old college professor contributes \$200 to her CREF Stock Account twice a month. She hopes to retire in 25 years with a retirement account worth \$800,000 or more. Her current account balance is \$15,000. Assuming that she will be able to earn an annual rate of 10.39 percent compounded semimonthly, will she be able to reach her retirement goal?

SOLUTION The professor's retirement account is a sinking fund, since it will be used to pay for future living expenses after she retires. The future value of her current account balance may be calculated using the compound interest formula. Since interest is compounded twice a month, $n = 24$.

$$\begin{aligned} A &= P\left(1 + \frac{r}{n}\right)^{nt} \\ &= 15{,}000\left(1 + \frac{0.1039}{24}\right)^{24(25)} \\ &= 15{,}000(13.3550) \\ &= 200{,}325.57 \end{aligned}$$

The current account balance will grow to \$200,325.57 over the next 25 years.

The future value of her future contributions may be calculated using the future value of an increasing annuity formula. For her scenario,

$m = 24 \cdot 25 = 600$ and $i = \frac{0.1039}{24} = 0.004329$.

$$\begin{aligned} FV &= PMT\,\frac{(1 + i)^m - 1}{i} \\ &= 200\,\frac{(1 + 0.004329)^{600} - 1}{0.004329} \\ &= 200(2853.9067) \\ &= 570{,}781.34 \end{aligned}$$

In 25 years, her future contributions will grow to \$570,781.34

As shown in the future value of an increasing annuity (with a nonzero balance) formula, her final account balance can be determined by summing the two amounts.

$$200{,}325.57 + 570{,}781.34 = 771{,}106.91$$

Her final account balance will be \$771,106.91. According to the scenario, she will be roughly \$29,000 short of her \$800,000 goal.

EXAMPLE 2

Using an Increasing Annuity to Plan for Retirement

In Example 1, the college professor ended up short of her investment goal. To meet her goal, she will need to either increase her payment amount or increase the number of payments.

(a) Determine the payment amount that will allow her to reach her goal in 25 years.

(b) Determine the number of \$200 payments that will allow her to reach her goal.

SOLUTION

(a) Since she is not increasing the length of her investment, her current balance will still grow to \$200,325.57. She needs to earn an additional \$599,674.43 from her future contributions in order to reach her \$800,000 goal. We can determine the payment amount by solving the future value of an increasing annuity formula for *PMT*.

$$FV = PMT\,\frac{(1+i)^m - 1}{i}$$

$$PMT = FV\frac{i}{(1+i)^m - 1}$$

$$\begin{aligned} PMT &= 599{,}674.43\left[\frac{0.004329}{(1+0.004329)^{600} - 1}\right] \\ &= 599{,}674.43(0.0003504) \\ &= 210.13 \end{aligned}$$

Increasing the payment amount by \$10.13 will increase the final account balance by \$28,910.07, slightly more than the \$28,893.09 we needed. A \$10.12 increase would increase the final account balance by \$28,881.53, slightly less than the \$28,893.09 we needed.

(b) Since she is increasing the amount of time her investment will earn interest, growth in both the current balance and in future contributions will be affected. We need the sum of her current and future investments to total \$800,000.

$$FV = P\left(1 + \frac{r}{n}\right)^{nt} + PMT\left[\frac{\left(1 + \frac{r}{n}\right)^{nt} - 1}{\frac{r}{n}}\right]$$

$$800{,}000 = 15{,}000(1.004329)^{24t} + 200\left[\frac{(1.004329)^{24t} - 1}{0.004329}\right]$$

$$800{,}000 = 15{,}000(1.1092)^{t} + 200\left[\frac{(1.1092)^{t} - 1}{0.004329}\right] \quad \text{Evaluate exponents}$$

$$800{,}000 = 15{,}000(1.1092)^{t} + 46{,}198.27[(1.1092)^{t} - 1] \quad \text{Divide 200 by 0.004329}$$

$$800{,}000 = 15{,}000(1.1092)^{t} + 46{,}198.27(1.1092)^{t} - 46{,}198.27 \quad \text{Distribute 46,198.27}$$

$$846{,}198.27 = 15{,}000(1.1092)^{t} + 46{,}198.27(1.1092)^{t} \quad \text{Add 46,198.27}$$

$$846{,}198.27 = 61{,}198.27(1.1092)^{t} \quad \text{Combine like terms}$$

$$13.8272 = (1.1092)^{t} \quad \text{Divide by 61,198.27}$$

$$\ln(13.8272) = t\ln(1.1092) \quad \text{Log both sides; Rule 7}$$

$$t = \frac{\ln(13.8272)}{\ln(1.1092)} \quad \text{Divide by ln (1.1092)}$$

$$t \approx 25.34 \quad \text{Evaluate on calculator}$$

She needs to make contributions for an additional 0.34 year (4.1 months). Since she makes contributions twice a month, she needs to make only nine more payments to reach her goal.

In Example 2, we saw how a small change in the payment amount or in the length of the investment may substantially increase the value of the investment. Likewise, seemingly small changes in the interest rate can have a dramatic impact on the growth of an investment. The magnifying factor is time. This phenomenon is commonly referred to as the **time value of money.** A dollar today is worth more than a dollar in the future, even after adjusting for inflation, because a dollar invested today can earn interest until the time the dollar in the future is received. By increasing her semimonthly payment by \$10.13 she contributed only an additional \$6078 to her retirement account. However, over the 25-year period, that additional investment grew to nearly \$29,000.

The future value of an increasing annuity formula is so applicable to everyday life that the TI-83 Plus has a built-in Time Value of Money Solver that allows us to solve annuity problems quickly and easily.

TECHNOLOGY TIP

Using the TVM Solver

1. Press the blue APPS button to bring up a list of available applications. Select `1:Finance...` from the list and press ENTER

2. Select `1:TVM Solver` from the list of available financial applications and press ENTER.

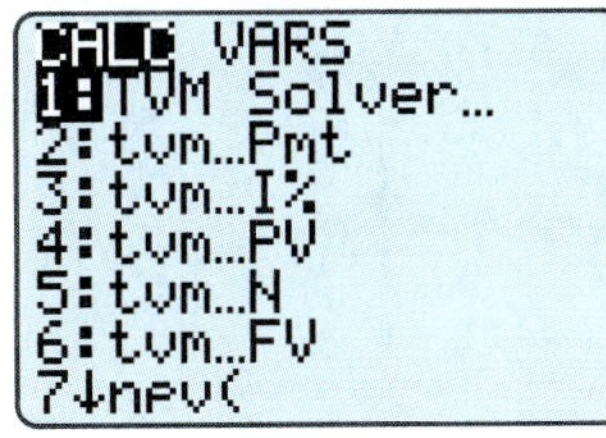

3. The TVM Solver displays several variables. `N` is the number of payments, `I%` is the annual interest rate (not the periodic rate) written as a percent, `PV` is the negative of the present value of the investment, `PMT` is the negative of the payment amount, `FV` is the future value of the investment, `P/Y` is the number of payments per year, `C/Y` is the number of compounding periods per year, and `PMT: END BEGIN` tells the calculator whether to calculate interest at the beginning or end of the compounding period.

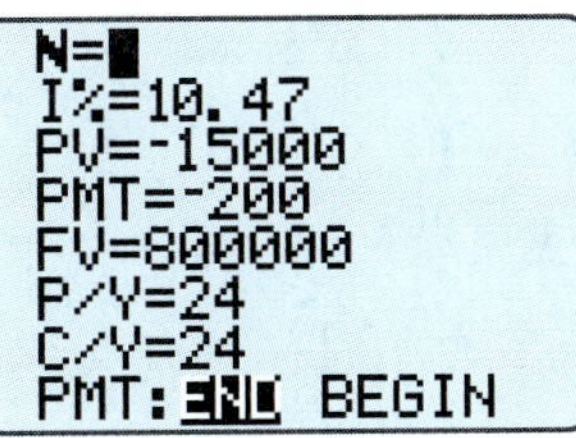

4. To solve for any quantity, place the cursor on the unknown quantity and press ALPHA then ENTER.

Since the TVM Solver is designed for decreasing annuities such as auto and home loans, we had to make the present value and payment amounts negative in order to use it with our increasing annuity problem. Admittedly, this is a bit counterintuitive. We will run through a couple of examples to allow you to practice using the TVM Solver.

EXAMPLE 3 Determining the Future Value of an Investment

You're contributing $100 per month to a retirement plan earning 6.31 percent compounded monthly. Your retirement account has a current balance of $2309.33. How long will it take for your investment to reach $10,000?

SOLUTION Entering each of the known values, we get the result shown in Figure 6.2.

FIGURE 6.2

Notice that the value of N is still 604.6393625 from a previous problem. We haven't solved for N yet. As soon as we do, the calculator will replace the value with the correct value of N. Solving for N, we get the result shown in Figure 6.3.

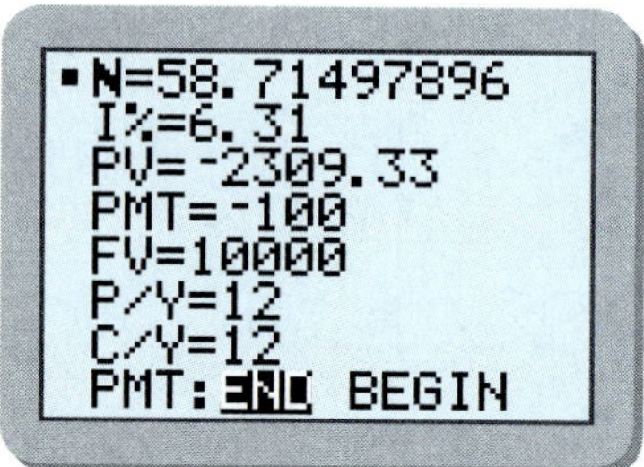

FIGURE 6.3

Notice that the calculator placed a black square to the left of N. This is the calculator's way of indicating that the value of N is a calculated value.

It will take 59 monthly payments for the account balance to reach $10,000. That is, it will take four years and eleven months for the account balance to reach $10,000.

EXAMPLE 4 Determining the Future Value of an Investment

Returning to the scenario in Example 3, what interest rate would you have to earn if you wanted your account balance to reach $10,000 in four years?

SOLUTION Four years of monthly payments equals 48 payments. Updating the TVM Solver, we have the result shown in Figure 6.4.

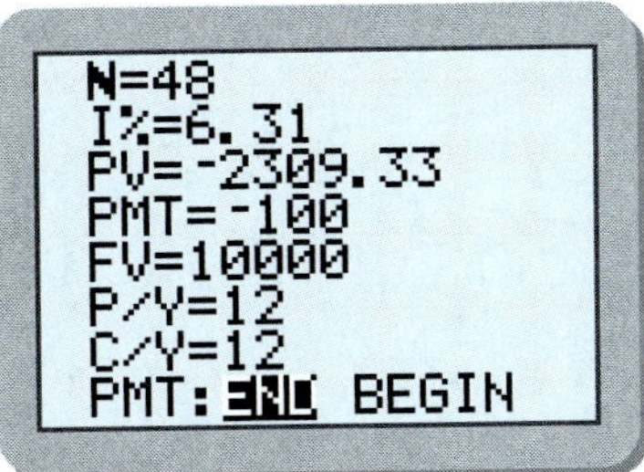

FIGURE 6.4

Solving for I%, we get the result shown in Figure 6.5.

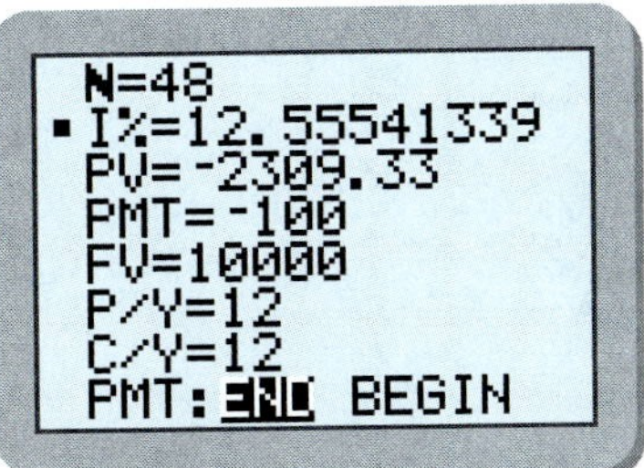

FIGURE 6.5

Notice that the calculator placed a black square to the left of the I%. This is the calculator's way of indicating that the value of I% is a calculated value.

You must earn an annual rate of 12.56 percent compounded monthly if you want the balance to grow to $10,000 in four years.

EXAMPLE 5

Comparing the Future Values of Two Investment Accounts

You plan on investing $50 per month for five years into one of two accounts. The first account pays 2.95 percent interest compounded daily. The second account pays 3.00 percent interest compounded quarterly. Which account will have the greatest value at the end of five years?

SOLUTION Notice that the compounding frequency and the number of payments are not equal, as in previous examples.

First Account:

Since interest is being compounded daily, C/Y=365.

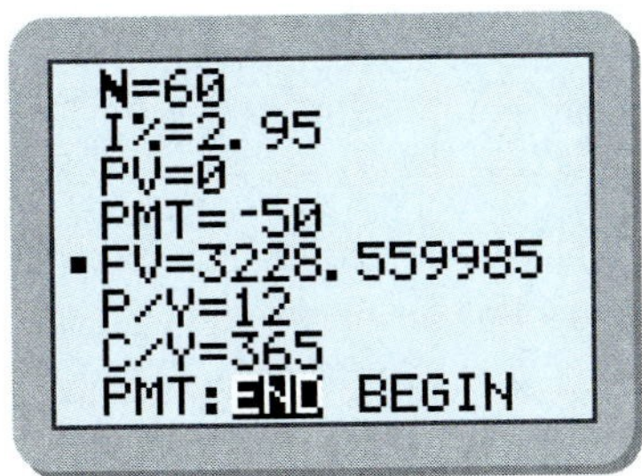

FIGURE 6.6

The future value will be \$3,228.56, as shown in Figure 6.6.

Second Account:

Since interest is being compounded quarterly, C/Y=4.

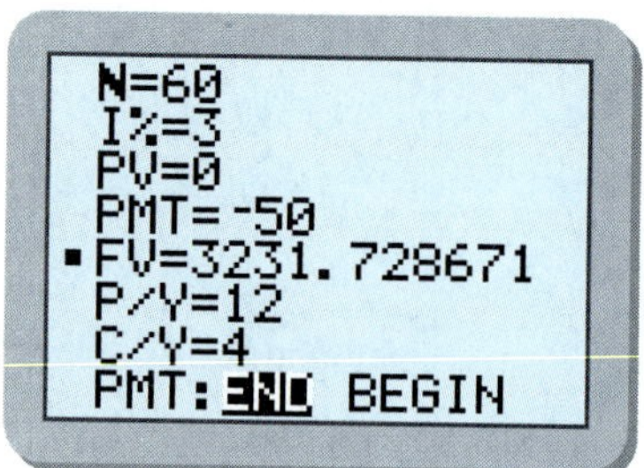

FIGURE 6.7

The future value of the account will be \$3,231.73, as shown in Figure 6.7. The second account will yield a higher return than the first account.

6.3 Summary

In this section, you learned how to use the future value of an increasing annuity formula to help you prepare for a comfortable retirement. You also learned how to use the TVM Solver to solve increasing annuity problems quickly.

6.3 Exercises

In Exercises 1–5, solve for the indicated value algebraically by using one of the future value of an increasing annuity formulas.

1. An investor opened a new account earning 6 percent interest compounded monthly. She plans to make a \$250 contribution each month. How long will it take for her account to be valued at \$10,000?

2. If an investor contributes \$100 a month to an account earning 4 percent interest compounded monthly, what will be the value of the account in five years?

3. The present value of an account earning 8 percent interest compounded monthly is \$2400. An investor needs the account to be worth \$4800 two years from now. How much must he contribute monthly?

4. How long will it take a \$2000 account balance earning 2 percent compounded monthly with monthly contributions of \$50 to grow to \$3000?

5. How much must you invest each month into an account earning 4 percent interest compounded monthly if you want the account balance to be at least \$2000 after two years?

In Exercises 6–15, retirement plan scenarios are set up using mutual funds, which, by their very nature, do not guarantee a particular rate of return. Mutual funds are used in these scenarios instead of savings accounts because they typically outperform savings accounts in the long run and are considered by many to be a better long-term investment. It is meaningful to look at the long-range performance of a mutual fund as an indicator of what its future performance might *be. Performance data are accurate as of August 31, 2002.*

Solve each of the increasing annuity problems algebraically. Then check your solution using the TVM Solver on your calculator.

6. **Mutual Funds** The CREF Global Equities Fund has earned an average of 8.37 percent since May 1, 1992. (**Source:** TIAA-CREF.) Assuming that the account will earn an annual rate of 8.37 percent compounded monthly, what will be the future value of the account in 10 years if $150 contributions are made monthly?

7. **Mutual Funds** The CREF Growth Fund has earned an average of 8.16 percent since April 29, 1994. (**Source:** TIAA-CREF.) Assuming that the account will earn an annual rate of 8.16 percent compounded monthly, what will be the future value of the account in 30 years if $200 contributions are made monthly?

8. **Mutual Funds** The CREF Equity Index Fund has earned an average of 11.33 percent since April 29, 1994. (**Source:** TIAA-CREF.) Assuming that the account will earn an annual rate of 11.33 percent compounded monthly, what will be the future value of the account in 25 years if $200 contributions are made monthly?

9. **Mutual Funds** The CREF Social Choice Fund has earned an average of 10.51 percent since March 1, 1990. (**Source:** TIAA-CREF.) Assuming that the account will earn an annual rate of 10.51 percent compounded monthly, what will be the future value of the account in 20 years if $300 contributions are made monthly?

10. **Mutual Funds** The CREF Bond Market Fund has earned an average of 8.00 percent since March 1, 1990. (**Source:** TIAA-CREF.) Assuming that the account will earn an annual rate of 8.00 percent compounded monthly, how many years will it take for the account to grow to $250,000 if $300 contributions are made monthly?

11. **Mutual Funds** The CREF Inflation-linked Bond Fund has earned an average of 6.88 percent since May 1, 1997. (**Source:** TIAA-CREF.) Assuming that the account will earn an annual rate of 6.88 percent compounded monthly, what will the future value of an account with a present value of $2500 be in 20 years if $300 contributions are made monthly?

12. **Mutual Funds** The TIAA Real Estate Account has earned an average of 8.11 percent since October 2, 1995. (**Source:** TIAA-CREF.) Assuming that the account will earn an annual rate of 8.11 percent compounded monthly, what will the future value of an account with a present value of $3200 be in 25 years if $125 contributions are made monthly?

13. **Mutual Funds** The CREF Money Market Account has earned an average of 5.51 percent since April 1, 1988. (**Source:** TIAA-CREF.) Assuming that the account will earn an annual rate of 5.51 percent compounded monthly, what will the future value of an account with a present value of $2000 be in 15 years if $100 contributions are made monthly?

14. **Mutual Funds** The Harbor Capital Appreciation Fund has earned an average of 13.16 percent since December 29, 1987. (**Source:** Harbor Funds.) Assuming that the account will earn an annual rate of 13.16 percent compounded monthly, what will the future value of an account with a present value of $6200 be in 30 years if $250 contributions are made monthly?

15. **Mutual Funds** The Large Cap Value Fund has earned an average of 11.38 percent since December 29, 1987. (**Source:** Harbor Funds.) Assuming that the account will earn an annual rate of 11.38 percent compounded monthly and has a present value of $6200, how long will it take to reach an account value of $124,000 if $250 contributions are made monthly?

In Exercises 16–29, solve each problem using the TVM Solver.

16. **Retirement Account** An investor hopes to increase his retirement account to $950,000 by the time he retires 10 years from now. The present value of the account is $250,259.12. He expects his investment to earn an annual rate of 13.16 percent compounded monthly. What must his monthly investment be in order for him to reach his goal?

17. **Retirement Account** An investor hopes to increase her retirement account to $1,250,000 by the time she retires 15 years from now. The present value of the account is $219,976.31. She expects her investment to earn an annual rate of 11.38 percent compounded monthly. What must her monthly investment be in order for her to reach her goal?

18. **Retirement Account** An investor hopes to increase her retirement account to \$1,000,000 by the time she retires 35 years from now. The present value of her account is \$2498.39. She expects her investment to earn an annual rate of 6.88 percent compounded monthly. What must her monthly investment be in order for her to reach her goal?

19. **Retirement Account** An investor hopes to increase his retirement account to \$500,000 by the time he retires two years from now. The present value of his account is \$425,524.21. He expects his investment to earn an annual rate of 5.51 percent compounded monthly. What must his monthly investment be in order for him to reach his goal?

20. **Retirement Account** An investor hopes to increase his retirement account to \$500,000 by the time he retires two years from now. The present value of his account is \$425,524.21. He can afford to make \$350 contributions monthly. Assuming that interest is compounded monthly, what interest rate must he earn in order for him to reach his goal?

21. **Retirement Account** An investor hopes to increase his retirement account to \$700,000 by the time he retires 11 years from now. The present value of his account is \$332,185.22. He can afford to make \$450 contributions monthly. Assuming that interest is compounded monthly, what interest rate must he earn in order for him to reach his goal?

22. **Retirement Account** An investor hopes to increase her retirement account to \$2,900,000 by the time she retires 21 years from now. The present value of her account is \$10,185.22. She can afford to make \$650 contributions monthly. Assuming that interest is compounded monthly, what interest rate must she earn in order for her to reach her goal?

23. **Retirement Account** An investor hopes to increase her retirement account to \$1,000,000 by the time she retires 44 years from now. The present value of her account is \$250. She can afford to make \$75 contributions monthly. Assuming that interest is compounded monthly, what interest rate must she earn in order for her to reach her goal?

24. **Buying a Home** A recent college graduate is saving up for the down payment on a home. She currently has \$2025 in a savings account earning an annual rate of 2.35 percent compounded *quarterly*. She contributes \$200 a month to the account. She anticipates that she will be able to afford a \$125,000 home. She intends to finance the home through the government FHA loan program, which requires a 5 percent down payment instead of the typical 20 percent down payment. How long will it take her to save up enough money to make a down payment that is 5 percent of the price of the home?

25. **Buying a Home** A grocery store manager is saving up for the down payment on a home. He and his wife currently have \$1200 in a savings account earning an annual rate of 3.25 percent compounded *quarterly*. They contribute \$500 a month to the account. They anticipate that they will be able to afford a \$195,000 home. They intend to finance the home through the government FHA loan program, which requires a 5 percent down payment instead of the typical 20 percent down payment. How long will it take them to save up enough money to make a down payment that is 5 percent of the price of the home?

26. **Buying a Home** A hotel catering manager is saving up for the down payment on a home. She and her husband currently have \$502.02 in a savings account earning an annual rate of 1.25 percent compounded monthly. She contributes her \$500.00 quarterly bonus to the account each quarter. They anticipate that they will be able to afford a \$105,000 home. How long will it take them to save up enough money to make a down payment that is 5 percent of the price of the home?

27. **Buying a Home** A steelworker is saving up for the down payment on his next home. He expects to make a profit of \$25,000 from the sale of his current home. He wants to move into a \$183,000 home and finance it with a conventional loan, which requires a 20 percent down payment. He currently has \$4200 in a savings account earning an annual rate of 2.33 percent compounded monthly. He contributes \$200 to the account each

time he gets his biweekly paycheck. Assuming that he will use the profit from the sale of his current home as part of the down payment, how long will it take him to save up enough money to make a down payment that is 20 percent of the price of the home?

28. **Buying a Home** An expert video-game programmer is saving up for the down payment on his next home. He expects to make a profit of \$20,000 from the sale of his current home. He wants to move into a \$525,000 home and finance it with a conventional loan, which requires a 20 percent down payment. He currently has \$24,200 in a conservative investment account earning an annual rate of 4.69 percent compounded daily. He contributes \$1,000 to the account monthly. How long will it take him to save up enough money to make a down payment that is 20 percent of the price of the home? Assume that he will use the profit from the sale of his first home as part of the down payment.

29. **Buying a Home** A freelance illustrator is saving up for the down payment on a condominium overlooking the ocean. She contributes \$15,000 semiannually to a savings account earning an annual rate of 3.89 percent compounded quarterly. (Artists and authors earning royalties on their work are typically paid twice a year.) She currently has \$15,290.23 in the account. Since she is uncertain of her future income, she doesn't want to finance any more than 25 percent of the price of the condominium. How long will it take her to save up enough money to make a 75 percent down payment on a \$110,000 condominium?

Exercises 30–41 are intended to challenge your ability to work with increasing annuities.

30. **Investment Value** Investment account A has a present value of \$1200 and is earning an annual rate of 6 percent interest compounded monthly. Investment account B has a present value of \$500 and is earning an annual rate of 6 percent interest compounded monthly. Monthly contributions of \$100 are made to investment account B.

    How long will it take for the two accounts to have the same value?

31. **Investment Value** Investment account A has a present value of \$8000 and is earning an annual rate of 8 percent interest compounded quarterly. Investment account B has a present value of \$2000 and is earning an annual rate of 6 percent interest compounded monthly. Monthly contributions of \$100 are made to investment account B.

    How long will it take for the two accounts to have the same value?

32. **Investment Value** A wise investor began investing \$50 a month in an account earning an annual rate of 4 percent interest compounded monthly on her 20th birthday. Her twin brother began investing \$100 a month into an account earning an annual rate of 4 percent compounded monthly 10 years later. How old will the twins be when the account values are equal?

33. **Investment Value** A wise investor began investing \$200 a month in an account earning an annual rate of 6 percent interest compounded monthly on his 18th birthday. His twin brother began investing \$200 a month into an account earning an annual rate of 12 percent compounded monthly 10 years later. How old will the twins be when the account values are equal?

34. **Solve the equation**

$$FV = PV(1 + i)^m + PMT\,\frac{(1 + i)^m - 1}{i}$$

    for PV. Simplify the result as much as possible.

35. Give an example of a case in which the result of Exercise 34 would be useful.

36. **Investment Value** An investment account earns 1 percent interest each *month*. (This is *not* the same as 1 percent interest compounded monthly.) If \$100 is contributed to the account monthly, what will be the value of the account after 10 payments are made? (Assume that deposits are made and interest is paid at the end of the month. An investment must be in the account for a full month before it earns interest.)

37. **Investment Value** An account earns 2 percent interest each *quarter*. (This is *not* the same as 2 percent interest compounded quarterly.) If \$200 is contributed to the account quarterly, what will be the value of the account after six payments are made? (Assume that deposits are made and interest is paid at the end of the quarter. An investment must be in the account for a full quarter before it earns interest.)

38. Investment Value An account earns 3 percent interest every six months. (This is *not* the same as 3 percent interest compounded semiannually.) If \$500 is contributed to the account every six months, what will be the value of the account after two years? (Assume that deposits are made and interest is paid at the end of the six-month period. An investment must be in the account for the full six months before it earns interest.)

39. Graph the functions

$$f(t) = \frac{100[(1.005)^{12t} - 1]}{0.005} \quad \text{and}$$

$$g(t) = \frac{30[(1.003)^{24t} - 1]}{0.003}$$

Find all points of intersection of the graphs. If f and g represent the value of two investment accounts t years from now, how should we interpret the meaning of the intersection point(s)?

40. Graph the functions

$$f(t) = \frac{100[(1.005)^{12t} - 1]}{0.005} \quad \text{and}$$

$$g(t) = \frac{60[(1.006)^{6t} - 1]}{0.006}$$

Find all points of intersection of the graphs. If f and g represent the value of two investment accounts t years from now, how should we interpret the meaning of the intersection point(s)?

41. Investment Value Two different investment accounts earn an annual rate of 6 percent interest compounded monthly over a period of 40 years. At the end of 40 years, which account will have a higher value?

Account A: \$100 monthly investments during the first 20 years and no investments during the last 20 years

Account B: No investments during the first 20 years and \$300 monthly investments during the last 20 years.

6.4 Present Value of a Decreasing Annuity

- Use the present value of a decreasing annuity formula to answer questions related to accounts with systematic payments, such as auto loans

GETTING STARTED A college student is planning on purchasing a two-year-old Honda Civic coupe. The average retail price of the used car is \$10,800. (**Source:** www.nadaguides.com.) The student has \$2000 for a down payment and wants to finance the balance of the loan over three years. She has been approved for a 5.24 percent loan by VirtualBank. (In September 2002, VirtualBank was identified by Bankrate.com as having one of the best auto loan rates nationally.) What will her monthly payments be? Including interest charges, how much will she actually pay for the car? Questions such as these can be answered by using the present value of a decreasing annuity formula.

In this section, we will discuss how to use the present value of a decreasing annuity formula as a tool in making financial decisions such as borrowing money for an auto or a home. We will also look at the power and peril of credit cards.

Amortization

A familiar decreasing annuity is the auto loan. Payments reduce the **principal** (amount owed) and cover interest charges. This process of paying off a loan is called **amortization.** An amortization table shows the lender and the loan recipient what portion of each payment goes to principal and what portion goes to interest. It also details how many payments are necessary to pay off the loan. Before signing any loan documents, wise consumers should make sure they understand the full cost of taking on a loan. The amortization table for the auto loan introduced at the start of this section is shown in Table 6.5. We generated the table using a spreadsheet.

TABLE 6.5

Payment Number	Principal (Present Value)	Interest (0.0044 × Principal)	Payment	New Balance (Principal + Interest − Payment)
1	$8,800.00	$38.43	$264.69	$8,573.74
2	$8,573.74	$37.44	$264.69	$8,346.49
3	$8,346.49	$36.45	$264.69	$8,118.24
4	$8,118.24	$35.45	$264.69	$7,889.00
5	$7,889.00	$34.45	$264.69	$7,658.76
6	$7,658.76	$33.44	$264.69	$7,427.51
7	$7,427.51	$32.43	$264.69	$7,195.26
8	$7,195.26	$31.42	$264.69	$6,961.99
9	$6,961.99	$30.40	$264.69	$6,727.70
10	$6,727.70	$29.38	$264.69	$6,492.38
11	$6,492.38	$28.35	$264.69	$6,256.04
12	$6,256.04	$27.32	$264.69	$6,018.67
13	$6,018.67	$26.28	$264.69	$5,780.26
14	$5,780.26	$25.24	$264.69	$5540.81
15	$5,540.81	$24.19	$264.69	$5,300.32
16	$5,300.32	$23.14	$264.69	$5,058.77
17	$5,058.77	$22.09	$264.69	$4,816.17
18	$4,816.17	$21.03	$264.69	$4,572.51
19	$4,572.51	$19.97	$264.69	$4,327.79

(Continued)

TABLE 6.5 *(Continued)*

Payment Number	Principal (Present Value)	Interest ($0.0044 \times$ Principal)	Payment	New Balance (Principal + Interest − Payment)
20	$4,327.79	$18.90	$264.69	$4,082.00
21	$4,082.00	$17.82	$264.69	$3,835.13
22	$3,835.13	$16.75	$264.69	$3,587.19
23	$3,587.19	$15.66	$264.69	$3,338.16
24	$3,338.16	$14.58	$264.69	$3,088.05
25	$3,088.05	$13.48	$264.69	$2,836.85
26	$2,836.85	$12.39	$264.69	$2,584.54
27	$2,584.54	$11.29	$264.69	$2,331.14
28	$2,331.14	$10.18	$264.69	$2,076.63
29	$2,076.63	$9.07	$264.69	$1,821.01
30	$1,821.01	$7.95	$264.69	$1,564.27
31	$1,564.27	$6.83	$264.69	$1,306.41
32	$1,306.41	$5.70	$264.69	$1,047.42
33	$1,047.42	$4.57	$264.69	$787.31
34	$787.31	$3.44	$264.69	$526.06
35	$526.06	$2.30	$264.69	$263.66
36	$263.66	$1.15	$264.81	$0.00

A graph of the loan balance as a function of the payment number is shown in Figure 6.8.

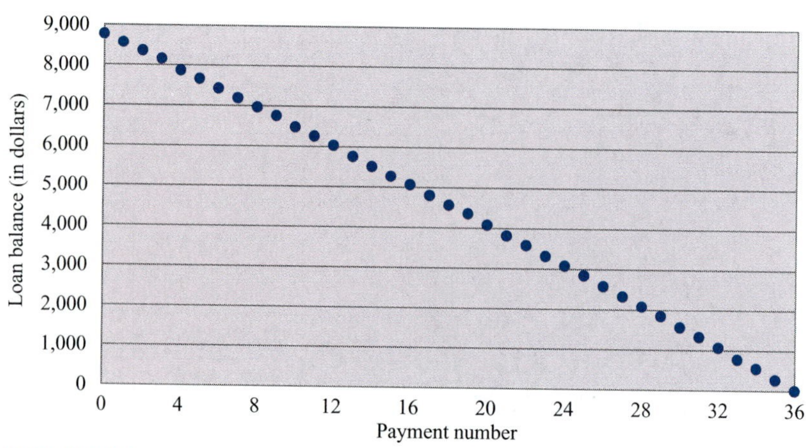

FIGURE 6.8

Since the interest rate is 5.24 percent compounded monthly, we needed to determine the monthly percentage rate in order to calculate the monthly interest.

$$\begin{aligned}\text{Monthly rate} &= \frac{\text{annual rate}}{12}\\ &= \frac{0.0524}{12}\\ &\approx 0.0044\end{aligned}$$

Monthly interest is about 0.44 percent of the principal. Since the first loan payment is not due until the month after the student acquired the loan, she will pay interest on the full loan amount in her first payment. The new principal balance ($8573.74) is calculated by adding the interest cost ($38.43) to the initial principal amount ($8800) and then subtracting the payment amount ($264.69). This process of calculating interest is repeated for the new principal balance. Since the principal balance has been reduced, the amount of interest due with the second payment is reduced to $37.44. By the 36th payment, interest is a mere $1.15. You may be wondering how we knew that the monthly payment should be $264.69. As you may have guessed, there is also a formula for decreasing annuities such as auto loans. The formula may be used to determine the amount of the monthly car payment.

PRESENT VALUE OF A DECREASING ANNUITY (WITH A FUTURE VALUE OF ZERO)

The present value PV of a decreasing annuity with a future value of zero is given by

$$PV = PMT\left[\frac{1-(1+i)^{-m}}{i}\right]$$

where $i = \frac{r}{n}$ is the periodic interest rate, $m = nt$ is the number of payments, and PMT is the payment amount.

The **present value** of an annuity is the amount of money that would need to be invested today in order to provide the desired number of equal periodic payments. We will demonstrate how the formula was developed by using an example. Suppose we want to withdraw $3000 per month from our retirement account for the next 25 years, beginning at the end of the month. The account is earning an annual rate of 6 percent interest compounded monthly. We want to know if we have enough money in the account to last us 25 years.

Since the payments will be withdrawn over a period of 25 years and will continue to earn interest until they are withdrawn, the present value of each payment is less than \$3000. We can calculate the present value of each payment by using a slightly modified version of the compound interest formula.

$$A = P\left(1 + \frac{r}{n}\right)^{nt}$$

$$A = PMT\left(1 + \frac{r}{n}\right)^{m} \qquad \text{Let } P = PMT \text{ and } m = nt$$

$$3000 = PMT\left(1 + \frac{0.06}{12}\right)^{m}$$

$$3000 = PMT(1 + 0.005)^{m}$$

$$3000 = PMT(1.005)^{m}$$

$$PMT = \frac{3000}{(1.005)^{m}}$$

We can calculate the present value of the first payment by substituting in $m = 1$.

$$PMT = \frac{3000}{(1.005)^{1}} = 2985.07$$

The present value of the first payment is \$2985.07, since it will earn interest for only a month before it is withdrawn. Calculating the present value of the second payment, we get

$$PMT = \frac{3000}{(1.005)^{2}} = 2970.22$$

The present value of the second payment is \$2970.22, since it will earn interest for two months before it is withdrawn. We can calculate the present value of any of the 300 payments.

$$PMT = \frac{3000}{(1.005)^{300}} = 671.90$$

The present value of the 300th payment is \$671.90. We can calculate the present value of the entire investment by adding up the present values of the individual payments.

$$PV = \frac{3000}{(1.005)^{300}} + \frac{3000}{(1.005)^{299}} + \cdots + \frac{3000}{(1.005)^{2}} + \frac{3000}{(1.005)^{1}}$$

Multiplying both sides of the equation by $(1.005)^{300}$, we get

$$(1.005)^{300} \cdot PV = (1.005)^{300}\left[\frac{3000}{(1.005)^{300}} + \frac{3000}{(1.005)^{299}} + \cdots + \frac{3000}{(1.005)^{2}} + \frac{3000}{(1.005)^{1}}\right]$$

$$(1.005)^{300} \cdot PV = 3000\left[\frac{(1.005)^{300}}{(1.005)^{300}} + \frac{(1.005)^{300}}{(1.005)^{299}} + \cdots + \frac{(1.005)^{300}}{(1.005)^{2}} + \frac{(1.005)^{300}}{(1.005)^{1}}\right] \qquad \text{Factor out 3000; Distribute } (1.005)^{300}$$

$$(1.005)^{300} \cdot PV = 3000[1 + (1.005)^{t} + \cdots + (1.005)^{298} + (1.005)^{299}] \qquad \text{Reduce quotients}$$

One amazing property of algebra is that $1 + x + x^2 + \cdots + x^{n-1} = \frac{x^n - 1}{x - 1}$. Consequently, $[1 + (1.005)^1 + \cdots + (1.005)^{298} + (1.005)^{299}]$ is equivalent to $\frac{1.005^{300} - 1}{1.005 - 1}$. Therefore, the equation may be rewritten as

$$(1.005)^{300} \cdot PV = 3000\left[\frac{(1.005)^{300} - 1}{1.005 - 1}\right]$$

$$PV = \frac{3000}{(1.005)^{300}}\left[\frac{(1.005)^{300} - 1}{0.005}\right] \quad \text{Divide both sides by } (1.005)^{300}$$

$$= 3000\left[\frac{(1.005)^{300} - 1}{(1.005)^{300}0.005}\right] \quad \text{Distribute } \frac{1}{(1.005)^{300}}$$

$$= 3000\left[\frac{1 - (1.005)^{-300}}{0.005}\right] \quad \text{Divide numerator by } (1.005)^{300}$$

Observe that this formula is of the form $PV = PMT\left[\frac{1 - (1 + i)^{-m}}{i}\right]$, that is, the present value of a decreasing annuity formula introduced earlier. Continuing with the computation, we get

$$PV = 3000(155.21)$$
$$= 465{,}620.59$$

If we have \$465,620.59 in our retirement account today, we will be able to withdraw \$3000 per month for the next 25 years. At the end of that time, our account balance will be zero.

EXAMPLE 1 Determining the Monthly Payment on an Auto Loan

A student intends to buy a two-year-old Honda Civic for \$10,800. She plans to put \$2000 down and finance the remaining balance on a 36-month loan charging 5.24 percent interest compounded monthly. What will be her monthly payment?

SOLUTION Since $10{,}800 - 2000 = 8800$, her initial loan amount is \$8800. Since payments are made monthly, her periodic interest rate is

$$i = \frac{0.0524}{12}$$
$$= 0.004367$$

She will make 36 payments, so $m = 36$. Substituting these values into the present value of a decreasing annuity formula, we get

$$PV = PMT\left[\frac{1 - (1 + i)^{-m}}{i}\right]$$

$$8800 = PMT\left[\frac{1 - (1 + 0.004367)^{-36}}{0.004367}\right]$$

$$8800 = PMT(33.24)$$

$$PMT = 264.69$$

Her monthly payment is \$264.69.

EXAMPLE 2

Determining the Loan Period on an Auto Loan

The boyfriend of the student in Example 1 wants her to buy a brand new $13,270 Civic coupe instead of the used model, even though she's told him that she can't afford more than a $264.69 monthly payment. He assures her that she can still make the same payment; she will just have to extend the length of her loan by a few months. How much longer will it take her to pay off the loan if she buys the new car instead of the used model?

SOLUTION She will still make the $2000 down payment, so she needs to finance only $11,270. The monthly periodic rate and payment amount remain the same. Substituting these values into the present value of a decreasing annuity formula, we get

$$PV = PMT\left[\frac{1-(1+i)^{-m}}{i}\right]$$

$$11{,}270 = 264.69\left[\frac{1-(1+0.004367)^{-m}}{0.004367}\right]$$

$$11{,}270 = 60{,}616.03[1-(1.004367)^{-m}] \quad \text{Divide 264.69 by 0.004367}$$

$$0.1859 = 1-(1.004367)^{-m} \quad \text{Divide both sides by 60,616.03}$$

$$-0.8141 = -(1.004367)^{-m} \quad \text{Subtract 1 from both sides}$$

$$0.8141 = (1.004367)^{-m} \quad \text{Multiply both sides by } -1$$

$$\ln(0.8141) = -m\ln(1.004367) \quad \text{Log both sides; Log Rule 7}$$

$$m = -\frac{\ln(0.8141)}{\ln(1.004367)} \quad \text{Divide both sides by } \ln(1.004367)$$

$$= 47.21 \quad \text{Evaluate on calculator}$$

She will have to make 48 payments to pay off the loan. (The last payment will be smaller than the first 47.) It will take her one year longer (48 months versus 36 months) to pay for the new car than to pay for the used car.

We can determine the actual amount paid for the car (including interest charges) by multiplying the payment amount by the number of payments. In Example 1, the total cost of the car was approximately

$$P = 2000 + 36(264.69)$$
$$= \$11{,}528.84$$

In Example 2, the total cost of the car was approximately

$$P = 2000 + 48(264.69)$$
$$= \$14{,}705.12$$

These are both approximations, since the final payment on a loan often differs slightly from the rest of the payments. This is due to the fact that we round payment amounts to the nearest cent instead of the nearest fraction of a penny. The final payment is used to compensate for the error introduced by rounding. In the case of Example 2, the final payment will be substantially different from the rest of the payments, since only 0.21 of a normal payment is necessary to pay off the loan.

Although we can use the present value of a decreasing annuity formula to determine the amount of any loan payment, one way to reduce the number of computations in future problems is to solve the present value of a decreasing annuity formula $PV = PMT\left[\frac{1-(1+i)^{-m}}{i}\right]$ for PMT.

$$PV = PMT\left[\frac{1-(1+i)^{-m}}{i}\right]$$

$$i(PV) = PMT[1-(1+i)^{-m}]$$ Multiply both sides by i

$$\frac{i(PV)}{1-(1+i)^{-m}} = PMT$$ Divide both sides by $1-(1+i)^{-m}$

This is referred to as the **amortization formula.**

AMORTIZATION

The payment PMT required to amortize (pay off) a debt of PV dollars is given by

$$PMT = \frac{i(PV)}{1-(1+i)^{-m}}$$

where $i = \frac{r}{n}$ is the periodic interest rate and $m = nt$ is the number of equal periodic payments.

EXAMPLE 3

Determining the Mortgage Payment for a Home Loan

Homeowners typically finance their homes using a 15-year or 30-year mortgage. Interest rates are typically lower for 15-year mortgages than for 30-year mortgages; however, the monthly payments are substantially larger. On September 4, 2002, mortgage rates hit their lowest point since the mid-1960s. The average 30-year fixed mortgage rate was 6.17 percent, and the average 15-year fixed mortgage rate was 5.62 percent. (**Source:** www.bankrate.com.) What would the mortgage payment be for a couple purchasing a \$233,750 home given that they made a 20 percent down payment? Calculate the payment amount for both mortgage options.

SOLUTION Since the couple is making a 20 percent down payment, only 80 percent of the purchase price will be financed. Since $0.8(233{,}750) = 187{,}000$, the amount financed is \$187,000.

15-year mortgage:

The monthly periodic rate is $i = \frac{0.0562}{12} = 0.004683$. The number of payments is $m = 15(12) = 180$. Applying the amortization formula, we have

$$\begin{aligned} PMT &= \frac{i(PV)}{1-(1+i)^{-m}} \\ &= \frac{0.004683(187{,}000)}{1-(1+0.004683)^{-180}} \\ &= \frac{875.78}{0.5687} \\ &= 1539.88 \end{aligned}$$

The monthly principal and interest payment for the 15-year mortgage is \$1539.88. (Although we round values before we write them down, we keep the actual values in our calculator. This ensures that our final answer is not adversely affected by round-off error.)

30-year mortgage:

The monthly periodic rate is $i = \frac{0.0617}{12} = 0.005142$. The number of payments is $m = 30(12) = 360$. Applying the amortization formula, we have

$$\begin{aligned} PMT &= \frac{i(PV)}{1-(1+i)^{-m}} \\ &= \frac{0.005142(187{,}000)}{1-(1+0.005142)^{-360}} \\ &= \frac{961.49}{0.8422} \\ &= 1141.68 \end{aligned}$$

The monthly principal and interest payment for the 30-year mortgage is \$1141.68.

The TVM Solver on our calculator may also be used to solve decreasing annuity problems; however, this time the value of *PMT* will be positive instead of negative. The present value amount will still be negative. We will illustrate how to use the TVM Solver for a decreasing annuity in the next example.

EXAMPLE 4

Determining How to Reduce the Length of a Loan

Suppose the couple in Example 3 chose the 30-year mortgage. If they increase their monthly payment to \$1200, how long will it take them to pay off the loan? If they want to pay off the loan in 20 years, how much should they pay each month?

SOLUTION Using the TVM Solver, we get the results shown in Figure 6.9.

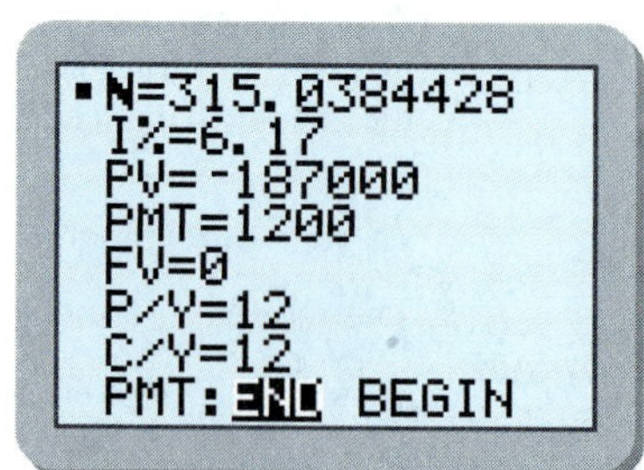

FIGURE 6.9

If they increase their monthly payment to \$1200, it will take about 315 monthly payments (26.25 years) for them to pay off the loan. The \$58.32 increase in their payment knocked more than three years off the loan.

Using the TVM Solver to solve for paying off the loan in 20 years, we get the results in Figure 6.10.

FIGURE 6.10

If they increase their payment to $1358.13, they'll be able to pay off the loan in 20 years.

Many first-time home buyers are surprised when they learn how much their home (including interest) actually costs. Truth-in-Lending laws require mortgage companies to disclose the total cost to home buyers before loan documents are signed.

The couple in Example 3 paid $1141.68 a month for 30 years. The total cost of their $233,750 home (including the $46,750 down payment) was $457,754.80, nearly double the value of their home! If they increased their payment to $1358.13, as shown in Example 4, their total home cost would be $372,701.20, an $85,053.60 savings. Fortunately, because of inflation, most homes appreciate over time. If housing prices increase by 2 percent annually, the $233,750 home will be worth over $423,000 thirty years later.

Sometimes we are interested in knowing when a decreasing annuity will reach a certain nonzero value. Since the interest rate on a 15-year mortgage is typically less than the rate on a 30-year mortgage, many homeowners refinance their homes with a 15-year mortgage once their principal balance reaches a certain level. The present value of a decreasing annuity (with a nonzero future value) formula may be used to address this situation.

PRESENT VALUE OF A DECREASING ANNUITY (WITH A NONZERO FUTURE VALUE)

The present value PV of a decreasing annuity with a nonzero future value FV is given by

$$PV = FV(1 + i)^{-m} + PMT\left[\frac{1 - (1 + i)^{-m}}{i}\right]$$

where $i = \frac{r}{n}$ is the periodic interest rate, $m = nt$ is the number of payments, and PMT is the payment amount.

EXAMPLE 5 **Determining When a Loan Will Reach a Designated Value**

A couple has purchased a home with a 30-year, \$150,000 mortgage charging 6 percent interest compounded monthly. Their monthly principal and interest payment is \$899.33. They want to refinance their home when the loan balance is \$75,000. How long will it take for the account value to drop below \$75,000?

SOLUTION The present value is \$150,000, and the future value is \$75,000. The loan rate is 6 percent, so $i = \frac{0.06}{12} = 0.005$. Since payments are made monthly, the total number of payments is given by $m = 12t$. We have

$$PV = FV(1+i)^{-m} + PMT\left[\frac{1-(1+i)^{-m}}{i}\right]$$

$$150{,}000 = 75{,}000(1+0.005)^{-12t} + 899.33\left[\frac{1-(1+0.005)^{-12t}}{0.005}\right]$$

$$150{,}000 = 75{,}000(0.9419)^{t} + 899.33\left[\frac{1-(0.9419)^{t}}{0.005}\right] \qquad \text{Evaluate exponents}$$

$$(0.005)150{,}000 = (0.005)\left\{75{,}000(0.9419)^{t} + 899.33\left[\frac{1-(0.9419)^{t}}{0.005}\right]\right\} \qquad \text{Multiply by 0.005}$$

$$750 = 375(0.9419)^{t} + 899.33[1-(0.9419)^{t}] \qquad \text{Distribute 0.005}$$

$$750 = 375(0.9419)^{t} + 899.33 - 899.33(0.9419)^{t} \qquad \text{Distribute 899.33}$$

$$-149.33 = 375(0.9419)^{t} - 899.33(0.9419)^{t} \qquad \text{Subtract 899.33}$$

$$-149.33 = -524.33(0.9419)^{t} \qquad \text{Group like terms}$$

$$0.2848 = (0.9419)^{t} \qquad \text{Divide by } -524.33$$

$$\ln(0.2848) = \ln(0.9419)^{t} \qquad \text{Log both sides}$$

$$\ln(0.2848) = t\ln(0.9419) \qquad \text{Log Rule 7}$$

$$t = \frac{\ln(0.2848)}{\ln(0.9419)} \qquad \text{Divide by } \ln(0.9419)$$

$$\approx 20.983\text{years} \qquad \text{Evaluate on calculator}$$

In 21 years (252 monthly payments), the account balance will drop below \$75,000.

Since we will solve multiple problems of this type in the exercises, it will be helpful to solve the present value of a decreasing annuity formula for m. This will make our computations in the exercises easier.

$$PV = FV(1+i)^{-m} + PMT\left[\frac{1-(1+i)^{-m}}{i}\right]$$

$$PV = FV(1+i)^{-m} + \frac{PMT}{i}[1-(1+i)^{-m}]$$

$$PV = FV(1+i)^{-m} + \frac{PMT}{i} - \frac{PMT}{i}(1+i)^{-m} \qquad \text{Distribute } \frac{PMT}{i}$$

$$PV - \frac{PMT}{i} = FV(1+i)^{-m} - \frac{PMT}{i}(1+i)^{-m} \qquad \text{Subtract } \frac{PMT}{i}$$

$$PV - \frac{PMT}{i} = \left(FV - \frac{PMT}{i}\right)(1+i)^{-m} \qquad \text{Group like terms}$$

$$\frac{PV - \frac{PMT}{i}}{FV - \frac{PMT}{i}} = (1+i)^{-m} \qquad \text{Divide by } \left(FV - \frac{PMT}{i}\right)$$

$$\frac{\frac{iPV - PMT}{i}}{\frac{iFV - PMT}{i}} = (1+i)^{-m} \qquad \text{Rewrite rational expressions}$$

$$\frac{iPV - PMT}{iFV - PMT} = (1+i)^{-m} \qquad \text{Rewrite rational expressions}$$

$$\ln\left(\frac{iPV - PMT}{iFV - PMT}\right) = \ln(1+i)^{-m} \qquad \text{Log both sides}$$

$$\ln\left(\frac{iPV - PMT}{iFV - PMT}\right) = -m\ln(1+i) \qquad \text{Log Rule 7}$$

$$m = \frac{\ln\left(\frac{iPV - PMT}{iFV - PMT}\right)}{-\ln(1+i)} \qquad \text{Divide both sides by } -\ln(1+i)$$

When using this form of the formula, it is important to remember that m represents the number of payments and, consequently, must be rounded to a whole number.

EXAMPLE 6

Determining When a Retirement Account Will Reach a Certain Value

After working hard for 45 years, an investor has accumulated \$450,000 in her retirement account. She has placed the money into an investment that guarantees a 5 percent interest rate compounded monthly. She plans to withdraw \$5000 per month. How many monthly withdrawals will she be able to make before the account value drops below \$100,000?

SOLUTION We have

$$\begin{aligned} i &= \frac{0.05}{12} \\ &= 0.004167 \end{aligned}$$

Using the modified form of the present value of a decreasing annuity formula, we get

$$\begin{aligned} m &= \frac{\ln\left(\frac{iPV - PMT}{iFV - PMT}\right)}{-\ln(1+i)} \\ &= \frac{\ln\left[\frac{0.004167(450{,}000) - 5000}{0.004167(100{,}000) - 5000}\right]}{-\ln(1 + 0.004167)} \end{aligned}$$

$$= \frac{\ln\left(\frac{-3125}{-4583.33}\right)}{-\ln(1 + 0.004167)}$$

$$= \frac{\ln(0.6818)}{-\ln(1.004167)}$$

$$= 92.11$$

She will be able to make 92 monthly withdrawals before the account balance drops below \$100,000.

Credit Cards

Many credit card companies target college students as a source of new customers. They offer free T-shirts, low introductory rates, and other gimmicks to attract potential patrons. Many college students need to develop a good credit history, and using a credit card responsibly can aid in that endeavor. Credit cards, however, are not annuities. The minimum payment required is typically a percentage of the account balance.

The author purchased a bunk bed (for his children) for \$532.22. He charged the purchase to his credit card. When he received his credit card statement, the minimum payment due was \$12.00, a mere 2.25 percent of the balance (Figure 6.11).

Furniture WHOLESALE Customer Service (Servicio al Cliente)

Days in Billing Cycle: 31 **Statement Date: 08/26/2002**

Account Summary

PAYMENT DUE DATE:	TOTAL MINIMUM PAYMENT DUE	NEW BALANCE	PAST DUE AMOUNT	AVAILABLE CREDIT
09/20/2002	\$12.00	\$532.22	\$0.00	\$2,468.00

To avoid additional finance charges, pay by the Payment Due date above: 1) the New Balances of your Regular and Reduced Rate Credit Plans, plus 2) the New Balance of any Same As Cash or Waived Finance Charge Credit Plan that has expired or that will expire in the current billing cycle, plus 3) any Minimum Payments due on an unexpired Same As Cash or Waived Finance Charge Credit Plan.

Transactions

Transaction Date	Transaction Detail	Promo. Type/Credit Plan	Amount
06/27/2002	Previous Balance		\$0.00
08/21/2002	Warehouse	Regular Purchase	\$532.22
08/26/2002	New Balance		\$532.22

Finance Charge Summary

Promotion Type/Credit Plan	Purchase Date	Promotion Expiration Date	Previous Balance	Average Daily Balance	Daily Periodic Rate	Corresponding APR	ANNUAL PERCENTAGE RATE (APR)	FINANCE CHARGES at Periodic Rate	Deferred FINANCE CHARGES	New Balance	Minimum Payment Due
Regular Purchase 00007-01	N/A	N/A	\$0.00	\$68.71	0.04287%	15.65%	15.65%	\$0.00	\$1.00	\$532.22	\$12.00

FIGURE 6.11

Suppose that he pays 2.25 percent of the balance (rounded up to the nearest dollar) every month. The card has a 15.65 percent annual percentage rate (APR). (APR is similar to annual percentage yield.)

For illustration purposes, we will assume that he makes his payment at the same time every month. Since the account balance will change once a month, we can convert the annual percentage rate into a monthly periodic rate. It appears that Costco calculates the APR by multiplying the daily periodic rate by 365. Therefore, we will calculate the monthly periodic rate by dividing the APR by 12.

$$\text{Monthly periodic rate} = \frac{0.1565}{12} \approx 0.0130$$

Using a spreadsheet, we set up an amortization table for his credit card. How long do you think it will take him to pay off the card? Five years? Ten years? Because of the size of the amortization table (Table 6.6), we will only show a few of the payments made over the next 21.25 years! If he makes only the minimum payment, he will pay \$569.41 in interest!

TABLE 6.6

Payment Number	Principal	Interest (1.30% of Principal)	Payment (2.25% of Principal rounded up)	New Balance (Principal + Interest − Payment)
1	\$532.22	\$6.94	\$12.00	\$527.16
2	\$527.16	\$6.88	\$12.00	\$522.04
3	\$522.04	\$6.81	\$12.00	\$516.84
4	\$516.84	\$6.74	\$12.00	\$511.58
5	\$511.58	\$6.67	\$12.00	\$506.26
7	\$500.86	\$6.53	\$12.00	\$495.39
9	\$489.85	\$6.39	\$12.00	\$484.24
10	\$484.24	\$6.32	\$11.00	\$479.56
50	\$315.22	\$4.11	\$8.00	\$311.33
100	\$175.97	\$2.29	\$4.00	\$174.26
200	\$39.38	\$0.51	\$1.00	\$38.90
254	\$1.60	\$0.02	\$1.00	\$0.62
255	\$0.62	\$0.01	\$0.63	\$0.00

Can we find a formula for the balance on a credit card with a minimum payment that is a percentage of the principal? Let i be the monthly periodic rate and $100r$ percent be the percentage of the account balance B that is required for the minimum payment. Then we have

Balance after first payment:

$$\begin{aligned} B_1 &= B + iB - rB \\ &= B(1 + i - r) \end{aligned}$$

Balance after second payment:

$$\begin{aligned} B_2 &= B(1 + i - r) + i[B(1 + i - r)] - r[B(1 + i - r)] \\ &= B(1 + i - r)(1 + i - r) \\ &= B(1 + i - r)^2 \end{aligned}$$

The pattern continues, and we get

Balance after nth payment:

$$B_n = B(1 + i - r)^n$$

ESTIMATE OF THE FUTURE BALANCE OF A CREDIT CARD

The balance on a credit card, B_n, after n minimum payments have been made may be estimated by

$$B_n = B(1 + i - r)^n$$

where B is the initial balance, i is the monthly periodic interest rate, and r is the percentage of the balance that is the minimum amount required to be paid.

This formula will give us an estimate of the balance after making n payments. It will not be precise, however, because many credit card companies impose additional rules on finance charges and payments, such as requiring a minimum \$0.50 finance charge or rounding up the minimum payment to the nearest dollar. One of the reasons they do so is to ensure that the debt is eventually paid off. Note that in the formula, $B_n \neq 0$ for any n. That is, the balance will never be paid off. The credit card companies also need to ensure that the interest they collect is sufficient to make billing the customer profitable.

EXAMPLE 7

Estimating the Future Balance of a Credit Card

A credit card has a \$532.22 balance and a 15.65 percent APR. The minimum payment required is 2.25 percent of the balance. Estimate the credit card balance after 50 payments have been made.

SOLUTION We have $B = 532.22$, $i = \dfrac{0.1565}{12} = 0.0130$, $r = 0.0225$, and $n = 50$.

$$\begin{aligned} B_{50} &= 532.22(1 + 0.0130 - 0.0225)^{50} \\ &= 532.22(0.9905)^{50} \\ &= 532.22(0.6218) \\ &= 330.92 \end{aligned}$$

We estimate that after 50 minimum payments, we will still owe \$330.92. Referring back to the amortization table presented earlier in this section, we see that the actual amount owed after the 50th payment was \$311.33. Even though the amortization table had the additional requirement of rounding up the minimum payment to the nearest dollar, the estimate was still close to the actual value.

6.4 Summary

In this section, you learned how to use the present value of a decreasing annuity formula as a tool in making financial decisions such as borrowing money to purchase an auto or a home. You also learned how to estimate a future credit card balance.

6.4 Exercises

In Exercises 1–20, use the present value of a decreasing annuity formula to solve the problem algebraically. Then check your work using the TVM Solver on your calculator. All quoted prices are accurate as of September 2002. Assume that interest is compounded monthly.

1. **Auto Loan** A new 2002 Chevrolet Tahoe 1500 LS has a manufacturer's suggested retail price (MSRP) of $32,954. (**Source:** www.car-prices-costs.com.) Determine the amount of the car payment if a buyer finances the entire retail price with an 8.99 percent, 48-month loan.

2. **Auto Loan** A new 2002 Volkswagen New Beetle GL has a manufacturer's suggested retail price (MSRP) of $15,900. (**Source:** www.car-prices-costs.com.) Determine the amount of the car payment if a buyer finances the entire retail price with a 3.95 percent, 48-month loan.

3. **Auto Loan** A new 2002 Chevrolet Corvette convertible has a manufacturer's suggested retail price (MSRP) of $48,205.00. (**Source:** www.car-prices-costs.com.) Determine the amount of the car payment if a buyer finances the entire retail price with a 6.99 percent, 60-month loan.

4. **Auto Loan** A new 2002 Toyota Camry Solara SE has a manufacturer's suggested retail price (MSRP) of $19,365. (**Source:** www.car-prices-costs.com.) Determine the amount of the car payment if a buyer finances the entire retail price with a 1.5 percent, 36-month loan.

5. **Auto Loan** A new 2002 Toyota Sequoia Limited 4 × 4 has a manufacturer's suggested retail price (MSRP) of $42,725. (**Source:** www.car-prices-costs.com.) A buyer gets a $7500 credit for her trade-in and finances the balance of the price at her local credit union with a 60-month, 5.99 percent loan. Determine the amount of the car payment.

6. **Auto Loan** A new 2002 Jaguar S-Type 4.0-Liter has a manufacturer's suggested retail price (MSRP) of $49,330. (**Source:** www.car-prices-costs.com.) A buyer gets a $18,500 credit for her trade-in and finances the balance of the price at her local credit union with a 48-month, 6.99 percent loan. Determine the amount of the car payment.

7. **Auto Loan** A new 2002 Jeep Grand Cherokee Laredo has a manufacturer's suggested retail price (MSRP) of $25,665. (**Source:** www.car-prices-costs.com.) A buyer gets an $8500 credit for her trade-in and finances the balance of the price at her local bank with a 72-month, 8.99 percent loan. Determine the amount of the car payment.

8. **Auto Loan** A new 2002 Ford Windstar LTD has a manufacturer's suggested retail price (MSRP) of $34,040. (**Source:** www.car-prices-costs.com.) A buyer gets a $17,200 credit for his trade-in and finances the balance of the price with a 60-month, 2.90 percent loan. Find the amount of the car payment.

9. **Auto Loan** A family wants to buy a 2002 Toyota Sequoia SR5 but wants to keep the monthly payments to $400 or less. If the family keeps the length of the loan to 48 months or less, the credit union will give them a loan with a 2.90 percent APR. The SUV retails for $31,256. (**Source:** www.car-prices-costs.com.) How much of a down payment does the family need to make?

10. **Auto Loan** A couple wants to buy a 2002 Ford Windstar LTD, but wants to keep the monthly payments to $450 or less. If

they keep the length of the loan to 36 months or less, the dealer will give them a loan with a 1.90 percent APR. The minivan retails for $34,040. (**Source:** www.car-prices-costs.com.) How much of a down payment does the couple need to make?

11. **Timeshare Condominiums** Worldmark, the Club by Trendwest, a popular timeshare condominium company, sells vacation credit packages to consumers. Buyers can cash in credits to stay in four-star accommodations around the globe. In 2002, a perpetual 6000-credit-per-year package cost $9300. In a sales presentation, a company representative indicated that with a 10 percent down payment, the company would finance the balance over seven years at a 13.9 percent interest rate. (**Source:** Worldmark, the Club.)

    If the buyer finances the package with the company, what will be the total cost of the package (including interest)?

12. **Timeshare Condominiums** After hearing the sales presentation referred to in Exercise 11, the author went to ebay.com and found a number of people selling their Worldmark, the Club memberships at a discount. One person offered the same 6000-credit package for $5100. However, the eBay seller indicated that the buyer would have to pay a $150 document fee to change the ownership of the package officially. (The author confirmed with Worldmark, the Club that memberships are transferable.) If the buyer finances the entire cost on his 9.90 percent APR credit card and makes $133.13 payments monthly, what is the actual cost of the package (including interest)?

13. **Home Mortgage** A 2608-square-foot, four-bedroom, two-bath home in Baker County, Florida, was advertised for $265,000 in September 2002. (**Source:** www.realtor.com.)

    What will the monthly payment on a 15-year, 7.25 percent mortgage be if the buyer makes a 20 percent down payment?

14. **Home Mortgage** A three-bedroom home on 41.6 acres of land in Kittitas County, Washington, was advertised for $220,000 in September 2002. (**Source:** www.realtor.com.)

    What will the monthly payment on a 30-year, 6.85 percent mortgage be if the buyer makes a 20 percent down payment?

15. **Home Mortgage** A three-bedroom home on five acres of land in Oregon was advertised for $89,900 in September 2002. (**Source:** www.realtor.com.)

    What will the monthly payment on a 15-year, 6.50 percent mortgage be if the buyer makes a 5 percent down payment?

16. **Home Mortgage** A two-bedroom, two-bath home in New York, New York, was advertised for $1,677,545 in September 2002. (**Source:** www.realtor.com.)

    What will the monthly payment on a 30-year, 7.75 percent mortgage be if the buyer makes a 20 percent down payment?

17. **Home Mortgage** A three-bedroom, two-bath home in Waukesha County, Wisconsin, was advertised for $249,900 in September 2002. (**Source:** www.realtor.com.)

    What will the monthly payment on a 20-year, 6.00 percent mortgage be if the buyer makes a 10 percent down payment?

18. **Home Mortgage** A 2992-square-foot, four-bedroom, 2.5-bath home in Lancaster County, Nebraska, was advertised for $149,900 in September 2002. (**Source:** www.realtor.com.)

    A couple has been prequalified for a 6.75 percent, 15-year VA loan. (VA loans, which may be obtained by military veterans, don't require a down payment.) The maximum monthly payment the couple can afford is $1000. Can the couple afford this home?

19. **Home Mortgage** A 3824-square-foot, six-bedroom, three-bath home in Queen Creek, Arizona, was advertised for $302,691 in February 2005. (**Source:** www.fultonhomes.com.)

    A family finances 80 percent of the purchase price with a 5.75 percent loan. If the family makes $1500 principal and interest payments monthly, how long will it take to pay off the loan?

20. **Home Mortgage** A three-bedroom, three-bath home on six acres in Rensselaer County, New York, was advertised for $895,000 in September 2002. (**Source:** www.realtor.com.)

    A retired executive has been prequalified for a 5.25 percent, 30-year loan. She intends to make a 20 percent down payment. The maximum monthly payment she can afford is $4000. Can she afford this home?

In Exercises 21–25, determine the amount of time it will take for the loan balance to reach the indicated value. Assume that interest is compounded monthly.

21. Loan Balance How long will it take for a 5 percent interest, $250,000 loan with a monthly payment of $1000 to reach a loan balance below $100,000?

22. Loan Balance How long will it take for an 8 percent interest, $100,000 loan with a monthly payment of $500 to reach a loan balance below $50,000?

23. Loan Balance How long will it take for a 9 percent interest, $13,000 loan with a monthly payment of $375 to have a loan balance below $5000?

24. Loan Balance How long will it take for an 18 percent interest, $9000 loan with a monthly payment of $400 to have a loan balance below $4000?

25. Loan Balance How long will it take for an 11 percent interest, $10,000 loan with a monthly payment of $200 to have a loan balance below $2000?

In Exercises 26–30, determine the amount of time it will take for the retirement account balance to reach the indicated value.

26. Retirement Account Balance How long will it take for a $300,000 retirement account earning an annual rate of 5 percent interest compounded annually to drop below $200,000 if $50,000 withdrawals are taken annually?

27. Retirement Account Balance How long will it take for a $190,000 retirement account earning 3 percent interest compounded annually to drop below $100,000 if $30,000 withdrawals are taken annually?

28. Retirement Account Balance How long will it take for a $800,000 retirement account earning 6 percent interest compounded annually to drop below $300,000 if $80,000 withdrawals are taken annually?

29. Retirement Account Balance How long will it take for a $400,000 retirement account earning an annual rate of 4 percent interest compounded monthly to drop below $50,000 if $5000 withdrawals are taken monthly?

30. Retirement Account Balance How long will it take for a $300,000 retirement account earning an annual rate of 6 percent interest compounded monthly to drop below $60,000 if $4000 withdrawals are taken monthly?

In Exercises 31–35, estimate the credit card balance after the indicated number of payments has been made. In each case, assume that the minimum payment is 2.25 percent of the current balance.

31. Credit Cards The initial balance on a credit card with an 18.9 percent APR is $2505.98. What will the estimated balance be after 20 payments have been made?

32. Credit Cards The initial balance on a credit card with a 21.9 percent APR is $1890.25. What will the estimated balance be after two years of payments have been made?

33. Credit Cards The initial balance on a credit card with a 15.67 percent APR is $632.10. What will the estimated balance be after six payments have been made? Create an amortization table to check the accuracy of your estimate.

34. Credit Cards The initial balance on a credit card with a 9.90 percent APR is $981.52. What will the estimated balance be after 36 payments have been made?

35. Credit Cards The initial balance on a credit card with a 13.35 percent APR is $1225.00. What will the estimated balance be after 15 payments have been made?

Exercises 36–40 are intended to challenge your understanding of decreasing annuities.

36. Retirement Planning An employee plans on working and contributing to her retirement account monthly for 44 years and then plans to live off her retirement savings for the next 25 years. She expects her retirement account to earn an annual rate of 8 percent compounded monthly. If she wants to make $3500 monthly withdrawals from her account during retirement, how much should she contribute to her retirement account monthly during her working years?

37. Retirement Planning An employee plans on working and contributing to his retirement account monthly for 30 years and then plans to live off his retirement savings for the next 20 years. He expects his retirement account to earn an annual rate of 6 percent compounded monthly. If he wants to make $3000 monthly withdrawals from his account during retirement, how much should he contribute to his retirement account monthly during his working years?

38. **Retirement Planning** An employee plans on working and contributing to his retirement account monthly for 40 years and then plans to live off his retirement savings for the next 20 years. He expects his retirement account to earn an annual rate of 6 percent compounded monthly. He can afford to make $300 monthly contributions to his retirement account. How much money can he withdraw from the account monthly during the 20 years of his retirement?

39. **Retirement Planning** An employee plans on working and contributing to her retirement account monthly for 30 years and then plans to live off her retirement savings for the next 35 years. She expects her retirement account to earn an annual rate of 7 percent compounded monthly. She can afford to make $600 monthly contributions to her retirement account. How much money can she withdraw from the account monthly during the 35 years of her retirement?

40. **Mortgage Payoff** A couple has a $200,000 mortgage with a 7 percent annual interest rate. If they pay $1500 a month for 120 months and then $2000 a month until the loan is paid off, how long will it take them to pay off the loan?

Chapter 6 Review Exercises

Section 6.1 *In Exercises 1–4, solve the exponential equation algebraically.*

1. $1.3^x = 1.69$
2. $3^x = 243$
3. $1.2^x = 1.44$
4. $(0.5)^x = 4$

In Exercises 5–6, solve the problems algebraically.

5. **Auto Fatalities** The percentage of auto accident fatalities that are not alcohol-related may be modeled by

$$F(t) = 49.96 + 5.112 \ln (t) \text{ percent}$$

where t is the number of years since the end of 1989. (**Source:** Modeled from *Statistical Abstract of the United States, 2001*, Table 1099, p. 688.) According to the model, when will less than 25 percent of auto accident fatalities be alcohol-related?

6. **Video-Game Industry** While much of the U.S. economy was struggling in fiscal year 2002, Electronic Arts Inc., which is known for its array of video-game software, earned more than $917 million in profits. Stockholders' equity in the company (total assets minus liabilities, preferred stock, and intangible assets) may be modeled by

$$S(t) = 566.9(1.225)^t \text{ million dollars}$$

where t is the number of years since fiscal year 1998. (**Source:** Modeled from Electronic Arts Inc Annual Report, March 31, 2002.) According to the model, how long will it take for stockholders' equity to reach $1 billion?

Section 6.2 *In Exercises 7–10, answer each of the questions using the simple and/or compound interest formulas. For each exercise, quoted rates are from Bankrate.com and are accurate as of August 2002.*

7. **Certificates of Deposit** NBC Bank offers a three-month CD paying an annual rate of 2.27 percent simple interest with a minimum investment of $5000. How much will the CD be worth at maturity if the minimum amount is invested?

8. **Certificates of Deposit** Beal Bank offers a three-month CD paying an annual rate of 2.23 percent compounded quarterly with a minimum investment of $1000. How much will the CD be worth at maturity if the minimum amount is invested?

9. **Certificates of Deposit** State Farm Bank offers a five-year CD paying an annual rate of 4.63 percent interest compounded daily. Stonebridge Bank offers a five-year CD

paying an annual rate of 4.65 percent compounded monthly. Assuming that the same amount of money is invested in each account, which account will be worth more at maturity?

10. **Certificates of Deposit** Providian National Bank offers a five-year CD paying an annual rate of 4.45 percent interest compounded daily. Eastern Savings Bank offers a five-year CD paying an annual rate of 4.46 percent compounded monthly. Assuming that the same amount of money is invested in each account, which account will be worth more at maturity?

In Exercises 11–12, calculate the annual percentage yield for each investment.

11. A savings account paying an annual rate of 2.57 percent compounded monthly.

12. A checking account paying an annual rate of 1.09 percent compounded daily.

In Exercises 13–14, determine how many years it will take for the two accounts to reach the same balance.

13. A $2000 investment in an account paying an annual rate of 2.75 percent compounded monthly and a $1800 investment in an account paying an annual rate of 3.02 percent compounded daily.

14. A $750 investment in an account paying an annual rate of 2.84 percent compounded quarterly and a $700 investment in an account paying an annual rate of 2.98 percent compounded daily.

In Exercises 15–16, convert the annual rate into a continuous rate or the continuous rate into an annual rate as appropriate.

15. Annual rate: 10.25 percent

16. Continuous rate: 3.23 percent

Section 6.3 *In Exercises 17–20, solve each of the increasing annuity problems algebraically. Check your solution using the TVM Solver on your calculator.*

17. **Mutual Funds** The Harbor Capital Appreciation fund earned an average of 11.51 percent in the 10-year period prior to August 31, 2002. (**Source:** Harbor Funds.) Assuming that the account will earn an annual rate of 11.51 percent compounded monthly, what will the future value of the account be in 10 years if $250 contributions are made monthly?

18. **Mutual Funds** The Harbor Growth fund earned an average of 4.14 percent annually in the 10-year period prior to August 31, 2002. (**Source:** Harbor Funds.) Assuming that the account will earn an annual rate of 4.14 percent compounded monthly, what will the future value of the account be in 10 years if $250 contributions are made monthly?

19. **Mutual Funds** The Harbor International fund earned an average of 10.05 percent in the 10-year period prior to August 31, 2002. (**Source:** Harbor Funds.) Assuming that the account has a current balance of $2320.00 and will earn an annual rate of 10.05 percent compounded monthly, what will the future value of the account be in 20 years if $300 contributions are made monthly?

20. **Mutual Funds** The Large Cap Value Fund earned an average of 10.12 percent in the 10-year period prior to August 31, 2002. (**Source:** Harbor Funds.) Assuming that the account has a current balance of $12,650.00 and will earn an annual rate of 10.12 percent compounded monthly, what will the future value of the account be in 30 years if $400 contributions are made monthly?

In Exercises 21–22, solve each problem using the TVM Solver.

21. **Retirement Savings** An investor hopes to increase his retirement account to $800,000 by the time he retires 15 years from now. The present value of the account is $35,289.19. He expects his investment to earn an annual rate of 9.89 percent compounded monthly. What must his monthly investment be in order for him to reach his goal?

22. **Home Buying** A rock star's ex-wife is saving up for a down payment on her next home. She wants to move into a $425,000 condominium in Paris with the former drummer of the band, and to finance it with a conventional loan requiring a 20 percent down payment. She currently has $34,820.21 in an offshore investment account earning an annual rate of 8.69 percent compounded daily. She contributes $3000 to the account monthly from her online sales of items containing her ex-husband's autograph. She wants to buy the condo within a year. Will she have enough money saved to make the down payment in 12 months?

Section 6.4 *In Exercises 23–26, use the present value of a decreasing annuity formula to solve the problem algebraically. Then check your work using the TVM Solver on your calculator. All quoted prices are accurate as of September 2002.*

23. Retirement Savings A retired couple has \$450,000 in an investment account earning an annual rate of 5 percent interest compounded monthly. Starting at the end of the month, they'll begin withdrawing \$4000 a month from the account. How long will it take for them to deplete all of the money in the account?

24. Auto Loan All used cars from Hunk-a-Heap Motors cost under \$1000. A young driver wants to buy a \$950 car he has nicknamed "the Bomb." He will pay 29 percent interest compounded daily on a two-year loan as a part of the dealer's "first-time buyer" package. What will his monthly payment be?

25. Investment Property A 4400-square-foot, ten-bedroom, six-bath duplex in Utah County, Utah, was advertised for \$309,000 in September 2002. (**Source:** www.realtor.com.)

A real estate investor has the required 20 percent down payment and is prequalified for a 6.25 percent, 15-year loan. She can afford a \$300 monthly payment and expects to bring in an additional \$2400 a month by renting out the rooms to local students. With her personal funds and the rent monies, will she have enough to make the monthly payment?

26. Investment Property A 1272-square-foot, three-bedroom, two-bath duplex in Utah County, Utah, was advertised for \$124,900 in September 2002. (**Source:** www.realtor.com.)

A real estate investor has the required 20 percent down payment and is prequalified for a 6.25 percent, 15-year loan. She can afford a \$300 monthly payment and expects to bring in an additional \$1000 a month by renting out the rooms to students at Brigham Young University and Utah Valley State College. With her personal funds and the rent monies, will she have enough to make the monthly payment?

In Exercises 27–28, estimate the credit card balance after the indicated number of payments has been made. In each case, assume that the minimum payment is 2.25 percent of the current balance.

27. The initial balance on a credit card with a 13.9 percent APR is \$3795.99. What will the estimated balance be after 15 payments have been made?

28. The initial balance on a credit card with a 10.9 percent APR is \$862.88. What will the estimated balance be after two years of payments have been made?

Make It Real

PROJECT 6

What to do

1. Find a home in your area that you would like to buy.
2. Open a savings account.
3. Calculate how much money you'll need for a 5 percent down payment on the house.
4. Calculate how long it will take you to save enough for the down payment if you contribute 5 percent of your monthly income (or $100 if you earn less than $2000 per month) to the savings account.
5. Find the best 30-year FHA mortgage rate offered in your area.
6. Find the monthly mortgage payment for the house from Step 1 using the rate from Step 5, assuming that you made a 5 percent down payment on the house.
7. Set a personal goal of getting a home mortgage within the next 10 years.

Where to look for data

Homes for Sale

www.realtor.com
www.homes.com
www.homebuilder.com
Local real estate guides (typically found outside supermarkets)

Mortgage Rates

www.bankrate.com
www.lendingtree.com
Wall Street Journal
Local banks and credit unions

Savings Account Rates

www.bankrate.com
Local banks and credit unions

Chapter 7

Sets and Probability

Although we typically try to control as many factors as possible, many things in life are influenced by an element of chance. If we understand basic probability concepts, we can estimate the likelihood of a certain event's occurrence. For example, did you know that in any class with 23 students or more, it is likely that two students in the class have the same birthday?

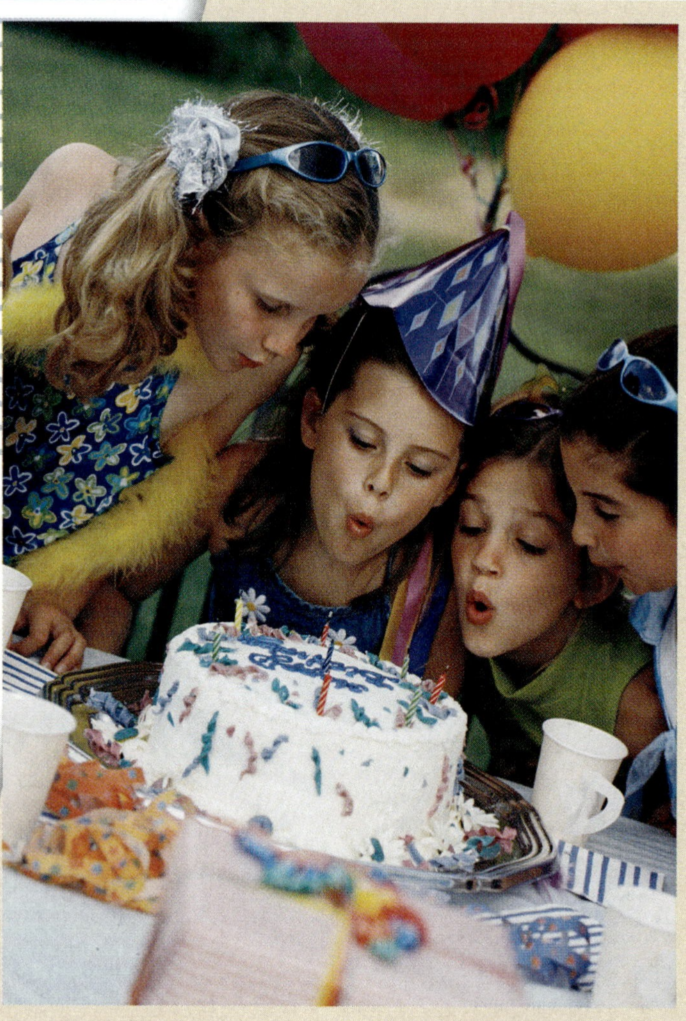

7.1 Introduction to Sets

- Use set notation to represent relationships between groups

GETTING STARTED Human resources departments have the arduous task of sorting through job applications and determining which applicants should be interviewed. One sorting strategy is to eliminate all applications that do not meet the advertised criteria for the job. The mathematical notion of sets is useful in developing the sorting process.

In this section, we will introduce set notation and Venn diagrams. We will show how sets can be used to clarify complex relationships.

In September 2002, Levi Strauss & Co. requested applications for the position of Director of Strategic Development in its San Francisco office. The director's duties included performing complex business analysis and providing a clear understanding of supply chain dynamics, the competition, and the company's marketplace performance to senior executives. In addition to other skills, the position required a bachelor's degree and a minimum of 10 years of business experience. (**Source:** www.levistrauss.com.)

Suppose the company's human resource department received applications from the nine candidates listed in Table 7.1.

TABLE 7.1

	Name	Education	Experience
1	Chen	M.B.A.	1 year internship at Bugle Boy
2	Smythe	High school diploma	12 years managing a clothing retail store
3	Rodriguez	B.S.	10 years in marketing; various positions
4	Lewis	B.A.	5 years market research and analysis
5	Black	B.S.	15 years teaching high school science
6	Nguyen	B.S.	11 years project management
7	Romero	M.B.A.	21 years marketing and management
8	Alexander	High school diploma	25 years in retail distribution
9	Farris	High school diploma	12 years as an actor/artist

The department wants to interview only those candidates who meet the company's minimum criteria. Candidates 1, 4, 5, and 9 lack the required business experience. Candidates 2, 8, and 9 lack the required educational background. The only candidates meeting both criteria are Candidates 3, 6, and 7.

An alternative way to represent the information is by using rectangles and circles in a **Venn diagram.** We will represent the set of all candidates with rectangle U, the **universal set.** We will represent the set of candidates with at least a bachelor's degree with circle A and the set of candidates with 10 or more years of business experience with circle B. Figure 7.1 represents the situation graphically.

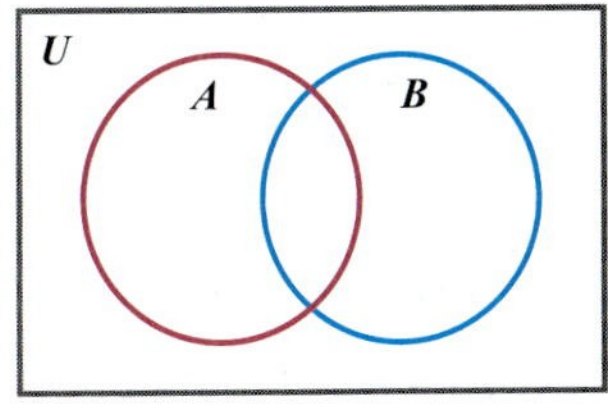

FIGURE 7.1

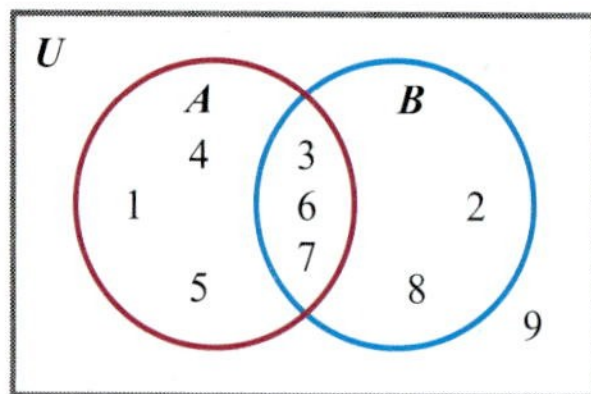

FIGURE 7.2

The set $A = \{1,3,4,5,6,7\}$. The set $B = \{2,3,6,7,8\}$. The entries inside the braces are called the **elements** of the set. We will now place the elements of each set in their appropriate location on the Venn diagram (Figure 7.2).

The candidates who meet the education requirement are in set A. The candidates who meet the business requirement are in set B. Candidates who meet both requirements are written in the **intersection of A and B,** the region that is common to both A and B. Candidate 9 doesn't meet either of the requirements and is written outside of both sets A and B.

EXAMPLE 1 Using a Venn Diagram to Sort Data

In September 2002, Electronic Arts advertised a Games Technical Support Representative position in Redwood Shores, California. Minimum qualifications included "avid computer and video gamer," "knowledge of Windows 95/98/Me Operating Systems in relation to the gaming environment (knowledge of MacOS and Windows XP is a plus)," and "self motivated and strong work ethics." **(Source:** jobs.ea.com.)

Use a Venn diagram to classify the hypothetical applicants in Table 7.2 according to their qualifications.

TABLE 7.2

	Name	Gaming Experience	Operating Systems	Work Ethic
1	Draxx	Video-game expert; plays 12 hours a day, knows 150 different games	Win 95/98/Me, Win XP, and MacOS	Fired from last two jobs for not coming to work
2	Argo	Six hours weekly on computer/video games for past 5 years	None	Successful athlete, average student, works in family business
3	Vogel	Five hours weekly on computer/video games for past 5 years	Win 95/98/Me	Eagle Scout, excellent student, successful athlete
4	Stewart	Risk, Monopoly, and other board games; hosts monthly bridge parties	None	Manages local retail store; active in a variety of professional organizations
5	Wilson	15 hours monthly on computer games	Win 95/98/Me, Win XP, and MacOS	Received outstanding employee of the year award

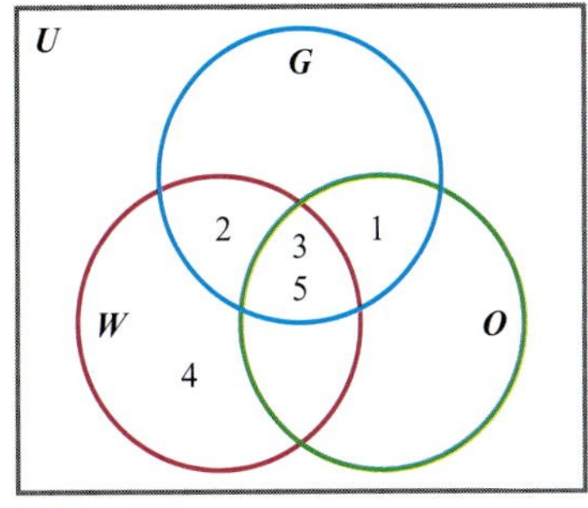

FIGURE 7.3

SOLUTION Let G be the set of applicants who are avid computer/video gamers, O be the set of applicants who know Win 95/98/Me, and W be the set of applicants with a strong work ethic. We'll consider anyone who plays computer games 15 or more hours a month to be an avid gamer. The Venn diagram is shown in Figure 7.3.

All of the applicants met at least one of the criteria; however, only applicants 3 and 5 met all three criteria.

To further our discussion of sets, it is necessary to develop some notation and terminology that will allow us to represent such concepts as the intersection of two sets symbolically.

SET TERMINOLOGY AND NOTATION

1. A set A is represented by separating its elements with commas and writing the list of elements within braces. For example, $A = \{1, 3, 5, 8\}$.
2. We write $x \in A$ and read "x is an element of A" if x is in A. For example, $3 \in A$.
3. We write $x \notin A$ and read "x is not an element of A" if x is not in A. For example, $4 \notin A$.
4. The set that contains no elements is called the **empty set** and is written $\varnothing$ or $\{\,\}$.
5. The set of all elements under consideration is called the **universal set** and is written U.
6. The set of all elements contained in U but not in A is called **A complement** and is written A'.
7. The set of all elements contained in A *and* in B is called the **intersection of A and B** and is written $A \cap B$. $A \cap B$ is read "A intersect B."
8. The set of all elements contained in A *or* in B *or* in both is called the **union of A and B** and is written $A \cup B$. $A \cup B$ is read "A union B."
9. A set A and a set B contain the same elements if and only if $A = B$.
10. If a set A is contained entirely in a set B, we say that "A is a **subset** of B" and write $A \subseteq B$.
11. If a set A is contained entirely in a set B, and $A \neq B$, then we say that "A is a **proper subset** of B" and write $A \subset B$.

EXAMPLE 2

Using Set Notation

Write each of the sentences using set notation, using Figure 7.4 as a reference.

(a) Two is an element of the intersection of G and W.
(b) The union of G and W contains the elements 1, 2, 3, 4, and 5.
(c) Four is not in O.

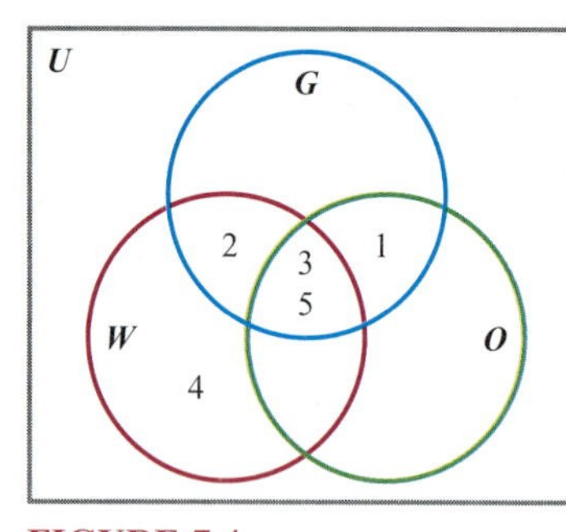

FIGURE 7.4

(d) Four is outside of G.
(e) Two is outside of the intersection of G and O.
(f) O is a proper subset of G.

SOLUTION

(a) $2 \in G \cap W$
(b) $G \cup W = \{1,2,3,4,5\}$
(c) $4 \notin O$ or $4 \in O'$
(d) $4 \in G'$ or $4 \notin G$
(e) $2 \notin (G \cap O)$ or $2 \in (G \cap O)'$
(f) $O \subset G$

EXAMPLE 3

Using a Venn Diagram to Interpret Set Notation

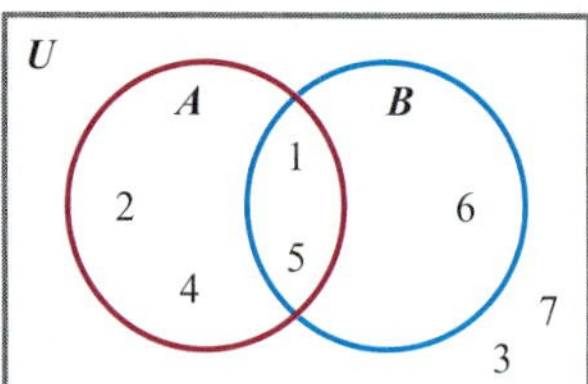

FIGURE 7.5

Use the Venn diagram in Figure 7.5 to find the elements in each of the sets. Then interpret the meaning of the results.

(a) $A \cap B$
(b) $A \cup B$
(c) U
(d) B'
(e) $A \cap B'$

SOLUTION

(a) $A \cap B = \{1, 5\}$ — Everything that is in both A and B
(b) $A \cup B = \{1, 2, 4, 5, 6\}$ — Everything that is in A or B or both
(c) $U = \{1, 2, 3, 4, 5, 6, 7\}$ — Everything in the rectangle
(d) $B' = \{2, 3, 4, 7\}$ — Everything not in B
(e) $A \cap B' = \{2, 4\}$ — Everything in A that's not in B

When two sets do not intersect, we call them **disjoint sets.** For example, the set of all licensed drivers and the set of unlicensed drivers are disjoint sets. The set of all blondes and the set of all people with Ph.D.'s are not disjoint sets because some blondes have Ph.D.'s.

When a set has a large number of elements, it is cumbersome to write down each element. Suppose C is a set containing all positive even numbers less than or equal to 100,000. Listing all 50,000 elements would require a visible effort and a lot of paper. Often it is much easier to define a set by a rule. In this case, **set-builder notation** is extremely useful.

SET-BUILDER NOTATION

The set A of all x such that x meets specific criteria may be written as

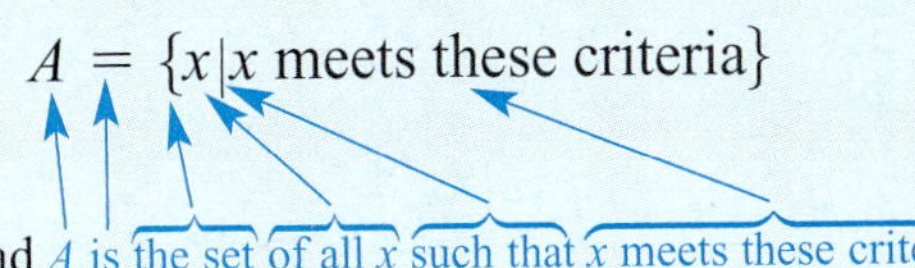

$A = \{x|x \text{ meets these criteria}\}$
and is read A is the set of all x such that x meets these criteria.

EXAMPLE 4

Using Set-Builder Notation

The set C is a set containing all positive even numbers less than or equal to 100,000. Write C using set-builder notation.

SOLUTION $C = \{x|x \text{ is a positive even number and } x \leq 100{,}000\}$; this is read "$C$ is the set of all x such that x is a positive even number and is less than or equal to 100,000."

EXAMPLE 5

Using Set Notation and a Venn Diagram

Many people buy life insurance so that their families will be financially secure in the event of the unexpected death of an income-earning family member. Life insurance premiums are determined in part by actuaries who analyze mortality (death) data. Term life insurance provides protection only for a specified period of time, whereas whole-life insurance provides coverage for an individual's whole life. Premiums for life insurance vary dramatically based upon the insured person's use of tobacco. In September 2002, a 30-year, $100,000 term life policy for a 20-year-old male nonsmoker cost $162 a year. The same policy for a 20-year-old male smoker cost $425 a year. (**Source:** Beneficial Life Insurance Company.)

A life insurance salesperson keeps a database of her customers on her home office computer. Her company has just reduced the rates for 20-40-year-old males who don't use tobacco. She plans on calling each of her 20-40-year-old male, nonsmoker clients to see if they want to increase their insurance coverage now that rates have dropped. She will use her database filter to create a list of the clients she needs to call. Use set notation to specify what search criteria she should use. Then draw a Venn diagram and shade the solution region.

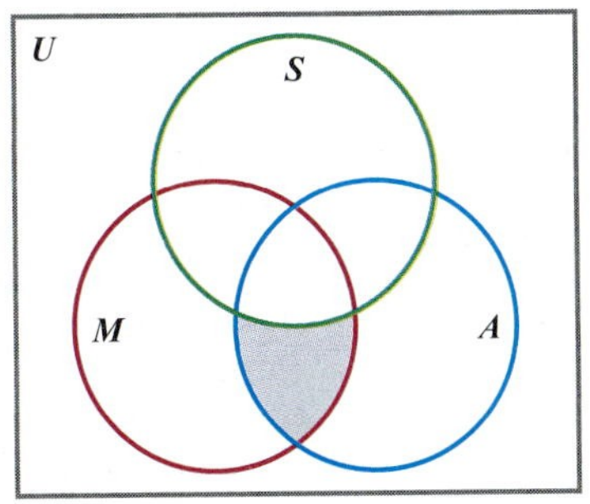

FIGURE 7.6

SOLUTION Let $S = \{x|x \text{ is a smoker}\}$, $M = \{x|x \text{ is a male}\}$, and $A = \{x|20 \leq x \leq 40 \text{ years old}\}$. The list of clients she wants to call is given by $(S' \cap M) \cap A$ and is shown as the shaded region in Figure 7.6.

EXAMPLE 6 **Interpreting Set Notation**

The results of a U.S. Department of Education study of instructional faculty and staff at two-year institutions is shown in Table 7.3.

TABLE 7.3

Employment Status and Years of Experience on Current Job	Instructional Faculty and Staff (in thousands)	Highest Credential Attained (percent)		
		Bachelor's or Less	Master's Degree	Ph.D. or First Professional
Total full time	94.9	17.5	63.7	18.8
Years of experience on current job				
Less than 10 years	46.0	21.6	60.8	17.6
10–19 years	25.8	19.2	63.6	17.3
20 or more years	23.2	7.5	69.6	22.9
Total part time	153.1	33.3	53.3	13.4
Years of experience on current job				
Less than 10 years	122.2	34.6	52.6	12.9
10–19 years	24.5	29.7	55.1	15.2
20 or more years	6.4	22.4	60.0	17.6

Source: U.S. Department of Education, National Center for Education Statistics, 1993 National Study of Postsecondary Faculty (NSOPF: 1993).

Let F be the group of full-time instructors and E be the group of people who have 20 or more years of teaching experience. Interpret the meaning of $F \cap E'$ and determine the number of people in that group.

SOLUTION E' is the group of people who have less than 20 years of teaching experience. $F \cap E'$ is the group of full-time faculty with less than 20 years of teaching experience. From the full-time section of the table, we see that 46.0 thousand instructors have less than 10 years' experience and 25.8 thousand have 10–19 years' experience. Since $F \cap E'$ is the union of these two groups, there are 71.8 thousand full-time instructors with less than 20 years of teaching experience.

7.1 Summary

In this section, you learned how to use set notation and Venn diagrams. You discovered how sets can be used in real life to clarify complex relationships.

7.1 Exercises

In Exercises 1–10, copy the Venn diagram given here and shade the region defined in each problem, if possible. Then list the elements of the shaded region.

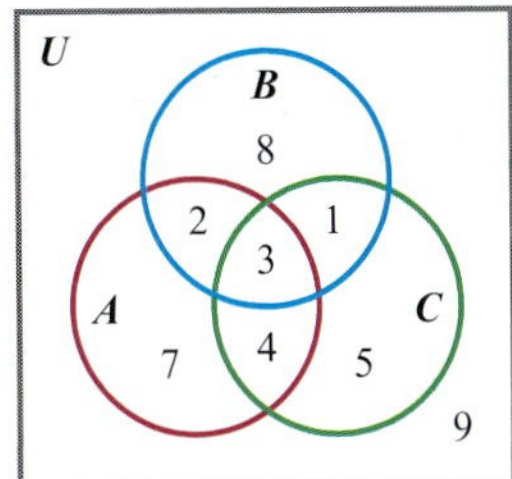

1. A'
2. $A \cap B$
3. $B \cup C$
4. $B' \cup C$
5. $(A \cap B)'$
6. $(A \cap B) \cup C$
7. $A \cap A'$
8. $(A \cap B) \cap C'$
9. $(A \cup B) \cap C'$
10. $((A \cap C) \cup (B \cap C))'$

In Exercises 11–15, let x represent a person.
$S = \{x | x \text{ is currently single}\}$,
$D = \{x | x \text{ never divorced}\}$, *and*
$M = \{x | x \text{ is or has been married}\}$.
Because of the complexity of multiple marriages, assume that each person in the universal set U has married at most once. Use set notation to represent the following groups of people.

11. Divorcees
12. Widows and widowers
13. People who never married
14. People who are currently married
15. Divorcees, widows, and widowers

In Exercises 16–18, use set notation to describe the relationship between A and B represented by the Venn diagram.

16.

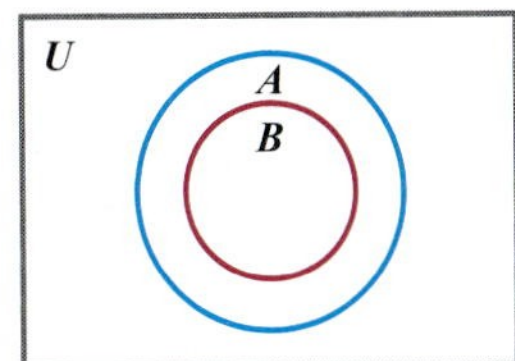

17.

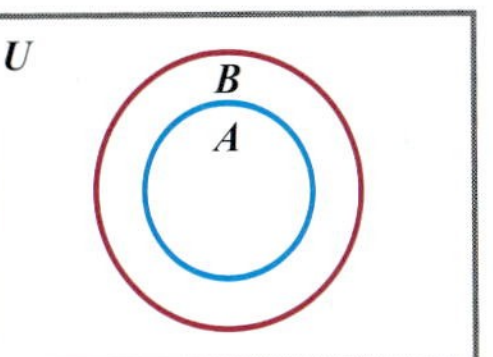

18.

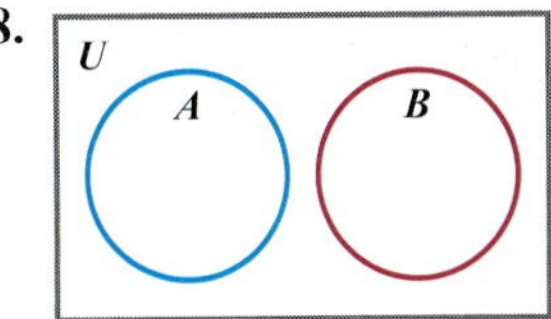

In Exercises 19–24, write the set using set-builder notation.

19. The set of all positive numbers less than 10
20. The set of all integers that are a multiple of 3
21. The set of numbers that are perfect squares (*Hint:* A perfect square is a number that is the product of two identical integer factors.)
22. The set of birth years for people who were 21 years old or younger at the end of 2006.
23. The set of numbers that are factors of 24
24. The set of negative integers

For Exercises 25–36, six people are categorized by their gender, household income, and employment status.

1: Male, \$67K, employed
2: Male, \$43K, employed
3: Female, \$52K, employed
4: Female, \$25K, unemployed
5: Male, \$31K, unemployed
6: Female, \$73K, unemployed

The group is further classified, with M being the set of all men, R being the set of people with household income more than \$50,000, and E being the set of employed people. Explain the real-life meaning of the set specified in each exercise. Then specify which people are elements of the set. (Hint: It may be helpful to draw a Venn diagram.)

25. $M \cap R$
26. $E \cup R$
27. $E' \cap R$
28. $(R \cap M)'$
29. $(E \cap M)' \cap R$

30. $(E' \cap R') \cup (M' \cap R)$

31. $(E' \cup R') \cap (M' \cup R)$

32. $(E \cup R') \cup (M' \cap R)$

33. $(E \cap R') \cup M$

34. $(E \cap R') \cap M$

35. $(M \cap R')'$

36. $(M \cup R') \cup E$

For Exercises 37–41, the results of a U.S. Department of Education study of instructional faculty and staff at two-year institutions are shown in the table. Let F be the group of full-time instructors and E be the group of people who have 20 or more years of teaching experience.

Employment Status and Years of Experience on Current Job	Instructional Faculty and Staff (in thousands)	Highest Credential Attained (percent)		
		Bachelor's or Less	Master's Degree	Ph.D. or First Professional
Total full time	94.9	17.5	63.7	18.8
Years of experience on current job				
Less than 10 years	46.0	21.6	60.8	17.6
10–19 years	25.8	19.2	63.6	17.3
20 or more years	23.2	7.5	69.6	22.9
Total part time	153.1	33.3	53.3	13.4
Years of experience on current job				
Less than 10 years	122.2	34.6	52.6	12.9
10–19 years	24.5	29.7	55.1	15.2
20 or more years	6.4	22.4	60.0	17.6

Source: U.S. Department of Education, National Center for Education Statistics, 1993 National Study of Postsecondary Faculty (NSOPF: 1993).

37. Interpret the meaning of $F \cup E$ and determine the number of people in that group.

38. Interpret the meaning of $F' \cap E$ and determine the number of people in that group.

39. Interpret the meaning of $F' \cap E'$ and determine the number of people in that group.

40. Interpret the meaning of $F' \cup E'$ and determine the number of people in that group.

41. Interpret the meaning of $(F \cap E')'$ and determine the number of people in that group.

Exercises 42–46 are intended to challenge your understanding of the concepts of sets and set notation.

42. Explain how the concepts of set intersection and union could be used by a business that tracks the type and number of products purchased by each customer.

43. DeMorgan's Law states that

$(A \cap B)' = A' \cup B'$ and $(A \cup B)' = A' \cap B'$.

For each statement, construct a Venn diagram for each side of the equal sign. Do you think DeMorgan's Law works for all sets A and B? Explain.

44. What is an equivalent way in which the statement $(A \cap B') \cup A'$ could be written?

45. What is an equivalent way in which the statement $(A \cap B') \cap (A' \cap B)$ could be written?

46. Are the two statements $(A \cap B') \cup B$ and $A \cap (B' \cup B)$ equivalent? Show the steps that lead to your conclusion.

7.2 Cardinality and the Addition and Multiplication Principles

- Count the number of items in a finite set

GETTING STARTED In the early 1940s, Marie Callender made a name for herself with her mouthwatering pies known for their light, flaky crusts and tasty fresh fruit fillings. Demand for her pies was so great that by 1948, the Callender family opened a wholesale bakery. The first Marie Callender's restaurant was opened in 1964, and by 2002, there were restaurants in more than 160 locations.

Marie Callender's offers seven pies called "Marie's Classics." Three of the pies list cream cheese as a main ingredient. Three of the pies list chocolate as an ingredient. Two pies don't list cream cheese or chocolate as an ingredient. (**Source:** www.mariecallender.com.) How many pies contain both cream cheese and chocolate? Questions such as this can be answered using the notion of cardinality.

In this section, we will show how to find the **cardinality of a set** (number of items in the set). We will also introduce the addition and multiplication principles for counting items in a set. We will use these concepts to analyze real-world data such as classifying Marie Callender's scrumptious pies.

CARDINALITY

If A is a set with a finite number of elements, then

$$n(A) = \text{the number of elements in } A$$

$n(A)$ is read "the cardinality of A."

Let U be the set of all Marie Callender's classic pies, C be the set of classic pies with cream cheese, and D be the set of classic pies with chocolate.

We have $n(U) = 7$, $n(C) = 3$, $n(D) = 3$, and $n((C \cup D)') = 2$. We want to find $n(C \cap D)$, the number of pies with both cream cheese and chocolate. To solve the problem, we need to know how to find the cardinality of a union and the cardinality of a complement.

CARDINALITY OF A UNION

If A and B are sets with a finite number of elements, then

$$n(A \cup B) = n(A) + n(B) - n(A \cap B)$$

Since the items in the intersection of A and B were "double counted" [once in $n(A)$ and once in $n(B)$], it was necessary to subtract $n(A \cap B)$.

CARDINALITY OF A COMPLEMENT

If U is a universal set with a finite number of elements and A is a subset of U, then

$$n(A') = n(U) - n(A) \text{ and } n(A) = n(U) - n(A')$$

From the cardinality of a complement formula, we know that

$$\begin{aligned} n(C \cup D) &= n(U) - n((C \cup D)') \\ &= 7 - 2 \\ &= 5 \end{aligned}$$

There are five pies with cream cheese or chocolate.

From the cardinality of a union formula, we know that

$$\begin{aligned} n(C \cup D) &= n(C) + n(D) - n(C \cap D) \\ 5 &= 3 + 3 - n(C \cap D) \\ -1 &= -n(C \cap D) \\ n(C \cap D) &= 1 \end{aligned}$$

One pie has both cream cheese and chocolate listed as ingredients. Each of the seven pies is identified in the Venn diagram in Figure 7.7.

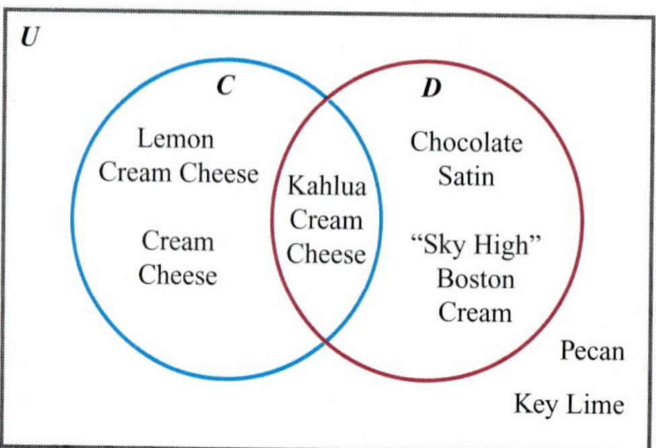

FIGURE 7.7

EXAMPLE 1

Finding the Cardinality of a Set

Wheaties has been known as "The Breakfast of Champions" since 1933. Packages featuring Babe Ruth, Mary Lou Retton, Michael Jordan, and Tiger Woods are among the top ten most popular Wheaties packages. Of the top ten packages, four packages feature people playing baseball and eight feature people in sports other than football. How many packages feature athletes not playing baseball or football?

SOLUTION Let U be the set of all top ten packages, B be the set of packages featuring baseball players, and F be the set of packages with football players. We must find $n((B \cup F)')$. We have $n(U) = 10$, $n(B) = 4$, and $n(F') = 8$.

We know that

$$\begin{aligned} n(F) &= n(U) - n(F') \\ &= 10 - 8 \\ &= 2 \end{aligned}$$

Since each athlete is featured playing only one sport, B and F are disjoint sets and $n(B \cap F) = 0$.

From the cardinality of a union formula, we know that

$$\begin{aligned} n(B \cup F) &= n(B) + n(F) - n(B \cap F) \\ &= 4 + 2 - 0 \\ &= 6 \end{aligned}$$

Six athletes played baseball or football.

From the cardinality of a complement formula, we know that

$$\begin{aligned} n((B \cup F)') &= n(U) - n(B \cup F) \\ &= 10 - 6 \\ &= 4 \end{aligned}$$

Four athletes playing sports other than baseball or football are featured. They include the Men's U.S. Gold Medal Hockey Team, Michael Jordan, Mary Lou Retton, and Tiger Woods.

Mary Lou Retton (featured in 1984) was the first female athlete to be a Wheaties spokesperson. Since then, the Women's U.S. Gold Medal Gymnastics Team, Amy Van Dyken (swimming), the Women's U.S. Gold Medal Hockey Team, and Sarah Hughes (figure skating) have joined the ranks of featured women.

Cartesian Products

Many sets are defined by linking two different quantities. The Cartesian product is used to maintain the linked relationship.

CARTESIAN PRODUCT

If A and B are sets with a finite number of elements, then the Cartesian product is

$$A \times B = \{(a, b) | a \in A \text{ and } b \in B\}$$

The elements of the set $A \times B$ are ordered pairs with the first value from set A and the second value from set B.

EXAMPLE 2

Determining the Outcomes of a Cartesian Product

In a game, you flip a coin and roll a die. What are the possible results?

SOLUTION

The coin has two possible outcomes: heads or tails.

The die has six possible outcomes: 1, 2, 3, 4, 5, 6.

Let $C = \{H, T\}$ be the set of possible outcomes for the coins and $D = \{1, 2, 3, 4, 5, 6\}$ be the set of possible outcomes for the die.

Then $C \times D = \left\{ \begin{matrix} (H, 1), & (H, 2), & (H, 3), & (H, 4), & (H, 5), & (H, 6), \\ (T, 1), & (T, 2), & (T, 3), & (T, 4), & (T, 5), & (T, 6), \end{matrix} \right\}$ is the set of possible results.

CARDINALITY OF A CARTESIAN PRODUCT

If A and B are sets with a finite number of elements, then

$$n(A \times B) = n(A) \cdot n(B)$$

From Example 2, we have $n(C) = 2$ and $n(D) = 6$. It follows that

$$\begin{aligned} n(C \times D) &= n(C) \cdot n(D) \\ &= 2 \cdot 6 \\ &= 12 \end{aligned}$$

This solution matches the result in Example 2.

EXAMPLE 3

Finding the Cardinality of a Cartesian Product

Two siblings have been accepted by Harvard, Yale, UCLA, and the University of Washington. The first sibling has also been accepted by the University of Michigan. How many different enrollment combinations are possible for the two siblings?

SOLUTION Let F be the number of enrollment choices for the first sibling and S be the number of enrollment choices for the second sibling. We have $n(F) = 5$ and $n(S) = 4$. So

$$\begin{aligned} n(F \times S) &= 5 \cdot 4 \\ &= 20 \end{aligned}$$

There are 20 possible enrollment combinations for the siblings.

Addition Principle

Marie Callender's classifies pies into five different categories: Famous Cream Pies, Callender's Double Cream Pies, Seasonal Fresh Fruit Pies, Marie's Classics, and Traditional Favorites. No pie is listed in more than one category (**Source:** www.mariecallender.com.) How many varieties of pie are there to pick from? Questions such as these may be answered using the Addition Principle.

ADDITION PRINCIPLE

When choosing an element from among n disjoint sets, the total number of elements to pick from is the *sum* of the cardinalities of each set.

There are four Famous Cream Pies, three Callender's Double Cream Pies, six Seasonal Fresh Fruit Pies (including the "no sugar added" varieties), seven Marie's Classics, and sixteen Traditional Favorites. The total number of pies to choose from is $4 + 3 + 6 + 7 + 16 = 36$.

EXAMPLE 4

Using the Addition Principle

As of September 10, 2002, Toyota of Puyallup had several used Toyota Tacoma SR5 trucks for sale, including four 2001 models, one 2000 model, and six 1999 models. (**Source:** ToyotaofPuyallup.com.) How many Toyota Tacoma trucks were for sale?

SOLUTION Since $4 + 1 + 6 = 11$, there were 11 Toyota Tacoma trucks for sale.

Multiplication Principle

The Royal Argosy dinner cruise is a popular attraction for visitors in the Seattle area. The cruise ship spends three hours on beautiful Elliot Bay while guests enjoy an exquisite dinner prepared by some of the area's premier chefs. The views of the Seattle skyline from the ship are phenomenal. The author and his wife dined on the Royal Argosy while celebrating their anniversary. They were given the following menu options:

Welcome Aboard Pre-Appetizer

Assorted Artisan Bread Basket

Starters

Hearts of Baby Romaine

Elliot's Dungeness Crab and Corn Bisque

Gathered Greens

Entrees

Macadamia Nut Halibut

Wild Chinook Salmon

New York Strip Loin

Filet Mignon Royal Argosy

Grilled Portobello

Desserts

Burnt Cream

Dark Chocolate Tower of Decadence

Mixed Berry Shortcake

Source: www.royalargosy.com.

How many different four-course meals can be ordered from the menu? Questions such as this can be answered using the Multiplication Principle.

MULTIPLICATION PRINCIPLE

The total number of possible selections that include one element from each of r sets is the *product* of the cardinalities of each set. That is, given sets $A_1, A_2, \ldots, A_r$ the total number of possible selections including exactly one element from each set is

$$n(A_1) \cdot n(A_2) \cdot \cdots \cdot n(A_r)$$

The dinner cruise offered one pre-appetizer, three starters, five entrees, and three desserts. The total number of possible dinner selections is $1 \cdot 3 \cdot 5 \cdot 3 = 45$.

EXAMPLE 5

Using the Multiplication Principle

Cars.com allows you to pick and price the features of your new car. In completing the online form for a 2002 Toyota Sequoia SR5 V8, potential buyers were given options in the following categories:

Styles (2 choices)
Drive type (2 choices)
Body color (9 choices)
Interior color (2 choices)
Option groups (3 choices)
Wheels (3 choices)
Tires (2 choices)
Audio (3 choices)
Other options (14 choices)

If one choice is made from each category, how many different Toyota Sequoias can be made?

SOLUTION Assuming that the sets are disjoint (as they appear to be), the multiplication principle may be used. There are $2 \cdot 2 \cdot 9 \cdot 2 \cdot 3 \cdot 3 \cdot 2 \cdot 3 \cdot 14 = 54{,}432$ different cars possible.

Decision Trees

Decision trees are a graphical representation of each of the possible outcomes that result from a series of choices. In a sense, they are a visual interpretation of the Multiplication Principle. They are most useful for a series of four or fewer choices from sets containing five or fewer elements. They can be used for larger sets, but they tend to become unwieldy. We will demonstrate how to use a decision tree with an example.

A student needs to enroll in three online classes next quarter: English, math, and art. There are two online English courses: Basic Composition and Basic Literature; there are three online math courses: Finite Math, Consumer Math, and Precalculus; and there are two online art courses: Art History and Design. We will construct the decision tree by making each of the choices in sequence. We must first choose an English class (see Figure 7.8).

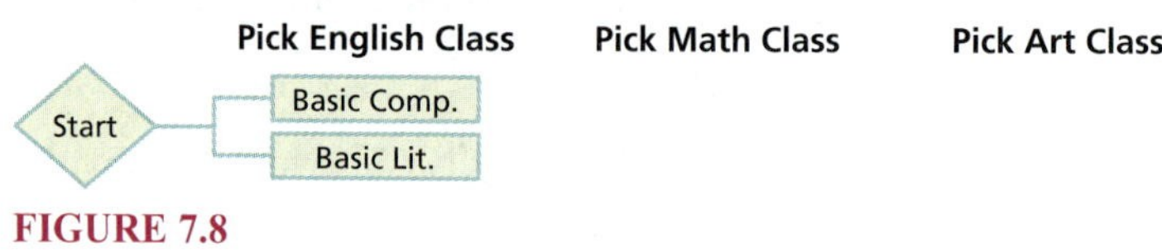

FIGURE 7.8

We will next choose a math class (see Figure 7.9).

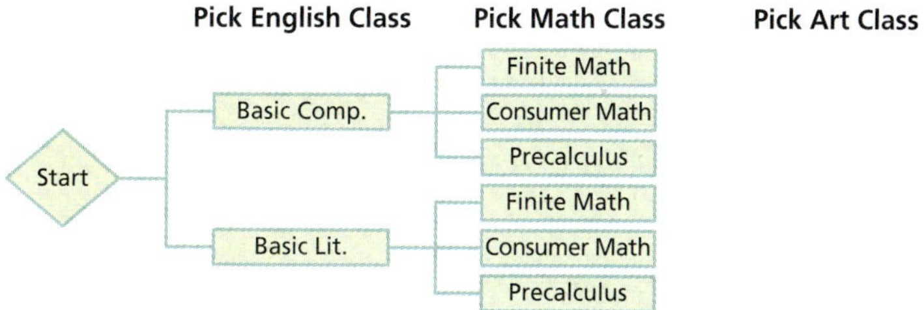

FIGURE 7.9

So far there are six different course schedules available to us. We will now pick the art class (see Figure 7.10).

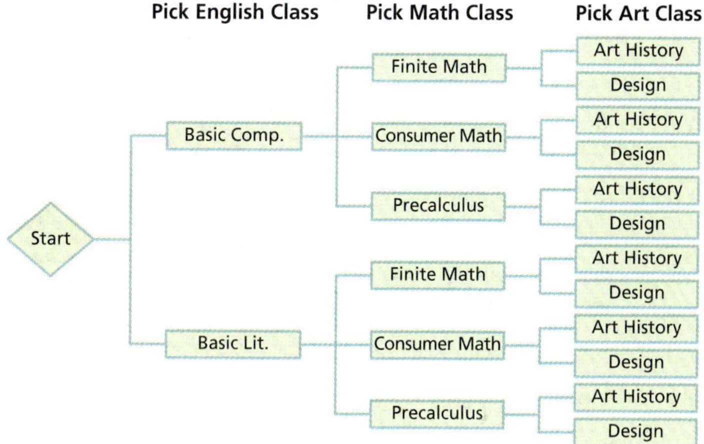

FIGURE 7.10

There are 12 schedules available to us. Reading from the decision tree, the first choice is Basic Composition, Finite Math, and Art History. The second choice is Basic Composition, Finite Math, and Design. The third choice is Basic Composition, Consumer Math, and Art History. The process of reading the diagram continues until you reach the final choice of Basic Literature, Precalculus, and Design.

7.2 Summary

In this section, you learned how to find the cardinality of a set and how to find a Cartesian product. You saw how the addition and multiplication principles are used in counting the items in a set. You finished the section by learning how to construct a decision tree.

7.2 Exercises

In Exercises 1–10, A and B are finite subsets of the universal set U with $n(U) = 15$, $n(A) = 10$, $n(A \cap B) = 4$, and $n(A \cup B) = 13$. Find the cardinality of each of the specified sets.

1. B
2. A'
3. B'
4. $A' \cup B'$
5. $A' \cap B'$
6. $(A \cup B)'$
7. $(A \cap B)'$
8. $A \cap B'$
9. $A' \cap B$
10. $A \cup B'$

In Exercises 11–15, calculate the cardinality of the Cartesian product $A \times B$ and write each of the elements of $A \times B$.

11. Let $A = \{\text{red, green, yellow}\}$ and $B = \{\text{apple, tomato}\}$.
12. Let $A = \{25, 36, 49\}$ and $B = \{x | x \text{ is a positive integer less than } 6\}$.
13. Let $A = \{\text{chocolate, vanilla}\}$ and $B = \{\text{ice cream, pie, shake, cookie}\}$.
14. Let $A = \{\text{a, e, i, o, u}\}$ and $B = \{\text{n, s, t}\}$.
15. Let $A = \{\text{Heads, Tails}\}$ and $B = \{x | x \text{ is a positive factor of } 6\}$.

In Exercises 16–25, use the Addition or Multiplication Principle as appropriate to determine the cardinality of the set.

16. **Online Class Schedules** Washington Online is a virtual campus of the Washington State two-year colleges. Students may enroll in a variety of courses at area colleges. For the fall quarter of 2002, South Seattle Community College offered three English courses, ten computer-related courses, and six other types of courses. (**Source:** www.waol.org.) How many schedule choices does a student who intends to enroll in one course from each category have?

17. **Online Class Schedules** For the fall quarter of 2002, Green River Community College offered two English courses, four social science courses, and four other types of online courses. (**Source:** www.waol.org.) How many schedule choices does a student who intends to enroll in one course from each category have?

18. **Ice Cream Varieties** The Tillamook Creamery Association, a farmer-owned cooperative in Oregon, operates the Tillamook Cheese creamery. Nearly a million tourists visit the creamery each year to watch the cheese makers at work and to partake of the rich Tillamook ice cream. On a hot summer day, the ice cream counter serves roughly 4000 ice cream cones.

    The creamery offers 4 types of cones (wafer cone, sugar cone, waffle cone, and waffle dish) and 45 varieties of ice cream, sherbet, and frozen yogurt. Cones may be served with one, two, or three scoops. (**Source:** www.tillamookcheese.com.) How many ice cream cone choices are there?

19. **Frozen Dessert Varieties** The Tillamook Cheese creamery sells 39 flavors of 13.5 percent butterfat ice cream, 3 varieties of frozen yogurt, 1 flavor of sherbet, and 2 varieties of reduced-fat ice cream. (**Source:** www.tillamookcheese.com.) How many different varieties of frozen dairy desserts does the company sell?

20. **Lodging and Dining** An executive is flying a potential business client to her Boston office. She has asked her administrative assistant to book a room for her client in a five-star hotel and to make dinner reservations for them at a top-rated restaurant. Her computer-savvy assistant surfs to www.find-luxury-hotels.com and pulls up a list of three five-star hotels in the Boston area: *Lenox Hotel, Fairmont Copley Plaza, and Millennium Bostonian.* He next goes to www.diningguide.net and finds two fine dining restaurants in downtown Boston: *Julien* and *Rowes Wharf Restaurant.* How many different dining/lodging options does the assistant have? Draw a decision tree showing the possible options.

21. **Lodging and Dining** If the administrative assistant from Exercise 20 includes 4-star and 4½-star hotels, there are 17 hotels to choose from. If he includes restaurants outside of the Boston downtown area, there are 8 fine dining restaurants to choose from. How many different dining/lodging options does he have?

22. **Menu Options** The Royal Argosy lunch cruise offers the following menu options:

Welcome Aboard Pre-Appetizer
Assorted Artisan Bread Basket

Starters
Elliot's Clam Chowder
Gathered Greens

Entrees
Hearts of Romaine & Alaskan Shrimp
Dungeness Crab Cakes
Wild Chinook Salmon
Herb Crusted Local Farm Chicken
Butternut Squash Ravioli

Desserts
Burnt Cream
Dark Chocolate Tower of Decadence
Warm Pineapple Upside-down Cake

Source: www.royalargosy.com.

How many different meals can be created from the menu, assuming that one item is selected from each category?

23. A calorie-conscious guest wants to forgo dessert on the lunch cruise in Exercise 22. How many meal options are there if dessert is omitted?

24. A calorie-conscious guest who has discovered that Burnt Cream (Crème Brulée) is the most exquisite dessert ever invented decides to enjoy a serving of Burnt Cream and start her diet tomorrow. Using the menu from Exercise 22, how many meal combinations are possible with Burnt Cream as a dessert?

25. Using the menu in Exercise 22, how many different dish choices (not meals) are there?

For Exercises 26–30, you are given the sets $R = \{\text{marble, block}\}$, $C = \{\text{black, red, blue}\}$, $S = \{\text{small, medium, large}\}$, *and* $M = \{\text{glass, wood, plastic}\}$. *Draw a decision tree to show all possible outcomes for the specified criteria.*

26. The outcomes with a single element from each of the sets R, C, and S.

27. The outcomes with a single element from each of the sets R, C, and M.

28. The outcomes with a single element from each of the sets R, S, and M.

29. The outcomes with a single element from each of the sets R, C, M, and S.

30. The outcomes with a single element from each of the sets R, C, M, and S that contain the elements *red* and *marble.*

For Exercises 31–35, refer to the accompanying table, which depicts the highest educational level attained by U.S. Air Force officers. Data were accurate as of May 2003.

Educational Level	Males (M)	Females (F)	Total
B.A./B.S. (B)	25,900	6,235	32,135
M.A./M.S. (S)	24,694	4,514	29,208
Ph.D. (D)	852	155	1,007
Professional Degree (P)	4,518	1,199	5,717
Unknown (N)	3,313	902	4,215
Total	59,277	13,005	72,282

Source: www.afpc.randolph.af.mil.

31. Calculate $n(D \cap F)$.

32. Calculate $n((M \cup F) \cap P)$.

33. Calculate $n(B' \cap F)$.

34. Calculate $n((P' \cap F) \cup (M \cap N'))$.

35. Calculate $n((N' \cap M) \cup F)$.

Exercises 36–40 are intended to challenge your understanding of the concepts covered in this section.

36. If $n(A \cap B) = n(A \cup B)$, what can you conclude about sets A and B?

37. If $n(A \cap B) \neq n(A \cup B)$, what can you conclude about sets A and B?

38. **Class Schedule** A student needs to select three courses for next semester: a math class, an English class, and a class from some other category. There are five math courses, four English courses, three P.E. courses, and four art courses available. How many different course schedules can be created?

39. **Sandwiches** Of the sandwiches listed on the McDonalds web site, nine contain beef, four contain chicken, and one includes fish.

Of the fourteen different sandwiches, seven contain lettuce and seven contain cheese. All sandwiches include a bun.
(**Source:** www.mcdonalds.com.)

How many different types of meat sandwiches could be made from the ingredients beef, chicken, fish, bun, lettuce, and cheese? (Assume that each sandwich contains only one type of meat and one bun. Additional ingredients are optional.)

40. Explain why the answer to Exercise 39 is less than 14.

7.3 Permutations and Combinations

- Use permutations and combinations to determine the number of possible outcomes of a series of choices

GETTING STARTED In an upset victory, Sarah Hughes won the women's figure skating gold medal in the 2002 Winter Olympics. Sarah, in fourth place after the short program, skated an impeccable long program that landed her the gold

medal. Since the games, she has reportedly received millions of dollars in endorsements and has been featured on the Wheaties cereal package. Fellow competitor Irina Slutskaya (Russia) won the silver medal, and teammate Michelle Kwan (United States) won the bronze. How many different ways could the top three skaters have placed? Questions such as these may be answered using permutations and combinations.

In this section, we will demonstrate how permutations and combinations can be used to count the number of possible results of an event. These concepts will lay the groundwork for our investigation into probability.

How many different ways could the three medalists have placed? Let's look at Table 7.4.

TABLE 7.4

	1	2	3	4	5	6
Gold	Hughes	Hughes	Slutskaya	Slutskaya	Kwan	Kwan
Silver	Slutskaya	Kwan	Hughes	Kwan	Hughes	Slutskaya
Bronze	Kwan	Slutskaya	Kwan	Hughes	Slutskaya	Hughes

There were six different ways the medalists could have placed. An alternative way to look at the ordering of the medalists is to look at the number of choices available for each medal and use the multiplication principle.

$$\overline{\text{Gold}} \quad \overline{\text{Silver}} \quad \overline{\text{Bronze}}$$

From the set $W = \{\text{Hughes, Slutskaya, Kwan}\}$, we have three choices for the gold medal. After choosing the gold medal winner, we have only two choices left for the silver medal. After choosing the silver medalist, only one choice remains for the bronze medal.

$$\underset{\text{Gold}}{\underline{3}} \cdot \underset{\text{Silver}}{\underline{2}} \cdot \underset{\text{Bronze}}{\underline{1}}$$

There are $3 \cdot 2 \cdot 1 = 6$ possible ways to order the elements of the set W.

How many different ways could the top five contestants have placed? Using the same argument, there are $5 \cdot 4 \cdot 3 \cdot 2 \cdot 1 = 120$ possible ways the top five contestants could have placed.

How many ways could the top 10 contestants have placed? There are $10 \cdot 9 \cdot 8 \cdot 7 \cdot 6 \cdot 5 \cdot 4 \cdot 3 \cdot 2 \cdot 1 = 3{,}628{,}800$ possible ways they could have placed. As you can see, it becomes increasingly difficult to write out the product as the number of different items in the set increases. To facilitate this process, we introduce **factorial** notation.

FACTORIAL

For nonnegative integers,

$$n! = n \cdot (n-1) \cdot (n-2) \cdot \cdots \cdot 2 \cdot 1$$

$n!$ is read "n factorial." By definition, $0! = 1$.

For example,

$$\begin{aligned} 10! &= 10 \cdot 9 \cdot 8 \cdot 7 \cdot 6 \cdot 5 \cdot 4 \cdot 3 \cdot 2 \cdot 1 \\ &= 3{,}628{,}800 \end{aligned}$$

Factorial notation is used extensively when determining the number of different ways to sequence a set of distinct items.

PERMUTATION OF n ITEMS

A **permutation of n items** is an ordered sequence of those items. The number of possible permutations of n items is

$$n! = n \cdot (n-1) \cdot (n-2) \cdot \cdots \cdot 2 \cdot 1$$

EXAMPLE 1

Counting the Number of Ways to Rank the Members of a Group

How many ways can a class of 15 students be ranked?

SOLUTION We must order the 15 students. There are 15! ways to rank them.

$$\begin{aligned} 15! &= 15 \cdot 14 \cdot 13 \cdot 12 \cdot 11 \cdot 10 \cdot 9 \cdot 8 \cdot 7 \cdot 6 \cdot 5 \cdot 4 \cdot 3 \cdot 2 \cdot 1 \\ &= 1{,}307{,}674{,}368{,}000 \end{aligned}$$

There are more than a trillion ways to rank order a class of 15 students!

Calculating $n!$ by hand can be cumbersome for even small values of n. The following Technology Tip details how to use a graphing calculator to calculate $n!$.

TECHNOLOGY TIP

Calculating n!

1. Type the value of n in the home screen.

```
15
```

2. Press MATH, use the blue arrows to move to the PRB menu, and select item 4:!. Press ENTER.

```
MATH NUM CPX PRB
1:rand
2:nPr
3:nCr
4:!
5:randInt(
6:randNorm(
7:randBin(
```

3. Press ENTER to display the result.

```
15!
  1.307674368E12
```

Returning to the Olympic figure skating event, we didn't know who would win medals in 2002, even after the short program. Michelle Kwan was favored to win, and Sarah Hughes was in fourth place. Out of the top 15 skaters, how many different ways could the contestants have won medals?

Let S be the set of the top 15 skaters at the end of the short program. We know that $n(S) = 15$. We still have only three medals to award. How many choices do we have for the gold medal? There are 15 contestants, so we have 15 choices. After selecting the gold medalist, how many choices do we have for the silver medal? We can choose from among the remaining 14 contestants. After selecting the gold and silver medalists, we have 13 contestants from which to pick the bronze medalist.

$$\underset{\text{Gold}}{\underline{15}} \cdot \underset{\text{Silver}}{\underline{14}} \cdot \underset{\text{Bronze}}{\underline{13}}$$

There are $15 \cdot 14 \cdot 13 = 2730$. possible ways to pick three elements from the set W. That is, there were 2730 different ways the medals could have been awarded to one of the top 15 short program contestants.

Observe that

$$\begin{aligned} 15 \cdot 14 \cdot 13 &= 15 \cdot 14 \cdot 13 \cdot \left(\frac{12 \cdot 11 \cdot 10 \cdot 9 \cdot 8 \cdot 7 \cdot 6 \cdot 5 \cdot 4 \cdot 3 \cdot 2 \cdot 1}{12 \cdot 11 \cdot 10 \cdot 9 \cdot 8 \cdot 7 \cdot 6 \cdot 5 \cdot 4 \cdot 3 \cdot 2 \cdot 1}\right) \\ &= \frac{15 \cdot 14 \cdot 13 \cdot 12 \cdot 11 \cdot 10 \cdot 9 \cdot 8 \cdot 7 \cdot 6 \cdot 5 \cdot 4 \cdot 3 \cdot 2 \cdot 1}{12 \cdot 11 \cdot 10 \cdot 9 \cdot 8 \cdot 7 \cdot 6 \cdot 5 \cdot 4 \cdot 3 \cdot 2 \cdot 1} \\ &= \frac{15!}{12!} \\ &= \frac{15!}{(15-3)!} \end{aligned}$$

We recognize the 15 as the number of unique elements in the set. We recognize the 3 as the number of medals to be awarded. Selecting 3 medalists from a group of 15 contestants is an example of a permutation of 15 items taken 3 at a time.

PERMUTATIONS OF n ITEMS TAKEN r AT A TIME

A **permutation of n items taken r at a time** is an ordered sequence of r items chosen from a set of n distinct items. The number of possible permutations of n items taken r at a time is

$$P(n, r) = \frac{n!}{(n-r)!}$$

EXAMPLE 2

Using Permutations in the Real World

In the early 1980s, Tommy Tutone made it big with the hit "867-5309 Jenny." The song tells of a guy trying to get up the nerve to call "Jenny," whose phone number he presumably found scrawled on a bathroom wall. The single hit number 4 on the charts in early 1982. (**Source:** musicfinder.yahoo.com.)

The number 8675309 is fascinating in its own right. First, no digit is repeated in the number. Second, the number is divisible only by 1 and itself.

How many different seven-digit phone numbers can be created without repeating any of the digits? Remember that phone numbers can't start with a 1 or a 0.

SOLUTION We must select each digit from the set $N = \{0, 1, 2, 3, 4, 5, 6, 7, 8, 9\}$. We know that $n(N) = 10$. Once a digit has been selected, it cannot be reused. We need seven digits to make the number.

$$\begin{aligned} P(n, r) &= \frac{n!}{(n-r)!} \\ P(10, 7) &= \frac{10!}{(10-7)!} \\ &= \frac{10!}{3!} \\ &= \frac{10 \cdot 9 \cdot 8 \cdot 7 \cdot 6 \cdot 5 \cdot 4 \cdot \cancel{3!}}{\cancel{3!}} \\ &= 604{,}800 \end{aligned}$$

There are 604,800 seven-digit numbers that don't repeat any digits. However, phone numbers can't start with a 0 or a 1, so we need to subtract all numbers that start with a 0 or a 1. Let's first consider the seven-digit numbers that start with a 0. We need to select six more digits from the remaining nine numbers.

$$\begin{aligned} P(n, r) &= \frac{n!}{(n-r)!} \\ P(9, 6) &= \frac{9!}{(9-6)!} \\ &= \frac{9!}{3!} \\ &= \frac{9 \cdot 8 \cdot 7 \cdot 6 \cdot 5 \cdot 4 \cdot \cancel{3!}}{\cancel{3!}} \\ &= 60{,}480 \end{aligned}$$

Similarly, there are 60,480 numbers that start with a 1. The number of seven-digit phone numbers is given by

$$604{,}800 - 2(60{,}480) = 483{,}840$$

Depending on what region of the country you live in, phone numbers beginning with certain other prefixes (e.g., 911, 611, or 411) also may not be used. The actual number of seven-digit phone numbers for a particular region will be less than the calculated 483,840.

Combinations

When the order in which items are selected from a set is unimportant, **combinations** are used instead of permutations. For example, suppose you plan on driving to a Norah Jones concert. You have five friends you would like to invite to go with you; however, your car can carry only three passengers in addition to you. How many different groupings of three friends do you have to pick from? For visual purposes, let's suppose your friends are named Alex, Ben, Carmen, Dina, and Ernesto. In Table 7.5, we represent the friends by the first initial of their names. If the order in which the group members are picked matters, there are $P(5, 3) = 60$ possible groupings (Table 7.5).

TABLE 7.5

ABC	ABD	ABE	ACD	ACE	ADE	BCD	BCE	BDE	CDE
ACB	ADB	AEB	ADC	AEC	AED	BDC	BEC	BED	CED
BAC	BAD	BAE	CAD	CAE	DAE	CBD	CBE	DBE	DCE
BCA	BDA	BEA	CDA	CEA	DEA	CDB	CEB	DEB	DEC
CAB	DAB	EAB	DAC	EAC	EAD	DBC	EBC	EBD	ECD
CBA	DBA	EBA	DCA	ECA	EDA	DCB	ECB	EDB	EDC

Notice that each group listed in the first column of the table contains Alex, Ben, and Carmen. If the order in which each friend is picked is unimportant, the six groups in the first column are equivalent. Likewise, the six groups in the second column are equivalent. The same is true for the remaining columns in the table. Therefore, there are $\frac{60}{6} = 10$ possible groupings if the order in which group members are picked is unimportant. There are 10 different combinations of three-person groups that can be picked from among the five friends.

COMBINATIONS OF *n* ITEMS TAKEN *r* AT A TIME

A **combination of *n* items taken *r* at a time** is an unordered sequence of r items chosen from a set of n distinct items. The number of possible combinations of n items taken r at a time is

$$C(n, r) = \frac{n!}{(n-r)! \cdot r!}$$

For all n and r, $C(n, r) \leq P(n, r)$.

The origin of the equation for calculating combinations is not immediately obvious. As we will show, $C(n, r) = \frac{P(n, r)}{r!}$.

$$\frac{P(n, r)}{r!} = \frac{\text{number of ways to pick } r \text{ things from a set of } n}{\text{number of ways to order a set of } r \text{ things}}$$

$$= \frac{\frac{n!}{(n-r)!}}{r!} \qquad \text{Definition of permutation}$$

$$= \frac{n!}{(n-r)!} \cdot \frac{1}{r!} \qquad \text{Division of fractions}$$

$$= \frac{n!}{(n-r)!r!} \qquad \text{Product of fractions}$$

$$= C(n, r)$$

For example, a soccer league official is asked to identify how many different matches among the eight teams in the league are possible. The official returns with the following chart:

$$\begin{bmatrix} \cancel{11} & 12 & 13 & 14 & 15 & 16 & 17 & 18 \\ 21 & \cancel{22} & 23 & 24 & 25 & 26 & 27 & 28 \\ 31 & 32 & \cancel{33} & 34 & 35 & 36 & 37 & 38 \\ 41 & 42 & 43 & \cancel{44} & 45 & 46 & 47 & 48 \\ 51 & 52 & 53 & 54 & \cancel{55} & 56 & 57 & 58 \\ 61 & 62 & 63 & 64 & 65 & \cancel{66} & 67 & 68 \\ 71 & 72 & 73 & 74 & 75 & 76 & \cancel{77} & 78 \\ 81 & 82 & 83 & 84 & 85 & 86 & 87 & \cancel{88} \end{bmatrix}$$

He explains that the number ij means that Team i plays Team j. He tells the league committee that there are 56 different matches possible, since a team can't play itself. One committee member points out that 12 and 21 mean the same thing, since Team 1 playing Team 2 is the same as Team 2 playing Team 1. She tells him that to be accurate, duplicate matches need to be eliminated.

$$\begin{bmatrix} \cancel{11} & \cancel{12} & \cancel{13} & \cancel{14} & \cancel{15} & \cancel{16} & \cancel{17} & \cancel{18} \\ 21 & \cancel{22} & \cancel{23} & \cancel{24} & \cancel{25} & \cancel{26} & \cancel{27} & \cancel{28} \\ 31 & 32 & \cancel{33} & \cancel{34} & \cancel{35} & \cancel{36} & \cancel{37} & \cancel{38} \\ 41 & 42 & 43 & \cancel{44} & \cancel{45} & \cancel{46} & \cancel{47} & \cancel{48} \\ 51 & 52 & 53 & 54 & \cancel{55} & \cancel{56} & \cancel{57} & \cancel{58} \\ 61 & 62 & 63 & 64 & 65 & \cancel{66} & \cancel{67} & \cancel{68} \\ 71 & 72 & 73 & 74 & 75 & 76 & \cancel{77} & \cancel{78} \\ 81 & 82 & 83 & 84 & 85 & 86 & 87 & \cancel{88} \end{bmatrix}$$

He eliminates the duplicate matches and reports that there are 28 unique matches possible. His result can easily be confirmed by calculating $C(8, 2)$.

$$\begin{aligned} C(n, r) &= \frac{n!}{(n-r)! \cdot r!} \\ C(8, 2) &= \frac{8!}{(8-2)! \cdot 2!} \\ &= \frac{8!}{6! \cdot 2!} \\ &= \frac{8 \cdot 7 \cdot \cancel{6!}}{\cancel{6!} \cdot 2!} \\ &= \frac{56}{2} \\ &= 28 \end{aligned}$$

Graphing calculators can quickly and easily compute both combinations and permutations, as detailed in the following Technology Tip.

TECHNOLOGY TIP

Calculating $C(n, r)$ and $P(n, r)$

1. Type the value of n in the home screen.

2. Press MATH, use the blue arrows to move to the PRB menu, and select item 2:nPr for permutations or 3:nCr for combinations.

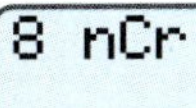

3. Type in the value of r and press ENTER to display the result.

8 nCr 2
28

EXAMPLE 3

Using Permutations in the Real World

A couple is expecting their third boy and is searching for a unique boy's name. The names of their first two boys contain five unique letters and start with *J*. Each name has a different vowel in the second and fourth positions and different consonants in the third and fifth positions. They want to maintain the pattern with their third son. How many names do they have to pick from?

SOLUTION There are 26 letters in the alphabet, 5 vowels and 21 consonants. (We will assume that y is a consonant.) The couple has used up *J*, so 20 consonants remain. Since the first letter is *J*, we need to pick only two vowels and two consonants. Does order matter? Since *Jared* is not the same as *Jerad*, order matters. The number of ways to pick two different vowels from the set of five vowels is given by

$$P(5, 2) = \frac{5!}{3!} = 20$$

The number of ways to pick two different consonants (excluding j) is

$$P(20, 2) = \frac{20!}{18!} = 380$$

The number of five-letter J names with alternating vowels and consonants is given by

$$P(5, 2) \cdot P(20, 2) = 20 \cdot 380$$
$$= 7600$$

There are 7600 such names.

An alternative way to look at the problem is to list the five slots and determine how many different letters may be used to fill each slot.

$$J\ \underline{5}\ \ \underline{20}\ \underline{4}\ \underline{19}$$
$$5 \cdot 20 \cdot 4 \cdot 19 = 7600$$

They have already used two names for their first two sons, so there are 7598 names remaining. When this chapter was written, the couple's top name candidates were *Jarin, Jaden, Jaxon,* and *Javin*.

The Multiplication Principle Revisited

To use permutations and combinations, we assume that no element can be picked twice. That is, once an item has been picked, it is eliminated from the selection pool. If an item is placed back into the selection pool after it is picked, the number of possible outcomes increases dramatically.

SELECTING r ITEMS FROM A SET OF n ITEMS (WITH REPLACEMENT)

If r items are chosen from a set of n items and each item is returned to the selection pool after it is picked, then the total number of possible groupings is n^r.

EXAMPLE 4

Counting the Number of Possible Phone Numbers

The North American Number Plan Administration (NANPA) is a third-party administrator that works with the telecommunications industry and oversees the assignment of area codes. On April 1, 2002, NANPA issued a press release indicating that three counties in Central Florida would be receiving a new overlay area code. According to the report, the 407 and 321 area codes were expected to be exhausted by first quarter 2004. (**Source:** NANPA.)

Certain phone prefixes are restricted and may not be used, most notably 000, 911, and 411. Suppose that there are a total of 300 restricted prefixes in Central Florida, including those prefixes starting with a 1 or a 0. (The phone company would not disclose to the author the exact number of restricted prefixes.) Assuming that all other phone numbers in the 407 and 321 area codes were used, how many phone numbers were there in Central Florida when the area codes were exhausted?

SOLUTION We will first consider the total number of possible seven-digit numbers and then subtract the phone numbers with restricted prefixes. We have seven slots to fill.

__ __ __ __ __ __ __

We may select each of the seven digits from the following 10 numbers: 0, 1, 2, 3, 4, 5, 6, 7, 8, 9. For the first slot, we have 10 choices, and likewise for the second through seventh slots.

$$\underline{10}\quad\underline{10}\quad\underline{10}\quad\underline{10}\quad\underline{10}\quad\underline{10}\quad\underline{10}$$

Multiplying the quantities together, we have $10^7 = 10{,}000{,}000$ possible seven-digit numbers. We will now eliminate the restricted numbers. For each restricted prefix, we have

$$\underline{1}\quad\underline{1}\quad\underline{1}\quad\underline{10}\quad\underline{10}\quad\underline{10}\quad\underline{10}$$

Multiplying the quantities together, there are 10,000 seven-digit numbers that start with any given prefix. We have 300 restricted prefixes, so we have

$$300 \cdot 10{,}000 = 3{,}000{,}000 \text{ restricted numbers}$$

Subtracting the quantity of restricted numbers from the total quantity of seven-digit numbers, we determine that there are 7,000,000 phone numbers in each area code, or a total of 14,000,000 numbers in the two area codes.

7.3 Summary

In this section, you learned how permutations and combinations can be used to count the number of possible results of an event. You also learned that permutations and combinations may be used only if an item is eliminated from the selection pool after it is picked.

7.3 Exercises

In Exercises 1–10, calculate the value of the expression without using a calculator.

1. $P(5, 2)$ **2.** $P(7, 5)$

3. $P(5, 5)$ **4.** $P(4, 1)$

5. $P(9, 0)$ **6.** $C(5, 2)$

7. $C(7, 5)$ **8.** $C(5, 5)$

9. $C(4, 1)$ **10.** $C(9, 0)$

In Exercises 11–20, calculate the value of the expression using a calculator.

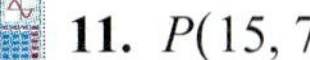

11. $P(15, 7)$ **12.** $P(20, 3)$

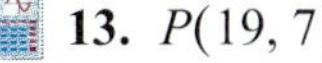

13. $P(19, 7)$ **14.** $P(39, 6)$

15. $P(369, 3)$ **16.** $C(15, 7)$

17. $C(20, 3)$

18. $C(19, 7)$

19. $C(39, 6)$ **20.** $C(369, 3)$

In Exercises 21–35, use permutations or combinations as appropriate to calculate the total number of possible outcomes.

21. **Lottery** In POWERBALL Lotto, five numbered white balls are drawn out of a drum of 49 white balls and one numbered red ball is drawn out of a drum of 42 red balls. To win the jackpot, you must match all five white balls (in any order) and the red ball. (**Source:** www.musl.com.) On September 7, 2002, the POWERBALL Lotto jackpot was $80 million. (**Source:** www.lotteryusa.com.) The probability of winning the jackpot is one divided by the total number of possible outcomes. How many outcomes are possible?

22. **License Plates** Some states use three letters followed by three numbers on their license plates. How many such license plates can be made if numbers and letters may be repeated?

23. **License Plates** How many license plates with three letters followed by three numbers can be made if no letters or numbers are repeated?

24. **Phone Numbers** Assuming that no phone numbers are prohibited, what is the maximum number of seven-digit phone numbers that could be issued in area code 360 if digits may be repeated?

25. **Phone Numbers** Assuming that no phone numbers are prohibited, what is the maximum number of seven-digit phone numbers with no repeating digits that could be issued in area code 360?

26. **Zip Codes** The U.S. Postal Service delivers mail based on a five-digit zip code. At most how many zip codes can begin with "98" if numbers may be repeated?

27. **Zip Codes** The U.S. Postal Service delivers mail based on a five-digit zip code. At most how many zip codes can begin with "98" if numbers may *not* be repeated?

28. **Soccer Matches** A local soccer league has 20 teams. How many different matches are possible?

29. **Class Schedules** A student needs to take a math class, two business classes, and a P.E. class. There are four math classes, seven business classes, and five P.E. classes that fit into her schedule. How many different class schedules does she have to pick from?

30. **Pizza Varieties** In 1995, Papa Aldo's Pizza and Murphy's Pizza merged into Papa Murphy's Pizza International. At the time of the merger, there were 140 stores. As of September 2002, there were 628 Papa Murphy's® Take 'N' Bake Pizza stores in 21 states. Ninety-eight percent of the stores are franchised. (**Source:** www.papamurphys.com.)

According to the company's web site, Papa Murphy's offers six different meat toppings, ten vegetable toppings, and four other toppings. Assuming that a topping is not used more than once on the same pizza, how many different types of three-topping pizzas can be made?

31. **Pizza Varieties** According to the company's web site, Papa Murphy's offers six different meat toppings, ten vegetable toppings, and four other toppings. How many different three-topping pizzas can be made if one topping is selected from each category?

32. **Pizza Varieties** According to the company's web site, Papa Murphy's offers six different meat toppings, ten vegetable toppings, and four other toppings. Assuming that a topping is not used more than once on the same pizza, how many different types of six-topping pizzas can be made?

33. **Pizza Varieties** According to the company's web site, Papa Murphy's offers six different meat toppings, ten vegetable toppings, and four other toppings. How many different types of six-topping pizzas can be made if two toppings are selected from each category?

34. **Radio Station Call Signs** Although there are 69 radio or TV stations with three-letter call signs, most stations are identified by a four-letter call sign. Stations to the west of the Mississippi River have call signs that begin with *K*, and stations to the east of the river have call signs that begin with *W*. (**Source:** www.earlyradiohistory.us.)

How many different four-letter radio station call signs can be made?

35. **Radio Station Call Signs** Radio station call signs were originally three letters long; however, the dramatic growth in the industry necessitated a move to a four-letter system in 1922. Four-letter call signs are required to begin with *K* or *W*. Furthermore, original regulations required that no letter could be repeated three times in a row. (**Source:** www.earlyradiohistory.us.)

How many four-letter call signs can be made without repeating a letter three times in a row?

Exercises 36–40 are intended to challenge your understanding of permutations and combinations.

36. How many three-digit numbers with no repeating digits are there that end with a 5, 7, or 9?

37. How many three-digit numbers are there with the property that each digit is greater than the digit to its right?

38. How many three-digit numbers are there with the property that each digit is less than or equal to the digit to its left?

39. A bag contains three types of coins: pennies, nickels, and dimes. The number of each type of coin is unknown; however, there is the same number of each type of coin (e.g., if there are five nickels, there are also five pennies and five dimes). There is a total of $3r$ coins in the bag. How many different ways are there to withdraw three coins from the bag whose combined value is 16 cents? (*Hint:* The answer will be in terms of r.) Assume that you can tell the difference between different coins of the same value.

40. A bag contains three types of coins: pennies, nickels, and dimes. The number of each type of coin is unknown; however, there is the same number of each type of coin (e.g., if there are five nickels, there are also five pennies and five dimes). There is a total of $3r$ coins in the bag. How many different ways are there to withdraw three coins from the bag whose combined value is *at most* 16 cents? (*Hint:* The answer will be in terms of r.) Assume that you can tell the difference between different coins of the same value.

7.4 Introduction to Probability

- Use probability terminology to describe real-life scenarios
- Use probability concepts to calculate probabilities

GETTING STARTED Many games include some element of chance. Players who understand the probability concepts associated with games of chance are better able to make wise decisions in their game play.

In this section, we will introduce the language and basic principles of probability. These will help us determine the likelihood of obtaining a desired result.

In discussing probability concepts, we use terms such as *experiment, trial, outcome,* and *sample space.* We begin by defining each term.

PROBABILITY TERMINOLOGY: EXPERIMENTS, TRIALS, OUTCOMES, AND SAMPLE SPACES

An **experiment** is an occurrence with an uncertain result. Each repetition of an experiment is called a **trial.** The possible results of each trial are called **outcomes.** The set of all possible outcomes of an experiment is called the **sample space** for the experiment.

To help you better understand the terminology, we will give several examples.

EXAMPLE 1

Using Probability Terminology

A coin is tossed. We are interested in the symbol on the face of the coin after it lands (heads or tails). Use probability terminology to describe the situation.

SOLUTION

Experiment: Toss a coin and observe the face of the coin after it lands

Trials: 1

Outcomes: Heads or tails

Sample space: $S = \{\text{heads, tails}\}$

EXAMPLE 2 Using Probability Terminology

Three people are picked from the audience at a U2 concert. We are interested in the gender of those selected, and we don't care about the order in which they are selected. That is, we consider picking a male first and then two females to be the same as picking two females first and then a male. Use probability terminology to describe the situation.

SOLUTION

Experiment: Randomly select three people from the U2 concert audience and observe their gender

Trials: 1

Outcomes: (M, M, M), (M, M, F), (M, F, F), (F, F, F)

Sample space: $S = \begin{Bmatrix} (\text{M, M, M}), (\text{M, M, F}), (\text{M, F, F}), \\ (\text{F, F, F}) \end{Bmatrix}$

If the order in which the people were selected was important to us, we would end up with a different sample space.

Sample space: $T = \begin{Bmatrix} (\text{M, M, M}), & (\text{M, M, F}), & (\text{M, F, M}), & (\text{F, M, M}), \\ (\text{M, F, F}), & (\text{F, M, F}), & (\text{F, F, M}), & (\text{F, F, F}) \end{Bmatrix}$

EXAMPLE 3 Using Probaility Terminology

In a game, a red die and a blue die are rolled. We're interested in the numbers shown on the dice. Use probability terminology to describe the situation.

SOLUTION

Experiment: Roll a red die and a blue die and observe the numbers shown

Trials: 1

Outcomes: (1, 1), (1, 2), (1, 3), . . . , (6, 3), (6, 4), (6, 5), (6, 6) (The first number is the red die value, and the second number is the blue die value.)

Sample space:

$$S = \begin{Bmatrix} (1,1) & (1,2) & (1,3) & (1,4) & (1,5) & (1,6) \\ (2,1) & (2,2) & (2,3) & (2,4) & (2,5) & (2,6) \\ (3,1) & (3,2) & (3,3) & (3,4) & (3,5) & (3,6) \\ (4,1) & (4,2) & (4,3) & (4,4) & (4,5) & (4,6) \\ (5,1) & (5,2) & (5,3) & (5,4) & (5,5) & (5,6) \\ (6,1) & (6,2) & (6,3) & (6,4) & (6,5) & (6,6) \end{Bmatrix}$$

EXAMPLE 4

Using Probability Terminology

In a game, two six-sided dice are rolled. We're interested in the sum of the numbers shown. Use probability terminology to describe the situation.

SOLUTION

Experiment: Roll two dice and observe the sum of the numbers shown

Trials: 1

Outcomes: 2, 3, 4, 5, 6, 7, 8, 9, 10, 11, 12

Sample space: $S = \{2, 3, 4, 5, 6, 7, 8, 9, 10, 11, 12\}$

EXAMPLE 5

Using Probability Terminology

Pick two coins from a purse containing nickels, dimes, and pennies. Use probability terminology to describe the situation.

SOLUTION

Experiment: Pick two coins from a purse with nickels, dimes, and pennies and observe the type of coin

Trials: 1

Outcomes: (nickel, nickel), (dime, nickel), (penny, nickel), (nickel, dime), (dime, dime), (penny, dime), (nickel, penny), (dime, penny), (penny, penny)

Sample space: $S = \left\{\begin{matrix} \text{(nickel, nickel)} & \text{(dime, nickel)} & \text{(penny, nickel)} \\ \text{(nickel, dime)} & \text{(dime, dime)} & \text{(penny, dime)} \\ \text{(nickel, penny)} & \text{(dime, penny)} & \text{(penny, penny)} \end{matrix}\right\}$

Often we are interested in a particular outcome because obtaining that outcome leads to some favorable result, such as scoring points, winning money, and so on. We use the terms *event* and *favorable outcomes* to describe such results.

MORE PROBABILITY TERMINOLOGY: EVENTS AND FAVORABLE OUTCOMES

A subset E of a sample space S that meets a specified criterion is called an **event**. The elements of E are called the **favorable outcomes.**

EXAMPLE 6

Using Probability Terminology

Three people are randomly picked from the audience of a U2 concert. The order in which the people are picked is important to us. [That is, we treat (M, M, F) and (M, F, M) as different elements, even though they both contain two males and one female.] We are interested in the outcomes that contain at least one woman. Use the terms *sample space, event,* and *favorable outcomes* to describe the situation.

SOLUTION

Sample space:

$$S = \begin{Bmatrix} (\text{M, M, M}), (\text{M, M, F}), (\text{M, F, M}), (\text{F, M, M}), (\text{M, F, F}), \\ (\text{F, M, F}), \quad (\text{F, F, M}), \quad (\text{F, F, F}) \end{Bmatrix}$$

$$\textit{Event: } E = \begin{Bmatrix} (\text{M, M, F}), (\text{M, F, M}), (\text{F, M, M}), (\text{M, F, F}), \\ (\text{F, M, F}), \quad (\text{F, F, M}), \quad (\text{F, F, F}) \end{Bmatrix}$$

Favorable outcomes:

(M, M, F), (M, F, M), (F, M, M), (M, F, F), (F, M, F), (F, F, M), (F, F, F)

EXAMPLE 7

Using Probability Terminology

Three people are randomly picked from a finite math class, and the gender of each person is recorded. We are interested in the outcomes that don't contain any men. Use the terms *sample space, event,* and *favorable outcomes* to describe the situation.

SOLUTION

Sample space:

$$S = \begin{Bmatrix} (\text{M, M, M}), (\text{M, M, F}), (\text{M, F, M}), (\text{F, M, M}), (\text{M, F, F}), \\ (\text{F, M, F}), \quad (\text{F, F, M}), \quad (\text{F, F, F}) \end{Bmatrix}$$

Event: $E = \{(\text{F, F, F})\}$

Favorable outcomes: (F, F, F)

EXAMPLE 8

Using Probabilty Terminology

In a game, a red die and a blue die are rolled. We observe the numbers shown on the pair of dice each time they are rolled. We want to know the dice pairs that have a sum of 7 or 11. Use the terms *sample space, event,* and *favorable outcomes* to describe the situation.

SOLUTION

Sample space:

$$S = \begin{Bmatrix} (1,1) & (1,2) & (1,3) & (1,4) & (1,5) & (1,6) \\ (2,1) & (2,2) & (2,3) & (2,4) & (2,5) & (2,6) \\ (3,1) & (3,2) & (3,3) & (3,4) & (3,5) & (3,6) \\ (4,1) & (4,2) & (4,3) & (4,4) & (4,5) & (4,6) \\ (5,1) & (5,2) & (5,3) & (5,4) & (5,5) & (5,6) \\ (6,1) & (6,2) & (6,3) & (6,4) & (6,5) & (6,6) \end{Bmatrix}$$

Event:

$E = \{(1, 6), (2, 5), (3, 4), (4, 3), (5, 2), (6, 1), (5, 6), (6, 5)\}$

Favorable outcomes:

(1, 6), (2, 5), (3, 4), (4, 3), (5, 2), (6, 1), (5, 6), (6, 5)

When playing a game where scoring points is based on random events such as rolling dice or picking cards, we are often interested in the likelihood of obtaining a favorable outcome. The notion of *probability* is used to measure the likelihood of success or failure.

BASIC PRINCIPLE OF PROBABILITY

Let a sample space S have equally likely outcomes. Let E be a subset of S. The **probability that event E occurs** is given by

$$P(E) = \frac{n(E)}{n(S)}$$

For any event, $0 \leq P(E) \leq 1$.

EXAMPLE 9

Calculating the Probability of an Event

A coin purse contains a penny, a nickel, a dime, a quarter, a 50-cent piece, and a silver dollar. What is the probability that a randomly selected coin's value is a multiple of 10?

SOLUTION Let S be the sample space and E be the set of coins whose value is a multiple of 10.

$$S = \{\text{penny, nickel, dime, quarter, 50-cent piece, silver dollar}\}$$

$$n(S) = 6$$

$$E = \{\text{dime, 50-cent piece, silver dollar}\}$$

$$n(E) = 3$$

$$\begin{aligned} P(E) &= \frac{n(E)}{n(S)} \\ &= \frac{3}{6} \\ &= 0.50 \end{aligned}$$

Assuming that there is an equally likely chance of picking any coin, there is a 50 percent chance that the selected coin's value will be a multiple of 10.

EXAMPLE 10

Calculating the Probability of an Event

In a game, a red die and a blue die are rolled. We observe the numbers shown on the pair of dice. What is the probability that the dice pair has a sum of 7 or 11?

SOLUTION

$$S = \begin{Bmatrix} (1,1) & (1,2) & (1,3) & (1,4) & (1,5) & (1,6) \\ (2,1) & (2,2) & (2,3) & (2,4) & (2,5) & (2,6) \\ (3,1) & (3,2) & (3,3) & (3,4) & (3,5) & (3,6) \\ (4,1) & (4,2) & (4,3) & (4,4) & (4,5) & (4,6) \\ (5,1) & (5,2) & (5,3) & (5,4) & (5,5) & (5,6) \\ (6,1) & (6,2) & (6,3) & (6,4) & (6,5) & (6,6) \end{Bmatrix}$$

$$n(S) = 36$$

Let the event E be the set of dice values that sum to 7 or 11.

$$E = \{(1, 6), (2, 5), (3, 4), (4, 3), (5, 2), (6, 1), (5, 6), (6, 5)\}$$
$$n(E) = 8$$

There are eight different dice rolls with a sum of 7 or 11.

$$\begin{aligned} P(E) &= \frac{n(E)}{n(S)} \\ &= \frac{8}{36} \\ &= 0.2222 \end{aligned}$$

There is a 22.22 percent chance that the sum of the dice will be 7 or 11.

Up until this point, we have used sample spaces with relatively few elements in each of our examples. When sample spaces become increasingly large, it is helpful to use the counting techniques covered in the previous sections. For Example 11, we are selecting five numbers (with replacement) from the set $\{1, 2, 3, 4, 5, 6\}$

EXAMPLE 11

Calculating the Probability of an Event

What is the probability that five dice rolled will all show the same number? (In the game of Yahtzee, this is the highest-scoring dice roll.)

SOLUTION

$$\begin{aligned} S &= \{x \mid x \text{ is a possible dice roll with 5 dice}\} \\ n(S) &= 6^5 \\ &= 7776 \end{aligned}$$

$$\begin{aligned} E &= \{x \mid x \text{ is a dice roll with dice all showing the same number}\} \\ n(E) &= P(6, 1) \\ &= 6 \end{aligned}$$

Pick one number from the possible six numbers: 1, 2, 3, 4, 5, 6

$$\begin{aligned} P(E) &= \frac{n(E)}{n(S)} \\ &= \frac{P(6, 1)}{6^5} \\ &= \frac{6}{6^5} \\ &= \frac{1}{6^4} \\ &= \frac{1}{1296} \\ &= 0.0007716 \\ &= 0.07716\% \end{aligned}$$

There is nearly a 0.08 percent chance that the dice will all show the same number.

EXAMPLE 12

Calculating the Probability of an Event

Four identical coins are tossed simultaneously. What is the probability that the faces of the coins will all be heads or all be tails when the coins land?

SOLUTION

$$
\begin{aligned}
S &= \{x \mid x \text{ is a possible coin toss with 4 coins}\} \\
n(S) &= 2^4 \\
&= 16 \\
E &= \{x \mid x \text{ is a coin toss showing all heads or all tails}\} \\
n(E) &= P(2, 1) && \text{From the two possible coin faces, we must pick one: heads or tails} \\
&= 2 && \text{One toss is all heads, the other toss is all tails} \\
P(E) &= \frac{n(E)}{n(S)} \\
&= \frac{2}{2^4} \\
&= \frac{2}{16} \\
&= 0.125 \\
&= 12.5\%
\end{aligned}
$$

There is a 12.5 percent probability that the coin toss will show either all heads or all tails.

Odds

In gaming situations, it is common to speak of the **odds** for or against the event instead of the probability of the event. Probability and odds are closely related, as detailed here.

PROBABILITY AND ODDS

If the probability of an event E is $P(E)$ with $0 < P(E) < 1$, then the

1. **Odds for *E*** are given by Odds for $E = \dfrac{P(E)}{P(E')}$
2. **Odds against *E*** are given by

$$
\begin{aligned}
\text{Odds against } E &= \frac{1}{\text{odds for } E} \\
&= \frac{1}{\dfrac{P(E)}{P(E')}} \\
&= \frac{P(E')}{P(E)}
\end{aligned}
$$

It is customary to represent odds as ratios of whole numbers.

In Example 12, we showed that the probability $P(E)$ that the faces of four tossed coins would match was 12.5 percent. Consequently, the probability that the faces do not all match is given by

$$\begin{aligned} P(E') &= 1 - P(E) \\ &= 1 - 0.125 \\ &= 0.875 \\ &= 87.5\% \end{aligned}$$

The odds for E are

$$\begin{aligned} \frac{P(E)}{P(E')} &= \frac{0.125}{0.875} \\ &= \frac{1}{7} \end{aligned}$$

If having all of the coin faces match is considered a win, then the odds of winning are 1 to 7. This is often written as 1:7 instead of $\frac{1}{7}$. The odds against winning are 7 to 1 (or 7:1). In other words, the odds of losing are 7:1.

EXAMPLE 13

Determining Odds

The probability of winning a dice game is 25 percent. What are the odds of winning and the odds of losing?

SOLUTION We have $P(E) = 0.25$. Therefore, $P(E') = 1 - 0.25 = 0.75$. The odds of winning are

$$\begin{aligned} \frac{P(E)}{P(E')} &= \frac{0.25}{0.75} \\ &= \frac{1}{3} \end{aligned}$$

The odds of winning are 1 to 3 (or 1:3). Consequently, the odds of not winning are 3 to 1. Therefore, the odds of losing are 3:1.

7.4 Summary

In this section, you learned basic probability terms and the basic principle of probability. You saw how these could be used to determine the likelihood of a favorable event. You also learned the relationship between probability and odds.

7.4 Exercises

In Exercises 1–10, use the terms experiment, trials, outcomes, *and* sample space *to describe the situation.*

1. A person is randomly picked from the population of Ellensburg, Washington, and the person's gender is noted.
2. A pair of dice is rolled, and the product of the numbers shown is noted.
3. A pair of dice is rolled, and the difference between the larger number and the smaller number is noted.
4. A coin is flipped, and the image on the side facing up is noted. The coin is flipped three times.
5. A bag of marbles contains five red, three white, and two blue marbles. Three marbles are pulled from the bag, and their color is noted.
6. A student enrolled in a five-credit math class is selected at random, and the number of credits on the student's class schedule is noted. (Assume that the number of credits cannot exceed 21.)
7. A student enrolled in a two-credit P.E. class is selected at random, and the number of credits on the student's class schedule is noted. (Assume that the number of credits cannot exceed 21.)
8. A player is selected at random from the high school football team, and the player's grade in school (i.e., ninth grade) is noted.
9. A grocery store classifies its breakfast cereals as Kellogg's, General Mills, or Other Brand. A box is selected at random, and its brand is noted. This process is repeated 10 times.
10. A clothing store catalogs its men's denim pants as Levi's, Bugle Boy, Calvin Klein, and Other Brands. A pair of denim pants is selected, and its brand is noted. This process is repeated 150 times.

In Exercises 11–20, identify the sample space for the experiment. Then write the specified events in set notation.

11. **Scholarship Applicants** Andrea (F), Miki (F), Miguel (M), Sara (F), and Sam (M) are scholarship contestants. (Their genders are noted in the parentheses.) Two of them are selected as finalists.
    (a) *Event:* At least one of the finalists is male.
    (b) *Event:* Both finalists are female.
    (c) *Event:* No more than one finalist is male.
12. **Scholarship Applicants** Julie (3.95), Dawn (4.00), Andres (3.53), Layla (3.83), Ari (3.71), and Saul (4.00) are scholarship candidates. (Their G.P.A.s are noted in the parentheses.) Two of them are selected as finalists.
    (a) *Event:* At least one of the finalists has a 4.0 G.P.A.
    (b) *Event:* Both finalists have above a 3.90 G.P.A.
    (c) *Event:* At least one finalist has a G.P.A. between 3.50 and 3.75.
13. **Game of Dice** In a game of dice, a red die and a blue die are rolled, and the numbers shown are noted.
    (a) *Event:* The sum of the numbers is even.
    (b) Event: The sum of the numbers is a multiple of 5.
    (c) *Event:* The sum of the numbers is not a multiple of 3.
14. **Game of Dice** In a game of dice, a red die and a blue die are rolled, and the numbers shown are noted.
    (a) *Event:* The numbers are the same.
    (b) *Event:* The numbers are both even.
    (c) *Event:* The difference between the larger number and the smaller number is exactly 2.
15. **True/False Test** A student takes a four-question true/false test, and her responses are noted. The correct answer to each question was *true*.
    (a) *Event:* She missed at most one question.
    (b) *Event:* She got at least two answers right.
    (c) *Event:* She got all of the questions wrong.
16. **True/False Test** A student takes a four-question true/false test, and his responses are noted. The correct answers to the questions (in order) were *TFFT.*
    (a) *Event:* He missed at most one question.
    (b) *Event:* He got at least three answers right.
    (c) *Event:* He got all of the questions wrong.
17. **House of Representatives** The two members of the U.S. House of Representatives from Idaho are selected in order of seniority, and their party affiliations (D, R, or I) are noted.
    (a) *Event:* More than half of them are Republicans (R).
    (b) *Event:* Democrats (D) outnumber Republicans (R).
    (c) *Event:* There are more non-Democrats (R and I) than Democrats (D).

18. **House of Representatives** The three members of the U.S. House of Representatives from Nebraska are selected in order of seniority, and their party affiliations (*D*, *R*, or *I*) are noted.

(a) *Event:* At least half of them are Republicans (*R*).
(b) *Event:* Democrats (*D*) outnumber Republicans (*R*).
(c) *Event:* There are fewer non-Democrats (*I* and *R*) than Democrats (*D*).

19. **House of Representatives** The four members of the U.S. House of Representatives from Mississippi are selected in order of seniority, and their genders are noted.

(a) *Event:* Females outnumber males.
(b) *Event:* The least senior representative is female.
(c) *Event:* The senior representative is male and the least senior representative is female.

20. **House of Representatives** The five members of the U.S. House of Representatives from Oregon are selected in order of seniority, and their genders are noted.

(a) *Event:* Females outnumber males.
(b) *Event:* The least senior representative is female.
(c) *Event:* The senior representative is male and the least senior representative is female.

In Exercises 21–30, assuming that all outcomes are equally likely, calculate the probability of each event. Then calculate the odds for and against each event. You may find it helpful to refer to your work in Exercises 11–20.

21. **Scholarship Applicants** Andrea (F), Miki (F), Miguel (M), Sara (F), and Sam (M) are scholarship contestants. (Their genders are noted in the parentheses.) Two of them are selected as finalists.

(a) *Event:* At least one of the finalists is male.
(b) *Event:* Both finalists are female.
(c) *Event:* No more than one finalist is male.

22. **Scholarship Applicants** Julie (3.95), Dawn (4.00), Andres (3.53), Layla (3.83), Ari (3.71), and Saul (4.00) are scholarship candidates. (Their G.P.A.s are noted in the parentheses.) Two of them are selected as finalists.

(a) *Event:* At least one of the finalists has a 4.0 G.P.A.
(b) *Event:* Both finalists have above a 3.90 G.P.A.
(c) *Event:* At least one finalist has a G.P.A. between 3.50 and 3.75.

23. **Game of Dice** In a game of dice, a red die and a blue die are rolled, and the numbers shown are noted.

(a) *Event:* The sum of the numbers is even.
(b) *Event:* The sum of the numbers is a multiple of 5.
(c) *Event:* The sum of the numbers is not a multiple of 3.

24. **Game of Dice** In a game of dice, a red die and a blue die are rolled, and the numbers shown are noted.

(a) *Event:* The numbers are the same.
(b) *Event:* The numbers are both even.
(c) *Event:* The difference between the larger number and the smaller number is exactly 2.

25. **True/False Test** A student takes a four-question true/false test, and her responses are noted. The correct answer to each question was true.

(a) *Event:* She missed at most one question.
(b) *Event:* She got at least two answers right.
(c) *Event:* She got all of the questions wrong.

26. **True/False Test** A student takes a four-question true/false test, and his responses are noted. The correct answers to the questions (in order) were *TFFT*.

(a) *Event:* He missed at most one question.
(b) *Event:* He got at least three answers right.
(c) *Event:* He got all of the questions wrong.

27. **House of Representatives** The two members of the U.S. House of Representatives from Idaho are selected in order of seniority, and their party affiliations (*D*, *R*, or *I*) are noted.

(a) *Event:* More than half of them are Republicans (*R*).
(b) *Event:* Democrats (*D*) outnumber Republicans (*R*).
(c) *Event:* There are more non-Democrats (*R* and *I*) than Democrats (*D*).

28. **House of Representatives** The three members of the U.S. House of Representatives from Nebraska are selected in order of seniority, and their party affiliations (*D*, *R*, or *I*) are noted.

(a) *Event:* At least half of them are Republicans (R).
(b) *Event:* Democrats (D) outnumber Republicans (R).
(c) *Event:* There are fewer non-Democrats (I and R) than Democrats (D).

29. **House of Representatives** The four members of the U.S. House of Representatives from Mississippi are selected in order of seniority, and their genders are noted.

(a) *Event:* Females outnumber males.
(b) *Event:* The least senior representative is female.
(c) *Event:* The senior representative is male and the least senior representative is female.

30. **House of Representatives** The five members of the U.S. House of Representatives from Oregon are selected in order of seniority, and their genders are noted.

(a) *Event:* Females outnumber males.
(b) *Event:* The least senior representative is female.
(c) *Event:* The senior representative is male and the least senior representative is female.

Exercises 31–35 are intended to challenge your understanding of basic probability concepts.

31. **Employment Status** In a five-person group, there are three women and two men. Two of the women and one of the men are employed. Assuming that all outcomes are equally likely, determine the probability of each event.

(a) *Event:* Two people selected from the group are both employed.
(b) *Event:* Two people selected from the group are both unemployed.
(c) *Event:* Two people selected from the group are employed women.

32. **Professions** In a thirty-person group, there are three doctors, five lawyers, and four teachers. Assuming that all outcomes are equally likely, determine the probability of each event.

(a) *Event:* Three people selected from the group are teachers.
(b) *Event:* Three people selected from the group are not lawyers.
(c) *Event:* At least one of three people selected from the group is a doctor.

33. **Dice Game** In a game of dice, three dice are rolled. The player wins if the sum *and* the product of the dice are even numbers. What is the probability that the player will win?

34. **Dice Game** In a game of dice, two dice are rolled. The player wins if the sum of the dice is greater than 5 *and* the product of the dice is less than 15. What is the probability that the player will win?

35. **Dice Game** In a game of dice, two dice are rolled. The player wins if the sum of the dice is less than 9 *and* the product of the dice is a multiple of 3. What is the probability that the player will win?

7.5 Basic Probability Concepts

- Use probability concepts to calculate probabilities

GETTING STARTED Political lobbyists representing various special interest groups seek to influence the decisions of members of Congress. Certain issues may receive support among Democrats, while other issues may receive support among women, regardless of their political affiliation. A lobbyist may use the concepts of probability to help him or her determine the likelihood of success with a particular issue.

In this section, we will discuss set operations for events and then introduce rules for the probability of a union and a complement. We will also introduce the idea of a probability distribution.

Often we are interested in the intersection, union, or complement of two events. As detailed here, the same set operations introduced earlier in the chapter may be used with events for a sample space.

SET OPERATIONS FOR EVENTS

Let E and F be events for a sample space S. Then

- $E \cap F$ is the set of outcomes in both E and F.
- $E \cup F$ is the set of all outcomes that are either in E or in F or in both E and F.
- E' is the set of all outcomes in S that are not in E.

EXAMPLE 1

Using and Interpreting Set Notation

A coin purse contains pennies, nickels, dimes, quarters, 50-cent pieces, and silver dollars. Pick a coin from a purse and note its value. Let E be the set of coins whose value is a multiple of 10. Let F be the set of coins whose value is a multiple of 25. Determine $E \cap F$, $E \cup F$, and E' and express in words the meaning of each set.

SOLUTION

$S = \{\text{penny, nickel, dime, quarter, 50-cent piece, silver dollar}\}$

$E = \{\text{dime, 50-cent piece, silver dollar}\}$ Coins whose value is a multiple of 10

$F = \{\text{quarter, 50-cent piece, silver dollar}\}$ Coins whose value is a multiple of 25

$E \cap F = \{\text{50-cent piece, silver dollar}\}$ Coins whose value is a multiple of 10 and 25

$E \cup F = \{\text{dime, quarter, 50-cent piece, silver dollar}\}$ Coins whose value is a multiple of 10 or 25

$E' = \{\text{penny, nickel, quarter}\}$ Coins whose value is not a multiple of 10

Not all events share elements in common. Events without any shared elements are **mutually exclusive events.**

MUTUALLY EXCLUSIVE EVENTS

Two events E and F are mutually exclusive if $E \cap F = \varnothing$.

EXAMPLE 2

Determining If Sets Are Mutually Exclusive

Pick a person from the crowd and note whether the person is a citizen or a noncitizen of the United States. Let E be the set of citizens and F be the set of noncitizens. Determine if the sets E and F are mutually exclusive.

SOLUTION

$S = \{\text{citizen, noncitizen}\}$

$E = \{\text{citizen}\}$ People who are citizens

$F = \{\text{noncitizen}\}$ People who are not citizens

$E \cap F = \varnothing$ People who are both citizens and not citizens

Since $E \cap F = \varnothing$, the events E and F are mutually exclusive.

When we talk about the probability of a union of two events E and F, we are looking for the likelihood that either event E **or** event F occurs. For example, a teenager may estimate that there is a 20 percent probability that Anastasia will ask him to the dance and a 40 percent probability that Marisol will ask him to the dance. The probability of the union of these two events is the probability that Anastasia **or** Marisol **or** both Anastasia and Marisol will ask him to the dance. To calculate the probability of a union, we use the probability of a union property.

PROBABILITY OF A UNION

Let E and F be events in a sample space S. Then

$$P(E \cup F) = P(E) + P(F) - P(E \cap F)$$

If the events E and F are mutually exclusive, then $P(E \cap F) = 0$, and the equation simplifies to

$$P(E \cup F) = P(E) + P(F)$$

EXAMPLE 3

Calculating the Probability of an Event

In 2002, the state of Missouri had nine representatives in the U.S. House of Representatives. Four of the representatives were Democrats, and two of the representatives were women. One of the representatives was a woman Democrat. (**Source:** www.clerk.house.gov.) If a political issue has strong support among both Democrats and women, and strong opposition from Republican men, what is the probability that a randomly selected representative from Missouri will support the issue?

SOLUTION The set $S = \{x|x \text{ is a representative from Missouri}\}$.

Let $E = \{x|x \text{ is a Democrat}\}$ and $F = \{x|x \text{ is a woman}\}$. We have $n(S) = 9$, $n(E) = 4$, $n(F) = 2$, and $n(E \cap F) = 1$.

We need to determine the probability that a representative is a woman **or** a Democrat. Recall that for an event E from a sample space S, $P(E) = \frac{n(E)}{n(S)}$.

$$\begin{aligned} P(E \cup F) &= P(E) + P(F) - P(E \cap F) \\ &= \frac{4}{9} + \frac{2}{9} - \frac{1}{9} \\ &= \frac{5}{9} \\ &= 0.5555 \end{aligned}$$

The probability that a representative is a Democrat or a woman is 55.55 percent. It is likely that the a randomly selected representative from the state of Missouri would support the issue.

EXAMPLE 4

Calculating the Probability of an Event

A coin is flipped three times, and the sequence of heads and tails is recorded. Let event E be the set of outcomes containing at least two heads. Let event F be the set of outcomes containing all heads or all tails. What is the probability that an outcome is in $E \cup F$?

SOLUTION

$S = \{\text{HHH, HHT, HTH, THH, HTT, THT, TTH, TTT}\}$

$n(S) = 8$

$E = \{\text{HHH, HHT, HTH, THH}\}$ The set of outcomes with at least two heads

$n(E) = 4$

$F = \{\text{HHH, TTT}\}$ The set of outcomes containing all heads or all tails

$n(F) = 2$

$E \cap F = \{\text{HHH}\}$ The set of outcomes with at least two heads and all heads or all tails

$n(E \cap F) = 1$

We must find $P(E \cup F)$.

$$\begin{aligned} P(E \cup F) &= P(E) + P(F) - P(E \cap F) \\ &= \frac{4}{8} + \frac{2}{8} - \frac{1}{8} \\ &= \frac{5}{8} \\ &= 0.6250 \end{aligned}$$

There is a 62.5 percent probability that an outcome is in $E \cup F$.

EXAMPLE 5

Calculating the Probability of an Event

Four men and two women sell high-quality kitchen knives. Three of them will be selected at random to work this weekend at a local fair. What is the probability that at most two men will have to work this weekend?

SOLUTION We must first determine the cardinality of our sample space S. The number of different groups of three people that can be selected from a group of six people is

$$\begin{aligned} n(S) &= C(6, 3) \\ &= \frac{6!}{3! \cdot 3!} \\ &= \frac{6 \cdot 5 \cdot 4}{3 \cdot 2 \cdot 1} \\ &= 20 \end{aligned}$$

Since there are only two women, at least one man will have to work this weekend. We will calculate the probability that a three-person group contains exactly one man and then calculate the probability that a three-person group contains exactly two men. The sum of these probabilities will give us the probability that at most two men will have to work this weekend.

Define event E to be the set of three-person groups containing exactly *one* man. The number of different groups of three people containing exactly one man is

$$n(E) = C(4, 1) \cdot C(2, 2)$$

Pick one out of four men. | Pick two out of two women.

$$\begin{aligned} n(E) &= C(4, 1) \cdot C(2, 2) \\ &= \frac{4!}{3!1!} \cdot \frac{2!}{2!0!} \\ &= 4 \cdot 1 \\ &= 4 \end{aligned}$$

There are four three-person groups containing exactly one man.

$$\begin{aligned} P(E) &= \frac{n(E)}{n(S)} \\ &= \frac{4}{20} \\ &= 0.20 \end{aligned}$$

There is a 20 percent probability that a three-person group will contain exactly one man.

Define event F to be the set of three-person groups containing exactly *two* men. The number of groups of three people containing exactly two men is

$$n(F) = C(4, 2) \cdot C(2, 1)$$

Pick two out of four men. | Pick one out of two women.

$$\begin{aligned} n(F) &= C(4, 2) \cdot C(2, 1) \\ &= \frac{4!}{2! \cdot 2!} \cdot \frac{2!}{1! \cdot 1!} \\ &= \frac{4 \cdot 3}{2 \cdot 1} \cdot \frac{2}{1} \\ &= 12 \end{aligned}$$

There are 12 three-person groups containing exactly two men.

$$\begin{aligned} P(F) &= \frac{n(F)}{n(S)} \\ &= \frac{12}{20} \\ &= 0.60 \end{aligned}$$

There is a 60 percent probability that a three-person group will contain exactly two men.

Since E and F are mutually exclusive, $P(E \cap F) = 0$. Therefore, the probability that at most one or two men will have to work this weekend is given by

$$\begin{aligned} P(E \cup F) &= P(E) + P(F) && \text{Since } E \text{ and } F \text{ are mutually exclusive} \\ &= 0.60 + 0.20 \\ &= 0.80 \end{aligned}$$

There is an 80 percent chance that at most two men will have to work this weekend.

The process of calculating probabilities can often be simplified by calculating the probability of a complement. The probability of a complement formula is derived by observing that there is a 100 percent probability of being in the sample space. That is,

$$\begin{aligned} P(S) &= \frac{n(S)}{n(S)} \\ &= 1 \end{aligned}$$

Note that $E \cup E' = S$. Since E and E' are mutually exclusive,

$$\begin{aligned} P(S) &= P(E \cup E') \\ 1 &= P(E) + P(E') && \text{Since } P(S) = 1 \\ P(E') &= 1 - P(E) \end{aligned}$$

PROBABILITY OF A COMPLEMENT

Let E be an event in a sample space S. Then

$$P(E') = 1 - P(E)$$

In Example 6, we will rework Example 5 using the probability of a complement formula.

EXAMPLE 6

Calculating the Probability of a Complement

Four men and two women sell high-quality kitchen knives. Three of them will be selected at random to work this weekend at a local fair. What is the probability that at most two men will have to work this weekend?

SOLUTION We must first determine the cardinality of our sample space S. The number of different groups of three people that can be selected from a group of six people is

$$\begin{aligned} n(S) &= C(6, 3) \\ &= \frac{6!}{3! \cdot 3!} \\ &= \frac{6 \cdot 5 \cdot 4}{3 \cdot 2 \cdot 1} \\ &= 20 \end{aligned}$$

Define event E to be the set of three-person groups containing exactly *three* men. The number of groups of three people containing exactly three men is

$$n(E) = C(4, 3) \cdot C(2, 0)$$

Pick three out of four men. Pick zero out of two women.

$$\begin{aligned} n(E) &= C(4, 3) \cdot C(2, 0) \\ &= \frac{4!}{3! \cdot 1!} \cdot \frac{2!}{2! \cdot 0!} \\ &= 4 \cdot 1 \\ &= 4 \end{aligned}$$

There are four three-person groups containing exactly three men.

$$\begin{aligned} P(E) &= \frac{n(E)}{n(S)} \\ &= \frac{4}{20} \\ &= 0.20 \end{aligned}$$

There is a 20 percent probability that a three-person group will contain exactly three men.

E' is the set of three-person groups not containing three men. In other words, it is the set of three-person groups containing at most two men.

$$\begin{aligned} P(E') &= 1 - P(E) \\ &= 1 - 0.20 \\ &= 0.80 \end{aligned}$$

There is an 80 percent chance that a three-person group will contain at most two men. This result is the same as that from Example 5 but required fewer computations.

EXAMPLE 7

Calculating the Probability of a Complement

Ten students are competing for a scholarship. Six of the students are women. What is the probability that a randomly selected application belongs to a man?

SOLUTION Since there are 10 applicants, $n(S) = 10$.

$$\begin{aligned} E &= \{x | x \text{ is a female applicant}\} \\ n(E) &= 6 \\ E' &= \{x | x \text{ is a male applicant}\} \\ n(E') &= n(S) - n(E) \\ &= 10 - 6 \\ &= 4 \\ P(E') &= \frac{n(E')}{n(S)} \\ &= \frac{4}{10} \\ &= 0.4 \end{aligned}$$

There is a 40 percent probability that the application belongs to a man.

When looking at the fairness of award processes, it is often helpful to look at the probability of obtaining an outcome that meets certain specified criteria. For example, suppose that the probability of a woman obtaining one of 10 identical scholarships is 55 percent. However, in actuality, men obtain all 10 of the scholarships. Since the actual result varies widely from our predicted result, we may want to analyze the awards process for gender bias.

EXAMPLE 8

Calculating the Probability of an Event

Ten students are competing for a scholarship. Exactly two of the men have a G.P.A. below 3.7. What is the probability that an applicant has a G.P.A above 3.7 or is a woman?

SOLUTION There are 10 students competing, so $n(S) = 10$. Let $E = \{x | x \text{ is a student who is a man}\}$ and $F = \{x | x \text{ is a student with a G.P.A below 3.7}\}$. The set of men with G.P.A.s below 3.7 is given by $E \cap F$. We know that $n(E \cap F) = 2$ and

$$\begin{aligned} P(E \cap F) &= \frac{n(E \cap F)}{n(S)} \\ &= \frac{2}{10} \\ &= 0.2 \end{aligned}$$

There is a 20 percent probability that a randomly selected applicant is a male with a G.P.A. below 3.7.

We know that

$$\begin{aligned}(E \cap F)' &= \{x|x \text{ is a student who is not a man with a G.P.A below 3.7}\} \\ &= \{x|x \text{ is a student who is a woman or has a G.P.A of 3.7 or better}\}\end{aligned}$$

$$\begin{aligned}P(E \cap F)' &= 1 - P(E \cap F) \\ &= 1 - 0.2 \\ &= 0.8\end{aligned}$$

There is an 80 percent probability that a randomly selected applicant is a woman or has a G.P.A of 3.7 or better.

To help us better understand our society, researchers often classify populations by demographic factors such as ethnic origin, religion, or gender. Doing so allows us to investigate the fairness of a variety of societal processes, such as education, employment, and home buying. Whenever a population is divided into a group of disjoint sets, a probability distribution may be used to determine the likelihood that a randomly selected element belongs to any specified set.

PROBABILITY DISTRIBUTION

Let *E, F, G, H, . . .* be mutually exclusive events whose union contains all outcomes in a sample space *S*. Then

$$P(E) + P(F) + P(G) + P(H) + \cdots = 1$$

The probabilities $P(E), P(F), P(G), P(H), \ldots$ make up a **probability distribution** for *S*.

EXAMPLE 9

Finding a Probabilty Distribution

A class of students contains twenty-five students classified by hair color: six blondes, four brunettes, two with red hair, seven with black hair, four with brown hair, and two with hair of other colors. Find a probability distribution for the class based on hair color.

SOLUTION Let $S = \{x|x \text{ is a student in the class}\}$, $E = \{x|x \text{ is a blonde}\}$, $F = \{x|x \text{ is a brunette}\}$, $G = \{x|x \text{ has red hair}\}$, $H = \{x|x \text{ has black hair}\}$, $I = \{x|x \text{ has brown hair}\}$, and $J = \{x|x \text{ has hair of another color}\}$. These events are mutually exclusive.

$$P(E) = \frac{6}{25} = 0.24 \qquad P(F) = \frac{4}{25} = 0.16 \qquad P(G) = \frac{2}{25} = 0.08$$

$$P(H) = \frac{7}{25} = 0.28 \qquad P(I) = \frac{4}{25} = 0.16 \qquad P(J) = \frac{2}{25} = 0.08$$

Observe that $0.24 + 0.16 + 0.08 + 0.28 + 0.16 + 0.08 = 1$. There is a 24 percent chance that a randomly selected student is blonde, a 16 percent chance that the student is brunette, an 8 percent chance that the student has red hair, a 28 percent chance that the student has black hair, a 16 percent chance that the student has brown hair, and an 8 percent chance that the student has hair of another color.

EXAMPLE 10

Finding a Probability Distribution

In 2002, the state of Colorado had six representatives in the U.S. House of Representatives, four Republicans and two Democrats. Only one of the representatives was female. (**Source:** www.clerk.house.gov.) Make a probability distribution based on political party. Then make another probability distribution based on gender.

SOLUTION Note that for both probability distributions, the sample space S is the set of six Colorado representatives.

$$E = \{x | x \text{ is a Republican}\}$$
$$F = \{x | x \text{ is a Democrat}\}$$

The events are mutually exclusive.

$$\begin{aligned} P(E) &= \frac{n(E)}{n(S)} \\ &= \frac{4}{6} \\ &= 0.6667 \end{aligned}$$

$$\begin{aligned} P(F) &= \frac{n(F)}{n(S)} \\ &= \frac{2}{6} \\ &= 0.3333 \end{aligned}$$

Observe that $0.6667 + 0.3333 = 1$.

There is a 66.67 percent probability that a randomly selected representative is a Republican and a 33.33 percent probability that the representative is a Democrat.

$$E = \{x | x \text{ is a male}\}$$
$$F = \{x | x \text{ is a female}\}$$

The events are mutually exclusive.

$$\begin{aligned} P(E) &= \frac{n(E)}{n(S)} \\ &= \frac{5}{6} \\ &= 0.8333 \end{aligned}$$

$$\begin{aligned} P(F) &= \frac{n(F)}{n(S)} \\ &= \frac{1}{6} \\ &= 0.1667 \end{aligned}$$

Observe that $0.8333 + 0.1667 = 1$.

There is an 83.33 percent probability that a randomly selected representative is male and a 16.67 percent probability that the representative is female.

7.5 Summary

In this section, you learned the rules for the probability of a union and a complement. You also learned how to find a probability distribution.

7.5 Exercises

In Exercises 1–5, determine $E \cap F$, $E \cup F$, and E' and express in words the meaning of each set.

1. Coin Purse A coin purse contains pennies, nickels, dimes, quarters, 50-cent pieces, and silver dollars. Pick a coin from the purse and note its value. Let E be the set of coins whose value is a multiple of 5. Let F be the set of coins whose value is a multiple of 50.

2. Recipe File A recipe file includes recipes for blueberry cobbler, apple crisp, oatmeal cookies, chocolate-covered strawberries, strawberry ice cream, chocolate ice cream, and German chocolate cake. Let E be the set of recipes containing fruit and F be the set of recipes containing chocolate.

3. Class Composition Let E be the students in a class who are on scholarship and F be the students in the class who have a cumulative G.P.A of 3.5 or higher.

4. Class Composition Some students are unsuccessful the first time they take a particular course but are able to successfully pass the course after retaking it. Let E be the group of students who are enrolled in this course for the first time and F be the group of students who have previously earned a grade below a C in this course.

5. United Nations Security Council In 2004, the United Nations Security Council consisted of Chile, China, France, Germany, Pakistan, Philippines, Romania, Russian Federation, Spain, United Kingdom, United States, Algeria, Angola, Benin, and Brazil. (**Source:** www.un.org.) Let E be the nations from Europe and F be the nations who are permanent members of the Security Council.

In Exercises 6–10, the sample space S is created by rolling two dice (one red and one white) and observing the numbers shown. Calculate $P(E \cup F)$ and $P(E \cap F)$.

6. $E = \left\{ x \mid x \text{ is a dice roll with an even sum} \right\}$

$F = \left\{ x \mid x \text{ is a dice roll with a sum less than 9} \right\}$

7. $E = \left\{ x \mid x \text{ is a dice roll with an odd sum} \right\}$

$F = \left\{ x \mid x \text{ is a dice roll with a sum less than 4} \right\}$

8. $E = \left\{ x \mid x \text{ is a dice roll with both numbers less than 5} \right\}$

$F = \left\{ x \mid x \text{ is a dice roll with both numbers even} \right\}$

9. $E = \left\{ x \mid x \text{ is a dice roll with both numbers more than 3} \right\}$

$F = \left\{ x \mid x \text{ is a dice roll with an even sum} \right\}$

10. $E = \left\{ x \mid x \text{ is a dice roll with both numbers equal} \right\}$

$F = \left\{ x \mid x \text{ is a dice roll with both numbers even} \right\}$

In Exercises 11–15, four women (Alicia, Britney, Carmela, and Dejanae) and two men (Ernesto and Frank) sell vegetable choppers at local fairs. Three of them will be selected at random to work this weekend. Find the probability of the specified event E.

11. $E = \left\{ x \mid x \text{ is a work group with more women than men} \right\}$

12. $E = \left\{ x \mid x \text{ is a work group with Alicia and Ernesto} \right\}$

13. $E = \left\{\begin{array}{l} x|x \text{ is a work group with} \\ \text{both men} \end{array}\right\}$

14. $E = \left\{\begin{array}{l} x|x \text{ is a work group with} \\ \text{at most one man} \end{array}\right\}$

15. $E = \left\{\begin{array}{l} x|x \text{ is a work group with} \\ \text{exactly two women} \end{array}\right\}$

In Exercises 16–20, three women (Alexis, Brianna, and Cali) and four men (Dakota, Eli, Fahim, and Gregorio) sell power blenders at local fairs. Four of them will be selected at random to work this weekend. Find the probability of the specified event E.

16. $E = \left\{\begin{array}{l} x|x \text{ is a work group with} \\ \text{Brianna and Dakota} \end{array}\right\}$

17. $E = \left\{\begin{array}{l} x|x \text{ is a work group without} \\ \text{Alexis and Fahim} \end{array}\right\}$

18. $E = \left\{\begin{array}{l} x|x \text{ is a work group with} \\ \text{all men} \end{array}\right\}$

19. $E = \left\{\begin{array}{l} x|x \text{ is a work group with} \\ \text{women and men} \end{array}\right\}$

20. $E = \left\{\begin{array}{l} x|x \text{ is a work group with} \\ \text{Cali and Eli} \end{array}\right\}$

21. Who would benefit from the results of Exercises 11–20? Explain.

In Exercises 22–30, find the probability distribution.

22. **House of Representatives** In 2002, the U.S. House of Representatives had 435 members. There were 223 Republicans, 209 Democrats, 1 Independent, and 2 vacancies. Find a probability distribution for the House of Representatives.

23. **Senate** In 2002, the U.S. Senate had 100 members, including 49 Republicans, 50 Democrats, and 1 Independent. Find a probability distribution for the Senate.

24. **Dice Game** A pair of dice is rolled. Find a probability distribution for the sum of the dice.

25. **Citizenship** A math class has 33 students, including 20 American citizens, 11 people with student visas, and 2 illegal aliens. Find a probability distribution for the legal status of the students in the class.

26. **Coin Collection** A box of twenty coins contains four quarters, five dimes, three nickels, and eight pennies. Find a probability distribution for the box of coins.

27. **Educational Status** A company with 40 employees tracks the highest earned degree of its employees. It has 13 employees with master's degrees, 17 employees with bachelor's degrees, and 10 employees with high school diplomas. Find a probability distribution for the educational status of the employees of the company.

28. **Dice Game** A six-sided die (numbered from one to six) and a four-sided die (numbered from one to four) are rolled. Find a probability distribution for the absolute value of the difference of the numbers shown on the faces of the dice.

29. **Dice Game** A six-sided die (numbered from one to six) and a four-sided die (numbered from one to four) are rolled. Find a probability distribution for the product of the numbers shown on the faces of the dice.

30. **Professions** A dinner party of sixteen guests includes three doctors, one dentist, one physical therapist, two teachers, three training specialists, three mechanics, a bartender, a preacher who gives weekly sermons, and a cook. The guests may be categorized into groups of those who treat the human body, those who teach people, those who fix mechanical things, and those who make or serve foodstuffs. Find a probability distribution for the employment categories.

Exercises 31–35 are intended to challenge your understanding of basic probability concepts.

31. **Mixed Groups** There are ten people in a courtroom, including six lawyers, three criminals, seven men, and one male lawyer who is a criminal. All of the women in the room, including the one who is a lawyer, aren't criminals. What is the probability that a person randomly selected from the group is a male lawyer?

32. **Doctor's Office** There are twenty people in a room, including six pregnant women, eleven sick people, two doctors, and three nurses. None of the medical personnel are sick, but one of them is pregnant. What is the probability that a person randomly selected from the group is a pregnant women who is sick?

33. **Dice Game** Two six-sided dice are rolled. Find a probability distribution for the product of the numbers shown minus the largest number shown on the faces of the dice.

34. **Dice Game** Three four-sided dice are rolled. Find a probability distribution for the difference between the largest and the smallest value shown on the faces of the dice.

35. **Dice Game** Based on the results of Exercise 34, is obtaining an even value or an odd value more likely?

Chapter 7 Review Exercises

Section 7.1 *In Exercises 1–4, copy the Venn diagram given here and shade the region defined in each problem, if possible. Then list the elements of the solution set.*

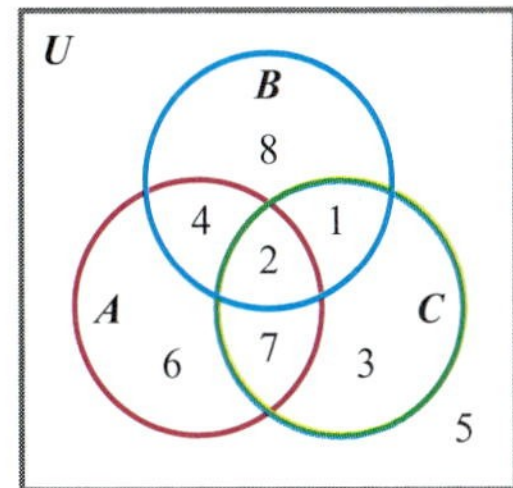

1. $(A \cap B)'$
2. $(A \cap B) \cup C$
3. $(A \cap B) \cap C'$
4. $(A \cup B) \cap C'$

For Exercises 5–8, six people are categorized by their gender, employment status, and highest academic degree as follows:

1: Male, employed, master's degree
2: Male, employed, high school diploma
3: Female, employed, bachelor's degree
4: Female, unemployed, bachelor's degree
5: Male, unemployed, Ph.D.
6: Female, unemployed, high school diploma

Further categorize the group by letting M be the set of all men, E be the set of employed people, and D be the set of people with a graduate degree (master's or Ph.D.). Explain the real-life meaning of the set specified in each problem. Then specify which people are elements of the set. (Hint: It may be helpful to draw a Venn diagram.)

5. $M \cap E$
6. $E \cup D$
7. $E' \cap D$
8. $(D \cap M)'$

Section 7.2 *In Exercises 9–12, A and B are finite subsets of the universal set U with $n(U) = 20$, $n(A) = 12$, $n(A \cap B) = 5$, and $n(A \cup B) = 17$. Find the cardinality of each of the specified sets.*

9. B'
10. $A' \cup B'$
11. $A' \cap B'$
12. $(A \cup B)'$

In Exercises 13–16, use the Addition or Multiplication Principle as appropriate to determine the cardinality of the set.

13. A restaurant offers 6 appetizers, 20 entrees, and 8 desserts. How many dish options are there?

14. A restaurant offers 6 appetizers, 20 entrees, and 8 desserts. How many three-course meal options are there?

15. A restaurant offers 7 appetizers, 25 entrees, and 4 desserts, including strawberry shortcake. How many three-course meal options are there with strawberry shortcake as dessert?

16. An office supply store offers three qualities of staplers: light-duty, medium-duty, and heavy-duty. The staplers come in grey, brown, or black. How many different types of staplers are there?

For Exercises 17–18, you are given the sets $R = \{$truck, sport utility vehicle$\}$, $C = \{$black, red, blue$\}$, $S = \{$two-wheel drive, four-wheel drive$\}$, and $M = \{$new, used$\}$. Draw a decision tree to show all possible outcomes for the specified criteria.

17. The outcomes with a single element from each of the sets R, C, and S.

18. The outcomes with a single element from each of the sets R, C, and M.

Section 7.3 *In Exercises 19–22, calculate the value of the expression without using a calculator.*

19. $P(9, 2)$ **20.** $P(15, 1)$

21. $C(6, 2)$ **22.** $C(10, 4)$

In Exercises 23–24, calculate the value of the expression using a calculator.

23. $P(25, 7)$ **24.** $C(25, 7)$

In Exercises 25–28, use permutations or combinations as appropriate to calculate the total number of possible outcomes.

Pizza Hut offers 4 different types of crust (Pan Pizza, Stuffed Crust Pizza, The Big New Yorker Pizza, and Thin 'n Crispy Pizza) and 12 toppings (Italian Sausage, Pepperoni, Beef Topping, Pork Topping, Ham, Chicken, Mushrooms, Green Peppers, Red Onions, Tomatoes, Black Olives, and Bacon). (**Source:** www.pizzahut.com.)

25. How many different types of one-topping pizzas can be made?

26. How many different types of three-topping Stuffed Crust Pizzas can be made?

27. How many different types of three-topping pizzas can be made?

28. How many different types of three-topping pizzas can be made with the meatless toppings?

Section 7.4 *In Exercises 29–30, use the terms* experiment, trials, outcomes, *and* sample space *to explain the situation.*

29. A sweepstakes contestant is randomly picked from a box of entries and the last digit of the entry number is noted. (Assume that entries may end in any digit.)

30. A pair of dice is rolled twice. Each time the sum of the dice is recorded.

In Exercises 31–32, identify the sample space for the experiment. Then write the specified event in set notation.

31. Adi (M), Berto (M), Marsha (F), Mariah (F), and Liz (F) are scholarship contestants. (Their genders are noted in the parentheses.) Two of them are selected as finalists.

(a) *Event:* At least one of the finalists is male.
(b) *Event:* Both finalists are female.
(c) *Event:* No more than one finalist is male.

32. Jorge (3.92), Shauna (3.97), Andrea (3.47), Mohammed (3.83), Teri (3.74), and Paul (4.00) are scholarship candidates. (Their G.P.A.s are noted in the parentheses.) Two of them are selected as finalists.

(a) *Event:* At least one of the finalists has a 4.0 G.P.A.
(b) *Event:* Both finalists have above a 3.90 G.P.A.
(c) *Event:* At least one finalist has a G.P.A. between 3.40 and 3.75.

Section 7.5 *In Exercises 33–36, three women (Alena, Betty, and Carmen) and two men (Don and Ernesto) deliver pizza. Three of them will be selected at random to work this weekend. Find the probability of the specified event.*

33. $E = \left\{ x \mid x \text{ is a group with more women than men} \right\}$

34. $E = \left\{ x \mid x \text{ is a group with Alena and Ernesto} \right\}$

35. $E = \left\{ x \mid x \text{ is a group with Carmen and at least one man} \right\}$

36. $E = \left\{ x \mid x \text{ is a group with Alena but not Carmen} \right\}$

Make It Real

PROJECT 7

Dice games, including Casino Craps, Pig, Bunco, Zilch, and Yahtzee©, among others, are popular diversions for people seeking a bit of entertainment. In this project, you'll make up a simple dice game.

What to do

1. Find a pair of dice (or a paradise!).
2. Determine how you will interpret each dice roll (sum of dice, product of dice, compare dice values, and so on).
3. List all outcomes for each dice roll.
4. Define three or more events that will be scored (e.g., sum of dice is 7).
5. Find the probability distribution for the dice game.
6. Design a scoring system based on the probabilities (low probability offers high points).
7. Name your game.

Example dice game

Bonkaroo

1. Roll two dice and sum their values.
2. *Sample space:* {2, 3, 4, . . . ,11, 12}
3. *Events:* {2}, {7}, {11}, {not 2, 7, 11}
4. There are 36 possible rolls with two dice. The ways to get each event are

   2: (1,1)

   7: (1, 6), (2, 5), (3, 4), (4, 3), (5, 2), (6, 1)

   11: (5, 6), (6, 5)

   not 2, 7, 11: the remaining 27 rolls

   Probability Distribution

   2: $\frac{1}{36} = 0.0278$

   7: $\frac{6}{36} = 0.1667$

   11: $\frac{2}{36} = 0.0556$

   not 2, 7, 11: $\frac{27}{36} = 0.75$
5. A 2 is worth 30 points, a 7 is worth 5 points, and an 11 is worth 15 points.
6. Players take turns rolling the dice.
7. The first player to score a total of 100 points and yell *Bonkaroo!* wins.

Chapter 8

Advanced Probability and Statistics

As diverse as we are as individuals, we still are remarkably similar to other people. Consider the physical feature of height. Ninety percent of Mexican American women are between 58 and 66 inches tall. (**Source:** www.halls.md.) If a woman is randomly selected from a large group of Mexican American women, how likely is it that she will be 70 inches tall? The statistical concept of a normal distribution may be used to calculate the probability of randomly selecting a woman of a particular height.

8.1 Conditional Probability
- Use conditional probability to determine if events are independent

8.2 Bayes' Theorem and Applications
- Apply Bayes' theorem to real-world applications
- Calculate the predictive-value positive and predictive-value negative of a test

8.3 Markov Chains
- Forecast future events with Markov chains

8.4 Random Variables and Expected Value
- Calculate the probability distribution of a random variable
- Draw a histogram
- Determine the expected value of a random variable

8.5 Measures of Central Tendency and Dispersion
- Calculate the mean, median, mode, and standard deviation of a data set

8.6 Normal Distributions
- Convert a normal distribution to a standard normal distribution
- Determine the area of a shaded region beneath a normal curve
- Calculate the probability of attaining a value within a designated number of standard deviations of the mean of a normal distribution

8.1 Conditional Probability

- Use conditional probability to determine if events are independent

GETTING STARTED In 2003, Washington state had nine representatives and two senators in Congress, all of whom were either Republicans or Democrats. Of the three female members of Congress, two were senators. The female representative in the House of Representatives was a Republican, and the female senators were Democrats. There were a total of eight congressional Democrats from Washington state. What is the probability that a randomly selected female member of Congress from Washington state is also a Democrat? What is the probability that a randomly selected member of the House of Representatives representing Washington state is female? Questions such as these can be answered using conditional probability.

In this section, we will show how conditional probability may be used when certain characteristics of a member of a set are known. We will also demonstrate how to create a probability tree and introduce the concept of independent and dependent events.

Let the sample space C represent the Washington state congressional delegation, the event D be the set of Washington state congressional Democrats, the event H be the set of Washington state Representatives, and the event F be the set of Washington state congressional women. Writing the cardinality of each set in the corresponding region on the Venn diagram yields Figure 8.1.

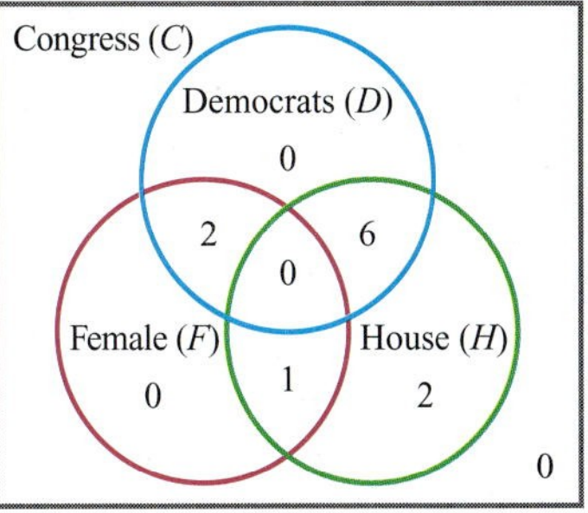

FIGURE 8.1

What is the probability that a randomly selected female member of Congress from Washington state is also a Democrat? In other words, what is the probability that a randomly selected member of Congress is a Democrat given that she is female? From the Venn diagram, we see that of the three women, two are Democrats. Since we know that the member of Congress is female, our sample space is restricted to the event F. We want to know $P(D)$, treating the event F as the entire sample space. In probability notation, we have

$$\begin{aligned} P(D) &= \frac{n(F \cap D)}{n(F)} \\ &= \frac{2}{3} \end{aligned}$$

Therefore, the probability that a randomly selected female member of Congress from Washington state is a Democrat is $\frac{2}{3}$, or 66.67 percent. The idea of restricting the sample space to a particular event is an important one. In the example, we restricted the sample space to the set of female members of Congress. We represent the notion of restricting the sample space to a particular set with conditional probability notation.

CONDITIONAL PROBABILITY

The probability that event A occurs given that event B occurs is given by

$$P(A|B) = \frac{P(A \cap B)}{P(B)}$$

provided that $P(B) \neq 0$.

This equation looks somewhat different from our earlier calculations. Will it yield the same result? Let's see. We want to find $P(D|F)$. Our sample space is the entire Washington state congressional delegation C.

$$\begin{aligned} P(D|F) &= \frac{P(D \cap F)}{P(F)} \\ &= \frac{\frac{n(D \cap F)}{n(C)}}{\frac{n(F)}{n(C)}} \\ &= \frac{\frac{2}{11}}{\frac{3}{11}} \\ &= \frac{2}{3} \end{aligned}$$

The result is the same as before.

EXAMPLE 1

Using Conditional Probability

The makeup of the Washington state congressional delegation is shown in Figure 8.2. What is the probability that a congressman is a member of the House of Representatives given that the congressman is a Democrat?

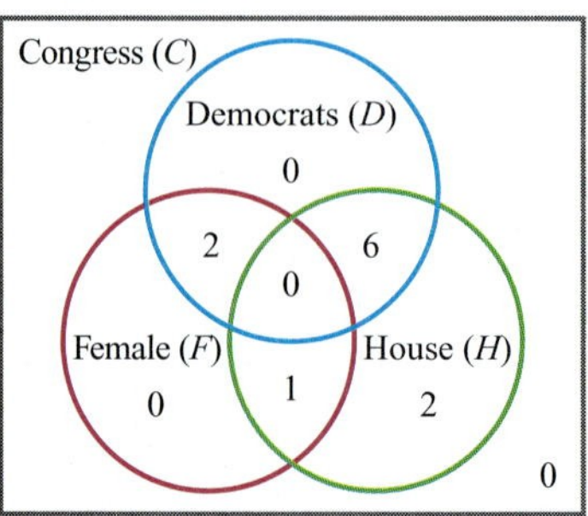

FIGURE 8.2

SOLUTION We are asked to find $P(H|D)$. From the conditional probability formula, we know that

$$\begin{aligned} P(H|D) &= \frac{P(H \cap D)}{P(D)} \\ &= \frac{\frac{n(H \cap D)}{n(C)}}{\frac{n(D)}{n(C)}} \\ &= \frac{n(H \cap D)}{n(D)} \\ &= \frac{6}{8} \\ &= 75\% \end{aligned}$$

Given that a member of Congress is a Democrat, there is a 75 percent probability that the member of Congress is a member of the House of Representatives.

EXAMPLE 2

Using Conditional Probability

An applied calculus class has 33 students enrolled, including 12 international students. There are 10 students in the class, 6 of whom are international students, who have Asian ancestors. What is the probability that a randomly selected student has Asian ancestors, given that the student is not an international student?

SOLUTION Let A be the set of students with Asian ancestors, I be the set of international students, and C be the set of students enrolled in the class. We are asked to find $P(A|I')$.

$$\begin{aligned} P(A|I') &= \frac{P(A \cap I')}{P(I')} \\ &= \frac{\frac{n(A \cap I')}{n(C)}}{\frac{n(I')}{n(C)}} \\ &= \frac{n(A \cap I')}{n(I')} \\ &= \frac{4}{21} \\ &\approx 19\% \end{aligned}$$

There is a 19 percent probability that a noninternational student has Asian ancestors.

EXAMPLE 3

Using Conditional Probability

Table 8.1 shows regional HIV/AIDS statistics for North America, the Caribbean, and the entire world as of the end of 2002. (**Source:** UNAIDS/WHO.)

TABLE 8.1

Region	People Living with HIV/AIDS (Infected Prior to 2002)	People Newly Infected with HIV in 2002	Total Number of People Infected with HIV/AIDS
North America	935,000	45,000	980,000
Caribbean	380,000	60,000	440,000
Worldwide	37,000,000	5,000,000	42,000,000

Is it more likely that a randomly selected person infected with HIV/AIDS lives in North America or lives in the Caribbean, given that the person was newly infected with HIV/AIDS in 2002?

SOLUTION Let C be the group of people with HIV/AIDS living in the Caribbean, let A be the group of people with HIV/AIDS living in North America, and let N be the group of people newly infected with HIV/AIDS. We are asked to compare $P(C|N)$ and $P(A|N)$

$$\begin{aligned} P(C|N) &= \frac{P(C \cap N)}{P(N)} \\ &= \frac{\frac{60{,}000}{42{,}000{,}000}}{\frac{5{,}000{,}000}{42{,}000{,}000}} \\ &= 1.2\% \end{aligned}$$

$$\begin{aligned} P(A|N) &= \frac{P(A \cap N)}{P(N)} \\ &= \frac{\frac{45{,}000}{42{,}000{,}000}}{\frac{5{,}000{,}000}{42{,}000{,}000}} \\ &= 0.9\% \end{aligned}$$

There is a 1.2 percent probability that a person infected with HIV/AIDS in 2002 lives in the Caribbean and a 0.9 percent probability that the person lives in North America. It is more likely that a person newly infected with HIV/AIDS in 2002 lives in the Caribbean than in North America.

We previously developed a formula to calculate the probability of the union of two events. In like manner, we can derive a formula for the probability of an intersection of two events. Recall that the conditional probability formula is given by $P(A|B) = \frac{P(A \cap B)}{P(B)}$. Since $P(B)$ is a real number, we can multiply both sides of the formula by $P(B)$. Doing so yields $P(A \cap B) = P(B) \cdot P(A|B)$. By a similar argument, $P(A \cap B) = P(A) \cdot P(B|A)$.

PRODUCT RULE OF PROBABILITY

Given events A and B,

$$P(A \cap B) = P(B) \cdot P(A|B) \text{ and } P(A \cap B) = P(A) \cdot P(B|A)$$

provided that $P(A) \neq 0$ and $P(B) \neq 0$. [If $P(A) = 0$ or $P(B) = 0$, then $P(A \cap B) = 0$.]

EXAMPLE 4

Using Conditional Probability in a Real-World Context

According to UNAIDS/WHO data, there is an 11.9 percent probability that a person with HIV/AIDS acquired the disease in 2002, given that the person lived in sub-Saharan Africa. There is a 70.0 percent probability that a person with HIV/AIDS in 2002 lived in sub-Saharan Africa. What is the probability that a randomly selected person who acquired HIV/AIDS in 2002 lived in sub-Saharan Africa?

SOLUTION Let S be the group of people with HIV/AIDS living in sub-Saharan Africa in 2002, and let N be the group of people newly infected with HIV in 2002. We are asked to find $P(S \cap N)$. We know $P(N|S) = 0.119$ and $P(S) = 0.700$.

$$\begin{aligned} P(S \cap N) &= P(S) \cdot P(N|S) \\ &= 0.700 \cdot 0.119 \\ &= 0.0833 \end{aligned}$$

There is an 8.33 percent probability that a randomly selected person with HIV/AIDS in 2002 acquired the disease in 2002 and lived in sub-Saharan Africa.

EXAMPLE 5

Finding the Probability of an Intersection of Sets

A couple has 11 grandchildren, including 5 girls and 6 boys. Two-thirds of the boys have curly hair. What is the probability that a randomly selected grandchild is a curly-haired boy?

SOLUTION Let B be the group of boys and C be the group of curly-haired grandchildren. We are asked to find $P(B \cap C)$. We know that $P(B) = \frac{6}{11}$ and $P(C|B) = \frac{2}{3}$.

$$\begin{aligned} P(B \cap C) &= P(B) \cdot P(C|B) \\ &= \frac{6}{11} \cdot \frac{2}{3} \\ &= \frac{4}{11} \\ &= 36.4\% \end{aligned}$$

There is a 36.4 percent probability that a randomly selected grandchild is a curly-haired boy.

Probability Trees

The product rule of probability may also be represented through the use of a *probability tree.* A **probability tree** is a decision tree with the probability of the event annotated on each branch of the tree. In Section 7.2, we introduced the scenario of a student who needed to enroll in three online classes: English, math, and art. The student could choose from two online English courses: Basic Composition and Basic Literature; three online math courses: Finite Math, Consumer Math, and Precalculus; and two online art courses: Art History and Design. Based on these options, we constructed the decision tree shown in Figure 8.3).

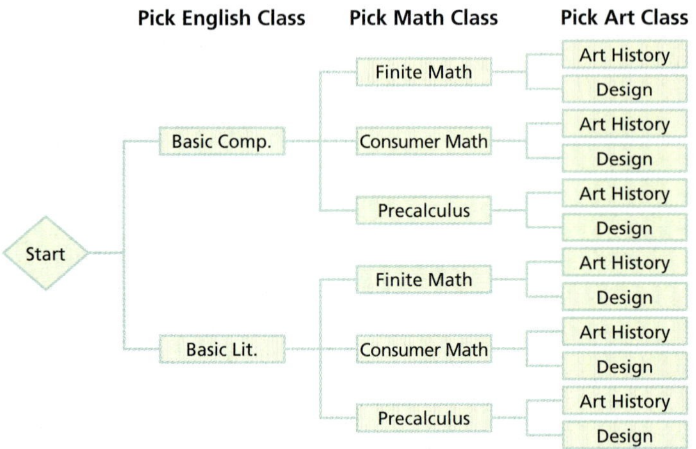

FIGURE 8.3

To convert the decision tree into a probability tree, we simply need to assign the probabilities to each event, as shown in Figure 8.4.

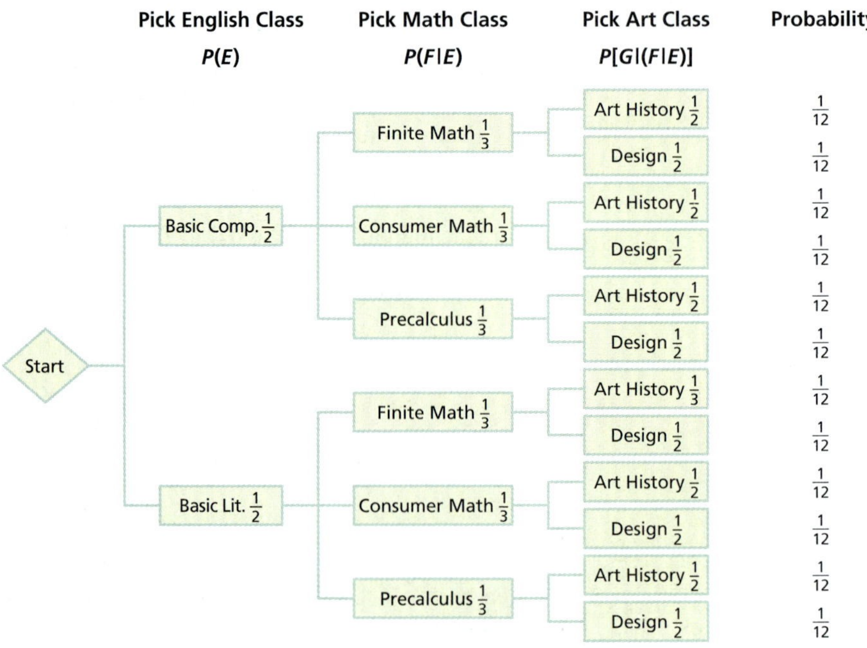

FIGURE 8.4

Since there are two English courses, the probability of randomly selecting a particular English course is $\frac{1}{2}$. Since there are three math courses, the probability of randomly selecting a particular math course is $\frac{1}{3}$. Likewise, since there are two art courses, the probability of randomly selecting a particular art course is $\frac{1}{2}$. Recall from our work in Chapter 7 that the total number of different three-course schedules is equal to the product of the number of choices for each course. That is, $2 \cdot 3 \cdot 2 = 12$. Since each of the schedules is different, the probability of randomly selecting one particular schedule is $\frac{1}{12}$. This same result may be obtained by multiplying the individual probabilities of each course selection as shown here. Assuming that the selection of any course in a given subject is equally likely, the probability of selecting any three-course schedule is $\frac{1}{12} = 8.33$ percent since $\frac{1}{2} \cdot \frac{1}{3} \cdot \frac{1}{2} = \frac{1}{12}$.

As just shown, the probability of choosing any of the 12 three-course schedules was the same for each schedule. However, if a choice alters the options for subsequent choices, the probabilities for different sequences of events often differ, as demonstrated in the following example.

EXAMPLE 6

Using a Probabilty Tree

In 2002, Missouri had nine members of the U.S. House of Representatives, five Republicans and four Democrats. Of the two women, one was a Democrat. (**Source:** clerk.house.gov.) Use a probability tree to determine the likelihood that a randomly selected representative is a female Democrat.

SOLUTION We will first choose the party of the representative and then select the gender (Figure 8.5).

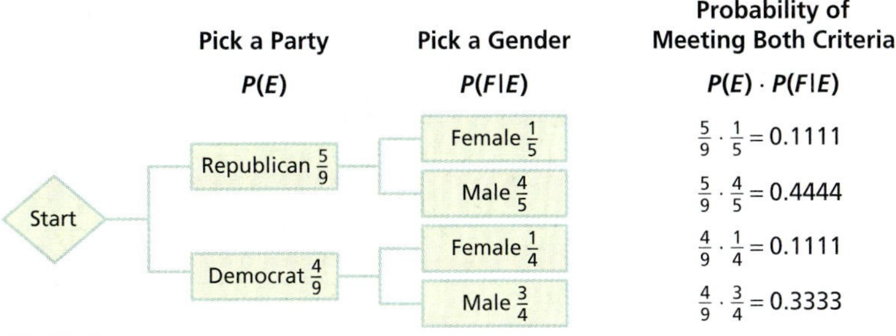

FIGURE 8.5

There is an 11.11 percent chance that a randomly selected Missouri representative is a female Democrat. It is most likely that a randomly selected Missouri representative is a male Republican (44.44 percent).

Independent Events

Observe that in the discussion of class schedules, $P(F|E) = P(F)$. That is, which English class was chosen did not alter the probability of choosing a given math class. However, in Example 6, $P(F|E) \neq P(F)$. That is, the probability of picking a male or female representative differed depending upon which political party was picked. If the probabilities of subsequent events are not altered by previous choices, the events are said to be **independent events.**

INDEPENDENT EVENTS

Events E and F are **independent events** if

$$P(E|F) = P(E) \text{ and } P(F|E) = P(F)$$

Events that are not independent are called **dependent events.** The events of selecting courses in the course schedule were independent events, whereas the events of selecting a representative's party affiliation and gender in Example 6 were dependent events.

EXAMPLE 7

Determining If Events Are Independent

The Federal Bureau of Investigation (FBI) collects crime statistics from cities across the nation and publishes the results regularly in the Uniform Crime Reports periodical. The crimes reported are murder, forcible rape, robbery, aggravated assault, burglary, larceny, and motor vehicle theft. In its December 16, 2002, report, the FBI published the crime statistics in Table 8.2.

TABLE 8.2 2002 Crime Statistics

City	Population (approx.)	Crimes
Anchorage, AK	260,300	6,325
Atlanta, GA	416,500	23,877
Boston, MA	589,100	17,463
Cleveland, OH	478,400	15,797
Green Bay, WI	102,300	1,585
Total	**1,846,600**	**65,047**

Source: FBI Uniform Crime Reports, January–June 2002, www.census.gov.

A family intends to move to one of the five cities. Assuming that it is equally likely that anyone in the selected city will be a crime victim and that crime rates are constant from year to year, determine if the likelihood of becoming a crime victim is independent of the city selected.

SOLUTION Let C be the event "the given city is selected" and V be the event "become a victim." We have the probability tree shown in Figure 8.6.

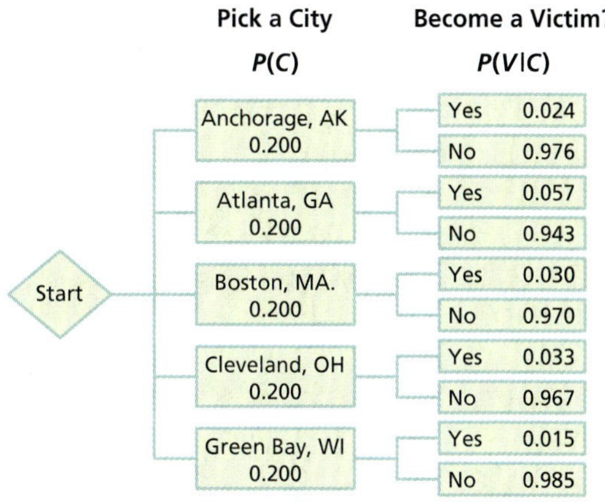

FIGURE 8.6

Based on the data for these five cities,

$$\begin{aligned} P(V) &= \frac{65{,}047}{1{,}846{,}600} \\ &= 0.0352 \\ &= 3.52\ \% \end{aligned}$$

If the crime rate is independent of the city, then each of the cities should have a 3.52 percent crime rate. However,

$$\begin{aligned} P(V|\text{Anchorage}) &= 2.4\% \\ P(V|\text{Atlanta}) &= 5.7\% \\ P(V|\text{Boston}) &= 3.0\% \\ P(V|\text{Cleveland}) &= 3.3\% \\ P(V|\text{Green Bay}) &= 1.5\% \end{aligned}$$

Therefore, the incidence of crime is dependent upon the city selected.

8.1 Summary

In this section, you learned how conditional probability may be used when certain characteristics of a member of a set are known. You also saw how to create a probability tree and became familiar with the concept of independent and dependent events.

8.1 Exercises

In Exercises 1–15, determine the indicated probability.

1. $P(A \cap B) = 0.45$ and $P(B) = 0.5$. Find $P(A|B)$.
2. $P(A \cap B) = 0.10$ and $P(B) = 0.2$. Find $P(A|B)$.
3. $P(A \cap B) = 0.01$ and $P(A) = 0.10$. Find $P(B|A)$.
4. $P(A \cap B) = 0.5$ and $P(A) = 0.5$. Find $P(B|A)$.
5. $P(A|B) = 0.42$ and $P(B) = 0.13$. Find $P(A \cap B)$.
6. $P(A|B) = 0.96$ and $P(B) = 0.32$. Find $P(A \cap B)$.
7. $P(B|A) = 0.21$ and $P(A) = 0.60$. Find $P(A \cap B)$.
8. $P(B|A) = 0.44$ and $P(A) = 0.25$. Find $P(A \cap B)$.
9. Find $P(F|D)$.

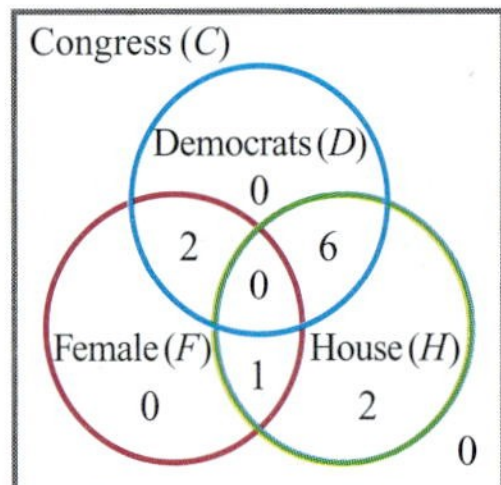

10. Find $P(D|F)$.

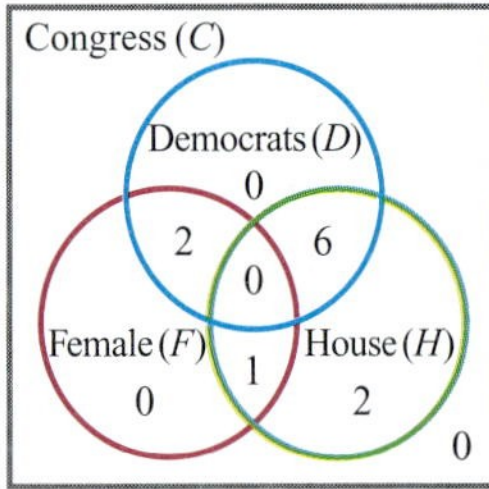

11. Find $P(D'|H)$.

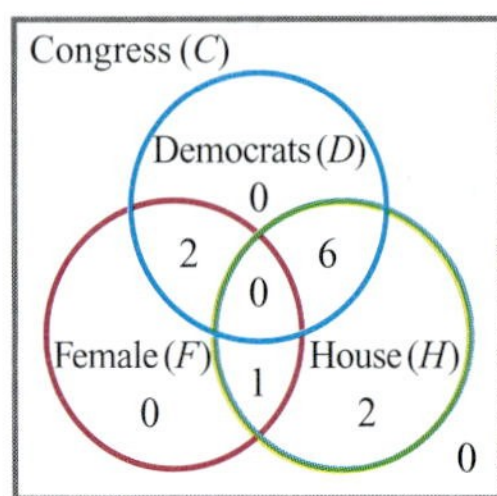

12. Find $P(D'|F')$.

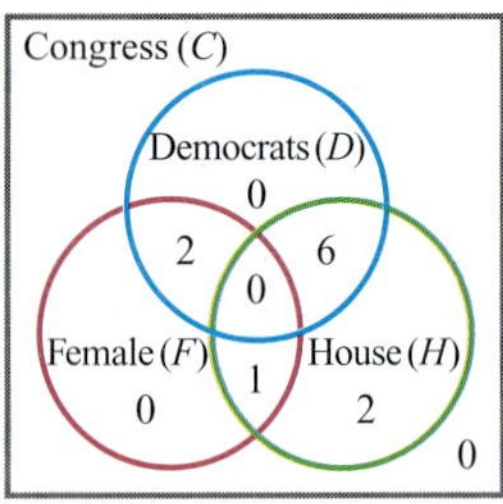

13. Find $P(H'|F)$.

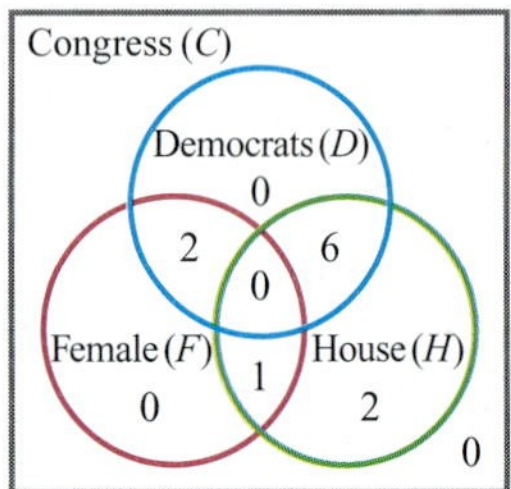

14. Find $P(F|H')$.

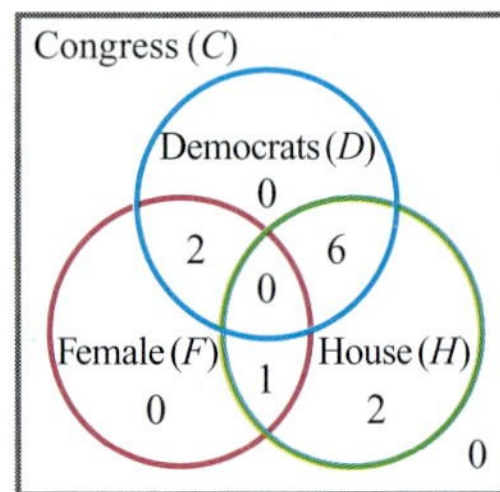

15. Find $P(H|F')$.

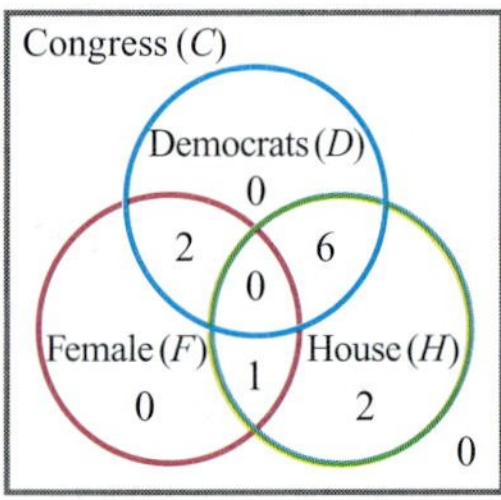

In Exercises 16–25, use the concept of conditional probability to answer the questions.

16. **Gumballs** A gumball machine contains 10 red, 12 yellow, and 8 blue gumballs. Assuming that all of the gumballs are equally likely to be dispensed when the machine handle is turned, what is the probability that the second gumball dispensed from the machine is blue, given that the first gumball dispensed was blue?

17. **Gumballs** A gumball machine contains 10 red, 12 yellow, and 8 blue gumballs. Assuming that all of the gumballs are equally likely to be dispensed when the machine handle is turned, what is the probability that the second gumball dispensed from the machine is yellow, given that the first gumball dispensed was red?

18. **Ethnic Background** A finite math class has 32 students enrolled, including 13 international students. There are 9 students, 6 of whom are international students, who have Hispanic ancestors. What is the probability that a randomly selected student has Hispanic ancestors, given that the student is not an international student?

19. **Employment Status** A finite math class has 28 students enrolled, including 15 women. There are 9 students, 6 of whom are women, who work full time. What is the probability that a randomly selected student doesn't work full time, given that the student is a woman?

20. **Demographics** There were 35 job candidates who submitted applications for a management position. Of the 35 applicants, 20 were men and 6 were from racial minorities. Exactly 2 men from a racial minority applied for the position. What is the probability that a randomly selected candidate is from a racial minority, given that the candidate is female?

21. **Demographics** There were 40 job candidates who submitted applications for a management position. Of the 40 applicants, 35 were women and 5 were from racial minorities. Exactly 2 men from a racial minority applied for the position. What is the probability that a randomly selected candidate is from a racial minority, given that the candidate is female?

22. **Demographics** Twenty people, including twelve African American men, applied for a management position at a recording studio. Five applicants, all African American men, were well qualified for the job. What is the probability that a person randomly selected for the position is not an African American man, given that the candidate is well qualified for the job?

23. **Demographics** Ninety people, including six women, applied for an advertising director position at an international shipping company. Twelve applicants, including five women, were well qualified for the job. What is the probability that a randomly selected person is a woman, given that she was well qualified?

24. **Discrimination** After being denied the position described in Exercise 23, a man sued the company, charging reverse discrimination. He argued that there was only a $\frac{6}{90} = 6.67$ percent probability that a woman would get the job and a 93.33 percent probability that a man would get the position. Considering the results of Exercise 23, how should the defense lawyer respond?

25. **Menu Choices** A menu offers 30 different items. There are 15 entrées, 3 of which are vegetarian dishes. When salads and other side dishes are included, there are a total of 10 vegetarian items on the menu. What is the likelihood that a randomly selected item chosen from the menu is not a vegetarian dish, given that the item is not an entrée?

In Exercises 26–30, use a probability tree to determine the solution.

26. **House of Representatives** In 2003, California had 53 members of the U.S. House of Representatives, 20 Republicans and 33 Democrats. Of the 17 female representatives, 1 was a Republican. (**Source:** clerk.house.gov.) What is the likelihood that a randomly selected representative is a female Democrat?

27. **House of Representatives** In 2003, California had 53 members of the U.S. House of Representatives, 20 Republicans and 33 Democrats. Of the 17 female representatives, 1 was a Republican. (**Source:** clerk.house.gov.) What is the likelihood that a randomly selected representative is a male Republican?

28. **Storage Media** Imation is a company that produces a variety of products, including CD-R and CD-RW storage media. According to its March 2003 product list, CD-R retail spindle packs are sold in packages containing 5, 10, 20, 25, 50, or 100 disks. CD-RW packs are sold in packages containing 1, 5, or 25 disks. (**Source:** www.imation.com.) What is the likelihood that a randomly selected item on the product list contains exactly 5 disks? Given that the selected disk is a CD-R disk, does the likelihood that it contains exactly 5 disks increase or decrease?

29. **Playing Cards** Suppose 3 cards are pulled from a 52-card deck containing 13 hearts, 13 diamonds, 13 clubs, and 13 spades. What is the probability that the third card drawn is a heart? Does the probability increase or decrease given that the first two cards drawn were hearts?

30. **Playing Cards** Suppose 3 cards are pulled from a 52-card deck containing 4 jacks. What is the probability that the third card drawn is a jack? Does the probability increase or decrease given that the first two cards drawn were jacks?

In Exercises 31–35, determine if the events E and F are independent or dependent.

31. Two coins are tossed.

$E =$ the first coin shows tails
$F =$ the second coin shows tails

32. Two dice are rolled.

$E =$ the first die shows a 5
$F =$ the second die shows a 1

33. Two coins are drawn from a bag containing five nickels, two dimes, and three pennies.

$E =$ the first coin is a dime
$F =$ the second coin is a nickel

34. Two coins are drawn from a bag containing five nickels, two dimes, and three pennies.

$E =$ the first coin is a dime
$F =$ the second coin is not a dime

35. A family has three children.

$E =$ the oldest child is a girl
$F =$ the second child is a girl

Exercises 36–38 are intended to challenge your understanding of probability concepts.

36. If $P(E|F) = P(F|E)$, what do you know about $P(E)$ and $P(F)$?

37. If $P(E|F) = P(F)$, are E and F independent? Explain.

38. Given $P(E \cap F) = P(F \cap G)$, show that $P(E \cap F \cap G') = P(E' \cap F \cap G)$.

8.2 Bayes' Theorem and Applications

- Apply Bayes' theorem to real-world applications
- Calculate the predictive-value positive and predictive-value negative of a test

GETTING STARTED One critical element of a disease control and prevention program is the ability to detect who has been infected by a disease. Blood screening and other tests are often used to identify infected people; however, sometimes these tests yield false-positive results. That is, the test indicates that the person is infected when in reality the person is not. Similarly, tests sometimes yield false-negative results. That is, the test indicates that the person is not infected when in reality the person is infected. One way to analyze the effectiveness of a test is by applying Bayes' theorem.

In this section, we will introduce Bayes' theorem and some of its applications. Bayes' theorem is used by the Centers for Disease Control and Prevention and others to investigate the accuracy of screening tests.

Our practical approach to Bayes' theorem differs from the approach taken by many other texts. We determined that it would be most beneficial for readers to learn how Bayes' theorem is actually used in medicine and industry. Many of the concepts covered herein may be applied by consumers in their personal lives.

When a disease detection test is administered, there are four possible outcomes: true positive, false positive, false negative, and true negative (see Table 8.3).

TABLE 8.3

Test Result	Actual Status: Disease Present	Actual Status: Disease Absent	Total
Positive	True positive (a)	False positive (b)	All positive tests ($a + b$)
Negative	False negative (c)	True negative (d)	All negative tests ($c + d$)
Total	All with disease ($a + c$)	All without disease ($b + d$)	Total tested ($a + b + c + d$)

Graphically, we may represent these results with a probability tree (Figure 8.7).

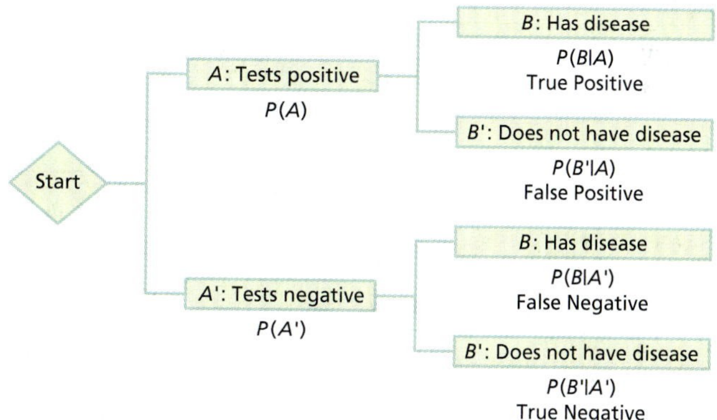

FIGURE 8.7

The probability that a test result will be positive when the test is administered to a person with the disease is referred to as the **sensitivity** of the test.

$$\text{Sensitivity} = \frac{a}{a + c} = \frac{\text{true positive}}{\text{all with disease}}$$

The probability that a test result will be negative when the test is administered to a person without the disease is referred to as the **specificity** of the test.

$$\text{Specificity} = \frac{d}{b + d} = \frac{\text{true negative}}{\text{all without disease}}$$

The probability that a person with a positive test result will actually have the disease is referred to as the **predictive-value positive (*PVP*)** of the test.

$$PVP = \frac{a}{a + b} = \frac{\text{true positive}}{\text{all with positive test}}$$

The probability that a person with a negative test result will not have the disease is referred to as the **predictive-value negative (*PVN*)** of the test.

$$PVN = \frac{d}{c + d} = \frac{\text{true negative}}{\text{all with negative test}}$$

The known percentage of people in the population that are infected is referred to as the **prevalence.**

EXAMPLE 1

Determining the Accuracy of a Screening Test

In June 1984, five companies were licensed to produce a test kit for detecting the HIV antibody. In March 1985, the first enzyme-linked immunosorbent assay (EIA) test kit was approved by the Food and Drug Administration.

The Centers for Disease Control and Prevention offer this case study in their 2003 *Screening for Antibody to the Human Immunodeficiency Virus* instructor's guide. A blood bank has an EIA sensitivity of 95.0 percent and an EIA specificity of 98.0 percent. Of the 1 million blood donors, 0.04 percent actually have the HIV antibody. (**Source:** www.cdc.gov.) Using this information, complete Table 8.4. Then calculate the *PVP* and *PVN* for the EIA test.

TABLE 8.4

	Actual Status		
Test Result	Disease Present	Disease Absent	Total
Positive			
Negative			
Total			

SOLUTION The number of people infected is given by

$$\begin{aligned}(\text{Prevalence})(\text{population}) &= 0.0004(1{,}000{,}000)\\ &= 400\end{aligned}$$

Therefore, the number of people not infected is given by

$$\begin{aligned}(1 - \text{prevalence})(\text{population}) &= (1 - 0.0004)(1{,}000{,}000)\\ &= (0.9996)(1{,}000{,}000)\\ &= 999{,}600\end{aligned}$$

(Alternatively, we could simply subtract the number infected from the total population: $1{,}000{,}000 - 400 = 999{,}600$.)

The number of true positives is given by

$$\begin{aligned}(\text{Sensitivity})(\text{infected}) &= 0.95(400)\\ &= 380\end{aligned}$$

The number of true negatives is given by

$$\begin{aligned}(\text{Specificity})(\text{not infected}) &= 0.98(999{,}600)\\ &= 979{,}608\end{aligned}$$

The number of false negatives is given by

$$\begin{aligned}\text{Infected} - \text{true positives} &= 400 - 380\\ &= 20\end{aligned}$$

The number of false positives is given by

$$\text{not infected} - \text{true negatives} = 999{,}600 - 979{,}608$$
$$= 19{,}992$$

The completed table is shown in Table 8.5.

TABLE 8.5

	Actual Status		
Test Result	Disease Present	Disease Absent	Total
Positive	380	19,992	20,372
Negative	20	979,608	979,628
Total	400	999,600	1,000,000

$$PVP = \frac{\text{true positive}}{\text{total positive}} = \frac{380}{20{,}372} = 0.018653 = 1.8653\%$$

$$PVN = \frac{\text{true negative}}{\text{total negative}} = \frac{979{,}608}{979{,}628} = 0.99998 = 99.998\%$$

There is about a 1.9 percent probability that a person with a positive test result actually has the disease. Consequently, we are highly skeptical of a positive test result. On the other hand, there is nearly a 100 percent probability that a person with a negative result doesn't have the disease. We are highly confident that if the test result is negative, the person does not have the disease.

EXAMPLE 2

Determining the Accuracy of E-mail Filtering Software

Between December 30, 2003, and January 2, 2004, the author received 49 e-mails. His Internet security software tagged 13 e-mails as possible spam. All of the tagged e-mails were in fact spam. There were 25 e-mails that were not spam. Determine the sensitivity, specificity, predictive-value positive, and predictive-value negative of the spam filter. Interpret the real-world significance of the results.

SOLUTION Data for the solution are given in Table 8.6.

TABLE 8.6

	Actual Status		
Test Result	Spam	Not Spam	Total
Tagged	13	0	13
Not Tagged	11	25	36
Total	24	25	49

$$\begin{aligned}\text{Sensitivity} &= \frac{\text{tagged spam}}{\text{total spam}}\\ &= \frac{13}{24}\\ &= 0.5417\\ &= 54.17\%\end{aligned}$$

$$\begin{aligned}\text{Specificity} &= \frac{\text{not spam and not tagged}}{\text{not spam}}\\ &= \frac{25}{25}\\ &= 1\\ &= 100\%\end{aligned}$$

$$\begin{aligned}PVP &= \frac{\text{tagged spam}}{\text{tagged}}\\ &= \frac{13}{13}\\ &= 1\\ &= 100\%\end{aligned} \qquad \begin{aligned}PVN &= \frac{\text{not spam and not tagged}}{\text{not tagged}}\\ &= \frac{25}{36}\\ &= 0.6944\\ &= 69.44\%\end{aligned}$$

Based on this sample, the spam filter appears to tag about 54 percent of all spam. If the filter tagged an e-mail as possible spam, it was spam. If the filter did not tag an e-mail as possible spam, there was a 69 percent chance that the e-mail was not spam.

According to IT industry experts, spam cost American businesses \$10 billion in 2003. As businesses have developed filtering software to block spam, however, they have run into a new challenge: false positives. Blocked legitimate e-mail cost U.S. businesses approximately \$3.5 billion in 2003. (**Source:** itmanagement.earthweb.com.)

An alternative approach to these types of problems is through the use of set notation. Recall that the product rule of probability states that for events A and B, $P(A \cap B) = P(B) \cdot P(A|B)$ and $P(A \cap B) = P(A) \cdot P(B|A)$ [for $P(A) \neq 0$ and $P(B) \neq 0$]. Furthermore, observe that $B = (B \cap A) \cup (B \cap A')$. ($B \cap A$ is the set of items in B that are also in A, whereas $B \cap A'$ is the set of items in B that are outside of A.) Therefore,

$$\begin{aligned}P(B) &= P[(B \cap A) \cup (B \cap A')]\\ &= P(B \cap A) + P(B \cap A') \text{ since } (B \cap A) \cap (B \cap A') = \varnothing && \text{Probability of a union}\\ &= P(B|A)P(A) + P(B|A')P(A') && \text{Product rule of probaility}\end{aligned}$$

We know that

$$\begin{aligned}P(A|B) &= \frac{P(A \cap B)}{P(B)} && \text{Definition of conditional probability}\\ &= \frac{P(B|A)P(A)}{P(B)} && \text{Product rule of probability}\\ &= \frac{P(B|A)P(A)}{P(B|A)P(A) + P(B|A')P(A')}\end{aligned}$$

This result is referred to as Bayes' theorem.

BAYES' THEOREM: BRIEF FORM

Let A and B be events. Then

$$P(A|B) = \frac{P(B|A)P(A)}{P(B|A)P(A) + P(B|A')P(A')}.$$

We can further understand Bayes' theorem by looking again at the corresponding probability tree (Figure 8.8).

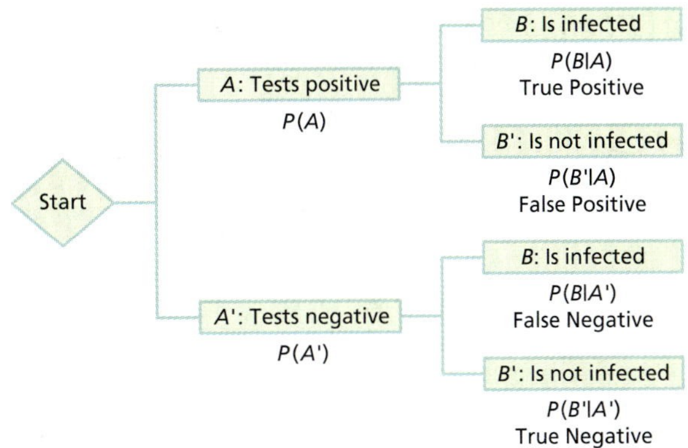

FIGURE 8.8

Bayes' theorem may be rewritten in terms of the probability tree.

BAYES' THEOREM: FROM A PROBABILITY TREE

Let A and B be events. Then

$$P(A|B) = \frac{\text{product of branch probabilities leading to } B \text{ through } A}{\text{sum of all branch probabilities leading to } B}$$

The numerator and denominator of the expression on the right-hand side of the equation in *Bayes' Theorem: From a Probability Tree* are equivalent to the numerator and denominator, respectively, of the expression on the right-hand side of the equation in Bayes' Theorem: Brief Form. That is, product of branch probabilities leading to B through $A = P(B|A)P(A)$ and sum of all branch probabilities leading to $B = P(B|A)P(A) + P(B|A')P(A')$. In Example 3, we investigate the scenario from Example 2 using the probability tree approach.

EXAMPLE 3

Applying Bayes' Theorem from a Probability Tree

Between December 30, 2003, and January 2, 2004, the author received 49 e-mails. His Internet security software tagged 13 e-mails as possible spam. All of the tagged e-mails were in fact spam. There were 25 e-mails that were not spam. What is the probability that a spam e-mail was tagged?

SOLUTION We use the given data to construct a probability tree (Figure 8.9).

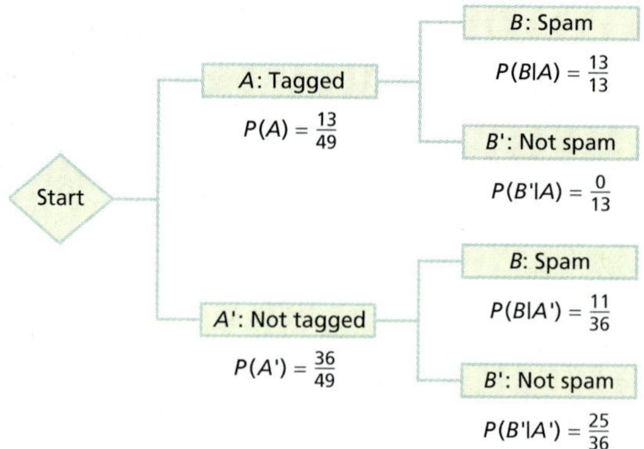

FIGURE 8.9

The probability that a spam e-mail was tagged is $P(A|B)$.

$$
\begin{aligned}
P(A|B) &= \frac{\text{product of branch probabilities leading to } B \text{ through } A}{\text{sum of all branch probabilities leading to } B} \\
&= \frac{\left(\frac{13}{49}\right)\left(\frac{13}{13}\right)}{\left(\frac{13}{49}\right)\left(\frac{13}{13}\right) + \left(\frac{36}{49}\right)\left(\frac{11}{36}\right)} \\
&= \frac{\left(\frac{13}{49}\right)}{\left(\frac{13}{49}\right) + \left(\frac{11}{49}\right)} = \frac{\left(\frac{13}{49}\right)}{\left(\frac{24}{49}\right)} \\
&= \frac{13}{24} \\
&= 0.5417
\end{aligned}
$$

There is a 54.17 percent probability that a spam e-mail was tagged.

EXAMPLE 4 Applying Bayes' Theorem in E-mail Filtering

Over a five-day period, a business's e-mail server received 2500 e-mails. Of these, 1100 were classified as spam and deleted before being delivered. Employees reported that 45 e-mails that were delivered were spam. Employees also reported that 10 legitimate e-mails that had been sent to the company were not delivered. (An IT investigation revealed that all 10 e-mails had been classified as spam when they reached the e-mail server.) What is the probability that a delivered e-mail was spam, and what is the probability that a legitimate e-mail was blocked?

SOLUTION We define the following sets:

$$
\begin{aligned}
B &= \{x|x \text{ is a blocked e-mail}\} \\
B' &= \{x|x \text{ is a delivered e-mail}\} \\
S &= \{x|x \text{ is spam}\} \\
S' &= \{x|x \text{ is not spam}\}
\end{aligned}
$$

We must determine $P(S|B')$ and $P(B|S')$. We have the results given in Table 8.7.

TABLE 8.7

	Actual Status		
Filter Result	Spam	Not Spam	Total
Blocked	1,100	10	1,110 $n(B)$
Delivered	45	1,345	1,390 $n(B')$
Total	1,145 $n(S)$	1,355 $n(S')$	2,500

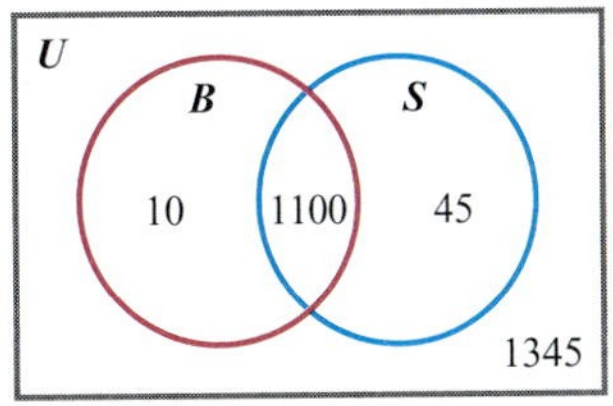

FIGURE 8.10

Using this information, we can generate a Venn diagram showing the cardinality of each subset (Figure 8.10).

Observe that

$$P(S) = \frac{1145}{2500} = 0.458 = 45.8\% \qquad P(S') = \frac{1355}{2500} = 0.542 = 54.2\%$$

$$P(B|S) = \frac{1100}{1145} = 0.961 = 96.1\% \quad \text{Sensitivity} \qquad P(B|S') = \frac{10}{1355} = 0.007 = 0.7\%$$

$$P(B'|S) = \frac{45}{1145} = 0.039 = 3.9\% \qquad P(B'|S') = \frac{1345}{1355} = 0.993 = 99.3\% \quad \text{Specificity}$$

We have $P(B|S') = 0.7\%$, so there is a 0.7 percent probability that the filter will block a legitimate e-mail. To find $P(S|B')$, we will use Bayes' theorem.

From Bayes' theorem, we know that

$$\begin{aligned} P(S|B') &= \frac{P(B'|S)P(S)}{P(B'|S)P(S) + P(B'|S')P(S')} \\ &= \frac{\left(\frac{45}{1145}\right)\left(\frac{1145}{2500}\right)}{\left(\frac{45}{1145}\right)\left(\frac{1145}{2500}\right) + \left(\frac{1345}{1355}\right)\left(\frac{1355}{2500}\right)} \\ &= \frac{\frac{45}{2500}}{\frac{45 + 1345}{2500}} = \frac{45}{1390} \\ &= 0.032 \\ &= 3.2\% \end{aligned}$$

There is a 3.2 percent probability that a delivered e-mail was spam.

Although it would have been quicker to calculate $P(S|B')$ from the Venn diagram in Example 4, we used Bayes' theorem to illustrate the relationship between sensitivity, specificity, and prevalence as related to the conditional probability.

$$\begin{aligned} P(S|B') &= \frac{P(B'|S)P(S)}{P(B'|S)P(S) + P(B'|S')P(S')} \\ &= \frac{(1 - \text{sensitivity})(\text{prevalence})}{(1 - \text{sensitivity})(\text{prevalence}) + (\text{specificity})(1 - \text{prevalence})} \end{aligned}$$

In fact, we can calculate any of the conditional probabilities simply by knowing the sensitivity, specificity, and prevalence of a test. Since we are often most interested in knowing the predictive-value positive and the predictive-value negative of a test, we will use Bayes' theorem to rewrite these conditional probabilities in terms of sensitivity, specificity, and prevalence.

Let $A = \{x|x \text{ yields a positive test}\}$ and $B = \{x|x \text{ is infected}\}$. Then sensitivity $= P(A|B)$, specificity $= P(A'|B')$, prevalence $= P(B)$, $PVP = P(B|A)$, and $PVN = P(B'|A')$. Applying Bayes' theorem,

$$\begin{aligned} PVP &= P(B|A) \\ &= \frac{P(A|B)P(B)}{P(A|B)P(B) + P(A|B')P(B')} && \text{Bayes' theorem} \\ &= \frac{P(A|B)P(B)}{P(A|B)P(B) + [1 - P(A'|B')]P(B')} && \text{Since } P(A|B') + P(A'|B') = 1 \\ &= \frac{P(A|B)P(B)}{P(A|B)P(B) + [1 - P(A'|B')][1 - P(B)]} && \text{Since } P(B) + P(B') = 1 \\ &= \frac{(\text{sensitivity})(\text{prevalence})}{(\text{sensitivity})(\text{prevalence}) + (1 - \text{specificity})(1 - \text{prevalence})} \end{aligned}$$

Similarly,

$$\begin{aligned} PVN &= P(B'|A') \\ &= \frac{P(A'|B')P(B')}{P(A'|B')P(B') + P(A'|B)P(B)} && \text{Bayes' theorem} \\ &= \frac{P(A'|B')[1 - P(B)]}{P(A'|B')[1 - P(B)] + P(A'|B)P(B)} && \text{Since } P(B) + P(B') = 1 \\ &= \frac{P(A'|B')[1 - P(B)]}{P(A'|B')[1 - P(B)] + [1 - P(A|B)]P(B)} && \text{Since } P(A|B) + P(A'|B') = 1 \\ &= \frac{(\text{specificity})(1 - \text{prevalence})}{(\text{specificity})(1 - \text{prevalence}) + (1 - \text{sensitivity})(\text{prevalence})} \end{aligned}$$

Since many medical tests report their effectiveness in terms of sensitivity and specificity, we will calculate the predictive-value positive and predictive-value negative using these terms in the next few examples.

PREDICTIVE-VALUE POSITIVE AND PREDICTIVE-VALUE NEGATIVE

Let $A = \{x | x \text{ yields a positive test}\}$ and $B = \{x | x \text{ is infected}\}$. Then sensitivity $= P(A|B)$, specificity $= P(A'|B')$, and prevalence $= P(B)$. The predictive-value positive and predictive-value negative of the test are given by

$$PVP = \frac{(\text{sensitivity})(\text{prevalence})}{(\text{sensitivity})(\text{prevalence}) + (1 - \text{specificity})(1 - \text{prevalence})}$$

and

$$PVN = \frac{(\text{specificity})(1 - \text{prevalence})}{(\text{specificity})(1 - \text{prevalence}) + (1 - \text{sensitivity})(\text{prevalence})}$$

EXAMPLE 5

Applying Bayes' Theorem in Cancer Detection

Prostate cancer is the second leading cause of cancer death among American men. The incidence of prostate cancer increases substantially for men over 50 years old. According to the National Cancer Institute's December 2003 *Screening for Prostate Cancer* report, the age-adjusted incidence for African American men is 275.3 per 100,000, and that for white males is 172.9 per 100,000. (**Source:** www.cancer.gov.)

One common test to detect prostate cancer is the prostate-specific antigen (PSA) test. Many professionals consider a PSA value of 4.0 to be positive for prostate cancer. Although the sensitivity and specificity of the test vary depending upon the age of the individual and the type of cancer, medical experts estimate that the sensitivity for detection of cancer occurring within the first four years after the test is 73 percent. The overall specificity is estimated to be 91 percent. (**Source:** www.ncbi.nlm.nih.gov.) Determine the *PVP* and *PVN* for African American and Caucasian men and interpret the meaning of the result.

SOLUTION The prevalence for African American men is 0.2753 percent and that for Caucasian men is 0.1729 percent. The sensitivity of the test is 73 percent, and the specificity is 91 percent.

African American Men

$$\begin{aligned} PVP &= \frac{(\text{sensitivity})(\text{prevalence})}{(\text{sensitivity})(\text{prevalence}) + (1 - \text{specificity})(1 - \text{prevalence})} \\ &= \frac{(0.73)(0.002753)}{(0.73)(0.002753) + (1 - 0.91)(1 - 0.002753)} \\ &= 0.02190 \\ &= 2.19\% \end{aligned}$$

Of the African American men with a PSA value of 4.0 or more, 2.19 percent will develop prostate cancer within 4 years of the test.

$$\begin{aligned} PVN &= \frac{(\text{specificity})(1 - \text{prevalence})}{(\text{specificity})(1 - \text{prevalence}) + (1 - \text{sensitivity})(\text{prevalence})} \\ &= \frac{(0.91)(1 - 0.002753)}{(0.91)(1 - 0.002753) + (1 - 0.73)(0.002753)} \\ &= 0.9992 \\ &= 99.92\% \end{aligned}$$

Of the African American men with a PSA value less than 4.0, 99.92 percent won't develop prostate cancer within 4 years of the test.

Caucasian Men

$$\begin{aligned} PVP &= \frac{(\text{sensitivity})(\text{prevalence})}{(\text{sensitivity})(\text{prevalence}) + (1 - \text{specificity})(1 - \text{prevalence})} \\ &= \frac{(0.73)(0.001729)}{(0.73)(0.001729) + (1 - 0.91)(1 - 0.001729)} \\ &= 0.01385 \\ &= 1.39\% \end{aligned}$$

Of the Caucasian men with a PSA value of 4.0 or more, 1.39 percent will develop prostate cancer within 4 years of the test.

$$\begin{aligned} PVN &= \frac{(\text{specificity})(1 - \text{prevalence})}{(\text{specificity})(1 - \text{prevalence}) + (1 - \text{sensitivity})(\text{prevalence})} \\ &= \frac{(0.91)(1 - 0.001729)}{(0.91)(1 - 0.001729) + (1 - 0.73)(0.001729)} \\ &= 0.9995 \\ &= 99.95\% \end{aligned}$$

Of the Caucasian men with a PSA value less than 4.0, 99.95 percent won't develop prostate cancer within 4 years of the test.

EXAMPLE 6

Applying Bayes' Theorem in Drug Testing

Screening for drug use by athletes has become a major issue in sports over the past several years. Laboratory urinalysis is the primary screening test used to detect drug use. Although the test is considered highly effective, the invasive nature of the urine collection procedure has concerned some privacy advocates. An alternative procedure, oral fluid analysis, shows promise as a less invasive yet highly effective drug screening test.

In a 2002 study of adult arrestees, it was estimated that the Intercept Oral Specimen Collection Device (IOSCD) was 95 percent sensitive and 98 percent specific for cocaine. (**Source:** www.criminology.fsu.edu.) In analyzing the effectiveness of the oral fluid collection method, it was assumed that the laboratory urinalysis approach accurately detected the use or nonuse of drugs.

Suppose that 8 percent of the athletes on a football team use cocaine. An athlete who tested positive for cocaine on the IOSCD test claims that he is not a drug user. If you were a lawyer assigned to represent the athlete, what analytical argument could you use to persuade officials that the athlete is telling the truth?

SOLUTION Since the athlete tested positive but claims not to be a drug user, we must determine the predictive-value positive.

$$\begin{aligned} PVP &= \frac{(\text{sensitivity})(\text{prevalence})}{(\text{sensitivity})(\text{prevalence}) + (1 - \text{specificity})(1 - \text{prevalence})} \\ &= \frac{(0.95)(0.08)}{(0.95)(0.08) + (1 - 0.98)(1 - 0.08)} \end{aligned}$$

$$= 0.8051$$
$$= 80.51\%$$

We could argue that although the test theoretically correctly identifies 95 percent of the cocaine users and 98 percent of non-cocaine users, the probability that our client is a user is about 81 percent.

We could further argue that the sensitivity and specificity estimates are imprecise. We could cite additional studies that showed different sensitivities and specificities. For example, a separate study showed that the IOSCD test had an 82 percent sensitivity and a 96 percent specificity for cocaine. **(Source:** www.criminology.fsu.edu.) Using these values, the PVP is substantially lower.

$$\begin{aligned} PVP &= \frac{(\text{sensitivity})(\text{prevalence})}{(\text{sensitivity})(\text{prevalence}) + (1 - \text{specificity})(1 - \text{prevalence})} \\ &= \frac{(0.82)(0.08)}{(0.82)(0.08) + (1 - 0.96)(1 - 0.08)} \\ &= 0.6406 \\ &= 64.06\% \end{aligned}$$

Using the results from the second study, the probability that our client is a cocaine user is about 64 percent. The varying probabilities call into question the reliability of the test results. Although the analytical argument may add strength to the athlete's case, other factors, such as the testimony of witnesses, the athlete's typical behavior, and so on, should be considered as well.

In the preceding examples, we used the brief form of Bayes' theorem. An expanded form of the theorem is more complex yet highly useful. Recall that the denominator of the rational expression in Bayes' theorem was calculated as follows:

$$\begin{aligned} P(B) &= P((B \cap A) \cup (B \cap A')) \\ &= P(B \cap A) + P(B \cap A') \text{ since } (B \cap A) \cap (B \cap A') = \varnothing && \text{Probability of a union} \\ &= P(A)P(B|A) + P(A')P(B|A') && \text{Product rule of probability} \end{aligned}$$

In the second step, we made the observation that the sets $(B \cap A)$ and $(B \cap A')$ were disjoint, yet their union made up the entirety of B (Figure 8.11).

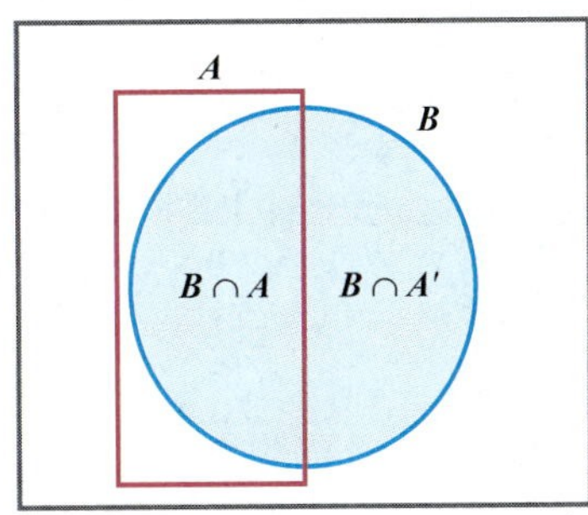

FIGURE 8.11

Although A and A' split B into only two regions, we may divide B into as many regions as we like. The division of a sample space into a group of disjoint sets whose union is the entire sample space is referred to as a **partition.** Figure 8.12 shows a partition of B into three disjoint sets.

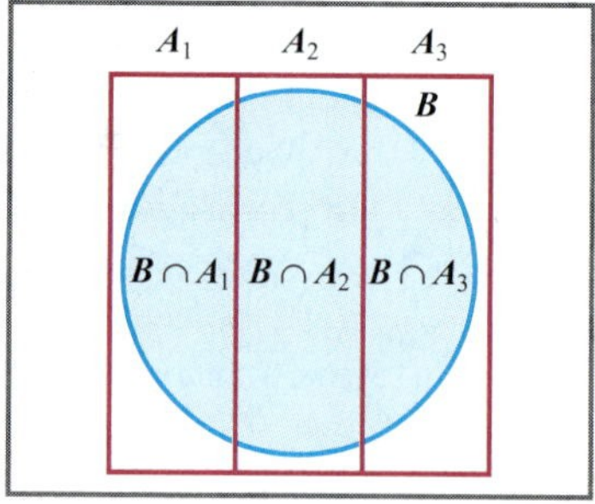

FIGURE 8.12

For the three-set partition of B, observe that $B = (B \cap A_1) \cup (B \cap A_2) \cup (B \cap A_3)$ and

$$\begin{aligned} P(B) &= P((B \cap A_1) \cup (B \cap A_2) \cup (B \cap A_3)) \\ &= P(B \cap A_1) + P(B \cap A_2) + P(B \cap A_3) && \text{Since } A_1, A_2, \text{ and } A_3 \text{ are disjoint} \\ &= P(B|A_1)P(A_1) + P(B|A_2)P(A_2) + P(B|A_3)P(A_3) && \text{Product rule of probability} \end{aligned}$$

The result leads us to the extended form of Bayes' theorem.

BAYES' THEOREM: EXTENDED FORM

Let events A_1, A_2, and A_3 form a partition of sample space B. Then

$$P(A_1|B) = \frac{P(B|A_1)P(A_1)}{P(B|A_1)P(A_1) + P(B|A_2)P(A_2) + P(B|A_3)P(A_3)}$$

A similar form would apply for subspaces with four or more partitions.

EXAMPLE 7

Applying Bayes' Theorem in Social Analysis

In 1999, 33.0 percent of live births in the United States were to unmarried mothers. Of mothers racially classified as Asian, 15.4 percent were unmarried. Of mothers racially classified as black, 68.9 percent were unmarried. Of mothers classified as other races, 27.2 percent were unmarried. Asian women made up 4.6 percent of the total number of women who gave birth in 1999, black women made up 15.3 percent, and women of other races made up 80.1 percent. (**Source:** *Statistical Abstract of the United States,* 2001, Table 827, p. 537.) What is the probability that an unmarried woman who gave birth in 1999 was Asian?

SOLUTION Define events $A_1 = \{x|x \text{ is a black woman}\}$, $A_2 = \{x|x \text{ is a black woman}\}$, $A_3 = \{x|x \text{ is a woman of another race}\}$, and $B = \{x|x \text{ is an unmarried mother}\}$. The events A_1, A_2, and A_3 form a partition of the racial classification. We know that $P(B) = 0.330$, $P(A_1) = 0.046$, $P(A_2) = 0.153$, $P(A_3) = 0.801$, $P(B|A_1) = 0.154$, $P(B|A_2) = 0.689$, and $P(B|A_3) = 0.272$. We are asked to find $P(A_1|B)$.

$$
\begin{aligned}
P(A_1|B) &= \frac{P(B|A_1)P(A_1)}{P(B|A_1)P(A_1) + P(B|A_2)P(A_2) + P(B|A_3)P(A_3)} \\
&= \frac{(0.154)(0.046)}{(0.154)(0.046) + (0.689)(0.153) + (0.272)(0.801)} \\
&= 0.021 \\
&= 2.1\%
\end{aligned}
$$

There is a 2.1 percent probability that a randomly selected unmarried woman who gave birth in 1999 was Asian.

Although we correctly applied Bayes' theorem in Example 7, we could have generated the result more quickly. Recall that the denominator of the expression is equal to $P(B)$. Since we knew that $P(B) = 0.33$, we could have calculated $P(A_1|B)$ as follows:

$$
\begin{aligned}
P(A_1|B) &= \frac{P(B|A_1)P(A_1)}{P(B|A_1)P(A_1) + P(B|A_2)P(A_2) + P(B|A_3)P(A_3)} \\
&= \frac{P(B|A_1)P(A_1)}{P(B)} = \frac{(0.154)(0.046)}{0.330} \\
&= 0.021 \\
&= 2.1\%
\end{aligned}
$$

Before diving into a problem with Bayes' theorem, it is wise to think through which approach will be most efficient. In the case of Example 7, simply applying the definition of conditional probability led us to our result more quickly. Bayes' theorem is most useful when we have incomplete information regarding the various probabilities.

EXAMPLE 8

Applying Bayes' Theorem in Educational Analysis

In an effort to assess the effectiveness of an intermediate algebra course in preparing students for college-level mathematics (precalculus, finite math, and contemporary math), the author collected the following statistics for Green River Community College. The statistics are based upon the number of students who passed the intermediate algebra course and enrolled in a college-level math course. Students are required to earn a grade of 2.0 or better in the intermediate algebra course, score well on a placement test, or obtain a waiver from a math instructor before they may enroll in the college-level math course. We restrict our analysis to students who took both intermediate algebra and a college level math course.

Of the students who earned a 2.0 or better in the intermediate algebra course on the first attempt, 80.8 percent earned a 2.0 or better in the college-level math course. Of the students who earned a 2.0 or better in intermediate algebra on the second attempt, 69.6 percent earned a 2.0 or better in the college-level math course. Of the students enrolled in the college-level math course, 96.1 percent earned a 2.0 or better in algebra on the first attempt and 3.9 percent earned a 2.0 or better in algebra on the second attempt. Further analyze the data and prepare a report for the Math Department.

SOLUTION We first acknowledge that the data won't give us a complete picture, since the sample does not include the students who took the intermediate algebra course but did not complete a college-level math course. However, our analysis will help us determine if the required 2.0 grade in intermediate algebra is a good predictor of performance in the college-level math course. We define events $A_1 = \{x|x \text{ is a student who earned 2.0 or better in algebra on the first attempt}\}$, $A_2 = \{x|x \text{ is a student who earned 2.0 or better in algebra on the second attempt}\}$, and $B = \{x|x \text{ is a student who earned 2.0 or better in a college-level math course}\}$. The events A_1 and A_2 form a partition of the population based on the number of times the student had to take the algebra course. We know that $P(B|A_1) = 0.808$, $P(B|A_2) = 0.696$, $P(A_1) = 0.961$, and $P(A_2) = 0.039$. We want to find answers to the following questions:

1. What is the probability that a student who earned a grade of 2.0 or better in a college-level math course earned a 2.0 or better in algebra on the first attempt? [Find $P(A_1|B)$.]

$$\begin{aligned} P(A_1|B) &= \frac{P(B|A_1)P(A_1)}{P(B|A_1)P(A_1) + P(B|A_2)P(A_2)} \\ &= \frac{(0.808)(0.961)}{(0.808)(0.961) + (0.696)(0.039)} \\ &= 0.966 \\ &= 96.6\% \end{aligned}$$

There is a 96.6 percent probability that a student who earned a grade of 2.0 or better in a college-level math course earned a 2.0 or better in intermediate algebra on the first attempt.

2. What is the probability that a student who earned a grade of 2.0 or better in a college-level math course earned a 2.0 or better in algebra on the second attempt? [Find $P(A_2|B)$.]

$$\begin{aligned} P(A_2|B) &= \frac{P(B|A_2)P(A_2)}{P(B|A_2)P(A_2) + P(B|A_1)P(A_1)} \\ &= \frac{(0.696)(0.039)}{(0.696)(0.039) + (0.808)(0.961)} \\ &= 0.034 \\ &= 3.4\% \end{aligned}$$

There is a 3.4 percent probability that a student who earned a grade of 2.0 or better in the college-level math course had to take intermediate algebra twice before obtaining a 2.0 in the algebra course.

The majority of students (96.6 percent) who did well in a college-level math course took the prerequisite intermediate algebra course only once. A small minority (3.4 percent) of those who did well in a college-level math course were required to take the intermediate algebra course twice. Students who had difficulty passing intermediate algebra on the first attempt did not do as well in the college-level math course. Our results indicate that 80.8 percent of the one-time algebra students passed the college-level math course, while only 69.6 percent of the repeating algebra students passed the college-level math course. These results could be shared with students during advising sessions to help them recognize the importance of putting forth their best effort in the intermediate algebra class the first time they enroll.

8.2 Summary

In this section, you learned how to apply Bayes' theorem in a variety of real-world applications ranging from drug test analysis to spam detection. You also discovered that Bayes' theorem is needed only when we have incomplete statistical information.

8.2 Exercises

In Exercises 1–10, calculate the prevalence, sensitivity, specificity, predictive-value positive, and predictive-value negative from the table. Then interpret the meaning of the results.

1.

	Actual Status	
Test Result	Disease Present	Disease Absent
Positive	125	18
Negative	20	822

2.

	Actual Status	
Test Result	Disease Present	Disease Absent
Positive	1,023	12
Negative	52	9,848

3.

	Actual Status	
Test Result	Spam	Not Spam
Positive	164	50
Negative	26	750

4.

	Actual Status	
Test Result	Spam	Not Spam
Positive	250	10
Negative	25	1,000

5.

	Will Pass Class	Will Fail Class
Studies	20	3
Doesn't Study	4	6

6.

	Will Pass Class	Will Fail Class
Studies	36	4
Doesn't Study	6	12

7.

	No Repairs Needed	Repairs Needed
Newer Home	56	14
Older Home	62	246

8.

	Has Children	Doesn't Have Children
Married	259	198
Single	123	326

9.

	Gets a Speeding Ticket	Doesn't Get a Speeding Ticket
Drives a Red Car	26	64
Drives a Non-red Car	39	189

10.

	Has Tongue Pierced	Doesn't Have Tongue Pierced
Has Tattoo	3	4
Doesn't Have Tattoo	5	22

In Exercises 11–20, complete the table by applying Bayes' theorem.

11.

	Percent
Prevalence	0.2
Sensitivity	98
Specificity	76
Predictive-value positive	
Predictive-value negative	

12.

	Percent
Prevalence	1
Sensitivity	87
Specificity	99
Predictive-value positive	
Predictive-value negative	

13.

	Percent
Prevalence	20
Sensitivity	50
Specificity	75
Predictive-value positive	
Predictive-value negative	

14.

	Percent
Prevalence	1.2
Sensitivity	88
Specificity	100
Predictive-value positive	
Predictive-value negative	

15.

	Percent
Prevalence	5
Sensitivity	85
Specificity	
Predictive-value positive	13
Predictive-value negative	

16.

	Percent
Prevalence	2
Sensitivity	60
Specificity	
Predictive-value positive	38
Predictive-value negative	

17.

	Percent
Prevalence	2
Sensitivity	92
Specificity	
Predictive-value positive	19
Predictive-value negative	

18.

	Percent
Prevalence	20
Sensitivity	
Specificity	62
Predictive-value positive	
Predictive-value negative	95

19.

	Percent
Prevalence	10
Sensitivity	
Specificity	98
Predictive-value positive	
Predictive-value negative	99

20.

	Percent
Prevalence	1
Sensitivity	
Specificity	50
Predictive-value positive	
Predictive-value negative	99

In Exercises 21–35, apply Bayes' theorem as appropriate in analyzing the real-world data.

21. **Skin Cancer Screening** A common screening test for skin cancer is the total cutaneous examination (TCE). One study cited in the American College of Preventative Medicine policy statement on skin cancer reports that the TCE has a 93.3 percent sensitivity, 97.8 percent specificity, predictive-value positive of 54 percent, and predictive-value negative of 99.8 percent when the examination is conducted by dermatologists. (**Source:** www.acpm.org.)

Interpret the meaning of sensitivity, specificity, predictive-value positive, and predictive-value negative in the context of this study.

22. **Breast Cancer Screening** The American Cancer Society and others recommend that women age 50 and older receive an annual mammogram, a breast cancer screening test. According to the American College of Preventative Medicine policy statement on breast cancer screening, the sensitivity of the mammogram is estimated to be 75 to 90 percent, and the specificity is estimated to be 90 to 95 percent. The predictive-value positive for women age 50 to 69 is estimated to be 60 to 80 percent. (**Source:** www.acpm.org.)

Explain how you could use these results when visiting a friend or relative who tested positive for breast cancer during a recent mammogram.

23. **Disease Detection** Invasive aspergillosis commonly occurs in people whose immune systems are compromised by illness or chemotherapy. In 2003, the Food and Drug Administration approved the Platelia Aspergillus EIA test as a disease detection tool. Based on clinical studies, the sensitivity of the test was estimated to be 81 percent and the specificity was estimated to be 89 percent. (**Source:** www.accessdata.fda.gov.) Since only a few thousand cases of the disease occur in the United States each year, the prevalence of the disease is estimated to be 0.002 percent.

Calculate and interpret the meaning of the predictive-value positive and the predictive-value negative for the Platelia Aspergillus EIA test.

24. **Birth Defect Screening** The incidence of Down's syndrome increases as a woman ages. When the mother is 35, roughly 1 out of 375 children born will have Down's syndrome. By age 45, the incidence increases to 1 out of 30 births. One screening test for Down's syndrome is a measurement of maternal serum alpha-fetoprotein (MSAFP). The MSAFP test has an estimated sensitivity of 25 percent and a specificity of about 90 percent. (**Source:** www.singhospi.com.)

Calculate the predictive-value positive and the predictive-value negative for MSAFP test for a 35-year-old and a 45-year-old woman. What conclusions can you draw from the results?

25. **Weather Forecasting** A meteorologist tracks the accuracy of her daily forecasts. She determines that if she forecasts rain, her prediction is accurate 65 percent of the time. If she forecasts no rain, her prediction is accurate 80 percent of the time. She typically forecasts rain

73 days out of a year. Today it is raining. What is the probability that her forecast for today's weather was accurate?

26. **Weather Forecasting** A meteorologist tracks the accuracy of his daily forecasts. He determines that if he forecasts rain, his prediction is accurate 69 percent of the time. If he forecasts no rain, his prediction is accurate 78 percent of the time. He typically forecasts rain 30 days out of a year. Today it is not raining. What is the probability that his forecast for today's weather was not accurate?

27. **Pregnancy Test** Signify® Card Pregnancy Test is a test used to detect human chorionic gonadotropin, an early indicator of pregnancy. The test boasts a 99.1 percent sensitivity and a 100 percent specificity. (**Source:** www.abbottdiagnostics.com.)

Out of a group of 100 women undergoing treatment for infertility, 4 people test positive for pregnancy. What is the probability that a woman with a positive test is actually pregnant? What is the probability that a woman with a negative test is actually pregnant?

28. **Out-of-Wedlock Births** In 1999, the percentage of live births to unmarried mothers in the United States reached a new high. Of mothers racially classified as Asian, 15.4 percent were unmarried. Of mothers racially classified as black, 68.9 percent were unmarried. Of mothers classified as other races, 27.2 percent were unmarried. Asian women made up 4.6 percent of the total number of women who gave birth in 1999, black women made up 15.3 percent, and women of other races made of 80.1 percent. (**Source:** *Statistical Abstract of the United States, 2001*, Table 827, p 537.) What is the probability that an unmarried woman who gave birth in 1999 was black?

29. **Law School Admission** A user-compiled online database of law school admission statistics seeks to help law school applicants determine their likelihood of acceptance. As of January 12, 2004, fourteen of the applicants to the law school at University of California-Berkeley had received word of their admission status. Of those with a Law School Data Assembly Service (LSDAS) GPA of 3.8 or higher, 90.9 percent were admitted to the school. Of those with an LSDAS GPA of less than 3.8, 33.3 percent were rejected. Eleven of the fourteen students had an LSDAS GPA of 3.8 or higher. (**Source:** www.lawschoolnumbers.com.)

What is the probability that an applicant who was rejected had an LSDAS GPA of 3.8 or above? What is the probability that an applicant who was accepted had an LSDAS GPA below 3.8?

30. **Law School Admission** A user-compiled online database of law school admission statistics seeks to help law school applicants determine their likelihood of acceptance. As of January 12, 2004, twelve of the applicants to the law school at the University of Notre Dame had received word of their admission status. Of those with an LSAT score of 163 or higher, 87.5 percent were admitted to the school. Of those with an LSAT score of less than 163, 75 percent were rejected. Eight of the twelve students scored 163 or higher on the LSAT. (**Source:** www.lawschoolnumbers.com.)

What is the probability that an applicant who was rejected had a LSAT score of 163 or higher? What is the probability that an applicant who was accepted scored below 163?

31. **Law School Admission** A user-compiled online database of law school admission statistics seeks to help law school applicants determine their likelihood of acceptance. As of January 12, 2004, 35 of the applicants to the law school at the University of Virginia had received word of their admission status: 77.1 percent were accepted, 11.4 percent were wait-listed, and 11.4 percent were rejected. (The percentages add up to 99.9 percent because of rounding.) Of those accepted, 100 percent had an LSAT score of 163 or higher. Of those wait-listed, 50 percent had an LSAT score of 163 or higher. Of those rejected, 0 percent had an LSAT score of 163 or higher. (**Source:** www.lawschoolnumbers.com.)

What is the probability that a student who had an LSAT score of 163 or higher was wait-listed?

32. **Mad Cow Disease** In December 2003, the first U.S. case of mad cow disease was discovered in Mabton, Washington. The international reaction was swift and decisive. More than 30 countries placed an immediate ban on beef from the United States.

The Prionics®-Check LIA is a test used to detect bovine spongiform encephalopathy (mad cow disease). Based on a study of 1400 cattle, the sensitivity and specificity of the test were both determined to be 100 percent. (**Source:** www.prionics.ch.)

How confident should consumers be that meat determined safe by the Prionics-Check LIA test is actually safe to eat? Explain.

33. **Ethnicity and Graduation Rates** The author determined the percentage of students taking courses at Green River Community College in 2000–2001 who graduated with an associate's degree by winter 2004. Treating the percentages as probabilities, he determined the following:

There was a 27.1 percent probability that a non-Latino/non-Hispanic student would earn a degree. There was a 17.3 percent probability that a Latino/Hispanic student would earn a degree. Latino/ Hispanic students made up 5.0 percent of the student population. According to this analysis, what is the probability that a student who earned a degree was Latino/Hispanic?

34. **Strep Throat Test** Signify Strep A is a test used for the detection of group A streptococcal antigens from a throat swab. The test has sensitivity of 96 percent and a specificity of 97.8 percent. (**Source:** www.abbottdiagnostics.com.) The incidence of strep throat is estimated to be 5 out of 1000 people. (**Source:** www.healthcentral.com.)

What is the probability that someone who tests negative for strep throat actually has the illness? What is the probability that someone who tests positive for strep throat doesn't have the illness?

35. **Infectious Mononucleosis Test** Signify Mono is a test used to detect infectious mononucleosis. The test boasts a 100 percent sensitivity and a 90.3 percent specificity. (**Source:** www.abbottdiagnostics.com.) The incidence of mononucleosis is estimated to be 8 out of 100,000 people. (**Source:** www.healthcentral.com.)

What is the probability that someone who tests negative for mono actually has the illness? What is the probability that someone who tests positive for mono doesn't have the illness?

Exercises 36–40 are intended to challenge your understanding of Bayes' theorem and its applications.

36. Assuming that the sensitivity and specificity of a disease screening test remain constant, what effect does an increase in the prevalence of the disease have on the predictive-value positive and predictive-value negative?

37. Let $A = \{x|x \text{ yields a positive test}\}$ and $B = \{x|x \text{ is infected}\}$. Then sensitivity $= P(A|B)$, specificity $= P(A'|B')$, and prevalence $= P(B)$. Write $P(A'|B)$ in terms of sensitivity, specificity, and/or prevalence.

38. **Drug Screening Test** A drug-use screening test is given to a group of 1000 drug users. The test indicates that all members of the group are drug users. As a result, the manufacturer claims that the screening test is 100 percent accurate. Do you agree with the manufacturer? Explain.

39. The sensitivity and specificity of a test are often estimated. What factors influence the accuracy of these estimates?

40. Let $B = \{x|x \text{ is a person living in Maine}\}$, $A_1 = \{x|x \text{ is a person with blue eyes}\}$, $A_2 = \{x|x \text{ is a person with brown eyes}\}$, and $A_3 = \{x|x \text{ is a person with green eyes}\}$. A student claims that $P(B) = P(B|A_1)P(A_1) + P(B|A_2)P(A_2) + P(B|A_3)P(A_3)$. Do you agree? Justify your answer.

8.3 Markov Chains

- Forecast future events with Markov chains

GETTING STARTED Business-savvy housing managers often analyze rental data to forecast occupancy rates in future years. A review of an apartment complex's lease records revealed that 85 percent of the 100 apartments vacated in 2005 had new occupants in 2006. Furthermore, 65 percent of the 400 apartments occupied in 2005 had the same occupants in 2006. (All occupants sign a one-year lease.)

Based on these results, the manager hypothesizes that the probability that a vacant apartment will be filled in the following year is 85 percent and the probability that a filled apartment will remain filled the following year is 65 percent. The housing manager can forecast the number of filled and vacant apartments in future years by using Markov chains.

In this section, we will demonstrate how to use Markov chains to address business concerns. Markov chains are frequently used to forecast consumer behavior. Markov chains are named for the Russian mathematician A. A. Markov, who developed many of the techniques covered in this section.

We summarize the apartment manager's hypothesis given at the beginning of this section in Table 8.8.

TABLE 8.8

		2006	
		Filled	Vacant
2005	Filled	0.65	0.35
	Vacant	0.85	0.15

Every apartment at the complex may be classified as either *filled* or *vacant*. These classifications are referred to as **states.** We are interested in the probability that an apartment *makes the transition* (moves) from one state to another. That is, what is the probability that a vacant apartment in 2005 will be filled in 2006? What is the probability that a filled apartment in 2005 will be vacant in 2006?

A **stochastic process** is a collection of possible states together with the transition probabilities between them. A square matrix with nonnegative entries is called a **stochastic matrix** or **transition matrix** if the sum of the entries in each row is 1. The probability table in Table 8.8 may be rewritten as a transition matrix.

Columns represent next state

$$\begin{array}{c} \\ S_1 \\ S_2 \end{array}\begin{array}{c} \begin{array}{cc} S_1 & S_2 \end{array} \\ \begin{bmatrix} 0.65 & 0.35 \\ 0.85 & 0.15 \end{bmatrix} \end{array} = P$$

Rows represent current state

In this matrix, state 1 = filled and state 2 = vacant. The matrix entry p_{ij} = the probability of moving from state i to state j. p_{ij} is called the **transition probability.** For example, $p_{12} = 0.35$ means that the probability that a filled apartment is vacant a year later is 35 percent. Similarly, $p_{22} = 0.15$ means that the probability that a vacant apartment remains vacant a year later is 15 percent.

MARKOV CHAIN

A finite Markov chain is a set of states $S_1, S_2, \ldots, S_n$ together with an $n \times n$ transition matrix P such that the entry p_{ij} of the matrix P is the probability of making the transition from S_i to S_j.

EXAMPLE 1 Interpreting the Meaning of the Entries of a Transition Matrix

Given the transition matrix $P = \begin{bmatrix} 0.25 & 0.45 & 0.30 \\ 0.00 & 1.00 & 0.00 \\ 0.75 & 0.00 & 0.25 \end{bmatrix}$, interpret the meaning of p_{12}, p_{22}, and p_{31}.

SOLUTION Since $p_{12} = 0.45$, the probability of making a transition from state S_1 to state S_2 is 45 percent. Similarly, the probability of moving from state S_2 to state S_2 is 100 percent, since $p_{22} = 1.00$. (That is, we never leave state 2 after we reach it.) Likewise, since $p_{31} = 0.75$, the probability of moving from state S_3 to state S_1 is 75 percent.

An alternative way to represent a transition matrix is through a **state transition diagram.** The diagram shows each of the states together with the probability of moving from one state to another. We have the transition matrix

$$P = \begin{bmatrix} 0.25 & 0.45 & 0.30 \\ 0.00 & 1.00 & 0.00 \\ 0.75 & 0.00 & 0.25 \end{bmatrix}$$

Since P is a 3×3 matrix, the Markov chain has three states: S_1, S_2, and S_3. We draw a circle for each state. For each transition with a nonzero probability p_{ij}, we draw an arrow from state S_i to state S_j (see Figure 8.13).

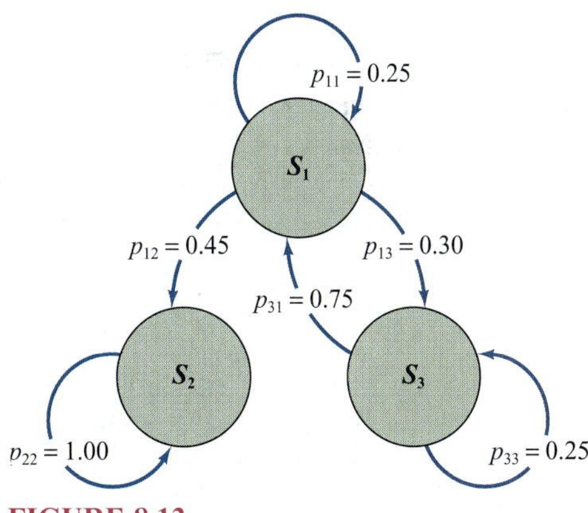

FIGURE 8.13

Observe that the only way to reach state S_2 from state S_3 is by passing through state S_1. Additionally, once state S_2 is reached, there is no leaving. We refer to S_2 as an **absorbing state.**

EXAMPLE 2

Drawing a State Transition Diagram

Draw the state transition diagram that corresponds with the given transition matrix. Identify which states, if any, are absorbing states.

$$P = \begin{bmatrix} 0.1 & 0.2 & 0.7 \\ 0.0 & 0.9 & 0.1 \\ 0.0 & 0.0 & 1.0 \end{bmatrix}$$

SOLUTION Using the transition matrix, we construct the state transition diagram (Figure 8.14).

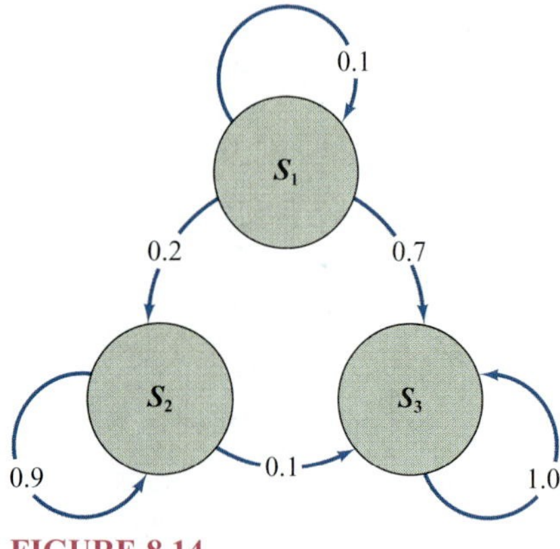

FIGURE 8.14

Observe that there are arrows leaving states S_1 and S_2; however, there are no arrows leaving state S_3. Thus, S_3 is the only absorbing state.

Powers of a Transition Matrix

In forecasting future events, we consider powers of the transition matrix P. Consider the apartment manager's transition matrix introduced earlier, $P = \begin{bmatrix} 0.65 & 0.35 \\ 0.85 & 0.15 \end{bmatrix}$. Since $p_{12} = 0.35$, there is a 35 percent probability of moving from state 1 to state 2 in exactly one step. In the context of the scenario, there is a 35 percent probability that a filled apartment will be vacant a year later. Now consider the matrix P^2.

$$\begin{aligned} P^2 &= P \cdot P \\ &= \begin{bmatrix} 0.65 & 0.35 \\ 0.85 & 0.15 \end{bmatrix} \begin{bmatrix} 0.65 & 0.35 \\ 0.85 & 0.15 \end{bmatrix} \\ &= \begin{bmatrix} 0.72 & 0.28 \\ 0.68 & 0.32 \end{bmatrix} \end{aligned}$$

What is the meaning of the entry p_{12} for the matrix P^2? This entry represents the probability of making a transition from state 1 to state 2 in exactly *two* steps. In this case, there is a 28 percent probability of making the transition from state 1 to state 2 in exactly *two* steps. In the context of the scenario, there is a 28 percent probability that a filled apartment will be vacant two years later.

What is the meaning of the entry p_{12} for the matrix P^3? This entry represents the probability of moving from state 1 to state 2 in exactly *three* steps.

$$\begin{aligned} P^3 &= P^2 \cdot P \\ &= \begin{bmatrix} 0.72 & 0.28 \\ 0.68 & 0.32 \end{bmatrix} \begin{bmatrix} 0.65 & 0.35 \\ 0.85 & 0.15 \end{bmatrix} \\ &= \begin{bmatrix} 0.706 & 0.294 \\ 0.714 & 0.286 \end{bmatrix} \end{aligned}$$

In this case, there is a 29.4 percent probability of moving from state 1 to state 2 in exactly *three* steps. In the context of the scenario, there is a 29.4 percent probability that a filled apartment will be vacant three years later. This pattern continues for any whole-number power of P.

POWERS OF A TRANSITION MATRIX

Let P be a transition matrix. Then the ijth entry of the matrix P^k is the probability of a transition from state i to state j in exactly k steps.

EXAMPLE 3

Forecasting Future Results with a Markov Chain

A business-savvy travel agent conducts an annual review of her financial records to look for trends in client spending patterns. She determines that 65 percent of her clients who booked a trip in 2003 also booked a trip in 2004. Additionally, 40 percent of her clients who did not book a trip in 2003 booked a trip in 2004. Use a Markov chain to forecast the spending behavior of her clients in 2005, in 2007, and in 2009.

SOLUTION We identify the states and write the transition matrix for the Markov chain.

State 1: Books a trip

State 2: Does not book a trip

$$P = \begin{bmatrix} 0.65 & 0.35 \\ 0.40 & 0.60 \end{bmatrix}$$

Since 2005 is two years away from 2003, we square the transition matrix.

$$\begin{aligned} P^2 &= \begin{bmatrix} 0.65 & 0.35 \\ 0.40 & 0.60 \end{bmatrix} \begin{bmatrix} 0.65 & 0.35 \\ 0.40 & 0.60 \end{bmatrix} \\ &= \begin{bmatrix} 0.56 & 0.44 \\ 0.50 & 0.50 \end{bmatrix} \end{aligned}$$

There is a 56 percent probability that a customer who books a trip in 2003 will book a trip in 2005. There is a 50 percent probability that a client who didn't book a trip in 2003 will book a trip in 2005.

Since 2007 is four years away from 2003, we raise the transition matrix to the fourth power.

$$P^4 = \begin{bmatrix} 0.65 & 0.35 \\ 0.40 & 0.60 \end{bmatrix}\begin{bmatrix} 0.65 & 0.35 \\ 0.40 & 0.60 \end{bmatrix}\begin{bmatrix} 0.65 & 0.35 \\ 0.40 & 0.60 \end{bmatrix}\begin{bmatrix} 0.65 & 0.35 \\ 0.40 & 0.60 \end{bmatrix}$$
$$= \begin{bmatrix} 0.54 & 0.46 \\ 0.53 & 0.47 \end{bmatrix}$$

There is a 54 percent probability that a customer who books a trip in 2003 will book a trip in 2007. There is a 53 percent probability that a client who didn't book a trip in 2003 will book a trip in 2007.

Since 2009 is six years away from 2003, we raise the transition matrix to the sixth power.

$$P^6 = \begin{bmatrix} 0.65 & 0.35 \\ 0.40 & 0.60 \end{bmatrix}\begin{bmatrix} 0.65 & 0.35 \\ 0.40 & 0.60 \end{bmatrix}\begin{bmatrix} 0.65 & 0.35 \\ 0.40 & 0.60 \end{bmatrix}\begin{bmatrix} 0.65 & 0.35 \\ 0.40 & 0.60 \end{bmatrix}\begin{bmatrix} 0.65 & 0.35 \\ 0.40 & 0.60 \end{bmatrix}\begin{bmatrix} 0.65 & 0.35 \\ 0.40 & 0.60 \end{bmatrix}$$
$$= \begin{bmatrix} 0.53 & 0.47 \\ 0.53 & 0.47 \end{bmatrix}$$

There is a 53 percent probability that a customer who books a trip in 2003 will book a trip in 2009. There is also a 53 percent probability that a client who didn't book a trip in 2003 will book a trip in 2009. In other words, there is a 53 percent probability that a randomly selected 2003 client will book a trip in 2009.

TECHNOLOGY TIP

Calculating Powers of a Square Matrix

1. Enter square matrix A into the calculator, using the Matrix Editor (see Section 2.2 for detailed steps). Then close the Matrix Editor.

```
MATRIX[A] 2 ×2
[ .65    .35 ]
[ .4     .6  ]

2,2=.6
```

2. Use the Matrix NAMES menu to place matrix A on the home screen.

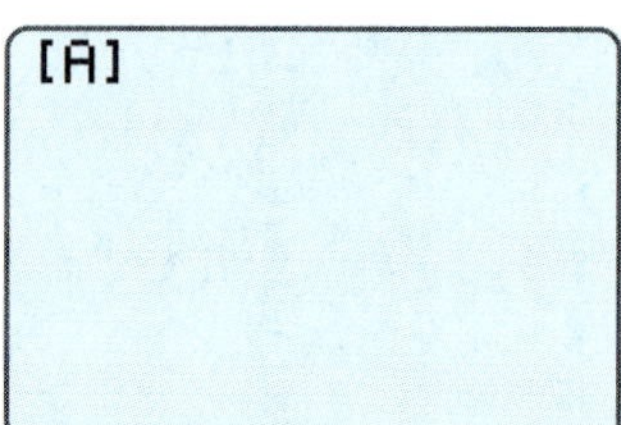

3. To calculate A^n, press the key sequence [^] [n], where n is the numerical value of the exponent.

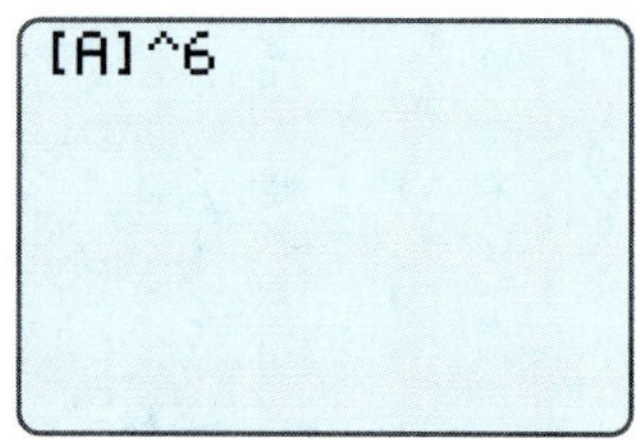

(Continued)

4. Press ENTER to display the result.

```
[A]^6
[[.5334472656 ...
 [.533203125  ...
```

5. To convert the matrix entries to rational numbers, press the key sequence MATH ENTER, ENTER. The calculator will convert the entries to rational numbers, if possible.

```
[A]^6
[[.5334472656 ...
 [.533203125  ...
Ans▸Frac
[[2185/4096 191...
 [273/512   239...
```

EXAMPLE 4

Forecasting Future Results with a Markov Chain

A survey of families belonging to the same church revealed the following. There was a 75 percent probability that children with mothers who were active in the church (attended church services weekly and volunteered service hours to the church) would be active in the church as adults. There was also a 75 percent probability that children with mothers who were less active in the church (did not attend church services weekly or did not volunteer service hours to the church) would be less active in the church as adults. (**Source:** Author's data.) What is the probability that the grandchild of an active mother is active in the church?

SOLUTION We identify the states and write the transition matrix for the Markov chain.

State 1: Active in the church

State 2: Less active in the church

$$P = \begin{bmatrix} 0.75 & 0.25 \\ 0.25 & 0.75 \end{bmatrix}$$

The time interval for this problem is one generation. Since we are interested in the active mother's grandchildren, we need to find P^2. (A grandchild is two generations away from its grandmother.)

$$P^2 = \begin{bmatrix} 0.75 & 0.25 \\ 0.25 & 0.75 \end{bmatrix}\begin{bmatrix} 0.75 & 0.25 \\ 0.25 & 0.75 \end{bmatrix}$$

$$= \begin{bmatrix} 0.625 & 0.375 \\ 0.375 & 0.625 \end{bmatrix}$$

There is a 62.5 percent probability that the grandchild of a woman who is active in the church will be active in the church.

State Distribution Vector

Returning to the apartment rental scenario given at the beginning of this section, we have a Markov chain with the transition matrix $P = \begin{bmatrix} 0.65 & 0.35 \\ 0.85 & 0.15 \end{bmatrix}$. We were told that in 2005, 400 of the 500 apartments were filled and the remaining 100 were vacant. In other words, 80 percent of the apartments were filled in 2005 and 20 percent were vacant. What percentage of the apartments will be filled in 2007?

$$\begin{aligned} \text{Rented in 2006} &= (\text{rented in 2005})(0.65) + (\text{vacant in 2005})(0.85) \\ &= (400)(0.65) + (100)(0.85) \\ &= 260 + 85 \\ &= 345 \end{aligned}$$

The manager forecasts that in 2006, 345 apartments will be filled and the remaining 155 will be vacant. In other words, 69 percent of the apartments will be filled and 31 percent will be vacant.

$$\begin{aligned} \text{Rented in 2007} &= (\text{rented in 2006})(0.65) + (\text{vacant in 2006})(0.85) \\ &= (345)(0.65) + (155)(0.85) \\ &= 224 + 132 \\ &= 356 \end{aligned}$$

The manager forecasts that in 2007, 356 apartments will be filled and the remaining 144 will be vacant. In other words, 71 percent of the apartments will be filled and 29 percent will be vacant.

Since 80 percent of the apartments were filled and 20 percent were vacant in 2005, we can alternatively represent the 2005 rental rates with the initial *state distribution vector* $D = [0.80 \quad 0.20]$. The expected percentage of apartments rented in 2006 is given by

$$DP = [\text{percentage filled} \quad \text{percentage vacant}]\begin{bmatrix} 0.65 & 0.35 \\ 0.85 & 0.15 \end{bmatrix}$$

$$\begin{aligned} &= [0.80 \quad 0.20]\begin{bmatrix} 0.65 & 0.35 \\ 0.85 & 0.15 \end{bmatrix} \\ &= [0.69 \quad 0.31] \end{aligned}$$

The expected percentage of apartments rented in 2007 is given by

$$\begin{aligned} (DP)P &= [0.69 \quad 0.31]\begin{bmatrix} 0.65 & 0.35 \\ 0.25 & 0.75 \end{bmatrix} \\ &= [0.71 \quad 0.29] \end{aligned}$$

These results agree with our earlier calculations.

STATE DISTRIBUTION VECTOR

A **state distribution vector** is a $1 \times n$ row matrix $D = [d_1 \ d_2 \ \ldots \ d_n]$ whose entries are nonnegative and add to 1. If D is the current probability distribution for a Markov chain with transition matrix P, then the state distribution vector given by DP is the next probability distribution.

EXAMPLE 5 Forecasting Future Results with a Markov Chain

According to market analysts, Palm Incorporated had 30.6 percent of the personal data assistant (PDA) market share in the third quarter of 2002. **(Source:** www.palmpower.com.) Assume that by the third quarter of the following year, 20 percent of current Palm users will move to a competing brand and 10 percent of users of competing brands will move to a Palm PDA.

According to these assumptions, what is Palm's expected market share in the third quarter of 2003, 2004, and 2005?

SOLUTION We define $S_1 =$ Palm user and $S_2 =$ other brand PDA user. The initial state distribution vector is $D_0 = [0.306 \quad 0.694]$. The transition matrix is $P = \begin{bmatrix} 0.80 & 0.20 \\ 0.10 & 0.90 \end{bmatrix}$.

$$\begin{aligned} D_1 &= D_0P \\ &= [0.306 \quad 0.694]\begin{bmatrix} 0.80 & 0.20 \\ 0.10 & 0.90 \end{bmatrix} \\ &= [0.314 \quad 0.686] \end{aligned}$$

By the third quarter of 2003, Palm is projected to have a 31.4 percent market share.

$$\begin{aligned} D_2 &= D_1P \\ &= [0.314 \quad 0.686]\begin{bmatrix} 0.80 & 0.20 \\ 0.10 & 0.90 \end{bmatrix} \\ &= [0.320 \quad 0.680] \end{aligned}$$

By the third quarter of 2004, Palm is projected to have a 32.0 percent market share.

$$\begin{aligned} D_3 &= D_2P \\ &= [0.320 \quad 0.680]\begin{bmatrix} 0.80 & 0.20 \\ 0.10 & 0.90 \end{bmatrix} \\ &= [0.324 \quad 0.676] \end{aligned}$$

By the third quarter of 2005, Palm is projected to have a 32.4 percent market share.

kth STATE DISTRIBUTION VECTOR

Let D_0 be the initial state distribution vector of a Markov chain with transition matrix P. The entry in row i and column j of P^k is the probability of moving from state i to state j in exactly k transitions. The *kth* **state distribution vector** is given by $D_k = D_0P^k$.

EXAMPLE 6 Finding State Distribution Vectors

Return to Example 5 and calculate the tenth, twentieth, and hundredth state distribution vectors. What do the results seem to infer?

SOLUTION

$$\begin{aligned} D_{10} &= D_0P^{10} \\ &= [0.306 \quad 0.694]\left(\begin{bmatrix} 0.80 & 0.20 \\ 0.10 & 0.90 \end{bmatrix}\right)^{10} \\ &= [0.306 \quad 0.694]\begin{bmatrix} 0.352 & 0.648 \\ 0.324 & 0.676 \end{bmatrix} \\ &= [0.333 \quad 0.667] \end{aligned}$$

By the third quarter of 2012, Palm is projected to have a 33.3 percent market share.

$$\begin{aligned} D_{20} &= D_0P^{20} \\ &= [0.306 \quad 0.694]\left(\begin{bmatrix} 0.80 & 0.20 \\ 0.10 & 0.90 \end{bmatrix}\right)^{20} \\ &= [0.306 \quad 0.694]\begin{bmatrix} 0.334 & 0.666 \\ 0.333 & 0.667 \end{bmatrix} \\ &= [0.333 \quad 0.667] \end{aligned}$$

By the third quarter of 2022, Palm is projected to have a 33.3 percent market share.

$$\begin{aligned} D_{100} &= D_0P^{100} \\ &= [0.306 \quad 0.694]\left(\begin{bmatrix} 0.80 & 0.20 \\ 0.10 & 0.90 \end{bmatrix}\right)^{100} \\ &= [0.306 \quad 0.694]\begin{bmatrix} 0.333 & 0.667 \\ 0.333 & 0.667 \end{bmatrix} \\ &= [0.333 \quad 0.667] \end{aligned}$$

By the third quarter of 2102, Palm is projected to have a 33.3 percent market share.

Observe that the state distribution vector remained constant for values of k beyond 10. That is, the projected market share did not change after 2012. We refer to the state distribution vector $D = [0.333 \quad 0.667]$ as a *steady state distribution vector* for the Markov chain, since multiplying this vector by P or any power of P leaves the vector unchanged.

STEADY STATE DISTRIBUTION VECTOR

A state distribution vector $D = [d_1 \; d_2 \; \ldots \; d_n]$ is a **steady state distribution vector** for a Markov chain with $n \times n$ transition matrix P if $DP = D$.

EXAMPLE 7 Finding a Steady State Distribution Vector

Find a steady state distribution vector (if it exists) for the Markov chain with transition matrix $P = \begin{bmatrix} 0.1 & 0.9 \\ 0.8 & 0.2 \end{bmatrix}$.

SOLUTION Define a steady state distribution vector $D = [d_1 \quad d_2]$. We know that

$$DP = D$$

$$[d_1 \quad d_2]\begin{bmatrix} 0.1 & 0.9 \\ 0.8 & 0.2 \end{bmatrix} = [d_1 \quad d_2]$$

$$[0.1d_1 + 0.8d_2 \quad 0.9d_1 + 0.2d_2] = [d_1 \quad d_2]$$

This result leads to the following system of equations:

$$0.1d_1 + 0.8d_2 = d_1$$
$$0.9d_1 + 0.2d_2 = d_2$$

Simplifying the system yields

$$-0.9d_1 + 0.8d_2 = 0$$
$$0.9d_1 - 0.8d_2 = 0$$

$$\left[\begin{array}{cc|c} -0.9 & 0.8 & 0 \\ 0.9 & -0.8 & 0 \end{array}\right]$$

$$\left[\begin{array}{cc|c} -0.9 & 0.8 & 0 \\ 0 & 0 & 0 \end{array}\right] \quad R_1 + R_2$$

The system appears to be dependent. However, since D is a state distribution vector, we know that $d_1 + d_2 = 1$. Adding this information to the system yields

$$\left[\begin{array}{cc|c} -0.9 & 0.8 & 0 \\ 1 & 1 & 1 \end{array}\right]$$

$$\left[\begin{array}{cc|c} 1 & 1 & 1 \\ -0.9 & 0.8 & 0 \end{array}\right] \quad \begin{array}{l} R_2 \\ R_1 \end{array}$$

$$\left[\begin{array}{cc|c} 1 & 1 & 1 \\ 0 & 17 & 9 \end{array}\right] \quad 9R_1 + 10R_2$$

$$\left[\begin{array}{cc|c} 1 & 1 & 1 \\ 0 & 1 & \frac{9}{17} \end{array}\right] \quad \frac{1}{17}R_2$$

$$\left[\begin{array}{cc|c} 1 & 0 & \frac{8}{17} \\ 0 & 1 & \frac{9}{17} \end{array}\right] \quad R_1 - R_2$$

The steady state distribution vector for the Markov chain with transition matrix $P = \begin{bmatrix} 0.1 & 0.9 \\ 0.8 & 0.2 \end{bmatrix}$ is $D = [\frac{8}{17} \quad \frac{9}{17}]$.

EXAMPLE 8

Finding a Steady State Distribution Vector

Find a steady state distribution vector (if it exists) for the Markov chain with transition matrix $P = \begin{bmatrix} 0.25 & 0.75 \\ 0.10 & 0.90 \end{bmatrix}$.

SOLUTION

$$DP = D$$

$$[d_1 \quad d_2]\begin{bmatrix} 0.25 & 0.75 \\ 0.10 & 0.90 \end{bmatrix} = [d_1 \quad d_2]$$

$$[0.25d_1 + 0.10d_2 \quad 0.75d_1 + 0.90d_2] = [d_1 \quad d_2]$$

This result leads to the following system of equations:

$$0.25d_1 + 0.10d_2 = d_1$$
$$0.75d_1 + 0.90d_2 = d_2$$

Simplifying the system yields

$$-0.75d_1 + 0.10d_2 = 0$$
$$0.75d_1 - 0.10d_2 = 0$$

$$\left[\begin{array}{cc|c} -0.75 & 0.10 & 0 \\ 0.75 & -0.10 & 0 \end{array}\right]$$

$$\left[\begin{array}{cc|c} -0.75 & 0.10 & 0 \\ 0 & 0 & 0 \end{array}\right] \quad R_1 + R_2$$

The system appears to be dependent. However, since D is a state distribution vector, we know that $d_1 + d_2 = 1$. Adding this information to the system yields

$$\left[\begin{array}{cc|c} -0.75 & 0.10 & 0 \\ 1 & 1 & 1 \end{array}\right]$$

$$\left[\begin{array}{cc|c} 1 & 1 & 1 \\ -0.75 & 0.10 & 0 \end{array}\right] \quad \begin{array}{l} R_2 \\ R_1 \end{array}$$

$$\left[\begin{array}{cc|c} 1 & 1 & 1 \\ 0 & 85 & 75 \end{array}\right] \quad 75R_1 + 100R_2$$

$$\left[\begin{array}{cc|c} 1 & 1 & 1 \\ 0 & 1 & \frac{75}{85} \end{array}\right] \quad \frac{1}{85}R_2$$

$$\left[\begin{array}{cc|c} 1 & 0 & \frac{10}{85} \\ 0 & 1 & \frac{75}{85} \end{array}\right] \quad R_1 - R_2$$

$$\left[\begin{array}{cc|c} 1 & 0 & \frac{2}{17} \\ 0 & 1 & \frac{15}{17} \end{array}\right] \quad \begin{array}{l} \text{Reduce fraction} \\ \text{Reduce fraction} \end{array}$$

The steady state distribution vector for the Markov chain with transition matrix $P = \begin{bmatrix} 0.25 & 0.75 \\ 0.10 & 0.90 \end{bmatrix}$ is $D = \left[\frac{2}{17} \ \frac{15}{17}\right]$.

EXAMPLE 9

Forecasting Genetic Makeup with a Markov Chain

The physical characteristics of a plant or animal's offspring are determined in large part by the genetic makeup of the parents. For some physical features, multiple genes from each parent determine the particular trait. For other physical features, a single gene from each parent determines the physical characteristic.

The North American Curly Horse is characterized by its curly coat. Whether the coat of a horse is curly or straight-haired is believed to be determined by a single gene from each parent. Mendelian genetics may be used to predict the coat of a foal based on its parents' genetic makeup.

Let C represent the curly-haired gene and c represent the straight-haired gene. Since curly hair is dominant, a horse with genetic makeup CC or Cc has curly hair. A horse with genetic makeup *cc* has straight hair. Horse breeders are often interested in predicting the genotype (genetic makeup) of a foal. If a mare of each genotype (*CC, Cc,* and *cc*) is bred with a stud of genotype *Cc*, the expected genotype of a foal may be determined by constructing a series of tables with column and row headings indicating the genotype of the parents and the entries in the table representing the expected genotype of the foal.

	C	*c*
C	*CC*	*Cc*
C	*CC*	*Cc*

	C	*c*
C	*CC*	*Cc*
c	*Cc*	*cc*

	C	*c*
c	*Cc*	*cc*
c	*Cc*	*cc*

Do the following:

1. Identify the genotype states.
2. Draw the transition diagram and its associated matrix.
3. Assuming that each mare of any future generation is bred with a stud of genotype *Cc,* determine the probability that a mare with genotype *CC* will have a great-grandchild with straight hair.
4. Suppose that the initial state distribution vector for the genotypes of a herd of horses is $D = [0.70 \quad 0.10 \quad 0.20]$. Determine the state distribution for the fourth generation.

SOLUTION Let $S_1 = CC$, $S_2 = Cc$, and $S_3 = cc$. From the tables, we can see that if a *CC* is bred with a *Cc*, there is a 50 percent probability that the foal will be a *CC* and a 50 percent probability that the foal will be a *Cc.* If a *Cc* is bred with a *Cc,* there is a 25 percent probability that the foal will be a *CC,* a 50 percent probability that the foal will be a *Cc*, and a 25 percent probability that the foal will be a *cc.* If a *cc* is bred with a *Cc*, there is a 50 percent probability that the foal will be a *Cc* and a 50 percent probability that the foal will be a *cc.* The preceding results yield the transition diagram in Figure 8.15 and its associated transition matrix.

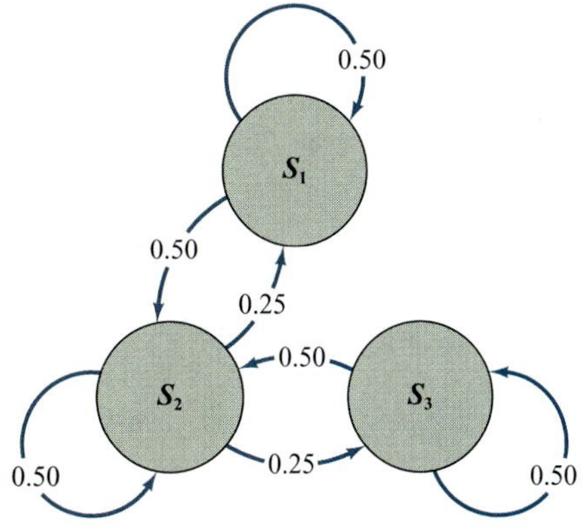

FIGURE 8.15

$$P = \begin{bmatrix} 0.50 & 0.50 & 0.00 \\ 0.25 & 0.50 & 0.25 \\ 0.00 & 0.50 & 0.50 \end{bmatrix}$$

To determine the probability that the great-grandchild of a *CC* mare is of genotype *cc*, we look at the p_{13} entry of the matrix P^3. (A great-grandchild is three generations away from its great-grandmother.)

$$P^3 = \begin{bmatrix} 0.50 & 0.50 & 0.00 \\ 0.25 & 0.50 & 0.25 \\ 0.00 & 0.50 & 0.50 \end{bmatrix}\begin{bmatrix} 0.50 & 0.50 & 0.00 \\ 0.25 & 0.50 & 0.25 \\ 0.00 & 0.50 & 0.50 \end{bmatrix}\begin{bmatrix} 0.50 & 0.50 & 0.00 \\ 0.25 & 0.50 & 0.25 \\ 0.00 & 0.50 & 0.50 \end{bmatrix}$$

$$= \begin{bmatrix} 0.31 & 0.50 & 0.19 \\ 0.25 & 0.50 & 0.25 \\ 0.19 & 0.50 & 0.31 \end{bmatrix}$$

There is a 19 percent probability that the great-grandchild of a mare of genotype *CC* will have straight hair, assuming that all mares of future generations are bred with a stud of genotype *Cc*.

The genotype state distribution for the fourth generation (three generations away from the first generation) is given by

$$DP^3 = [0.70 \quad 0.10 \quad 0.20]\begin{bmatrix} 0.31 & 0.50 & 0.19 \\ 0.25 & 0.50 & 0.25 \\ 0.19 & 0.50 & 0.31 \end{bmatrix}$$

$$= [0.28 \quad 0.50 \quad 0.22]$$

Of the fourth generation of horses, 28 percent are expected to have genotype *CC*, 50 percent are expected to have genotype *Cc*, and 22 percent are expected to have genotype *cc*.

8.3 Summary

In this section, you learned how to use Markov chains to forecast future results. You saw that the long-range behavior of a Markov chain often stabilizes over time.

8.3 Exercises

In Exercises 1–5, interpret the meaning of p_{13}, p_{21}, and p_{33} in the given transition matrix.

1. $P = \begin{bmatrix} 1 & 0 & 0 \\ 0 & 0.2 & 0.8 \\ 0 & 0.3 & 0.7 \end{bmatrix}$

2. $P = \begin{bmatrix} 0.7 & 0.0 & 0.3 \\ 0.1 & 0.4 & 0.5 \\ 0.2 & 0.8 & 0.0 \end{bmatrix}$

3. $P = \begin{bmatrix} 0.0 & 0.9 & 0.1 \\ 0.2 & 0.0 & 0.8 \\ 0.4 & 0.6 & 0.0 \end{bmatrix}$

4. $P = \begin{bmatrix} 0.1 & 0.2 & 0.7 \\ 0.2 & 0.3 & 0.5 \\ 0.3 & 0.4 & 0.3 \end{bmatrix}$

5. $P = \begin{bmatrix} 1 & 0 & 0 \\ 0 & 1 & 0 \\ 0.5 & 0 & 0.5 \end{bmatrix}$

In Exercises 6–15, draw the state transition diagram that corresponds to the given transition matrix. Identify which states, if any, are absorbing states.

6. $P = \begin{bmatrix} 0 & 1 \\ 1 & 0 \end{bmatrix}$

7. $P = \begin{bmatrix} 1 & 0 \\ 0 & 1 \end{bmatrix}$

8. $P = \begin{bmatrix} 0.1 & 0.9 \\ 0.4 & 0.6 \end{bmatrix}$ **9.** $P = \begin{bmatrix} 1.00 & 0.00 \\ 0.05 & 0.95 \end{bmatrix}$

10. $P = \begin{bmatrix} 1 & 0 & 0 \\ 0 & 0.2 & 0.8 \\ 0 & 0.3 & 0.7 \end{bmatrix}$

11. $P = \begin{bmatrix} 0.7 & 0.0 & 0.3 \\ 0.1 & 0.4 & 0.5 \\ 0.2 & 0.8 & 0.0 \end{bmatrix}$

12. $P = \begin{bmatrix} 0.0 & 0.9 & 0.1 \\ 0.2 & 0.0 & 0.8 \\ 0.4 & 0.6 & 0.0 \end{bmatrix}$

13. $P = \begin{bmatrix} 0.1 & 0.2 & 0.7 \\ 0.2 & 0.3 & 0.5 \\ 0.3 & 0.4 & 0.3 \end{bmatrix}$

14. $P = \begin{bmatrix} 0 & 1 & 0 & 0 \\ 0 & 0 & 1 & 0 \\ 0 & 0 & 0 & 1 \\ 1 & 0 & 0 & 0 \end{bmatrix}$

15. $P = \begin{bmatrix} 0 & 0.5 & 0.5 & 0 \\ 0 & 0 & 0.5 & 0.5 \\ 0 & 0 & 1 & 0 \\ 0 & 0 & 0 & 1 \end{bmatrix}$

In Exercises 16–20, write the transition matrix P that corresponds to the state transition diagram.

16.

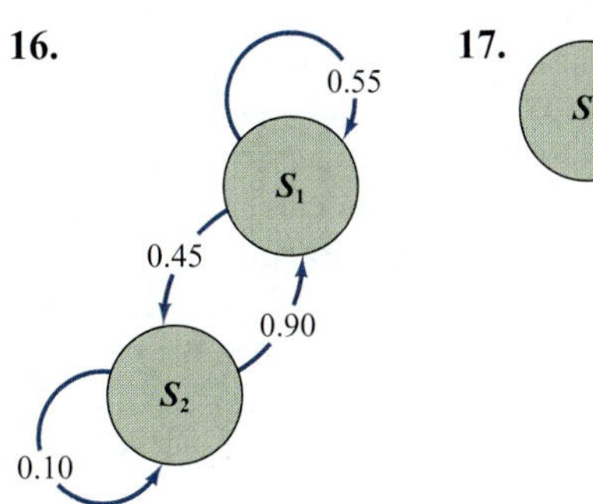

17.

1.00
S_1 S_2
1.00

18.

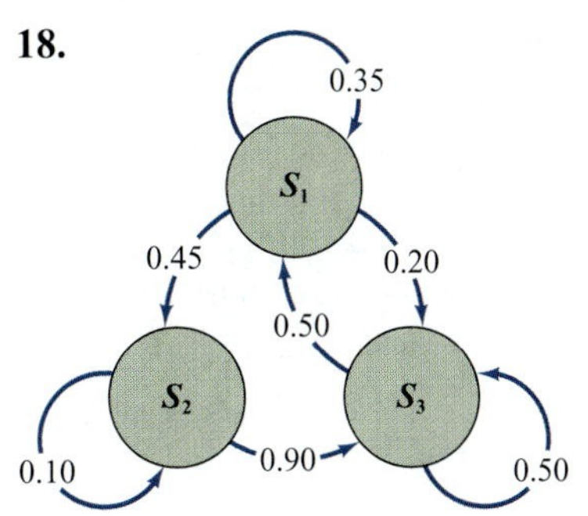

19.

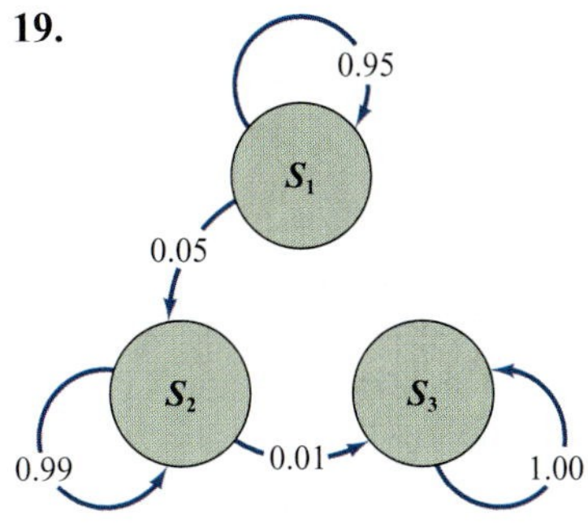

20.

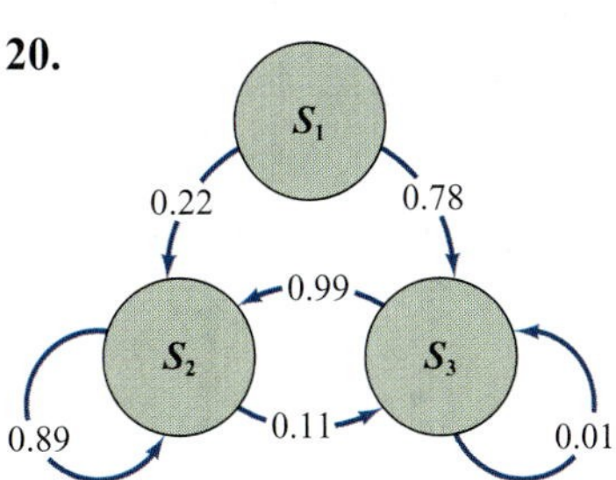

In Exercises 21–30, find a steady state distribution vector (if it exists) for each transition matrix.

21. $P = \begin{bmatrix} 0.2 & 0.8 \\ 0.3 & 0.7 \end{bmatrix}$ **22.** $P = \begin{bmatrix} 0.5 & 0.5 \\ 0.1 & 0.9 \end{bmatrix}$

23. $P = \begin{bmatrix} 0.1 & 0.9 \\ 0.4 & 0.6 \end{bmatrix}$ **24.** $P = \begin{bmatrix} 1.00 & 0.00 \\ 0.05 & 0.95 \end{bmatrix}$

25. $P = \begin{bmatrix} 0.64 & 0.36 \\ 0.25 & 0.75 \end{bmatrix}$

26. $P = \begin{bmatrix} 0.7 & 0.0 & 0.3 \\ 0.1 & 0.4 & 0.5 \\ 0.2 & 0.8 & 0.0 \end{bmatrix}$

27. $P = \begin{bmatrix} 0.0 & 0.9 & 0.1 \\ 0.2 & 0.0 & 0.8 \\ 0.4 & 0.6 & 0.0 \end{bmatrix}$

28. $P = \begin{bmatrix} 1.0 & 0.0 & 0.0 \\ 0.0 & 0.5 & 0.5 \\ 0.0 & 0.7 & 0.3 \end{bmatrix}$

29. $P = \begin{bmatrix} 0 & 1 & 0 \\ 0 & 0 & 1 \\ 1 & 0 & 0 \end{bmatrix}$

30. $P = \begin{bmatrix} 0 & 0.5 & 0.5 & 0 \\ 0 & 0 & 0.5 & 0.5 \\ 0 & 0 & 1 & 0 \\ 0 & 0 & 0 & 1 \end{bmatrix}$

In Exercises 31–35, calculate and interpret the meaning of P^2, P^5, and P^7 given a Markov chain with transition matrix P.

31. $P = \begin{bmatrix} 0.12 & 0.00 & 0.88 \\ 0.88 & 0.12 & 0.00 \\ 0.00 & 0.88 & 0.12 \end{bmatrix}$

32. $P = \begin{bmatrix} 0.5 & 0.5 & 0.0 \\ 0.0 & 0.5 & 0.5 \\ 0.0 & 0.0 & 1.0 \end{bmatrix}$

33. $P = \begin{bmatrix} 0.50 & 0.25 & 0.25 \\ 0.25 & 0.50 & 0.25 \\ 0.25 & 0.25 & 0.50 \end{bmatrix}$

34. $P = \begin{bmatrix} 0 & 1 & 0 \\ 0 & 0 & 1 \\ 1 & 0 & 0 \end{bmatrix}$

35. $P = \begin{bmatrix} 0.33 & 0.67 & 0.00 \\ 0.10 & 0.90 & 0.00 \\ 0.00 & 0.00 & 1.00 \end{bmatrix}$

In Exercises 36–40, determine the answer to each question by calculating the appropriate power of the transition matrix.

36. Family Business In forecasting future workforce needs, the head of a small company estimates that there is a 20 percent probability that an employee who is working for the company today will not be working for the company one year from now. There is a 15 percent probability that an employee who used to work for the company will be employed by the company one year from now. Assuming that these estimates are accurate for future years, what is the probability that a current employee will not be employed by the company three years from now?

37. Coin Toss Game A unique coin toss game is played as follows. To start the game, a coin is tossed and allowed to fall upon a tabletop. The face of the coin (heads or tails) is noted. The process is then repeated, with scoring as follows: If the coin shows heads or matches the previous coin toss, a win is counted. Otherwise, a loss is counted. Is it possible for a player to lose 25 rounds in a row? Explain. Then calculate the probability that a person who loses the first round will win the 25th round.

38. Deterioration of Theater Films Each time a movie is shown in a theater, the actual film is exposed to the possibility of being scratched or of accumulating specks of dust. Consequently, the quality of the projected film image deteriorates over time.

Suppose that there is a 0.2 percent probability that an undamaged film frame will be damaged during the next showing of the film and that there is a 100 percent probability that a damaged film frame will remain damaged during the next showing of the film. What is the probability that an undamaged film frame will have become damaged after 200 additional showings of the film?

39. College Enrollment The educational attainment of a student's parents has a dramatic effect on the educational attainment of the student. According to a 2001 Department of Education study, the percentage of high school graduates enrolled in college varied dramatically based upon the educational attainment of the parents. If the student's parents' highest level of education was a high school diploma or less, 59 percent of the students enrolled in college. If the parents' highest level of education was some college, 75 percent of the students enrolled in college. If the parents' highest level of education was a bachelor's degree or higher, 93 percent of the students enrolled in college. (**Source:** Special Analysis 2001, *Students Whose Parents Did Not Go to College: Postsecondary Access, Persistence, and Attainment,* U.S. Department of Education.) Based on these results and others presented within the study, we combine the last two percentages and estimate that 86 percent of students whose parents attended college also enrolled in college.

We define state 1 as "attended college" and state 2 as "did not attend college." Treating each of the percentages as a probability, what is the probability that the grandchild of a person who attended college also went to college?

40. College Enrollment Referring to the scenario given in Exercise 39, explain whether or not a person's educational attainment substantially affects the likelihood that a great-great-grandchild of that person will enroll in college. (*Hint:* A great-great-grandchild is four generations away from its great-great-grandparent.)

In Exercises 41–46, do the following: (i) identify the states of the Markov chain, (ii) draw the transition diagram, (iii) write the transition matrix P, (iv) identify the initial state distribution vector D, and (v) calculate DP and interpret the meaning of the result.

41. **Software Market Share** According to market researchers, Mercury Interactive had a 55.2 percent market share of the automated software quality tools market in 2002. (**Source:** www.mercuryinteractive.com.) Assume that the probability that a Mercury Interactive customer will change to a competitor's program is 15 percent and that the probability that a competitor's customer will change to a Mercury Interactive program is 20 percent.

42. **Banking Deposit Market Share** According to market researchers, J. P. Morgan Chase had a 22.52 percent market share of the banking deposit market as of January 30, 2003. (**Source:** www.snl.com.) Assume that the probability that a J. P. Morgan Chase customer will transfer all of his or her deposits to a competitor's bank is 10 percent and that the probability that a competitor's customer will transfer all of his or her deposits to a J. P. Morgan Chase bank is 20 percent.

43. **Stock Prices** The closing share price of Apple Computer Corporation fluctuated during the 20 trading days in January 2004. On 11 of the trading days, the share price was higher than the preceding day's closing price. On the remaining trading days, the share price was lower than the preceding day's closing price. Further analysis of the data revealed that the probability of moving from a high price to a higher price on the next day was 45 percent. Similarly, the probability of moving from a low price to a lower price on the next day was 25 percent. On the first trading day of 2004, the share price was lower than on the preceding trading day.

44. **Employment and Mortality Rates** At the 10-year reunion of a high school graduating class, it is discovered that 94 percent of the graduates are employed, 5 percent of the graduates are unemployed, and 1 percent of the graduates are dead. Assume that the probability that an unemployed graduate will become employed by the 20-year reunion is 84 percent, and the probability that the graduate will be dead is 5 percent. The probability that an employed graduate will become unemployed by the 20-year reunion is 8 percent, and the probability that the graduate will be dead is 2 percent.

45. **Workout Schedules** In an effort to ensure that a fitness club has an adequate balance of cardiovascular and weight-training equipment, a manager monitors the workout behavior of the club's clients. The manager determines that a client who does weight training today is expected to do one of the following tomorrow:

Weight training (10 percent probability)
Cardiovascular (70 percent probability)
Not work out (20 percent probability)

A client who does cardiovascular training today is expected to do one of the following tomorrow:

Weight training (60 percent probability)
Cardiovascular (20 percent probability)
Not work out (20 percent probability)

A client who does not work out today is expected to do one of the following tomorrow:

Weight training (40 percent probability)
Cardiovascular (50 percent probability)
Not work out (10 percent probability)

Today 30 percent of the club's members did weight training, 25 percent of the members did cardiovascular training, and 45 percent of the members did not work out at the club.

46. **Gambling** A gambler predicts that if she wins the current game of poker, the probability that she will win the next game is 55 percent. She predicts that if she loses the current game of poker, there is a 40 percent probability that she will win the next game. So far this evening she has won 30 percent of her poker games.

In Exercises 47–55, use the Markov chain to answer the questions.

47. **Unemployment Rate** Unemployment rates fluctuate over time. In 1992, 7.5 percent of civilians 16 years old and older were unemployed. In 1996, 5.4 percent were unemployed. In 2000, 4.0 percent were unemployed. (**Source:** *Statistical Abstract of the United States, 2001*, Table 567, p. 367.) Based on these data, the unemployment rate may be modeled by a Markov chain with transition matrix

$$P = \begin{bmatrix} 0.671 & 0.329 \\ 0.004 & 0.996 \end{bmatrix}$$

and states S_1 = unemployed and S_2 = employed. The time interval for the matrix is four years. (That is, there are four years between transitions.) The initial state distribution vector is $D_0 = [0.075 \quad 0.925]$.

(a) Calculate and interpret the meaning of D_1 and D_2.

(b) Compare the results of part (a) with the reported unemployment data. Does the model seem to fit the data well? Explain.

(c) Use the model to predict the unemployment rate in 2004.

(d) According to this model, what percentage of the population is expected to be unemployed in the long run?

48. **Union Membership—Public Sector** Union membership rates tend to fluctuate over time. In 1990, 36.5 percent of the public-sector workers were members of a union. In 1995, 37.7 percent were union members. In 2000, 37.5 percent were union members. (**Source:** *Statistical Abstract of the United States, 2001*, Table 637, p. 411.) Based on these data, union membership rates for public-sector workers may be modeled by a Markov chain with transition matrix

$$P = \begin{bmatrix} 0.271 & 0.729 \\ 0.438 & 0.562 \end{bmatrix}$$

and states S_1 = union member and S_2 = not a union member. The time interval for the matrix is five years. (That is, there are five years between transitions.) The initial state distribution vector is $D_0 = [0.365 \quad 0.635]$.

(a) Calculate and interpret the meaning of D_1 and D_2.

(b) Compare the results of part (a) with the reported union membership data. Does the model seem to fit the data well? Explain.

(c) Use the model to predict the union membership rate in 2005.

(d) According to this model, what percentage of public-sector workers will be members of a union in the long run?

(e) How could union leaders use the results of part (d)?

49. **Union Membership—Private Sector** In 1990, 11.9 percent of the private-sector workers were members of a union. In 1995, 10.3 percent were union members. In 2000, 9.0 percent were union members. (**Source:** *Statistical Abstract of the United States, 2001*, Table 637, p. 411.) Based on these data, union membership rates for private-sector workers may be modeled by a Markov chain with transition matrix

$$P = \begin{bmatrix} 0.819 & 0.181 \\ 0.006 & 0.994 \end{bmatrix}$$

and states S_1 = union member and S_2 = not a union member. The time interval for the matrix is five years. (That is, there are five years between transitions.) The initial state distribution vector is $D_0 = [0.119 \quad 0.881]$.

(a) Calculate and interpret the meaning of D_1 and D_2.

(b) Compare the results of part (a) with the reported union membership data. Does the model seem to fit the data well? Explain.

(c) Use the model to predict the union membership rate in 2005.

(d) According to this model, what percentage of private-sector workers will be members of a union in the long run?

(e) How could union leaders use the results of part (d)?

50. **African Population** In 1990, humans living in Africa made up 11.9 percent of the world's population. By 2000, the percentage had increased to 13.2 percent. According to Census Bureau estimates, Africans are expected to make up 14.4 percent of the world's population by 2010. (**Source:** *Statistical Abstract of the United States, 2001*, Table 1325, p. 829.) Based on these data, the percentage of humans living in Africa may be modeled by a Markov chain with transition matrix

$$P = \begin{bmatrix} 0.945 & 0.055 \\ 0.022 & 0.978 \end{bmatrix}$$

and states S_1 = living in Africa and S_2 = not living in Africa. The time interval for the matrix is 10 years. (That is, there are 10 years between transitions.) The initial state distribution vector is $D_0 = [0.119 \quad 0.881]$.

(a) Use the model to calculate the percentage of the human population expected to be living in Africa in 2050.

(b) The Census Bureau projects that by 2050, Africans will make up 20.3 percent of the world's human population. Does this estimate agree with the Markov chain model? Explain.

(c) According to the model, what percentage of the world's population is projected to live in Africa in the long run?

51. **Child Abuse and Neglect** In 1996, nearly 956,000 children in the United States were victims of child abuse. The majority of these children were neglected by their parents or guardians. In 1996, 51.6 percent of abused children were neglected. In 1997, 54.6 percent were neglected, and in 1998, 53.5 percent were neglected. (**Source:** *Statistical Abstract of the United States, 2001*, Table 328, p. 199.) Based on these data, the percentage of abused children who were neglected may be modeled by a Markov chain with transition matrix

$$P = \begin{bmatrix} 0.369 & 0.631 \\ 0.735 & 0.265 \end{bmatrix}$$

and states S_1 = neglected and S_2 = not neglected.

(a) According to the model, what percentage of abused children were neglected in 1999?
(b) According to the model, what percentage of abused children will be neglected in the long run?
(c) What can you do to help reduce child abuse? (*Hint:* Visit www.preventchildabuse.org.)

52. **Supermarkets with ATMs** In 1992, 28 percent of supermarkets had automatic teller machines. The percentage had increased to 60 percent by 1996 and to 64 percent by 2000. (**Source:** *Statistical Abstract of the United States, 2001*, Table 1033, p. 651.) Based on these data, the percentage of supermarkets with ATMs may be modeled by a Markov chain with transition matrix

$$P = \begin{bmatrix} 0.690 & 0.310 \\ 0.565 & 0.435 \end{bmatrix}$$

and states S_1 = has ATM and S_2 = doesn't have ATM. The time interval between transitions is four years.

(a) Does the model fit the data well? Justify your answer.
(b) Do the transition probabilities seem to make sense in the context of the problem?
(c) According to the model, will the percentage of supermarkets with ATM machines ever reach 80 percent? Justify your answer.
(d) Do you think that this model is an accurate predictor of the percentage of supermarkets with ATMs? Justify your answer.

53. **Supermarkets with a Pharmacy** In 1992, 18 percent of supermarkets had a pharmacy. The percentage had increased to 26 percent by 1996 and to 32 percent by 2000. (**Source:** *Statistical Abstract of the United States, 2001*, Table 1033, p. 651.) Based on these data, the percentage of supermarkets with a pharmacy may be modeled by a Markov chain with transition matrix

$$P = \begin{bmatrix} 0.875 & 0.125 \\ 0.125 & 0.875 \end{bmatrix}$$

and states S_1 = has pharmacy and S_2 = doesn't have pharmacy. The time interval between transitions is four years.

(a) Does the model fit the data well? Justify your answer.
(b) Do the transition probabilities seem to make sense in the context of the problem?
(c) According to the model, what is the maximum predicted percentage of supermarkets that will have a pharmacy?
(d) Do you think that this model is an accurate predictor of the percentage of supermarkets with a pharmacy? Justify your answer.

54. **Supermarkets with a Deli** In 1992, 78 percent of supermarkets had a deli. The percentage had increased to 80 percent by 1996 and to 81 percent by 2000. (**Source:** *Statistical Abstract of the United States, 2001*, Table 1033, p. 651.) Based on these data, the percentage of supermarkets with a deli may be modeled by a Markov chain with transition matrix

$$P = \begin{bmatrix} 0.91 & 0.09 \\ 0.41 & 0.59 \end{bmatrix}$$

and states S_1 = has deli and S_2 = doesn't have deli. The time interval between transitions is four years.

(a) Does the model fit the data well? Justify your answer.
(b) Do the transition probabilities seem to make sense in the context of the problem?
(c) According to the model, what is the maximum predicted percentage of supermarkets that will have a deli?

55. **Supermarkets with Service Fish Departments** In 1992, 41 percent of supermarkets had a service fish department. The percentage had increased to 46 percent by 1996

and then decreased to 45 percent by 2000. (**Source:** *Statistical Abstract of the United States, 2001*, Table 1033, p. 651.) Based on these data, the percentage of supermarkets with a service fish department may be modeled by a Markov chain with transition matrix

$$P = \begin{bmatrix} 0.342 & 0.658 \\ 0.542 & 0.458 \end{bmatrix}$$

and states $S_1 =$ has service fish department and $S_2 =$ doesn't have service fish department. The time interval between transitions is four years.

(a) Does the model fit the data well? Justify your answer.

(b) Do the transition probabilities seem to make sense in the context of the problem?

(c) According to the model, what percentage of supermarkets will have a service fish department in the long run?

Exercises 56–61 are intended to challenge your understanding of Markov chains.

56. Find three distinct steady state vectors for the transition matrix

$$P = \begin{bmatrix} 0.2 & 0.8 & 0.0 \\ 0.1 & 0.9 & 0.0 \\ 0.0 & 0.0 & 1.0 \end{bmatrix}$$

57. The vector $D = [d_1 \quad d_2 \quad d_3]$ is a steady state distribution vector for the transition matrix

$$P = \begin{bmatrix} 1 & 0 & 0 \\ 0.2 & 0.6 & 0.2 \\ 0 & 0 & 1 \end{bmatrix}$$

Determine the value of and/or relationship between d_1, d_2, and d_3. (*Hint:* We're looking for something more than just $d_1 + d_2 + d_3 = 1$.)

58. A Markov chain with a 5×5 transition matrix P has the property that its first four states are absorbing states. The probability of making a transition from the fifth state S_5 to state S_i is $ai\%$, where a is a real number. Write the numeric values of the transition matrix P.

59. Given the transition matrix

$$P = \begin{bmatrix} 0 & 0 & 1 \\ 1 & 0 & 0 \\ 0 & 1 & 0 \end{bmatrix}$$

determine for which values of k $P^k = P$. Show the work that leads to your conclusion.

60. Consider a Markov chain with transition matrix

$$P = \begin{bmatrix} 0 & 0 & 1 \\ 1 & 0 & 0 \\ 0 & 1 & 0 \end{bmatrix}$$

and initial state distribution vector $[0.1 \ 0.3 \ 0.6]$. List all possible state distribution vectors for the Markov chain. What will the state probability distribution be after 291 transitions? (*Hint:* Refer to the results of Exercise 59.)

61. Suppose that a Markov chain with transition matrix $P = \begin{bmatrix} a & 1-a \\ b & 1-b \end{bmatrix}$ has state distribution vectors $d_0 = [x \quad 1-x]$, $d_1 = [y \quad 1-y]$, and $d_2 = [z \quad 1-z]$. Show that if $x \neq y$, $0 < x \leq 1$, and $0 < y \leq 1$, then $a = \dfrac{xz + y - y^2 - z}{x - y}$ and $b = \dfrac{xz - y^2}{x - y}$.

8.4 Random Variables and Expected Value

- Calculate the probability distribution of a random variable
- Draw a histogram
- Determine the expected value of a random variable

GETTING STARTED Many decision-making processes involve a certain element of chance. People often take risks after weighing the likelihood of a negative consequence. By using mathematical methods, we can quantify risk and make better-informed decisions.

In this section, we will demonstrate how to calculate the probability distribution of a *random variable*, draw a *histogram*, and determine the *expected value of a random variable.* These techniques provide important inputs into any decision-making process.

Random Variables and Histograms

RANDOM VARIABLE

A **random variable** is a function that assigns a number to each outcome in a sample space.

The term *random variable* is an odd name for this function, since it is neither random nor a variable. However, since the term is considered standard, we too will use it. Capital letters X and Y are often used to represent random variables.

Consider a simple coin toss. When a tossed coin lands, it will result in one of two outcomes: heads or tails. We may define a random variable X by assigning a value of 0 to *tails* and a value of 1 to *heads.* When $X = 0$, we know that a *tail* is displayed. When $X = 1$, we know that a *head* is displayed.

Random variables are not limited to experiments with just two possible outcomes. In many games of dice, we are interested in the sum of the face values of two dice. There are 36 outcomes in the sample space. We can define a random variable $X =$ sum of the face values of two six-sided dice. The value of the random variable is obtained by counting the dots on the face of each die (Figure 8.16).

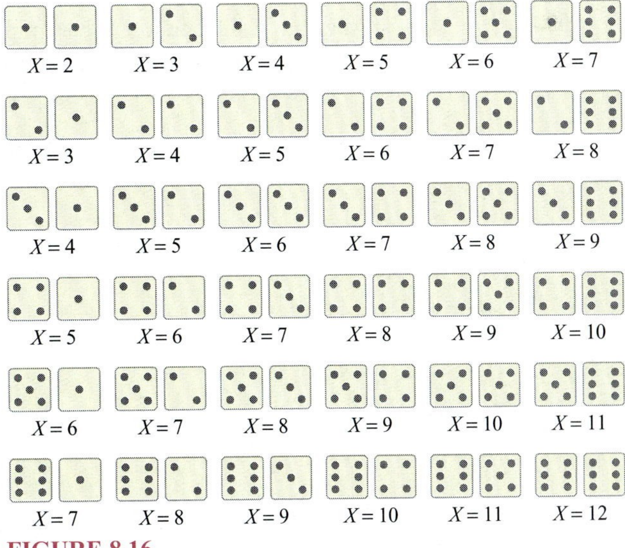

FIGURE 8.16

Some of the random variable values occur frequently (e.g., $X = 7$), while others occur only once (e.g., $X = 2$). We identify the number of times each value occurs in a **frequency distribution** (Table 8.9).

TABLE 8.9

x	2	3	4	5	6	7	8	9	10	11	12
Frequency	1	2	3	4	5	6	5	4	3	2	1

In games of chance, we are interested in the probability of a particular value occurring. To determine the probability of an event, we divide the frequency of the event by the total number of possible outcomes. We summarize all of the probabilities with a probability distribution.

PROBABILITY DISTRIBUTION OF A RANDOM VARIABLE

The **probability distribution of a random variable X** is denoted by

$$P(X = x) = p(x)$$

and represents the collection of probabilities for each possible value of x.

In the case of the dice sum, the random variable X may take on any of the whole-number values between 2 and 12. To calculate $P(X = x)$, we count the number of times X takes on the value of x and divide it by the total number of outcomes in the sample space (Table 8.10).

TABLE 8.10

x	2	3	4	5	6	7	8	9	10	11	12
$P(X = x)$	$\frac{1}{36}$	$\frac{2}{36}$	$\frac{3}{36}$	$\frac{4}{36}$	$\frac{5}{36}$	$\frac{6}{36}$	$\frac{5}{36}$	$\frac{4}{36}$	$\frac{3}{36}$	$\frac{2}{36}$	$\frac{1}{36}$

Since $P(X = x)$ is a probability distribution, the values in the table will always sum to 1. The graph of a probability distribution is referred to as a **histogram.** A histogram is a bar graph. The width of each bar is equal to 1 unit, and the height of each bar is the probability. The histogram for the random variable X is shown in Figure 8.17.

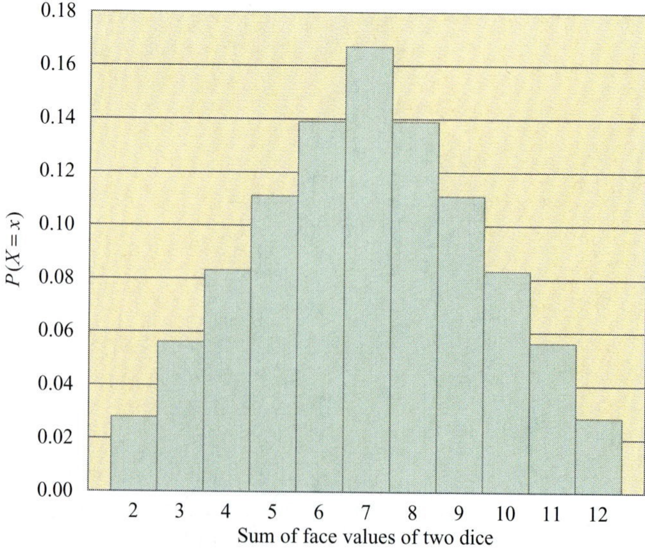

FIGURE 8.17

This particular histogram is symmetric about $X = 7$. We see visually that of all possible outcomes, that $X = 7$ is most likely and $X = 2$ or $X = 12$ is least likely.

EXAMPLE 1

Finding a Probabilty Distribution and Its Associated Histogram

The random variable Y is defined to be the absolute value of the difference in the face values of two dice. Determine the probability distribution for Y and draw the associated histogram.

SOLUTION The random variable Y may take on any of the whole-number values between 0 and 5. The frequency of each outcome varies as shown in Figure 8.18.

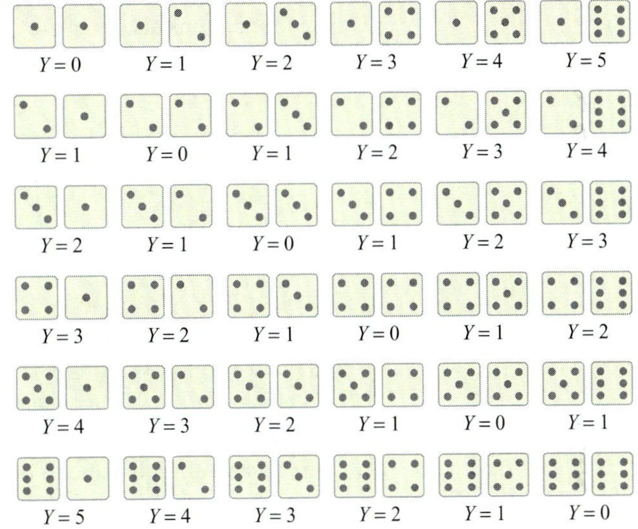

FIGURE 8.18

The probability distribution of Y is shown in Table 8.11 and the histogram in Figure 8.19.

TABLE 8.11

y	0	1	2	3	4	5
$P(Y = y)$	$\frac{6}{36}$	$\frac{10}{36}$	$\frac{8}{36}$	$\frac{6}{36}$	$\frac{4}{36}$	$\frac{2}{36}$

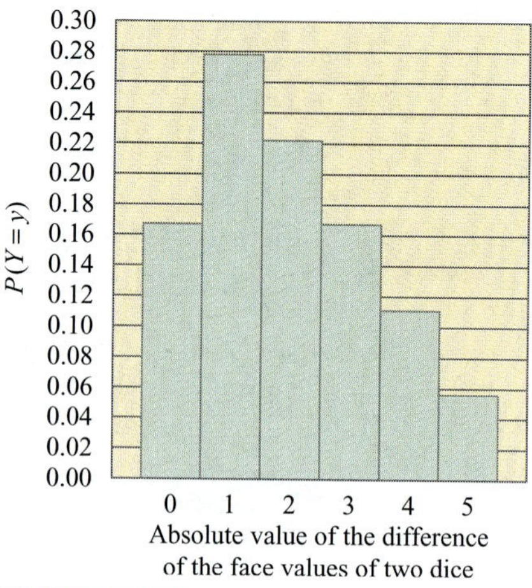

FIGURE 8.19

Unlike the previous histogram, the histogram in Figure 8.19 is not symmetric. The outcome $X = 1$ is the most likely, and the outcome $X = 5$ is the least likely.

The TI-83 Plus may be used to draw the histogram of a probability distribution, as illustrated in the following Technology Tip.

TECHNOLOGY TIP

Drawing a Histogram

1. Press STAT ENTER to open the Statistics Editor. Then enter the values of the random variable in L1 and the frequency of each value in L2.

L1	L2	L3
0	6	
1	10	
2	8	
3	6	
4	4	
5	2	

L3(1) =

2. Use the blue arrow to move the cursor to the L3 heading. Press the key sequence 2nd, 2, ÷, and the number of outcomes in the sample space. (This operation will divide the values in List 2 by the number of outcomes in the sample space.)

L1	L2	L3
0	6	
1	10	
2	8	
3	6	
4	4	
5	2	

L3 =L2/36

3. Press ENTER to display the probability distribution of the random variable in List 3.

L1	L2	L3
0	6	.16667
1	10	.27778
2	8	.22222
3	6	.16667
4	4	.11111
5	2	.05556

L3(7) =

(Continued)

4. Press the key sequence 2nd, Y=, ENTER to open Plot 1. Select the entries depicted in the figure to the right to set up a histogram with random variable values from List 1 and probabilities from List 3.

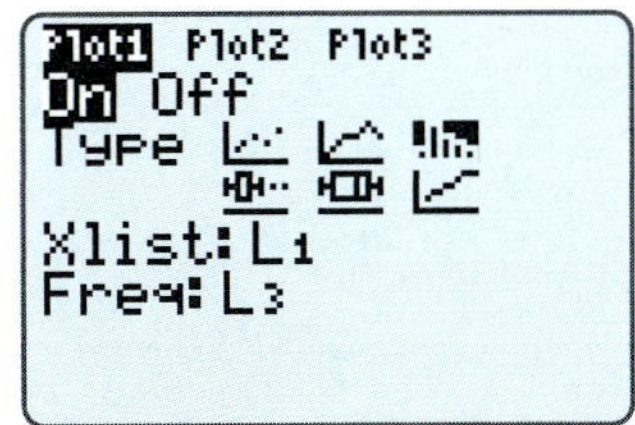

5. Press WINDOW and adjust the values. `Xmin` is the smallest value of the random variable, `Xmax` is the largest value, and `Xscl` is the difference between consecutive values of the random variable. `Ymin` is 0, and `Ymax` should be a value slightly larger than the greatest probability value in the distribution.

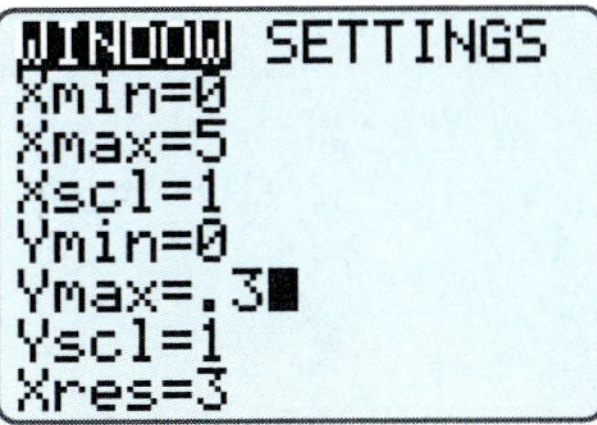

6. Press GRAPH to display the histogram.

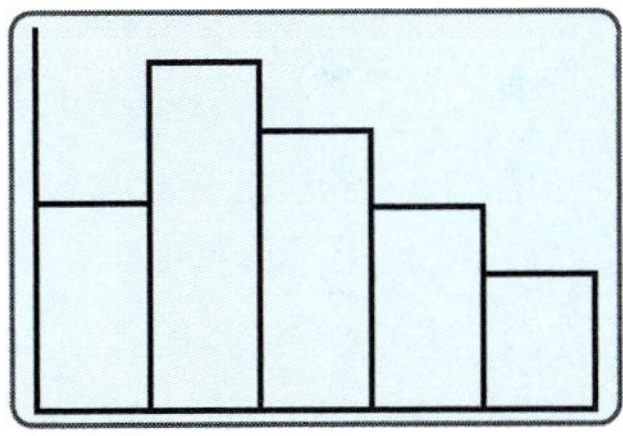

Expected Value

If we were to roll a pair of dice a large number of times, what would be the average value of the random variable? In Example 1, we found the probability distribution for Y to be that shown in Table 8.12.

TABLE 8.12

y	0	1	2	3	4	5
$P(Y = y)$	$\frac{6}{36}$	$\frac{10}{36}$	$\frac{8}{36}$	$\frac{6}{36}$	$\frac{4}{36}$	$\frac{2}{36}$

To determine the average value of the random variable, we multiply each value of the random variable by its associated probability and sum the results. The end result is referred to as the **expected value** $E(Y)$ of the random variable.

$$\begin{aligned} E(Y) &= 0\left(\frac{6}{36}\right) + 1\left(\frac{10}{36}\right) + 2\left(\frac{8}{36}\right) + 3\left(\frac{6}{36}\right) + 4\left(\frac{4}{36}\right) + 5\left(\frac{2}{36}\right) \\ &= 0 + \frac{10}{36} + \frac{16}{36} + \frac{18}{36} + \frac{16}{36} + \frac{10}{36} \\ &= \frac{70}{36} \\ &= 1.944 \end{aligned}$$

Although the value 1.944 cannot occur in a single dice roll, it is expected that the long-run average of the values will be 1.944.

EXPECTED VALUE OF A RANDOM VARIABLE

The **expected value** of a random variable X with probability distribution

x_i	x_1	x_2	...	x_n
$P(X = x_i)$	p_1	p_2	...	p_n

is given by

$$E(x) = x_1p_1 + x_2p_2 + \cdots + x_np_n$$

where p_i is the probability that x_i occurs.

EXAMPLE 2 Finding a Probability Distribution

A group of 14 applicants to Berkeley Law School in 2003 reported LSAT scores between 169 and 179. Of these applicants, 12 were accepted to the law school and 2 were rejected. (**Source:** www.lawschoolnumbers.com.) A random sample of 5 applicants is selected from the 14-applicant pool.

Let $X =$ the number of applicants in the sample who were rejected. Find the probability distribution of X and determine the expected number of rejected applicants in the sample.

SOLUTION The number of ways to select 5 applicants from a pool of 14 applicants is

$$\begin{aligned} C(14, 5) &= \frac{14!}{9!5!} \\ &= 2002 \end{aligned}$$

Since only 2 applicants were rejected, any sample will contain either 0, 1, or 2 rejected applicants. The number of samples with 0 rejected applicants is given by $C(12, 5)\ C(2, 0)$, the number with 1 rejected applicant is given by $C(12, 4)\ C(2, 1)$, and the number with 2 rejected applicants is given by $C(12, 3)\ C(2, 2)$. Therefore,

$$\begin{aligned} p_0 &= \frac{C(12, 5)\ C(2, 0)}{C(14, 5)} \\ &= \frac{792 \cdot 1}{2002} \\ &= 0.40 \end{aligned} \qquad \begin{aligned} p_1 &= \frac{C(12, 4)\ C(2, 1)}{C(14, 5)} \\ &= \frac{495 \cdot 2}{2002} \\ &= 0.49 \end{aligned}$$

$$\begin{aligned} p_2 &= \frac{C(12, 3)\ C(2, 2)}{C(14, 5)} \\ &= \frac{220 \cdot 1}{2002} \\ &= 0.11 \end{aligned}$$

The probability distribution of X is given by

x	0	1	2
$P(X = x)$	0.40	0.49	0.11

The expected value of X is given by

$$\begin{aligned} E(X) &= 0(0.40) + 1(0.49) + 2(0.11) \\ &= 0.71 \end{aligned}$$

If a random sample of 5 applicants was repeatedly selected from the 14-applicant pool, we expect that the average number of applicants in the sample who were rejected would be 0.71.

EXAMPLE 3

Using a Probability Distribution in a Real-World Context

The average premium for homeowner's insurance nationwide in 1999 was \$487. In that year, 8.63 percent of homeowners filed payable insurance claims. The average payout was \$3773. (**Source:** www.iii.org.) Assuming that there was an 8.63 percent probability that an insured homeowner received a payment for damages in 1999, determine the expected return of a homeowner who had a policy in 1999.

SOLUTION In 1999, there was an 8.63 percent probability that a homeowner received a payment. The average payment amount was \$3773. However, since the insured owner had to pay a \$487 premium, the net return was \$3286 (\$3773 − \$487).

In 1999, there was a 91.37 percent probability that a homeowner did **not** receive a payment. For these homeowners, the net return was −\$487.

Let $X =$ the average net insurance return. The probability distribution of X is given by

x	−487	3,286
$P(X = x)$	0.9137	0.0863

The expected value is given by

$$\begin{aligned} E(X) &= -487(0.9137) + 3286(0.0863) \\ &= -\$161.39 \end{aligned}$$

If the insurance payment data remained constant over a number of years, the average net loss to an insured homeowner is expected to be \$161.39 per year.

EXAMPLE 4

Using a Probability Distribution in a Real-World Context

The National Thoroughbred Racing Association publishes wagering information for people interested in betting on a horse race. The association reports that a \$2.00 winning ticket on an entry with 10 to 1 odds will pay \$22.00 to the winner. (**Source:** www.ntra.com.) A gambler repeatedly wagers on entries with 10 to 1 odds. In the long run, what is her expected net return per ticket?

SOLUTION Let $X =$ the net return. If the gambler wins, her net gain is \$20, since the ticket cost \$2. If she loses, her net loss is \$2. The probability distribution of X is given by

x	−2	20
$P(X = x)$	$\frac{10}{11}$	$\frac{1}{11}$

since the odds of losing are 10 to 1. The expected value is given by

$$E(X) = -2\left(\frac{10}{11}\right) + 20\left(\frac{1}{11}\right)$$
$$= 0$$

In the long run, she should expect to earn \$0 per entry.

A game with an expected value of \$0 is referred to as a **fair game.** The payoff amount for the wager in Example 4 is considered to be fair, since the expected value of the net return was \$0. If the value had been negative, the game would have been in favor of the racetrack. If the value had been positive, the game would have been in favor of the person placing the wager.

EXAMPLE 5

Using a Probability Distribution to Make Informed Decisions

A newlywed couple debating who should do the dishes decides to settle the dispute by rolling a pair of dice. He proposes that if none of the face values are odd, he will do the dishes. Otherwise, she must do them. Should she accept his proposal?

SOLUTION We must determine who is favored by the game. We define the random variable X to be the number of dice with an odd face value. The random variable may take on the value of 0, 1, or 2. He wins if $X = 0$, and she wins if $X = 1$ or $X = 2$. Since each die has three odd faces and three even faces, there is a 50 percent probability that a die face will show an odd number and a 50 percent probability that the die face will show an even number. We have

$$P(\text{both odd}) = (0.5)(0.5) = 0.25$$
$$P(\text{first odd, second even}) = (0.5)(0.5) = 0.25$$
$$P(\text{first even, second odd}) = (0.5)(0.5) = 0.25$$
$$P(\text{both even}) = (0.5)(0.5) = 0.25$$

Therefore,

x	0	1	2
$P(X = x)$	0.25	0.50	0.25

We can see from the table that there is a 75 percent probability that she will win. The expected value of the game is

$$E(X) = 0(0.25) + 1(0.5) + 2(0.25)$$
$$= 1$$

She should readily accept his offer, since she is likely to win the game and thus not have to do the dishes.

EXAMPLE 6

Using a Probability Distribution to Make Informed Decisions

In February 2004, the collision and comprehensive loss insurance coverage offered by Progressive Max Insurance Company on a 1990 Toyota pickup truck (driven by a 35-year-old male with an accident-free history) was \$105 per six-month period. (**Source:** www.progressive.com.) This quote required a \$250 deductible. The estimated value of the vehicle was \$2000. Assuming that there is a 0.2 percent probability that the vehicle will sustain a collision or a comprehensive loss during the six-month period, determine whether or not the driver should carry collision and comprehensive loss insurance coverage.

SOLUTION If an insured driver is in a crash, his total loss will include the six-month premium and the \$250 deductible (\$105 + \$250 = \$365). Using this result together with the additional data given in the problem, we construct Table 8.13.

TABLE 8.13

	Insurance Coverage	No Insurance Coverage
Collision or comprehensive loss ($p = 0.002$)	−\$365	−\$2,000
No collision or comprehensive loss ($p = 0.998$)	−\$105	\$0

We will first determine the expected loss for a driver with insurance coverage.

$$\begin{aligned} E(X) &= 0.002(-365) + 0.998(-105) \\ &= -105.52 \end{aligned}$$

The expected six-month loss for a driver with comprehensive and collision coverage is \$105.52.

We will next determine the expected loss for a driver without insurance coverage.

$$\begin{aligned} E(X) &= 0.002(-2000) + 0.998(0) \\ &= -4 \end{aligned}$$

The expected six-month loss for the driver without collision and comprehensive coverage is \$4.00. Since \$4.00 < \$105.52, the driver may want to decline collision and comprehensive loss coverage, since the expected loss is lower when collision and comprehensive loss coverage is not carried.

8.4 Summary

In this section you learned how to calculate the probability distribution of a random variable, draw a histogram, and determine the expected value of a random variable. You used these concepts in a variety of consumer decision-making applications.

8.4 Exercises

In Exercises 1–10, determine the probability distribution for the random variable X and draw its associated histogram.

1. Two coins are tossed.

$X =$ the number of heads shown

2. Five cards are pulled from a deck of cards containing 13 hearts, 13 clubs, 13 diamonds, and 13 spades.

$X =$ the number of cards that are hearts

3. Five cards are pulled from a deck of cards containing 4 kings and 48 other cards.

$X =$ the number of cards that are kings

4. Four marbles are randomly selected from a bucket of marbles containing five red marbles, three blue marbles, and two green marbles.

$X =$ the number of green marbles

5. A sample of 6 batteries is pulled from an order of 1000 batteries with a known defect rate of 1.3 percent.

$X =$ the number of defective batteries in the sample

6. Three six-sided dice are rolled.

$X =$ the number of dice showing an even number

7. A sample of 10 law school applicants is randomly selected from a pool of 500 applicants. The law school has a known acceptance rate of 65 percent.

$X =$ the number of applicants in the sample who were accepted

8. Three people are playing the game Rock-Paper-Scissors. (At each round of the game, each person indicates either Rock, Paper, or Scissors.)

$X =$ the number of Rocks

9. Four marbles are randomly selected from a bucket of marbles containing three blue marbles, three yellow marbles, two red marbles, and one green marble.

$X =$ the number of red marbles

10. Five people are randomly selected from a group containing ten convicts and four police officers.

$X =$ the difference between the number of police officers and convicts

In Exercises 11–23, apply the concepts of probability distribution and expected value, as appropriate.

11. Raffle Tickets A campus club sells 250 raffle tickets at a price of \$2 each. The raffle offers one \$50 prize, two \$25 prizes, and five \$10 prizes. What is the expected value of a raffle ticket?

12. Raffle Tickets A campus club sells 500 raffle tickets at a price of \$2 each. The raffle offers one \$100 prize, two \$50 prizes, and five \$25 prizes. What is the expected value of a raffle ticket?

13. Income Tax Returns Based on data from 1988 through 1998, the probability of an income tax return being examined by an Internal Revenue Service tax auditor is estimated to be 0.44 percent. Based on the data from the same time period, the average tax penalty for audited income tax returns was \$2707. (**Source:** *Statistical Abstract of the United States, 2001*, Table 471, p. 313.)

For an additional \$25 fee, an accountant offers to pay any tax penalties of his clients who are audited by the IRS. Would you recommend the audit-guarantee service? Justify your answer.

14. Income Tax Returns An accountant determines that 0.20 percent of the income tax returns she processes are audited by the IRS. The average tax penalty for her audited returns is \$1240. She guarantees her clients that she will pay any tax penalties on audited returns. How much must she charge for the guarantee service in order to break even?

15. Test-Taking Strategy Each question on a multiple-choice test has four possible choices. A correct answer earns 1 point, an incorrect answer loses 0.5 point, and a blank answer earns 0 points. If the student can't eliminate any of the choices for a particular question, should he guess? Explain.

16. Test-Taking Strategy Each question on a multiple-choice test has five possible choices. A correct answer earns 1 point, an incorrect answer loses 0.25 point, and a blank answer earns 0 points. If a student can eliminate one of the incorrect choices on a question and then guesses from among the remaining choices, what is the expected value of the points earned on that question?

17. Test-Taking Strategy Each question on a multiple-choice test has four possible choices. A correct answer earns 1 point, an incorrect answer loses 0.5 point, and a blank answer earns 0 points. A student determines that she can eliminate one of the incorrect choices on a question. If she guesses from among the remaining three choices, what is the expected value of the points earned on that question?

18. Projected Earnings A student works two part-time jobs. She typically works 30 hours per month as a waitress, earning an average of \$8.00 per hour, and 50 hours per month as a tutor, earning \$6.50 per hour. What are her expected average earnings per hour during a given month?

19. Auto Insurance Policy Choices In 2004, United Services Automobile Association charged the author \$89.85 to insure his minivan against collision and comprehensive loss for a period of six months. (**Source:** USAA insurance policy.) The author estimated the value of the minivan to be \$5300 and the probability of suffering a collision or comprehensive loss in the next six months to be 0.1 percent. Should the author continue to buy collision and comprehensive coverage on the minivan? Justify your answer.

20. Auto Insurance Policy Choices An automobile insurance policy's comprehensive and collision coverage costs \$620 for a six-month period. The value of the insured vehicle is estimated to be \$23,400. The insured driver estimates that there is a 1 percent probability that he will destroy his car in a collision in the next six months. Should he continue to buy collision and comprehensive loss coverage? Justify your answer.

21. Homeowners' Insurance In 2000, the average consumer expenditure on homeowners' insurance was \$508 per year. In that year, 7.72 percent of insured homeowners filed payable claims. The average payout was \$4168 per claim. (**Source:** www.iii.org.) Assuming that the consumer insurance expenditure, filing frequency, and payout amount remain constant for future years, determine the expected value of a homeowners' insurance policy and interpret the meaning of the result.

22. Lottery Winnings The Florida Mega Money lottery game sells tickets for \$1 each. The estimated probability of winning various cash prizes is shown in the following table.

Prize Amount	Probability of Winning
\$500,000	$\frac{1}{2,986,523}$
\$1,200	$\frac{1}{142,216}$
\$350	$\frac{1}{18,667}$
\$50	$\frac{1}{890}$
\$25	$\frac{1}{639}$
\$3	$\frac{1}{77}$
\$2	$\frac{1}{31}$

Source: www.flalottery.com.

What is the expected value of a Mega Money lottery ticket?

23. Lottery Winnings On February 28, 2004, the value of the Florida Lotto jackpot was estimated to be \$3 million. Lottery tickets cost \$1 each. The estimated probability of winning various cash prizes is shown in the following table.

Prize Amount	Probability of Winning
Jackpot	$\frac{1}{22,957,481}$
\$5,000	$\frac{1}{81,411}$
\$70	$\frac{1}{1,417}$
\$5	$\frac{1}{72}$

Source: www.flalottery.com.

What is the expected value of a Florida Lotto ticket when the jackpot is valued at \$3 million? \$5 million? \$10 million?

In Exercises 24–30, calculate the expected rate of change in the share price of the given company. Assume that the probability of an increase or decrease in the share price is equal to the percentage of time that the share price actually increased or decreased, respectively.

24. Investment Performance The share price of American Eagle Outfitters, Inc., varied between February 2 and March 3, 2004. The price increased on 71.4 percent of the trading days and decreased on 28.6 percent of the trading days. On days when the price increased, the average increase was \$0.45 per day. On days when the price decreased, the average change in price was −\$0.28 per day. Based on these data, what was the expected change in the share price on any day in the given time period?

25. **Investment Performance** The share price of Big Dog Holdings, Inc., varied between February 2 and March 3, 2004. The price increased on 47.6 percent of the trading days, remained the same 19.1 percent of the trading days, and decreased on 33.3 percent of the trading days. On days when the price increased, the average increase was \$0.16 per day. On days when the price decreased, the average change in price was −\$0.14 per day. Based on these data, what was the expected change in the share price on any day in the given time period?

26. **Investment Performance** The share price of Pepsico, Inc., varied between 1977 and 2002. The annual closing price increased during 68 percent of the years and decreased during 32 percent of the years. In years when the share price increased, the average annual increase was \$3.28 per year. In years when the share price decreased, the average annual change in price was −\$1.95 per year. Based on these data, what was the expected change in the share price in any year in the given time period?

27. **Investment Performance** The share price of Scholastic Corp. varied between 1992 and 2002. The annual closing price increased during 72.7 percent of the years and decreased during 27.3 percent of the years. In years when the share price increased, the average annual increase was \$8.81 per year. In years when the share price decreased, the average annual change in price was −\$11.50 per year. Based on these data, what was the expected change in the share price in any year in the given time period?

28. **Investment Performance** The share price of Gap, Inc., varied between February 2 and March 3, 2004. The price increased on 42.9 percent of the trading days and decreased on 57.1 percent of the trading days. On days when the price increased, the average increase was \$0.44 per day. On days when the price decreased, the average change in price was −\$0.15 per day. Based on these data, what was the expected change in the share price on any day in the given time period?

29. **Investment Performance** The share price of the Coca Cola Company varied between February 2 and March 3, 2004. The price increased on 33.3 percent of the trading days, remained constant on 4.8 percent of the trading days, and decreased on 61.9 percent of the trading days. On days when the price increased, the average increase was \$0.47 per day. On days when the price decreased, the average change in price was −\$0.29 per day. Based on these data, what was the expected rate of change in the share price on any day in the given time period?

30. **Investment Performance** The share price of the Coca Cola Company varied between 1965 and 2002. The annual closing price increased during 75 percent of the years and decreased during 25 percent of the years. In years when the share price increased, the average annual increase was \$2.53 per year. In years when the share price decreased, the average annual change in price was −\$2.50 per year. Based on these data, what was the expected change in the share price in any year in the given time period?

Exercises 31–35 are intended to challenge your understanding of probability distributions and expected values of random variables.

31. **Life Insurance Premiums** Term life insurance is used to insure a person's life for a fixed period of time. In February 2004, First Colony Life advertised a \$500,000 30-year term life insurance policy for a 36-year-old woman in excellent health. The \$440 annual premium guaranteed that if the woman died before February 2034, her designated beneficiaries would receive \$500,000. (**Source:** vitalterm.lifelinkcorp.com.)

The woman anticipates that there is a 1.5 percent probability she will die before her policy expires. Assuming that the total cost of the policy is $C = 440t$, where t is the number of years she pays on the policy, what is the expected value of the life insurance policy? (*Hint:* The expected value will be in terms of t.)

32. Refer to the expected value in Exercise 31. Determine for which whole-number values of t the expected value of the life insurance policy will be positive and interpret the meaning of the result.

33. **Dice Roll** A fair six-sided die has six equally likely outcomes: 1, 2, 3, 4, 5, 6. A fair four-sided die has four equally likely outcomes: 1, 2, 3, 4. The two dice are rolled. A random variable X is defined to be the larger of the face values shown on the two dice. Determine the expected value of X.

34. **Insurance Rates** An insurance company estimates that the probability that a client will file a legitimate claim in a given year is 4 percent. The company predicts that the average payout for each legitimate claim will be \$2600. In order to be profitable, the company must earn an average of \$200 per client per year, after deducting the amount of money spent on claims. How much should the company charge policyholders for the insurance?

35. **Insurance Rates** An insurance company estimates that the probability that a client will file a legitimate claim in a given year is 3 percent. In order to be profitable, the company must earn an average of \$200 per client per year, after deducting the amount of money spent on claims. The company is currently charging each client \$350 per year. In order to reach its profitability goal, what level does it need to keep its average payout below?

8.5 Measures of Central Tendency and Dispersion

- Calculate the mean, median, mode, and standard deviation of a data set

GETTING STARTED Every year, the U.S. government makes direct payments to the agriculture industry. The amount of money paid to the industry varies widely from year to year. By using a variety of statistical measures, we can anticipate future agricultural funding requirements. In this section, we will investigate an area of mathematics referred to as *statistics*. We will introduce the statistical measures of *mean*, *median*, *mode*, and *standard deviation* and apply them to real-life scenarios such as the agricultural payments.

Measures of Central Tendency

Table 8.14 shows the amount of money the government paid directly to the agriculture industry between 1990 and 2000.

TABLE 8.14 Direct Government Payments to Agriculture Industry

Year (t)	Payments (P) (billions of dollars)	Year (t)	Payments (P) (billions of dollars)
1990	9.3	1996	7.3
1991	8.2	1997	7.5
1992	9.2	1998	12.4
1993	13.4	1999	21.5
1994	7.9	2000	22.9
1995	7.3		

Source: *Statistical Abstract of the United States, 2001*, Table 807, p. 527.

What is the typical amount of money that the government pays to the agriculture industry in a given year? We can answer this question using each of the three common **measures of central tendency** that are used in statistical analysis: mean, median, and mode. The mean finds the "typical" payment by dividing the sum of all payments by the number of payments. Many people refer to this as the *average* payment.

MEAN

The *mean*, $\bar{x}$, (read "x bar") of a group of n numbers $x_1, x_2, \ldots, x_n$ is given by

$$\bar{x} = \frac{x_1 + x_2 + \cdots + x_n}{n}$$

To calculate the mean of the payments, we add up the payment amounts and divide by the number of years.

$$\begin{aligned}\bar{x} &= \frac{9.3 + 8.2 + 9.2 + 13.4 + 7.9 + 7.3 + 7.3 + 7.5 + 12.4 + 21.5 + 22.9}{11} \\ &= \frac{126.9}{11} \\ &\approx 11.5 \text{ billion dollars per year}\end{aligned}$$

Over the 11-year period, the mean (or average) annual payment amount was \$11.5 billion.

The median finds the "typical" payment by giving the middle value of all payments listed in order from smallest to largest.

MEDIAN

The *median* of a group of n numbers $x_1, x_2, \ldots, x_n$ is found by listing the numbers in numerical order and locating the value(s) in the center of the list. If n is odd, there will be a single value at the center of the list. If n is even, there will be two values at the center of the list, and the median is the mean of these two values.

To calculate the median of the payments, we must first list the payment amounts in numerical order.

7.3, 7.3, 7.5, 7.9, 8.2, 9.2, 9.3, 12.4, 13.4, 21.5, 22.9

↑ Median (9.2)

Since there are an odd number of values in the list, the centermost value will be the median. The median amount of money spent by the government annually on direct agricultural payments is \$9.2 billion.

The mode finds the "typical" payment by choosing the payment amount that occurs most often.

MODE

The *mode* of a group of n numbers $x_1, x_2, \ldots, x_n$ is the value that is repeated most often. If all values are repeated the same number of times, there is no mode.

To calculate the mode of the payments, we identify the payment amount that is repeated most often. All of the payment amounts occur exactly once with the exception of 7.3, which occurs twice. The mode of the payments is $7.3 billion.

Which of the measures of central tendency is the most useful? It depends. For the agricultural payment data, we have

$$\text{Mean} = 11.5$$
$$\text{Median} = 9.2$$
$$\text{Mode} = 7.3$$

Although the payment amount has ranged between $7.3 and $22.9 billion, the average amount of the annual payment is $11.5 billion. If we expect that future payments will oscillate within this range, the mean may be a good predictor for future payment amounts.

Observe that the payment amounts in 1999 and 2000 are substantially larger than the payments in all of the other years. If we believe that this dramatic increase in payment size is an anomaly and not indicative of a future trend, then the median may be the best measure to use. Unlike the mean, the median is not dramatically affected by a data value that is substantially different from the other values in the data set.

The mode for the agricultural payments was 7.3. This value occurred twice, while all other values occurred only once. If this value had appeared substantially more often (say 50 percent of all data values were 7.3), we might have concluded that the mode would be the best predictor of future payment amounts. However, since 7.3 occurred only one more time than all of the other data values, we do not expect it to be a good predictor of future payment amounts.

EXAMPLE 1

Calculating and Interpreting Measures of Central Tendency

The amount of money spent on aerobic shoes in the United States between 1991 and 2000 is shown in Table 8.15.

TABLE 8.15 Spending on Aerobic Shoes

Year (t)	Sales (S) (millions of dollars)
1991	600
1992	590
1993	500
1994	356
1995	372
1996	401
1997	380
1998	334
1999	275
2000	272

Source: *Statistical Abstract of the United States, 2001*, Table 1251, p. 766.

Calculate the mean, median, and mode for the annual amount of money spent on aerobic shoe sales and interpret the meaning of each result.

SOLUTION

$$\bar{x} = \frac{600 + 590 + 500 + 356 + 372 + 401 + 380 + 334 + 275 + 272}{10}$$

$$= \frac{4080}{10}$$

$$= 408$$

From 1991 through 2000, the mean (average) annual amount of money spent on aerobic shoes was $408 million.

To find the median, we sequence the sales values in numerical order and seek to find the centermost value.

$$272, 275, 334, 356, \underset{\uparrow}{372}, \underset{\uparrow}{380}, 401, 500, 590, 600$$

Since there is an even number of terms in the sequence, there are two values occupying the centermost positions. To determine the median, we find the mean of these two values.

$$\text{Median} = \frac{372 + 380}{2}$$

$$= 376$$

In five of the ten years, the amount of money spent on aerobic shoe sales was less than $376 million. In the remaining five years, the amount of money spent was more than $376 million.

Since all of the data values occur with the same frequency, there is no mode.

EXAMPLE 2 Calculating Measures of Central Tendency

In 2003–2004, the Scottsdale Community College Mathematics Department had 15 full-time faculty members. For each member of the department, we recorded the number of years that the member had been employed by the college. We summarized our results in the frequency distribution in Table 8.16.

TABLE 8.16

Years Employed (x)	2	4	7	8	9	14	15	20	21	29
Frequency (number of math faculty)	3	2	2	1	1	1	2	1	1	1

Source: Scottsdale Community College General Catalog and Student Handbook.

From the frequency distribution, we see that three members had been employed for two years, two members had been employed for four years, two members had been employed for seven years, and so on.

Calculate the mean, median, and mode for the number of years a department member had been employed by the college and interpret the meaning of each result. Then draw a histogram of the frequency distribution, indicating the location of the median and mean.

SOLUTION Calculating the mean from a frequency distribution is somewhat different from calculating the mean from a data set. The frequency value tells us the number of times we must write the corresponding value of x (e.g., since the frequency of 2 is 3, we write the 2 three times in the numerator sum).

$$\bar{x} = \frac{(2 + 2 + 2) + (4 + 4) + (7 + 7) + 8 + 9 + 14 + (15 + 15) + 20 + 21 + 29}{15}$$

$$= \frac{159}{15}$$

$$= 10.6 \text{ years}$$

The mean amount of time a member of the department had been employed by the college was 10.6 years. Does this signify that half of the members had worked fewer than 10.6 years and the other half of the members had worked more? No. We can see from the frequency distribution that more than half (9 of the 15 members) had worked fewer than 10.6 years.

The outlying value of 29 is largely responsible for increasing the size of the mean. The value 29 is 8 years away from the next closest data value (21), while all of the other values in the data set are within 3 units of the next closest data value. If the value 29 had been removed from the data set, the mean would have been 9.3. (This is 1.3 units lower than the mean of the entire data set.) In contrast, if one of the values of 7 had been removed from the data set, the mean would have been 10.9. (This is 0.3 unit higher than the mean of the entire data set.)

If we want to split the math faculty into two similarly sized groups based upon seniority (the number of years at the college), we need to use the median. Listing the data values in numerical order yields

$$2, 2, 2, 4, 4, 7, 7, 8, 9, 14, 15, 15, 20, 21, 29$$

The eighth term is the middle value and is equal to 8. That is, the median is 8. Seven of the faculty members had worked at the college for fewer than eight years, and seven of the faculty members had worked at the college for more than eight years. To create two similarly sized groups, we could add the one math faculty member who had worked at the college for eight years to either of the groups.

The mode of the number of years math faculty had worked at the college is 2, since this is the value with the largest frequency in our frequency distribution. If the math faculty were broken into groups based upon the year they were hired, the largest group would be those who had worked at the college for two years.

A histogram of the frequency distribution is shown in Figure 8.20.

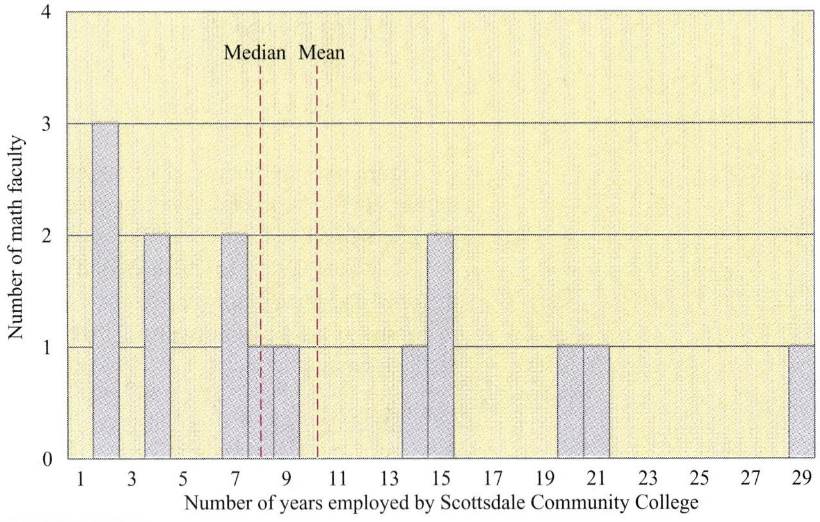

FIGURE 8.20

Of the three measures of central tendency, the mean is used the most often. It is easy to compute, and it tends to be a very reliable measure. That is, repeated samples taken from the same population are likely to have similar means. However, the mean is easily influenced by extreme values, as was shown in Example 2.

The median, on the other hand, is not significantly influenced by extreme values. However, the median is often difficult to calculate for large data sets, since calculating the median requires sequencing the values in numerical order. The sorting function of a spreadsheet is often used to sequence the values of a large data set.

The mode is easy to find but may not yield a unique value for each data set. For example, the data set $\{2, 2, 3, 3, 4, 4, 5, 6\}$ has mode 2, 3, and 4! The values 2, 3, and 4 each occur twice, while the other values in the list occur only once.

Measures of Dispersion

While measures of central tendency identify the "middle" or "average" of a data set, **measures of dispersion** identify how "spread out" the data items are. For example, consider the sets of data in Table 8.17.

TABLE 8.17

List A	List B
4	1
5	1
6	1
6	1
7	7
7	7
8	12
8	12
9	14
10	14
Mean = 7	Mean = 7
Median = 7	Median = 7

Both data sets have the same mean and median. However, the data sets have very different frequency distributions, as shown in the histograms in Figures 8.21 and 8.22).

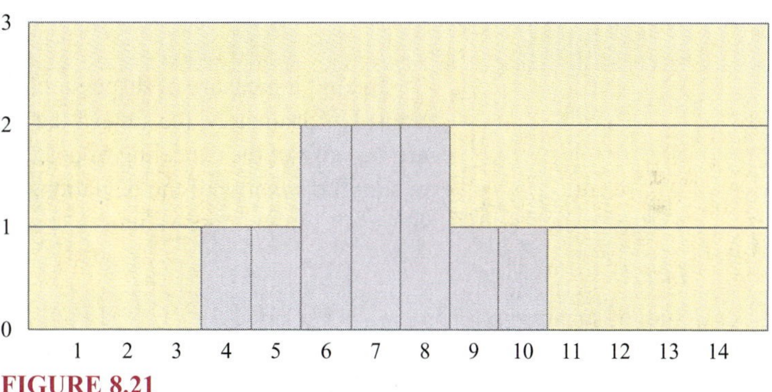

FIGURE 8.21

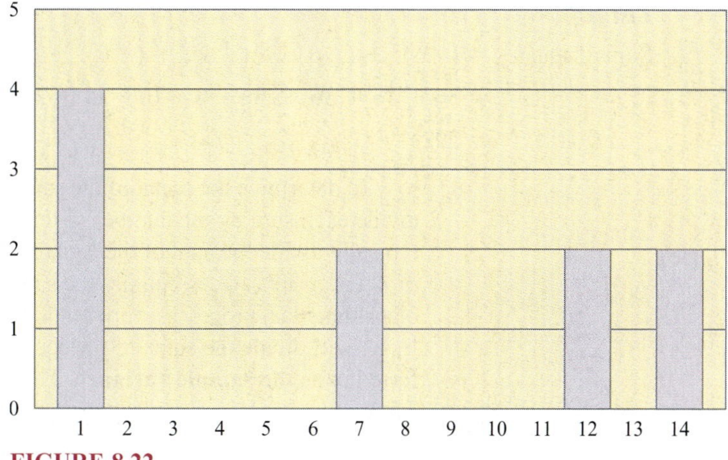

FIGURE 8.22

The data values from List A are clustered close to the mean. On the other hand, most of the data values from List B are far away from the mean. We can quantify the *dispersion* of the data by looking at the *deviations* from the mean. A **deviation from the mean** is the difference found by subtracting the mean from each number in the sample (Table 8.18).

TABLE 8.18

List A	Deviation from the Mean	List B	Deviation from the Mean
4	−3	1	−6
5	−2	1	−6
6	−1	1	−6
6	−1	1	−6
7	0	7	0
7	0	7	0
8	1	12	5
8	1	12	5
9	2	14	7
10	3	14	7
Mean = 7		Mean = 7	

To find a measure of dispersion, we might try finding the mean of the deviations. Unfortunately, the sum of the deviations from the mean will always equal zero because of the canceling of positive and negative values. To offset this effect, we sum the *squares* of the deviations. The data set that is more widely dispersed will result in the larger value.

List A:

$$\begin{aligned}\text{Sum of squares} &= (-3)^2 + (-2)^2 + (-1)^2 + (-1)^2 + (0)^2 + (0)^2 + (1)^2 + (1)^2 + (2)^2 + (3)^2 \\ &= 9 + 4 + 1 + 1 + 0 + 0 + 1 + 1 + 4 + 9 \\ &= 30\end{aligned}$$

List B:

$$\begin{aligned}\text{Sum of squares} &= (-6)^2 + (-6)^2 + (-6)^2 + (-6)^2 + (0)^2 + (0)^2 + (5)^2 + (5)^2 + (7)^2 + (7)^2 \\ &= 36 + 36 + 36 + 36 + 0 + 0 + 25 + 25 + 49 + 49 \\ &= 292\end{aligned}$$

To determine the mean of the sum of the squares of the deviations, we would divide each sum by the number of terms in the data set, n. However, statisticians typically divide the sum of the squares by $n - 1$ instead of by n. For large values of n, the results are essentially the same. (Using $n - 1$ instead of n makes some other statistical processes work out nicely.) Since $n = 10$ for both List A and List B, we will divide the sums of the squares of the deviations by 9. The result is referred to as the **sample variance.** The symbol s^2 is used to represent the sample variance.

$$\text{List A sample variance} = \frac{30}{9}$$

$$s^2 = 3.333$$

$$\text{List B sample variance} = \frac{292}{9}$$

$$s^2 = 32.44$$

It is often convenient to write the sample variance using *sigma notation*. The Greek letter sigma (Σ) is used to represent the concept of summing a number of values. The expression

$$(x_1 - \bar{x})^2 + (x_2 - \bar{x})^2 + \cdots + (x_n - \bar{x})^2$$

may be written as $\sum_{i=1}^{n}(x_i - \bar{x})^2$. The variable i is an index value that moves through whole-number values beginning at 1 and continuing until it reaches n. Each time the index value changes, another term is added to the summation. For example,

$$\begin{aligned}
\sum_{i=1}^{n}(x_i - \bar{x})^2 &= (x_1 - \bar{x})^2 + \sum_{i=2}^{n}(x_i - \bar{x})^2 \\
&= (x_1 - \bar{x})^2 + (x_2 - \bar{x})^2 + \sum_{i=3}^{n}(x_i - \bar{x})^2 \\
&= (x_1 - \bar{x})^2 + (x_2 - \bar{x})^2 + (x_3 - \bar{x})^2 + \sum_{i=4}^{n}(x_i - \bar{x})^2 \\
&\vdots \\
&= (x_1 - \bar{x})^2 + (x_2 - \bar{x})^2 + \cdots + (x_{n-1} - \bar{x})^2 + \sum_{i=n}^{n}(x_i - \bar{x})^2 \\
&= (x_1 - \bar{x})^2 + (x_2 - \bar{x})^2 + \cdots + (x_{n-1} - \bar{x})^2 + (x_n - \bar{x})^2
\end{aligned}$$

Summation notation may feel cumbersome at first; however, as you practice, you will become more comfortable with it. The sample variance is typically written using summation notation.

SAMPLE VARIANCE

The *sample variance*, s^2, of a group of n numbers $x_1, x_2, \ldots, x_n$ with mean $\bar{x}$ is given by

$$s^2 = \frac{\sum_{i=1}^{n}(x_i - \bar{x})^2}{n - 1}$$

The sample variance may be calculated more easily by using the equivalent form

$$s^2 = \frac{\left[\sum_{i=1}^{n}(x_i)^2\right] - n(\bar{x})^2}{n - 1}$$

Converting from $s^2 = \frac{\sum_{i=1}^{n}(x_i - \bar{x})^2}{n-1}$ to $s^2 = \frac{\left[\sum_{i=1}^{n}(x_i)^2\right] - n(\bar{x})^2}{n-1}$ is not immediately obvious. We will demonstrate the equivalency of the two expressions using $n = 2$. A similar approach may be used to show that the two expressions are equivalent for any value of n. For $n = 2$, the sample variance equation may be written as follows:

$$
\begin{aligned}
s^2 &= \frac{\sum_{i=1}^{2}(x_i - \bar{x})^2}{2-1} \\
&= \frac{(x_1 - \bar{x})^2 + (x_2 - \bar{x})^2}{1} && \text{Convert from sigma notation} \\
&= (x_1 - \bar{x})^2 + (x_2 - \bar{x})^2 \\
&= (x_1 - \bar{x})(x_1 - \bar{x}) + (x_2 - \bar{x})(x_2 - \bar{x}) \\
&= [(x_1)^2 - 2x_1\bar{x} + (\bar{x})^2] + [(x_2)^2 - 2x_2\bar{x} + (\bar{x})^2] && \text{Calculate products} \\
&= (x_1)^2 + (x_2)^2 - 2x_1\bar{x} - 2x_2\bar{x} + 2(\bar{x})^2 && \text{Reorder the terms} \\
&= (x_1)^2 + (x_2)^2 - 2\bar{x}(x_1 + x_2) + 2(\bar{x})^2 && \text{Factor out } 2\bar{x} \\
&= (x_1)^2 + (x_2)^2 - 2\bar{x}(2\bar{x}) + 2(\bar{x})^2 && \text{Since } \bar{x} = \frac{x_1 + x_2}{2} \Rightarrow x_1 + x_2 = 2\bar{x} \\
&= \left[\sum_{i=1}^{2}(x_i)^2\right] - 4(\bar{x})^2 + 2(\bar{x})^2 && \text{Rewrite with sigma notation} \\
&= \left[\sum_{i=1}^{2}(x_i)^2\right] - 2(\bar{x})^2 && \text{Group like terms} \\
&= \frac{\left[\sum_{i=1}^{2}(x_i)^2\right] - 2(\bar{x})^2}{2-1}
\end{aligned}
$$

This alternative form of writing the sample variance is a useful shortcut for calculating the sample variance.

Since the sample variance was derived from the square of the deviations from the mean, we must calculate the square root of the sample variance to return to the same units as the data. The result is called the **sample standard deviation.**

SAMPLE STANDARD DEVIATION

The *sample standard deviation*, s, of a group of n numbers $x_1, x_2, \ldots, x_n$ with mean $\bar{x}$ is given by

$$s = \sqrt{\frac{\left[\sum_{i=1}^{n}(x_i)^2\right] - n(\bar{x})^2}{n-1}}$$

The sample standard deviation is the most common measure of dispersion. If the data points tend to be bunched up close to the mean, the standard deviation will be relatively small. If the data points are spread out from the mean, the standard deviation will be relatively large.

Earlier in our discussion, we derived the following results for List A and List B:

$$\text{List A sample variance} = \frac{30}{9}$$
$$s^2 = 3.333$$
$$\text{List B sample variance} = \frac{292}{9}$$
$$s^2 = 32.44$$

It follows that

$$\text{List A sample standard deviation} = \sqrt{\frac{30}{9}}$$
$$s = 1.826$$
$$\text{List B sample standard deviation} = \sqrt{\frac{292}{9}}$$
$$s = 5.696$$

Note that the more widely dispersed data set has a much larger standard deviation.

The Russian mathematician Pafnuty L. Chebyshev (1821–1894) discovered a fascinating relationship between the sample standard deviation of a data set and the elements of the data set. This relationship is referred to as Chebyshev's rule.

CHEBYSHEV'S RULE

For any set of data, at least $\frac{k^2 - 1}{k^2}$ data points fall within k standard deviations of the mean. That is, at least $\frac{k^2 - 1}{k^2}$ data points lie in the interval $[\bar{x} - ks, \bar{x} + ks]$.

According to Chebyshev's Rule, at least $\frac{3}{4}$ of the data points of any data set fall within two sample standard deviations of the mean. For List A, we have

$$\bar{x} - 2s = 7 - 2(1.826) = 3.348 \quad \text{and} \quad \bar{x} + 2s = 7 + 2(1.826) = 10.652$$

So at least 75 percent $\left(\frac{3}{4}\right)$ of the data points of List A lie within the interval $[3.348, 10.652]$. In fact, all 10 of the data points of List A $\{4, 5, 6, 6, 7, 7, 8, 8, 9, 10\}$ lie within this interval.

For List B, we have

$$\bar{x} - 2s = 7 - 2(5.696) = -4.392 \quad \text{and} \quad \bar{x} + 2s = 7 + 2(5.696) = 18.392$$

So at least 75 percent $\left(\frac{3}{4}\right)$ of the data points of List B lie within the interval $[-4.392, 18.392]$. In fact, all 10 of the data points of List B $\{1, 1, 1, 1, 7, 7, 12, 12, 14, 14\}$ lie within this interval.

EXAMPLE 3 Finding the Mean and Standard Deviation

From 1965 through 1999, there were 3753 successful space launches worldwide (Table 8.19). (A successful space launch entails attainment of Earth orbit or Earth escape.)

TABLE 8.19 Worldwide Space Launches

Time Period (t)	Successful Launches (N)
1965–1969	586
1970–1974	555
1975–1979	607
1980–1984	605
1985–1989	550
1990–1994	466
1995–1999	384

Source: *Statistical Abstract of the United States, 2001*, Table 793, p. 519.

Determine the mean and sample standard deviation for the number of successful launches in a five-year period. How many of the data points lie within one sample standard deviation of the mean?

SOLUTION

$$\bar{x} = \frac{586 + 555 + 607 + 605 + 550 + 466 + 384}{7}$$

$$= \frac{3753}{7}$$

$$\approx 536$$

An average of 536 successful space launches occurred during each five-year period.

To calculate the sample standard deviation, we will first calculate the component parts of the equation and then substitute in the simplified results. We first determine $\sum_{i=1}^{n}(x_i)^2$.

$$\sum_{i=1}^{7}(x_i)^2 = (586)^2 + (555)^2 + (607)^2 + (605)^2 + (550)^2 + (466)^2 + (384)^2$$

$$= 2{,}053{,}007$$

We next find $n(\bar{x})^2$.

$$7(\bar{x})^2 = 7\left(\frac{3753}{7}\right)^2$$

$$\approx 2{,}012{,}144$$

Substituting these results into the sample standard deviation equation yields

$$\begin{aligned} s &= \sqrt{\frac{\left[\sum_{i=1}^{7}(x_i)^2\right] - 7(\bar{x})^2}{7-1}} \\ &= \sqrt{\frac{2{,}053{,}007 - 2{,}012{,}144}{6}} \\ &= \sqrt{\frac{40{,}863}{6}} \\ &\approx 83 \end{aligned}$$

The sample standard deviation (rounded to the nearest whole number) is equal to 83 successful launches.

To determine the interval of values within one standard deviation of the mean, we calculate the following:

$$\begin{aligned} \bar{x} - s &= 536 - 83 \\ &= 453 \end{aligned} \qquad \text{and} \qquad \begin{aligned} \bar{x} + s &= 536 + 83 \\ &= 619 \end{aligned}$$

All data points within one standard deviation of the mean lie in the interval [453, 619]. Six of the seven points lie within this interval. Only the value 384 is not within one standard deviation of the mean.

The TI-83 Plus quickly and easily calculates the mean, median, and standard deviation of a data set, as shown in the following Technology Tip.

TECHNOLOGY TIP

Calculating the Mean, Median, and Standard Deviation

1. Press STAT ENTER to open the Statistics Editor. Enter the values from the list in L1.

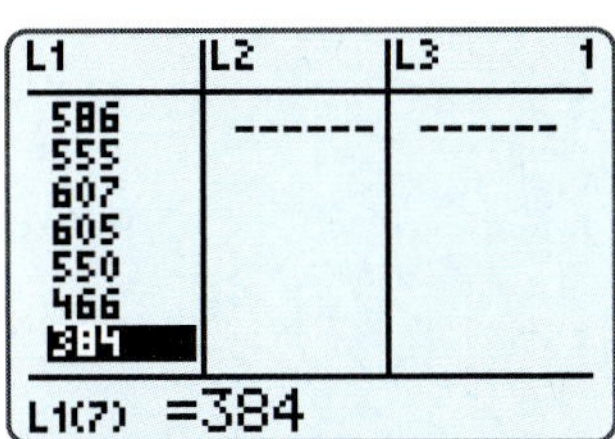

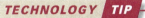

2. Press STAT and use the blue arrows to move to the CALC menu.

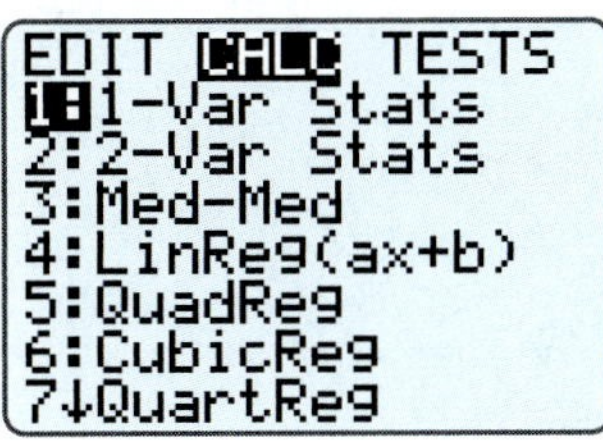

(Continued)

3. Press ENTER to place the command `1-Var Stats` on the home screen and press ENTER again to calculate statistics for the data set. The mean of the data set is $\bar{x} \approx 536$. The sample standard deviation (given by $Sx = 82.52\ldots$) is $s \approx 83$.

```
1-Var Stats
 x̄=536.1428571
 Σx=3753
 Σx²=2053007
 Sx=82.5256093
 σx=76.40386784
↓n=7
```

4. To find the median, press the blue down arrow to display the rest of the statistical results. The median is shown as `Med=555`. The median is 555.

```
1-Var Stats
↑n=7
minX=384
Q1=466
Med=555
Q3=605
maxX=607
```

The technology tip also shows the **population standard deviation** σ. The Greek letter σ is read "sigma." It is the lowercase form of Σ. The population standard deviation is the square root of the **population variance.** The population variance is given by

$$\sigma^2 = \frac{\left[\sum_{i=1}^{n}(x_i)^2\right] - n(\bar{x})^2}{n}$$

and the population standard deviation is given by

$$\sigma = \sqrt{\frac{\left[\sum_{i=1}^{n}(x_i)^2\right] - n(\bar{x})^2}{n}}$$

These equations are nearly identical to those used to calculate the sample variance and sample standard deviation. The only difference is that the denominator of the rational expression is n instead of $n - 1$. We will use the population standard deviation extensively as we discuss normal distributions in the next section.

EXAMPLE 4

Finding the Mean and Standard Deviation

On September 25, 2004, travelocity.com advertised a variety of fares for a one-way flight on December 25, 2004, from Phoenix, Arizona, to Orlando, Florida, as shown by the frequency distribution in Table 8.20.

TABLE 8.20

Fare	Frequency
\$152	5
\$159	1
\$184	1
\$197	1
\$210	2
\$242	3
\$254	1
\$284	1
\$301	2
\$307	1
\$314	1
\$322	1

Source: www.travelocity.com.

Calculate the mean and the sample standard deviation for the data set. Then determine the number of data values within two standard deviations of the mean.

SOLUTION We can determine the total number of flights by summing the frequencies.

$$\text{Total number of flights} = 5 + 1 + 1 + 1 + 2 + 3 + 1 + 1 + 2 + 1 + 1 + 1$$
$$= 20$$

Because of the relatively large size of the data set, we will use technology to calculate the mean and standard deviation, as demonstrated in the previous Technology Tip (see Figure 8.23).

```
1-Var Stats
 x̄=226.45
 Σx=4529
 Σx²=1100261
 Sx=62.68927131
 σx=61.1019435
↓n=20
```

FIGURE 8.23

We see that $\bar{x} = 226.45$. The mean (average) ticket price is \$226.45. The sample standard deviation is \$62.69. The equation $\Sigma x = 4529$ means that the sum of the 20 fares is \$4529. The equation $n = 20$ indicates that 20 data points have been entered into the calculator. (It is good practice to count the number of data points manually as well to ensure that you didn't inadvertently omit entering one of the data points.)

We are asked to find the number of data points that are within two standard deviations of the mean.

$$\bar{x} - 2s = 226.45 - 2(62.69) = 226.45 - 125.38 = 101.07$$

$$\bar{x} + 2s = 226.45 + 2(62.69) = 226.45 + 125.38 = 351.83$$

Since all 20 fares are between $101.07 and $351.83, all of the fares lie within two standard deviations of the mean.

8.5 Summary

In this section, you learned how to use the statistical measures of mean, median, mode, and standard deviation in analyzing a variety of real-life data sets. You also discovered that Chebyshev's rule is useful for estimating the dispersion of the points in any data set.

8.5 Exercises

In Exercises 1–15, calculate the mean, median, and mode for each data set and interpret the meaning of each result. Create a frequency distribution for each data set. Then draw a histogram of the frequency distribution, indicating the location of the median and mean.

1. **University of Oklahoma Football** As of the end of the 2003 season, 16 football players had scored 30 or more touchdowns during their careers at the University of Oklahoma. The number of touchdowns scored was as follows: 57, 53, 51, 43, 41, 37, 37, 36, 35, 35, 34, 32, 32, 31, 30, and 30. (**Source:** www.soonerstats.com.)

2. **Space Shuttle Columbia** The space shuttle Columbia flew 30 missions between 1984 and 2001; however, the number of missions flown each year varied as shown in the following table.

Year	Number of Missions
1984	2
1985	4
1986	0
1987	0
1988	1
1989	2
1990	2
1991	2
1992	2
1993	2
1994	2
1995	2
1996	0
1997	2
1998	2
1999	2
2000	1
2001	2

Source: www.pao.ksc.nasa.gov.

The space shuttle Columbia and her crew were lost during entry on February 1, 2003, 16 minutes before the scheduled landing time.

3. **Yale University Football Ranking** In the early 1900s, Yale University was a football powerhouse. The following table shows how one ranking system rated the Yale team from 1900 to 1909.

Year	Ranking
1900	1
1901	4
1902	2
1903	3
1904	7
1905	2
1906	4
1907	6
1908	9
1909	4

Source: www.kiko13.com.

4. **Death Rates** Death rates in the United States between 1950 and 1995 have fluctuated as shown in the following table.

Year	Deaths (per 1000 people)
1950	9.6
1955	9.3
1960	9.5
1965	9.4
1970	9.5
1975	8.8
1980	8.8
1985	8.8
1990	8.6
1995	8.8

Source: *Statistical Abstract of the United States, 2001*, Table 68, p. 59.

5. **Births to Teenagers** Between 1990 and 1999, the percentage of all births that were to teenage mothers ranged from 12.3 percent to 13.1 percent, as shown in the following table.

Year	Births to Teenagers (percent of all births)
1990	12.8
1991	12.9
1992	12.7
1993	12.8
1994	13.1
1995	13.1
1996	12.9
1997	12.7
1998	12.5
1999	12.3

Source: *Statistical Abstract of the United States, 2001*, Table 76, p. 63.

6. **Births to Unmarried Women** Between 1990 and 1999, the percentage of all births that were to unmarried women ranged from 26.6 percent to 33.0 percent, as shown in the following table.

Year	Births to Unmarried Mothers (percent of all births)
1990	26.6
1991	28.0
1992	30.1
1993	31.0
1994	32.6
1995	32.2
1996	32.4
1997	32.4
1998	32.8
1999	33.0

Source: *Statistical Abstract of the United States, 2001*, Table 76, p. 63.

7. **Tropical Storms and Hurricanes** Between 1980 and 2000, the number of tropical storms and hurricanes in the North Atlantic ranged between 4 and 19 per year, as shown in the following table.

Year	Tropical Storms and Hurricanes
1980	11
1981	11
1982	5
1983	4
1984	12
1985	11
1986	6
1987	7
1988	12
1989	11
1990	14
1991	8
1992	7
1993	8
1994	7
1995	19
1996	13
1997	7
1998	14
1999	12
2000	15

Source: *Statistical Abstract of the United States, 2001*, Table 368, p. 221.

8. **U.S. Marines** Between 1980 and 2000, the number of U.S. Marines ranged between 173 and 200 thousand, as shown in the following table.

Year	U.S. Marines (thousands)
1980	188
1981	191
1982	192
1983	194
1984	196
1985	198
1986	199
1987	200
1988	197
1989	197
1990	197
1991	194
1992	185
1993	178
1994	174
1995	175
1996	175
1997	174
1998	173
1999	173
2000	173

Source: *Statistical Abstract of the United States, 2001*, Table 500, p. 329.

9. **Homes of More than 2400 Square Feet** Between 1985 and 1999, the percentage of new homes built that had 2400 or more square feet ranged from 17 percent to 34 percent, as shown in the following table.

Year	Percentage of New Homes with 2400+ Square Feet (percent)
1985	17
1986	18
1987	21
1988	25
1989	26
1990	29
1991	28
1992	29
1993	29
1994	28
1995	28
1996	30
1997	31
1998	32
1999	34

Source: *Statistical Abstract of the United States, 2001*, Table 938, p. 597.

10. **Homes of Less than 1200 Square Feet** Between 1985 and 1999, the percentage of new homes built that had 1200 or less square feet ranged from 7 percent to 20 percent, as shown in the following table.

Year	Percentage of New Homes with Less than 1200 Square Feet (percent)
1985	20
1986	17
1987	14
1988	12
1989	13
1990	11
1991	12
1992	10
1993	9
1994	9
1995	10
1996	9
1997	8
1998	7
1999	7

Source: *Statistical Abstract of the United States, 2001*, Table 938, p. 597.

11. **Fuel Production** Between 1990 and 1999, the U.S. percentage of the world's fuel production ranged from 19 percent to 23 percent, as shown in the following table.

Year	U.S. Percentage of World Fuel Production (percent)
1990	19
1991	20
1992	20
1993	19
1994	21
1995	20
1996	21
1997	21
1998	22
1999	23

Source: *Statistical Abstract of the United States, 2001*, Table 872, p. 559.

12. **Milled Rice Production** Between 1990 and 2000, the U.S. percentage of the world's milled rice production ranged from 1.4 percent to 1.8 percent, as shown in the following table.

Year	U.S. Percentage of World Milled Rice Production (percent)
1990	1.5
1991	1.4
1992	1.6
1993	1.5
1994	1.8
1995	1.5
1996	1.4
1997	1.5
1998	1.5
1999	1.6
2000	1.5

Source: *Statistical Abstract of the United States, 2001*, Table 821, p. 534.

13. **Citrus Fruit Consumption** Between 1980 and 2000, the per capita consumption of citrus fruit in the United States ranged from 19 to 28 pounds, as shown in the following table.

Year	Per Capita Citrus Fruit Consumption (pounds)
1980	26
1981	24
1982	23
1983	28
1984	23
1985	22
1986	24
1987	24
1988	25
1989	24
1990	21
1991	19
1992	24
1993	26
1994	25
1995	24
1996	25
1997	27
1998	27
1999	21
2000	26

Source: *Statistical Abstract of the United States, 2001*, Table 832, p. 539.

14. **Potato Consumption** Between 1980 and 1999, the per capita consumption of potatoes in the United States ranged from 46 to 51 pounds, as shown in the following table.

Year	Per Capita Potato Consumption (pounds)
1980	51
1981	46
1982	47
1983	50
1984	48
1985	46
1986	49
1987	48
1988	50
1989	50
1990	47
1991	50
1992	49
1993	51
1994	50
1995	50
1996	51
1997	49
1998	48
1999	48

Source: *Statistical Abstract of the United States, 2001*, Table 832, p. 539.

15. **Top Ten Movies** In 2003, the movie with the highest gross theater income was *The Lord of the Rings: Return of the King*, distributed by New Line studio. The following table shows the number of New Line movies that ranked in the 10 top-grossing movies for each of the given years.

Year	Number of New Line Movies in the 10 Top Grossing Movies
1993	0
1994	2
1995	1
1996	0
1997	0
1998	1
1999	1
2000	0
2001	2
2002	2
2003	2

Source: www.boxofficemojo.com.

In Exercises 16–30, calculate the mean and the sample standard deviation for each data set. Then determine the percentage of data points within two standard deviations of the mean. (You may want to apply Chebyshev's rule, as appropriate, to substantiate your results.)

16. **Home Prices** On March 29, 2004, Realtor.com advertised five homes in Ivins, Utah, in the \$200,000–\$225,000 price range. The amount of square footage offered by each home was 3400, 3400, 2738, 4332, and 4332 square feet, respectively. (**Source:** www.realtor.com.)

17. **Home Prices** On March 29, 2004, Realtor.com advertised four homes in the Bronx and in the city of Yonkers, New York, in the \$200,000–225,000 price range. The amount of square footage offered by each home was 1600, 1500, 1116, and 1497 square feet, respectively. (**Source:** www.realtor.com.)

18. **Air Fares** On April 2, 2004, travelocity.com advertised a variety of fares for a one-way flight on April 8, 2004, from Seattle to Phoenix, as shown by the following frequency distribution.

Fare	Frequency
$93	4
$94	6
$101	3
$104	4
$154	1
$162	1
$164	3

Source: www.travelocity.com.

19. **Arnold Schwarzenegger Movies** Between 1991 and 2003, Arnold Schwarzenegger was featured in at least 10 movies. The gross income from theaters for each movie ranged from $34.6 million to $146.3 million, as shown in the following table.

Movie	Gross Income (millions)
Terminator 3: Rise of the Machines	$143.7
Collateral Damage	$40.1
The 6th Day	$34.6
End of Days	$66.9
Batman & Robin	$107.3
Jingle All the Way	$60.6
Eraser	$101.3
Junior	$36.8
True Lies	$146.3
Last Action Hero	$50.0

Source: www.boxofficemojo.com.

20. **Drew Barrymore Movies** Between 1998 and 2003, Drew Barrymore was featured in at least 12 movies. The gross income from theaters for each movie ranged from $0.5 million to $125.3 million, as shown in the following table.

Movie	Gross Income (millions)
Duplex	$9.7
Charlie's Angels 2: Full Throttle	$100.8
Confessions of a Dangerous Mind	$16.0
Riding in Cars with Boys	$30.2
Freddy Got Fingered	$14.3
Donnie Darko	$0.5
Charlie's Angels	$125.3
Titan A.E. (voice)	$22.8
Never Been Kissed	$55.5
Home Fries	$10.5
Ever After: A Cinderella Story	$65.7
The Wedding Singer	$80.2

Source: www.boxofficemojo.com.

21. **Birth Rates** Birth rates in the United States between 1950 and 1995 have varied widely, as shown in the following table.

Year	Births (per 1,000 people)
1950	24.1
1955	25.0
1960	23.7
1965	19.4
1970	18.4
1975	14.6
1980	15.9
1985	15.8
1990	16.7
1995	14.8

Source: *Statistical Abstract of the United States, 2001,* Table 68, p. 59.

22. **Infant Death Rates** Infant death rates in the United States declined dramatically between 1950 and 1995, as shown in the following table.

Year	Infant Deaths (per 1,000 live births)
1950	29.2
1955	26.4
1960	26.0
1965	24.7
1970	20.0
1975	16.1
1980	12.6
1985	10.6
1990	9.2
1995	7.6

Source: *Statistical Abstract of the United States, 2001*, Table 68, p. 59.

23. **Divorce Rates** Divorce rates in the United States between 1950 and 1995 have fluctuated as shown in the following table.

Year	Divorces (per 1,000 people)
1950	2.6
1955	2.3
1960	2.2
1965	2.5
1970	3.5
1975	4.8
1980	5.2
1985	5.0
1990	4.7
1995	4.1

Source: *Statistical Abstract of the United States, 2001*, Table 68, p. 59.

24. **Marriage Rates** Marriage rates in the United States between 1950 and 1995 have fluctuated as shown in the following table.

Year	Marriages (per 1,000 people)
1950	11.1
1955	9.3
1960	8.5
1965	9.3
1970	10.6
1975	10.0
1980	10.6
1985	10.1
1990	9.8
1995	7.6

Source: *Statistical Abstract of the United States, 2001*, Table 68, p. 59.

25. **Women's Basketball Champions** On April 6, 2004, the University of Connecticut women's basketball team defeated Tennessee and won the national title for the third year in a row. In the final six games leading up to the championship, Connecticut beat its opponents by notable point margins. The following table shows the team played and the difference between the winning and losing scores for each of these games.

Team	Difference Between Winning and Losing Scores
Penn	36
Auburn	26
U.C. Santa Barbara	8
Penn State	17
Minnesota	9
Tennessee	9

Source: www.uconnhuskies.com.

26. **Air Pollution** Between 1990 and 1999, the national ambient air carbon monoxide concentration decreased, as shown in the following table.

Year	Carbon Monoxide Pollution (in parts per million)
1990	5.8
1991	5.7
1992	5.3
1993	5.0
1994	5.1
1995	4.6
1996	4.3
1997	3.9
1998	3.8
1999	3.7

Source: *Statistical Abstract of the United States, 2001*, Table 355, p. 216.

27. **Tornadoes** Between 1980 and 1997, the number of tornadoes ranged between 656 and 1298 per year, as shown in the following table.

Year	Tornadoes	Year	Tornadoes
1980	866	1989	856
1981	783	1990	1,133
1982	1,046	1991	1,132
1983	931	1992	1,298
1984	907	1993	1,176
1985	684	1994	1,082
1986	764	1995	1,235
1987	656	1996	1,170
1988	702	1997	1,148

Source: *Statistical Abstract of the United States, 2001*, Table 368, p. 221.

28. **Tropical Storms and Hurricanes** Between 1980 and 2000, the number of tropical storms and hurricanes in the North Atlantic ranged between 4 and 19 per year, as shown in the following table.

Year	Tropical Storms and Hurricanes
1980	11
1981	11
1982	5
1983	4
1984	12
1985	11
1986	6
1987	7
1988	12
1989	11
1990	14
1991	8
1992	7
1993	8
1994	7
1995	19
1996	13
1997	7
1998	14
1999	12
2000	15

Source: *Statistical Abstract of the United States, 2001*, Table 368, p. 221.

29. **Real Networks Net Revenue** The net revenue of Real Networks, Inc., between 1999 and 2003 is shown in the following table.

Year	Net Revenue (thousands of dollars)
1999	131,242
2000	241,538
2001	188,905
2002	182,679
2003	202,377

Source: Real Networks Inc 2003 Report, p. 14.

30. **Education and Earnings** The average annual earnings of full-time, male workers in 1999 is shown in the following table based on their level of educational attainment.

Educational Attainment	Average Earnings
Less than 9th grade	$18,743
9th–12th grade (no diploma)	$18,908
High school graduate (or equivalent)	$30,414
Some college (no degree)	$33,614
Associate degree	$40,047
Bachelor's degree	$57,706
Master's degree	$68,367
Professional	$120,352
Doctorate	$97,357

Source: *Statistical Abstract of the United States, 2001*, Table 677, p. 441.

Exercises 31–35 are intended to challenge your understanding of measures of central tendency and dispersion.

31. **Starting Salaries for Law Graduates** A professional football player with a multimillion-dollar contract pursues a law degree in the off-season. In the year after the player obtains his degree, which measure of first-year salary of new law school graduates will be most accurate: mean, median, or mode? Explain.

32. Describe the relationship between the mean of a data set and the expected value of a random variable.

33. **Dice Game Activity** Roll a pair of dice 10 times, each time noting the sum of the face values of the dice. Calculate the mean, median, mode, and standard deviation for the data set. Do you think the results would be close to the same if the dice were rolled 100 times? 1000 times? Explain.

34. Using a spreadsheet to simulate the rolling of dice, the author "rolled" a pair of dice 1000 times and summed their face values. The frequency of each event is recorded in the following table.

Sum of Face Values	Frequency
2	31
3	49
4	90
5	111
6	127
7	160
8	136
9	106
10	96
11	57
12	37

Calculate the mean, median, mode, and standard deviation for the data set. Do these results support the conclusions you made in Exercise 33? Explain.

35. Two data sets each have the following statistics:

$$\text{Mean} = 4$$
$$\text{Median} = 4$$
$$\text{Mode} = \textit{none}$$
$$\text{Standard deviation} = 1$$

Are the two data sets identical? Justify your answer.

8.6 Normal Distributions

- Convert a normal distribution to a standard normal distribution
- Determine the area of a shaded region beneath a normal curve
- Calculate the probability of attaining a value within a designated number of standard deviations of the mean of a normal distribution

GETTING STARTED Many colleges use placement tests such as the SAT as a factor in determining scholarship recipients. They assume that students who outperform their peers on the SAT are likely to excel in their academic studies at the college. Many large data sets representing human characteristics or performance, such as SAT scores, may be modeled by *normal distributions.*

In this section we will introduce normal distributions. Normal distributions are one of the most common types of probability distributions occurring in the real world. Normal distributions are *continuous* probability distributions, since the data values of the distribution may assume any one of an infinite number of values within a given interval. Recall that not all probability distributions are continuous. Consider the histogram of the discrete probability distribution for the sum of two dice given in Section 8.4 (Figure 8.24). The sum of two dice may take on only whole-number values between 2 and 12. It is impossible for the sum of two dice to equal any other value (e.g., 3.463).

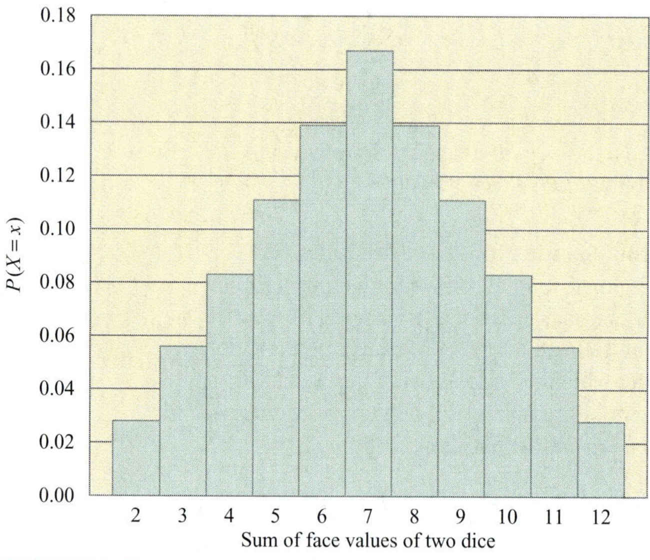

FIGURE 8.24

This distribution is a *discrete* probability distribution, since there are a finite number of possible outcomes (in this case, 11 outcomes).

In contrast, the birth weight of a boy is a *continuous distribution* because the boy's weight may take on any number of values within a given interval. Is it possible for a newborn boy to weigh 6.21347 pounds? Absolutely! Although a weight scale may be unable to detect the difference between a newborn weighing 6.21000 pounds and a newborn weighing 6.21347 pounds, the fact remains that the newborns have different weights. If we were to draw the graph of the continuous distribution for the weights of newborn boys, it would look something like the continuous curve shown in Figure 8.25.

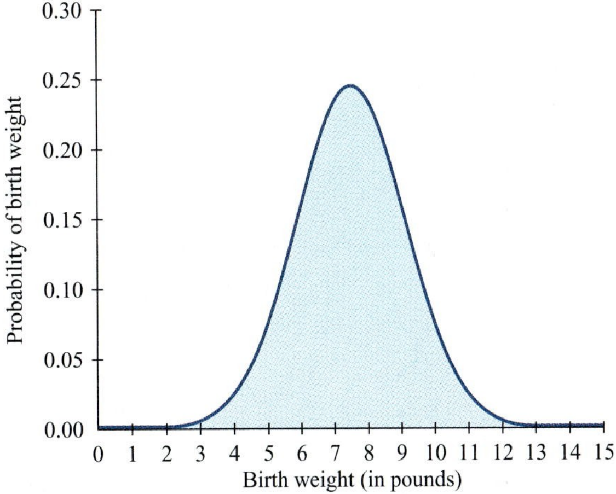

FIGURE 8.25

As shown in the graph, most newborn boys weigh between 5.5 and 9.5 pounds. The median birth weight of a newborn boy is about 7.5 pounds. (**Source:** National Center for Health Statistics growth chart at www.cdc.gov.)

Although discrete probability distributions are not continuous distributions, continuous curves are often used to model discrete probability distributions. The normal curve is one of the most common types of continuous curves used to model discrete data sets.

NORMAL DISTRIBUTION

A **normal distribution** is a continuous probability distribution with the following properties:

1. The graph of the distribution is a bell-shaped curve with a horizontal asymptote at the horizontal axis. (The graph is called a *normal curve.*)
2. The mean, median, and mode of the distribution are equal to one another.
3. The graph of the distribution is symmetric about the vertical line that passes through the population mean μ. (μ is the Greek letter mu.)
4. The distribution may be completely determined by knowing the population standard deviation σ and population mean μ of the distribution.
5. The area below the graph of the distribution and above the horizontal axis is equal to 1.

Figures 8.26 to 8.28 show the graphs of three different normal distributions. Each of the distributions has a mean of $\mu = 0$; however, they have different standard deviations. Notice that as the standard deviation σ gets larger, the normal curve becomes flatter.

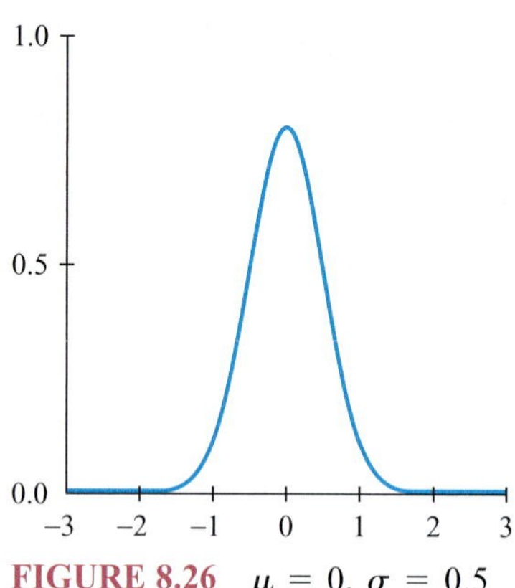

FIGURE 8.26 $\mu = 0, \sigma = 0.5$

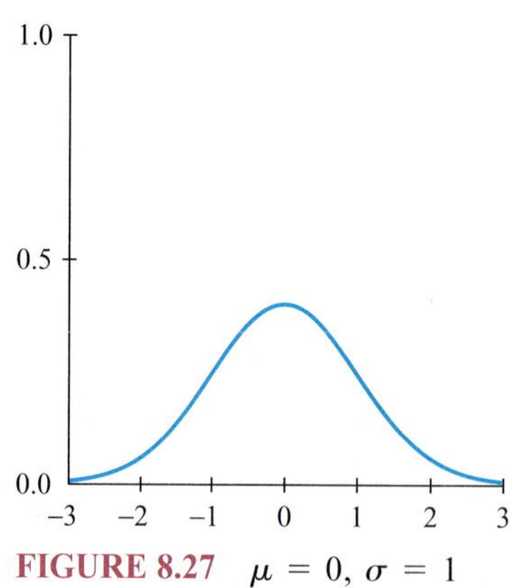

FIGURE 8.27 $\mu = 0, \sigma = 1$

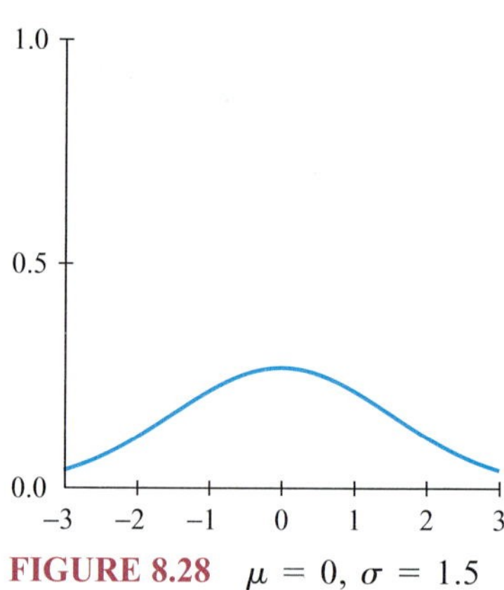

FIGURE 8.28 $\mu = 0, \sigma = 1.5$

The graph shown in Figure 8.27 is the normal curve for the *standard normal distribution.*

STANDARD NORMAL DISTRIBUTION

The **standard normal distribution** is the normal distribution with $\mu = 0$ and $\sigma = 1$.

Any normal distribution may be converted to a standard normal distribution by calculating the *z-scores* for the distribution. The **z-score** for a particular value x is given by $z = \frac{x - \mu}{\sigma}$. The z-score shows how many standard deviations x is away from the mean. For example, suppose a normal distribution has $\mu = 1.2$ and $\sigma = 0.4$. The z-score for $x = 2.2$ is

$$\begin{aligned} z &= \frac{2.2 - 1.2}{0.4} \\ &= \frac{1.0}{0.4} \\ &= 2.5 \end{aligned}$$

$x = 2.2$ is 2.5 standard deviations away from the mean. We can easily confirm the result by adding 2.5 standard deviations to the mean.

$$\begin{aligned} \mu + 2.5\sigma &= 1.2 + 2.5(0.4) \\ &= 1.2 + 1.0 \\ &= 2.2 \end{aligned}$$

EXAMPLE 1

Finding the z-Score for a Normal Distribution

Determine the z-score for $x = 15$ given a normal distribution with $\mu = 20$ and $\sigma = 3$.

SOLUTION

$$z = \frac{x - \mu}{\sigma}$$
$$= \frac{15 - 20}{3}$$
$$= -\frac{5}{3}$$
$$\approx -1.67$$

The value $x = 15$ is approximately 1.67 standard deviations below the mean.

Area Under a Normal Curve

All normal distributions have the amazing property that the area of the shaded region between the normal curve and the horizontal axis on the interval $[a, b]$ is equal to the probability that a randomly selected data value will fall within the interval $[a, b]$. Consequently, the probability that a z-score lies within n standard deviations of the mean is determined by the area of the shaded region between the standard normal curve and the horizontal axis on the interval $[-n, n]$. Since any normal distribution may be converted to a standard normal distribution, the area under the standard normal curve on an interval $[0, z]$ is typically listed in a table displaying various values of z (Table 8.21). By using the table, we can quickly determine the probability that a z-score lies within n standard deviations of the mean.

TABLE 8.21

The area A under the standard normal curve on the interval $[0, z]$ is given in the table.

z	A	z	A	z	A	z	A	z	A	z	A	z	A	z	A
0.00	0.0000	0.21	0.0832	0.42	0.1628	0.63	0.2357	0.84	0.2995	1.05	0.3531	1.26	0.3962	1.47	0.4292
0.01	0.0040	0.22	0.0871	0.43	0.1664	0.64	0.2389	0.85	0.3023	1.06	0.3554	1.27	0.3980	1.48	0.4306
0.02	0.0080	0.23	0.0910	0.44	0.1700	0.65	0.2422	0.86	0.3051	1.07	0.3577	1.28	0.3997	1.49	0.4319
0.03	0.0120	0.24	0.0948	0.45	0.1736	0.66	0.2454	0.87	0.3078	1.08	0.3599	1.29	0.4015	1.50	0.4332
0.04	0.0160	0.25	0.0987	0.46	0.1772	0.67	0.2486	0.88	0.3106	1.09	0.3621	1.30	0.4032	1.51	0.4345
0.05	0.0199	0.26	0.1026	0.47	0.1808	0.68	0.2517	0.89	0.3133	1.10	0.3643	1.31	0.4049	1.52	0.4357
0.06	0.0239	0.27	0.1064	0.48	0.1844	0.69	0.2549	0.90	0.3159	1.11	0.3665	1.32	0.4066	1.53	0.4370
0.07	0.0279	0.28	0.1103	0.49	0.1879	0.70	0.2580	0.91	0.3186	1.12	0.3686	1.33	0.4082	1.54	0.4382
0.08	0.0319	0.29	0.1141	0.50	0.1915	0.71	0.2611	0.92	0.3212	1.13	0.3708	1.34	0.4099	1.55	0.4394
0.09	0.0359	0.30	0.1179	0.51	0.1950	0.72	0.2642	0.93	0.3238	1.14	0.3729	1.35	0.4115	1.56	0.4406
0.10	0.0398	0.31	0.1217	0.52	0.1985	0.73	0.2673	0.94	0.3264	1.15	0.3749	1.36	0.4131	1.57	0.4418
0.11	0.0438	0.32	0.1255	0.53	0.2019	0.74	0.2704	0.95	0.3289	1.16	0.3770	1.37	0.4147	1.58	0.4429
0.12	0.0478	0.33	0.1293	0.54	0.2054	0.75	0.2734	0.96	0.3315	1.17	0.3790	1.38	0.4162	1.59	0.4441
0.13	0.0517	0.34	0.1331	0.55	0.2088	0.76	0.2764	0.97	0.3340	1.18	0.3810	1.39	0.4177	1.60	0.4452
0.14	0.0557	0.35	0.1368	0.56	0.2123	0.77	0.2794	0.98	0.3365	1.19	0.3830	1.40	0.4192	1.61	0.4463
0.15	0.0596	0.36	0.1406	0.57	0.2157	0.78	0.2823	0.99	0.3389	1.20	0.3849	1.41	0.4207	1.62	0.4474
0.16	0.0636	0.37	0.1443	0.58	0.2190	0.79	0.2852	1.00	0.3413	1.21	0.3869	1.42	0.4222	1.63	0.4484
0.17	0.0675	0.38	0.1480	0.59	0.2224	0.80	0.2881	1.01	0.3438	1.22	0.3888	1.43	0.4236	1.64	0.4495
0.18	0.0714	0.39	0.1517	0.60	0.2257	0.81	0.2910	1.02	0.3461	1.23	0.3907	1.44	0.4251	1.65	0.4505
0.19	0.0753	0.40	0.1554	0.61	0.2291	0.82	0.2939	1.03	0.3485	1.24	0.3925	1.45	0.4265	1.66	0.4515
0.20	0.0793	0.41	0.1591	0.62	0.2324	0.83	0.2967	1.04	0.3508	1.25	0.3944	1.46	0.4279	1.67	0.4525

(Continued)

(Continued)

z	A	z	A	z	A	z	A	z	A	z	A	z	A	z	A
1.68	0.4535	1.97	0.4756	2.26	0.4881	2.55	0.4946	2.84	0.4977	3.13	0.4991	3.42	0.4997	3.71	0.4999
1.69	0.4545	1.98	0.4761	2.27	0.4884	2.56	0.4948	2.85	0.4978	3.14	0.4992	3.43	0.4997	3.72	0.4999
1.70	0.4554	1.99	0.4767	2.28	0.4887	2.57	0.4949	2.86	0.4979	3.15	0.4992	3.44	0.4997	3.73	0.4999
1.71	0.4564	2.00	0.4772	2.29	0.4890	2.58	0.4951	2.87	0.4979	3.16	0.4992	3.45	0.4997	3.74	0.4999
1.72	0.4573	2.01	0.4778	2.30	0.4893	2.59	0.4952	2.88	0.4980	3.17	0.4992	3.46	0.4997	3.75	0.4999
1.73	0.4582	2.02	0.4783	2.31	0.4896	2.60	0.4953	2.89	0.4981	3.18	0.4993	3.47	0.4997	3.76	0.4999
1.74	0.4591	2.03	0.4788	2.32	0.4898	2.61	0.4955	2.90	0.4981	3.19	0.4993	3.48	0.4997	3.77	0.4999
1.75	0.4599	2.04	0.4793	2.33	0.4901	2.62	0.4956	2.91	0.4982	3.20	0.4993	3.49	0.4998	3.78	0.4999
1.76	0.4608	2.05	0.4798	2.34	0.4904	2.63	0.4957	2.92	0.4982	3.21	0.4993	3.50	0.4998	3.79	0.4999
1.77	0.4616	2.06	0.4803	2.35	0.4906	2.64	0.4959	2.93	0.4983	3.22	0.4994	3.51	0.4998	3.80	0.4999
1.78	0.4625	2.07	0.4808	2.36	0.4909	2.65	0.4960	2.94	0.4984	3.23	0.4994	3.52	0.4998	3.81	0.4999
1.79	0.4633	2.08	0.4812	2.37	0.4911	2.66	0.4961	2.95	0.4984	3.24	0.4994	3.53	0.4998	3.82	0.4999
1.80	0.4641	2.09	0.4817	2.38	0.4913	2.67	0.4962	2.96	0.4985	3.25	0.4994	3.54	0.4998	3.83	0.4999
1.81	0.4649	2.10	0.4821	2.39	0.4916	2.68	0.4963	2.97	0.4985	3.26	0.4994	3.55	0.4998	3.84	0.4999
1.82	0.4656	2.11	0.4826	2.40	0.4918	2.69	0.4964	2.98	0.4986	3.27	0.4995	3.56	0.4998	3.85	0.4999
1.83	0.4664	2.12	0.4830	2.41	0.4920	2.70	0.4965	2.99	0.4986	3.28	0.4995	3.57	0.4998	3.86	0.4999
1.84	0.4671	2.13	0.4834	2.42	0.4922	2.71	0.4966	3.00	0.4987	3.29	0.4995	3.58	0.4998	3.87	0.4999
1.85	0.4678	2.14	0.4838	2.43	0.4925	2.72	0.4967	3.01	0.4987	3.30	0.4995	3.59	0.4998	3.88	0.4999
1.86	0.4686	2.15	0.4842	2.44	0.4927	2.73	0.4968	3.02	0.4987	3.31	0.4995	3.60	0.4998	3.89	0.4999
1.87	0.4693	2.16	0.4846	2.45	0.4929	2.74	0.4969	3.03	0.4988	3.32	0.4995	3.61	0.4998	3.90	0.5000
1.88	0.4699	2.17	0.4850	2.46	0.4931	2.75	0.4970	3.04	0.4988	3.33	0.4996	3.62	0.4999	3.91	0.5000
1.89	0.4706	2.18	0.4854	2.47	0.4932	2.76	0.4971	3.05	0.4989	3.34	0.4996	3.63	0.4999	3.92	0.5000
1.90	0.4713	2.19	0.4857	2.48	0.4934	2.77	0.4972	3.06	0.4989	3.35	0.4996	3.64	0.4999	3.93	0.5000
1.91	0.4719	2.20	0.4861	2.49	0.4936	2.78	0.4973	3.07	0.4989	3.36	0.4996	3.65	0.4999	3.94	0.5000
1.92	0.4726	2.21	0.4864	2.50	0.4938	2.79	0.4974	3.08	0.4990	3.37	0.4996	3.66	0.4999	3.95	0.5000
1.93	0.4732	2.22	0.4868	2.51	0.4940	2.80	0.4974	3.09	0.4990	3.38	0.4996	3.67	0.4999	3.96	0.5000
1.94	0.4738	2.23	0.4871	2.52	0.4941	2.81	0.4975	3.10	0.4990	3.39	0.4997	3.68	0.4999	3.97	0.5000
1.95	0.4744	2.24	0.4875	2.53	0.4943	2.82	0.4976	3.11	0.4991	3.40	0.4997	3.69	0.4999	3.98	0.5000
1.96	0.4750	2.25	0.4878	2.54	0.4945	2.83	0.4977	3.12	0.4991	3.41	0.4997	3.70	0.4999	3.99	0.5000

EXAMPLE 2

Finding and Interpreting the Meaning of the Area Under a Standard Normal Curve

Use Table 8.21 to determine the area beneath the standard normal curve on the interval $[0, 1.49]$. Then determine the probability that a randomly selected data value will fall within 1.49 standard deviations of the mean.

SOLUTION Referring to Table 8.21, we see that when $z = 1.49$, $A = 0.4319$. Since the standard normal distribution is symmetric about $x = 0$, the area between the curve and the horizontal axis on the interval $[-1.49, 1.49]$ is equal to $2A$. Therefore, the probability that a randomly selected z-score is within 1.49 standard deviations of the mean is

$$
\begin{aligned}
P(-1.49 \leq z \leq 1.49) &= 2A \\
&= 2(0.4319) \\
&= 0.8638
\end{aligned}
$$

There is an 86.38 percent likelihood that a randomly selected z-score lies within 1.49 standard deviations of the mean. We represent this graphically by shading the corresponding region of the normal curve (Figure 8.29).

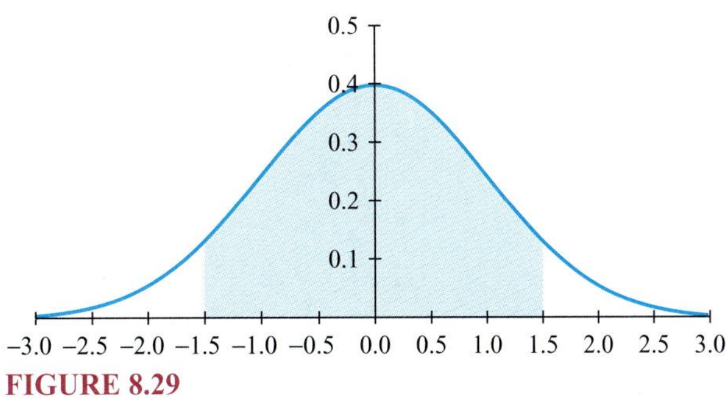

FIGURE 8.29

The built-in features of the TI-83 Plus calculator eliminate the need to refer to a z-score table. The calculator can draw and calculate the area of any region beneath a normal curve, as detailed in the following two Technology Tips.

TECHNOLOGY TIP

Calculating the Area Beneath a Normal Curve

1. Press [2nd] [VARS] to open the `DISTR` menu.

2. Move the cursor to `2:normalcdf(` and press [ENTER]. The normalcdf function requires the following four parameters: a, b, μ, and σ. (a and b are the starting and ending values of the interval, respectively.)

```
normalcdf(-1.49,
1.49,0,1)
```

3. After entering the parameters, press [ENTER] to display the area beneath the normal curve on the interval $[a, b]$. The area beneath the standard normal curve on $[-1.49, 1.49]$ is 0.8638.

```
normalcdf(-1.49,
1.49,0,1)
              .8637757036
```

TECHNOLOGY TIP

Drawing the Area Beneath a Normal Curve

1. Press WINDOW and set `Xmin` to $\mu - 3\sigma$ and `Xmax` to $\mu + 3\sigma$. For example, if $\mu = 4$ and $\sigma = 3$, then `Xmin=-5` and `Xmax=13`.

```
WINDOW
Xmin=-5
Xmax=13
Xscl=1
Ymin=-.1
Ymax=.2
Yscl=1
Xres=1
```

2. Press 2nd VARS to open the `DISTR` menu and move the cursor to the `DRAW` menu. Select the `1:ShadeNorm(` function and press ENTER.

```
DISTR DRAW
1:ShadeNorm(
2:Shade_t(
3:Shadeχ²(
4:ShadeF(
```

3. The ShadeNorm function requires the following four parameters: a, b, μ, and σ. To shade the region within one standard deviation of the mean, we let $a = \mu - \sigma$ and $b = \mu + \sigma$. In this case, $a = 1$ and $b = 7$.

```
ShadeNorm(1,7,4,
3)
```

4. Press ENTER to display the shaded region.

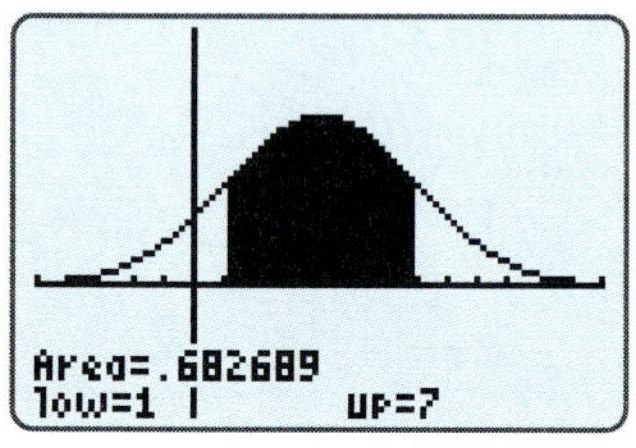

EXAMPLE 3

Using Technology to Analyze a Normal Distribution

Using a graphing calculator, determine the probability that a randomly selected value from a normal distribution with $\mu = 2.32$ and $\sigma = 0.48$ lies within 1.5 standard deviations of the mean. Then draw the shaded region of the normal distribution.

SOLUTION We must first determine $a = \mu - 1.5\sigma$ and $b = \mu + 1.5\sigma$.

$$\begin{aligned} a &= \mu - 1.5\sigma \\ &= 2.32 - 1.5(0.48) \\ &= 1.60 \end{aligned} \qquad \begin{aligned} b &= \mu + 1.5\sigma \\ &= 2.32 + 1.5(0.48) \\ &= 3.04 \end{aligned}$$

Using the normalcdf function on the TI-83 Plus yields the result shown in Figure 8.30.

FIGURE 8.30

The probability that a randomly selected value lies within 1.5 standard deviations of the mean is 86.64 percent.

In order to set the window, we find the values three standard deviations away from the mean. Doing so will ensure that the majority of the normal curve will appear in the window, even though only a portion of the region beneath the curve will be shaded.

$$\begin{aligned} \text{Xmin} &= \mu - 3\sigma \\ &= 2.32 - 3(0.48) \\ &= 0.88 \end{aligned} \qquad \text{and} \qquad \begin{aligned} \text{Xmax} &= \mu + 3\sigma \\ &= 2.32 + 3(0.48) \\ &= 3.76 \end{aligned}$$

Using the ShadeNorm function, we draw the shaded region (see Figure 8.31).

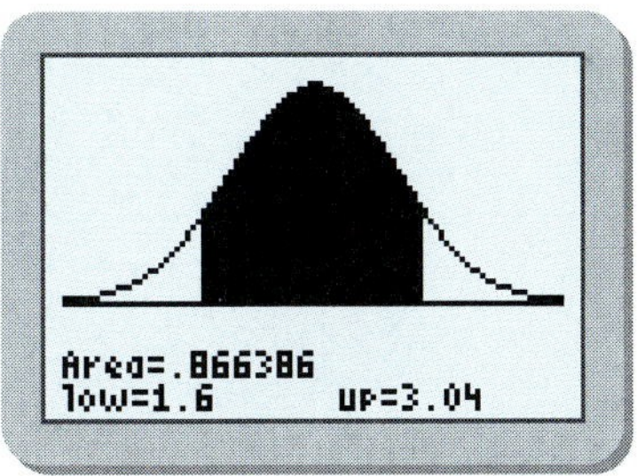

FIGURE 8.31

In Example 3, we drew the graph of the normal distribution over the interval $[\mu - 3\sigma, \mu + 3\sigma]$. From the image we see that nearly the entire graph of the normal distribution was contained within this domain. In fact, for all normal distributions, the majority of values will fall within three standard deviations of the mean.

EMPIRICAL RULE

All normal distributions have the following properties.

- Approximately 68.3 percent of values lie within one standard deviation of the mean.
- Approximately 95.4 percent of values lie within two standard deviations of the mean.
- Approximately 99.7 percent of values lie within three standard deviations of the mean.

EXAMPLE 4

Finding the Area Under a Standard Normal Curve

What percentage of the total area under the standard normal curve lies between the z-scores of $z = 1.94$ and $z = 3.08$?

SOLUTION From the z-score table, we see that the area below the curve on the interval $[0, 1.94]$ is 0.4738. The area below the curve on the interval $[0, 3.08]$ is 0.4990. Since we are interested in the area below the curve on $[1.94, 3.08]$, we calculate the difference between the two values.

$$0.4990 - 0.4738 = 0.0252$$

Therefore, 2.52 percent of the total area below the standard normal curve is between the z-scores of $z = 1.94$ and $z = 3.08$.

EXAMPLE 5

Using a Normal Distribution to Forecast Unknown Results

An IQ test has a mean of 100 and a standard deviation of 15. Assuming that test scores are normally distributed, what percentage of the population is expected to score above 110?

SOLUTION We first compute the z-score.

$$\begin{aligned} z &= \frac{x - \mu}{\sigma} \\ &= \frac{110 - 100}{15} \\ &= \frac{10}{15} \\ &\approx 0.67 \end{aligned}$$

The score 110 is 0.67 standard deviation above the mean. Referring to the z-score table, we see that the area between 0 and 0.67 is 0.2486. Since 50 percent of all scores are above the mean, the percentage of the population expected to score above 110 is given by

$$0.5000 - 0.2486 = 0.2514$$

Therefore, 25.14 percent of the population is expected to score above 110 on the IQ test.

EXAMPLE 6

Using a Normal Distribution to Forecast Unknown Results

In 2003, the mean SAT verbal score for college-bound seniors was 507 with a standard deviation of 111. (**Source:** www.collegeboard.com.) Assuming that the scores are normally distributed, what score would a student have to earn to be in the top 10 percent of all test takers?

SOLUTION We are looking for a test score x and a z-score z such that

$$A = 0.40 \qquad \text{and} \qquad z = \frac{x - 507}{111}$$

(We pick $A = 0.40$ because we already know that 50 percent of the scores lie below the mean. We are looking for a score above the mean that has 40 percent of the scores above the mean below it. The resulting score will have a total of 90 percent of the scores below it, since 50 percent + 40 percent = 90 percent.)

Looking in the z-score table, we see that when $z = 1.28, A = 0.3997$, and when $z = 1.29, A = 0.4015$. Since the first value of A is closer to 0.40, we pick $z = 1.28$.

$$1.28 = \frac{x - 507}{111}$$

$$142.08 = x - 507$$

$$x = 649.08$$

$$\approx 650 \qquad \text{Since an SAT score must be a multiple of 10}$$

Assuming that the scores are normally distributed, a student earning 650 or better would be in the top 10 percent of college-bound seniors who took the test.

EXAMPLE 7

Using a Normal Distribution to Forecast Unknown Results

During a well-baby exam, a nurse informed a mother that her 16-month-old son was at the 95th percentile for height (that is, 95 percent of toddlers his age were shorter than he was). The average height of a 16-month-old toddler is about 31.5 inches. (**Source:** www.cdc.gov.) The woman's son was 33.5 inches tall. Approximate the value of σ for the height of a 16-month-old boy.

SOLUTION We know that 50 percent of the heights lie below the 31.5-inch mean. We need to first determine what z-score yields a value of $A = 0.45$, since 50 percent + 45 percent = 95 percent. From the z-score table, we see that when $z = 1.64$, $A = 0.4495$, and when $z = 1.65$, $A = 0.4505$. We estimate that when $z = 1.645$, $A = 0.4500$. We know that $z = \frac{x - \mu}{\sigma}$. Therefore,

$$z = \frac{x - \mu}{\sigma}$$

$$1.645 = \frac{33.5 - 31.5}{\sigma}$$

$$1.645 = \frac{2}{\sigma}$$

$$\sigma = \frac{2}{1.645}$$

$$\sigma = 1.216 \text{ inches}$$

The population standard deviation σ for the height of a 16-month-old boy is about 1.2 inches.

8.6 Summary

In this section, you learned how to use normal distributions. You saw that any normal distribution may be converted into a standard normal distribution by calculating the corresponding z-scores. You also learned how to use a z-score table.

8.6 Exercises

In Exercises 1–10, calculate the z-score for the given values of x, μ, and σ.

1. $x = 4, \mu = 5, \sigma = 1.1$
2. $x = 34, \mu = 25, \sigma = 4$
3. $x = -2, \mu = 0, \sigma = 0.9$
4. $x = 124, \mu = 111, \sigma = 10.5$
5. $x = 0.21, \mu = 0.42, \sigma = 0.84$
6. $x = 22, \mu = 49, \sigma = 10$
7. $x = 0.94, \mu = 0.49, \sigma = 0.17$
8. $x = 1.57, \mu = 1.34, \sigma = 0.32$
9. $x = -0.29, \mu = -1.19, \sigma = 0.76$
10. $x = 789, \mu = 256, \sigma = 121$

In Exercises 11–20, use the z-score table to determine the area beneath the standard normal curve on the interval $[0, z]$. Then determine the probability that a randomly selected data value will fall within z standard deviations of the mean.

11. $z = 2.13$
12. $z = 1.50$
13. $z = 2.39$
14. $z = 3.29$
15. $z = 0.21$
16. $z = 1.78$
17. $z = 0.67$
18. $z = 1.35$
19. $z = 0.42$
20. $z = 3.61$

In Exercises 21–30, use a graphing calculator to shade the region within one standard deviation of the mean for a normal distribution with the indicated values of μ and σ.

21. $\mu = 5.2, \sigma = 2.6$
22. $\mu = 0.2, \sigma = 0.6$
23. $\mu = 9.8, \sigma = 3.5$
24. $\mu = 5.7, \sigma = 0.1$
25. $\mu = 7.9, \sigma = 2.7$
26. $\mu = 5.8, \sigma = 9.7$
27. $\mu = -1.2, \sigma = 1.7$
28. $\mu = -0.3, \sigma = 1.5$
29. $\mu = -1.7, \sigma = 7.4$
30. $\mu = 9.1, \sigma = 0.2$

In Exercises 31–35, determine the percentage of the total area between the normal curve and the horizontal axis that is between the given z-scores.

31. $z = 0.21$ and $z = 1.25$
32. $z = 1.98$ and $z = 3.96$
33. $z = -1.44$ and $z = 1.45$
34. $z = -0.25$ and $z = 2.97$
35. $z = 2.99$ and $z = 3.81$

In Exercises 36–45, apply normal distribution concepts to answer the questions.

36. **SAT Math Scores** In 2003, the mean SAT math score for college-bound seniors was 519 with a standard deviation of 115. (**Source:** www.collegeboard.com.) Assuming that the scores are normally distributed, what score would a student have to earn to be in the top 10 percent of all test takers on the math portion of the exam? (Round your answer to the nearest multiple of 10.)

37. **SAT Math Scores—Utah** In 2003, the mean SAT math score for college-bound Utah women was 543 points with a standard deviation of 105 points. (**Source:** www.collegeboard.com.) Assuming that the scores are normally distributed, what score would a woman have to earn to be in the top 10 percent of all female test takers in Utah on the math portion of the exam? (Round your answer to the nearest multiple of 10.)

38. **Quality Control: Canned Tuna** Four cans of Bumble Bee® Chunk Light Tuna were weighed on a kitchen scale. The weight of the cans varied from 185 to 195 grams. Assuming that the weight of the cans is normally distributed with $\mu = 190$ grams and $\sigma = 3.54$, what is the probability than a can of tuna weighs less than 180 grams?

39. **Quality Control: Canned Beans** Five cans of S&W® Cut Green Beans were weighed on a kitchen scale. The weight of the cans varied from 440 to 455 grams. Assuming that the weight of the cans is normally distributed with $\mu = 448$ grams and $\sigma = 5.10$, what is the probability that a can of green beans weighs less than 440 grams?

40. **Quality Control: Boxed Pudding** Six boxes of Hain Pure Foods® Super Pudding™ were weighed on a kitchen scale. The weight of the boxes varied from 80 to 85 grams. Assuming that the weight of the boxes is normally distributed with $\mu = 81.3$ grams and $\sigma = 1.97$, what is the probability that a box of pudding weighs more than 89 grams?

41. **Quality Control: Canned Albacore** Six cans of Kirkland Signature™ Solid White Albacore were weighed on a kitchen scale. The weight of the cans varied from 185 to 187 grams. Assuming that the weight of the cans is normally distributed with $\mu = 185.7$ grams and $\sigma = 0.94$, what is the probability than a can of albacore weighs more than 190 grams?

42. Exercises 38–41 dealt with the weight variations of packaged foodstuffs. Explain how a manufacturer could use analytical results such as those presented in the indicated exercises to improve its business.

43. **Weight of a Boy** According to a growth chart published by the Centers for Disease Control, the average weight of a 24-month-old boy is approximately 27.5 pounds. The chart further indicates that 5 percent of all 24-month-old boys weigh less than 23.5 pounds. (**Source:** www.cdc.gov.) Assuming that the weights of 24-month-old boys are normally distributed, what weight range is within two standard deviations of the mean?

44. **Weight of a Girl** According to a growth chart published by the Centers for Disease Control, the average weight of a 36-month-old girl is approximately 30.5 pounds. The chart further indicates that 90 percent of all 36-month-old girls weigh less than 36 pounds. (**Source:** www.cdc.gov.) Assuming that the weights of 36-month-old girls are normally distributed, what is the standard deviation for the normal distribution?

45. **Well-Baby Exam** During a well-baby exam, pediatricians typically compare a patient's measurements with growth charts such as those published by the CDC. Why do you think that pediatricians would be interested in comparing a child to charts showing weight-for-age and height-for-age percentiles?

Exercises 46–50 are intended to challenge your understanding of normal distributions.

46. Two different normal distributions have the same mean μ; however, the standard deviation of one of the distributions is σ_1, while the standard deviation of the other is σ_2. What will be the relationship between the z-scores of each of the distributions for a given value of x?

47. Two different normal distributions have the same standard deviation σ; however, the mean of the first distribution is μ_1, while the mean of the second distribution is μ_2. What is the relationship between the z-scores of the distributions for a given value of x?

48. The equation of a graph of a normal curve is given by

$$f(x) = \frac{1}{\sigma\sqrt{2\pi}}e^{-\frac{1}{2}\left(\frac{x-\mu}{\sigma}\right)^2}$$

The graph reaches a maximum value at $x = \mu$. If $f(\mu) = k$, then determine the value of σ in terms of k.

49. Will the sum of two normal distributions be a normal distribution? Show the work leading to your conclusion.

50. Give the equation of a normal distribution with $\sigma = 1$ but $\mu \neq 0$. What is the relationship between the graph of this distribution and the standard normal curve?

Chapter 8 Review Exercises

Section 8.1 *In Exercises 1–3, determine the indicated probability.*

1. $P(A \cap B) = 0.25$ and $P(B) = 0.35$. Find $P(A|B)$.

2. Find $P(H'|D')$.

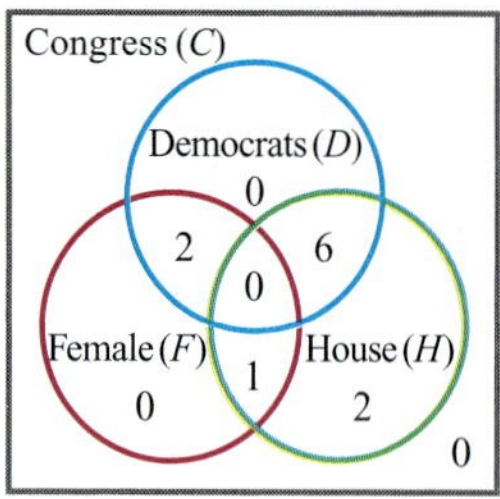

3. Twenty-two students, including twelve seniors, entered a school art contest. Five students, three of whom were seniors, won awards. What is the probability that a student won an award, given that the student was a senior?

In Exercises 4–5, use a probability tree to determine the solution.

4. In 2003, California had 55 members of the U.S. Congress, including 2 senators and 53 representatives. Of the 35 Democrats, 2 were senators. (**Source:** clerk.house.gov.) What is the likelihood that a randomly selected member of Congress is a Democratic representative?

5. In 2003, California had 55 members of the U.S. Congress, including 2 senators and 53 representatives. Of the 35 Democrats, 2 were senators. (**Source:** clerk.house.gov.) What is the likelihood that a randomly selected member of Congress is *not* a Democratic representative?

In Exercises 6–7, determine if the events E and F are independent or dependent.

6. Two coins are pulled from a bag containing one nickel, one dime, and one penny.

   $E =$ the first coin is a dime
   $F =$ the second coin is a nickel

7. Two coins are pulled from a bag containing five nickels and one dime.

   $E =$ the first coin is a nickel
   $F =$ the second coin is a dime

Section 8.2 *In Exercises 8–9, complete the table by applying Bayes' theorem.*

8.

	Percent
Prevalence	5
Sensitivity	92
Specificity	76
Predictive-value positive	
Predictive-value negative	

9.

	Percent
Prevalence	0.1
Sensitivity	25
Specificity	99
Predictive-value positive	
Predictive-value negative	

In Exercises 10–11, apply Bayes' theorem to the real-life application.

10. **Hepatitis B Test** ACON® HbsAg is a one-step hepatitis B surface antigen test strip produced by ACON Laboratories. According to the test package insert, the test has a sensitivity of at least 99 percent and a specificity of 97 percent. (**Source:** www.aconlab.com.)

The incidence of chronic hepatitis B infection is estimated to be 1 out of 100 people in the United States. What is the probability that someone who tests negative for hepatitis B actually has the illness?

11. **Hepatitis C Test** ACON HCV Rapid Test Device (Serum/Plasma) is a one-step hepatitis C virus test device produced by ACON Laboratories. According to the package insert, the test was compared with a leading commercial HCV test that used a different method. The comparison yielded the following results.

		Other HCV Method	
		Positive HCV	Negative HCV
ACON HCV	Positive	145	2
	Negative	0	144

Source: www.aconlab.com.

Calculate and interpret the meaning of the sensitivity, specificity, predictive-value positive, and predictive value negative for the ACON HCV test.

Section 8.3 *In Exercises 12–13, draw the state transition diagram that corresponds with the given transition matrix. Identify which states, if any, are absorbing states.*

12. $P = \begin{bmatrix} 0 & 1 & 0 \\ 1 & 0 & 0 \\ 0.5 & 0.25 & 0.25 \end{bmatrix}$

13. $P = \begin{bmatrix} 0 & 0.1 & 0.9 \\ 0.3 & 0.7 & 0 \\ 0.8 & 0 & 0.2 \end{bmatrix}$

In Exercises 14–15, write the transition matrix P that corresponds with the state transition diagram.

14.

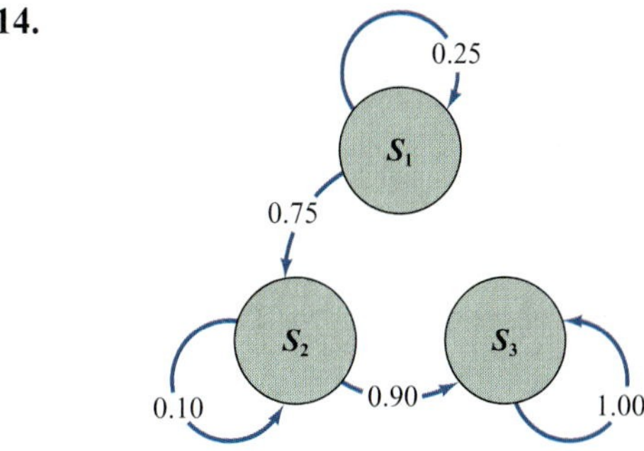

15.

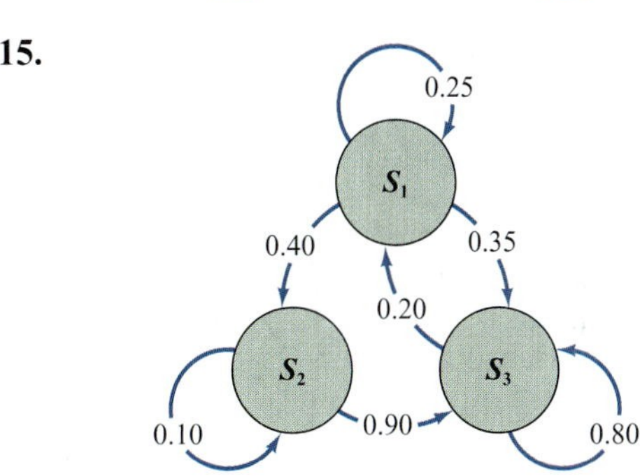

In Exercises 16–17, do the following: (i) identify the states of the Markov chain, (ii) draw the transition diagram, (iii) write the transition matrix P, (iv) identify the initial state distribution vector D, and (v) calculate DP and interpret the meaning of the result.

16. **Market Share** Assume that an advertising company has 15 percent of the market share for advertising. Suppose that the probability that a client will change to a competitor is 55 percent and that the probability that a competitor's client will change to the advertising company is 45 percent.

17. **Market Share** Suppose that two toothpaste brands dominate the market. The first brand, A, has 38 percent of the market share. The second brand, B, has 32 percent of the market share. The probability that a Brand A customer will become a brand B customer by the end of the year is 20 percent. The probability that a Brand A customer will become a customer of another non-B brand by the end of the year is 12 percent. The probability that a customer of Brand B will become a Brand A customer by the end of the year is 18 percent. The probability that a Brand B customer will become a customer of another non-A brand by the end of the year is 22 percent. There is a 50 percent probability that a customer using a brand other than Brand A or Brand B will remain with her or his current brand. There is a 20 percent probability that the customers will become a Brand A customer by the end of the year.

Section 8.4

In Exercises 18–19, determine the probability distribution for the random variable X and draw its associated histogram.

18. A coin is tossed three times in a row

$X =$ the number of heads in the longest consecutive sequence of heads

19. Three six-sided dice are tossed.

$X =$ the number of different values shown

In Exercises 20–22, apply the concepts of probability distribution and expected value, as appropriate.

20. Raffle Tickets A campus club sells 420 raffle tickets at a price of $2 each. The raffle offers one $100 prize, two $50 prizes, and ten $10 prizes. What is the expected value of a raffle ticket?

21. True/False Tests A true/false test is scored as follows: A correct answer adds 5 points to the overall score. An incorrect answer deducts 5 points from the overall score. A blank answer adds 0 points to the overall score.

A person taking the test doesn't know the answer to one of the questions. Should she guess? Justify your answer.

22. Income Tax Returns An accountant determines that 0.10 percent of the income tax returns she processes are audited by the IRS. The average tax penalty for her audited returns is $1650. She guarantees her clients that she will pay any tax penalties on audited returns. How much must she charge for the guarantee service in order to break even?

Section 8.5

In Exercises 23–25, calculate the mean, median, mode, and standard deviation and interpret the meaning of the results.

23. Women's Basketball Champions On April 6, 2004, the University of Connecticut women's basketball team defeated Tennessee and won the national title for the third year in a row. In the final six games leading up to the championship, Connecticut outscored each of its opponents. The table shows the number of points scored by Connecticut in each of its final six games.

Losing Team	Points Scored by Connecticut
Penn	91
Auburn	79
U.C. Santa Barbara	63
Penn State	66
Minnesota	67
Tennessee	70

Source: www.uconnhuskies.com.

24. Golf Champion Up until 2004, Phil Mickelson had been classified by some as the best player never to win a major tournament. That changed at the 2004 Augusta National golf tournament. As a result of a remarkable performance on the last nine holes of the fourth round, Phil Mickelson surpassed his opponents to clinch his first major victory. The table details Mickelson's performance in the fourth round.

Difference Between Score and Par For the Hole	Frequency
−1	6
0	9
1	3

Source: www.golf.com.

25. Weekend Theater Sales The movie *My Big Fat Greek Wedding* was somewhat unknown when it was first introduced in April 2002; however, it steadily grew in popularity among the public. The table shows the gross weekend theater sales of the movie from some of its top-grossing weekends.

Weekend in 2002	Gross Weekend Theater Sales
Aug. 23–25	$7,261,842
Aug. 30–Sept. 2	$14,809,546
Sept. 6–8	$10,372,316
Sept. 13–15	$10,772,146
Sept. 20–22	$9,748,969
Sept. 27–29	$9,434,602
Oct. 4–6	$8,223,801
Oct. 11–13	$8,453,159
Oct. 18–20	$7,145,309
Oct. 25–27	$6,209,500

Source: www.boxofficemojo.com.

Section 8.6 *In Exercises 26–27, calculate the z-score for the given values of* x, μ, and σ.

26. $x = 12.1$, $\mu = 10.9$, $\sigma = 6.8$

27. $x = 9.4$, $\mu = 9.5$, $\sigma = 0.6$

In Exercises 28–29, use the z-score table to determine the percentage of values of a normal distribution lying within the indicated number of standard deviations of the mean.

28. $z = 2.13$

29. $z = 0.25$

In Exercises 30–31, use a graphing calculator to shade the region within two standard deviations of the mean for a normal distribution with the indicated values of μ *and* σ.

30. $\mu = 4.5$, $\sigma = 1.5$

31. $\mu = 5.22$, $\sigma = 0.11$

In Exercises 32–33, determine the percentage of the total area between the normal curve and the horizontal axis that is between the given z-scores.

32. $z = -0.91$ and $z = 1.42$

33. $z = 1.00$ and $z = 2.00$

In Exercise 34, apply normal distribution concepts to answer the question.

34. **Fresh Whole Carrot Weight** A particular bag of fresh whole carrots contains carrots ranging in weight from 55 grams to 110 grams. Assuming that the individual weights of the carrots used by the producer to fill each bag of carrots are normally distributed with $\mu = 76$ and $\sigma = 18.4$, what percentage of carrots in the bag are likely to weigh 110 grams or more?

Make It Real

PROJECT 8

What to do

1. Collect a set of 10 or more data points from an area of personal interest.
2. Calculate the mean, median, mode, and standard deviation for the data set.
3. Explain the real-world meaning of each of the measures of central tendency.
4. Describe how your results could be used to improve business or society.

Example project

Age of Grandchild	Frequency
1	1
2	1
3	1
4	1
5	1
7	2
8	1
10	2

Mean: 5.7

The average age of a grandchild is 5.7 years.

Median: 6

Five grandchildren are less than six years old and five grandchildren are more than six years old.

Mode: 7 and 10

The most common age of a grandchild is seven years old and ten years old.
Standard deviation: 3.199

If the grandparents plan to give each grandchild $1000 in the year the child turns 18, they could use these results for financial planning. From the mode, they can determine the years in which the most money will be required. From the mean, they can determine that the average age of a grandchild will be 18 in about 12 years. The median tells them that half of the grandchildren will turn 18 in the next 12 years.

Answers to Odd-Numbered Exercises 

CHAPTER 1

Section 1.1 *(page 12)*

1. Your weight is a function of your age since at any instant in time you have only one weight.

3. Temperature is a function of the time of day.

5. The number of salmon in a catch is not a function of the number of fish caught.

7. $C(4) = 159.80$
The total cost of four pairs of shoes is $159.80.

9. $H(2) = 56$
Two seconds after he jumped, the cliff diver is 56 feet above the water.

11. $E(4) = 0.06$
In the fourth quarter since December 1999, shares in the tortilla company earned $0.06 per share.

13. $P(4) \approx 18.72$
On November 8, 2001, the closing stock price of the computer company was approximately $18.72.

15. It appears that near $x = 1$, the graph goes vertical. If it does so, a vertical line drawn at that point would touch the graph in multiple locations. However, if the graph doesn't actually go vertical near $x = 1$, then it is a function. One drawback of reading a graph is that it is sometimes difficult to tell if the graph goes vertical or not.

17. The graph is a function.

19. The graph is a function.

21.

23.

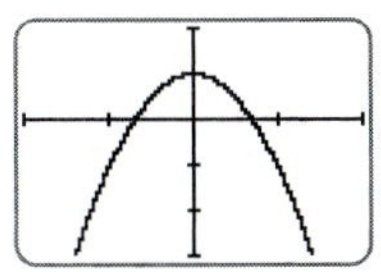

25. $y \approx 4$; $y = 4$

27. $y \approx 2$; y is undefined.

29. The domain is all real numbers.

31. The domain is all real numbers.

33. The domain is all real numbers except $t = 1$. That is, $\{t \mid t \neq 1\}$.

35. The domain is all real numbers greater than or equal to -1. That is $\{a \mid a \geq -1\}$.

37. The domain is the set of real numbers greater than or equal to -3. That is, $\{x \mid x \geq -3\}$.

39. The domain is the set of whole numbers. That is, $\{n \mid n \text{ is a whole number}\}$.

41. Using our current calendar system, the domain of the function is the set of whole numbers between 1 and the current year.

43. The domain is all real numbers except $x = -1$ and $x = 1$.

45. Yes. Even though the domain value $x = 1$ is listed twice in the table, it is linked with the same range value both times, $y = 6$. Similarly, $x = 4$ is listed twice in the table, but each domain value is linked to the range value $y = 2$.

Section 1.2 *(page 27)*

1. $m = -1$ **3.** $m = -0.2$ **5.** $m = 0$

7. y-intercept: $(0, 10)$
x-intercept: $(-2, 0)$

9. y-intercept: $(0, 11)$
x-intercept: $(-5.5, 0)$

11. y-intercept: $(0, -4)$
x-intercept: $\left(\frac{4}{3}, 0\right)$

13. The slope-intercept form of the line is $y = -x + 7$. The standard form of the line is $x + y = 7$. A point-slope form of the line is $y - 5 = -1(x - 2)$.

15. The slope-intercept form of the line is $y = -0.2x + 3.64$. The standard form of the line is $5x + 25y = 91$. A point-slope form of the line is $y - 3.4 = -0.2(x - 1.2)$.

17. The slope-intercept form of the line is $y = 0x + 2$ and is commonly written as $y = 2$. The standard form of the line is $0x + y = 2$ and is also often written as $y = 2$. A point-slope form of the line is $y - 2 = 0(x + 2)$.

19. $y = 4x - 2$

21. $y - 4 = 0.5(x - 2)$

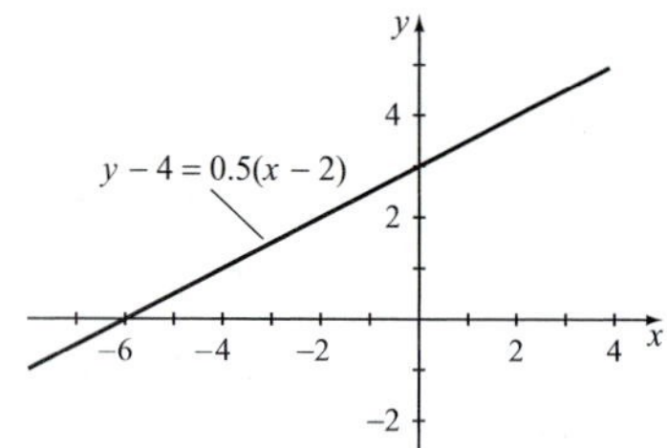

23. $2x - 3y = 5$

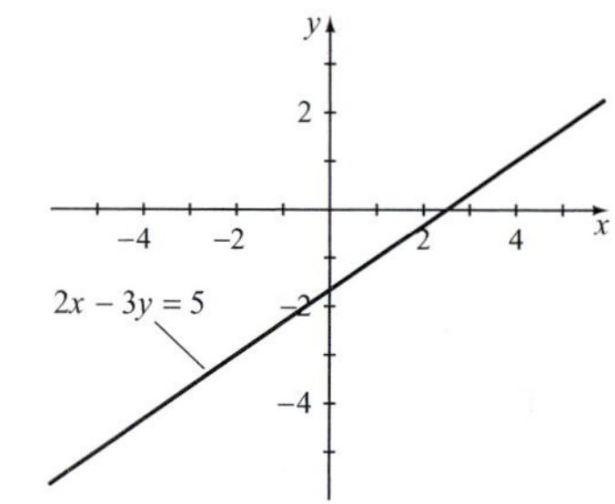

25. $y = -\frac{2}{3}x + \frac{4}{3}$

27. The slope-intercept form of the line is $y = \frac{2}{3}x + 3$.

29. The equation of the line is $x = 3$.

31. The data represents a linear function.

$$m = 993.2 \text{ year 2000 dollars per year}$$

Between 1989 and 1999, the U.S. average personal income (in year 2000 dollars) increased by an average of $993.20 per year.

33. The data represents a linear function.

$$m = 1183.01 \text{ dollars per month}$$

Between September 2001 and October 2001, the employee's take-home pay increased at a rate of $1183.01 per month.

35. The data represents a linear function.

$$m = \$0.0375 \text{ per pound}$$

It costs an average of $0.0375 per pound to dispose of clean wood.

37. In order to consume 8 grams of fiber, you would need to eat 3 servings $\left(2\frac{1}{4}\text{ cups}\right)$ of Wheaties along with the large banana.

39. The equation of the vertical line is $x = 4$.

41. The slope-intercept form is $y = mx + b$, and the point-slope form is $y - y_1 = m(x - x_1)$. For a vertical line, the slope m is undefined and thus may not be substituted into either of the two forms.

43. Vertical lines are not functions, since they fail the Vertical Line Test. However, all nonvertical lines are functions.

45. $x = 3$

47. The x-intercept is $\left(\frac{c}{a}, 0\right)$.

The y-intercept is $\left(0, \frac{c}{b}\right)$.

The slope of the line is $m = -\frac{a}{b}$.

49. b must equal 23. **51.** (a, b)

Section 1.3 *(page 45)*

1. (a) The scatter plot shows that the data are near-linear.

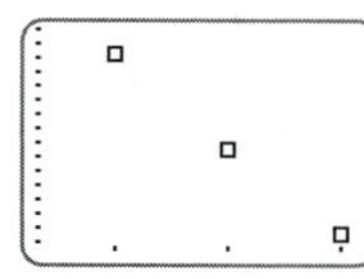

(b) The equation of the line of best fit is $y = -6.290x + 110.9$.

(c) $m = -6.290$ means that the Harbor Capital Appreciation Fund share price is dropping at a rate of \$6.29 per month.

The y-intercept means that in month 0 of 2000, the fund price was \$110.94. Since the months of 2000 begin with 1, not 0, the y-intercept does not represent the price at the end of January. It could, however, be interpreted as being the price at the end of December 1999.

(d) This model is a useful tool to show the trend in the stock price between October and December 2000. Since stock prices tend to be volatile, we are somewhat skeptical of the accuracy of data values outside of that domain. (Answers may vary.)

3. (a) The scatter plot shows that the data are near-linear.

(b) The equation of the line of best fit is $y = 1165.9x + 80{,}887$.

(c) $m = 1165.9$ means that Washington State public university enrollment is increasing by about 1166 students per year.

The y-intercept means that, according to the model, Washington State public university enrollment was 80,887 in 1990.

(d) The model fits the data extremely well, as shown by the graph of the line of best fit.

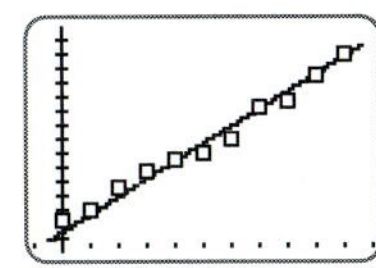

This model could be used by Washington State legislators and university administrators in budgeting and strategic planning. (Answers may vary).

5. (a) The scatter plot shows that although the data are not linear, they are nonincreasing.

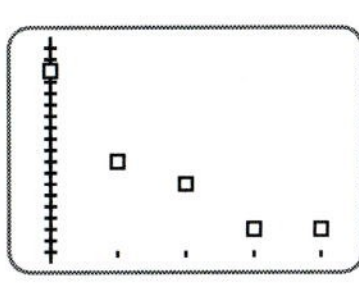

(b) The equation of the line of best fit is $y = -3.4x + 39.6$.

(c) $m = -3.4$ means that the per capita income ranking of North Carolina is changing at a rate of 3.4 places per year. That is, the state is moving up in the rankings by about 3 places per year.

The y-intercept means that in 1995 (year 0), North Carolina was ranked 40th out of the 50 states. (Only positive whole number rankings make sense.)

(d) This model is not a highly accurate representation of the data, as shown by the graph of the line of best fit, so it should be used with caution.

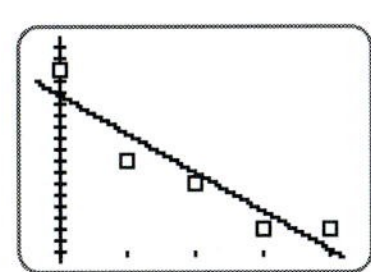

However, an incumbent government official could use the model in a 2000 reelection campaign as evidence that the state's economy had improved during his or her tenure in office. The official might also use the model to claim that the trend of improvement will continue if he or she is reelected. (Answers may vary.)

7. (a)

$$C(x) = \begin{cases} 15.25 & 0 \le x \le 320 \\ 0.04274x + 1.036 & 340 \le x \le 760 \\ 0.04409x & 780 \le x \end{cases}$$

For values of $x \ge 780$, $C(x)$ is directly proportional to x.

(b) \$15.25

(c) \$23.26

(d) The cost per pound is lowest when 780 or more pounds of trash are disposed of. As a construction company, we would try to keep the weight of our trash deliveries at or above 780 pounds. (Answers may vary.)

9. **(a)** $C(t) = \begin{cases} 29.99 & 0 \le t \le 200 \\ 0.4t - 50.01 & t > 200 \end{cases}$

(b) For the Qwest plan, we have $C(300) = \$82.49$

For the Sprint plan, we have $C(300) = \$69.99$

The Sprint plan is the best deal for a customer who uses 300 anytime minutes.

11. **(a)** Let t be the production year of a Toyota Land Cruiser 4-Wheel Drive and V be the value of the vehicle in 2001.

$$V = 4505t - 8{,}966{,}350$$

(b) \$7610

(c) The model substantially underestimated the value of a 1992 Land Cruiser.

13. **(a)** Let F be the number of grams of fat in x Chef Salads. We have $F = 8x$.

The number of fat grams is directly proportional to the number of Chef Salads.

(b) less than 10 salads

(c) Up to 7.75 salads

(d) Up to 5.875 salads

15. Let s be the number of segments and C be the maximum total cost of the ticket (in dollars) from Seattle to Phoenix. We have

$$C = 10s + 222$$

Note that $s \ge 1$.

17. a is the price of a cup of coffee, b is the price of a bagel, and c is the price of a muffin.

19. The correlation coefficient $r = 0$ indicates that the model doesn't fit the data well.

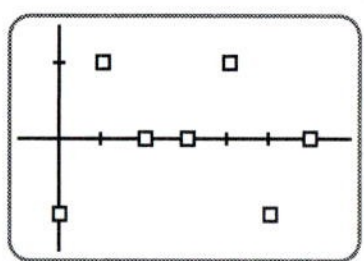

In addition, we can see from the scatter plot that the model doesn't fit well.

21. The y-intercept form of a line passing through the origin is

$$y = mx$$

This implies that y is directly proportional to x.

If the linear model $y = mx$ fits the original data well, it is likely that the dependent variable of the original data is directly proportional to the independent variable. However, if the linear model $y = mx$ does not fit the data well, the dependent and independent variables of the original data are not directly proportional.

23. The line of best fit is $y = 1$. This model fits the table of data perfectly.

```
LinReg
 y=ax+b
 a=0
 b=1
 r²=
 r=
```

However, from the calculator display, we see that r is undefined.

The correlation coefficient is formally defined as

$$r = \frac{n(\Sigma xy) - (\Sigma x)(\Sigma y)}{\sqrt{n(\Sigma x^2) - (\Sigma x)^2}\sqrt{n(\Sigma y^2) - (\Sigma y)^2}}$$

$$r = \frac{0}{0}$$

Because we can't divide by zero and there is a zero in the denominator of the expression, the correlation coefficient r is undefined.

25. $y = \begin{cases} -2x + 10 & -2 \le x \le 4 \\ 0.5x & 4 < x \le 10 \end{cases}$

CHAPTER 1 REVIEW EXERCISES

Section 1.1 *(page 49)*

1. $C(2) = 99.90$
The cost to buy two pairs of shoes is \$99.90.

3. $H(2) = 36$
The cliff diver is 36 feet above the water 2 seconds after he jumps from a 100-foot cliff.

5. The domain is all real numbers.

7. The domain is all real numbers except $r = 1$ and $r = -1$.

9. At $t = 2$, $P \approx 21.03$. The stock price of the computer company at the end of the day two days after December 16, 2001, was about \$21.03.

11. The graph represents a function.

Section 1.2 *(page 50)*

13. $m = \dfrac{2}{3}$

15. $m = -11$

17. The y-intercept is $(0, 18)$.
The x-intercept is $(6, 0)$.

19. $y = -1x + 7$
In standard form, $x + y = 7$.

Section 1.3 *(page 50)*

21. **(a)** Let x be the number of large orders of French fries consumed. The amount of fat consumed is given by $F(x) = 29x$ grams.
(b) Let y be the number of Big N' Tasty sandwiches consumed. The amount of fat consumed is given by $G(y) = 32y$ grams.
(c) Only one combination meal may be eaten.

23. **(a)** Let t be the production year of the Mercedes-Benz Roadster two-door SL500 and $v(t)$ be the value of the car in 2001.

$$v(t) = 7312.5t - 14{,}558{,}975$$

(b) \$58,712.50
(c) The linear model was extremely effective at accurately predicting the value of the 1999 Mercedes-Benz Roadster two-door SL500. The \$212.50 difference between the predicted value of the model and the NADA guide average value was negligible.

CHAPTER 2

Section 2.1 *(page 73)*

1. (2, 1)

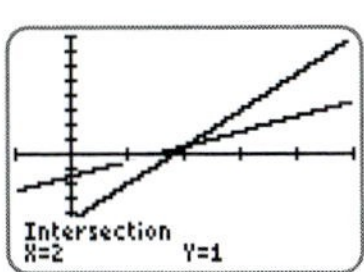

3. $(-0.2118, 5)$

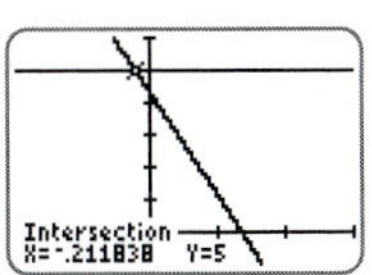

5. The lines are parallel and thus do not intersect. So, a solution does not exist.

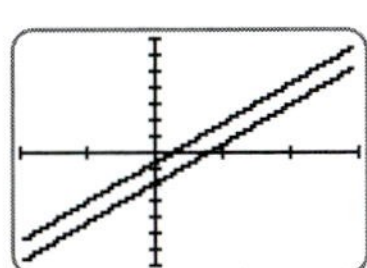

7. $(0, -9)$

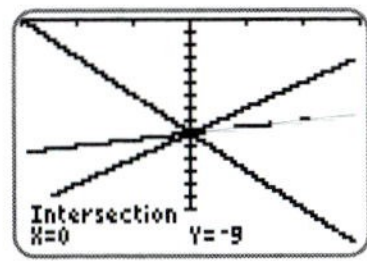

9. The lines do not intersect at a common point. So, a solution does not exist.

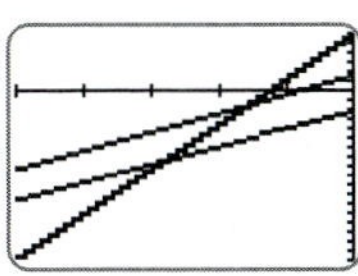

11. $x = 2, y = 1$ **13.** $x \approx -0.2118, y = 5$

15. There is no solution. **17.** $x = 0, y = -9$

19. The three lines do not intersect at a common point, so there is no solution.

21. By early 1996 ($t = 1.354$), the diabetes infection rates had become equal.

23. Graphing the three models simultaneously yields the following:

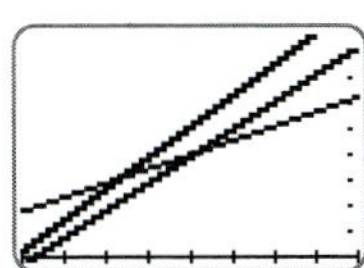

Since the functions do not intersect at a common point, the cost of the plans will never be equal.

25. At the end of 2022.

27. Near the end of 1996, the sports team industry and independent artist industry payrolls were equal. Since the original models were based upon just two years of data (1998 and 1999), we are uncertain as to whether the payroll of each industry can, in general, be modeled well by a linear function.

If the linear model fits additional data points well near $t = 0$, $t = -2.170$ is likely to be a good estimate for the time when the payrolls of both industries were equal.

29. According to the model, the per capita consumption of fish exceeded the per capita consumption of chicken before the latter part of 1956, 14.20 years before the end of 1970. Since the value $t = -14.20$ is so far outside of the data set ($0 \le t \le 29$), we are somewhat skeptical of this result.

31. $x = 2, y = 1, z = 5$ **33.** $x = 1, y = 0, z = 0$

35. Any point of the form $\left(-\frac{9}{8}t - \frac{45}{4}, \frac{1}{8}t + \frac{45}{4}, t\right)$ is a solution.

37. $x = 9, y = 6, z = -4$

39. $x = 1, y = 0, z = -1$

41. $x = 1, y = 0, z = 1, w = 0$

43. The company has $100 left over after paying the dye supplier for 250 bottles of sky blue dye and paying the lease on the building. It is mathematically possible for it to generate sufficient revenue to pay the dye supplier and the building lease.

45. No. Whenever the number of variables exceeds the number of equations, there are 0 or infinitely many solutions.

Section 2.2 *(page 89)*

1. 2×3 3. 1×3 5. 1×1

7. $$\begin{aligned} x + 3y &= 12 \\ 4x + y &= 7 \\ -x + 6y &= 9 \end{aligned}$$

9. $$\begin{aligned} 3x - y + 9z &= 13 \\ -3x + y &= -31 \end{aligned}$$

11. The matrix is not in reduced row echelon form because the leading 1 in R_2 is to the *right* of the leading 1 in R_3.

13. The matrix is not in reduced row echelon form because the leading entry in R_2 is not a 1.

15. The matrix is in reduced row echelon form.

17. The matrix is not in reduced row echelon form because the leading 1 in R_1 is to the *right* of the leading 1 in R_2.

19. The matrix is not in reduced row echelon form because the leading 1 in R_2 is not the only nonzero entry in its corresponding column.

21. $\left[\begin{array}{cc|c} 2 & 5 & 2 \\ 3 & -5 & 3 \end{array}\right]$ reduces to $\left[\begin{array}{cc|c} 1 & 0 & 1 \\ 0 & 1 & 0 \end{array}\right]$

The solution is $x = 1$ and $y = 0$.

23. $\left[\begin{array}{ccc|c} -2 & 6 & 4 & 10 \\ 4 & -12 & 2 & -20 \\ 3 & 4 & -1 & 11 \end{array}\right]$ reduces to

$\left[\begin{array}{ccc|c} 1 & 0 & 0 & 1 \\ 0 & 1 & 0 & 2 \\ 0 & 0 & 1 & 0 \end{array}\right]$

So $x = 1$, $y = 2$, and $z = 0$.

25. $\left[\begin{array}{ccc|c} 3 & -2 & 1 & 6 \\ 11 & -20 & -1 & 0 \\ 0 & 1 & 1 & 3 \end{array}\right]$ reduces to $\left[\begin{array}{ccc|c} 1 & 0 & 0 & 2 \\ 0 & 1 & 0 & 1 \\ 0 & 0 & 1 & 2 \end{array}\right]$

So $x = 2$, $y = 1$, and $z = 2$.

27. $\left[\begin{array}{cc|c} 1 & -1 & -5 \\ 9 & 1 & 25 \\ 29 & 1 & 65 \end{array}\right]$ reduces to $\left[\begin{array}{cc|c} 1 & 0 & 2 \\ 0 & 1 & 7 \\ 0 & 0 & 0 \end{array}\right]$

The solution is $x = 2$ and $y = 7$.

29. $\left[\begin{array}{cc|c} 2 & -4 & 16 \\ 9 & 1 & -4 \\ -3 & 6 & -24 \end{array}\right]$ reduces to $\left[\begin{array}{cc|c} 1 & 0 & 0 \\ 0 & 1 & -4 \\ 0 & 0 & 0 \end{array}\right]$

The solution is $x = 0$ and $y = -4$.

31. $\left[\begin{array}{cc|c} 2 & -2 & 0 \\ 3 & 1 & 4 \\ 5 & -1 & 5 \end{array}\right]$ reduces to $\left[\begin{array}{cc|c} 1 & 0 & 0 \\ 0 & 1 & 0 \\ 0 & 0 & 1 \end{array}\right]$

The system is inconsistent because $0x + 0y \neq 1$.

33. $\left[\begin{array}{ccc|c} 1 & 1 & 1 & 6 \\ 2 & -1 & 1 & 3 \end{array}\right]$ reduces to $\left[\begin{array}{ccc|c} 1 & 0 & \frac{2}{3} & 3 \\ 0 & 1 & \frac{1}{3} & 3 \end{array}\right]$

We have $x + \frac{2}{3}z = 3$ and $y + \frac{1}{3}z = 3$. This is a dependent system. All solutions are of the form

$$\begin{aligned} x &= -\frac{2}{3}t + 3 \\ y &= -\frac{1}{3}t + 3 \\ z &= t \end{aligned}$$

for any real number t.

35. $\left[\begin{array}{ccc|c} 3 & 5 & 1 & 6 \\ 6 & 10 & 2 & 10 \end{array}\right]$ reduces to $\left[\begin{array}{ccc|c} 1 & \frac{5}{3} & \frac{1}{3} & 0 \\ 0 & 0 & 0 & 1 \end{array}\right]$

This system is inconsistent because $0x + 0y + 0z \neq 2$.

37. $\left[\begin{array}{ccc|c} 1 & 1 & 1 & 6 \\ 2 & -1 & 1 & 3 \\ 4 & -2 & 3 & 9 \end{array}\right]$ reduces to $\left[\begin{array}{ccc|c} 1 & 0 & 0 & 1 \\ 0 & 1 & 0 & 2 \\ 0 & 0 & 1 & 3 \end{array}\right]$

The solution is $x = 1$, $y = 2$, and $z = 3$.

39. $\left[\begin{array}{ccc|c} 3 & 2 & 3 & 6 \\ 2 & -5 & 1 & -11 \\ 4 & 2 & 3 & 3 \end{array}\right]$ reduces to

$\left[\begin{array}{ccc|c} 1 & 0 & 0 & -3 \\ 0 & 1 & 0 & \frac{30}{17} \\ 0 & 0 & 1 & \frac{65}{17} \end{array}\right]$

The solution is $x = -3$, $y = \frac{30}{17}$, and $z = \frac{65}{17}$.

41. 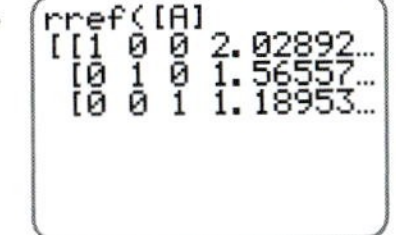

$x = 2.0289, y = 1.5656, z = 1.1895$

43. 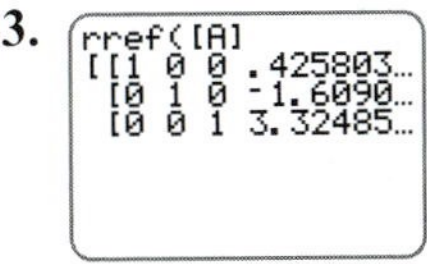

$x = 0.42580, y = -1.6090, z = 3.3249$

45. 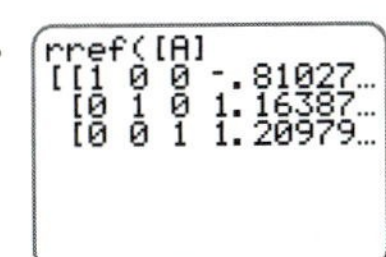

$x = -0.81028, y = 1.1639, z = 1.2098$

47. 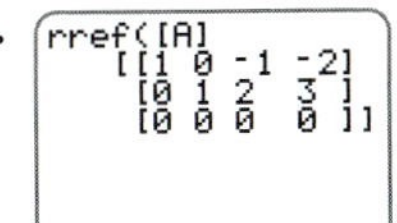

The system is dependent. We have $x - z = -2$ and $y + 2z = 3$. All solutions are of the form

$$x = t - 2$$
$$y = -2t + 3$$
$$z = t$$

for any real number t.

49. 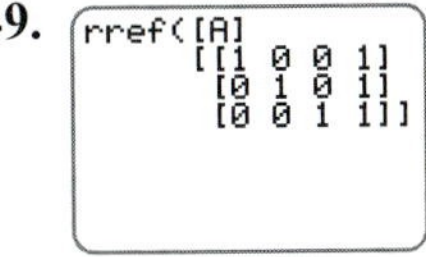

$x = 1, y = 1, z = 1$

Section 2.3 *(page 99)*

1. (a) We have $S = 7.62t + 284$, where S is the projected number of students enrolled in second grade t years after 1998.

   We have $T = 9.99t + 284$, where T is the projected number of students enrolled in third grade t years after 1998.

   (b) $S = T$ when $t = 0$

   (c) According to the models, the same number of students (284) was projected to be enrolled in second grade and third grade in 1998. The model projections slightly underestimated the actual second grade enrollment and overestimated the actual third grade enrollment.

   (d) The model results could be used for strategic planning in the allocation of classrooms and teachers. The projections would also be helpful in determining when to build a new elementary school. According to the models, the number of students enrolled in second grade is expected to increase at a rate of about 8 students per year, while third grade enrollment is expected to increase at an approximate rate of 10 students per year. (Answers will vary.)

3. (a) We have $C = 0.9818t + 23.74$, where C is the annual per capita consumption of chicken (in pounds) t years after 1970.

   We have $F = 0.1372t + 11.72$, where F is the annual per capita consumption of fish and shellfish (in pounds) t years after 1970.

   (b) $C = F$ when $t \approx -14.2$

   (c) According to the models, the per capita consumption of chicken and of fish and shellfish was equal in late 1956 (14.2 years before the end of 1970). At that time, the annual per capita consumption was approximately 9.8 pounds.

   We are somewhat skeptical of this result and question the accuracy of the interpretations, since the value $t = -14.2$ is so far outside of the original data set. For the original data set, $0 \leq t \leq 29$.

   (d) Despite the apparent inaccuracy of the interpretation of the intersection point, the models do have practical value. The annual per capita chicken consumption is increasing at a rate of 0.98 pound per year. The annual per capita fish and shellfish consumption is increasing at a rate of 0.14 pound per year. Poultry farmers and seafood producers could use these figures as a baseline when attempting to determine the effects of nationwide advertising campaigns. For example, suppose that poultry farmers launch a nationwide chicken advertising campaign at the start of 2007 and, at the end of 2007, the annual per capita chicken consumption has increased by 2 pounds. Since the model only projects a 0.98-pound increase, the producers conclude that the advertising campaign may have had an effect on chicken consumption. However, other variables would have to be analyzed to ensure that the increase was in fact due to the advertising and not some other market condition. (Answers will vary.)

5. (a) We have $N = 2.969t + 24.23$, where N is the number of people (in millions) who attended a National League baseball game t years after 1994.

We have $A = 1.689t + 24.83$, where A is the number of people (in millions) who attended an American League baseball game t years after 1994.

(b) $A = N$ when $t \approx 0.4688$

(c) According to the models, in mid-1995 (0.47 year after the end of 1994), the baseball game attendance of the two different leagues was equal. We are skeptical of this result for a couple of reasons. First, the linear models did not fit the data well. Second, the baseball season doesn't last the entire year. Therefore, we must be careful about drawing conclusions based on decimal estimates from the model. Round to the nearest year, we could estimate that annual baseball game attendance at games in the two leagues was equal in 1994 or 1995.

(d) Because of the relative inaccuracy of the models, we should be wary of any conclusions drawn as a result. The slopes of the two functions, $m_N = 2.969$ and $m_A = 1.689$, show that National League game attendance increased by roughly 3.0 million people per year and American League game attendance increased by roughly 1.7 million people per year. Representatives of Major League Baseball could use these results as an impetus for research to determine why National League game attendance is growing faster than American League game attendance. (Answers will vary.)

7. (a) The percentage of births to unmarried women in the United States may be modeled by

$$U = \frac{5}{6}t + 18$$

where t is the number of years since the end of 1980.

The percentage of births to unmarried women in the United Kingdom may be modeled by

$$K = \frac{13}{9}t + 12$$

where t is the number of years since the end of 1980.

(b) $U = K$ when $t \approx 9.82$

(c) According to the models, the percentage of all births in the United States and the United Kingdom that were to unmarried women were equal near the end of 1990 ($t = 9.82$). At that time, approximately 26 percent of all births were to unmarried women.

(d) Although the percentage of births to unmarried women in the United States exceeded the percentage of births to unmarried women in the United Kingdom by 6 percentage points at the end of 1980, by the end of 1998 the percentage of births to unmarried women in the United States trailed the percentage of births to unmarried women in the United Kingdom by 5 percentage points. Assuming that both countries have the goal of limiting the number of births to unmarried women, the countries could compare their policies and pregnancy planning services in an effort to determine why the United States had greater success than the United Kingdom. (Answers will vary.)

9. (a) The cost (in cents) to send a 1-ounce letter in the United States t years after 1985 may be modeled by $L = 0.7409t + 23.12$.

The cost (in cents) to send a postcard in the United States t years after 1985 may be modeled by $P = 0.3912t + 14.81$.

(b) $L = P$ when $t \approx -23.76$

(c) The models indicate that the cost to send a 1-ounce letter and the cost to send a postcard were equal in early 1962 ($t = -23.76$). We are skeptical of these results because $t = -23.76$ lies so far outside of the original data set ($0 \le t \le 16$).

Additionally, in our lifetime the cost of sending a letter has always exceeded the cost of sending a postcard.

(d) Postal service companies (e.g., the U.S. Postal Service, UPS, FedEx) could use the model to help determine future postage rates. The postage cost of a 1-ounce letter has increased by approximately 3 cents every four years, while the postage cost of a postcard has increased by about 2 cents every five years. (Answers may vary.)

11. (a) The average annual earnings (in dollars) of a paper manufacturing employee t years after the end of 1995 may be modeled by $E = 1335.1t + 39{,}408$.

The average annual earnings (in dollars) of a printing and publishing employee t years after the end of 1995 may be modeled by $P = 1644.7t + 34{,}351$

(b) $E = P$ when $t \approx 16.333$

(c) According to the models, in early 2012 $(t = 16.333)$, the annual earnings of paper manufacturing employees and of printing and publishing employees are expected to be equal. Although the results make sense in their real-world context and the models fit the data very well, the solution $t = 16.333$ is so far outside of the original data set $(0 \leq t \leq 4)$ that we must be cautious in using the result.

(d) If current trends continue, printing and publishing employees will eventually make more than paper manufacturing employees. Printers and publishing companies could use this information to show the strength of their industry as they recruit prospective employees. (Answers will vary.)

13. (a) The number of clothing and accessories stores t years after the end of 1998 may be modeled by $C(t) = -929t + 152{,}603$

The number of food and beverage stores t years after the end of 1998 may be modeled by $F(t) = 3854t + 147{,}652$

(b) $C = F$ when $t \approx 1.035$

(c) According to the models, shortly after the end of 1999, the number of food and beverage stores surpassed the number of clothing and accessories stores.

(d) Entrepreneurs and investors could use this information to forecast the growth in the two industries. The fact that the number of food and beverage stores is increasing while the number of clothing and accessories stores is decreasing would be important to consider when making an investment decision. (Answers may vary.)

15. (a) The number of men's clothing stores t years after the end of 1998 may be modeled by $M(t) = -416t + 11{,}861$

The number of children's clothing stores t years after the end of 1998 may be modeled by $C(t) = 168t + 5165$

(b) $M = C$ when $t \approx 11.47$

(c) According to the models, roughly 6 months into 2010, the number of men's clothing stores and the number of children's clothing stores will be equal. The fact that the intersection point is so far outside of the original data set calls into question the accuracy of the projection.

(d) Clothing producers could use these models to determine in what specialty area to focus their production efforts. (Answers may vary.)

17. (a) The retail sales (in millions of dollars) of building materials and supplies stores t years after the end of 1992 may be modeled by $B(t) = 14{,}675.0t + 157{,}819$

The retail sales (in millions of dollars) of department stores t years after the end of 1992 may be modeled by $D(t) = 6889.85t + 187{,}399$

(b) $B = D$ when $t \approx 3.80$

(c) According to the models, roughly 10 months into 1996, the annual retail sales of building materials and supplies stores and the annual retail sales of department stores was equal.

(d) Investors could use these models in deciding in which industries to invest. Although the department store sales initially exceeded the building supplies sales, department store sales were increasing at a much slower rate than building supplies store sales. As a result, building supplies store sales ultimately surpassed department store sales. (Answers may vary.)

19. He should invest \$818.18 in the Bond Market account, \$272.73 in the Growth account, and \$1909.09 in the Stock account.

21. It should buy 30 Corollas, 10 Camrys, and 15 Prius cars.

23. The quadratic model for the Johnson & Johnson cost of goods sold is

$$C(t) = 103t^2 + 315t + 8539$$

where t is the number of years after 1999 and C is in millions of dollars.

25. The quadratic model for the Gatorade/Tropicana North America net sales is

$$R(t) = -107t^2 + 496t + 3452$$

where t is the number of years after 1999 and R is in millions of dollars.

27. In order to use up all of the buns, patties, and cheese slices, the family should make 60 hamburgers, 16 cheeseburgers, and 28 double cheeseburgers.

29. The breeder should purchase 9 bags of venison meal and rice dog food, 0 bags of natural dog food, and 4 bags of lamb dog food.

31. In order to use up all of the supplies, the company should assemble 20 compact, 40 standard, and 30 deluxe first-aid kits.

33. If 1000 floor tickets, 5000 lower tickets, and 6000 upper tickets are sold, the desired revenue is obtained.

35. In order to use all of the available labor, the company should produce 40 rockers, 80 armchairs, and 40 benches.

37. There are 14 pennies, 22 nickels, 22 dimes, and 6 quarters.

39. We pick $x_4 = 150$ and $x_4 = 200$ and generate a table of values showing each solution.

x_4	x_1	x_2	x_3
150	100	50	0
200	150	100	50

(Answers may vary.)

CHAPTER 2 REVIEW EXERCISES

Section 2.1 (page 106)

1. (1, 7)

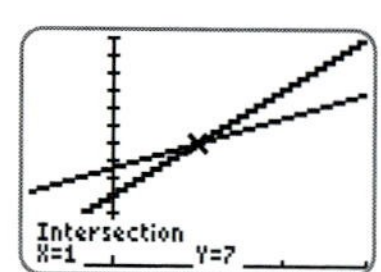

3. (5, −6)

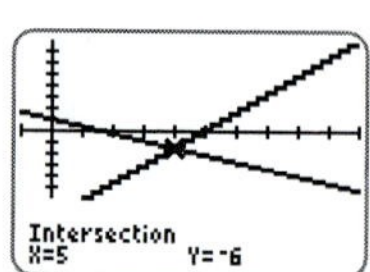

5. The lines are parallel and do not intersect. So, there is no solution.

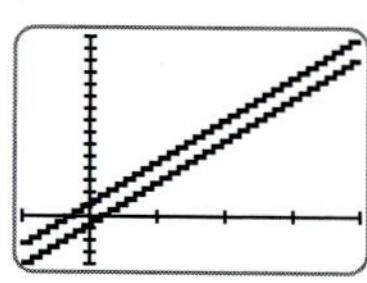

7. (1, 0, 2)

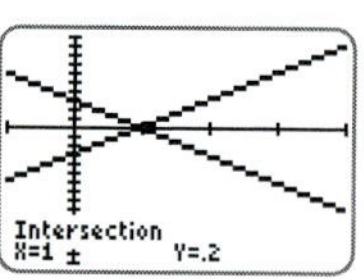

9. $x = 1, y = 7$ **11.** $x = 5, y = -6$

13. There is no solution. **15.** $x = 1, y = 0.2$

Section 2.2 (page 107)

17. The leading entry in row 2 is not the only nonzero entry in its corresponding column. Also, the leading entry in row 3 is not to the right of the leading entry in the row above it.

19. All rows of zeros are not at the bottom of the matrix. Also the leading entry in row 3 is not to the right of the leading entry in the row above it.

21. Reduced row echelon form

23. Reduced row echelon form

25. $\left[\begin{array}{cc|c} 2.1 & -1 & -8 \\ 3.4 & 1 & 8 \end{array}\right]$ reduces to $\left[\begin{array}{cc|c} 1 & 0 & 0 \\ 0 & 1 & 8 \end{array}\right]$

$x = 0, y = 8$

27. $\left[\begin{array}{cc|c} 4 & -8 & -16 \\ 1 & 1 & 5 \end{array}\right]$ reduces to $\left[\begin{array}{cc|c} 1 & 0 & 2 \\ 0 & 1 & 3 \end{array}\right]$

$x = 2, y = 3$

29. $\left[\begin{array}{ccc|c} 1 & 1 & 1 & 6 \\ 2 & -1 & 1 & 3 \\ 3 & 0 & 2 & 9 \end{array}\right]$ reduces to $\left[\begin{array}{ccc|c} 1 & 0 & \frac{2}{3} & 3 \\ 0 & 1 & \frac{1}{3} & 3 \\ 0 & 0 & 0 & 0 \end{array}\right]$

$x = -\frac{2}{3}z + 3$ and $y = -\frac{1}{3}z + 3$

Let $z = t$. Then $x = -\frac{2}{3}t + 3$ and $y = -\frac{1}{3}t + 3$.

31. $\left[\begin{array}{ccc|c} 1 & -1 & 3 & 5 \\ 2 & 4 & 1 & -1 \end{array}\right]$ reduces to

$\left[\begin{array}{ccc|c} 1 & 0 & \frac{13}{6} & \frac{19}{6} \\ 0 & 1 & -\frac{5}{6} & -\frac{11}{6} \end{array}\right]$

Let $z = t$. Then $x = -\frac{13}{6}t + \frac{19}{6}$ and $y = \frac{5}{6}t - \frac{11}{6}$.

33. $\left[\begin{array}{ccc|c} 1 & 0 & 0 & 0.3289 \\ 0 & 1 & 0 & -0.7230 \\ 0 & 0 & 1 & 0.5717 \end{array}\right]$

35. $\left[\begin{array}{ccc|c} 1 & 0 & 0.5066 & 0.8054 \\ 0 & 1 & -0.6303 & -0.3195 \\ 0 & 0 & 0 & 0 \end{array}\right]$

Section 2.3 *(page 107)*

37. (a) $K(t) = \frac{13}{9}t + 12$

$I(t) = \frac{4}{3}t + 40$

(b) $I = K$ when $t = 252$

(c) According to the model, the number of births to unmarried mothers in Iceland and the United Kingdom will be the same 252 years after the end of 1980. According to the model, 376 percent of births will be to unmarried mothers. This doesn't make sense, since the maximum percentage of births that can be to unmarried mothers is 100 percent.

(d) The models could be used by sociologists and government policymakers to forecast the unmarried birth rate in years subsequent to 1998. However, as shown in part (c), the results should be used with caution.

39. (a) $M(t) = 153.72t + 27{,}110$
$W(t) = 461.28t + 1283.0$

(b) $M = W$ when $t = 83.97$

(c) According to the models, NCAA basketball attendance at men's and women's games will be equal approximately 84 years after the end of 1985. In that year, attendance is projected to be 40,019 thousand for both men's and women's basketball. The result makes sense, but the point of intersection is so far outside of the original data set that we are skeptical of the result.

(d) The models could be used by the NCAA as a baseline in monitoring fan support of men's and women's basketball programs.

41. The quadratic model is
$S(t) = 29.9t^2 + 76.6t + 178.4$

According to the model, in 2004 Starbucks will generate $4639 million in sales.

43. The company should assemble 15 compact, 35 standard, and 25 deluxe first-aid kits.

CHAPTER 3

Section 3.1 *(page 121)*

1. $A + B = \begin{bmatrix} -4 & 6 \\ 10 & 12 \\ -4 & 3 \end{bmatrix}$

3. Since B and C do not have the same dimensions, they may not be added together.

5. $D + C = \begin{bmatrix} 3 & 0 & 4 \\ -1 & 9 & 8 \\ 14 & 6 & 2 \end{bmatrix}$

7. $3B = \begin{bmatrix} -15 & 0 \\ 21 & 24 \\ -27 & 3 \end{bmatrix}$

9. $-2C + 2D = \begin{bmatrix} -2 & -12 & 8 \\ 2 & 2 & 12 \\ 8 & 20 & 4 \end{bmatrix}$

11.

```
1.2[A]
[[1.44   7.56 ....
 [-10.92 5.04 2...
 [1.08   -2.4 ....
```

$1.2A = \begin{bmatrix} 1.44 & 7.56 & 0.48 \\ -10.92 & 5.04 & 2.04 \\ 1.08 & -2.4 & 0.36 \end{bmatrix}$

(*Hint:* Use the blue arrow keys on the calculator to display the third column of the matrix.)

13.

```
1.2[A]-2.3[B]
[[-1.78  8.25   ...
 [-4.48  -7.61  ...
 [-20.08 -22.18...
```

$1.2A - 2.3B = \begin{bmatrix} -1.78 & 8.25 & -0.44 \\ -4.48 & -7.61 & -14.29 \\ -20.08 & -22.18 & -4.24 \end{bmatrix}$

15.

```
-1.1[A]+2.9[B]
[[2.74  -7.8  ....
 [1.89  11.33 1...
 [25.69 27.14 5...
```

$-1.1A + 2.9B = \begin{bmatrix} 2.74 & -7.8 & 0.72 \\ 1.89 & 11.33 & 18.72 \\ 25.69 & 27.14 & 5.47 \end{bmatrix}$

17.

-8.7[A]+8.7[B]
[[1.74 -57.42 ...
[54.81 11.31 ...
[72.21 92.22 ...

$$-8.7A + 8.7B = \begin{bmatrix} 1.74 & -57.42 & 0 \\ 54.81 & 11.31 & 46.98 \\ 72.21 & 92.22 & 14.79 \end{bmatrix}$$

19. 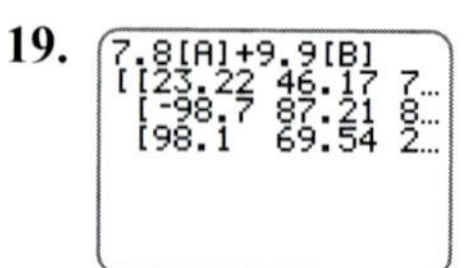

$$7.8A + 9.9B = \begin{bmatrix} 23.22 & 46.17 & 7.08 \\ -98.7 & 87.21 & 83.55 \\ 98.1 & 69.54 & 22.14 \end{bmatrix}$$

21.

Combined	LSL	RB	MJ
Consumer loan	2,700	0	2,950
Car loan	0	27,700	0
Credit card	0	8,400	0
Mortgage	0	82,500	0

23.

2005–2006 Pay Scale	240 Credits ($)	300 Credits ($)	360 Credits ($)
Level 1	40,208	44,341	48,475
Level 2	42,365	46,498	50,631
Level 3	44,521	48,654	52,787
Level 4	46,678	50,811	54,943

25. The 2006 prices are projected to be as follows:

	Albertsons	A&W	Henry Weinhards
Root beer, 6-pack	$1.79	$1.88	$5.06
Cream soda, 12-pack	$3.58	$3.76	$10.13
Club soda, 2 liters	$1.11		

27. Average Markup

	Accord	Civic
2000 model	$2,300	$1,875
2001 model	$2,400	$1,975

29. Surplus Energy

Years Since 1960	Natural Gas (quadrillion BTUs)	Coal (quadrillion BTUs)
0	0.27	0.98
10	−0.13	2.34
20	−0.48	3.18
30	−0.94	3.21
40	−3.59	0.25

31.

Surplus Energy

Years Since 1960	Nuclear Electric Power (quadrillion BTUs)	Coal (quadrillion BTUs)
0	0	0.98
10	0	2.34
20	0	3.18
30	0	3.21
40	0	0.25

33. Total Renewable Energy Consumption 1997–1999

Renewable Energy Type	Energy Consumed (quadrillion BTUs)
Conventional hydroelectric power	10.91
Geothermal energy	1.00
Biomass	9.48
Solar energy	0.22
Wind energy	0.10

35. Difference in Energy Consumption

Renewable Energy Type	Energy Consumed (quadrillion BTUs)
Conventional hydroelectric power	−0.22
Geothermal energy	0.00
Biomass	0.35
Solar energy	0.01
Wind energy	0.01

The amount of conventional hydroelectric power consumed in 1999 was less than the three-year average. The amount of geothermal energy consumed in 1999 was equal to the three-year average. The amount of biomass, solar, and wind energy consumed in 1999 exceeded the three-year average.

37. $B = \begin{bmatrix} -2 & 74 & -4 \\ 27 & -20 & -50 \\ -38 & 1 & \frac{11}{3} \end{bmatrix}$

39. $B = \begin{bmatrix} \frac{3}{2} & -11 & 1 \\ -\frac{15}{4} & 3 & 8 \\ \frac{11}{2} & 0 & -\frac{7}{12} \end{bmatrix}$

Section 3.2 *(page 138)*

1. [0] **3.** $A + B = \begin{bmatrix} 5 & 0 \\ 0 & 0 \end{bmatrix}$

5. $4A = \begin{bmatrix} 12 & 16 \\ -4 & -8 \end{bmatrix}$

7. $4A - 2B = \begin{bmatrix} 8 & 24 \\ -6 & -12 \end{bmatrix}$

9. $AB = \begin{bmatrix} 10 & -4 \\ -4 & 0 \end{bmatrix}$

11. We cannot calculate CD, since C is a 3×2 matrix and D is a 3×3 matrix.

13. $CA = \begin{bmatrix} 20 & 24 \\ 17 & 22 \\ 18 & 22 \end{bmatrix}$

15. $CB = \begin{bmatrix} 20 & -24 \\ 13 & -22 \\ 17 & -22 \end{bmatrix}$

17. $B^{-1} = \begin{bmatrix} \frac{1}{4} & \frac{1}{2} \\ -\frac{1}{8} & \frac{1}{4} \end{bmatrix}$

19. $AA^{-1} = \begin{bmatrix} 1 & 0 \\ 0 & 1 \end{bmatrix}$

21. The matrix is invertible because $\det(A) = 1$.

$$A^{-1} = \begin{bmatrix} 2 & -3 \\ -1 & 2 \end{bmatrix}$$

23. The matrix is singular because $\det(C) = 0$.

25. The matrix is invertible because $\det(E) = 1.1$.

$$E^{-1} = \begin{bmatrix} \frac{34}{11} & \frac{7}{11} \\ -\frac{40}{11} & -\frac{5}{11} \end{bmatrix}$$

27. The matrix is singular because $\det(B) = 0$.

29. The matrix is invertible because $\det(D) = 1.99$.

$$D^{-1} = \begin{bmatrix} \frac{100}{199} & \frac{110}{199} \\ -\frac{90}{199} & \frac{100}{199} \end{bmatrix}$$

31. From 1995 through 1999, the total amount of money spent on employee wages and salaries was 109,129,532 thousand dollars, or roughly $109 billion.

33. The first smoothie has 68 mg vitamin C, 53 mg calcium, and 7.5 g fiber. The second smoothie has 115.5 mg vitamin C, 86 mg calcium, and 7.2 g fiber.

35. The tropical fruit smoothie has 74.1 mg vitamin C, 41 mg calcium, and 13.8 g fiber. The pina colada smoothie has 34.1 mg vitamin C, 28 mg calcium, and 11.4 g fiber.

37. $RP = \begin{bmatrix} -52 & -7 \end{bmatrix}$

From 1996 to 1998, the number of military personnel dropped by 52 thousand and the number of army personnel dropped by 7 thousand.

$$PC = \begin{bmatrix} 565 \\ 520 \end{bmatrix}$$

In 1996, the number of military personnel not in the army was 565 thousand. In 1998, the number of military personnel not in the army was 520 thousand.

39. The accumulated sales for 1999–2001 was $22,792.2 million.

The accumulated cost of goods sold for 1999–2001 was $10,780.6 million.

The accumulated profit for 1999–2001 was $12,011.6 million.

41. $A = \begin{bmatrix} 0 & 0 \\ 0 & 0 \end{bmatrix}$, $B = \begin{bmatrix} 1 & 1 \\ 2 & 2 \end{bmatrix}$, and $C = \begin{bmatrix} 1 & 0 \\ 1 & 0 \end{bmatrix}$

(Answers may vary.)

43. We have

$$AA^{-1} = \begin{bmatrix} a & 0 & 0 \\ 0 & b & 0 \\ 0 & 0 & c \end{bmatrix} \begin{bmatrix} \frac{1}{a} & 0 & 0 \\ 0 & \frac{1}{b} & 0 \\ 0 & 0 & \frac{1}{c} \end{bmatrix}$$

$$= \begin{bmatrix} 1 & 0 & 0 \\ 0 & 1 & 0 \\ 0 & 0 & 1 \end{bmatrix}$$

$$= I$$

and

$$A^{-1}A = \begin{bmatrix} \frac{1}{a} & 0 & 0 \\ 0 & \frac{1}{b} & 0 \\ 0 & 0 & \frac{1}{c} \end{bmatrix} \begin{bmatrix} a & 0 & 0 \\ 0 & b & 0 \\ 0 & 0 & c \end{bmatrix}$$

$$= \begin{bmatrix} 1 & 0 & 0 \\ 0 & 1 & 0 \\ 0 & 0 & 1 \end{bmatrix}$$

$$= I$$

Therefore, the matrix A^{-1} is the inverse of A. A is singular if a, b, or c is equal to zero.

45. Let $A = \begin{bmatrix} a & b & c \\ b & d & e \\ c & e & f \end{bmatrix}$. Then

$$A^2 = \begin{bmatrix} a & b & c \\ b & d & e \\ c & e & f \end{bmatrix} \begin{bmatrix} a & b & c \\ b & d & e \\ c & e & f \end{bmatrix}$$

$$= \begin{bmatrix} a^2 + b^2 + c^2 & ab + bd + ce & ac + be + cf \\ ab + bd + ce & b^2 + d^2 + e^2 & bc + de + ef \\ ac + be + cf & bc + de + ef & c^2 + e^2 + f^2 \end{bmatrix}$$

The matrix A^2 is symmetric.

Section 3.3 (page 157)

1. $A^{-1} = \begin{bmatrix} 1.2 & -1.4 & -0.2 \\ -0.4 & 0.8 & -0.1 \\ 0.2 & -0.4 & 0.3 \end{bmatrix}$

3. $A^{-1} = \begin{bmatrix} 0 & 2 & -1 \\ -1 & 5 & -2 \\ 0.5 & -4 & 2 \end{bmatrix}$

5. $A^{-1} = \begin{bmatrix} 0 & 0.2 & -0.8 \\ 1 & -1.2 & 4.8 \\ -0.5 & 0.6 & -1.9 \end{bmatrix}$

7. A is singular.

9. A is singular.

11. $\begin{bmatrix} 1 & 1 & 1 \\ 2 & -1 & 1 \\ 4 & -2 & 3 \end{bmatrix} \begin{bmatrix} x \\ y \\ z \end{bmatrix} = \begin{bmatrix} 6 \\ 3 \\ 9 \end{bmatrix}$; $x = 1, y = 2, z = 3$

13. $\begin{bmatrix} 3 & 2 & 3 \\ 2 & -5 & 1 \\ 4 & 2 & 3 \end{bmatrix} \begin{bmatrix} x \\ y \\ z \end{bmatrix} = \begin{bmatrix} 6 \\ -11 \\ 3 \end{bmatrix}$; $x = -3, y = \frac{30}{17}$, $z = \frac{65}{17}$

15. $\begin{bmatrix} 1 & 1 & 1 \\ 2 & -1 & 1 \\ 4 & -2 & 3 \end{bmatrix}\begin{bmatrix} x \\ y \\ z \end{bmatrix} = \begin{bmatrix} 3 \\ 13 \\ 28 \end{bmatrix}$; $x = 4$, $y = -3$, $z = 2$

17. $\begin{bmatrix} 3 & -1 & 2 \\ 4 & 3 & 5 \\ 6 & 8 & 4 \end{bmatrix}\begin{bmatrix} x \\ y \\ z \end{bmatrix} = \begin{bmatrix} 12 \\ 45 \\ 64 \end{bmatrix}$; $x = 2$, $y = 4$, $z = 5$

19. $\begin{bmatrix} 2 & 5 & 3 \\ 1 & 4 & 2 \\ 0 & 2 & 4 \end{bmatrix}\begin{bmatrix} x \\ y \\ z \end{bmatrix} = \begin{bmatrix} 0 \\ -1 \\ 2 \end{bmatrix}$; $x = 1$, $y = -1$, $z = 1$

21. The matrix A is singular.

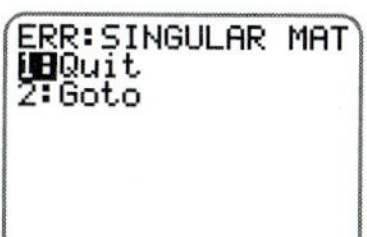

23.

```
[C]⁻¹
[[0 0            ...
 [0 .2380952381 ...
 [2 -.571428571 ...
```

$$C^{-1} \approx \begin{bmatrix} 0 & 0 & 3.333 \\ 0 & 0.2381 & -1.429 \\ 2 & -0.5714 & -19.24 \end{bmatrix}$$

25.

```
[E]⁻¹
[[-1.458333333 ...
 [-.25         ...
 [-.125        ...
 [.8333333333  ...
```

$$E^{-1} \approx \begin{bmatrix} -1.458 & 0.9722 & 0.09722 & 0.05556 \\ -0.25 & 0.5 & -0.25 & 0 \\ -0.125 & 0.08333 & 0.2083 & -0.1667 \\ 0.8333 & -0.5556 & -0.05556 & 0.1111 \end{bmatrix}$$

27.

```
[B]⁻¹
[[-.0667481663 ...
 [.104400978   ...
 [-.0596577017 ...
```

$$B^{-1} \approx \begin{bmatrix} -0.06675 & 0.1858 & -0.1296 \\ 0.1044 & -0.03423 & 0.1002 \\ -0.05966 & 0.01956 & 0.2285 \end{bmatrix}$$

29. The matrix D is singular.

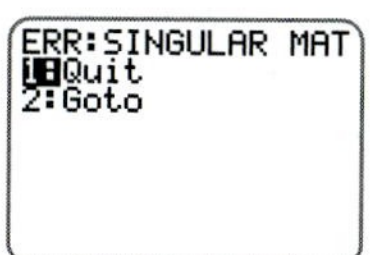

31.

	Protein Blend	Low-Fat Mix	Low-Carb Mix
Protein	23.03	21.47	16.02
Carbohydrates	38.99	40.37	25.12
Fat	70.47	67.95	77.60

33. To earn a 6% return, \$1539.17 should be invested in the Money Market fund and \$460.83 should be invested in the Large Cap Value fund.

To earn an 8% return, \$924.73 should be invested in the Money Market fund and \$1075.27 should be invested in the Large Cap Value fund.

To earn a 10% return, \$310.29 should be invested in the Money Market fund and \$1689.71 should be invested in the Large Cap Value fund.

35. If a student has earned 40.91 credits or less, it is mathematically possible for the student to earn a cumulative GPA of 3.5 by the time he completes 90 credits. If the student has earned more than 40.91 credits, it is mathematically impossible for the student to earn a cumulative GPA of 3.5 credits by the time he completes 90 credits. Only the student with 30 credits could conceivably increase his GPA to a 3.5 by the time he earns 90 credits.

37. $A^{-1} = \begin{bmatrix} 1 & \frac{1}{2}a & -\frac{1}{2}a & -\frac{1}{2}a \\ 0 & \frac{1}{2} & \frac{1}{2} & -\frac{1}{2} \\ 0 & \frac{1}{2} & -\frac{1}{2} & \frac{1}{2} \\ 0 & -\frac{1}{2} & \frac{1}{2} & \frac{1}{2} \end{bmatrix}$

39. The statement is true.

41. The statement is false.

43. We have $AA^{-1} = I$ and $BB^{-1} = I$. If both A and B are $n \times n$ matrices, AB is defined. We must show that if AB is defined, $(AB)(AB)^{-1} = I$.

$$\begin{aligned}(AB)(AB)^{-1} &= (AB)(B^{-1}A^{-1}) \\ &= A(BB^{-1})A^{-1} \\ &= AIA^{-1} \\ &= AA^{-1} \\ &= I\end{aligned}$$

Thus, if A and B are invertible, then AB is invertible.

45. $A = \begin{bmatrix} 1 & 0 & 0 \\ 0 & 1 & 0 \\ 0 & 0 & 1 \end{bmatrix}$ and $A = \begin{bmatrix} 1 & 0 & 0 \\ 0 & 0 & 1 \\ 0 & 1 & 0 \end{bmatrix}$

Section 3.4 *(page 165)*

1. t_{11} means that 0.2 unit of Sector A is required to produce 1 unit in Sector A. t_{12} means that 0.5 unit of Sector A is required to produce 1 unit in Sector B. t_{21} means that 0.4 unit of Sector B is required to produce 1 unit in Sector A. t_{22} means that 0.8 unit of Sector B is required to produce 1 unit in Sector B.

3. t_{12} means that 0.01 unit of oil products is required to produce 1 unit of mining products. t_{21} means that 0.00 units of mining products are required to produce 1 unit of oil products. t_{23} means that 0.00 units of mining products are required to produce 1 unit of support products. t_{31} means that 0.02 unit of support products is required to produce 1 unit of oil products.

5. From the technology matrix, we see that \$0.10 of agriculture products and \$0.30 of manufacturing products are required to produce \$1 of manufacturing products.

7. $I - T = \begin{bmatrix} 0.8 & -0.1 \\ -0.6 & 0.7 \end{bmatrix}$

$(I - T)^{-1} = \begin{bmatrix} 1.4 & 0.2 \\ 1.2 & 1.6 \end{bmatrix}$

9. \$408 million of agriculture products and \$864 million of manufacturing products are required to meet the internal and external demands of the economy.

11. Roughly \$69.9 billion of textile products and \$54.1 billion of apparel products are required to meet the internal and external demands of the economy.

13. Roughly \$431.4 billion of vehicle products and \$163.4 billion of transportation products are required to meet the internal and external demands of the economy.

15. Roughly \$131.7 billion of metals, \$232.5 billion of products, and \$232.6 billion of machinery are required to meet the internal and external demands of the economy.

17. Roughly \$143.3 billion of computer products, \$284.6 billion of management, and \$431.2 billion of support are required to meet the internal and external demands of the economy.

19. Roughly \$507.0 billion of bank products, \$224.5 billion of securities, \$455.5 billion of insurance products, and \$71.6 billion of funds products are required to meet the internal and external demands of the economy.

21. The sum of each column is greater than 1. This implies that it costs more than \$1 of input to produce \$1 of output in each of the given industries. An economy with input costs exceeding output value cannot survive.

23. Since output in Sectors B and C doesn't require any input from Sector A, the strike should not affect production in Sectors B and C.

25. The sum of each column must be positive and less than 1.

CHAPTER 3 REVIEW EXERCISES

Section 3.1 *(page 167)*

1. $A + B = \begin{bmatrix} 7 & 9 \\ 0 & -8 \\ 9 & 9 \end{bmatrix}$

3. $2A = \begin{bmatrix} 4 & 6 \\ -6 & -8 \\ 0 & 14 \end{bmatrix}$

5. $2A + 3B = \begin{bmatrix} 19 & 24 \\ 3 & -20 \\ 27 & 20 \end{bmatrix}$

7. $5.3B = \begin{bmatrix} 18.02 & -1.59 & -2.12 \\ -20.14 & 29.68 & 38.16 \\ 11.66 & 13.78 & 10.6 \end{bmatrix}$

9. $4.1A + 0.1B = \begin{bmatrix} 5.26 & 24.98 & -1.68 \\ -37.69 & 17.78 & 7.69 \\ 3.91 & 12.56 & 13.73 \end{bmatrix}$

11. Average Markup

	Celica	MR2 Spyder
2000 model	$2275	$2550
2001 model	$2400	$2625

Section 3.2 *(page 168)*

13. $BA = \begin{bmatrix} 25 & 42 \\ 12 & 20 \end{bmatrix}$

15. $DC = \begin{bmatrix} 30 & 27 \\ -10 & -8 \\ 16 & 17 \end{bmatrix}$

17. AC is undefined.

19. $B^{-1} = \begin{bmatrix} -1 & 2.5 \\ 1 & -2 \end{bmatrix}$

21. $\det(A) = 36$

Since $\det(A) \neq 0$, the matrix is invertible.

23. $B^{-1} = \begin{bmatrix} \frac{1}{2} & -\frac{31}{32} & -\frac{901}{80} \\ 0 & \frac{5}{16} & \frac{27}{8} \\ 0 & 0 & -2 \end{bmatrix}$

Section 3.3 *(page 168)*

25. $A^{-1} = \begin{bmatrix} 1 & -\frac{5}{3} & \frac{1}{6} \\ 2 & -4 & \frac{1}{2} \\ -1 & 2 & 0 \end{bmatrix}$

27. $A^{-1} = \begin{bmatrix} \frac{5}{3} & -\frac{8}{3} & -1 \\ -\frac{14}{3} & \frac{23}{3} & 2 \\ \frac{10}{3} & -\frac{16}{3} & -1 \end{bmatrix}$

29. $x = 8, y = -6, z = 12$

31. To earn a 7% return, $1661.31 should be invested in the Growth Fund and $338.69 in the Capital Appreciation Fund.

To earn a 9% return, $1077.37 should be invested in the Growth Fund and $922.63 in the Capital Appreciation Fund.

To earn an 11% return, $493.43 should be invested in the Growth Fund and $1506.57 in the Capital Appreciation Fund.

33. Her flower cost is $18.41 for Type 1, $24.94 for Type 2, and $25.81 for Type 3.

The retail price is $27.62 for Type 1, $37.41 for Type 2, and $38.72 for Type 3.

Section 3.4 *(page 169)*

35. Roughly $231.5 billion of machine products and $431.3 billion of auto products are required to meet the internal and external demands of the economy.

CHAPTER 4

Section 4.1 *(page 183)*

1. $2x + y \leq 6$; $P = (2, 4)$

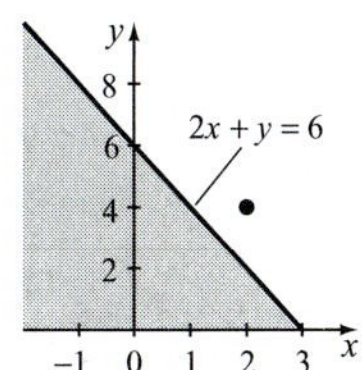

The point $(2, 4)$ is not in the solution region.

3. $x + 5y \leq 10$; $P = (0, 0)$

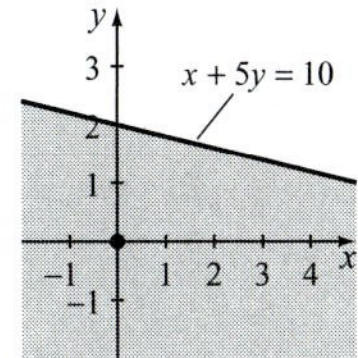

The point $(0, 0)$ is in the solution region.

5. $-2x + 4y \geq -2$; $P = (1, 2)$

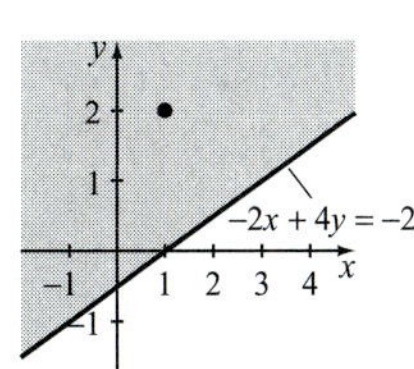

The point $(1, 2)$ is in the solution region.

7. $5x - 4y \le 0$; $P = (1, 0)$

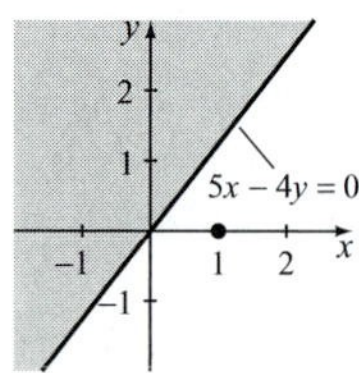

The point $(1, 0)$ is not in the solution region.

9. $2x - y \ge 8$; $P = (-3, 2)$

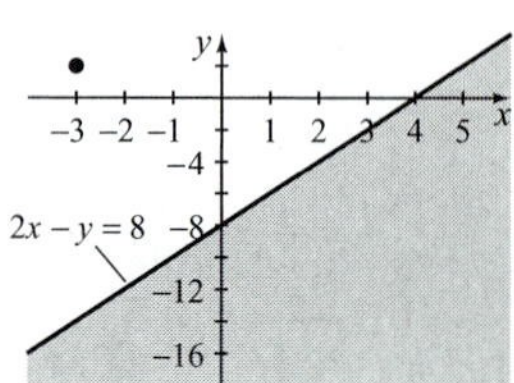

The point $(-3, 2)$ is not in the solution region.

11. $-4x + y \ge 2$
$-2x + y \ge 1$
$x \le 0$

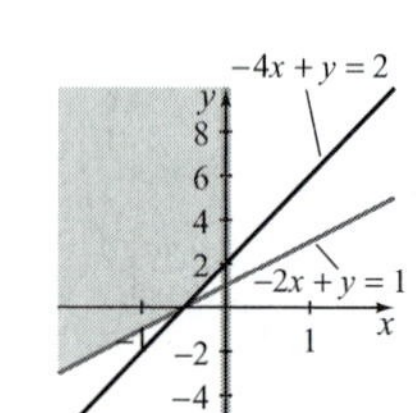

13. $-2x + 6y \le 8$
$4x - 12y \le -6$

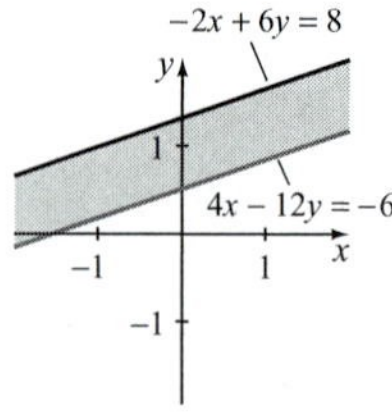

15. $3x - 2y \le 4$
$11x - 20y \ge 2$

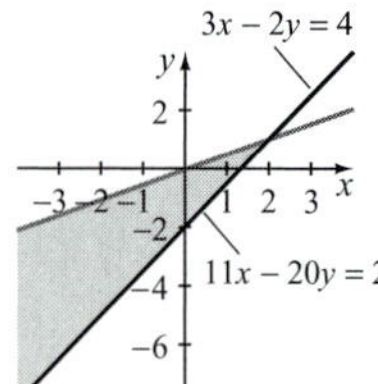

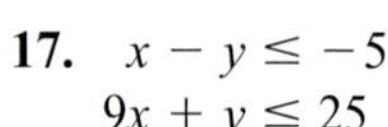

17. $x - y \le -5$
$9x + y \le 25$

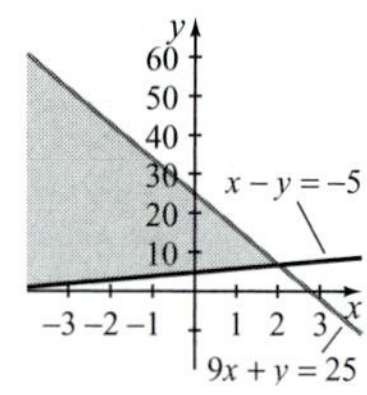

19. $6x + 2y \le 10$
$-x - 2y \ge -5$
$x \ge 0$
$y \ge 0$

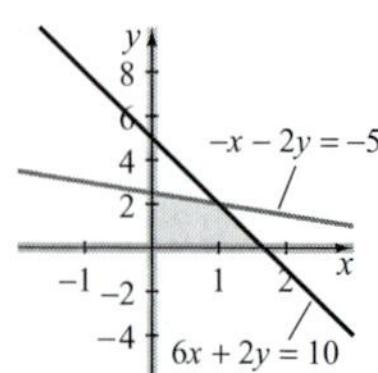

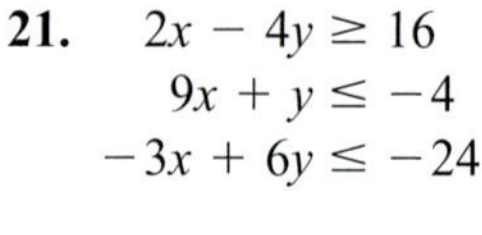

21. $2x - 4y \ge 16$
$9x + y \le -4$
$-3x + 6y \le -24$

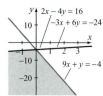

Note: The first and third linear equations describe the same line.

23. $8x - y \ge 3$
$x + 2y \le 11$
$9x + y \le 14$

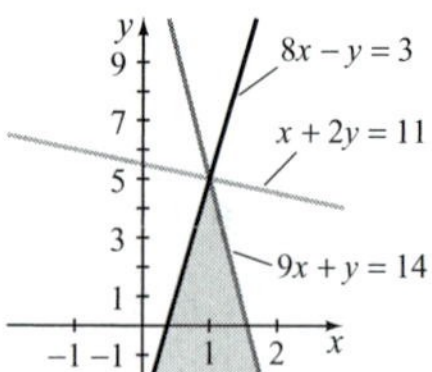

25. $-5x + y \ge 0$
$2x + y \le 4$
$y \le 1$
$x \le 1$

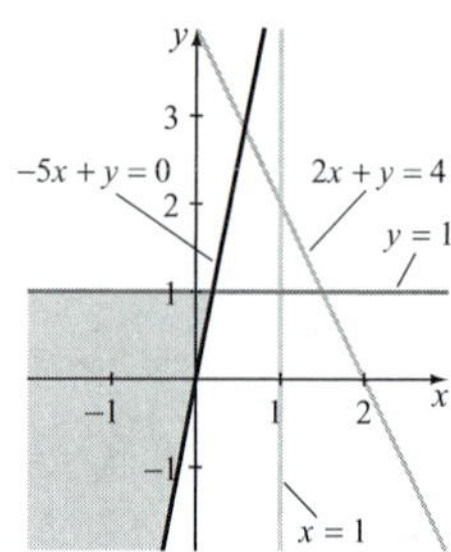

27. Let x be the number of granola bars and y be the number of cereal bars. We have

Sodium: $130x + 65y \le 2400$

Fat: $3.5x + 1.5y \le 80$

and the corresponding solution region for nonnegative x and nonnegative y

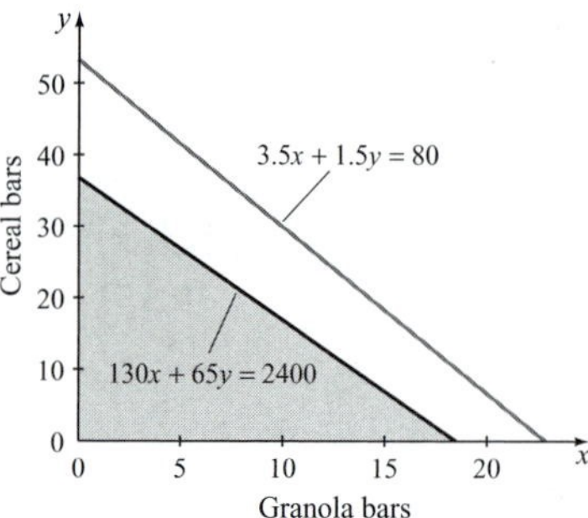

Any serving combination contained within the solution region would meet the dietary guidelines. For example, 8 granola bars and 20 cereal bars is a solution.

29. Let x be the number of hours spent managing the copy center and y be the number of hours spent designing brochures. We have

Hours worked: $x + y \le 50$

Minimum copy center hours: $x \ge 35$

Maximum copy center hours: $x \le 45$

Earnings: $900 + 25y \ge 1100$

Nonnegative constraints: $x \ge 0, y \ge 0$

and the corresponding solution region

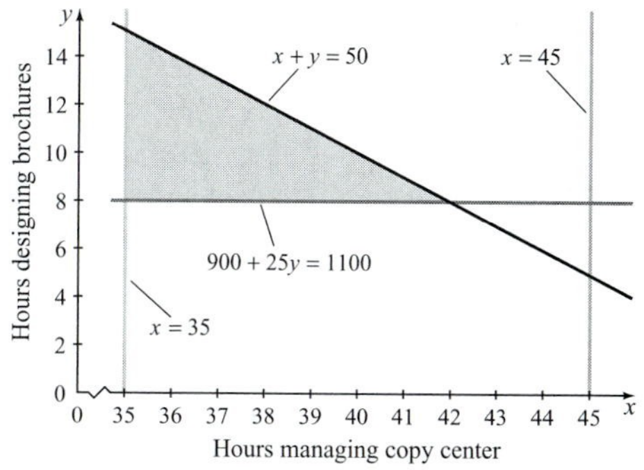

If the employee is able to consistently work 42 hours or less each week managing the copy center, he will be able to meet his goals. However, any week that the man works more than 42 hours at the copy center, he will be unable to meet his weekly income goal unless he exceeds his 50-hour-per-week workload limit.

31. $$\begin{aligned} 2x + 3y &\le 6 \\ -2x + 4y &\ge 4 \\ -5x + y &\le 15 \\ x &\le 5 \\ y &\ge 2 \end{aligned}$$

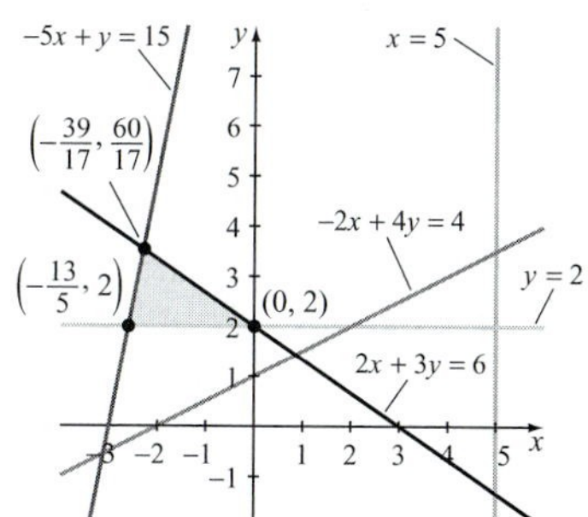

33. $$\begin{aligned} -x + y &\le 0 \\ -x - y &\ge -4 \\ y &\ge 2 \end{aligned}$$

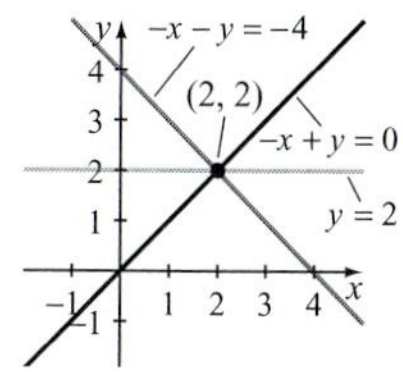

There is a single solution to the system of inequalities: $(2, 2)$. This may be loosely referred to as the "corner point" of the region.

35. $$\begin{aligned} x &\ge 1 \\ y &\le 3 \\ y &\ge \frac{2}{3}x - \frac{1}{3} \\ y &\ge 1 \end{aligned}$$

37. $$\begin{aligned} y &\le -\frac{1}{4}x + 5 \\ y &\le -4x + 20 \\ x &\ge 0 \\ x &\le 5 \end{aligned}$$

39. $$\begin{aligned} x &\ge 0 \\ x &\le 0 \\ y &\ge 0 \\ y &\le 0 \end{aligned}$$

The solution region is bounded and consists of a single solution: $(0, 0)$. This may be loosely referred to as a corner point.

Section 4.2 *(page 199)*

1. The optimal solution is $(0.6, 1.6)$, and the optimal value is 13.

3. The optimal solution is $(0, 16)$, and the optimal value is 16.

5. The optimal solution is $(2, 4)$, and the optimal value is 42.

7. The optimal solution is $(3, 0)$, and the optimal value is 6.

9. The feasible region is empty, so the linear programming problem has no solution.

11. The optimal solution is $(0, 1)$, and the optimal value is 7.

13. The optimal solution is $(2, 4)$, and the optimal value is 58.

15. The optimal solution is $(4, 5)$, and the optimal value is 54.

17. The optimal solution is $(5, 0)$, and the optimal value is 100.

19. Since we may make y as big as we like and still remain in the feasible region, z will never attain a maximum value. The linear programming problem has no solution.

21. Let x be the number of cartons of medium peaches and y be the number of cartons of small peaches. The wholesale cost is the objective function and is given by $z = 9x + 10y$. We are to minimize the function subject to the following constraints:

Minimum demand: $60x + 70y \ge 420$

Maximum demand: $60x + 70y \le 630$

Nonnegative constraints: $x \ge 0,\ y \ge 0$

The optimal solution is $(0, 6)$ with optimal value $z = 60$. The minimum wholesale cost occurs when six cartons of small peaches are purchased. The cost is \$60.

23. Let x be the number of shares in the Stock account and y be the number of shares in the Growth account.

$$\text{Minimize} \quad P = 6x + 7y$$

$$\text{Subject to} \quad \begin{cases} 174x + 55y \geq 2000 \\ x \geq 0, y \geq 0 \end{cases}$$

The optimal solution is $(11.49, 0)$ with optimal value $P = 68.94$. In order to minimize risk, he should invest all of his money in the Stock account.

25. Let x be the number of bags of venison dog food and y be the number of bags of lamb dog food.

$$\text{Minimize} \quad P = 21.99x + 22.99y$$

$$\text{Subject to} \quad \begin{cases} 2x + 3y \geq 30 \\ 4x - 9y \geq 0 \\ x \geq 0, y \geq 0 \end{cases}$$

The optimal solution is $(9, 4)$ with optimal value $P = 289.87$. Buying 9 bags of venison dog food and 4 bags of lamb dog food yields the minimum cost.

27. Let x be the number of rockers and y be the number of benches produced.

$$\text{Maximize} \quad P = 314x + 57y$$

$$\text{Subject to} \quad \begin{cases} 4x + y \leq 360 \\ 7x + 3y \leq 730 \\ x + y \leq 150 \\ x \geq 0, y \geq 0 \end{cases}$$

The optimal solution is $(90, 0)$ with optimal value $P = 28{,}260$. Producing 90 rockers and 0 benches will yield the maximum profit of \$28,260.

29. Let x be the number of 29-passenger vehicles and y be the number of 49-passenger vehicles.

$$\text{Minimize} \quad P = 1000x + 1800y$$

$$\text{Subject to} \quad \begin{cases} 27x + 45y \geq 135 \\ 2x + 4y \leq 16 \\ x \geq 0, y \geq 0 \end{cases}$$

The optimal solution is $(5, 0)$ with optimal value $P = 5000$. In order to minimize transportation costs, the PTA should charter five 29-passenger vehicles. (Only 10 chaperones will be needed for the trip.)

31. Let x be the number of *CSI* spots and y be the number of *Law and Order* spots.

$$\text{Maximize} \quad P = 14{,}834x + 10{,}557y$$

$$\text{Subject to} \quad \begin{cases} x \geq 10 \\ y \geq 10 \\ 31x + 25y \leq 870 \end{cases}$$

The optimal solution is $(20, 10)$ with optimal value $P = 402{,}250$. In order to maximize the number of viewers seeing the commercial spot, the company should purchase 20 spots on *CSI* and 10 spots on *Law and Order*. The number of viewers is projected to be 402,250,000. (*Note:* This calculation counts a viewer each time a person sees the spot. That is, if someone views both shows and sees all 30 spots, that person is counted as 30 viewers.)

33. Let x be the number of shares in Harley-Davidson, Inc., and y be the number of shares in Polaris Industries, Inc.

$$\text{Maximize} \quad P = 2.56x + 2.50y$$

$$\text{Subject to} \quad \begin{cases} 62.7x + 50.5y \leq 4000 \\ 0.40x + 0.92y \geq 60 \\ x \geq 0, y \geq 0 \end{cases}$$

The optimal solution is $(0, 79.2)$ with optimal value $P = 198$. In order to maximize earnings, the investor should purchase 79.2 shares of Polaris Industries. (Because of round-off error, this will cost only \$3999.60. If we round the required number of shares to 79.208, the cost is \$4000.00.)

35. Let x be the number of shares of Polaris Industries and y be the number of shares of Winnebago.

$$\text{Maximize} \quad P = 0.92x + 0.20y$$

$$\text{Subject to} \quad \begin{cases} 50.50x + 33.42y \leq 17{,}000 \\ 2.5x + 1.8y \geq 900 \\ x \geq 0, y \geq 0 \end{cases}$$

The optimal solution is $(71.0, 401.4)$ with optimal value $P = 145.60$. In order to maximize dividends, the investor should buy 71.0 shares of Polaris Industries and 401.4 shares of Winnebago Industries. (Because of round-off error, this actually results in an investment of \$17,000.29. If we went out more decimal places on the share amounts, we could get a more accurate answer. The investor may choose to simply work with the rounded amounts instead of worrying about the few extra cents.)

37. Yes. For example, the feasible region described by the following constraints contains exactly one point: $(0, 0)$.

$$x + y \leq 0$$
$$x \geq 0, y \geq 0$$

The maximum and minimum value of any objective function will be zero.

39. Yes. For example,

Minimize $z = x + y$

Subject to $\{y \geq x$

This feasible region has no corner points. In fact, any linear programming problem with exactly one inequality as a constraint will not have any corner points.

Section 4.3 *(page 230)*

1. The problem is a standard maximization problem.

3. The problem is not a standard maximization problem. The objective function is not a linear function.

5. The problem is not a standard maximization problem. It is missing the nonnegative constraints: $x \geq 0, y \geq 0, z \geq 0$.

7. The problem is not a standard maximization problem. The constant term of the first linear constraint is negative (-24).

9. The problem is not a standard maximization problem. It is missing the nonnegative constraints: $x \geq 0, y \geq 0, z \geq 0$.

11. The corner point $(3, 4)$ yields the maximum objective function value of 59. The feasible region is shown in the following graph.

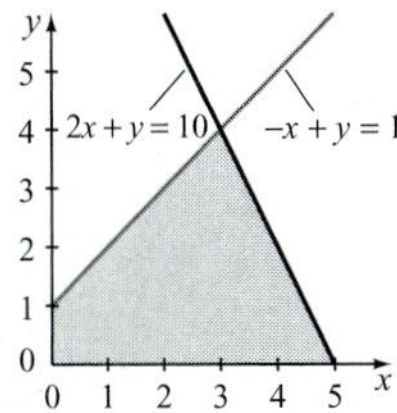

Evaluating the objective function $P = 9x + 8y$ at each of the corner points yields

x	y	P
0	0	0
0	1	8
3	4	59
5	0	45

13. The corner point $(4, 0)$ yields the maximum objective function value of 20. The feasible region is shown in the following graph.

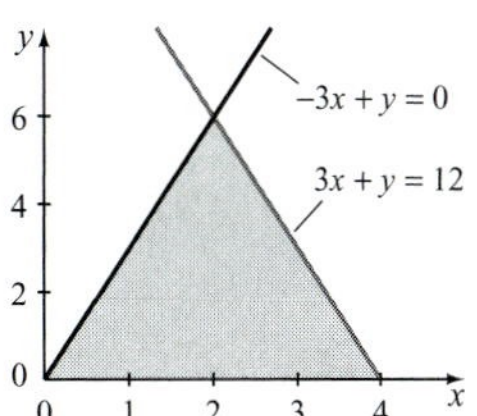

Evaluating the objective function $P = 5x - 2y$ at each of the corner points yields

x	y	P
0	0	0
2	6	−2
4	0	20

15. The corner point $(2, 4)$ yields the maximum objective function value of 24.The feasible region is shown in the following graph.

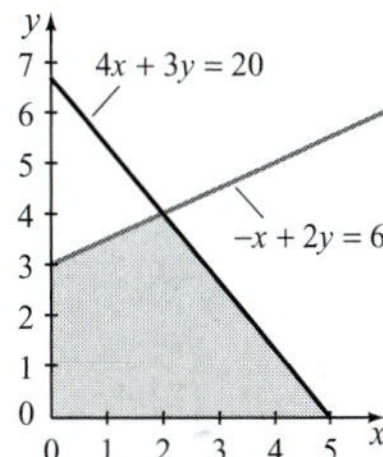

Evaluating the objective function $P = 4x + 4y$ at each of the corner points yields

x	y	P
0	0	0
0	3	12
2	4	24
5	0	20

17. The corner point $(2, 3)$ yields the maximum objective function value of 27. The feasible region is shown in the following graph.

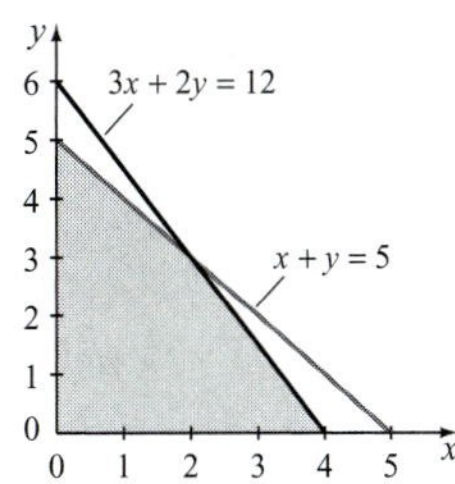

Evaluating the objective function $P = 6x + 5y$ at each of the corner points yields

x	y	P
0	0	0
0	5	25
2	3	27
4	0	24

19. The corner point $(5, 0)$ yields the maximum objective function value of 35. The feasible region is shown in the following graph.

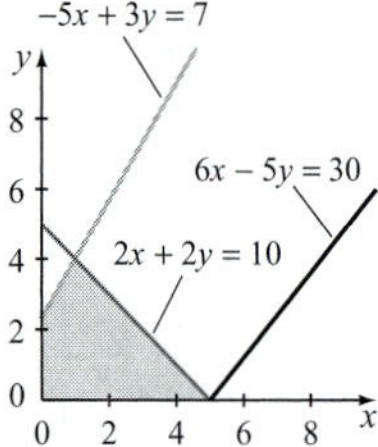

Evaluating the objective function $P = 7x - 5y$ at each of the corner points yields

x	y	P
0	0	0
0	$\frac{7}{3}$	$\frac{-35}{3}$
1	4	-13
5	0	35

21. The solution $(0, 0, 60)$ yields a maximum objective function value of 360.

23. The solution $(0, 0, 40)$ yields a maximum objective function value of 480.

25. The solution $(45, 0, 0)$ yields a maximum objective function value of 225.

27. The solution $(0, 0, 4.6)$ yields a maximum objective function value of 46.

29. The solution $(0, 4, 0)$ yields a maximum objective function value of 4.

31. Let x be the number of batches of Cranberry Cooler, y be the number of batches of Nancy's Party Punch, and z be the number of batches of Jimmy Wallbanger. We are to

Maximize $P = 1x + 9y + 1z$

Subject to
$$\begin{cases} \text{Cranberry: } 4x + 32y \le 128 \\ \text{Lemon-lime: } 2x + 6z \le 64 \\ \text{Nonnegative: } x \ge 0, y \ge 0, z \ge 0 \end{cases}$$

We can make the maximum $46\frac{2}{3}$ servings by making 4 batches of Nancy's Party Punch and $10\frac{2}{3}$ batches of Jimmy Wallbanger.

33. Let x be the number of hamburgers, y be the number of cheeseburgers, and z be the number of double cheeseburgers. We are to

Maximize $P = x + y + 2z$

Subject to
$$\begin{cases} \text{Beef patties: } x + y + 2z \le 132 \\ \text{Buns: } x + y + z \le 104 \\ \text{Cheese slices: } y + 2z \le 72 \\ \text{Nonnegative: } x \ge 0, y \ge 0, z \ge 0 \end{cases}$$

The optimal solution is $(60, 0, 36)$. The maximum number of beef patties that can be used is 132 patties. This occurs when 60 hamburgers, no cheeseburgers, and 36 double cheeseburgers are made.

35. Let x be the number of 50-packs, y be the number of 100-packs, and z be the number of 600-packs. We are to

Maximize the revenue $R = 50x + 100y + 600z$

Subject to
$$\begin{cases} \text{Quantity: } 50x + 100y + 600z \le 1000 \\ \text{Budget: } 10x + 18y + 96z \le 175 \\ \text{Nonnegative: } x \ge 0, y \ge 0, z \ge 0 \end{cases}$$

We also have the additional constraints $x \le 17$, $y \le 9$, and $z \le 1$. The optimal solution is $(0, 4, 1)$ with optimal value $R = 1000$. She should order no 50-packs, four 100-packs, and one 600-pack. Her maximum revenue will be $1000.

37. **(a)** The corner points are $(0, 0)$, $(0, 4)$, $(2, 6)$, $(4, 5)$, and $(6, 0)$.

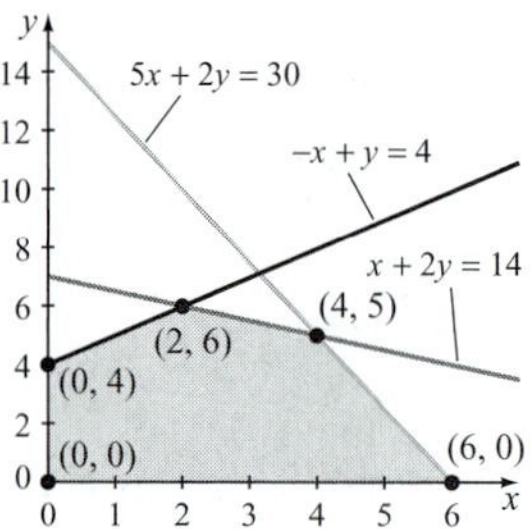

(b) The optimal solution is $(4, 5)$ with optimal value $z = 36$.

(c) The basic feasible solution $(0, 0)$ is the first corner point of the region.

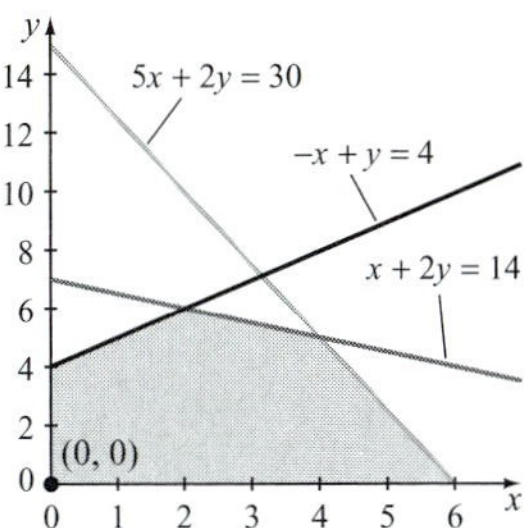

The feasible solution associated with the second tableau is (0, 4).

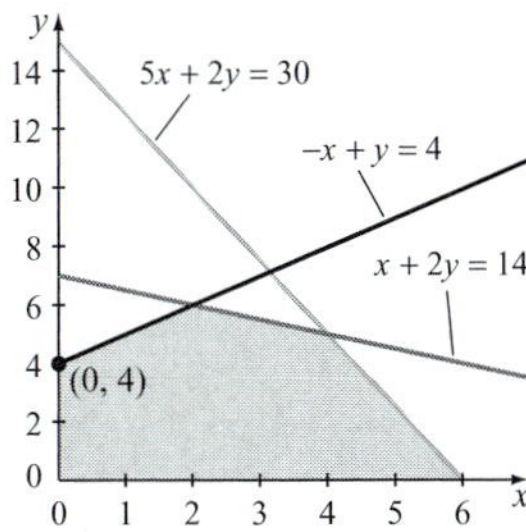

The feasible solution associated with the third tableau is (2,6).

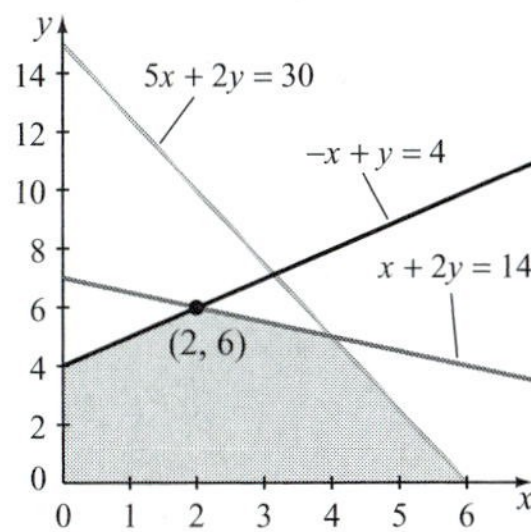

The feasible solution associated with the final tableau is (4,5).

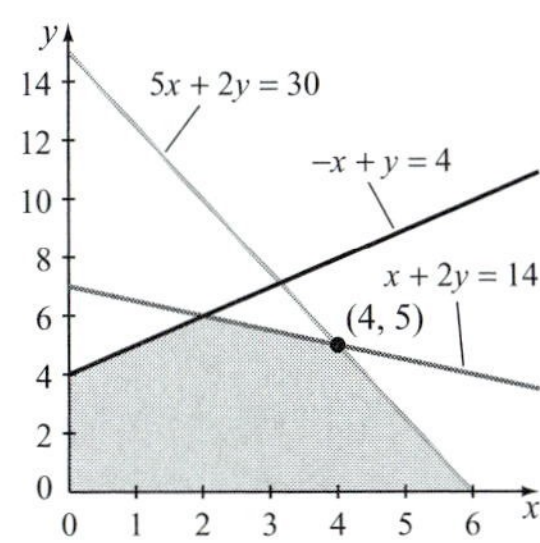

An alternative correct solution moves from (0, 0) to (6, 0) to (4, 5). This occurs if you pick the x-column as the pivot column in the initial simplex tableau.

(d) The simplex method starts at (0, 0) and moves from corner point to corner point along the border of the feasible region while simultaneously increasing the value of the objective function.

Section 4.4 *(page 249)*

1. The problem is a standard minimization problem.

3. The objective function $P = 6xy$ is not of the form $P = ax + by$. Therefore, the problem is not a standard minimization problem.

5. The required nonnegative constraints $x \geq 0$, $y \geq 0$, and $z \geq 0$ are not given. Therefore, the problem is not a standard minimization problem.

7. The constraint $-6x - y + 4z \geq -24$ is not of the form $Ax + By + Cz + \cdots \geq M$ with M nonnegative. Therefore, the problem is not a standard minimization problem.

9. The required nonnegative constraints $x \geq 0$ and $y \geq 0$ are not given. Therefore, the problem is not a standard minimization problem.

11. $A^T = \begin{bmatrix} 2 & 5 & 8 \\ 2 & 9 & 1 \\ 1 & 7 & 0 \\ 4 & 3 & 1 \end{bmatrix}$ **13.** $A^T = \begin{bmatrix} 1 & 6 \\ 2 & 8 \\ 4 & 1 \end{bmatrix}$

15. $A^T = \begin{bmatrix} 9 & 7 \\ 3 & 2 \\ -1 & -2 \\ 4 & 1 \end{bmatrix}$

17. $A^T = \begin{bmatrix} 9 & -3 & 4 & 5 \\ 2 & -2 & 0 & 6 \\ 18 & 6 & 0 & 1 \end{bmatrix}$

19. $A^T = \begin{bmatrix} 0 & -5 & 8 & -1 & 4 \\ 2 & 4 & 1 & 2 & 2 \\ 1 & 7 & 0 & 1 & 1 \end{bmatrix}$

21. (i) The dual problem is

$$\text{Maximize} \quad P = 6x + 11y$$

$$\text{Subject to} \quad \begin{cases} 3x + 2y \leq 1 \\ x + 3y \leq 2 \\ x \geq 0, y \geq 0 \end{cases}$$

(ii) The final simplex tableau for the dual problem is

$$\begin{array}{ccccc|c} x & y & s & t & P & \\ \hline 1.5 & 1 & 0.5 & 0 & 0 & 0.5 \\ -3.5 & 0 & -1.5 & 1 & 0 & 0.5 \\ \hline 10.5 & 0 & 5.5 & 0 & 1 & 5.5 \end{array}$$

(iii) Reading from the s and t columns in the bottom row, we see that the objective function of the *minimization* problem has the optimal solution $x = 5.5$ and $y = 0$ with optimal value $P = 5.5$.

(iv)

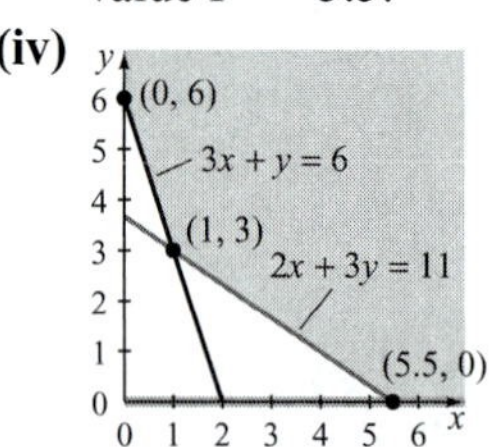

x	y	$P = x + 2y$
0	6	6
1	3	7
5.5	0	5.5

23. (i) The dual problem is

$$\text{Maximize} \quad P = 30x + 15y$$

$$\text{Subject to} \quad \begin{cases} 5x + 10y \le 9 \\ 2x - 5y \le 6 \\ x \ge 0, y \ge 0 \end{cases}$$

(ii) The final simplex tableau for the dual problem is

$$\begin{array}{c} \begin{matrix} x & y & s & t & P \end{matrix} \\ \left[\begin{array}{ccccc|c} 1 & 2 & 0.2 & 0 & 0 & 1.8 \\ 0 & -9 & -0.4 & 1 & 0 & 2.4 \\ \hline 0 & 45 & 6 & 0 & 1 & 54 \end{array}\right] \end{array}$$

(iii) Reading from the s and t columns in the bottom row, we see that the objective function of the *minimization* problem has the optimal solution $x = 6$ and $y = 0$ with optimal value $P = 54$.

(iv)

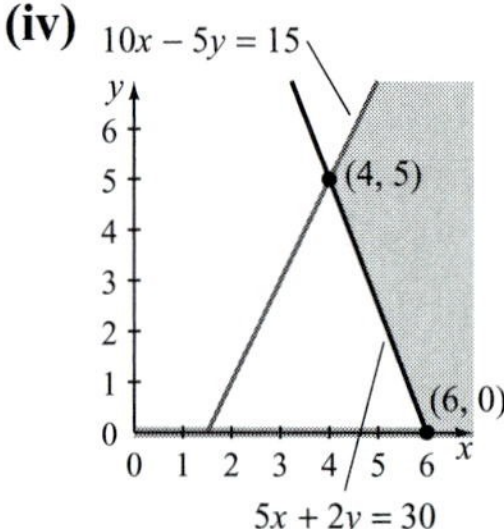

x	y	$P = 9x + 6y$
4	5	66
6	0	54

25. (i) The dual problem is

$$\text{Maximize} \quad P = 10x + 25y$$

$$\text{Subject to} \quad \begin{cases} -5x + 5y \le 5 \\ 2x + y \le 20 \\ x \ge 0, y \ge 0 \end{cases}$$

(ii) The final simplex tableau for the dual problem is

$$\begin{array}{c} \begin{matrix} x & y & s & t & P \end{matrix} \\ \left[\begin{array}{ccccc|c} 0 & 1 & \frac{2}{15} & \frac{1}{3} & 0 & \frac{22}{3} \\ 1 & 0 & -\frac{1}{15} & \frac{1}{3} & 0 & \frac{19}{3} \\ \hline 0 & 0 & \frac{8}{3} & \frac{35}{3} & 1 & \frac{740}{3} \end{array}\right] \end{array}$$

(iii) Reading from the s and t columns in the bottom row, we see that the objective function of the *minimization* problem has the optimal solution $x = \frac{8}{3}$ and $y = \frac{35}{3}$ with optimal value $P = \frac{740}{3} = 246\frac{2}{3}$.

(iv)

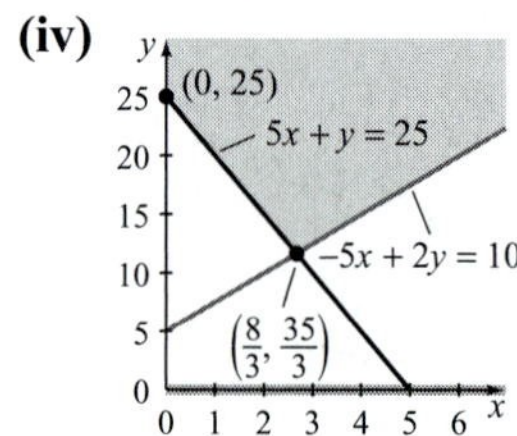

x	y	$P = 5x + 20y$
0	25	500
$\frac{8}{3} = 2\frac{2}{3}$	$\frac{35}{3} = 11\frac{2}{3}$	$\frac{740}{3} = 246\frac{2}{3}$

27. (i) The dual problem is

$$\text{Maximize} \quad P = 5x + 11y$$

$$\text{Subject to} \quad \begin{cases} x + 3y + 4z \le 3 \\ x + 2y - z \le 2 \\ x \ge 0, y \ge 0, z \ge 0 \end{cases}$$

(ii) One final simplex tableau for the dual problem is

$$\begin{array}{c} \begin{matrix} x & y & z & s & t & P \end{matrix} \\ \left[\begin{array}{cccccc|c} -\frac{1}{11} & 0 & 1 & \frac{2}{11} & -\frac{3}{11} & 0 & 0 \\ \frac{5}{11} & 1 & 0 & \frac{1}{11} & \frac{4}{11} & 0 & 1 \\ \hline 0 & 0 & 0 & 1 & 4 & 1 & 11 \end{array}\right] \end{array}$$

(iii) Reading from the s and t columns in the bottom row, we see that the objective function of the *minimization* problem has the optimal solution $x = 1$ and $y = 4$ with optimal value $P = 11$.

(iv) 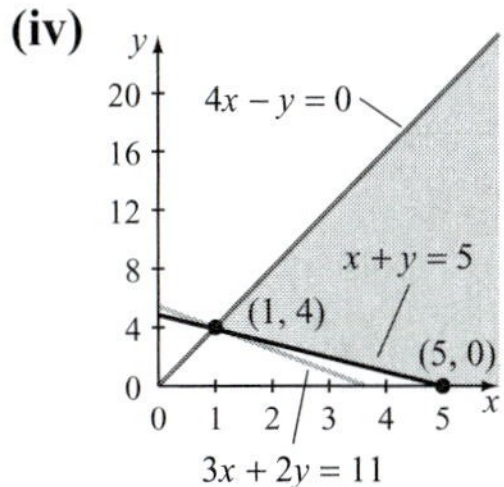

x	y	$P = 3x + 2y$
1	4	11
5	0	15

29. (i) The dual problem is

$$\text{Maximize} \quad P = 4x + 2y + 6z$$

$$\text{Subject to} \quad \begin{cases} x + y + 2z \le 6 \\ x - y - z \le 5 \\ x \ge 0, y \ge 0, z \ge 0 \end{cases}$$

(ii) The final simplex tableau for the dual problem is

$$\begin{array}{cccccc|c} x & y & z & s & t & P & \\ 0 & \frac{2}{3} & 1 & \frac{1}{3} & -\frac{1}{3} & 0 & \frac{1}{3} \\ 1 & -\frac{1}{3} & 0 & \frac{1}{3} & \frac{2}{3} & 0 & \frac{16}{3} \\ \hline 0 & \frac{2}{3} & 0 & \frac{10}{3} & \frac{2}{3} & 1 & \frac{70}{3} \end{array}$$

(iii) Reading from the s and t columns in the bottom row, we see that the objective function of the *minimization* problem has the optimal solution $x = \frac{10}{3}$ and $y = \frac{2}{3}$ with optimal value $P = \frac{70}{3}$.

(iv) 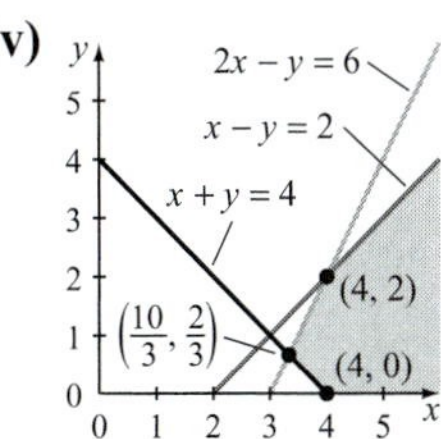

x	y	$P = 6x + 5y$
$\frac{10}{3}$	$\frac{2}{3}$	$\frac{70}{3} = 23\frac{1}{3}$
4	0	24
4	2	34

31. Let x be the number of bags of chicken dog food, y be the number of bags of lamb dog food, and z be the number of bags of turkey dog food. We must

$$\text{Minimize} \quad P = 8x + 8y + 9z$$

$$\text{Subject to} \quad \begin{cases} x + y + z \ge 15 \\ -2x + y - z \ge 0 \\ x \ge 0, y \ge 0, z \ge 0 \end{cases}$$

The optimal solution is $x = 0$, $y = 15$, $z = 0$ with optimal value $P = 120$. The breeder should buy 15 bags of the lamb variety of dog food to obtain the minimum cost of $120. (An alternative optimal solution is $x = 5$, $y = 10$, $z = 0$, and $P = 120$.)

33. Let x be the number of bags of Nature's Recipe brand dog food, y be the number of bags of Nutro Max brand dog food, and z be the number of bags of PETsMART brand dog food. We must

$$\text{Minimize} \quad P = 22x + 13y + 23z$$

$$\text{Subject to} \quad \begin{cases} 8x + 7y + 12z \ge 70 \\ -20x + 7y \ge 0 \\ -16x + 7y \ge 0 \\ x \ge 0, y \ge 0, z \ge 0 \end{cases}$$

The optimal solution is $x = 0$, $y = 10$, $z = 0$ with optimal value $P = 130$. The optimal solution indicates the breeder should buy 10 bags of the Nutro Max brand of dog food to obtain a minimum cost of $130.

35. Let x be the number of cases shipped from Monroe to Birmingham, y be the number of cases shipped from Monroe to Scottsboro, z be the number of cases shipped from Shelbyville to Birmingham, and w be the number of cases shipped from Shelbyville to Scottsboro.

We need to minimize the shipment cost function, $C = 0.68x + 0.65y + 0.57z + 0.32w$, subject to

$$x + y \ge 600$$
$$w + z \ge 400$$
$$x + z \ge 700$$
$$y + w \ge 300$$
$$x \ge 0, y \ge 0, z \ge 0, w \ge 0$$

The optimal solution is $x = 600$, $y = 0$, $z = 100$, $w = 300$ and $P = 561$. In order to minimize shipping costs, 600 cases should be shipped from Monroe to Birmingham, 0 cases should be shipped from Monroe to Scottsboro, 100 cases should be shipped from Shelbyville to Birmingham, and 300 cases should be shipped from Shelbyville to Scottsboro. The minimal shipping cost is \$561.

37. Let x be the number of cases shipped from Monroe to Calhoun, y be the number of cases shipped from Monroe to Scottsboro, z be the number of cases shipped from Shelbyville to Calhoun, and w be the number of cases shipped from Shelbyville to Scottsboro.

We need to minimize the shipment cost function, $C = 0.39x + 0.65y + 0.50z + 0.32w$, subject to

$$\begin{aligned} x + y &\geq 400 \\ w + z &\geq 200 \\ x + z &\geq 300 \\ y + w &\geq 300 \\ x \geq 0, y \geq 0, z &\geq 0, w \geq 0 \end{aligned}$$

The optimal solution is $x = 300$, $y = 100$, $z = 0$, $w = 200$ with optimal value $P = 246$. In order to minimize shipping costs, 300 cases should be shipped from Monroe to Calhoun, 100 cases should be shipped from Monroe to Scottsboro, 0 cases should be shipped from Shelbyville to Calhoun, and 200 cases should be shipped from Shelbyville to Scottsboro. The minimal shipping cost is \$246.

39. Let x be the number of times they went shopping and y be the number of times they went to a baseball game. We must

$$\text{Minimize } P = 200x + 250y$$

$$\text{Subject to } \begin{cases} 3x + 4y \geq 10 \\ 11x + 14y \geq 36 \\ x \geq 2 \\ x \geq 0, y \geq 0 \end{cases}$$

The optimal solution is $x = 2$ and $y = 1$ with optimal value $P = 650$.

Section 4.5 (page 272)

1.
$$\begin{aligned} 2x + y + z + s &= 10 \\ x + z + t &= 3 \\ x - 2z - u &= 1 \\ -4x - 2y - 3z + P &= 0 \end{aligned}$$

3.
$$\begin{aligned} -12x + 8y + 2z + s &= 15 \\ 8x + 5y - 2z + t &= 10 \\ 3x - 2y + 6z - u &= 5 \\ -3x + y - 9z + P &= 0 \end{aligned}$$

5.
$$\begin{aligned} 4x - 3y - 2z + s &= 12 \\ -2x + 4y + 6z + t &= 12 \\ 3x - 2y + 4z - u &= 12 \\ -10x - 9y - z + P &= 0 \end{aligned}$$

7.
$$\begin{array}{c} \\ * \\ \\ * \\ \\ \end{array} \begin{array}{c} \begin{array}{cccccc} x & y & s & t & u & P \end{array} \\ \left[\begin{array}{cccccc|c} 9 & 4 & -1 & 0 & 0 & 0 & 10 \\ -2 & 5 & 0 & 1 & 0 & 0 & 8 \\ 1 & 1 & 0 & 0 & -1 & 0 & 12 \\ \hline 5 & 4 & 0 & 0 & 0 & 1 & 0 \end{array}\right] \end{array}$$

9.
$$\begin{array}{c} \\ * \\ \\ * \\ \\ \end{array} \begin{array}{c} \begin{array}{cccccc} x & y & s & t & u & P \end{array} \\ \left[\begin{array}{cccccc|c} 0 & 6 & -2 & 0 & 0 & 0 & 16 \\ 5 & 1 & 0 & 1 & 0 & 0 & 1 \\ 0 & 3 & 0 & 4 & -4 & 0 & 6 \\ \hline 0 & 10 & 0 & 5 & 0 & 3 & 9 \end{array}\right] \end{array}$$

11. The optimal solution is $(5, 0)$ with optimal value $P = 10$. The points corresponding to each of the tableaus are shown in the figure below.

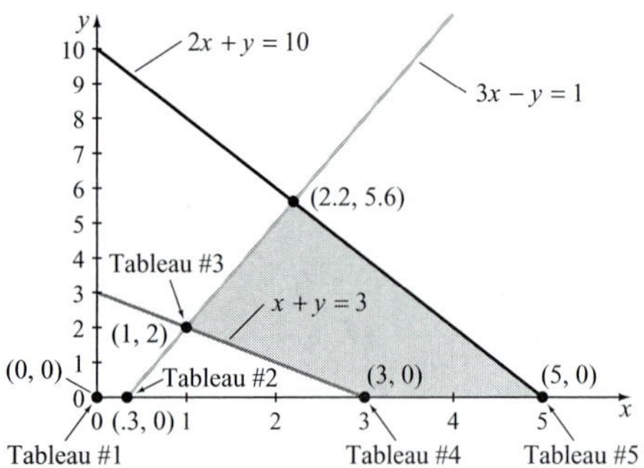

13. The optimal solution is $(2, 4)$ with optimal value $P = -22$. The points corresponding to each of the tableaus are shown in the figure below.

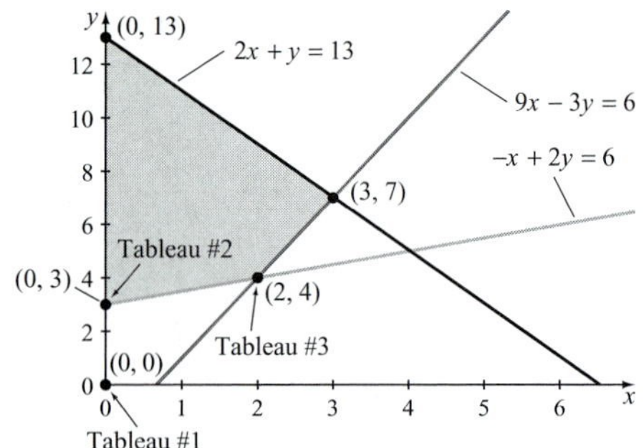

15. The optimal solution is $(0, 6)$ with optimal value $P = 24$. The points corresponding to each of the tableaus are shown in the figure below.

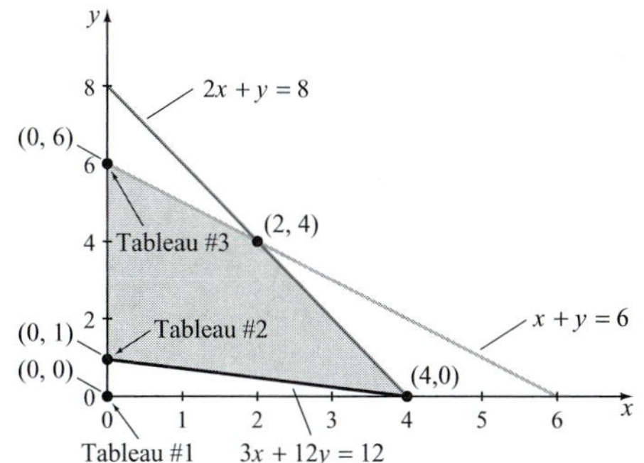

17. The optimal solution is $(2, 4)$ with optimal value $C = -30$. Since $C = -P$, P has the optimal value $P = 30$. The points corresponding to each of the tableaus are shown in the figure below.

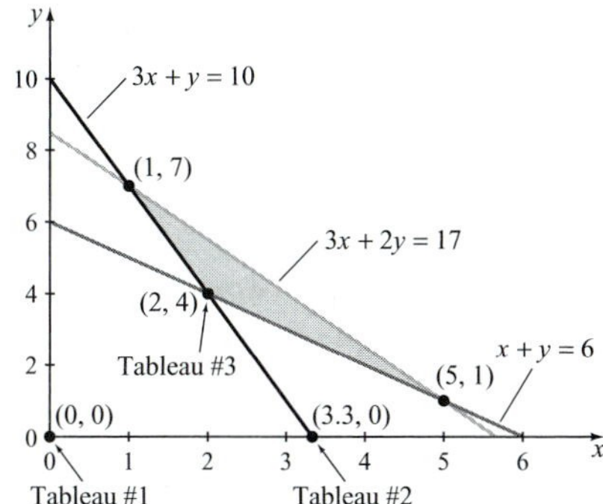

19. The feasible region is unbounded and the objective function C cannot be maximized. Consequently, the objective function $P = -C$ cannot be minimized. There is no solution to the linear programming problem.

The points corresponding to each of the tableaus are shown in the figure below.

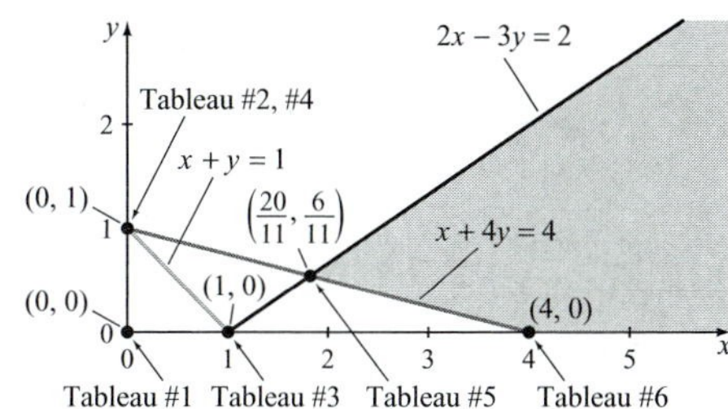

21. The optimal solution is $(5, 5, 0)$ with optimal value $P = 25$.

23. The optimal solution is $(0, 0, 8)$ with optimal value $P = 8$.

25. The optimal solution is $(0, 0, 6)$ with optimal value $P = 12$.

27. The optimal solution is $(7, 0, 0)$ with optimal value $P = 42$.

29. The optimal solution for $C = x + y - 4z$ is $(0, 0, 4)$ with optimal value $C = -16$. The optimal solution for $P = -x - y + 4z$ is also $(0, 0, 4)$ with optimal value $P = 16$.

31. The feasible region is unbounded and there is no optimal solution for $C = -9x + 3y + 3z$. Thus, $P = 9x - 3y - 3z$ does not have an optimal solution.

33. The optimal solution for $C = -6x + 2y - 4z$ is $(0, 0, 1)$ with optimal value $C = -4$. Therefore, the optimal solution for $P = 6x - 2y + 4z$ is $(0, 0, 1)$ with optimal value $P = 4$.

35. The optimal solution is $(0, \frac{22}{3}, 1)$ with optimal value $P = 24$.

37. Let x be the number of 50-packs, y be the number of 100-packs, and z be the number of 600-packs. We are to

Maximize $P = 1.50(50x + 100y + 600z)$
$= 75x + 150y + 900z$

Subject to
$$\begin{cases} 50x + 100y + 600z \geq 900 & \text{At least 900 batteries must be purchased} \\ 10x + 18y + 96z \leq 150 & \text{At most \$150 may be spent} \\ z \leq 1 & \text{At most one 600-pack may be purchased} \\ x \geq 0,\ y \geq 0,\ z \geq 0 \end{cases}$$

The optimal solution is $(0, 3, 1)$ with optimal value $P = 1350$. In order to maximize revenue, she should buy three 100-packs and one 600-pack. Her maximum revenue will be \$1350.

39. Let x be the number of batches of rice pudding, y be the number of batches of tapioca pudding, and z be the number of batches of vanilla pudding. We want to

Maximize $P = 24x + 18y + 12z$

Subject to
$$\begin{cases} 12x + 12y + 6z \leq 108 & \text{At most 108 cups of milk may be used} \\ 1.5x + 1.5y + 1.5z \leq 150 & \text{At most 150 cups of sugar may be used} \\ 9x + 9y + 6z \leq 84 & \text{At most 84 eggs may be used} \\ x + y \geq 2 & \text{The sum of the batches of rice pudding and tapioca pudding must be at least two} \\ x \geq 0,\ y \geq 0,\ z \geq 0 \end{cases}$$

The optimal solution is $(9, 0, 0)$ with optimal value $P = 216$. In order to maximize the number of servings, the deli should make 9 batches of rice pudding. No other pudding variety needs to be made.

41. Let x be the number of $\frac{1}{2}$-cup servings of pinto beans, y be the number of $\frac{1}{2}$-cup servings of kidney beans, and z be the number of $\frac{1}{2}$-cup servings of black beans. We want to *minimize*

$$P = y$$

This is equivalent to *maximizing*

$$C = -y$$

We have the following constraints:

$x + y + z \le 6$	At most 6 1/2-cup servings
$7x + 11y + 2z \ge 39$	At least 39 grams of fiber is needed
$6x + 8y + 6z \ge 42$	At least 42 grams of protein is needed

$x \ge 0, y \ge 0, z \ge 0$

The optimal solution for $C = -y$ is $(0, 3, 3)$ with optimal value $C = -3$. The optimal solution for $P = y$ is $(0, 3, 3)$ with optimal value $P = 3$.

We should use three $\frac{1}{2}$-cup servings of kidney beans and three $\frac{1}{2}$-cup servings of black beans. In other words, we should use 1.5 cups of kidney beans and 1.5 cups of black beans.

43. Let x be the number of students transported from the North region to School 1, y be the number of students transported from the North region to School 2, z be the number of students transported from the South region to School 1, and w be the number of students transported from the South region to School 2.

The total transportation cost is given by

$$P = 120x + 180y + 100z + 150w.$$

We are to *minimize* this amount. In other words, we are to

Maximize $C = -P$.

Subject to

$x + z \le 500$	School 1 holds at most 500 students
$y + w \le 420$	School 2 holds at most 420 students
$x + y \ge 400$	The North region has at least 400 students
$z + w \ge 430$	The South region has at least 430 students
$x \ge 0, y \ge 0, z \ge 0, w \ge 0$	

The optimal solution for $C = -120x - 180y - 100z - 150w$ is $(400, 0, 100, 330)$ with optimal value $C = -107500$. The optimal solution for $P = 120x + 180y + 100z + 150w$ is $(400, 0, 100, 330)$ with optimal value $P = 107{,}500$.

The minimum transportation cost is \$107,500, which occurs when 400 children are bused from the North region to School 1, 100 children are bused from the South region to School 1, and 330 children are bused from the South region to School 2.

45. Let x be the number of desks shipped from Phoenix to Kingman, y be the number of desks shipped from Phoenix to Flagstaff, z be the number of desks shipped from Las Vegas to Kingman, and w be the number of desks shipped from Las Vegas to Flagstaff.

The total transportation cost is given by

$$P = 1.85x + 1.45y + 1.05z + 2.51w$$

We are to *minimize* this amount. In other words, we are to

Maximize $C = -P$.

Subject to

$x + y \le 500$	Phoenix can ship at most 500 desks
$z + w \le 200$	Las Vegas can ship at most 200 desks
$x + z \ge 250$	Kingman needs at least 250 desks
$y + w \ge 400$	Flagstaff needs at least 400 desks
$x \ge 0, y \ge 0, z \ge 0, w \ge 0$	

The optimal solution for $C = -1.85x - 1.45y - 1.05z - 2.51w$ is $(50, 400, 200, 0)$ with optimal value $C = -882.5$. The optimal solution for $P = 1.85x + 1.45y + 1.05z + 2.51w$ is $(50, 400, 200, 0)$ with optimal value $P = 882.5$.

The minimum transportation cost is \$882.50, which occurs when 50 desks are shipped from Phoenix to Kingman, 400 desks are shipped from

Phoenix to Flagstaff, and 200 desks are shipped from Las Vegas to Kingman.

47. The optimal solution for $C = -x - y - z - w$ is $(4, 0, 2, 3)$ with optimal value $C = -9$. The optimal solution for $P = x + y + z + w$ is $(4, 0, 2, 3)$ with optimal value $P = 9$.

49.

$$\text{Maximize} \quad P = x$$

$$\text{Subject to} \quad \begin{cases} x + y + z \leq 2 \\ x + y + z \geq 5 \\ x + 2y \leq 0 \\ x \geq 0, y \geq 0, z \geq 0 \end{cases}$$

There are no values of x, y, and z that can simultaneously satisfy the first two constraints. If the sum of the variables is less than 2, the first constraint is satisfied but the second constraint is not. If the sum of the variables is greater than 5, the second constraint is satisfied but the first constraint is not. Thus the feasible region is empty. (Answers will vary).

CHAPTER 4 REVIEW EXERCISES

Section 4.1 *(page 276)*

1. $P = (2, 9)$ is a solution.

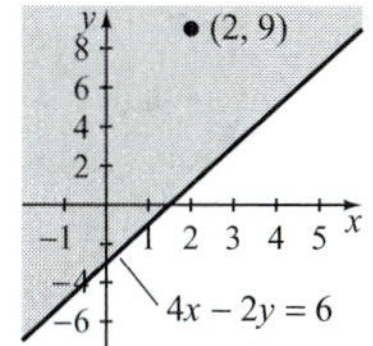

3. $P = (8, 9)$ is a solution.

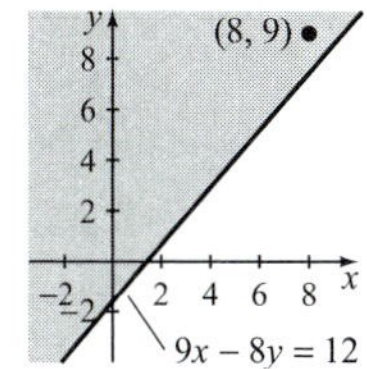

5. **7.**

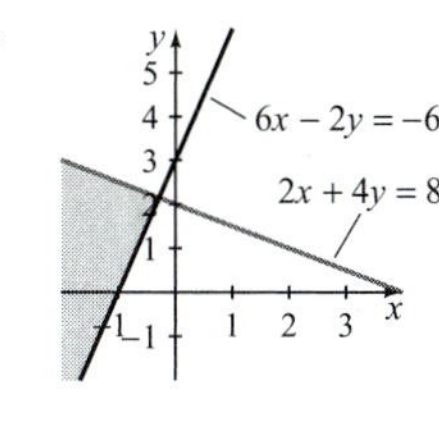

9. Let x be the number of hours she spends working in the copy center and y be the number of hours she spends designing brochures. We have the following constraints.

$x + y \leq 50$	She works at most 50 hours per week
$x \geq 40$	She must work at least 40 hours in the copy center
$x \leq 50$	She may work at most 50 hours in the copy center
$30y \geq 200$	She must earn at least \$200 making brochures
$x \geq 0, y \geq 0$	

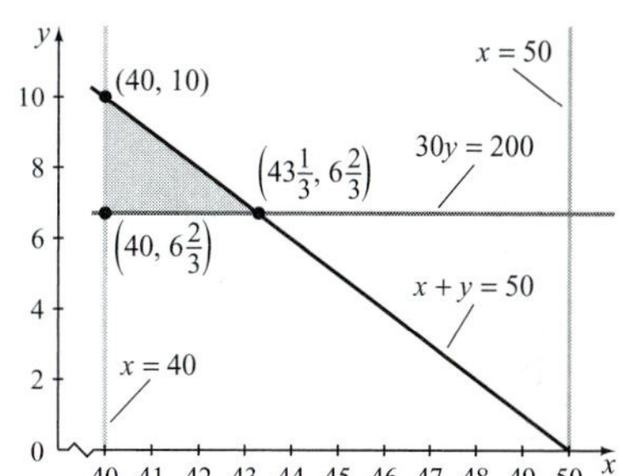

In order to meet her income and workload goals, she can work at most 43 hours and 20 minutes a week at the copy center. Since she has no control over the amount of time she works at the copy center, some weeks she will meet her goals and other weeks she will not. In the weeks she doesn't meet her goals, she will have to either work extra hours or live on less money. The struggle between standard of living and quality of life is a dilemma for many workers.

Section 4.2 *(page 277)*

11. The optimal solution is $(2, 4)$ with optimal value $z = 34$.

13. The optimal solution is $\left(2\frac{2}{3}, 0\right)$ with optimal value $z = 13\frac{1}{3}$.

Section 4.3 *(page 277)*

15. The problem is not a standard maximization problem because it is missing the nonnegative constraints $x \geq 0$ and $y \geq 0$. Additionally, the second constraint $-x + -y \leq -20$ is not of the form $Ax + By \leq M$ with M nonnegative.

17. The optimal solution is $(3, 4)$ with optimal value $P = 34$. We verify our solution graphically.

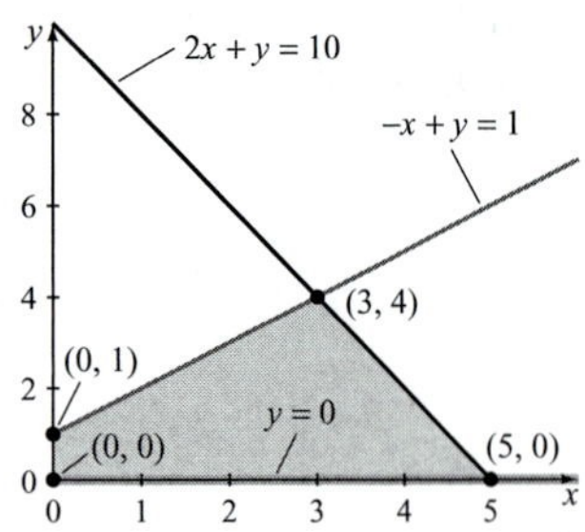

We evaluate the objective function $P = -2x + 10y$ at each of the corner points.

x	y	$P = -2x + 10y$
0	0	0
0	1	10
3	4	34
5	0	−10

19. The optimal solution is $(60, 0, 0)$ with optimal value $P = 360$.

21. Let x be the number of 100-packs, y be the number of 600-packs, and z be the number of 1200-packs. She earns \$100 for each 100-pack sold, \$600 for each 600-pack sold, and \$1200 for each 1200-pack sold. She must

Maximize $P = 100x + 600y + 1200z$

Subject to
$$\begin{cases} 100x + 600y + 1200z \leq 1800 & \text{She wants at most 1800 batteries} \\ 18x + 96y + 180z \leq 300 & \text{She wants to spend at most \$300} \\ x \geq 0, y \geq 0, z \geq 0 \end{cases}$$

Furthermore, to increase the likelihood that we will get a whole-number solution, we determine the maximum number of each size pack we could buy with the \$300 and add these as additional constraints.

$x \leq 16$ She can buy at most sixteen \$18 100-packs with \$300

$y \leq 3$ She can buy at most three \$96 600-packs with \$300

$z \leq 1$ She can buy at most one \$180 1200-pack with \$300

The optimal solution is $(0, 1, 1)$ with optimal value $P = 1800$. She should order one 1200-pack and one 600-pack in order to get the maximum revenue of \$1800.

Section 4.4 (page 278)

23. This is a standard minimization problem.

25. The second constraint is not of the form $Ax + By \geq M$ with M nonnegative and cannot be made into the correct form by multiplying by -1. Thus the problem is not a standard minimization problem.

27. $A^T = \begin{bmatrix} 12 & 14 & 16 \\ 10 & 12 & 14 \\ 8 & 10 & 12 \end{bmatrix}$

29. **(i)** The dual problem is

Maximize $P = 20x + 11y$

Subject to $\begin{cases} 4x + 2y \leq 4 \\ 5x + 3y \leq 2 \\ x \geq 0, y \geq 0 \end{cases}$

(ii) The final tableau is

$$\begin{array}{cccccc} x & y & s & t & P & \\ \end{array}$$
$$\left[\begin{array}{ccccc|c} 0 & -2 & 5 & -4 & 0 & 12 \\ 5 & 3 & 0 & 1 & 0 & 2 \\ \hline 0 & 1 & 0 & 4 & 1 & 8 \end{array}\right]$$

(iii) Reading from the s and t columns in the bottom row, we see that the objective function of the *minimization* problem has the optimal solution $x = 0$ and $y = 4$ with optimal value $P = 8$.

(iv)

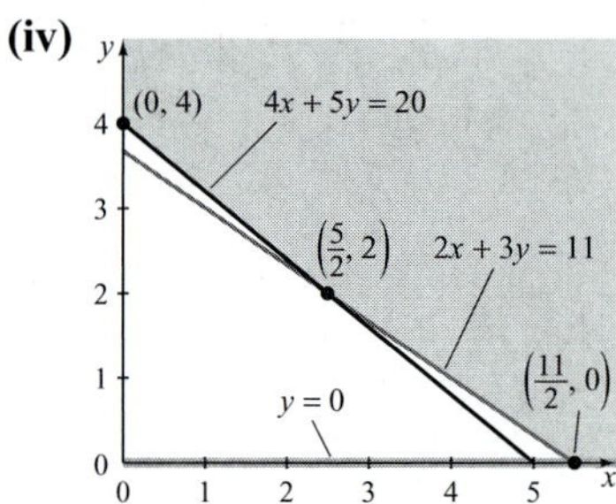

x	y	$P = 4x + 2y$
0	4	8
$\frac{5}{2}$	2	14
$\frac{11}{2}$	0	22

31. **(i)** The dual problem is

$$\text{Maximize} \quad P = 26x + 25y$$
$$\text{Subject to} \quad \begin{cases} 5x + 10y \le 2 \\ 2x - \quad 5y \le 3 \\ x \ge 0, y \ge 0 \end{cases}$$

(ii) The final tableau for the dual problem is

$$\begin{array}{ccccc|c} x & y & s & t & P & \\ 5 & 10 & 1 & 0 & 0 & 2 \\ 0 & -45 & -2 & 5 & 0 & 11 \\ \hline 0 & 27 & \frac{26}{5} & 0 & 1 & \frac{52}{5} \end{array}$$

(iii) Reading from the s and t columns in the bottom row, we see that the objective function of the *minimization* problem has the optimal solution $x = \frac{26}{5} = 5.2$ and $y = 0$ with optimal value $P = \frac{52}{5} = 10.4$.

(iv)

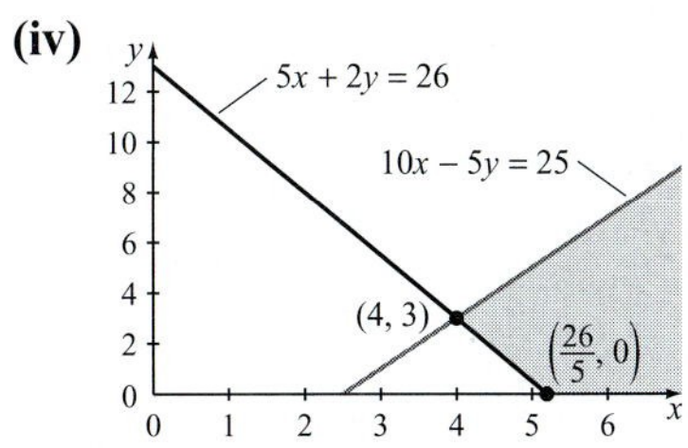

x	y	$P = 2x + 3y$
4	3	17
$\frac{26}{5}$	0	$\frac{52}{5} = 10.4$

33. Let x be the number of bags of Authority brand dog food, y be the number of bags of Bil-Jac brand dog food, and z be the number of bags of Iams brand dog food.

We must

$$\text{Minimize} \quad P = 20x + 19y + 9z$$
$$\text{Subject to} \quad \begin{cases} 33x + 18y + 8z \ge 330 & \text{At least 330 pounds of dog food is needed} \\ 0.66x - 0.54y - 0.32z \ge 0 & \text{At least 30\% protein} \\ 0.18y + 0.08z \ge 0 & \text{At least 3\% fiber} \\ x \ge 0, y \ge 0, z \ge 0 \end{cases}$$

The optimal solution is $(10, 0, 0)$ with optimal value $P = 200$. In order to minimize costs, the kennel should buy 10 bags of the Authority brand dog food. Using the rounded prices, we calculate that the cost will be \$200.

Section 4.5 *(page 279)*

35.
$$\begin{aligned} 2x + 4y + z + s &= 10 \\ x + 2z + t &= 3 \\ 3x - 2z - u &= 1 \\ -2x - 3y - z + P &= 0 \end{aligned}$$

37. The optimal solution is $(1, 0, 2)$ with optimal value $P = 24$.

CHAPTER 5

Section 5.1 *(page 299)*

1. Concave up, y-intercept $(0, 1)$, vertex $(1, 0)$

3. Concave up, y-intercept $(0, 0)$, vertex $(-0.5, -0.75)$

5. Concave up, y-intercept $(0, 2.1)$, vertex $(0.25, 1.925)$

7. $y = -x^2 + 4x - 1$ **9.** $y = x^2 - 10x + 5$

11. **(a)** $W(t) = 36.92t^2 - 43.60t + 2072$

(b)

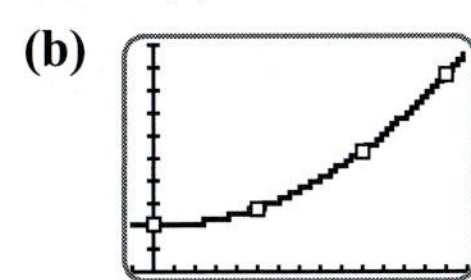

(c) As shown in the scatter plot, the model appears to fit the data well. It seems reasonable that women's basketball game attendance will continue to increase; consequently, we believe this model is good.

13. **(a)** $D(t) = 896.67t^2 + 723.33t - 540$

(b)

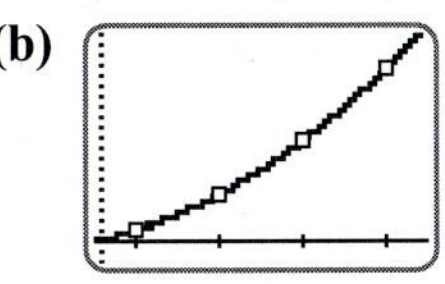

(c) As shown in the scatter plot, the model appears to fit the data well. Since the scatter plot fits the data well and we expect that DVD shipments will continue to increase, we believe this model is a good model.

15. **(a)** $C(v) = -0.03550v^2 + 85.451v - 47{,}197$

(b)

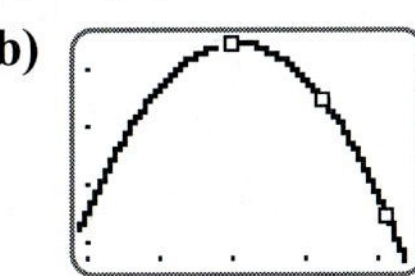

(c) As shown in the scatter plot, the model fits the data perfectly. Although the scatter plot fits the data perfectly, we believe that the model has some definite limitations. According to the model, as the number of blank VHS tapes drops below 1195, the number of CD-R disks will also begin to decrease. This doesn't seem reasonable. We anticipate that CD-R disk sales will continue to increase even as VHS tape sales decrease.

17. (a) (See part d.)

(b) Since the data set consists of three nonlinear points, a quadratic model will fit the data perfectly.

(c) The quadratic model is
$I(t) = -10.31t^2 + 972.4t + 22{,}530$

(d)

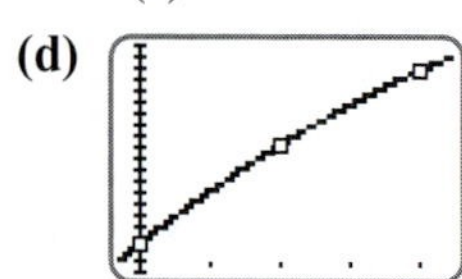

(e) Since the model fits the data and since we expect personal incomes to continue to increase, we anticipate that the model will be relatively accurate in forecasting per capita personal incomes between 1980 and 2010.

19. (a)

(b) The scatter plot does not seem to resemble a parabola. Consequently, we do not think that a quadratic function will fit the data well.

21. (a) (See part d.)

(b) The scatter plot roughly resembles a parabola, so we will model the data with a quadratic function.

(c) The quadratic model is
$M(t) = 0.02628t^2 + 0.3340t + 47.46$

(d)

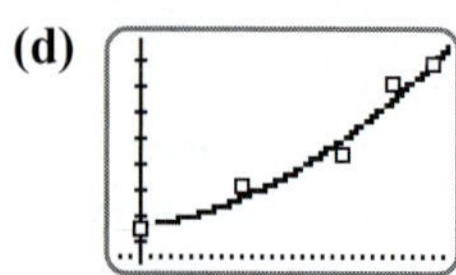

(e) Since the model fits the data and since it seems reasonable to assume that the upward trend in conventional mortgages will continue, we anticipate that the model will be relatively accurate in forecasting the percentage of new privately owned one-family houses financed with a conventional mortgage between 1977 and 2005. (Note: 2005 is arbitrarily selected as a year *near* the last data point.)

(f) When $t \approx 5.4$, $M(t) = 50$. Since t is the number of years since the end of 1970, we estimate that in mid-1976 ($t = 5.4$), 50 percent of mortgages were conventional mortgages.

23. (a)

(b) The scatter plot does not seem to resemble a parabola. Consequently, we do not think that a quadratic function will fit the data well.

25. (a) (See part d.)

(b) The scatter plot seems to resemble a portion of a parabola. We think that a quadratic model will fit the data.

(c) The quadratic model is
$H(t) = -0.01293t^2 + 1.398t + 57.60$

(d)

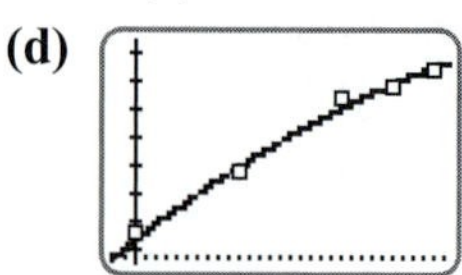

(e) We believe the model is a good model because it seems reasonable to assume that the percentage of homes with a garage will continue to increase.

(f) Since $H(33.6) \approx 90$, we estimate that in mid-2004 ($t = 33.6$), 90 percent of new homes had garages.

27. (a) (See part d.)

(b) Although the scatter plot appears to be near-linear, we can still construct a quadratic model. When quadratic models are used to model near-linear functions, the value of a is relatively close to zero. (If we had our choice of which type of model to construct, we would construct a linear model because it is simpler.)

(c) The quadratic model is
$P(t) = -0.02143t^2 + 8.279t + 68.27$

(d)

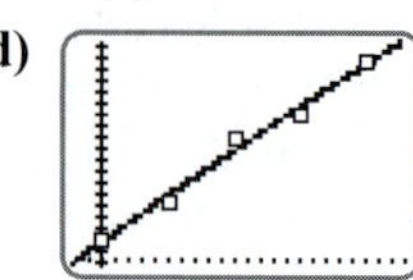

(e) Since the model fits the scatter plot fairly well, and since home sales prices typically increase over time, we believe that our model will relatively accurately predict future home prices in the northeastern United States.

29. (a) (See part d.)
(b) The scatter plot looks concave up. Consequently, we expect that a parabola will fit the data.
(c) The quadratic model is $P(t) = 0.0417t^2 + 0.06833t + 2.455$
(d)

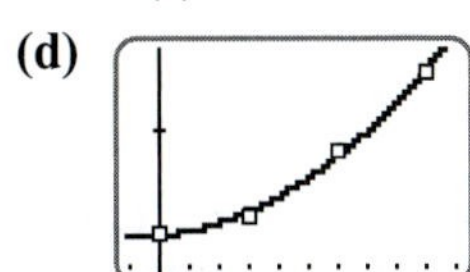

(e) Since the model fits the scatter plot fairly well, and since auto leasing appears to be increasing in popularity, we believe that our model will relatively accurately predict the percentage of households leasing vehicles.
(f) Since $P(7) \approx 5$, we estimate that at the end of 1996, 5 percent of households were leasing vehicles.

31. (a) (See part d.)
(b) The scatter plot looks like a concave up parabola, so we believe that a quadratic model will fit the data well.
(c) The quadratic model is $S(t) = 29.23t^2 + 79.33t + 177.4$
(d)

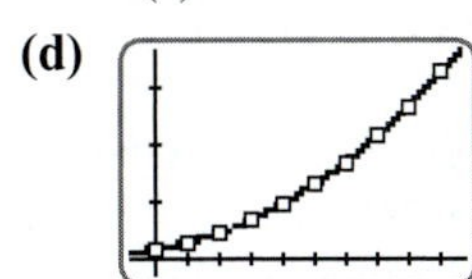

(e) Since the model fits the data extremely well, and since we expect Starbucks sales to continue to increase, we believe we have a good model.
(f) Since $S(11.6) \approx 5000$, we expect that midway through fiscal year 2005 ($t = 11.6$), Starbucks sales will reach $5000 million.

33. (a) (See part d.)
(b) The scatter plot looks more or less concave up and increasing. A quadratic model will probably be a good choice for the data set.
(c) The quadratic model is $P(t) = 0.0007346t^2 + 0.009701t + 0.7803$
(d)

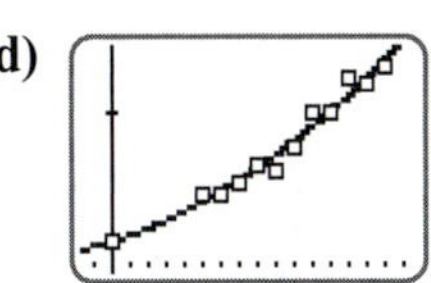

(e) Since the model fits the data relatively well, and since we expect the price of chicken to continue to increase (because of inflation), we believe we have a good model.

35. (a) (See part d.)
(b) The scatter plot is generally concave up and increasing. We believe that a quadratic model will fit the data well.
(c) The quadratic model is $S(t) = 0.9195t^2 + 23.25t + 2652$
(d)

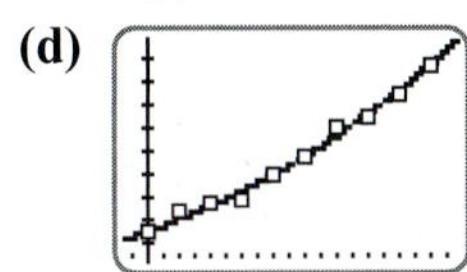

(e) Since the model fits the data relatively well, and since it seems reasonable to anticipate that private college enrollment will continue to increase, we believe that the model is a good model.

37. (a) (See part d.)
(b) The scatter plot appears concave up and increasing. A quadratic model may fit the data well.
(c) The quadratic model is $P(t) = 123.73t^2 + 445.58t + 22{,}847$
(d)

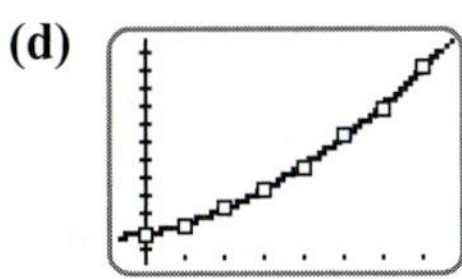

(e) Since the model fits the data relatively well, and since it seems reasonable to anticipate that personal incomes will continue to increase, we believe the model is a good model.

39. (a) (See part d.)
(b) The scatter plot appears near-linear. Nevertheless, we can come up with a decent quadratic model for the data.
(c) The quadratic model is $P(t) = -14.470t^2 + 883.18t + 17{,}620$
(d)

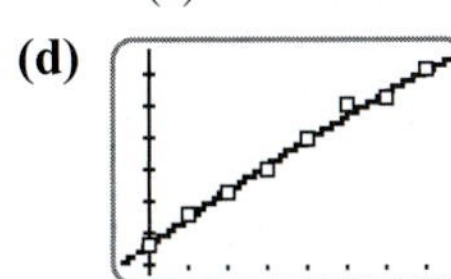

(e) Since the model fits the data relatively well, and since it seems reasonable to anticipate that personal incomes will continue to increase, we believe that the model is a good model.

41. (a) (See part d.)
(b) The scatter plot appears slightly concave down and increasing. A quadratic model should fit the data well.
(c) The quadratic model is
$P(t) = -32.71t^2 + 778.1t + 2341$
(d)

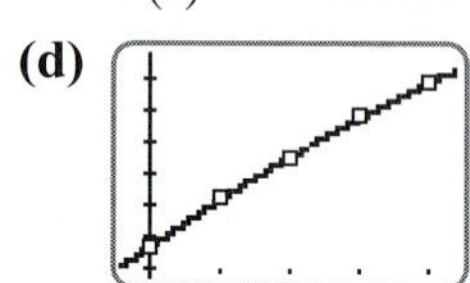

(e) Since the model fits the data well, and since it seems reasonable that the company's profits will continue to increase, we believe that the model is a good model.
(f) Since $P(6.5) \approx 6000$, we estimate that midway through 2004 ($t = 6.5$), Johnson & Johnson profits reached \$6000 million.

43. (a) (See part d.)
(b) The scatter plot is concave up and increasing. A quadratic function may fit the data well.
(c) The quadratic model is
$F(t) = 12.66t^2 + 48.52t + 2102$
(d)

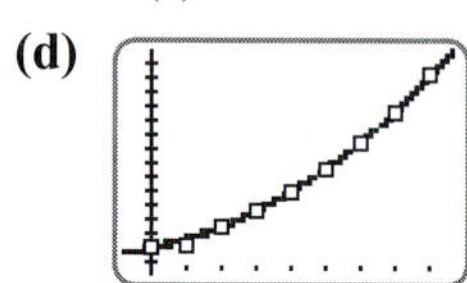

(e) Since the model fits the data well, and since it seems reasonable that the payroll will continue to grow, we believe that the model is a good model.

45. (a) (See part d.)
(b) The scatter plot appears more or less concave down and increasing. A quadratic model may fit the data well.
(c) The quadratic model is
$F(t) = -0.04079t^2 + 1.453t + 401.6$
(d)

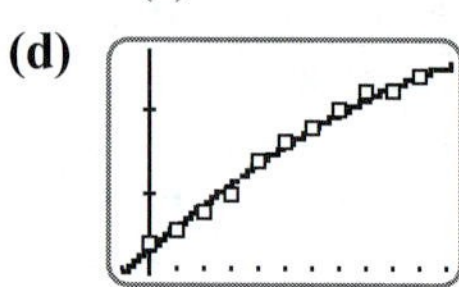

(e) Since the model fits the data well, and since it seems reasonable that the number of private-school teachers will continue to increase, we believe that the model is a good model.

47. (a) (See part d.)
(b) The scatter plot appears more or less concave up and increasing. A quadratic model may fit the data well.
(c) The quadratic model is
$W(t) = 0.003197t^2 + 0.1333t + 4.606$
(d)

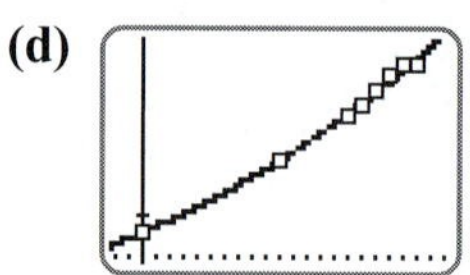

(e) Since the model fits the data well, and since it seems reasonable that wages will continue to increase, we believe that the model is a good model.

49. (a) (See part d.)
(b) The scatter plot appears concave up. A quadratic model may fit the data well.
(c) The quadratic model is
$A(t) = 9.396t^2 - 11.58t + 1063$
(d)

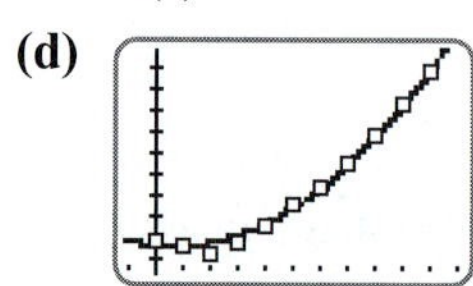

(e) Since the model fits the data well, and since it seems reasonable that advertising expenditures will continue to increase, we believe that the model is a good model.
(f) Since $A(10.6) \approx 2000$, we estimate that in mid-2001 ($t = 10.6$), billboard advertising expenditures reached \$2000 million (\$2 billion).

51. (Answers may vary.)
$y = ax(x - 4)$
$y = x^2 - 4x$ when $a = 1$
$y = -x^2 + 4x$ when $a = -1$

53. No. Although any set of three *nonlinear* points defines a unique quadratic function, the three points given all lie on the same line, $y = 2x + 1$.

55. Since the function is nonnegative for all values of x, the graph of the parabola must be entirely above the x axis. Consequently, the graph must be concave up. (If it were concave down, it would cross the x axis.) Since the graph is concave up, $a > 0$.

Section 5.2 *(page 323)*

1. The quartic function that best fits the data is given by $S(t) = -0.53891t^4 + 15.050t^3 - 103.86t^2 + 48.261t + 12{,}393$. $S(16) = 12{,}904$ means that there were 12,904 thousand students enrolled in high school in 2000.

3. The quadratic function that best fits the data is $R(t) = 1.874t^2 - 1.425t + 2165$.
$R(21.5) \approx 3000$ means that the number of public elementary and secondary teachers reached 3 million (3000 thousand) in mid-2002 ($t = 21.5$).

5. The cubic function that best fits the data is $C(t) = -0.5278t^3 + 5.905t^2 - 4.234t + 4.048$.
$C(7) \approx 83$ means that 83 percent of public school classrooms had Internet access in 2001.
$C(8) \approx 78$ means that 78 percent of public school classrooms had Internet access in 2002. The 2001 estimate seems reasonable, since we expect Internet access in schools to become increasingly common. The 2002 estimate does not seem reasonable, since it forecasts a decline in Internet access.

7. The cubic model that best fits the data is $C(t) = 0.1563t^3 - 2.268t^2 + 14.84t + 164.0$.
$C(15) \approx 404$ means that there will be 404 million debt cards by the end of 2005. Our model estimate is substantially higher than the consulting firm's 270 million card projection.

9. The quartic function that best fits the data is given by $S(t) = 4.807t^4 - 98.56t^3 + 541.3t^2 + 2936t + 10{,}610$. $S(25) \approx 760{,}000$ means that 760 billion shares were traded in 2005. Because of the rapid growth in the market in recent years, this estimate seems reasonable. However, the recession in the early 2000s may have affected the number of shares sold.

11. The quartic function that best fits the data is $W(t) = 100t^4 - 1518.5t^3 + 8244.4t^2 - 17{,}294t + 42{,}667$. $W(6.5) \approx 40{,}000$, meaning that in mid-1999 ($t = 6.5$), the average annual wage per worker for computer and equipment retailers first reached \$40,000.

13. The quadratic function of best fit is $W(t) = 1421.4t^2 - 2321.4t + 56{,}429$.
$W(7.6) \approx 120{,}000$, meaning that in mid-2000 ($t = 7.6$), the average wage per worker in the prepackaged software development industry reached \$120,000.

15. The cubic model that best fits the software wholesaler's wages is $W(t) = 113.89t^3 - 251.19t^2 + 277.78t + 52{,}571$. We estimate that at the end of 2002, the average wage of both software wholesalers and software retailers was \$145,000.

17. The quartic model that best fits the data is $W(t) = 51.515t^4 - 701.52t^3 + 3119.7t^2 - 2450.4t + 41{,}563$. $W(7) \approx 60{,}000$, meaning that the television broadcasting average wage will first exceed \$60,000 at the end of 1999. This estimate may be a bit high, but it is not unreasonable. Since $t = 7$ is close to the end of the data set ($t = 6$) and \$60,000 is somewhat close to \$54,600, we are comfortable with the estimate.

19. The cubic model that best fits the data is $W(t) = -0.005860t^3 + 0.2430t^2 - 0.06623t + 6.541$. $C(32) \approx 61.2$, meaning that 61.2 percent of TV homes had cable at the end of 2002. We would generally expect the percentage of homes with cable to continue to increase, as it has over the 29 years of the data set. For this reason, the estimate seems unreasonable. However, if people are replacing cable with satellite dishes or some other technology, it is possible that the percentage of homes with cable could decline.

21. A polynomial of degree n may have at most $n - 1$ bends.

23. $f(x) = f(-x)$

25. $y = x^4 - 2x^3 - x^2 + 2x$

Section 5.3 *(page 343)*

1. The graph is decreasing and concave up. The y-intercept is (0, 4).

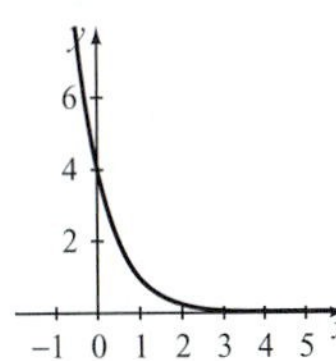

3. The graph is increasing and concave up. The y-intercept is (0, 0.5).

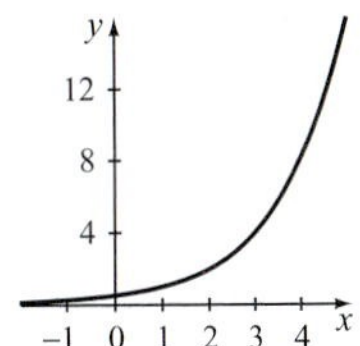

5. The graph is increasing and concave up. The y-intercept is (0, 0.4).

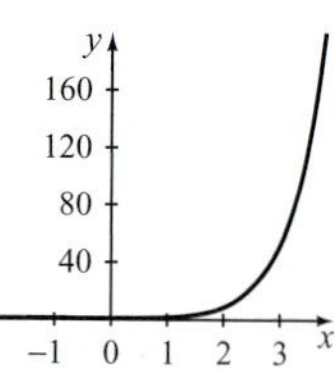

7. The graph is decreasing and concave down. The y-intercept is $(0, -1.2)$.

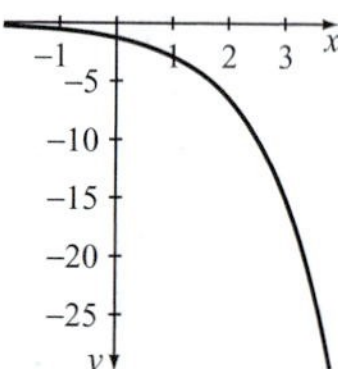

9. The graph is decreasing and concave up. The y-intercept is (0, 3).

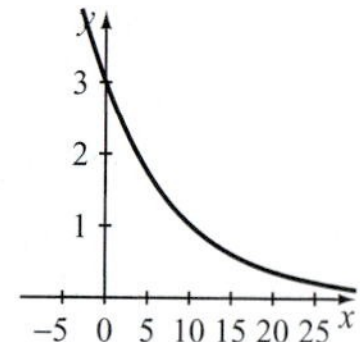

11. $y = 2(3)^x$ 13. $y = 5(2)^x$

15. $y = 64\left(\frac{1}{2}\right)^x$ 17. $y = 0.25(4)^x$

19. $y = 256\left(\frac{1}{2}\right)^x$

21. $I(t) = 82.67(1.061)^t$ (rounded model)
$I(25) = 364.5$ (using unrounded model)
At the end of 2005, dental prices are expected to be 364.5 percent of the 1984 price.

23. $I(t) = 106.1(0.9657)^t$ (rounded model)
$I(25) = 44.4$ (using unrounded model)
At the end of 2005, the price of television set is expected to be 44.4 percent of the 1984 price.

25. $I(t) = 86.75(1.051)^t$ (rounded model)
$I(25) = 302.7$ (using unrounded model)
At the end of 2005, the price of admission to an entertainment venue is expected to be 302.7 percent of its 1984 price.

27. $B(t) = 235(1.0232)^t$
where B is the balance (in dollars) and t is the number of years from now. In just over 2 years and 8 months ($t \approx 2.7$), the balance is projected to reach \$250.

29. $p(t) = 59.90(1.28)^t$
where p is the Dave Matthews Band concert tickets price and t is the number of years since September 2004. A concert ticket that costs \$59.90 in September 2004 is expected to cost \$98.14 in September 2006.

31. $x = 8.827$ 33. $x = 14.21$ 35. $x = 2$

37. (a) The predicted value of the investment is given by $V(t) = 2000(1.0867)^t + 1000(1.0693)^t$ where V is the investment value (in dollars) and t is the number of years from now.
(b) $V(20) = \$14{,}368.66$
Before changing the amount of money invested in each account, the investor should consider her risk tolerance. (Answers may vary.)

39. (a) Growth account: \$699.40
Inflation-linked Bond account: \$1163.20
(b) The combined value of the accounts is projected to double by June 2017 ($t = 14.43$).

Section 5.4 *(page 357)*

1. $y = 2$ 3. $y = 6$ 5. $y = -2$

7. $y = \log_4(x)$ is concave down and increasing.

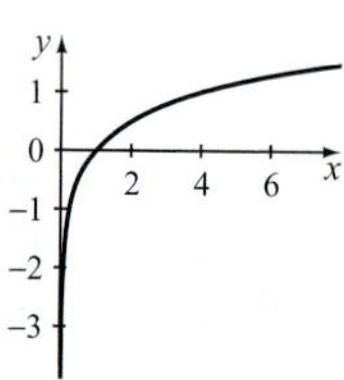

9. $y = \log_{0.7} x$ is concave up and decreasing.

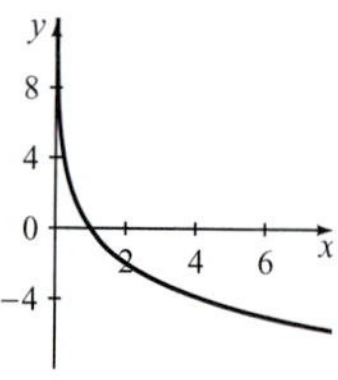

11. $x = 6$ 13. $x = -1$ 15. $x = 3$

17. $x = \frac{1}{16}$ 19. $x = 243$ 21. $4\log(2)$

23. $\log(2x^4)$ 25. $\log(jam)$

27. $\log\left(\frac{1}{324}x^{-16}\right)$ **29.** $-\log(x)$

31. $2\ln(x)$ **33.** $2\ln(x)$ **35.** $\ln(27)$

37. $\ln\left(\frac{1}{64}\right)$ **39.** $\ln(729x^5)$

41. $T(i) = -73.69 + 16.71\ln(i)$
$T(125) = 7.0$
According to the model, the Consumer Price Index reached 125 seven years after 1980. That is, at the end of 1987, the price of dental services were 125 percent of their 1984 price.

43. $T(i) = 129.8 - 27.76\ln(i)$
$T(125) = -4.2$
According to the model, the Consumer Price Index was at 125 4.2 years before the end of 1980. That is, in late 1976, the price of a television set was 125 percent of its 1984 price.

45. $T(i) = -88.48 + 19.84\ln(i)$
$T(125) = 7.3$
According to the model, the Consumer Price Index was at 125 7.3 years after the end of 1980. That is, in early 1988, the price of admission to an entertainment venue was 125 percent of its 1984 price.

47. We know that $y = \log_b(x)$ is equivalent to $x = b^y$. From the definition of $y = \log_b(x)$, we also know that b is positive.

We know that raising a positive number to any power yields a positive number. So b^y will always be positive. But $x = b^y$, so x must always be positive. Hence, the domain of $y = \log_b(x)$ is all positive real numbers.

49. No.

Section 5.5 *(page 370)*

1. $S(t) = 8.451t^3 - 111.9t^2 + 470.0t + 965.0$

Since $S(12) \approx 5099$, we forecast that gaming hardware sales will be \$5099 million in 2002.

3. $C(t) = 0.3214t^2 + 0.2500t + 37.36$
Since $C(8) \approx 60$, we forecast that the 2005–06 tuition cost per credit will be \$60.

5. $S(t) = 1165.9t + 80{,}887$

Since $S(20) \approx 104{,}200$, we forecast that the 2010 public university enrollment will be 104,200.

7. $C(t) = \dfrac{1106}{1 + 0.1259e^{0.5491t}}$

Since $S(13) \approx 7$, we forecast that there will be 7 pediatric AIDS cases in the United States in 2005.

9. $A(t) = \dfrac{38.56}{1 + 14.52e^{-0.8032t}} + 16$

Since $A(10) \approx 54$, we forecast that there were 54 air carrier accidents in 2002.

11. $C(t) = 0.60(1.03)^t$ dollars, where t is the number of years from now.

13. $P(t) = 3t + 87$ dollars, where t is the number of years from now.

15. The logistic model for monthly product sales is

$$S(t) = \frac{200.67}{1 + 25.85e^{-0.4973t}}$$

17. The logistic model for the data is

$$P(t) = \frac{6930}{1 + 166.2e^{-0.4019t}} + 2600$$

The maximum projected population of the city is $6930 + 2600 = 9530$. According to the logistic model, the 2003 population was 6258, since $P(13) \approx 6258$.

The exponential model that best fits the aligned data is

$$P(t) = 67.59(1.377)^t + 2600$$

According to the exponential model, the 2003 population was 6910, since $P(13) \approx 6910$. Although this estimate is still below the actual 2003 population (7480), it is substantially better than the logistic model estimate. (Answers may vary.)

19.

$$T(x) = \begin{cases} 0.1x & \text{if } 0 \le x \le 7000 \\ 0.15x - 350 & \text{if } 7000 < x \le 28{,}400 \\ 0.25x - 3190 & \text{if } 28{,}400 < x \le 68{,}800 \end{cases}$$

21. In addition to looking at how well the model fits the data, we must consider the expected future behavior of the thing being modeled. Since the model is being used to forecast future events, we must make sure that the model's prediction seems reasonable based on what we know about the thing being modeled. (Answers may vary.)

23. Businesses can benefit financially by accurately predicting the future. Although mathematical models are imperfect at forecasting the future, they offer an educated guess as to what might happen. In short, they increase the likelihood of an accurate prediction. (Answers may vary.)

25. Because of the dramatic variability in the data set, we don't recognize any trends or patterns in the scatter plot. None of the standard mathematical models will fit this data set well. Our best guess might be to identify the range of y values and state that we would expect future data values to fall within that range. (Answers may vary.)

CHAPTER 5 REVIEW EXERCISES

Section 5.1 *(page 375)*

1. The quadratic model that best fits the data is

$$A(t) = 42.293t^2 + 158.32t + 6574.5$$

Since $A(12) \approx 14{,}564$, we estimate that \$14,564 million was spent on magazine advertising in 2002.

3. Early in the second quarter of 2001 ($t \approx 10.3$).

5. $A(t) = 34.992t^2 + 94.909t + 8963.2$

Since $A(11.8) \approx 15{,}000$, we estimate that \$15 billion was spent on Yellow Pages advertising in the one-year period that ended in late 2002 ($t = 11.8$).

Section 5.2 *(page 376)*

7. $A(t) = -32.085t^3 + 656.16t^2 - 1640.7t + 31{,}914$

Since $A(11) \approx 50{,}557$, we estimate that \$50,557 million was spent on newspaper advertising in 2001 ($t = 11$).

9. $A(t) = 10.222t^4 - 221.51t^3 + 1616.4t^2 - 2487.8t + 26{,}671$

Since $A(11.0) \approx 50{,}000$, we estimate that \$50 billion was spent on broadcast TV advertising in 2001 ($t = 11$). Broadcast TV annual advertising expenditures beyond 2001 are expected to exceed \$50 billion.

11. $E(t) = 0.001228t^3 - 0.03338t^2 + 0.6439t + 9.844$

Since $E(23) \approx 21.94$, we estimate that average hourly earnings of a manufacturing industry employee in Michigan in 2003 ($t = 23$) was \$21.94.

13. The manufacturing wage in Florida is roughly 65 percent of the Michigan manufacturing wage. Thus labor costs in Florida will be substantially less than labor costs in Michigan. For this reason, a new manufacturing business may prefer to start up in Florida as opposed to Michigan. However, additional factors such as business tax rate, infrastructure, availability and cost of raw materials, etc., should also be considered. (Answers may vary.)

Section 5.3 *(page 379)*

15. The function $y = -0.3(2.8)^x$ is concave down and decreasing. The y-intercept is $(0, -0.3)$.

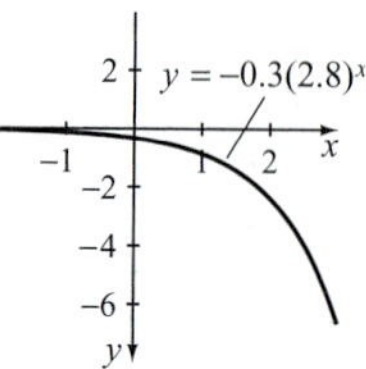

17. $y = 9(2)^x$

19. $N(t) = 22{,}925(1.0727)^t$
$N(10) \approx 46{,}248$ means that there will be 46,248 McDonalds restaurants in 2007.

21. $E(t) = 4500(1.023)^t$
$E(5) \approx 5042$ means that five years from now, my monthly household expenses are expected to be \$5042.

23. $x \approx 1.723$

Section 5.4 *(page 379)*

25. $y = -1$

27. The graph of $y = \log_{0.4}(x)$ is decreasing and concave up.

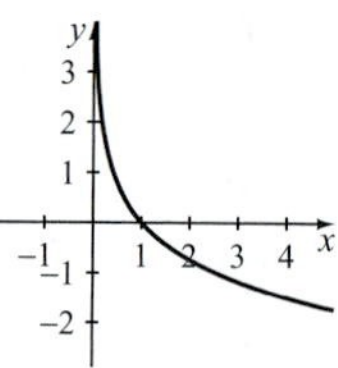

29. $x = 4$

31. $x = 0.04$

33. $\log(2x^2)$

35. $-\log(3)$

37. $F(t) = 49.96 + 5.112 \ln t$. Since $F(10) \approx 61.7$, we estimate that in 1999, 61.7 percent of auto accident fatalities were not alcohol-related.

Section 5.5 *(page 380)*

39. $P(t) = \dfrac{57.81}{1 + 0.5443e^{0.3900t}}$

CHAPTER 6

Section 6.1 *(page 389)*

1. $x = 4$ **3.** $x = 2$

5. $x \approx 1.585$ **7.** $t \approx -0.614$

9. $t \approx 0.5195$

11. $t \approx 24$; At the end of 2012 ($t = 24$), the number of congregations is expected to reach 15,000.

13. $t \approx 36.47$; By mid-2015 ($t = 36.47$), the amount of land in farms is projected to drop to 910 million acres.

15. $t \approx 20.28$; By early 2001 ($t = 20.28$), the annual per capita bottled water consumption was projected to reach 20 gallons.

17. $t \approx 2.59$; In mid-1993, the number of people who considered themselves nondenominational Christians was expected to equal the number of people who consider themselves Lutherans. At that time, both groups were expected to have 9228 thousand American adult members.

19. $t \approx 76.9$; The populations of Ohio and Illinois are projected to be equal by late 2072 ($t = 76.9$). At that time, the population of each state is expected to be 14,052 thousand people.

21. $x \approx 0.7510$.

23. The salaries of public elementary and secondary school teachers and public university assistant professors are projected to be equal in early 1995 ($t = 4.07$) and in early 2035 ($t = 44.1$).

25. At the end of 2003 ($t = 4$), the number of unit cases sold is expected to equal the number of dollars of net operating revenue.

27. $t \approx 9.298$; It will take more than 9 years for the investment value to reach \$1000.

29. **(a)** The account values are expected to be equal after about 4 years and 4 months ($t = 4.323$).
(b) The investments will each be valued at \$697.36, with a combined value of \$1394.72.
(c) The combined value of the investments after 30 years will be \$5113.30.
(d) \$5080.07
(e) (Answers may vary.)

31. Yes. Although the combined value of the accounts initially decreases, the combined value increases after the end of the second year.

33. $x \approx 0.7304$

35. The equation will have a solution for all $a \leq 0$.

37. The equation has at least one solution: $x = 0$.

39. When $a = b$, the graphs intersect in infinitely many places.

Section 6.2 *(page 402)*

1. \$612.50 **3.** \$5137.50 **5.** \$632.16

7. Even though the stated rate of ING DIRECT (4.8%) was higher than the stated rate of Ascencia Bank (4.7%), the difference in compounding methods made Ascencia Bank the better value. ING DIRECT yielded \$620 at maturity while Ascencia Bank yielded \$632.16.

9. At maturity, the accounts will have the same value.

11. $APY = 3.55$ percent **13.** $APY = 3.49$ percent

15. $APY = 2.38$ percent

17. The account values will never be equal to each other in the future.

19. The account values will be equal in 22.3 years.

21. Annual rate: 7.50 percent

23. Continuous rate: 3.24 percent

25. Annual rate: 11.65 percent

27. Continuous rate: 15.87 percent

29. 735,777.99 percent **31.** 8478.33 percent

33. **(a)** 4.13 percent per year
(b) A loaf of white bread will cost \$2.06 in 2008.

35. **(a)** 4.33 percent per year
(b) In 2003 ($t = 19.37$)

37. 9.05 percent **39.** 4.16 percent

41. 4.55 percent **43.** 5.89 percent

45. $r = 0$ percent or $r = 200$ percent

47. 791.6 percent

Section 6.3 *(page 414)*

1. About 3 years ($t \approx 3.0$)

3. \$76.55 monthly **5.** \$80.19 monthly

7. \$307,934.23 **9.** \$243,470.19

11. \$163,879.49 **13.** \$32,460.06

15. About 13.5 years ($t \approx 13.5$)

17. \$100.00 per month **19.** \$988.56 per month

21. 5.66 percent **23.** 10.91 percent

25. It will take them 17 months to save up the \$9750 down payment, assuming that interest is prorated between quarters. If interest is not prorated between quarters, they will have to wait until the end of the quarter (18 months). At that time, they will have \$10,469.34.

27. 72 weeks **29.** 2.5 years

31. 7.529 years (assuming that interest is prorated)

33. When $t = 30.26$, the account values are expected to be equal. The twins will be 48 years old and about 3 months into their 49th year when the accounts reach the same value.

35. (Answers may vary.) **37.** \$1261.62

39. The graphs intersect at (0, 0) and (56.104041, 554,549.98). Each account started out with the same initial value: \$0. Shortly after the 56th year of the investments, both accounts simultaneously attained the same value: \$554,549.98.

41. Account A

Section 6.4 (page 433)

1. \$819.91 **3.** \$954.29 **5.** \$680.83

7. \$309.32 **9.** \$13,148.39

11. Adding in the \$930 down payment, the total cost for the timeshare is \$14,607.02.

13. \$1935.27 **15.** \$743.97 **17.** \$1611.33

19. About 26 years ($t \approx 25.9$)

21. With \$1000 monthly payments, the loan balance will never reach \$100,000. This is because the payments don't cover the interest on the loan.

23. 2 years and 3 months (27 payments)

25. 4 years and 9 months (57 payments)

27. 4 years (four annual payments)

29. 7 years (84 monthly payments)

31. \$2188.51 **33.** \$597.13 **35.** \$1031.83

37. \$416.86 per month **39.** \$4676.32 per month

CHAPTER 6 REVIEW EXERCISES

Section 6.1 (page 436)

1. $x = 2$ **3.** $x = 2$

5. In 2123 and any year thereafter

Section 6.2 (page 436)

7. \$5028.38 **9.** Stonebridge Bank

11. $APY = 2.60$ percent

13. 38 years and 7 months ($t \approx 38.6$)

15. Continuous rate: 9.76 percent

Section 6.3 (page 437)

17. \$55,883.41 **19.** \$246,464.02

21. \$1572.91 monthly

Section 6.4 (page 438)

23. 152 months (12 years and 8 months)

25. Yes. **27.** \$3219.72

CHAPTER 7

Section 7.1 (page 447)

1. $A' = \{1, 5, 8, 9\}$

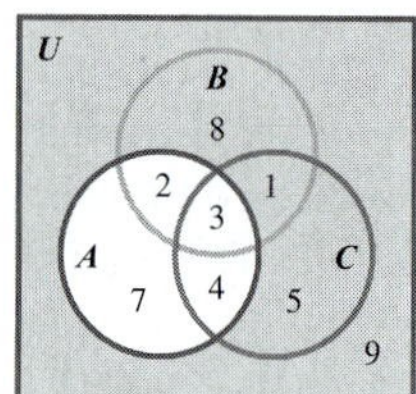

3. $B \cup C = \{1, 2, 3, 4, 5, 8\}$

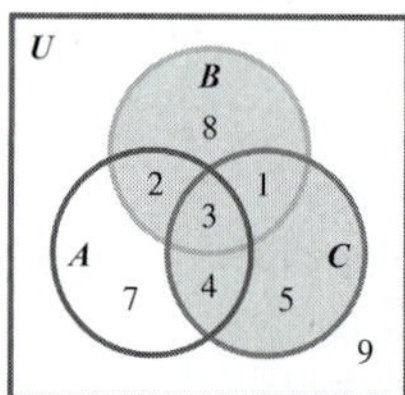

5. $(A \cap B)' = \{1, 4, 5, 7, 8, 9\}$

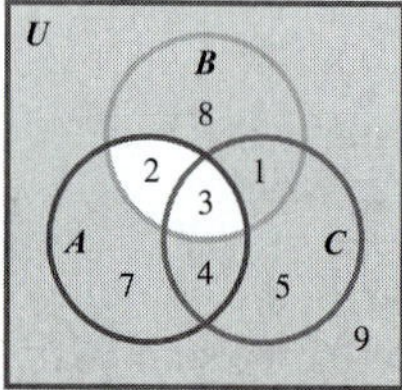

7. $A \cap A' = \{\}$

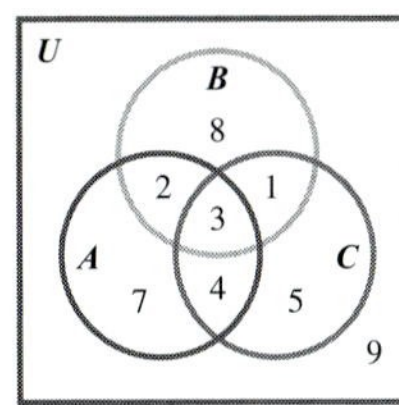

9. $(A \cup B) \cap C' = \{2, 7, 8\}$

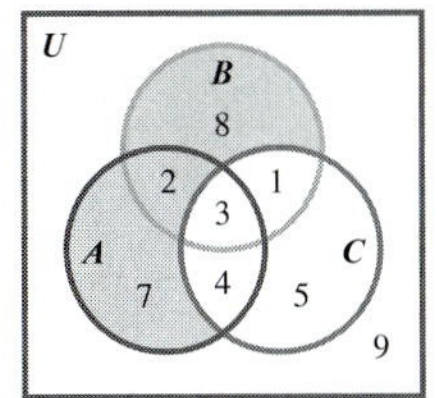

11. D' **13.** M' **15.** $S \cap M$ **17.** $A \subset B$

19. $\{x | 0 < x < 10\}$

21. $\{x | x = k^2$ for some integer $k\}$

23. $\{x | x$ is a factor of $24\}$

The following Venn diagram was used for Exercises 25–36.

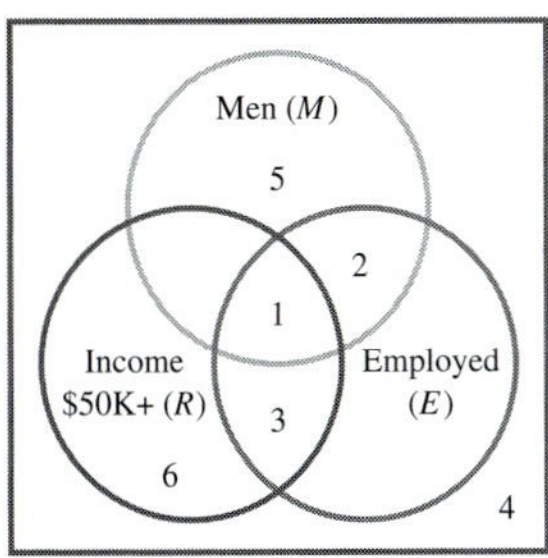

25. $M \cap R = \{1\}$; This is the set of men with a household income above \$50,000.

27. $E' \cap R = \{6\}$; This is the set of unemployed people with a household income above \$50,000.

29. $(E \cap M)' \cap R = \{3, 6\}$; This is the set of people with a household income above \$50,000 who aren't employed men.

31. $(E' \cup R') \cap (M' \cup R) = \{4, 6\}$; This is the set of people who are unemployed or have a household income not greater than \$50,000 who are also women or have a household income above \$50,000. In simpler terms, it is the set of people who are unemployed women.

33. $(E \cap R') \cup M = \{1, 2, 5\}$; This is the set of people who are employed with a household income not greater than \$50,000 together with all of the men.

35. $(M \cap R')' = \{1, 3, 4, 6\}$; This is the set of people who are not men with a household income of at most \$50,000. In other terms, it is the set of men who have a household income greater than \$50,000 together with the set of all women.

37. $F \cup E$ is the group of full-time instructors together with the group of instructors with 20 or more years of teaching experience. There are 94.9 thousand full-time instructors. There are 6.4 thousand part-time instructors with 20 or more years of teaching experience. There are a total of 101.3 thousand people in $F \cup E$.

39. $F' \cap E'$ is the set of part-time instructors with less than 20 years of teaching experience. There are 122.2 thousand part-time instructors with less than 10 years' experience and 24.5 thousand part-time instructors with 10–19 years' experience. There are a total of 146.7 thousand people in $F' \cap E'$.

41. $(F \cap E')'$ is the group of people who aren't full-time instructors with less than 20 years of teaching experience. In other terms, it is the set of all part-time instructors together with the full-time instructors with 20 or more years of teaching experience. There are 153.1 thousand part-time instructors, and there are 23.2 thousand full-time instructors with 20 or more years of teaching experience. There are 176.3 thousand people in $(F \cap E')'$. (*Note:* Because of the rounding in the table, you may also correctly end up with 176.2 thousand people.)

43. $(A \cap B)'$

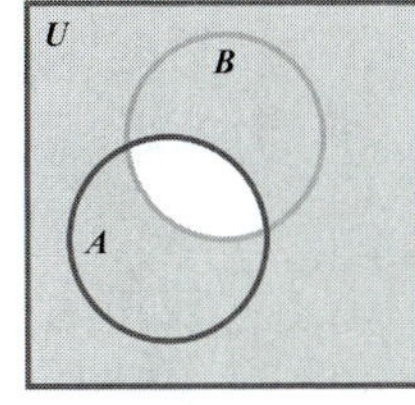

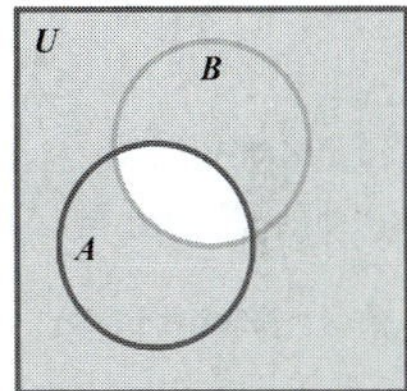

Since both statements yield the same Venn diagram, the statements are equivalent. That is,

$$(A \cap B)' = A' \cup B'$$

$(A \cup B)'$

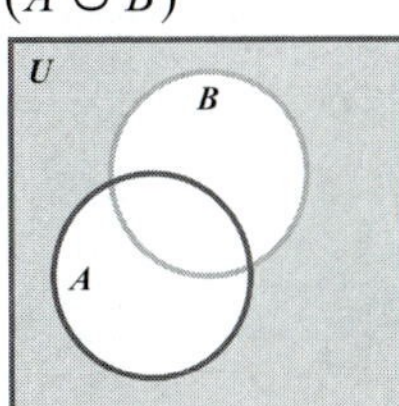

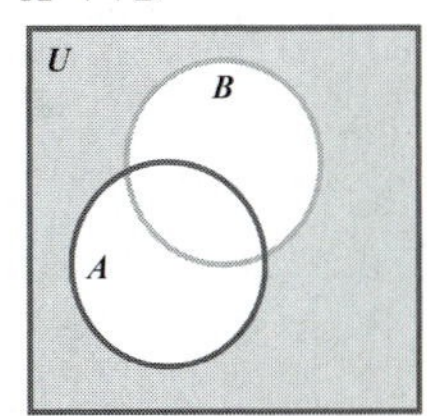

Since both statements yield the same Venn diagram, the statements are equivalent. That is,

$$(A \cup B)' = A' \cap B'$$

Since A and B represented arbitrary sets, DeMorgan's Law is true for any sets A and B.

45. $(A \cap B') \cap (A' \cap B) = \varnothing$

Section 7.2 *(page 456)*

1. $n(B) = 7$ **3.** $n(B') = 8$

5. $n(A' \cap B') = 2$ **7.** $n((A \cap B)') = 11$

9. $n(A' \cap B) = 3$

11. $n(A \times B) = 6$

$$A \times B = \left\{\begin{array}{l}\text{red apple, green apple,}\\ \text{yellow apple, red tomato,}\\ \text{green tomato, yellow tomato}\end{array}\right\}$$

13. $n(A \times B) = 8$

$$A \times B = \left\{\begin{array}{l}\text{chocolate ice cream, chocolate pie,}\\ \text{chocolate shake, chocolate cookie,}\\ \text{vanilla ice cream, vanilla pie,}\\ \text{vanilla shake, vanilla cookie}\end{array}\right\}$$

15. $n(A \times B) = 8$

$$A \times B = \left\{\begin{array}{l}(H, 1), (H, 2), (H, 3), (H, 6),\\ (T, 1), (T, 2), (T, 3), (T, 6)\end{array}\right\}$$

17. 32 schedules **19.** 45 frozen dairy desserts

21. 136 different dining/lodging options

23. 10 different meal options

25. 11 dish choices

27. The outcomes with a single element from each of the sets R, C, and M.

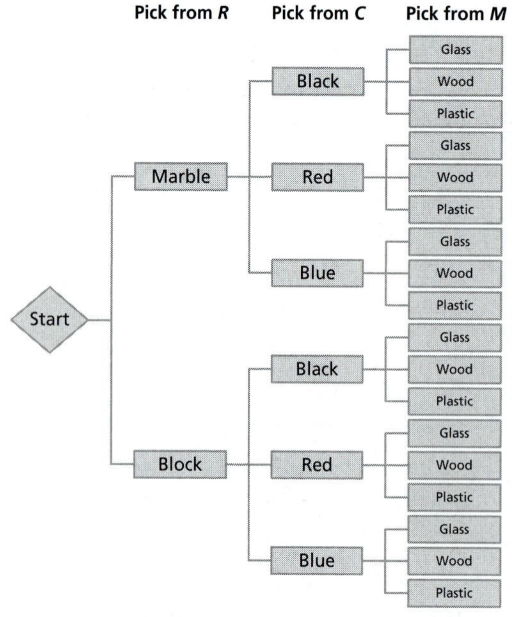

29. The outcomes with a single element from each of the sets R, C, M, and S.

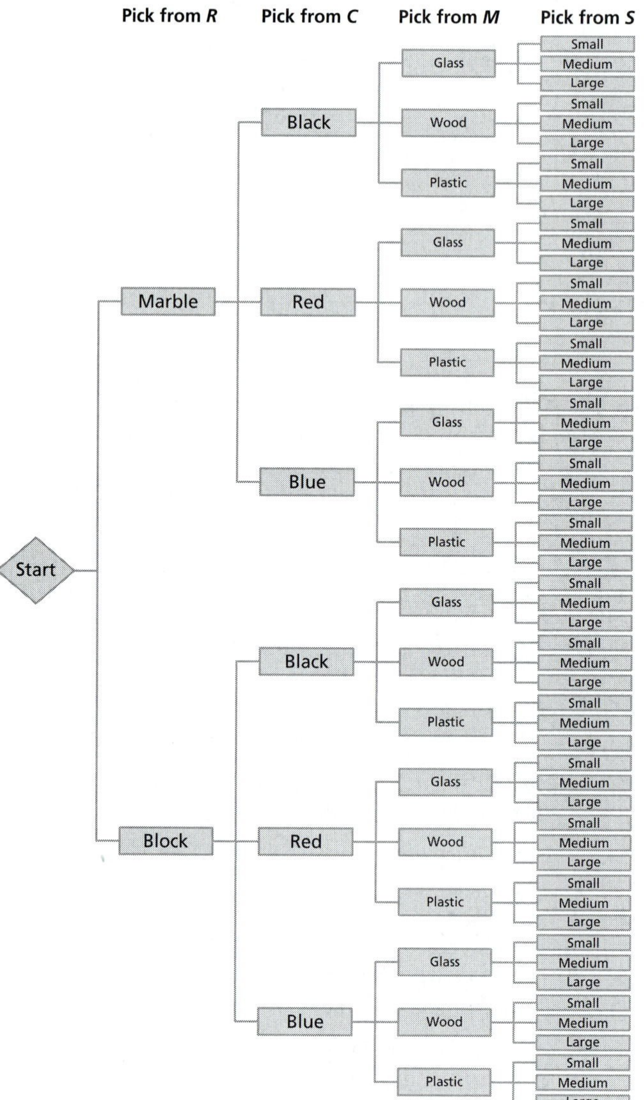

31. $n(D \cap F) = 155$ **33.** $n(B' \cap F) = 6770$

35. $n((N' \cap M) \cup F) = 68{,}969$

37. $A \neq B$ **39.** 12 sandwiches

Section 7.3 *(page 467)*

1. $P(5, 2) = 20$ **3.** $P(5, 5) = 120$

5. $P(9, 0) = 1$ **7.** $C(7, 5) = 21$

9. $C(4, 1) = 4$

11. $P(15, 7) = 32{,}432{,}400$

13. $P(19, 7) = 253{,}955{,}520$

15. $P(369, 3) = 49{,}835{,}664$

17. $C(20, 3) = 1140$

19. $C(39, 6) = 3{,}262{,}623$

21. 80,089,128

23. 11,232,000 license plates

25. 604,800 phone numbers

27. 720 zip codes

29. 420 schedules

31. 240 pizzas

33. 4050 pizzas

35. 35,048 call signs

37. 120 numbers

39. r^3 ways

Section 7.4 *(page 477)*

1. Experiment: Pick a person from Ellensburg and note the gender
Trial: 1
Outcomes: M, F
Sample space: $\{M, F\}$

3. Experiment: Roll a pair of dice and calculate the difference between the larger number and the smaller number
Trial: 1
Outcomes: 0, 1, 2, 3, 4, 5
Sample space: {0, 1, 2, 3, 4, 5}

5. Experiment: Pick three marbles from a bag and note their color
Trial: 1
Outcomes: RRR, RRW, RRB, RWW, RBB, RBW, WWW, WWB, WBB
Sample space:
$$\left\{\begin{matrix}\text{RRR, RRW, RRB, RWW, RBB,} \\ \text{RBW, WWW, WWB, WBB}\end{matrix}\right\}$$

7. Experiment: Pick a student in a two-credit class and note the number of credits on the student's class schedule
Trial: 1
Outcomes: 2, 3, 4, . . . , 20, 21
Sample space: {2, 3, 4, . . . , 20, 21}

9. Experiment: Pick a box of cereal and note its brand
Trial: 10
Outcomes: Kellogg's, General Mills, Other Brand
Sample space: {Kellogg's, General Mills, Other Brand}

11. The sample space is {(Miguel, Andrea), (Miguel, Miki), (Miguel, Sara), (Miguel, Sam), (Sam, Andrea), (Sam, Miki), (Sam, Sara), (Andrea, Miki), (Andrea, Sara), (Miki, Sara)}

(a) *Event:* At least one of the finalists is male. {(Miguel, Andrea), (Miguel, Miki), (Miguel, Sara), (Miguel, Sam), (Sam, Andrea), (Sam, Miki), (Sam, Sara)}
(b) *Event:* Both finalists are female. {(Andrea, Miki), (Andrea, Sara), (Miki, Sara)}
(c) *Event:* No more than one finalist is male. {(Miguel, Andrea), (Miguel, Miki), (Miguel, Sara), (Sam, Andrea), (Sam, Miki), (Sam, Sara), (Andrea, Miki), (Andrea, Sara), (Miki, Sara)}

13.

$$S = \left\{\begin{matrix}(1,1) & (1,2) & (1,3) & (1,4) & (1,5) & (1,6) \\ (2,1) & (2,2) & (2,3) & (2,4) & (2,5) & (2,6) \\ (3,1) & (3,2) & (3,3) & (3,4) & (3,5) & (3,6) \\ (4,1) & (4,2) & (4,3) & (4,4) & (4,5) & (4,6) \\ (5,1) & (5,2) & (5,3) & (5,4) & (5,5) & (5,6) \\ (6,1) & (6,2) & (6,3) & (6,4) & (6,5) & (6,6)\end{matrix}\right\}$$

(a) $$E = \left\{\begin{matrix}(1,1) & (1,3) & (1,5) \\ (2,2) & (2,4) & (2,6) \\ (3,1) & (3,3) & (3,5) \\ (4,2) & (4,4) & (4,6) \\ (5,1) & (5,3) & (5,5) \\ (6,2) & (6,4) & (6,6)\end{matrix}\right\}$$

(b) $$E = \left\{\begin{matrix}(1,4) & (2,3) & (3,2) & (4,1) \\ (4,6) & (5,5) & (6,4) & \end{matrix}\right\}$$

(c) $$E = \left\{\begin{matrix}(1,1) & (1,3) & (1,4) & (1,6) \\ (2,2) & (2,3) & (2,5) & (2,6) \\ (3,1) & (3,2) & (3,4) & (3,5) \\ (4,1) & (4,3) & (4,4) & (4,6) \\ (5,2) & (5,3) & (5,5) & (5,6) \\ (6,1) & (6,2) & (6,4) & (6,5)\end{matrix}\right\}$$

15. S = {TTTT, TTTF, TTFT, TFTT, FTTT, TTFF, TFTF, FTTF, TFFT, FTFT, FFTT, TFFF, FTFF, FFTF, FFFT, FFFF}

(a) E = {TTTT, TTTF, TTFT, TFTT, FTTT}
(b) E = {TTTT, TTTF, TTFT, TFTT, FTTT, TTFF, TFTF, FTTF, TFFT, FTFT, FFTT}
(c) E = {FFFF}

17. $S = \{DD, DR, RD, DI, ID, IR, RI, II, RR\}$

(a) $E = \{RR\}$
(b) $E = \{DD, DI, ID\}$
(c) $E = \{IR, RI, II, RR\}$

19. $S = \{$FFFF, FFFM, FFMF, FMFF, MFFF, FFMM, FMFM, MFFM, MFMF, MMFF, FMMF, FMMM, MFMM, MMFM, MMMF, MMMM$\}$

(a) $E = \{$FFFF, FFFM, FFMF, FMFF, MFFF$\}$
(b) $E = \{$FFFF, FFMF, FMFF, MFFF, MFMF, MMFF, FMMF, MMMF$\}$
(c) $E = \{$MFFF, MFMF, MMFF, MMMF$\}$

21. $n(S) = 10$

(a) $P(E) = 0.7$
Odds for E: 7:3
Odds against E: 3:7
(b) $P(E) = 0.3$
Odds for E: 3:7
Odds against E: 7:3
(c) $P(E) = 0.9$
Odds for E: 9:1
Odds against E: 1:9

23. $N(S) = 36$

(a) $P(E) = 0.5$
Odds for E: 1:1
Odds against E: 1:1
(b) $P(E) = \frac{7}{36} \approx 0.1944$
Odds for E: 7:29
Odds against E: 29:7
(c) $P(E) = \frac{2}{3} \approx 0.6667$
Odds for E: 2:1
Odds against E: 1:2

25. $n(S) = 16$

(a) $P(E) = 0.3125$
Odds for E: 5:11
Odds against E: 11:5
(b) $P(E) = 0.6875$
Odds for E: 11:5
Odds against E: 5:11
(c) $P(E) = 0.0625$
Odds for E: 1:15
Odds against E: 15:1

27. $n(S) = 9$

(a) $P(E) = \frac{1}{9} \approx 0.1111$
Odds for E: 1:8
Odds against E: 8:1
(b) $P(E) = \frac{3}{9} \approx 0.3333$
Odds for E: 3:6
Odds against E: 6:3
(c) $P(E) = \frac{4}{9} \approx 0.4444$
Odds for E: 4:5
Odds against E: 5:4

29. $n(S) = 16$

(a) $P(E) = \frac{5}{16} = 0.3125$
Odds for E: 5:11
Odds against E: 11:5
(b) $P(E) = \frac{8}{16} = 0.5$
Odds for E: 1:1
Odds against E: 1:1
(c) $P(E) = \frac{4}{16} = 0.25$
Odds for E: 1:3
Odds against E: 3:1

31. **(a)** $P(E) = \frac{6}{20} = 0.3$
(b) $P(E) = \frac{2}{20} = 0.1$
(c) $P(E) = \frac{2}{20} = 0.1$

*In our solution, we treated 35 and 53 as different groups even though the same two people belonged to each group. If you had treated 35 and 53 as the same group (combination versus permutation), the sample space would have had 10 elements.

The probability for each of the events would remain the same.

33. $P(E) = \frac{108}{216} = 0.50$ **35.** $P(E) = \frac{12}{36} \approx 0.33$

Section 7.5 *(page 489)*

1. $E \cap F = \{$50-cent piece, silver dollar$\}$
$E \cap F$ is the set of coins whose value is both a multiple of 5 and a multiple of 50.
$E \cup F = \left\{\begin{matrix}\text{nickel, dime, quarter,} \\ \text{50-cent piece, silver dollar}\end{matrix}\right\}$
$E \cup F$ is the set of coins whose value is a multiple of 5 or a multiple of 50.
$E' = \{$penny$\}$
E' is the set of coins whose value is *not* a multiple of 5.

3. $E \cap F = \left\{\begin{array}{l} x|x \text{ is a student on} \\ \text{scholarship with a} \\ \text{cumulative GPA of} \\ \text{at least 3.5} \end{array}\right\}$

$E \cap F$ is the set of students on scholarship who have a cumulative GPA of 3.5 or higher.

$E \cup F = \left\{\begin{array}{l} x|x \text{ is a student who is} \\ \text{on scholarship or has a} \\ \text{cumulative GPA of at} \\ \text{least 3.5} \end{array}\right\}$

$E \cup F$ is the set of students on scholarship together with the set of students who have a cumulative GPA of 3.5 or higher.

$E' = \left\{\begin{array}{l} x|x \text{ is a student who is} \\ \text{not on scholarship} \end{array}\right\}$

E' is the set of students who are not on scholarship.

5. $E \cap F = \left\{\begin{array}{l} x|x \text{ is a European nation} \\ \text{that is a permanent member} \\ \text{of the UN Security Council} \end{array}\right\}$

$E \cap F$ is the set of European nations that are permanent members of the security council.

$E \cup F = \left\{\begin{array}{l} x|x \text{ is a European nation or} \\ \text{is a permanent member} \\ \text{of the UN Security Council} \end{array}\right\}$

$E \cup F$ is the set of European nations on the Security Council together with the permanent members of the Security Council.

$E' = \left\{\begin{array}{l} x|x \text{ is a non-European nation} \\ \text{on the UN Security Council} \end{array}\right\}$

E' is the set of all non-European nations on the UN Security Council.

7. $P(E \cup F) = \frac{19}{36} \approx 0.528$ and

$P(E \cap F) = \frac{2}{36} \approx 0.056$

9. $P(E \cup F) = \frac{22}{36} \approx 0.611$ and

$P(E \cap F) = \frac{5}{36} \approx 0.139$

11. $P(E) = 0.8$

13. $P(E) = 0.2$

15. $P(E) = 0.6$

17. $P(E) = \frac{1}{7} \approx 0.143$

19. $P(E) = \frac{34}{35} \approx 0.971$

21. An employer who was interested in giving all of her employees an equal opportunity to work could track the work assignments of each employee and periodically compare the work assignment to the various probabilities. For example, if the probability of having an all-women work team was 20 percent and several months of employee data showed that 50 percent of the work teams contained only women, the employer might be concerned that she is being biased against men. (Answers may vary.)

23. $E = \{x|x \text{ is a Republican}\}$
$F = \{x|x \text{ is a Democrat}\}$
$G = \{x|x \text{ is an Independent}\}$

$$P(E) = 0.49$$
$$P(F) = 0.50$$
$$P(G) = 0.01$$

25. $E = \{x|x \text{ is an America citizen}\}$
$F = \{x|x \text{ has a student visa}\}$
$G = \{x|x \text{ is an illegal alien}\}$

$$P(E) = \frac{20}{33} \approx 0.606$$
$$P(F) = \frac{11}{33} \approx 0.333$$
$$P(G) = \frac{2}{33} \approx 0.061$$

27. $E = \{x|x \text{ has a master's degree}\}$
$F = \{x|x \text{ has a bachelor's degree}\}$
$G = \{x|x \text{ has a high school diploma}\}$

$$P(E) = \frac{13}{40} = 0.325$$
$$P(F) = \frac{17}{40} = 0.425$$
$$P(G) = \frac{10}{40} = 0.25$$

*Note: The company is tracking the *highest* earned degree of each employee. Although someone with a master's degree also has a bachelor's degree and high school diploma, that person is counted only in the category of his or her highest degree.

29. The elements of the sample space that yield each product are indicated here, together with the probability of obtaining the product:

$1: (1, 1) \Rightarrow P(E) = \frac{1}{24}$

$2: (2, 1), (1, 2) \Rightarrow P(E) = \frac{1}{12}$

$3: (3, 1), (1, 3) \Rightarrow P(E) = \frac{1}{12}$

$4: (4, 1), (1, 4), (2, 2) \Rightarrow P(E) = \frac{1}{8}$

$5: (5, 1) \Rightarrow P(E) = \frac{1}{24}$

$6: (6, 1), (2, 3), (3, 2) \Rightarrow P(E) = \frac{1}{8}$

$8: (2, 4), (4, 2) \Rightarrow P(E) = \frac{1}{12}$

$9: (3, 3) \Rightarrow P(E) = \frac{1}{24}$

$10: (5, 2) \Rightarrow P(E) = \frac{1}{24}$

$12: (6, 2), (3, 4), (4, 3) \Rightarrow P(E) = \frac{1}{8}$

$15: (5, 3) \Rightarrow P(E) = \frac{1}{24}$

$16: (4, 4) \Rightarrow P(E) = \frac{1}{24}$

$18: (6, 3) \Rightarrow P(E) = \frac{1}{24}$

$20: (5, 4) \Rightarrow P(E) = \frac{1}{24}$

$24: (6, 4) \Rightarrow P(E) = \frac{1}{24}$

31. 50%

33. The elements of the sample space that yield each value are indicated here, together with the probability of obtaining the value:

$0: (1, 1), (1, 2), (1, 3), (1, 4), (1, 5), (1, 6), (2, 1), (3, 1), (4, 1), (5, 1), (6, 1) \Rightarrow P(E) = \frac{11}{36} \approx 0.306$

$2: (2, 2) \Rightarrow P(E) = \frac{1}{36} \approx 0.028$

$3: (2, 3), (3, 2) \Rightarrow P(E) = \frac{2}{36} \approx 0.056$

$4: (2, 4), (4, 2) \Rightarrow P(E) = \frac{2}{36} \approx 0.056$

$5: (2, 5), (5, 2) \Rightarrow P(E) = \frac{2}{36} \approx 0.056$

$6: (2, 6), (6, 2), (3, 3) \Rightarrow P(E) = \frac{3}{36} \approx 0.083$

$8: (3, 4), (4, 3) \Rightarrow P(E) = \frac{2}{36} \approx 0.056$

$10: (5, 3), (3, 5) \Rightarrow P(E) = \frac{2}{36} \approx 0.056$

$12: (6, 3), (3, 6), (4, 4) \Rightarrow P(E) = \frac{3}{36} \approx 0.083$

$15: (5, 4), (4, 5) \Rightarrow P(E) = \frac{2}{36} \approx 0.056$

$18: (6, 4), (4, 6) \Rightarrow P(E) = \frac{2}{36} \approx 0.056$

$20: (5, 5) \Rightarrow P(E) = \frac{1}{36} \approx 0.028$

$24: (6, 5), (5, 6) \Rightarrow P(E) = \frac{2}{36} \approx 0.056$

$30: (6, 6) \Rightarrow P(E) = \frac{1}{36} \approx 0.028$

35. Obtaining an odd value is more likely than obtaining an even value.

CHAPTER 7 REVIEW EXERCISES

Section 7.1 *(page 491)*

1. $(A \cap B)' = \{1, 3, 5, 6, 7, 8\}$

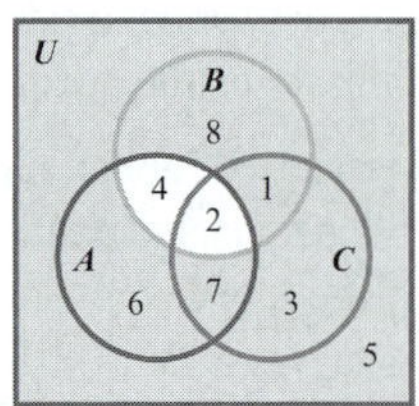

3. $(A \cap B) \cap C' = \{4\}$

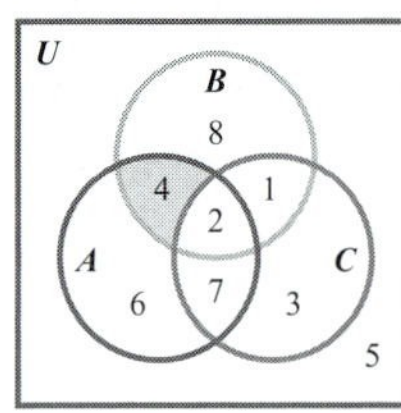

5. $M \cap E$ is the set of employed men. $M \cap E = \{1, 2\}$

7. $E' \cap D$ is the set of unemployed people with graduate degrees. $E' \cap D = \{5\}$

Section 7.2 *(page 491)*

9. $n(B') = 10$

11. $n(A' \cap B') = 3$

13. 34 dishes

15. 175 three-course meals with strawberry shortcake

17. The outcomes with a single element from each of the sets R, C, and S.

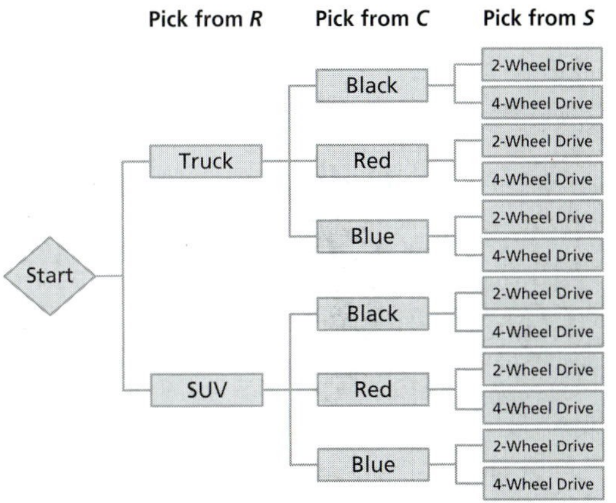

Section 7.3 *(page 492)*

19. $P(9, 2) = 72$ **21.** $C(6, 2) = 15$

23. $P(25, 7) = 2{,}422{,}728{,}000$

25. 48 one-topping pizzas

27. 880 three-topping pizzas

Section 7.4 *(page 492)*

29. **Experiment:** Pick an entry and note the last number of the entry
Trial: 1
Outcomes: 0, 1, 2, 3, 4, 5, 6, 7, 8, 9
Sample space: {0, 1, 2, 3, 4, 5, 6, 7, 8, 9}

31. $S = \left\{\begin{array}{l}(\text{Adi, Berto}), (\text{Adi, Marsha}),\\ (\text{Adi, Mariah}), (\text{Adi, Liz}),\\ (\text{Berto, Marsha}), (\text{Berto, Mariah}),\\ (\text{Berto, Liz}), (\text{Marsha, Mariah}),\\ (\text{Marsha, Liz}), (\text{Mariah, Liz})\end{array}\right\}$

(a) *Event:* At least one of the finalists is male.

$E = \left\{\begin{array}{l}(\text{Adi, Berto}), (\text{Adi, Marsha}),\\ (\text{Adi, Mariah}), (\text{Adi, Liz}),\\ (\text{Berto, Marsha}), (\text{Berto, Mariah}),\\ (\text{Berto, Liz})\end{array}\right\}$

(b) *Event:* Both finalists are female.

$E = \left\{\begin{array}{l}(\text{Marsha, Mariah}),\\ (\text{Marsha, Liz}), (\text{Mariah, liz})\end{array}\right\}$

(c) *Event:* No more than one finalist is male.

$E = \left\{\begin{array}{l}(\text{Adi, Marsha}), (\text{Adi, Mariah}),\\ (\text{Adi, Liz}), (\text{Berto, Marsha}),\\ (\text{Berto, Mariah}), (\text{Berto, Liz}),\\ (\text{Marsha, Mariah}), (\text{Marsha, Liz}),\\ (\text{Mariah, Liz})\end{array}\right\}$

Section 7.5 *(page 492)*

33. $P(E) = \frac{7}{10} = 0.7$ **35.** $P(E) = \frac{5}{10} = 0.5$

CHAPTER 8

Section 8.1 *(page 504)*

1. 0.9 **3.** 0.1 **5.** 0.0546 **7.** 0.126

9. 0.25 **11.** $0.3\overline{3}$ **13.** $0.6\overline{6}$ **15.** 1

17. 0.414 **19.** 0.6 **21.** 0.0857

23. $0.41\overline{6}$ **25.** $0.53\overline{3}$ **27.** 0.358

29. 0.25; decreases to 0.22 **31.** Independent

33. Dependent **35.** Independent **37.** No

Section 8.2 *(page 521)*

1.

Prevalence	14.7%
Sensitivity	86.2%
Specificity	97.9%
PVP	87.4%
PVN	97.6%

Of the entire population, 14.7% have the disease. Of those who have the disease, 86.2% test positive. Of those without the disease, 97.9% test negative. Of those who test positive, 87.4% have the disease. Of those who test negative, 97.6% don't have the disease.

3.

Prevalence	19.2%
Sensitivity	86.3%
Specificity	93.8%
PVP	76.6%
PVN	96.6%

Of all e-mails, 19.2% are spam. Of those that are spam, 86.3% test positive. Of those that are not spam, 93.8% test negative. Of those that test positive, 76.6% are spam. Of those that test negative, 96.6% aren't spam.

5.

Prevalence	72.7%
Sensitivity	83.3%
Specificity	66.7%
PVP	87.0%
PVN	60.0%

Of the entire population, 72.7% pass the class. Of those who pass the class, 83.3% study. Of those who fail the class, 66.7% don't study. Of those who study, 87.0% pass the class. Of those who don't study, 60.0% don't pass the class.

7. | | |
|---|---|
| Prevalence | 31.2% |
| Sensitivity | 47.5% |
| Specificity | 94.6% |
| *PVP* | 80.0% |
| *PVN* | 79.9% |

Of all the homes, 31.2% don't need repairs. Of those that don't need repairs, 47.5% are newer homes. Of those that need repairs, 94.6% are older homes. Of those that are newer homes, 80.0% don't need repairs. Of those that are older homes, 79.9% need repairs.

9. | | |
|---|---|
| Prevalence | 20.4% |
| Sensitivity | 40.0% |
| Specificity | 74.7% |
| *PVP* | 28.9% |
| *PVN* | 82.9% |

Of the entire population, 20.4% got a speeding ticket. Of those who got a speeding ticket, 40.0% drive a red car. Of those who didn't get a speeding ticket, 74.7% don't drive a red car. Of those who drive a red car, 28.9% got a speeding ticket. Of those who don't drive a red car, 82.9% didn't get a speeding ticket.

11. | | |
|---|---|
| Prevalence | 0.2% |
| Sensitivity | 98% |
| Specificity | 76% |
| *PVP* | **0.81%** |
| *PVN* | **99.99%** |

13. | | |
|---|---|
| Prevalence | 20% |
| Sensitivity | 50% |
| Specificity | 75% |
| *PVP* | **33.3%** |
| *PVN* | **85.7%** |

15. | | |
|---|---|
| Prevalence | 5% |
| Sensitivity | 85% |
| Specificity | **70%** |
| *PVP* | 13% |
| *PVN* | **98.9%** |

17. | | |
|---|---|
| Prevalence | 2% |
| Sensitivity | 92% |
| Specificity | **92.0%** |
| *PVP* | 19% |
| *PVN* | **99.8%** |

19. | | |
|---|---|
| Prevalence | 10% |
| Sensitivity | **91.1%** |
| Specificity | 98% |
| *PVP* | **83.5%** |
| *PVN* | 99% |

21. 93.3% sensitivity: If a person has skin cancer, there is a 93.3% chance that the test will be positive. 97.8% specificity: If a person doesn't have skin cancer, there is a 97.8% chance that the test will be negative. 54% *PVP*: If a test is positive, there is a 54% chance that the person has skin cancer.
99.98% *PVN*: If a test is negative, there is a 99.8% chance that the person has no skin cancer.

23. 0.0147% *PVP*: If the test is positive, there is a 0.0147% chance that the person has invasive aspergillosis. 99.9996% *PVN*: If the test is negative, there is a 99.9996% chance that the person does not have invasive aspergillosis.

25. 44.8%

27. 100%; 0.037%

29. 50.1%; 16.7%

31. 6.9%

33. 3.3%

35. 0.00%; 99.92%

37. 1 − sensitivity

39. Accuracy of the tools; having a positive way of identifying people who have the disease for which you are testing; size of the sample (answers may vary)

Section 8.3 *(page 538)*

1. $p_{13} = 0$ means that the probability of moving from state 1 to state 3 is 0%.
$p_{21} = 0$ means that the probability of moving from state 2 to state 1 is 0%.
$p_{33} = 0.7$ means that the probability of moving from state 3 to state 3 is 70%.

3. $p_{13} = 0.1$ means that the probability of moving from state 1 to state 3 is 10%.
$p_{21} = 0.2$ means that the probability of moving from state 2 to state 1 is 20%.
$p_{33} = 0.0$ means that the probability of moving from state 3 to state 3 is 0%.

5. $p_{13} = 0$ means that the probability of moving from state 1 to state 3 is 0%.
$p_{21} = 0$ means that the probability of moving from state 2 to state 1 is 0%.
$p_{33} = 0.5$ means that the probability of moving from state 3 to state 3 is 50%.

7. S_1 and S_2 are absorbing states.

$$P = \begin{bmatrix} 1 & 0 \\ 0 & 1 \end{bmatrix}$$

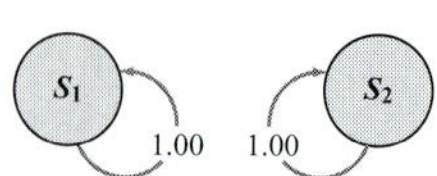

9. S_1 is an absorbing state.

$$P = \begin{bmatrix} 1.00 & 0.00 \\ 0.05 & 0.95 \end{bmatrix}$$

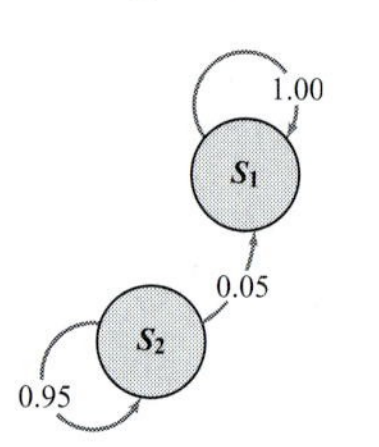

11. $P = \begin{bmatrix} 0.7 & 0.0 & 0.3 \\ 0.1 & 0.4 & 0.5 \\ 0.2 & 0.8 & 0.0 \end{bmatrix}$

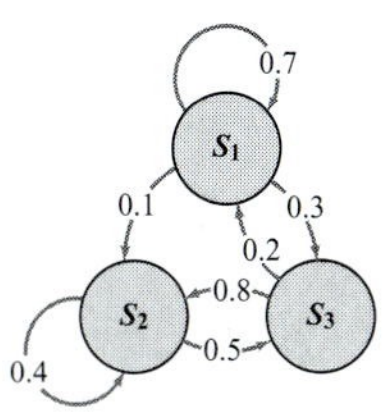

13. $P = \begin{bmatrix} 0.1 & 0.2 & 0.7 \\ 0.2 & 0.3 & 0.5 \\ 0.3 & 0.4 & 0.3 \end{bmatrix}$

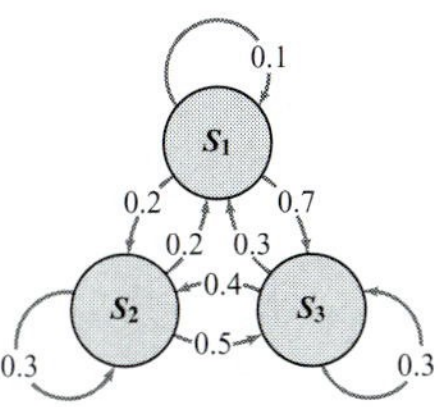

15. S_3 and S_4 are absorbing states.

$$P = \begin{bmatrix} 0 & 0.5 & 0.5 & 0 \\ 0 & 0 & 0.5 & 0.5 \\ 0 & 0 & 1 & 0 \\ 0 & 0 & 0 & 1 \end{bmatrix}$$

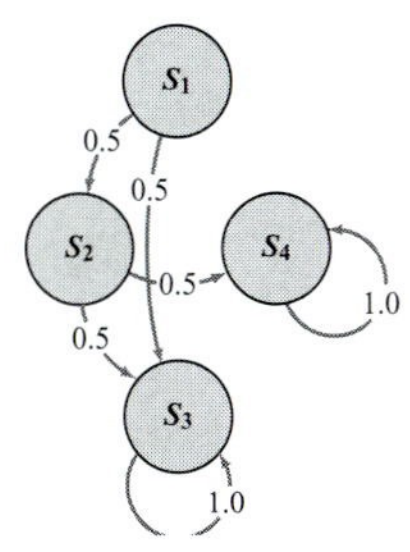

17. $P = \begin{bmatrix} 0 & 1 \\ 1 & 0 \end{bmatrix}$

19. $P = \begin{bmatrix} 0.95 & 0.05 & 0.00 \\ 0.00 & 0.99 & 0.01 \\ 0.00 & 0.00 & 1.00 \end{bmatrix}$

21. $D = \begin{bmatrix} \frac{3}{11} & \frac{8}{11} \end{bmatrix}$

23. $D = \begin{bmatrix} \frac{4}{13} & \frac{9}{13} \end{bmatrix}$

25. $D = \begin{bmatrix} \frac{25}{61} & \frac{36}{61} \end{bmatrix}$

27. $D = \begin{bmatrix} \frac{26}{115} & \frac{48}{115} & \frac{41}{115} \end{bmatrix}$

29. $D = \begin{bmatrix} \frac{1}{3} & \frac{1}{3} & \frac{1}{3} \end{bmatrix}$

31. $P^2 = \begin{bmatrix} 0.0144 & 0.7744 & 0.2112 \\ 0.2112 & 0.0144 & 0.7744 \\ 0.7744 & 0.2112 & 0.0144 \end{bmatrix}$

$P^5 = \begin{bmatrix} 0.0982 & 0.5411 & 0.3607 \\ 0.3607 & 0.0982 & 0.5411 \\ 0.5411 & 0.3607 & 0.0982 \end{bmatrix}$

$P^7 = \begin{bmatrix} 0.3950 & 0.1600 & 0.4450 \\ 0.4450 & 0.3950 & 0.1600 \\ 0.1600 & 0.4450 & 0.3950 \end{bmatrix}$

For the matrix P^k, the entry p_{ij} is the probability of moving from state i to state j in exactly k steps.

33. $P^2 = \begin{bmatrix} 0.3750 & 0.3125 & 0.3125 \\ 0.3125 & 0.3750 & 0.3125 \\ 0.3125 & 0.3125 & 0.3750 \end{bmatrix}$

$P^5 = \begin{bmatrix} 0.3340 & 0.3330 & 0.3330 \\ 0.3330 & 0.3340 & 0.3330 \\ 0.3330 & 0.3330 & 0.3340 \end{bmatrix}$

$P^7 = \begin{bmatrix} 0.3334 & 0.3333 & 0.3333 \\ 0.3333 & 0.3334 & 0.3333 \\ 0.3333 & 0.3333 & 0.3334 \end{bmatrix}$

For the matrix P^k, the entry p_{ij} is the probability of moving from state i to state j in exactly k steps.

35. $P^2 = \begin{bmatrix} 0.1759 & 0.8241 & 0 \\ 0.1230 & 0.877 & 0 \\ 0 & 0 & 1 \end{bmatrix}$

$P^5 = \begin{bmatrix} 0.1304 & 0.8696 & 0 \\ 0.1298 & 0.8702 & 0 \\ 0 & 0 & 1 \end{bmatrix}$

$P^7 = \begin{bmatrix} 0.1299 & 0.8701 & 0 \\ 0.1299 & 0.8701 & 0 \\ 0 & 0 & 1 \end{bmatrix}$

For the matrix P^k, the entry p_{ij} is the probability of moving from state i to state j in exactly k steps.

37. It is not possible for a player to lose 25 rounds in a row. The possible outcomes for any sequence of three coin tosses are

$$\begin{Bmatrix} \text{HHH, HHT, HTH, HTT,} \\ \text{THH, THT, TTH, TTT} \end{Bmatrix}$$

The outcomes correspond with the indicated win-loss sequences: win-win, win-lose, lose-win, lose-win, win-win, win-lose, win-win, win-win. Every outcome of three consecutive coin tosses will result in at least one win.

The probability that a player who lost the first round won the 25th round is 75%.

39. 82%

41. **(i)** Let S_1 = Mercury Interactive customer and S_2 = competitor's customer.

(ii)

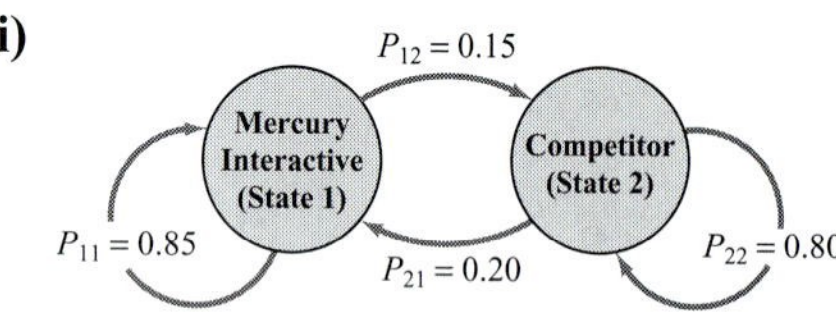

(iii) $P = \begin{bmatrix} 0.85 & 0.15 \\ 0.20 & 0.80 \end{bmatrix}$

(iv) $D = [0.552 \quad 0.448]$

(v) $DP = [0.5588 \quad 0.4412]$

We estimate that Mercury Interactive had a 55.9% market share of the automated software quality tools market in 2003.

43. **(i)** Let S_1 = price higher than previous day and S_2 = price lower than previous day.

(ii)

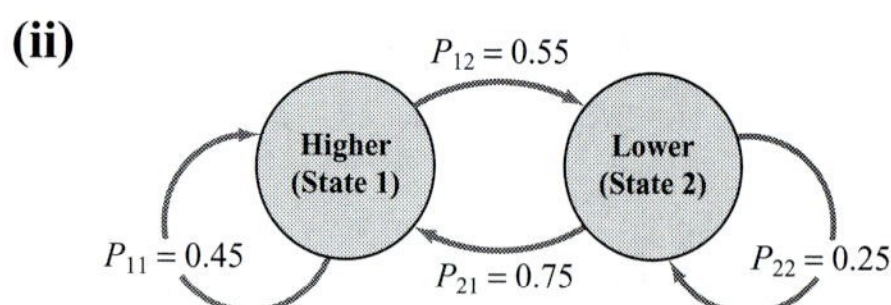

(iii) $P = \begin{bmatrix} 0.45 & 0.55 \\ 0.75 & 0.25 \end{bmatrix}$

(iv) $D = \begin{bmatrix} \frac{11}{20} & \frac{9}{20} \end{bmatrix}$

(v) $DP = [0.585 \quad 0.415]$

According to the Markov chain model, we estimate that at the end of the next 20 trading days, 58.5% of the share prices will have been higher than the previous day's price and 41.5% of the share prices will have been lower than the previous day's price.

45. **(i)** Let S_1 = weight training, S_2 = cardiovascular, and S_3 = not work out.

(ii)

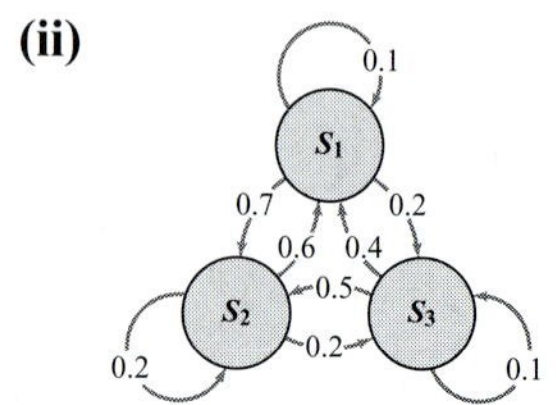

(iii) $P = \begin{bmatrix} 0.1 & 0.7 & 0.2 \\ 0.6 & 0.2 & 0.2 \\ 0.4 & 0.5 & 0.1 \end{bmatrix}$

(iv) $D = [0.30 \quad 0.25 \quad 0.45]$

(v) $DP = [0.36 \quad 0.485 \quad 0.155]$

According to the Markov chain model, we estimate that tomorrow 36% of the members will do weight lifting, 48.5% will do cardiovascular exercises, and 15.5% won't work out.

47. **(a)** $D_1 = [0.054 \quad 0.946]$

5.4% of civilians 16 years old or older were unemployed and 94.6% were employed in 1996.

$D_2 = [0.040 \quad 0.960]$

4.0% of civilians 16 years old or older were unemployed and 96.0% were employed in 2000.

(b) The model yielded the same results as the raw data. The model fits the data well.

(c) $D_3 = [0.031 \quad 0.969]$

According to the model, 3.1% of civilians 16 years old and older were unemployed in 2004.

(d) The steady state distribution vector is

$$D = [0.012 \quad 0.988]$$

The model forecasts that the long-term unemployment rate will be 1.2%.

49. **(a)** $D_1 = [0.103 \quad 0.897]$

10.3% of private-sector workers were union members in 1995 and 89.7% were not.

$D_2 = [0.090 \quad 0.910]$

9.0% of private-sector workers were union members in 2000 and 91.0% were not.

(b) The model yielded the same results as the raw data. The model fits the data well.

(c) $D_3 = [0.079 \quad 0.921]$

According to the model, 7.9% of private-sector workers were union members in 2005 and 92.1% were not.

(d) The steady state distribution vector is

$$D = [0.032 \quad 0.968]$$

The model forecasts that the long-term percentage of private-sector workers who are union members will be 3.2%.

(e) According to the model, union enrollment is forecasted to decline dramatically. In response to these results, union leaders could increase recruitment efforts and implement additional worker-friendly policies to increase membership. (Answers may vary.)

51. (a) $D_3 = [0.539 \quad 0.461]$
The percentage of abused children who were neglected in 1999 is estimated to be 53.9%.

(b) 53.8%

(c) Reducing child abuse may seem like a daunting task. Nevertheless, there is much that can be done by each of us. On a broad scale, any initiatives that strengthen positive family relationships will help to reduce child abuse.

Prevent Child Abuse America has indicated that supporting children and parents in your family and community reduces the likelihood of child abuse and neglect. Simple things such as offering to babysit; donating clothing, furniture, and toys to other families; being kind to new parents; and being a good neighbor go a long way in helping to prevent child abuse. (**Source:** www.preventchildabuse.org.) (Answers may vary.)

53. (a) Yes. The model fits the data perfectly.
$D_1 = [0.26 \quad 0.74]$ and $D_2 = [0.32 \quad 0.68]$

(b) The transition matrix indicates that a supermarket with a pharmacy has a 12.5% probability of *not* having a pharmacy four years later. It seems reasonable that a supermarket would replace an unprofitable service, such as a deli, bakery, or bank. The remaining probabilities in the matrix all seem reasonable.

(c) 50%.

(d) This model seems reasonable. There will always be supermarkets for which an in-store pharmacy will not make good business sense. We don't have any information that leads us to believe that more than 50% of supermarkets will have pharmacies in the future. (Answers may vary.)

55. (a) Yes. The model fits the data perfectly.
$D_1 = [0.460 \quad 0.540]$ and
$D_2 = [0.450 \quad 0.550]$

(b) The transition matrix indicates that a supermarket with a service fish department has a 65.8% probability of *not* having a service fish department four years later. This seems a bit higher than we would expect. Likewise, the transition matrix indicates that there is a 54.2% probability that a store without a service fish department will have such a department four years later. In light of the high probability of service fish departments going out of business within four years, it seems unlikely that there would be a high percentage of stores without service fish departments that were clamoring to add a service fish department.

(c) 45.2%

57. $d_2 = 0$ and $d_3 = 1 - d_1$

59. Observe that

$$P^2 = \begin{bmatrix} 0 & 1 & 0 \\ 0 & 0 & 1 \\ 1 & 0 & 0 \end{bmatrix}$$

$$P^3 = \begin{bmatrix} 1 & 0 & 0 \\ 0 & 1 & 0 \\ 0 & 0 & 1 \end{bmatrix}$$

Since P^3 is the identity matrix

$$\begin{aligned} P^4 &= P^3 \cdot P \\ &= IP \\ &= P \end{aligned}$$

Consider $k = 3n + 1$.

$$\begin{aligned} P^k &= P^{3n+1} \\ &= P^{3n} \cdot P \\ &= (P^3)^n P \\ &= (I)^n P \\ &= IP \\ &= P \end{aligned}$$

In general, if $k = 3n + 1$ (for nonnegative integer values of n) then $P^k = P$.

61.

$$\begin{aligned} d_1 &= d_0 P \\ &= [x \quad 1 - x]\begin{bmatrix} a & 1 - a \\ b & 1 - b \end{bmatrix} \\ &= [ax + (1 - x)b \quad (1 - a)x + (1 - b)(1 - x)] \end{aligned}$$

Since $d_1 = [y \quad 1 - y]$, we know

$$y = ax + b(1 - x)$$

$$\begin{aligned} d_2 &= d_1 P \\ &= [y \quad 1 - y]\begin{bmatrix} a & 1 - a \\ b & 1 - b \end{bmatrix} \\ &= [ay + (1 - y)b \quad (1 - a)y + (1 - b)(1 - y)] \end{aligned}$$

Since $d_2 = [z \quad 1 - z]$, we know

$$z = ay + b(1 - y)$$

We have a system of equations in a and b.

$$\begin{aligned} ax + b(1 - x) &= y \\ ay + b(1 - y) &= z \end{aligned}$$

Writing the system as an augmented matrix yields

$$\left[\begin{array}{cc|c} x & 1-x & y \\ y & 1-y & z \end{array}\right]$$

$$\left[\begin{array}{cc|c} x & 1-x & y \\ 0 & x-y & xz-y^2 \end{array}\right] \quad xR_2 - yR_1$$

$$\left[\begin{array}{cc|c} x^2-xy & 0 & x(xz+y-y^2-z) \\ 0 & x-y & xz-y^2 \end{array}\right] \quad (x-y)R_1-(1-x)R_2$$

$$\left[\begin{array}{cc|c} 1 & 0 & \dfrac{xz+y-y^2-z}{x-y} \\ 0 & 1 & \dfrac{xz-y^2}{x-y} \end{array}\right] \quad \begin{array}{l} \frac{1}{x^2-xy}R_1 \\ \frac{1}{x-y}R_2 \end{array}$$

Thus $a = \dfrac{xz+y-y^2-z}{x-y}$ and $b = \dfrac{xz-y^2}{x-y}$.

Section 8.4 *(page 554)*

1.

X	$P(X)$
0	$\frac{1}{4}$
1	$\frac{1}{2}$
2	$\frac{1}{4}$

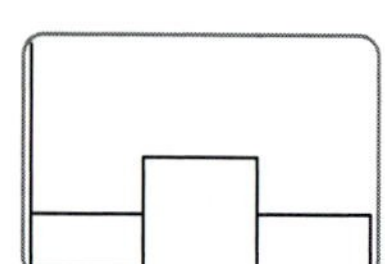

3.

X	$P(X)$
0	0.659
1	0.299
2	0.040
3	0.001
4	0.00002

5.

X	$P(X)$
0	0.9243
1	0.07342
2	0.00224
3	3.3×10^{-5}
4	2.5×10^{-7}
5	9×10^{-10}
6	1×10^{-12}

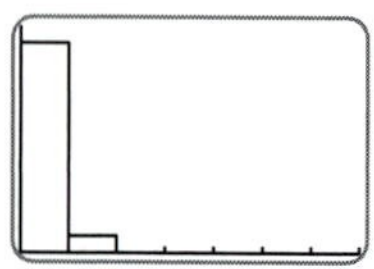

7.

X	$P(X)$
0	2.3×10^{-5}
1	4.6×10^{-4}
2	0.004
3	0.020
4	0.068
5	0.154
6	0.240
7	0.254
8	0.175
9	0.071
10	0.013

9.

X	$P(X)$
0	$\frac{35}{126} = 0.278$
1	$\frac{70}{126} = 0.556$
2	$\frac{21}{126} = 0.167$

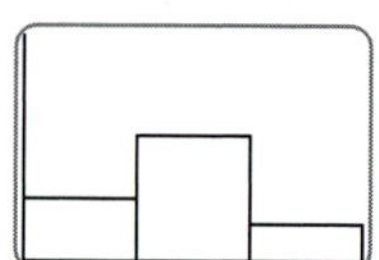

11. X = net amount won

X	$P(X)$
-2	$\frac{242}{250} = 0.968$
48	$\frac{1}{250} = 0.004$
23	$\frac{2}{250} = 0.008$
8	$\frac{5}{250} = 0.020$

$$\begin{aligned} E(X) &= (-2)(0.968) + (48)(0.004) \\ &\quad + (23)(0.008) + (8)(0.020) \\ &= -1.40 \end{aligned}$$

The expected value is $-\$1.40$.

13. *With audit protection*
X = cost

X	$P(X)$
25	0.9956
0	0.0044

$E(X) = 24.89$

Without audit protection
X = cost

X	$P(X)$
0	0.9956
2,707	0.0044

$E(X) = 11.91$

Since the expected cost without audit protection is less than the expected cost with audit protection, we would not recommend the audit-guarantee service. However, since the audit protection is relatively cheap, you may choose to purchase it anyway to eliminate the possibility of being hit with a large tax penalty.

15. X = points awarded if you guess

X	$P(X)$
1	$\frac{1}{4} = 0.25$
-0.5	$\frac{3}{4} = 0.75$

$E(X) = -0.125$

No. If he guesses, we expect he will lose 0.125 point. If he doesn't guess, no points are lost.

17. X = points awarded if she guesses

X	$P(X)$
1	$\frac{1}{3} \approx 0.333$
-0.5	$\frac{2}{3} \approx 0.667$

$E(X) = 0$

The expected value is 0 points.

19. *With insurance*
X = cost to author

X	$P(X)$
-89.95	0.001
-89.95	0.999

$E(X) = -89.95$

Without insurance
X = cost to author

X	$P(X)$
$-5{,}300$	0.001
0	0.999

$E(x) = -5.30$

No. The expected value of not having insurance is less than the expected value of having insurance. However, if the author wants to eliminate the risk of a \$5300 loss, he may want to buy the insurance anyway.

21. X = net income from insurance

X	$P(X)$
-508	0.9228
3,660	0.0772

$E(X) = -186.23$

The expected value is $-\$186.23$. Since the homeowner considers any payout as a gain, the long-term average annual cost for "peace of mind" is expected to be \$186.23.

23. *\$3 million jackpot*
$E(X) \approx -0.69$

\$5 million jackpot
$E(X) \approx -0.60$

\$10 million jackpot
$E(X) \approx -0.38$

25. \$0.03 per day

27. \$3.27 per year

29. $-\$0.02$ per day

31. $E(X) = -5502 - 6.6t$

33. $3.91\overline{6}$

35. $x \leq 5000$

Section 8.5 *(page 572)*

1. Mean: 38.38
Median: 35.5
Mode: 30, 32, 35, 37

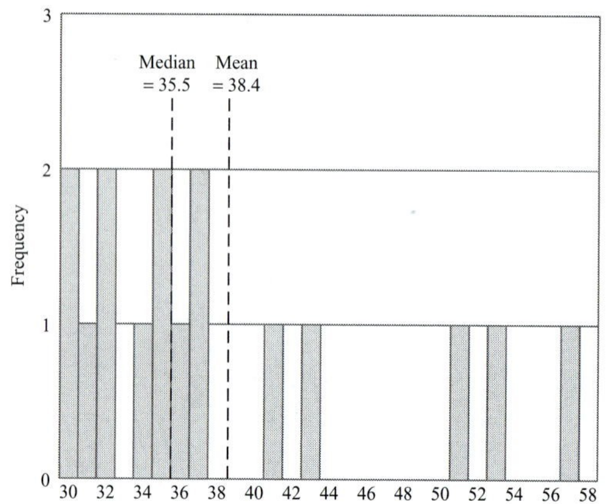

Of those players scoring more than 30 touchdowns, the mean number of touchdowns scored was 38.38. Half of the players had scored more than 35.5 touchdowns and half of the players had scored fewer than 35.5 touchdowns. The number of touchdowns scored most frequently were 30, 32, 35, and 37.

3. Mean: 4.2
Median: 4
Mode: 4

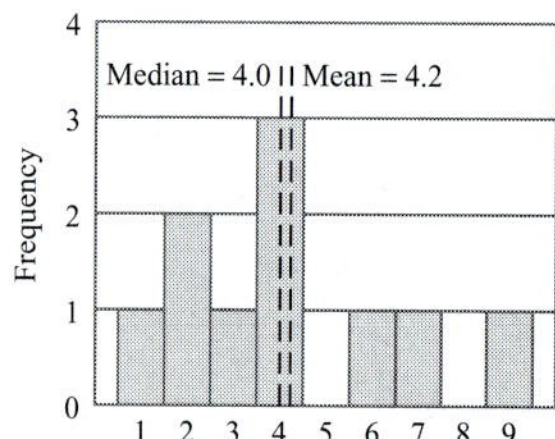

Between 1900 and 1909, the mean rank of the Yale football team was 4.2. Placing its rankings in numerical order, the middlemost ranking was 4. The most frequent ranking was also 4.

5. Mean: 12.78
Median: 12.8
Mode: 12.7, 12.8, 12.9, 13.1

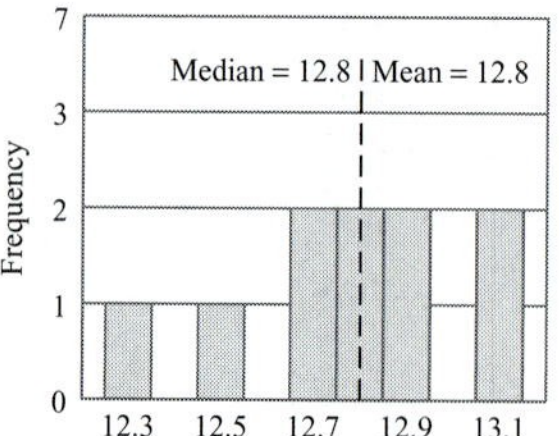

Between 1990 and 1999, the mean percentage of births to teenage mothers was 12.78%. Placing the percentages in numerical order, the middlemost percentage was 12.8%. The percentages occurring most frequently were 12.7%, 12.8%, 12.9%, and 13.1%.

7. Mean: 10.19
Median: 11
Mode: 7, 11

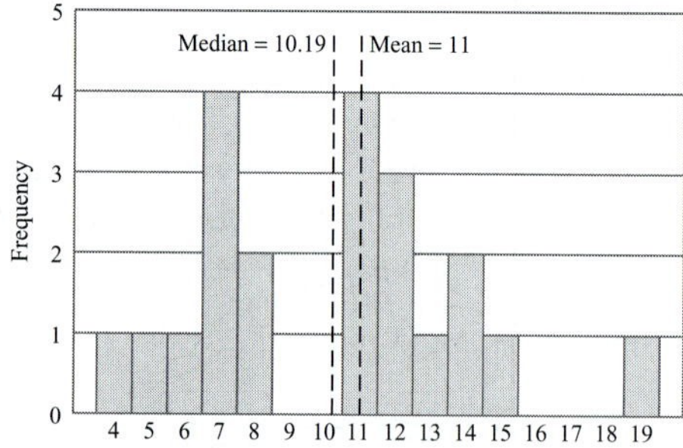

Between 1980 and 2000, the mean number of tropical storms and hurricanes each year was 10.19. Placing the number of tropical storms and hurricanes in numerical order, the middlemost number was 11. The number of tropical storms and hurricanes occurring most often in a year was 7 and 11.

9. Mean: 27
Median: 28
Mode: 28, 29

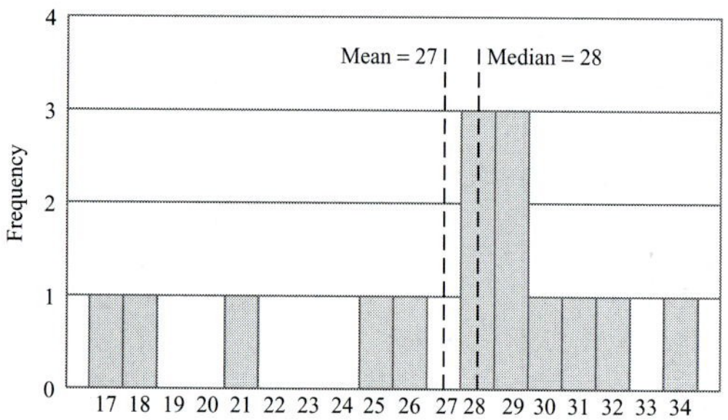

Between 1990 and 1999, the mean percentage of new homes with 2400 or more square feet was 27%. Placing the percentages in numerical order, the middlemost percentage was 28%. The percentages occurring most often were 28% and 29%.

11. Mean: 20.6
Median: 20.5
Mode: 20, 21

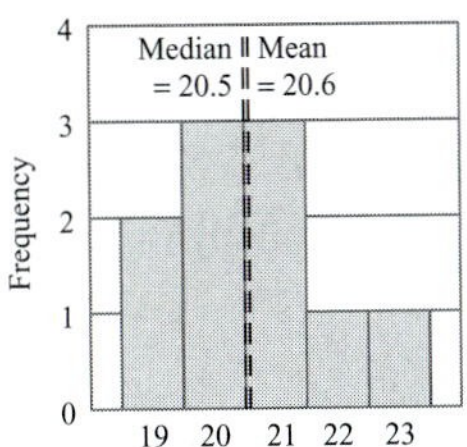

Between 1990 and 1999, the mean U.S. percentage of worldwide fuel production was 20.6%. Placing the percentages in numerical order, the middlemost percentage was 20.5%. The percentages occurring most often were 20% and 21%.

13. Mean: 24.19
Median: 24
Mode: 24

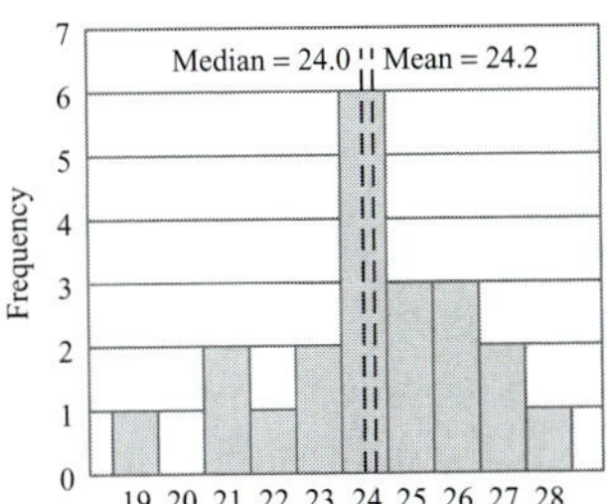

Between 1980 and 2000, the mean per capita citrus fruit consumption was 24.19%. Placing the percentages in numerical order, the middlemost percentage was 24%. The percentage occurring most often was 24%.

15. Mean: 1
Median: 1
Mode: 0, 2

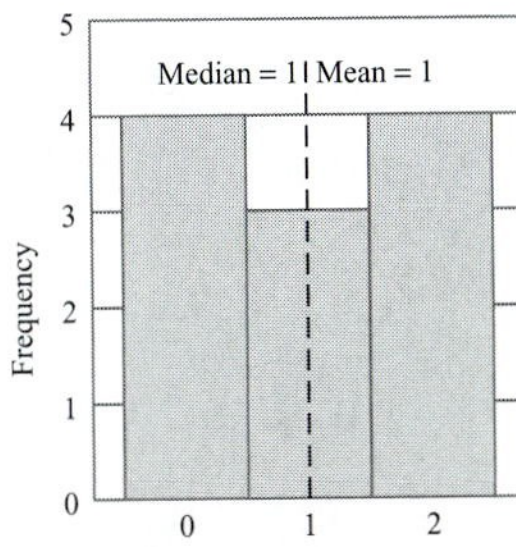

Between 1993 and 2003, the mean number of top ten grossing movies of New Line studio was 1. Placing the number of top ten grossing movies in numerical order, the middlemost number was 1. The number of top ten grossing movies that occurred most often was 0 and 2.

17. $\bar{x} = 1428.25$ square feet
$s = 213.6$ square feet
100% within 2 standard deviations

19. $\bar{x} = 78.76$ million dollars
$s = 42.95$ million dollars
100% within 2 standard deviations

21. $\bar{x} = 18.84$ births per 1000 people
$s = 4.03$ births per 1000 people
100% within 2 standard deviations

23. $\bar{x} = 3.69$ divorces per 1000 people
$s = 1.21$ divorces per 1000 people
100% within 2 standard deviations

25. $\bar{x} = 17.5$ points
$s = 11.4$ points
100% within 2 standard deviations

27. $\bar{x} = 976.1$ tornadoes
$s = 206.0$ tornadoes
100% within 2 standard deviations

29. $\bar{x} \approx 189{,}348$ thousand dollars
$s \approx 39{,}732$ thousand dollars
100% within 2 standard deviations

31. Median. The mean will be skewed because of the huge salary the football player makes.

33. Answers will vary.

35. No. The sets {3, 4, 5} and {3, 3, 4, 4, 5, 5} both satisfy the requirements but are different sets.

Section 8.6 *(page 592)*

1. -0.91 **3.** -2.22 **5.** -0.25

7. 2.65 **9.** 1.18

11. 0.4834, 0.9668 **13.** 0.4916, 0.9832

15. 0.0832, 0.1664 (from z-score table)

17. 0.2486, 0.4972 (from z-score table)

19. 0.1628, 0.3256 (from z-score table)

21. $\mu = 5.2, \sigma = 2.6$

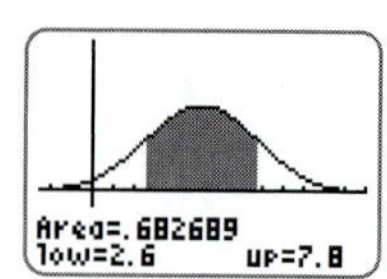

23. $\mu = 9.8, \sigma = 3.5$

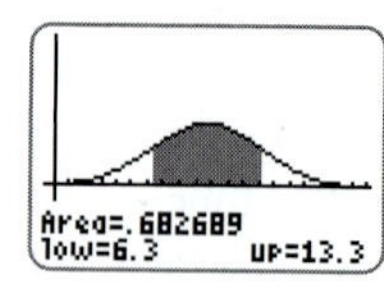

25. $\mu = 7.9, \sigma = 2.7$

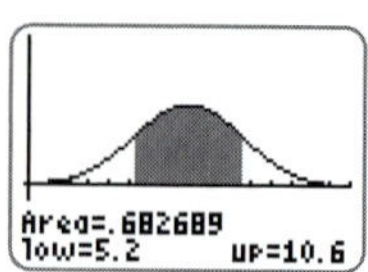

27. $\mu = -1.2, \sigma = 1.7$

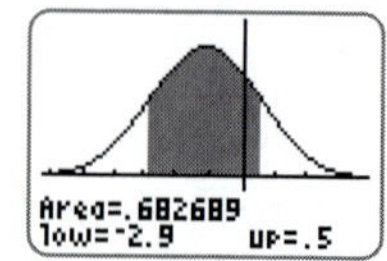

29. $\mu = -1.7, \sigma = 7.4$

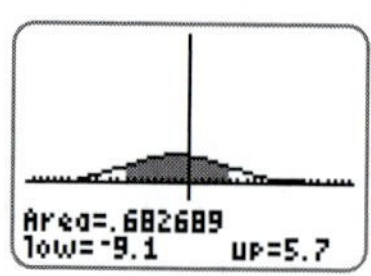

31. 31.12% **33.** 85.15% **35.** 0.13%

37. $x \geq 680$ **39.** ≈ 0.0582 **41.** ≈ 0.0000

43. 22.7 pounds to 32.3 pounds

45. The doctor would want to know that the baby is growing at a rate that is not so unusual as to be a problem.

47. $z_1 = z_2 + \dfrac{\mu_2 - \mu_1}{\sigma}$

49. The graphs of the normal distributions $f(x) = \dfrac{1}{\sqrt{2\pi}} e^{-1/2(x-4)^2}$ and $g(x) = \dfrac{1}{\sqrt{2\pi}} e^{-1/2(x-8)^2}$ are shown in the figure below.

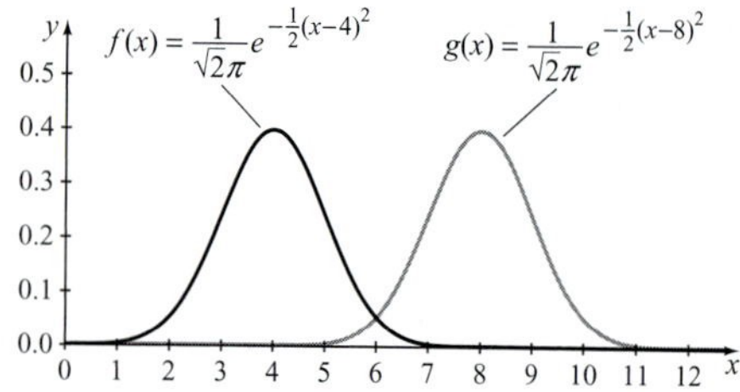

The graph of $h(x) = f(x) + g(x)$ (shown below) is clearly not a normal curve.

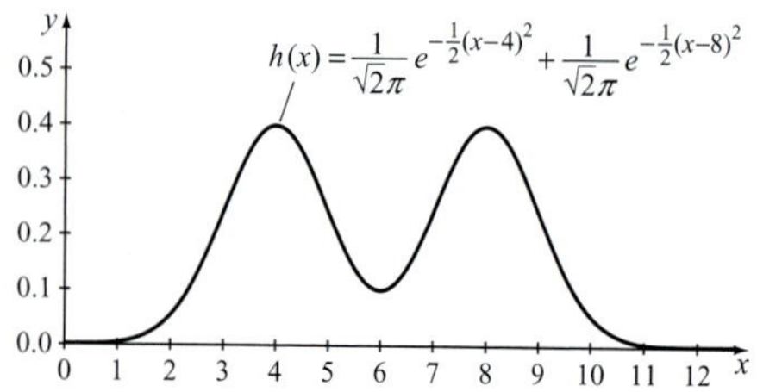

Therefore, the sum of two normal distributions is not guaranteed to be a normal distribution.

CHAPTER 8 REVIEW EXERCISES

Section 8.1 *(page 594)*

1. $P(A|B) = 0.714$

3. Let S be the set of seniors and W be the set of winners. $P(W|S) = 0.25$

5. 40.0%

7. The events are dependent.

Section 8.2 *(page 594)*

9.

	%
Prevalence	0.1
Sensitivity	25
Specificity	99
Predictive-value positive	**2.44**
Predictive-value negative	**99.9**

11. Sensitivity = 100%
There is a 100% probability that a person who tested positive with the other HCV method also tested positive with ACON HCV.

Specificity = 98.6%
There is a 98.6% probability that a person who tested negative with the other HCV method also tested negative with ACON HCV.

Predictive-value positive = 98.6%
There is a 98.6% probability that a person who tested positive with ACON HCV also tested positive with the other HCV method.

Predictive-value negative = 100%
There is a 100% probability that a person who tested negative with ACON HCV also tested negative with the other HCV method.

Section 8.3 *(page 595)*

13. $P = \begin{bmatrix} 0 & 0.1 & 0.9 \\ 0.3 & 0.7 & 0 \\ 0.8 & 0 & 0.2 \end{bmatrix}$

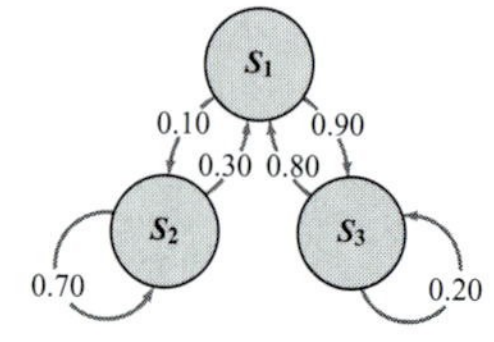

15. $P = \begin{bmatrix} 0.25 & 0.40 & 0.35 \\ 0 & 0.10 & 0.90 \\ 0.20 & 0 & 0.80 \end{bmatrix}$

17. (i) Let S_1 = Brand A; S_2 = Brand B, and S_3 = other brand.

(ii)

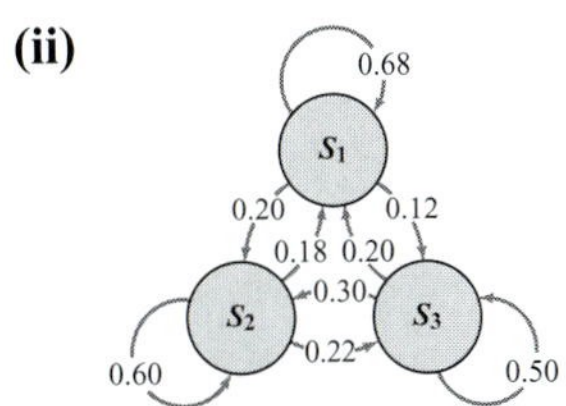

(iii) $P = \begin{bmatrix} 0.68 & 0.20 & 0.12 \\ 0.18 & 0.60 & 0.22 \\ 0.20 & 0.30 & 0.50 \end{bmatrix}$

(iv) $D = [0.38 \quad 0.32 \quad 0.30]$

(v) $DP = [0.376 \quad 0.358 \quad 0.266]$

At the end of the year, Brand A is expected to have 37.6% of the market share; Brand B, 35.8% of the market share; and other brands, 26.6% of the market share.

Section 8.4 (page 596)

19.

X	$P(X)$
1	$\frac{6}{216} \approx 0.028$
2	$\frac{90}{216} \approx 0.417$
3	$\frac{120}{216} \approx 0.556$

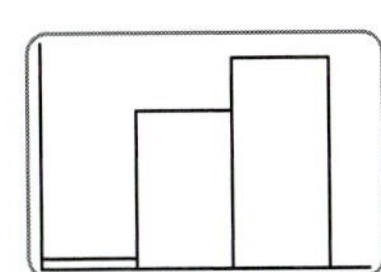

21. X = points awarded if she guesses

X	$P(X)$
−5	0.50
5	0.50

$E(X) = 0$

The expected value if she guesses is 0 points. If she doesn't guess, she will earn 0 points. According to the expected value, if she were to guess on every question, we would expect her overall score to be near 0. From a probability standpoint, either option is a good choice, since guessing and not guessing yield the same expected value. We would encourage her to think through the problem thoroughly, look back at other problems for clues, and then take the risk and guess.

Section 8.5 (page 596)

23. Mean = 72.67

The average number of points per game was 72.67 points.

Median = 68.50

When the six scores are lined up in numerical order, the average (mean) of the middle two values is 68.50 points.

Mode = none

No score was repeated.

Standard deviation = 10.5

Based on Chebyshev's rule, we anticipate that at least 75% of the scores lie within 21 points (2 standard deviations) of the mean.

25. Mean = 9,243,119

The average gross sales per weekend was $9,243,119.

Median = 8,943,880.5

When the 10 sales values are lined up in numerical order, the average (mean) of the middle two values is $8,943,880.50.

Mode = none

No weekend sales amount was repeated.

Standard deviation = 2,451,367.52

Based on Chebyshev's rule, we anticipate that at least 75% of the weekend sales lie within $4,902,735.04 (2 standard deviations) of the mean.

Section 8.6 (page 597)

27. $z = -0.1667$

29. 19.74%

31.

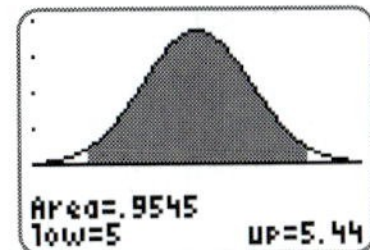

33. 13.59%

Index

Index of Businesses, Products, and Associations

Derivative Rules

For differentiable functions $f(x)$, $g(x)$, and $u = h(x)$ and constants n and c, the following rules apply:

Constant Rule $\frac{d}{dx}(c) = 0$	
Power Rule $\frac{d}{dx}(x^n) = nx^{n-1}$	**Generalized Power Rule** $\frac{d}{dx}(u^n) = nu^{n-1}u'$
Constant Multiple Rule $\frac{d}{dx}[c \cdot f(x)] = c \cdot f'(x)$	
Sum and Difference Rule $\frac{d}{dx}[f(x) \pm g(x)] = f'(x) \pm g'(x)$	
Product Rule $\frac{d}{dx}[f(x) \cdot g(x)] = f'(x)g(x) + g'(x)f(x)$	
Quotient Rule $\frac{d}{dx}\left[\frac{f(x)}{g(x)}\right] = \frac{f'(x)g(x) - g'(x)f(x)}{[g(x)]^2}$	
Exponential Rule $\frac{d}{dx}[b^x] = \ln(b) \cdot b^x$	**Generalized Exponential Rule** $\frac{d}{dx}[b^u] = \ln(b) \cdot b^u u'$
Logarithmic Rule $\frac{d}{dx}[\log_b x] = \frac{1}{\ln(b) \cdot x}$	**Generalized Logarithmic Rule** $\frac{d}{dx}[\log_b u] = \frac{1}{\ln(b) \cdot u} u'$
Chain Rule $\frac{d}{dx}\{f[g(x)]\} = f'[g(x)]g'(x)$	**Chain Rule (Alternate Form)** $\frac{dy}{dx} = \frac{dy}{dt} \cdot \frac{dt}{dx}$

Integral Rules

For integrable functions $f(x)$, $g(x)$, $u(x)$, and $v(x)$ and constants n, b, c, and C, the following rules apply:

Power Rule $\int x^n\,dx = \frac{x^{n+1}}{n+1} + C$ for $n \neq -1$
Rule for $\frac{1}{x}$ $\int \frac{1}{x}\,dx = \ln\lvert x\rvert + C$
Constant Multiple Rule $\int [c \cdot f(x)]\,dx = c \cdot \int f(x)\,dx$
Sum and Difference Rule $\int [f(x) \pm g(x)]\,dx = \int f(x)\,dx \pm \int g(x)\,dx$
Exponential Rule $\int b^x\,dx = \frac{b^x}{\ln(b)} + C$
Integration by Parts $\int u\,dv = uv - \int v\,du$

Common Functions	Equation
Constant	$y = c$
Linear	$y = mx + b$
Quadratic	$y = ax^2 + bx + c$
Cubic	$y = ax^3 + bx^2 + cx + d$
Quartic	$y = ax^4 + bx^3 + cx^2 + dx + k$
Power	$y = x^n$
Exponential	$y = ab^x$
Logarithmic	$y = \log_b(x)$

(Continued)